Introduction to Electric Circuits

Grown-ups never understand anything for themselves, and it is tiresome for children to be always and forever explaining things to them.

The Little Prince, Antoine de Saint– Exupery

Dedicated to our children
Christine Joy Dorf, Renée Dorf Schafer
and
Lance Evan, Mari Susann,
Jessica Lynn and Sarah Marie Svoboda

Introduction to
Electric
Circuits

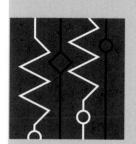

Third Edition

Richard C. Dorf
University of California, Davis

James A. Svoboda
Clarkson University

JOHN WILEY & SONS, Inc.
New York • Chichester • Brisbane • Toronto • Singapore

About the Cover Photo

On June 29, 1995 the U.S. Space Shuttle Atlantis docked with Russia's Mir Space Station orbiting 245 miles above the earth. The docking maneuver required control of the Space Shuttle to achieve alignment with the Mir station within 3 inches and 2 degrees of the assigned position. The photo shows Atlantis's aft cargo bay and Spacelab science module visible through a window on the station. A 70mm camera, carried into space by the STS-71 crew aboard Atlantis, was used to expose the image. The linkup enabled the seven crew members to visit Mir, and it allowed the three Mir-18 crew members, in space since March of 1995, access to Spacelab. That module was quite busy with tests and data collection involving the three until Atlantis brought them home on July 7, 1995.

Photo courtesy of National Aeronautics and Space Administration.

ACQUISITIONS EDITOR	Steven Elliot
MARKETING MANAGER	Debra Riegert
FREELANCE PRODUCTION MANAGER	Charlotte Hyland
PRODUCTION COORDINATION	HRS/Electronic Text Management
MANUFACTURING MANAGER	Mark Cirillo
SENIOR ILLUSTRATION COORDINATOR	Edward Starr
ILLUSTRATION COORDINATION	HRS/Electronic Text Management
ILLUSTRATIONS	Wellington Studios
COVER DESIGNER	David Levy

This book was set in Garamond Light by Progressive Information Technologies and printed and bound by *RR Donnelley/Willard*. The cover was printed by *Phoenix Color*.

Recognizing the importance of preserving what has been written, it is a policy of John Wiley & Sons, Inc. to have books of enduring value published in the United States printed on acid-free paper, and we exert our best efforts to that end.

The paper in this book was manufactured by a mill whose forest management programs include sustained yield harvesting of its timberlands. Sustained yield harvesting principles ensure that the number of trees cut each year does not exceed the amount of new growth.

ISBN 0-471-12702-7

Printed in the United States of America

10 9 8 7 6 5 4 3 2

ABOUT THE AUTHORS

Richard C. Dorf, professor of electrical and computer engineering at the University of California, Davis, teaches graduate and undergraduate courses in electrical engineering in the fields of circuits and control systems. He earned a Ph.D. in electrical engineering from the U.S. Naval Postgraduate School, an M.S. from the University of Colorado, and a B.S. from Clarkson University. Highly concerned with the discipline of electrical engineering and its wide value to social and economic needs, he has written and lectured internationally on the contributions and advances in electrical engineering.

Professor Dorf has extensive experience with education and industry and is professionally active in the fields of robotics, automation, electric circuits, and communications. He has served as a visiting professor at the University of Edinburgh, Scotland; the Massachusetts Institute of Technology; Stanford University, and the University of California, Berkeley.

A Fellow of the Institute of Electrical and Electronic Engineers, Dr. Dorf is widely known to the profession for his *Modern Control Systems,* 7th Edition (Addison-Wesley, 1995) and The International Encyclopedia of Robotics (Wiley, 1988). Dr. Dorf is also the coauthor of *Circuits, Devices and Systems* (with Ralph Smith), 5th Edition (Wiley, 1992). Dr. Dorf edited the widely used Electrical Engineering Handbook (CRC Press and IEEE Press) published in 1993.

James A. Svoboda is an associate professor of electrical and computer engineering at Clarkson University where he teaches courses on topics such as circuits, electronics and computer programming. He earned a Ph.D. in electrical engineering from the University of Wisconsin, Madison, an M.S. from the University of Colorado and a B.S. from General Motors Institute.

Sophomore Circuits is one of Professor Svoboda's favorite courses. He has taught this course to 2000 undergraduates at Clarkson University over the last 16 years. In 1986, he received Clarkson University's Distinguished Teaching Award.

Professor Svoboda has written several research papers describing the advantages of using nullors to model electric circuits for computer analysis. He is interested in the way that technology affects engineering education and has developed several software packages for use in Sophomore Circuits.

Professor Svoboda's email address is svoboda@sun.soe.clarkson.edu. His spot on the internet is located at http://sunspot.ece.clarkson.edu:1050/~svoboda/.

PREFACE

The central theme of *Introduction to Electric Circuits* is the concept that electric circuits are a part of the basic fabric of modern technology. Given this theme, we endeavor to show how the analysis and design of electric circuits is inseparably intertwined with the ability of the engineer to design complex electronic, communication, computer, and control systems as well as consumer products.

Approach

This book is designed for a one- to three-semester course in electric circuits or linear circuit analysis. The presentation is geared to readers who are being exposed to the basic concepts of electric circuits for the first time, and the scope of the work is broad. Students should come to the course with a basic knowledge of differential and integral calculus.

The emphasis on circuit design is enhanced in this edition by a design challenge that is introduced at the start of each chapter. The design of circuits is a unique advantage of this book.

The text is designed with maximum flexibility in mind. A flowchart immediately following this preface demonstrates alternative chapter organizations that can accommodate different course outlines without disrupting continuity.

Since this book is designed for introductory courses, a major effort has been made to provide the history as well as the current motivation for each of the topics. Thus, circuits are shown to be the results of real invention and the answers to real needs in industry, the office, and the home. Although the tools of electric circuit analysis may be partially abstract, electric circuits are the building blocks of modern society. The analysis and design of electric circuits are critical skills for all engineers.

Changes in the Third Edition

The third edition provides a revised introduction to operational amplifiers, based on the simplest model, the ideal op amp, a topic covered in Chapter 6. With this clearer and readable introduction, students can be more prepared for the study of related topics, such as filter circuits. Consistent with the theme of the book, this edition contains expanded coverage of the design of op amp circuits. As noted above, however, chapters can be rearranged to accommodate different course outlines.

In addition, extensive revisions of Chapters 13 and 14—Frequency Response and Laplace transform, respectively—will enable the reader to more readily grasp these important topics.

The Design Challenge is introduced at the beginning of each chapter, and the Design Challenge Solution is provided at the end of the chapter. Each challenge solution requires the knowledge provided in each chapter and thus motivates the reader by illustrating the necessity of the chapter's content.

 Mathematical Toolboxes are used, where appropriate, to summarize mathematical tools that are needed as the reader proceeds through a chapter.

Revised optional PSpice sections and problems, denoted in the text with a computer symbol, are integrated throughout the text as appropriate. An extensive appendix serves as a tutorial to the PSpice program.

The problems at the end of each chapter are reorganized to correspond in number to the section of the chapter. Thus, for example, P10.4-1, is the first problem based on section 10.4.

A new section is added to each chapter called *Verification Examples* where we illustrate techniques that can be used to verify the correctness of solutions provided by computers or other persons. This provides excellent practice for the reader in the verification of results. At the end of each chapter we provide several *Verification Problems* that challenge the reader to accomplish the verification process. These verification methods provide an antidote to the all-too-frequent phenomenon of Garbage In-Garbage Out (GIGO).

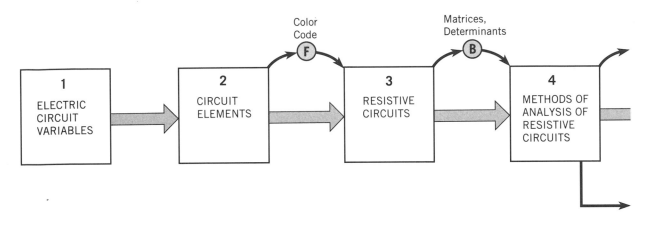

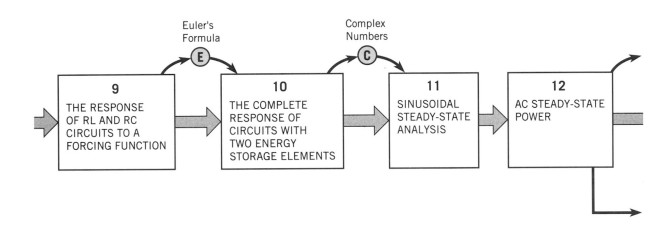

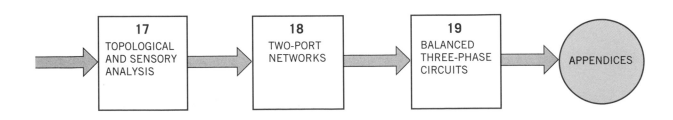

Alternative Chapter Organizations

This flowchart illustrates the alternative chapter organizations that can accommodate different course outlines without disrupting continuity.

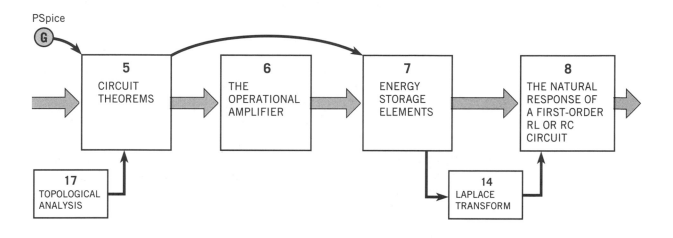

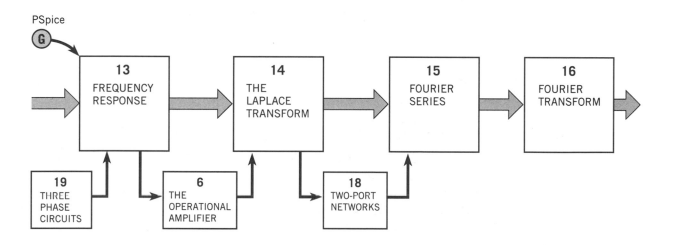

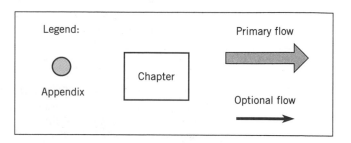

Chapter Features

Preview   Each chapter opens with a preview outlining the content and objective of the chapter.

Design Challenge   Following the preview, the reader is challenged by a design problem that can only be solved by utilizing the material presented in the chapter. The student is asked to consider the challenge and then proceed to read the content of the chapter.

Design Challenge Solution   At the end of each chapter the Design Challenge is solved utilizing the chapter's methods and approach. The design method utilizes a step-by-step methodology for solving design problems. The steps of the methodology are (1) state the problem, (2) define the situation, (3) state the goal, (4) generate a plan to solve the problem, and (5) take action using the plan.

Historical Vignettes   After a chapter preview, the first section of each chapter is dedicated to one aspect of electrical engineering history and current practice. This enables the reader to witness the past and current motivations for the development of modern electrical engineering devices and methods as well as to grasp the excitement of engineering in the 1990s.

Illustrative Examples   Because this book is oriented toward gaining expertise in problem solving, we have included more than 200 illustrative examples. Here, as in the end-of-chapter problems, the derivation of results is precise from a mathematical standpoint, but the solution of practical problems is emphasized.

Summary and Glossary   Each chapter includes a summary and a glossary of terms and concepts at the end of the chapter.

Exercises and Problems   The exercises and problems are organized into five types

- Exercises provide the students with factual review; answers are provided immediately following the exercise.
- Problems serve as extensions of the Illustrative Examples within the chapter; answers to selected problems are provided in the text, and other answers can be found in the Instructor's Manual.

- Verification problems challenge the reader to consider a problem with a stated set of results. The reader must then verify the correctness of the results provided in the problem statement.
- Design problems, written with ABET accreditation standards in mind, provide practice in applying the material to interesting design situations. For example, students design a car airbag ignition circuit and an electric fence to deflect sharks from the beach. These open-ended problems vary in level of sophistication.
- Optional PSpice problems encourage computer use.

There are more than 300 exercises and 800 homework problems. The reader should first consider the exercises and then attempt selected problems. Answers are liberally provided to the readers, enabling a check of the solutions. Answers are available for all exercises.

Contents

The book begins with the development of electric circuit elements and the electric variables used to describe them. Then resistive circuits are studied to provide an in-depth introduction to the concept of a circuit and its analysis. Chapter 6 focuses on the widely used electronic element, the operational amplifier. The large number of useful circuits that can be constructed by using the operational amplifier with resistors and capacitors is demonstrated.

Next the many useful theorems and principles developed for insightful analysis of electric circuits are considered. Then the discussion moves to a description of energy storage elements—the inductor and the capacitor—which are widely used in industrial and office systems and in consumer products, such as the television receiver or the automobile.

Chapters 8 and 9 begin the study of the time response of electric circuits that incorporate energy storage elements. I also introduce the concept of switched circuits, which play a very important role in modern circuits. Chapter 10 is concerned with the study of the time response of circuits with two or more energy storage elements, which are described by differential equations.

In Chapter 11, the discussion focuses on the steady-state behavior of circuits with sinusoidal sources and develops the concepts of the phasor transform and impedance.

Chapter 12 is concerned with the study of ac steady-state power. In this chapter I also introduce the electric transformer, another important circuit device.

Chapter 13 provides an introduction to the concept of frequency response of a circuit and the utility of Bode diagrams. Then in Chapter 14 the study of the Laplace transform and its use in the analysis of electric circuits are considered.

Chapter 15 provides an introduction to the concept of describing signals by the Fourier series and their use in circuit analysis and design. In Chapter 16 the Fourier transform and the very important concept of a signal spectrum are examined.

Chapter 17 describes the topological analysis of a circuit and the use of graphs for nodal, mesh, and state variable analysis. Chapter 18 is concerned with the description of two-port networks, the T-to-TT transformation, and the interconnection of circuits. Finally, in Chapter 19 balanced three-phase circuits and their general utility in modern power systems are studied.

The appendices provide a review of matrices, determinants, and complex numbers. Two new appendices describe Euler's formula and the standard resistor color code. Finally, an extensive appendix describes the analysis and system program, PSpice, which can be used to obtain the value of many variables in a circuit and to verify a solution obtained analytically.

Supplements

An Instructor's Manual Written by the text's authors and double checked by a team of content experts, the Instructor's Manual includes sample course syllabi, answers to all exercises and problems in the text, and selected solutions suitable for use as transparency masters.

 PSpice

(A free copy of PSpice is available from the Microsim Corporation. Class instructors can receive complimentary evaluation versions for both the IBM-PC and Macintosh by submitting a request on educational letterhead to Product Marketing Department, MicroSim Corporation, 20 Fairbanks, Irvine, CA 92718.)

Acknowledgments

We are grateful to many people whose efforts have gone into the making of this textbook. We gratefully acknowledge the contribution of Joseph Tront in the development of the original content used in Appendix G on PSpice. We wish to thank reviewers of the Third Edition, especially Gary Ybarra of Duke University, who provided many useful suggestions for improvement. We specifically thank the following individuals:

Jonny Anderson, University of Washington
Daniel Bukofzer, California State University, Fresno
Mauro Caputo, Hofstra University
Yu Chang, Union College
Sherif Embabi, Texas A&M University
Dennis Fitzgerald, California State Polytechnic, Pomona
Gary Ford, University of California — Davis
Prashant Gandhi, Villanova University
Charles Gauder, University of Dayton
Ed Gerber, Drexel University
Victor Gerez, Montana State University
Paul Gordy, Tidewater Community College
Cliff Griggs, Rose Hulman Institute
Nazli Gundes, University of California, Davis
Robert Herrick, Purdue University
DeVerls Humphreys, Brigham Young University
Aziz Inan, University of Portland
Ashok Iyer, University of Nevada, Las Vegas
J. Michael Jacob, Purdue University
Rich Johnson, Lawrence Tech University
Ravindra Joshi, Old Dominion University
Gerald Kane, University of Tulsa
Steven Kaprielian, Lafayette College
Richard Klafter, Temple University
Joseph Kozikowski, Villanova University
Robert Laramore, Purdue University
Haniph Latchman, University of Florida
Gerald Lemay, University of Massachusetts — Dartmouth
Gary Lipton, Lafayette College
Steve McFee, McGill University
Robert Miller, Virginia Tech
David Moffatt, Ohio State University

Paul Murray, Mississippi State University
Burkes Oakley, University of Illinois, Urbana-Champaign
Anthony J. A. Oxtoby, Purdue University
Albert Peng, Central Michigan University
Bill Perkins, University of Illinois, Urbana-Champaign
Kouros Saiedpazouki, Villanova University
Edgar Sanchez, Texas A&M University
Thomas Schubert, University of San Diego
David Skitek, University of Missouri, Kansas City
Charles Smith, University of Mississippi
Jerry Suran, University of California, Davis
Arthur Sutton, California State Polytechnic University
William Sutton, George Mason University
Xiao-Bang Xu, Clemson University
Mark Yoder, Rose Hulman University

Richard C. Dorf
James Svoboda

3RD ED. © 1996

TABLE OF CONTENTS

Chapter 6

THE OPERATIONAL AMPLIFIER 225

Chapter 7

ENERGY STORAGE ELEMENTS 285

Chapter 12

AC STEADY-STATE POWER 569

Chapter 13

FREQUENCY RESPONSE 539

Chapter 14

THE LAPLACE TRANSFORM   711

Chapter 15

FOURIER SERIES   771

Chapter 16

FOURIER TRANSFORM   809

Chapter 17

TOPOLOGICAL AND SENSITIVITY ANALYSIS   839

CHAPTER 1

ELECTRIC CIRCUIT VARIABLES

PREVIEW

From the beginning of recorded time, humans have explored the electrical phenomena they have experienced in everyday life. As scientists developed the knowledge of electrical charge, they formulated the laws of electricity as we know them today.

In this chapter we illustrate the thinking process underlying the design of an electric circuit and briefly review the history of electrical science up to the late 1800s. With the knowledge of electricity available by the turn of the century, scientists and engineers analyzed and built electric circuits. In this chapter we explore how electric elements may be described and analyzed in terms of the variables charge, current, voltage, power, and energy.

The design of electric circuits is the process of combining electric elements to provide desired values of these circuit variables. We now illustrate the design process in the context of a jet valve controller circuit.

1-1 DESIGN CHALLENGE

JET VALVE CONTROLLER

Often we need a circuit that provides energy to a device such as a pump or a valve. It is necessary to determine the required current and voltage provided to the device so that it will operate for a desired time period. Here we consider a jet valve controller that requires 40 mJ of energy for 1 minute of operation. The energy will be supplied to the jet valve controller by another element, i.e., a battery. We need to draw the circuit model of this jet valve controller and its energy supply. Let us proceed to consider simple electric circuits and describe the voltage and current in terms of the energy delivered to an element such as a jet valve controller. Then, at the end of the chapter we will determine the current and voltage required to provide 40 mJ of energy for 1 minute of operation and describe the necessary battery.

We will return to the design challenge for the jet valve controller in Section 1-10 after briefly reviewing the history of electrical science and examining the circuit variables in detail.

1-2 ‖ THE EARLY HISTORY OF ELECTRICAL SCIENCE

Electricity is a natural phenomenon controlled for the purposes of humankind. With this phenomenon we have developed communications, lighting, and computing devices.

Electricity is the physical phenomenon arising from the existence and interaction of electric charge.

2

Prehistoric people experienced the properties of magnetite—permanently magnetized pieces of ore, often called lodestones. These magnetic stones were strong enough to lift pieces of iron.

The philosopher Thales of Miletus (640–546 B.C.) is thought to have been the first person who observed the electrical properties of amber. He noted that when amber was rubbed it acquired the ability to pick up light objects such as straw and dry grass. He also experimented with the lodestone and knew of its power to attract iron. By the thirteenth century, floating magnets were used for compasses.

William Gilbert of England published the book *De Magnete* in 1600, and it represented the greatest forward step in the study of electricity and magnetism up to that time. In his studies, he found a long list of materials that could be electrified. Gilbert also presented a procedure for analyzing physical phenomena by a series of experiments, which we now call the scientific method.

Following Gilbert's lead, Robert Boyle published his many experimental results in 1675, as shown in Figure 1-1. Boyle was one of the early experimenters with electricity in a vacuum.

Otto von Guericke (1602–1686) built an electrical generator and reported it in his *Experimenta Nova* of 1672. This device, shown in Figure 1-2, was a sulfur globe on a shaft that could be turned on its bearings. When the shaft was turned with a dry hand held on the globe, an electrical charge gathered on the globe's surface. Guericke also noted small sparks when the globe was discharged.

A major advance in electrical science was made in Leyden, Holland, in 1746, when Pieter van Musschenbroek introduced a jar that served as a storage apparatus for static electricity. The jar was coated inside and out with tinfoil, and a metallic rod was attached to the inner foil lining and passed through the lid. As shown in Figure 1-3, Leyden jars were gathered in groups (called batteries) and arranged with multiple connections, thereby further improving the discharge energy.

People have always watched but few have analyzed that great display of power present in the electrical discharge in the sky called lightning and illustrated by Figure 1-4. In the late 1740s, Benjamin Franklin developed the theory that there are two kinds of charge, positive and negative. With this concept of charge, Franklin developed his famous kite experiment in June 1752 and his innovation, the lightning rod, for draining the electrical

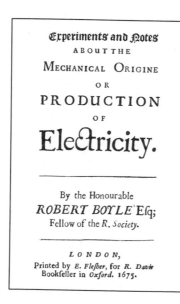

Figure 1-1 Title page of Robert Boyle's book on electrical science, published in 1675. Courtesy of the Institution of Electrical Engineers, London.

Figure 1-2 In his studies of attraction and gravitation, Guericke devised the first electrical generator. When a hand was held on a sulfur ball revolving in its frame, the ball attracted paper, feathers, chaff, and other light objects. Courtesy of Burndy Library.

charge from the clouds. Franklin, shown in Figure 1-5, was the first great American electrical scientist.

In 1767 Joseph Priestly published the first book on the history of electricity; the title page is shown in Figure 1-6. Twenty years later, Professor Luigi Galvani of Bologna, Italy, carried out a series of experiments with a frog's legs. Galvani noted that the leg of a dead

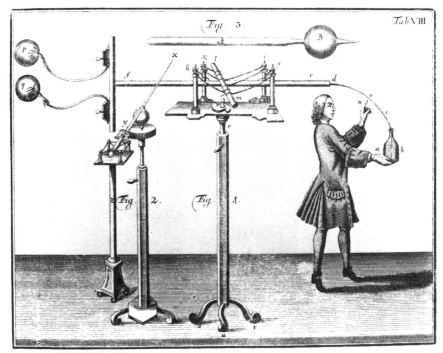

Figure 1-3 Illustration of a Leyden jar for retaining electrical charges. The charges developed from the rotating glass globes (left) were transmitted through the central conductor and led down a wire to the bottle, which was partly filled with water. With the bottle held in one hand, the other hand completed the circuit—with a resulting shock. Courtesy of Burndy Library.

Figure 1-4 A display of lightning. Courtesy of the National Severe Storms Laboratory.

Figure 1-5 Benjamin Franklin and a section of the lightning rod erected in his Philadelphia home in September 1752. Divergence of the balls indicated a charged cloud overhead. With this apparatus Franklin discovered that most clouds were negatively charged and that "'tis the earth that strikes into the clouds, and not the clouds that strike into the earth." Courtesy of Burndy Library.

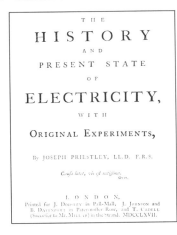

Figure 1-6 Title page of the book on electrical science by Joseph Priestley (1767). Courtesy of the Institution of Electrical Engineers, London.

frog would twitch when dissected with a metal scalpel, and he published his findings in 1791. Galvani found that the frog's legs twitched when subjected to an electrical discharge, as shown in Figure 1-7.

Alessandro Volta of Padua, Italy, shown in Figure 1-8, recognized that the twitch was caused by two dissimilar metals that were moist and touching at one end and in contact with the frog's leg nerves at the other end. He went on to construct an electrochemical

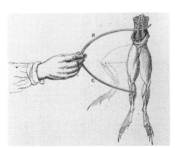

Figure 1-7 Luigi Galvani in 1786 noticed that a frog's leg would twitch if touched by two dissimilar metals, copper and zinc. Courtesy of Burndy Library.

Figure 1-8 Alessandro Volta. Courtesy of Burndy Library.

pile, as illustrated in Figure 1-9, that consisted of pairs of zinc and silver disks separated by brine-soaked cloth or paper. This voltaic pile could cause the sensation of current flowing through a person who placed a hand at either end of the pile.

Volta, with the invention of his pile, or electric battery, in 1800, was able to show a steady current in a closed circuit. Volta was remembered 54 years after his death when the electromotive force was officially named the *volt*.

The foundation of electrodynamics was laid by André-Marie Ampère. The former studies of electricity he called electrostatics, to highlight the difference. During the 1820s he defined the electric current and developed means of measuring it. Ampère, shown in

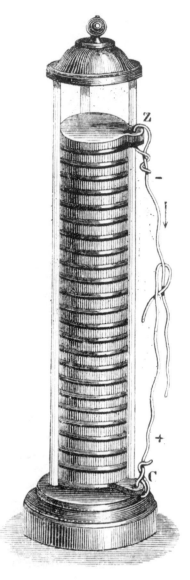

Figure 1-9 The voltaic pile, which was a series of successions of three conducting substances—a plate of silver, a plate of zinc, and a similar piece of spongy matter impregnated with a saline solution—repeated 30 or 40 times. Courtesy of Burndy Library.

Figure 1-10 André-Marie Ampère. Courtesy of the Library of Congress.

Figure 1-10, was honored by having the unit of electric current, the ampere, named after him in 1881.

A paper published by James Prescott Joule in 1841 claimed the discovery of the relationship between a current and the heat or energy produced, which we call Joule's law. Joule is shown in Figure 1-11, and the unit of energy is called the joule in his honor.

The theories of electrodynamics were stated in mathematical terms by James Clerk Maxwell, a Scottish mathematical physicist, in papers published between 1855 and 1864. His famous book *Treatise on Electricity and Magnetism* was published in 1873. Maxwell in his student days is shown in Figure 1-12.

The major events in electrical science and engineering are summarized in Table 1-1.

Figure 1-11 James Prescott
Joule (1818–1889). Courtesy
of the Smithsonian Institution.

Figure 1-12 Maxwell in 1855 as a student at
Cambridge University, England. Courtesy of
Burndy Library.

Table 1-1
Major Events in Electrical Science and Engineering

1600	William Gilbert published *De Magnete*.
1672	Otto von Guericke published *Experimenta Nova*.
1675	Robert Boyle published *Production of Electricity*.
1746	Demonstration of the Leyden jar in Holland.
1750	Benjamin Franklin invented the lightning conductor.
1767	Joseph Priestley published *The Present State of Electricity*.
1786	Luigi Galvani observed electrical convulsion in the legs of dead frogs.

1800	Alessandro Volta announced the voltaic pile.
1801	Henry Moyes was the first to observe an electric arc between carbon rods.
1820	Hans Oersted discovered the deflection of a magnetic needle by current in a wire.
1821	Michael Faraday produced magnetic rotation of a conductor and magnet—the first electric motor.
1825	André-Marie Ampère defined electrodynamics.
1828	Joseph Henry produced silk-covered wire and more powerful electromagnets.
1831	Michael Faraday discovered electromagnetic induction and carried out experiments with an iron ring and core. He also experimented with a magnet and a rotating disk.
1836	Samuel Morse devised a simple relay.
1836	Electric light from batteries shown at the Paris Opéra.
1841	James Joule stated the relation between current and energy produced.
1843	Morse transmitted telegraph signals from Baltimore to Washington, D.C.
1850	First channel telegraph cable laid from England to France.
1858	Atlantic telegraph cable completed and first message sent.
1861	Western Union established telegraph service from New York to San Francisco.
1863	James Clerk Maxwell determined the ohm.
1873	Maxwell published *Treatise on Electricity and Magnetism*.
1875	Alexander Graham Bell invented the telephone.
1877	Thomas Edison invented the carbon telephone transmitter.
1877	Edison Electric Light Company formed.
1881	First hydropower station brought into use at Niagara, New York.
1881	Edison constructed the first electric power station at Pearl Street, New York.
1883	Overhead trolley electric railways started at Portrush and Richmond, Virginia.
1884	Philadelphia electrical exhibition.
1884	American Institute of Electrical Engineers (AIEE) formed as professional society for electrical engineers.
1885	Organization of the American Telephone and Telegraph Company.
1886	H. Hollerith introduced his tabulating machine.
1897	J. J. Thomson discovered the electron.
1899	Guglielmo Marconi transmitted radio signals from South Foreland to Wimereux, England.
1904	John Ambrose Fleming invented the thermionic diode.
1906	Lee De Forest invented the triode.
1912	Institute of Radio Engineers (IRE) formed as professional society for radio engineers.
1915	Commercial telephone service established from New York to San Francisco.
1927	Television established experimentally.
1933	Edwin Armstrong demonstrated FM radio transmission.
1936	Boulder Dam hydroelectric scheme with 115,000-horsepower turbines.
1946	Electronic digital computer ENIAC.
1948	William Shockley, John Bardeen, and Walter Brattain produced the first practical transistor.
1958	First voice transmission from a satellite. Laser invented.
1959	Invention of integrated circuit by Jack Kilby and Robert Noyce.
1963	Institute of Electrical and Electronics Engineers formed as merger of AIEE and IRE.
1980	First fiber optic cable installed in Chicago.
1987	Superconductivity demonstrated at 95 K.

1-3 ELECTRIC CIRCUITS AND CURRENT FLOW

The outstanding characteristics of electricity when compared with other power sources are its mobility and flexibility. Electrical energy can be moved to any point along a couple of wires and, depending on the user's requirements, converted to light, heat, or motion.

An **electric circuit** or electric network is an interconnection of electrical elements linked together in a closed path so that an electric current may continuously flow.

Consider a simple circuit consisting of two well-known electrical elements, a battery and a resistor, as shown in Figure 1-13. Each element is represented by the two-terminal element shown in Figure 1-14. Elements are sometimes called devices. Terminals are sometimes called nodes.

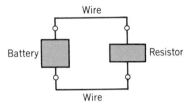

Figure 1-13 A simple circuit.

Figure 1-14 A general two-terminal electrical element with terminals a and b.

The basic element shown in Figure 1-14 has two terminals, cannot be subdivided into other elements, and can be described mathematically in terms of the electrical variables, voltage and current. Wires interconnect the elements to close the circuit so that a continuous current may flow. A more complex set of elements, with a particular terminal interconnection, is shown in Figure 1-15.

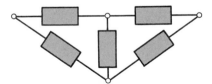

Figure 1-15 A circuit with five elements.

A *current* may flow in an electric circuit. *Current is the time rate of change of charge past a given point.* Charge is the intrinsic property of matter responsible for electric phenomena. The quantity of charge q can be expressed in terms of the charge on one electron, which is -1.602×10^{-19} coulombs. Thus, -1 coulomb is the charge on 6.24×10^{18} electrons. The current flows through a specified area A and is defined by the electric charge passing through the area per unit of time. Thus, define q as the charge expressed in coulombs (C).

Charge is the quantity of electricity responsible for electric phenomena.

Then we can express current as

$$i = \frac{dq}{dt} \tag{1-1}$$

The unit of current is the ampere (A); an ampere is 1 coulomb per second.

Current is the time rate of flow of electric charge past a given point.

Note that throughout this chapter we use a lowercase letter, such as q, to denote a variable that is a function of time, $q(t)$. We use an uppercase letter, such as Q, to represent a constant.

The flow of current is conventionally represented as a flow of positive charges. This convention was initiated by Benjamin Franklin. Of course, we now know that charge flow in metal conductors results from electrons with a negative charge. Nevertheless, we will conceive of current as the flow of positive charge, according to accepted convention.

Figure 1-16 shows the notation that we use to describe a current. There are two parts to this notation: a value (perhaps represented by a variable name) and an assigned direction. As a matter of vocabulary, we say that a current exists *in* or *through* an element. Figure 1-16 shows that there are two ways to assign the direction of the current through an

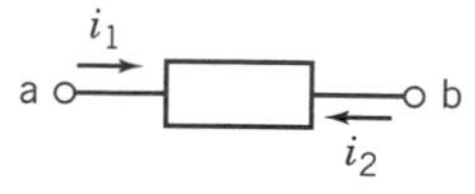

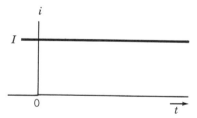

Figure 1-16 Current flow in a circuit element.

element. The current i_1 is the rate of flow of electric charge from terminal a to terminal b. On the other hand, the current i_2 is the flow of electric charge from terminal b to terminal a. The currents i_1 and i_2 are similar but different. They are the same size but have different directions. Therefore, i_2 is the negative of i_1 and

$$i_1 = -i_2$$

We always associate an arrow with a current to denote its direction. A complete description of current requires both a value (which can be positive or negative) and a direction (indicated by an arrow).

If the current flowing through an element is constant, we represent it by the constant I, as shown in Figure 1-17. A constant current is called a *direct current* (dc).

Figure 1-17 A direct current of magnitude I.

A **direct current** (dc) is a current of constant magnitude.

A time-varying current $i(t)$ can take many forms, such as a ramp, a sinusoid, or an exponential, as shown in Figure 1-18. The sinusoidal current is called an alternating current, or ac.

If the charge q is known, the current i is readily found using Eq. 1-1. Alternatively, if the current i is known, the charge q is readily calculated. Note that from Eq. 1-1 we obtain

$$q = \int_{-\infty}^{t} i \, d\tau$$

$$= \int_{0}^{t} i \, d\tau + q(0) \tag{1-2}$$

where $q(0)$ is the charge at $t = 0$.

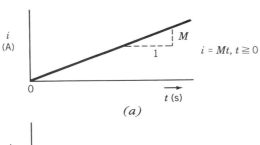

(a)

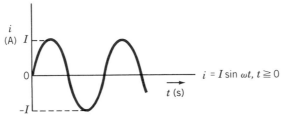

(b)

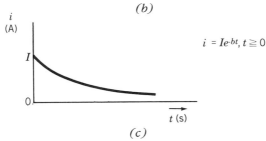

(c)

Figure 1-18 (a) A ramp with a slope M. (b) A sinusoid. (c) An exponential. I is a constant. The current i is zero for $t < 0$.

Example 1-1

Find the current flowing past terminal a of an element when the charge that has entered the element is

$$q = 12t \quad \text{C}$$

where t is the time in seconds.

Solution

Recall that the unit of charge is coulombs, C. Then the current, from Eq. 1-1, is

$$i = \frac{dq}{dt} = 12 \text{ A}$$

where the unit of current is amperes, A.

Example 1-2

Find the charge that has entered the terminal of an element by time t when the current is

$$i = Mt, \qquad t \geq 0$$

as shown in Figure 1-18a, and M is a constant. Assume that the charge is zero at $t = 0$ ($q(0) = 0$).

Solution

Using Eq. 1-2, we have

$$q = \int_0^t M\tau \, d\tau$$

$$= M\frac{t^2}{2} \text{ C}$$

Example 1-3

Find the charge that has entered the terminal of an element from $t = 0$ s to $t = 3$ s when the current is as shown in Figure 1-19.

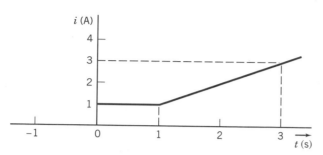

Figure 1-19 Current waveform for Example 1-3.

Solution

From Figure 1-19 we can describe $i(t)$ as

$$i(t) = \begin{cases} 0 & t < 0 \\ 1 & 0 \le t \le 1 \\ t & t > 1 \end{cases}$$

Using Eq. 1-2, we have

$$q = \int_0^3 i(t)\, dt = \int_0^1 1\, dt + \int_1^3 t\, dt$$

$$= 1t \Big|_0^1 + \frac{t^2}{2} \Big|_1^3$$

$$= 1 + \frac{1}{2}(9 - 1)$$

$$= 5\ \mathrm{C}$$

Alternatively, we note that integration of $i(t)$ from $t = 0$ to $t = 3$ s simply requires the calculation of the area under the curve shown in Figure 1-19. Then, we have

$$q = 1 + 2 \times 2$$
$$= 5\ \mathrm{C}$$

Example 1-4

Find the charge, $q(t)$, and sketch its waveform when the current entering a terminal of an element is as shown in Figure 1-20. Assume that $q(0) = 0$.

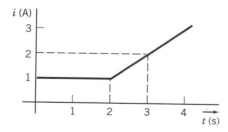

Figure 1-20 Current waveform for Example 1-4.

Solution

From Figure 1-20, we can describe $i(t)$ as

$$i(t) = \begin{cases} 0 & t < 0 \\ 1 & 0 \le t \le 2 \\ t - 1 & t > 2 \end{cases}$$

Using Eq. 1-2, we have

$$q(t) = \int_0^t i(\tau) \, d\tau$$

Hence, when $0 \leq t \leq 2$, we have

$$q = \int_0^t 1 \, d\tau = t \ \text{C}$$

When $t \geq 2$, we obtain

$$q = \int_0^t i(\tau) \, d\tau = \int_0^2 1 \, dt + \int_2^t (\tau - 1) \, d\tau$$

$$= t \Big|_0^2 + \frac{\tau^2}{2} \Big|_2^t - \tau \Big|_2^t$$

$$= \frac{t^2}{2} - t + 2 \ \text{C}$$

The sketch of $q(t)$ is shown in Figure 1-21. Note that $q(t)$ is a continuous function of time even though $i(t)$ has a discontinuity at $t = 0$.

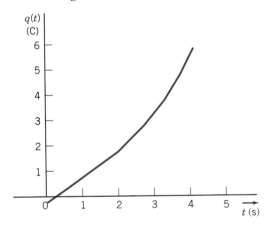

Figure 1-21 Sketch of $q(t)$ for Example 1-4.

EXERCISE 1-1

Find the charge that has entered an element by time t when $i = 8t^2 - 4t$ A, $t \geq 0$. Assume $q(0) = 0$.

Answer: $q(t) = \dfrac{8}{3} t^3 - 2t^2$ C

EXERCISE 1-2

The total charge that has entered an element is $q = q_0 \, e^{-2t}$, $t \geq 0$. Find the current. Note that q_0 is the charge at $t = 0$.

Answer: $i = -2q_0 e^{-2t}$ A
$= -2q$ A

EXERCISE 1-3

Find the charge that has entered a terminal between $t = 0$ and $t = 5$ s when $i = 10t^2$ A, $t \geq 0$.
Answer: 416.67 C

EXERCISE 1-4

The total charge $q(t)$, in coulombs, that has entered the terminal of an element is

$$q(t) = \begin{cases} 0 & t < 0 \\ 2t & 0 \leq t \leq 2 \\ 3 + e^{-2(t-2)} & t > 2 \end{cases}$$

Find the current $i(t)$ and sketch its waveform for $t \geq 0$.
Answer:

$$i(t) = \begin{cases} 0 & t < 0 \\ 2 & 0 < t \leq 2 \\ -2e^{-2(t-2)} & t > 2 \end{cases}$$

1-4 SYSTEMS OF UNITS

In representing a circuit and its elements, we must define a consistent system of units for the quantities occurring in the circuit. At the 1960 meeting of the General Conference of Weights and Measures, the representatives modernized the metric system and created the Système International d'Unités, commonly called SI units.

SI is *Système International d'Unités* or the International System of Units.

The fundamental, or base, units of SI are shown in Table 1-2. Symbols for units that represent proper (persons') names are capitalized; the others are not. Periods are not used after the symbols, and the symbols do not take on plural forms. The derived units for other

Table 1-2
SI Base Units

Quantity	SI Unit	
	Name	Symbol
Length	meter	m
Mass	kilogram	kg
Time	second	s
Electric current	ampere	A
Thermodynamic temperature	kelvin	K
Amount of substance	mole	mol
Luminous intensity	candela	cd

Table 1-3

Derived Units in SI

Quantity	Unit Name	Formula	Symbol
Acceleration—linear	meter per second per second	m/s^2	
Velocity—linear	meter per second	m/s	
Frequency	hertz	s^{-1}	Hz
Force	newton	$kg \cdot m/s^2$	N
Pressure or stress	pascal	N/m^2	Pa
Density	kilogram per cubic meter	kg/m^3	
Energy or work	joule	$N \cdot m$	J
Power	watt	J/s	W
Electric charge	coulomb	$A \cdot s$	C
Electric potential	volt	W/A	V
Electric resistance	ohm	V/A	Ω
Electric conductance	siemens	A/V	S
Electric capacitance	farad	C/V	F
Magnetic flux	weber	$V \cdot s$	Wb
Inductance	henry	Wb/A	H

physical quantities are obtained by combining the fundamental units. Table 1-3 shows the more common derived units along with their formulas in terms of the fundamental units and/or preceding derived units. Symbols are shown for the units that have them.

The basic units such as length in meters (m), time in seconds (s), and current in amperes (A) can be used to obtain the derived units. Then, for example, we have the unit for charge (C) derived from the product of current and time ($A \cdot s$). The fundamental unit for energy is the joule (J), which is force times distance or $N \cdot m$.

The great advantage of the SI system is that it incorporates a decimal system for relating large or smaller quantities to the basic unit. The powers of 10 are represented by standard prefixes given in Table 1-4. An example of the common use of a prefix is the centimeter (cm), which is 0.01 meter.

The decimal multiplier must always accompany the appropriate units and is never written by itself. Thus, we may write 2500 W as 2.5 kW. Similarly, we write 0.012 A as 12 mA.

Table 1-4

SI Prefixes

Multiple	Prefix	Symbol
10^{12}	tera	T
10^9	giga	G
10^6	mega	M
10^3	kilo	k
10^{-2}	centi	c
10^{-3}	milli	m
10^{-6}	micro	μ
10^{-9}	nano	n
10^{-12}	pico	p
10^{-15}	femto	f

Example 1-5

A mass of 150 grams experiences a force of 100 newtons. Find the energy or work expended if the mass moves 10 centimeters. Also, find the power if the mass completes its move in 1 millisecond.

Solution

The energy is found as

$$
\begin{aligned}
\text{Energy} &= \text{force} \times \text{distance} \\
&= 100 \times 0.1 \\
&= 10 \text{ J}
\end{aligned}
$$

Note that we used the distance in units of meters. The power is found from

$$
\text{Power} = \frac{\text{energy}}{\text{time period in seconds}}
$$

where the time period is 10^{-3} s. Thus,

$$
\text{Power} = \frac{10}{10^{-3}} = 10^4 \text{ W}
$$

EXERCISE 1-5

Find the number of picoseconds in 10 milliseconds.
Answer: 10^{10}

EXERCISE 1-6

A constant current of 4 kA flows through an element. What is the charge that has passed through the element in the first millisecond?
Answer: 4 C

EXERCISE 1-7

Find the work done by a constant force of 10 mN applied to a mass of 150 grams for a distance of 50 cm.
Answer: 5 mJ

EXERCISE 1-8

A current of 5 μA flows through a wire. (a) How many coulombs of charge have passed through the wire in 10 seconds? (b) How many coulombs would pass through the wire in 2 years?
Answer: (a) 5×10^{-5} C
(b) 315.4 C

1-5 | VOLTAGE

The basic variables in an electrical circuit are current and voltage. These variables describe the flow of charge through the elements of a circuit and the energy required to cause charge to flow. Figure 1-22 shows the notation that we use to describe a voltage. There are two parts to this notation: a value (perhaps represented by a variable name) and an assigned direction. The value of a voltage may be positive or negative. The direction of a voltage is given by its polarities $(+,-)$. As a matter of vocabulary, we say that a voltage exists *across* an element. Figure 1-22 shows that there are two ways to label the voltage across an element. The voltage v_{ba} is proportional to the work required to move a positive charge from terminal b to terminal a. On the other hand, the voltage v_{ab} is proportional to the work required to move a positive charge from terminal a to terminal b. We sometimes read v_{ba} as "the voltage at terminal b with respect to terminal a." Similarly, v_{ab} can be read as "the voltage at terminal a with respect to terminal b." Alternatively, we sometimes say that v_{ab} is the voltage drop from terminal a to terminal b. The voltages v_{ab} and v_{ba} are similar but different. They have the same magnitude but different directions. This means that

$$v_{ab} = -v_{ba}$$

When considering v_{ba}, terminal b is called the "+ terminal" and terminal a is called the "− terminal." On ther other hand, when talking about v_{ab}, terminal a is called the "+ terminal" and terminal b is called the "− terminal."

The **voltage** across an element is the work (energy) required to move a unit positive charge from the − terminal to the + terminal. The unit of voltage is the volt, V.

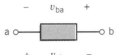

Figure 1-22 Voltage across a circuit element.

We call the current a *through variable* and the voltage an *across variable* because they respectively flow through and appear across the element. *Power* is the product of the through and across variables. For example, in a translational system, $p = $ force $\times$ velocity. In an electrical system

(through) (across)

$$p = iv \qquad i = dq/dt$$

Power measures the rate at which energy is transformed, dw/dt.

Then,

$$v = \frac{p}{i} = \frac{dw/dt}{dq/dt}$$

We may now write the equation for the voltage across the element as

$$v = \frac{dw}{dq} \tag{1-3}$$

where v is voltage, w is energy (or work), and q is charge. A charge of 1 coulomb delivers an energy of 1 joule as it moves through a voltage of 1 volt.

Energy is the capacity to perform work.

For incremental changes in w and q we may rewrite Equation (1-3) as

$$v = \frac{\Delta w}{\Delta q}$$

where Δw is the incremental change in energy.

Example 1-6

A charge of 5 kC passes through an element, and the energy required is 20 MJ. Find the voltage across the element.

Solution

$$v = \frac{\Delta w}{\Delta q} = \frac{20 \times 10^6}{5 \times 10^3} = 4 \text{ kV}$$

Example 1-7

A constant current of 2 amperes flows through an element. The energy to move the current for 1 second is 10 joules. Find the voltage across the element.

Solution

First, find the charge transferred from

$$i = \frac{\Delta q}{\Delta t}$$

or $\Delta q = i \cdot \Delta t = 2 \cdot 1 = 2$ C. Then

$$v = \frac{\Delta w}{\Delta q} = \frac{10}{2} = 5 \text{ V}$$

Example 1-8

The average current in a typical lightning thunderbolt is 2×10^4 A and its typical duration is 0.1s (Williams 1988). The voltage between the clouds and the ground is 5×10^8 V. Determine the total charge transmitted to the earth and the energy released.

Solution

The total charge is

$$Q = \int_0^{0.1} i(t)\, dt$$

$$= \int_0^{0.1} 2 \times 10^4\, dt = 2 \times 10^3 \text{ C}$$

Then the total energy released is

$$w = Q \cdot v = 2 \times 10^3 \times 5 \times 10^8 = 10^{12} \text{ J}$$
$$= 1 \text{ TJ}$$

EXERCISE 1-9

Find the energy required to move 2 coulombs of charge through a constant voltage of 4 volts.
Answer: 8 J

EXERCISE 1-10

A current of $i = 10$ A is delivered to an element for 5 s. Find the energy required to maintain a voltage of 10 volts.
Answer: 500 J

EXERCISE 1-11

A 110-V electric light bulb is connected to the terminals of a bank of storage batteries producing 110 V. The current through the bulb is 6/11 A. Find the energy delivered to the bulb during a 1-s period.
Answer: 60 J

1-6 POWER AND ENERGY

The power and energy delivered to an element are of great importance. For example, the useful output of an electric light bulb can be expressed in terms of power. We know that a 300-watt bulb delivers more light than a 100-watt bulb.

Power is the time rate of expending or absorbing energy.

Thus, we have the equation

$$p = \frac{dw}{dt} \tag{1-4}$$

where p is power in watts, w is energy in joules, and t is time in seconds. The power associated with the current flow through an element is

$$
\begin{aligned}
p &= \frac{dw}{dt} \\
&= \frac{dw}{dq} \cdot \frac{dq}{dt} \\
&= v \cdot i
\end{aligned}
\tag{1-5}
$$

From Eq. 1-5 we see that the power is simply the product of the voltage across an element times the current through the element. The power has units of watts. Clearly, the sign convention we have adopted tells whether the power is delivered to (absorbed by) the element or extracted from (supplied by) the element.

Two circuit variables correspond to each element of a circuit: a voltage and a current. Figure 1-23 shows that there are two different ways to arrange the directions of this current and voltage. In Figure 1-23*a*, the current is directed from the + terminal of the voltage to the − terminal. In contrast, in Figure 1-23*b*, the assigned direction of the current is directed from the − terminal of the voltage to the + terminal.

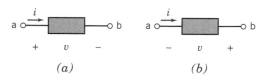

Figure 1-23 (*a*) The passive convention is used for element voltage and current. (*b*) The passive convention is not used.

First consider Figure 1-23*a*. For this situation, the assigned direction of the current is directed from the + terminal of the voltage direction to the − terminal, which is called "the passive convention." In the passive convention, the voltage indicates the work required to move a positive charge in the direction indicated by the current. Accordingly, the power calculated by multiplying the element voltage by the element current

$$p = v\,i \tag{1-6}$$

is the power **absorbed** by the element. (This power is also called "the power dissipated by the element" and "the power delivered *to* the element.") The power absorbed by an element can be either positive or negative. This will depend on the values of the element voltage and current.

Next, consider Figure 1-23*b*. Here the passive convention has not been used. Instead, the direction of the current is from the − terminal of the voltage to the + terminal. In this case, the voltage indicates the work required to move a positive charge in the direction opposite to the direction indicated by the current. Accordingly, the power calculated by multiplying the element voltage by the element current when the passive convention is not used is

$$p = vi$$

and is the power **supplied** by the element. (This power is also called "the power delivered *by* the element.") The power supplied by an element can be either positive or negative. This will depend on the values of the element voltage and current.

The power absorbed by an element and the power supplied by that same element are related by

power absorbed = −power supplied

The rules for the passive convention are summarized in Table 1-5. When the element voltage and current adhere to the passive convention, the energy absorbed by an element can be determined from Eq. 1-4 by rewriting it as

$$dw = p\,dt \tag{1-7}$$

On integrating, we have

$$w = \int_{-\infty}^{t} p\,d\tau \tag{1-8}$$

If the element only receives power for $t \geq t_0$ and we let $t_0 = 0$, then we have

$$w = \int_{0}^{t} p\,d\tau \tag{1-9}$$

Table 1-5

Power Absorbed or Supplied by an Element

Power Absorbed by an Element	Power Supplied by an Element
Since the reference directions of v and i adhere to the passive convention, the power $$p = vi$$ is the power absorbed by the element.	Since the reference directions of v and i do not adhere to the passive convention, the power $$p = vi$$ is the power supplied by the element.

Example 1-9

Let us consider the element shown in Figure 1-23a when $v = 4$ V and $i = 10$ A. Find the power absorbed by the element and the energy absorbed over a 10-s interval.

Solution

The power absorbed by the element is

$$p = vi$$
$$= 4 \cdot 10$$
$$= 40 \text{ W}$$

The energy absorbed by the element is

$$w = \int_0^{10} p \, dt$$
$$= \int_0^{10} 40 \, dt$$
$$= 40 \cdot 10$$
$$= 400 \text{ J}$$

Example 1-10

Consider the element shown in Figure 1-24. The current i and voltage v_{ab} adhere to the passive convention so the power *absorbed* by this element is

$$power\ absorbed = i \cdot v_{ab}$$
$$= 2 \cdot (-4)$$
$$= -8 \text{ W}$$

The current i and voltage v_{ba} do not adhere to the passive convention, so the power *supplied* by this element is

$$power\ supplied = i \cdot v_{ba}$$
$$= 2 \cdot (4)$$
$$= 8 \text{ W}$$

As expected

$$power\ absorbed = -\ power\ supplied$$

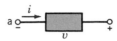

$i = 2$ A

$v_{ba} = 4$ V

a ○ ─── b

$+ \quad v_{ab} = -4$ V $\quad -$

Figure 1-24 The element considered in Example 1-10.

Now let us consider an example when the passive convention is not used. Then $p = vi$ is the power supplied by the element.

Example 1-11

Consider the circuit shown in Figure 1-25 with $v = 8e^{-t}$ V and $i = 20e^{-t}$ A for $t \geq 0$. Find the power supplied by this element and the energy supplied by the element over the first second of operation. We assume that v and i are zero for $t < 0$.

Figure 1-25 An element with the current flowing into the terminal with a negative voltage sign.

Solution

The power supplied is

$$
\begin{aligned}
p &= vi \\
&= (8e^{-t})(20e^{-t}) \\
&= 160e^{-2t} \text{ W}
\end{aligned}
$$

This element is providing energy to the charge flowing through it.

The energy supplied during the first second is

$$
\begin{aligned}
w &= \int_0^1 p\ dt \\
&= \int_0^1 (160e^{-2t})\ dt \\
&= 160 \left. \frac{e^{-2t}}{-2} \right|_0^1 \\
&= \frac{160}{-2}(e^{-2} - 1) \\
&= 80(1 - e^{-2}) \\
&= 69.2 \text{ J}
\end{aligned}
$$

This element is said to be supplying energy.

Example 1-12

An experimenter in a lab assumes that an element is absorbing power as shown in Figure 1-26. She then measures the voltage and current and determines they are $v = +12$ V and $i = -2$ A. Determine if the element is absorbing or supplying energy.

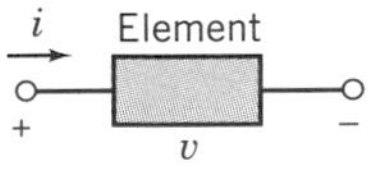

Figure 1-26 This element adheres to the passive convention, so $p = vi$ is the power absorbed by the element.

Solution

The power absorbed by the element is

$$p = vi$$
$$= 12 \cdot (-2)$$
$$= -24 \text{ W}$$

Let's change the current direction as shown in Figure 1-27. Then $i_1 = 2$ A and $v = 24$ V. Since i_1 and v do not adhere to the passive convention $p = i_1 \cdot v = 24$ W is the power supplied by the element.

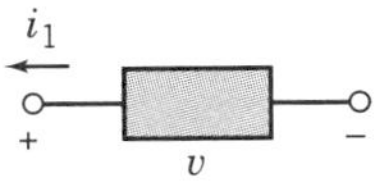

Figure 1-27 The passive convention is not used, so $p = vi$ is the power supplied by the element.

An alternative approach is to note that a negative power absorbed is equivalent to a positive power delivered. Then, if $p = -24$ W absorbed, we have $p = +24$ W delivered.

EXERCISE 1-12

Find the power and the energy for the first 10 seconds of operation of the element shown in Figure 1-23a when $v = 10$ V and $i = 20$ A.
Answer: $p = 200$ W, $w = 2$ kJ

EXERCISE 1-13

Find the power and energy supplied during the first 10 seconds of operation for the element shown in Figure 1-23 when $v = 50e^{-10t}$ V and $i = 5e^{-10t}$ A. The circuit begins operation at $t = 0$.
Answer: $p = 250e^{-20t}$ W, $w = 12.5$ J

EXERCISE 1-14

A hydroelectric power plant can deliver power to remote users. A representation of the power plant is shown in Figure 1-23a. If $v = 100$ kV and $i = 120$ A, find the power supplied by the hydroelectric plant and the energy supplied each day.
Answer: $p = 12$ MW, $w = 1.04$ TJ

1-7 VOLTMETERS AND AMMETERS

Measurements of dc current and voltage are made with direct-reading (analog) or digital meters, as shown in Figure 1-28. A direct-reading meter has an indicating pointer whose angular deflection depends on the magnitude of the variable it is measuring. A digital meter displays a set of digits indicating the measured variable value.

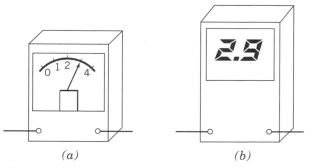

Figure 1-28 *(a)* A direct-reading (analog) meter and *(b)* a digital meter.

An ideal ammeter measures the current flowing through its terminals, as shown in Figure 1-29*a*, and has zero voltage, v_m, across its terminals. An ideal voltmeter measures the voltage across its terminals, as shown in Figure 1-29*b*, and has a terminal current i_m equal to zero. Practical measuring instruments only approximate the ideal conditions. For a practical ammeter the voltage across its terminals is negligibly small. Similarly, the current into the terminal of a voltmeter is usually negligible.

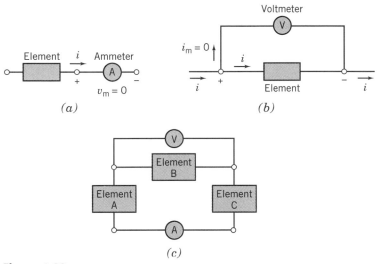

Figure 1-29 *(a)* Ideal ammeter and *(b)* ideal voltmeter. *(c)* A circuit with an ammeter and a voltmeter.

A circuit incorporating three elements, a voltmeter, and an ammeter is shown in Figure 1-29*c*. The voltmeter measures the voltage across element B, and the ammeter measures the current through elements A and C.

1-8 CIRCUIT DESIGN

Design is a purposeful activity in which a designer has in mind an idea about a desired outcome. It is the process of originating circuits and predicting how these circuits will fulfill objectives. Engineering design is the process of producing a set of descriptions of a circuit that satisfy a set of performance requirements and constraints.

The design process can be considered to incorporate three phases: analysis, synthesis, and evaluation. The first task is to diagnose, define, and prepare—that is, to understand the problem and produce an explicit statement of goals. The second task involves finding plausible solutions. The third task concerns judging the validity of solutions relative to the goals and selecting among alternatives. A cycle is implied in which the solution is revised and improved by reexamining the analysis. These three phases form the basis of a framework for planning, organizing, and evolving design projects.

Design is the process of creating a circuit to satisfy a set of goals.

This design process is used in this text and is illustrated by Design Challenges included in each chapter. The design method used here explicitly follows the procedure illustrated in Figure 1-30.

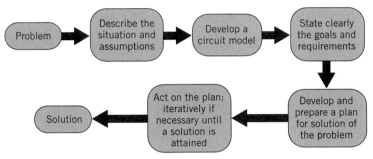

Figure 1-30 The design procedure used for design challenge solutions at the end of each chapter.

1-9 VERIFICATION EXAMPLE

Engineers are frequently called upon to verify that a solution to a problem is indeed correct. For example, proposed solutions to design problems must be checked to confirm if all of the specifications have been satisfied. Computer output must be reviewed to guard against data entry errors. Claims made by vendors must be examined critically.

Engineering students are also called upon to verify the correctness of their work. For example, occasionally just a little time remains at the end of an exam. It is useful to be able to identify quickly those solutions that need more work.

This text includes some examples, called verification examples, that illustrate techniques that are useful for checking the solutions of the kinds of problems discussed in that chapter. At the end of each chapter there are some problems, called verification problems, that provide an opportunity to practice these techniques.

Let's proceed to demonstrate the verification of a calculation provided in a laboratory report.

A laboratory report states the measured values of v and i for the circuit element shown in Figure 1-31 are -5 V and -2 A, respectively. The report states that the power absorbed by the element is 10 W. Verify this statement.

Solution
The circuit shown in Figure 1-31 adheres to the passive sign convention.

Then, the power absorbed is

$$p = vi$$

Figure 1-31 A circuit element with a measured current and voltage.

Substituting v and i, we have

$$P = (-5)(-2)$$
$$= 10 \text{ W}$$

Thus, we have verified that the circuit element is absorbing 10 W.

1-10 DESIGN CHALLENGE SOLUTION

JET VALVE CONTROLLER

Problem

A small, experimental space rocket uses a two-element circuit, as shown in Figure 1D-1, to control a jet valve from point of liftoff at $t = 0$ until expiration of the rocket after 1 minute. The energy that must be supplied by element 1 for the 1-minute period is 40 mJ. Element 1 is a battery to be selected.

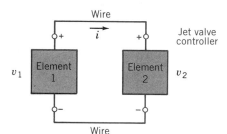

Figure 1D-1 The circuit to control a jet valve for a space rocket.

It is known that $i(t) = De^{-t/60}$ mA for $t \geq 0$, and the voltage across the second element is $v_2(t) = Be^{-t/60}$ V for $t \geq 0$. The maximum magnitude of the current, D, is limited to 1 mA. Determine the required constants D and B, and describe the required battery.

Define the Situation, State the Assumptions, and Develop a Model

1 The current enters the plus terminal of the second element.
2 The current leaves the plus terminal of the first element.
3 The wires are perfect and have no effect on the circuit (they do not absorb energy).
4 The model of the circuit, as shown in Figure 1D-1, assumes that the voltage across the two elements is equal, that is, $v_1 = v_2$.
5 The battery voltage v_1 is $v_1 = Be^{-t/60}$ V where B is the initial voltage of the battery that will discharge exponentially as it supplies energy to the valve.
6 The circuit operates from $t = 0$ to $t = 60$ s.
7 The current is limited, so D $\leq$ 1mA.

The Goal

Determine the energy supplied by the first element for the 1-minute period and then select the constants D and B. Describe the battery selected.

Generate a Plan

First, find $v_1(t)$ and $i(t)$ and then obtain the power, $p_1(t)$, supplied by the first element. Then, using $p_1(t)$, find the energy supplied for the first 60 s.

Prepare the Plan

Goal	Equation	Need	Information
The energy w_1 for the first 60 s	$w_1 = \displaystyle\int_0^{60} p_1(t)\, dt$	$p_1(t)$	v_1 and i known except for constants D and B

Take Action Using the Plan

First, we need $p_1(t)$, so we first calculate

$$p_1(t) = iv_1$$
$$= (De^{-t/60} \times 10^{-3}\ \text{A})\,(Be^{-t/60}\ \text{V})$$
$$= DBe^{-t/30} \times 10^{-3}\ \text{W}$$
$$= DBe^{-t/30}\ \text{mW}$$

Second, we need to find w_1 for the first 60 s as

$$w_1 = \int_0^{60} (DBe^{-t/30} \times 10^{-3})\,dt$$

$$= \frac{DB \times 10^{-3} e^{-t/30}}{-1/30}\bigg|_0^{60}$$

$$= -30DB \times 10^{-3}(e^{-2} - 1)$$

$$= 25.9DB \times 10^{-3}\ \text{J}$$

Since we require $w_1 = 40$ mJ,

$$40 = 25.9DB$$

Next select the limiting value, D = 1, to get

$$B = \frac{40}{(25.9)(1)}$$

$$= 1.54\ \text{V}$$

Thus, we select a 2-V battery so that the magnitude of the current is less than 1 A.

SUMMARY

The uses of electric power are diverse and very important to modern societies. However, electrical science developed slowly over the centuries with many studies concerning the nature of charge. As scientists became aware of the ability to store and control charge, they formulated the idea of a circuit.

A circuit consists of electrical elements linked together in a closed path so that an electric current may flow. The current flowing in the circuit is the time rate of change of the charge.

The SI units are used by today's engineers and scientists. Using decimal prefixes, we may simply express electrical quantities with a wide range of magnitudes.

The voltage across an element is the work required to move a unit of charge through the element. When the passive convention is used to assign the reference directions, the product of the element current and the element voltage gives the power absorbed by the element.

TERMS AND CONCEPTS

Charge   The fundamental unit of matter responsible for electric phenomena.

Electric Circuit   Interconnection of electrical elements in a closed path.

Current   The rate of flow of electric charge. Time rate of change of charge, $i = dq/dt$.

Design   Creating a circuit to satisfy a set of goals.

Direct Current   Current of constant magnitude.

Electricity   Physical phenomena arising from the existence and interaction of electric charge.

Energy   Capacity to perform work.

Power   Energy per unit period of time, $p = dw/dt$.

SI   Système International d'Unités; the International System of Units.

Voltage   Energy required to move a positive charge of 1 coulomb through an element, $v = dw/dq$.

REFERENCES   Chapter 1

Atherton, W.A. *From Compass to Computer,* San Francisco Press, San Francisco, 1984.

Dunsheath, P. *A History of Electrical Engineering,* Faber and Faber, London, 1962.

Metzger, T.L. "Electric Rockets," *Discover,* March 1989, pp. 18–22.

Meyer, H.W. *A History of Electricity and Magnetism,* Burndy Library, Norwalk, Conn., 1972.

Williams, E.R. "The Electrification of Thunderstorms," *Scientific American,* November 1988, pp. 88–99.

Williams, L.P. "André-Marie Ampère," *Scientific American,* January 1989, pp. 90–97.

PROBLEMS

Section 1-3   Electric Circuits and Current Flow

P 1.3-1   A wire carries a constant current of 10 mA. How many coulombs pass a cross section of the wire in 20 s?
Answer: $q = 0.2$ C

P 1.3-2   The total charge that has accumulated on the positive plate of a capacitor varies with time, as shown in Figure P 1.3-2. The charge follows $q = At^2$ for the first 4 s. (a) Find the value of current at $t = 2, 6, 9, 12, 16$ s. (b) Sketch the current $i(t)$ as a function of time. (c) What is the total charge that accumulates on the capacitor after 16 s? (d) What is the average current between $t = 0$ and $t = 16$ s?

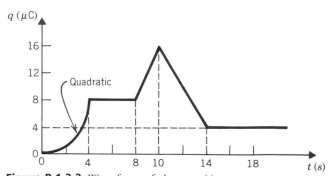

Figure P 1.3-2 Waveform of charge $q(t)$.

P 1.3-3   The current entering the positive terminal of a device varies with time as shown in Figure P 1.3-3. The current is zero at 12 ms and 18 ms. What total amount of charge has passed through the device at $t =$ (a) 5 ms, (b) 10 ms, (c) 15 ms, and (d) 25 ms?

Answer: (a) $q(5 \text{ ms}) = 30 \ \mu\text{C}$
(b) $q(10 \text{ ms}) = 59 \ \mu\text{C}$
(c) $q(15 \text{ ms}) = 58 \ \mu\text{C}$
(d) $q(25 \text{ ms}) = 86 \ \mu\text{C}$

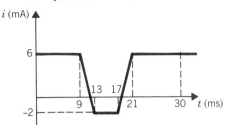

Figure P 1.3-3 Waveform of current $i(t)$.

P 1.3-4 The charge entering the terminal of a device is given by $q(t) = 2k_1 t + k_2 t^2$ C. If $i(0) = 4$ and $i(3) = -4$, find k_1 and k_2.

Answer: $k_1 = 2, \ k_2 = -4/3$

P 1.3-5 In a closed electric circuit, the number of electrons that pass a given point is known to be 10 billion per second. Find the corresponding current flow in amperes.

P 1.3-6 The current flowing in an element is

$$i(t) = \begin{cases} 2 - \dfrac{t}{2} & 0 \le t \le 1 \\ 2 & -3 \le t < 0 \\ \dfrac{1}{2} & \text{otherwise} \end{cases}$$

(a) Sketch $i(t)$ versus time.
(b) Determine the charge entering the element from $t = 0$ to $t = 4$ s.

P 1.3-7 An electroplating bath, as shown in Figure P 1.3-7, is used to plate silver uniformly onto objects such as kitchenware and plates. A current of 600 A flows for 20 minutes, and each coulomb transports 1.118 mg of silver. What is the weight of silver deposited in grams?

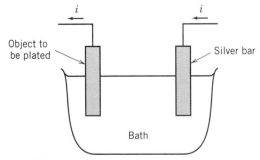

Figure P 1.3-7 An electroplating bath.

Section 1-5 Voltage

P 1.5-1 The current in a wire is given by $i(t) = 12 \sin 2\pi t$ amps for $t > 0$ and $i(t) = 0$ for $t < 0$. (a) Find the total charge passing a cross section of the wire between $t = 0$ and $t = 1/8$ s. (b) If the same current enters the positive terminal of an element whose voltage is given by

$$v = 4 \int_0^t i \, d\tau \text{ volts}$$

find the power delivered to the element.

Answer: (a) $q(t) = -0.559$ C
(b) $p = (24/\pi)(\cos 2\pi t - 1) \sin 2\pi t$ W

P 1.5-2 Modern technology has produced a small 1.5-volt alkaline battery with a nominal stored energy of 150 joules. For how many days will it power a calculator that draws a 2-mA current? Can you see why automatic shutoff is a good idea?

P 1.5-3 An electric range has a constant current of 10 A entering the positive voltage terminal with a voltage of 110 V. The range is operated for 2 hours. (a) Find the charge in coulombs that passes through the range. (b) Find the power absorbed by the range. (c) If electric energy costs 6 cents per kilowatt-hour, determine the cost of operating the range for 2 hours.

P 1.5-4 A walker's cassette tape player uses four AA batteries in series to provide 6 V to the player circuit. The four alkaline battery cells store a total of 200 watt-seconds of energy. If the cassette player is drawing a constant 10 mA from the battery pack, how long will the cassette operate at normal power?

Section 1-6 Power and Energy

P 1.6-1 The charge entering the positive terminal of an element is $q(t) = 10e^{-2t} \ \mu\text{C}$, where t is in seconds. It is determined that $v = 5 \, di/dt$, where v is in volts and i is in amperes. Find (a) the power delivered to the element, $p(t)$, and (b) the energy supplied to the element between 0 and 2 s.

P 1.6-2 A circuit element with a constant voltage of 4 V across it dissipates 4 J of energy in 2 minutes. What is the current through the element?

P 1.6-3 When a battery is being charged, the circuit of Figure 1-22 applies. The current entering the positive terminal of a 10-volt battery varies linearly from 3 to 9 mA between $t = 0$ and $t = 15$ minutes. (a) How much charge passes through the battery during the first 10 minutes? (b) What is the power absorbed by the battery at $t = 5$ minutes and at $t = 10$ minutes? (c) How much energy is supplied to the battery during the entire 15 minutes?

Answer: (a) 3.0 C
(b) 50 mW, 70 mW
(c) 54 J

P 1.6-4 A large storage battery is required for fishing and trolling boats that run on electric motors. One such battery, the Die-Hard, provides 675 A at 12 V for 30 s to start a big boat moving. Once the boat is moving, the battery is able to provide 20 A at 11 V for 200 minutes. (a) Calculate the power provided during the starting period and the trolling period. (b) Calculate the energy provided over the total of the startup and trolling periods.

P 1.6-5 The energy w absorbed by a two-terminal device is shown in Figure P 1.6-5 as a function of time. If the voltage across the device is $v(t) = 12 \cos \pi t$ V, where t is in ms, find the current entering the positive terminal at $t = 1, 3, 6$ ms for the circuit of Figure 1-22.

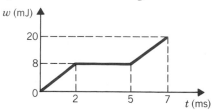

Figure P 1.6-5 Waveform of energy $w(t)$.

P 1.6-6 The current through and voltage across an element vary with time as shown in Figure P 1.6-6. Sketch the power delivered to the element for $t > 0$. What is the total energy delivered to the element between $t = 0$ and $t = 25$ s? The element is absorbing power as shown in Figure 1-22.

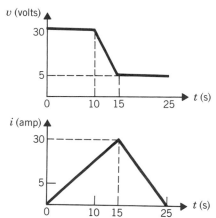

Figure P 1.6-6 (a) Voltage v(t) and (b) current i(t) for an element.

P 1.6-7 An automobile battery is charged with a constant current of 2 A for 5 hours. The terminal voltage of the battery is $v = 11 + 0.5t$ V for $t > 0$, where t is in hours. (a) Find the energy delivered to the battery during the 5 hours and sketch $w(t)$. (b) If electric energy costs 10 cents/kWh, find the cost of charging the battery for 5 hours.

Answer: (b) 1.23 cents

P 1.6-8 The circuit shown in Figure P 1.6-8a has a current source i as shown in Figure P 1.6-8b. The resulting voltage across the circuit is shown in Figure P 1.6-8c. (a) Determine the power $p(t)$ and the energy $w(t)$ absorbed by the element. (b) Sketch $p(t)$ and $w(t)$.

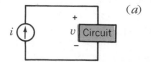

(a)

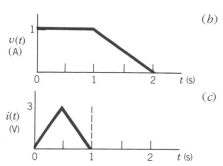

Figure P 1.6-8 (a) Element and its (b) current and (c) voltage.

P 1.6-9 Find the power p supplied by the element shown in Figure P 1.6-9 when $v = 10$ V and $i = 12$ mA.
Answer: $p = 120$ mW

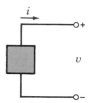

Figure P 1.6-9 An element.

P 1.6-10 A battery is delivering power to an automobile starter. (a) When the current is $i = 10e^{-t}$ A and $v = 12e^{-t}$ V, find the power supplied by the source. (b) Find the energy $w(t)$ delivered by the source to the starter.

P 1.6-11 An element has a current i entering the terminal with a positive voltage v. Using measuring devices, we find that $i = 4e^{-50t}$ mA and $v = 10 - 20e^{-50t}$ V for $t \geq 0$. (a) How much power is absorbed by the element at $t = 10$ ms? (b) How much energy is absorbed by the element in the interval $0 \leq t \leq \infty$?

P 1.6-12 A 12-V car battery is connected so that it supplies power to car headlights while the engine is disconnected. (a) Find the power delivered by the car battery when the current is 1 A. (b) Find the power absorbed by the headlights when the current is 1 A. (c) Find the energy absorbed by the headlights over an interval of 10 minutes.

P 1.6-13 The voltage across a battery is a constant 15 V. The current through the battery (entering the positive terminal) is $i(t) = 10 - 5 \sin 2\pi t$ mA. (a) Find the power $p(t)$ and sketch it in the interval $0 \leq t \leq 1$ s. (b) Calculate the energy received by the battery in the interval $0.5 \leq t \leq 1$ s. This battery is being charged by the current. Interpret the power $p(t)$ delivered to the battery.

P 1.6-14 The voltage across an element is $v(t) = at$ V for $t > 0$. The current entering the positive terminal is $i(t) = be^{-t}$ A for $t > 0$. Find the expression for the energy $w(t)$ for $t > 0$, assuming $w(0) = 0$.

P 1.6-15 Calculate the power absorbed or supplied by each element in Figure P 1.6-15. State in each case whether power is absorbed or supplied.

Figure P 1.6-15

P 1.6-16 Neglecting losses, determine the power that can be developed from Niagara Falls, which has an average height of 168 feet and over which the water flows at 500,000 tons per minute.
Answer: 3.8 GW

P 1.6-17 The battery of a flashlight develops 3 V and the current through the bulb is 200 mA. What power is absorbed by the bulb? Find the energy absorbed by the bulb in a 5-minute period.

P 1.6-18 A Die-Hard 12-V automobile battery can deliver 2×10^6 J over a 10-hour interval. What is the current through the battery?
Answer: 4.63 A

P 1.6-19 A large thunderbolt has a current of 160 A and transfers 10^{20} electrons in 0.1 second. Determine the power generated when the voltage from cloud to ground is 100 kV. (Williams 1988)

P 1.6-20 Sixty years after it was conceived, the electromagnetic launcher known as the railgun is finally coming of age (Metzger 1989). A railgun launches a projectile with a powerful pulse of electric current. The railgun projectile completes the circuit between two parallel rails of conductive material, creating a magnetic field that pushes against the projectile and slings it forward. It continues to accelerate for its entire ride down the launcher. The greater the current or the longer the rails, the higher the speed the projectile can reach.

One experimental railgun uses 14,000 12-V experimental high-current batteries in series to generate 168 kV. They can produce a 5-s jolt of 2.5 MA. Calculate the energy generated and determine whether a 10-gram projectile can be launched to 10 km above the earth.

VERIFICATION PROBLEMS

VP 1-1 A laboratory report states the measured values of v and i are 10 V and -3 A, respectively, for the circuit shown in Figure VP 1-1. The report states the power absorbed by the element is 30 W. Do you agree? Why or why not?

Figure VP 1-1 An element and its current and voltage.

VP 1-2 An electron beam is carrying 10^{15} electrons per second and is accelerated by passing through a voltage of 10 kV. The design engineer for this TV picture tube states that this circuit will result in a beam power of 16 W. Check this calculation and, if necessary, determine the required voltage to achieve 16 W.

DESIGN PROBLEMS

DP 1-1 A current source $i(t)$ is applied to a circuit as shown in Figure DP 1-1a. The current waveform and the power waveform are shown in Figure DP 1-1b. Determine a suitable $v(t)$. Also, calculate and show the resulting energy waveform.

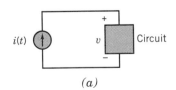

(a)

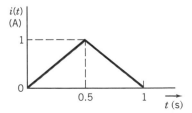

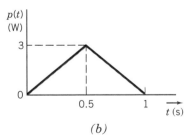

(b)

Figure DP 1-1 (a) Circuit and its (b) current and (c) power.

DP 1-2 Reconsider the Design Challenge Solution when the jet valve operates from $t = 0$ to $t = 4$ minutes when the rocket expires. The required energy supplied by element 1 for the 4-minute period is 35 mJ. Determine suitable values for D and B.

CHAPTER 2

CIRCUIT ELEMENTS

PREVIEW

Electric circuits consist of interconnections of circuit elements. In this chapter we first consider the great example of Thomas A. Edison, who may be called the first electrical engineer. In the 1880s Edison demonstrated his electric lighting system, and electrical engineering began to emerge as a profession.

Circuit elements, such as Edison's electric lamp, may be represented by a model of their behavior described in terms of the terminal current and voltage of each element. In this chapter we consider how electrical elements may be described in order to interconnect them in a circuit and then describe the circuit's overall behavior.

We describe how the resistance of an element depends on the material used in its construction. Then we consider the linear model for the voltage and current of the resistor as developed by Georg Ohm in 1827. In addition we describe the linear model for voltage and current sources. Finally, we describe the models for switches and transducers.

2-1 DESIGN CHALLENGE

TEMPERATURE SENSOR

Currents can be measured easily using ammeters. A temperature sensor such as Analog Devices' AD590 can be used to measure temperature by converting temperature to current [Dorf 1993]. Figure 2D-1 shows a symbol used to represent a temperature sensor. In order for this sensor to operate properly, the voltage v must satisfy the condition

$$4 \text{ volts} \leq v \leq 30 \text{ volts}$$

When this condition is satisfied, the current i, in μA, is numerically equal to the temperature T, in °K. The phrase "numerically equal" indicates that the two variables have the same value but different units.

$$i = k \cdot T \quad \text{where } k = 1 \frac{\mu\text{A}}{°\text{K}}$$

The goal is to design a circuit using the AD590 to measure the temperature of a container of water. In addition to the AD590 and an ammeter, several power supplies and an assortment of standard 2% resistors are available. The power supplies are voltage sources. Power supplies having voltages of 10, 12, 15, 18, or 24 volts are available.

Figure 2D-1 A temperature sensor.

The solution to this problem will have to wait until we know more about the devices used to build electric circuits. We will return to this problem at the end of the chapter.

2-2 | THOMAS A. EDISON—THE FIRST ELECTRICAL ENGINEER

Electrical energy can be converted into heat, light, and mechanical energy. Nature exhibits the first two conversions in lightning. When the battery was invented in 1800, it was expected that the continuous current available could be used to produce the familiar spark. This continuous current was ultimately used in the arc light for illumination.

During the first two-thirds of the nineteenth century, gas was used to provide illumination and was the marvel of the age. By 1870 there were 390 companies manufacturing gas for lighting in the United States.

In 1878, in Ohio, Charles F. Brush invented an electric arc lighting system. Later, a search was undertaken to find a suitable light that would give less glare than an arc light. Thus, research was centered on finding a material that could be heated to incandescence without burning when an electric current was passed through it.

Thomas A. Edison (1847–1931) realized in 1878 that the development of an electric lighting device was of great commercial importance. His success was both a technical and a commercial achievement. Edison presided over a well-staffed laboratory in Menlo Park, New Jersey, for many years. In an interview, Edison stated:

I have an idea that I can make the electric light available for all common uses, and supply it at a trifling cost, compared with that of gas. There is no difficulty about dividing up the electric currents and using small quantities at different points. The trouble is in finding a candle that will give a pleasant light, not too intense, which can be turned on or off as easily as gas [McMahon 1984].

Edison was on the right track, seeking a direct replacement for the gas light that would be easily replaceable and easy to control. As we may recall, Edison eventually found the "ideal" filament for his light bulb in the form of a carbonized thread in 1879. Other materials that could be used for a thin high-resistance thread were later substituted. Edison also undertook a program to improve the vacuum in a light bulb, thus reducing the effects of occluding gases and improving the efficiency of the filament and bulb. Edison is shown in Figure 2-1 with several of his lamps.

With his early engineering insight, Edison saw that a complete lighting system was required, as he stated:

The problem then that I undertook to solve was . . . the production of the multifarious apparatus, methods, and devices, each adapted for use with every other, and all forming a comprehensive system [McMahon 1984].

By 1882 the system had been conceived, designed, patented, and tested at Menlo Park. The equipment for the lighting system was manufactured by the Edison companies. In his prime and during the development of his electric lighting system, Edison depended on investment bankers for funds. By 1882 he had built a system with 12,843 light bulbs within a few blocks of Wall Street, New York. How consistent he was in his approach can be seen by the way he finally marketed the electric light. To emphasize to customers that they

Figure 2-1 Thomas A. Edison in the laboratory at Menlo Park, New Jersey. He is shown with his Edison lamps, discovered in 1879. Courtesy of Edison National Historical Site.

Figure 2-2 Edison's tower of lights, shown at the Philadelphia Exhibition, 1884. Courtesy of Edison National Historical Site.

were buying an old familiar product—light—and not a new unfamiliar product—electricity—the bills were for light-hours rather than kilowatts. Edison exhibited his lights at the Philadelphia Exhibition in 1884, as shown in Figure 2-2. Today, nearly one-quarter of the electricity sold in the United States is devoted to lighting uses.

The great Electrical Exhibition of 1884 provided the site for the first annual meeting of a new U.S. professional society, the American Institute of Electrical Engineers. Edison, among others, called for a college course of study in electrical engineering. By the mid-1880s such courses had begun at the Massachusetts Institute of Technology and Cornell University. By 1890, electrical engineering as a profession and an academic discipline had begun with Edison as role model and leader and then progressed to the formation of a professional society and a college course of study.

2-3 ‖ ENGINEERING AND LINEAR MODELS

One view of engineering describes it as the activity of problem solving under constraints. A somewhat expanded definition follows:

Engineering combines the study of mathematics and natural and social sciences to direct the forces of nature for the benefit of humankind.

The art of engineering is to take a bright idea and, using money, materials, knowledgeable people, and a regard for the environment, produce something the buyer wants at an affordable price.

The ultimate objective of engineering work is the design and production of specific items, often called hardware. Many such devices incorporate electric circuits. The engineer, in the accomplishment of a task,

1 Analyzes the problem.
2 Synthesizes a solution.
3 Evaluates the results and, possibly, resynthesizes a solution.

Another way of looking at the task is to say that the engineer must

1 Understand the problem.
2 Devise a plan to solve the problem.
3 Carry out the plan.
4 Look back and check the solution obtained.
5 Based on the solution, possibly reformulate the initial problem and solve it again.

This method is illustrated by the design challenge solution appearing at the end of each chapter.

Engineers use *models* to represent the elements of an electric circuit. We generate models for manufactured elements and devices in order to manipulate parameters and establish bounds on the devices' operating characteristics. Intuition demands a simple model and ease of solution. This may lead to simplification in the form of aggregations of variables and properties of a device or circuit. Although the model serves to illuminate the real thing, it is *not* the real thing. Thus, a model is constructed to facilitate understanding and enhance prediction. In our work we will construct models of elements and then interconnect them to form a *circuit model*.

A **model** is an object or pattern of objects or an equation that represents an element or circuit.

An automobile ignition circuit is shown in Figure 2-3*a*, and the circuit model of the ignition circuit is shown in Figure 2-3*b*.

The idealized models of electric devices are precisely defined. It is important to distinguish between actual devices and their idealized models, which we call circuit elements.

The goal of circuit analysis is to predict the quantitative electrical behavior of physical circuits. Its aim is to predict and to explain the terminal voltages and terminal currents of the circuit elements and thus the overall operation of the circuit.

A device or element is *linear* if the element's excitation and response satisfy certain properties. Consider the element shown in Figure 2-4. The excitation is the current i, and the response is the voltage v. When the element is subjected to a current i_1, it provides a response v_1. Furthermore, when the element is subjected to a current i_2, it provides a response v_2. For a linear circuit, it is necessary that the excitation $i_1 + i_2$ result in a response $v_1 + v_2$. This is usually called the *principle of superposition*.

Furthermore, it is necessary that the magnitude scale factor be preserved for a linear element. If the element is subjected to an excitation of ki, where k is a constant multiplier, then it is necessary that the response of a linear device be equal to kv. This is called the *property of homogeneity*. A circuit is linear if, and only if, the properties of superposition and homogeneity are satisfied for all excitations and responses.

A **linear element** satisfies the properties of superposition and homogeneity.

Let us restate mathematically the two required properties of a linear circuit, using the arrow notation to imply the transition from excitation to response:

$$i \longrightarrow v$$

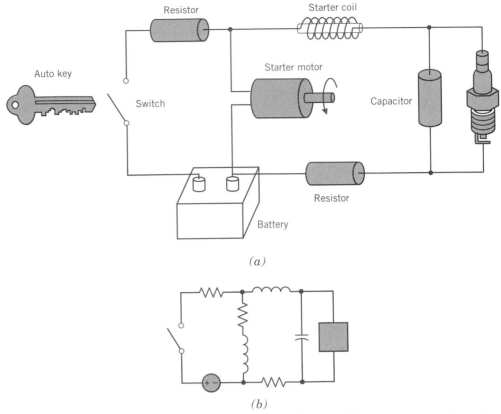

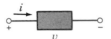

Figure 2-3 *(a)* An automobile ignition circuit and *(b)* a model of the ignition circuit for starting a car.

Figure 2-4 An element with an excitation current i and a response v.

Then we may state the two properties required as follows.

Superposition:

$$i_1 \longrightarrow v_1$$
$$i_2 \longrightarrow v_2$$
$$\text{then} \quad i_1 + i_2 \longrightarrow v_1 + v_2 \tag{2-1}$$

Homogeneity:

$$i \longrightarrow v$$
$$\text{then} \quad ki \longrightarrow kv \tag{2-2}$$

A device that does not satisfy the superposition and homogeneity principles is said to be nonlinear.

Example 2-1

Consider the element represented by the relationship between current and voltage as

$$v = Ri$$

Determine whether this device is linear.

Solution

Since

$$v_1 = Ri_1$$

and

$$v_2 = Ri_2$$

then

$$v_1 + v_2 = Ri_1 + Ri_2$$
$$= R(i_1 + i_2)$$

Thus, it satisfies the property of superposition. Since

$$v_1 = Ri_1$$

we have for an excitation $i_2 = ki_1$

$$v_2 = Ri_2$$
$$= Rki_1$$

Therefore,

$$v_2 = kv_1$$

satisfies the principle of homogeneity. Because the element satisfies the properties of superposition and homogeneity, it is linear.

Example 2-2

Now let us consider an element represented by the relationship between current and voltage:

$$v = i^2$$

Determine whether this device is linear.

Solution

Since

$$v_1 = i_1^2 \quad \checkmark$$

and

$$v_2 = i_2^2 \quad \checkmark$$

then

$$v_1 + v_2 = i_1^2 + i_2^2 \quad \checkmark$$

Also, we have

$$(i_1 + i_2)^2 = i_1^2 + 2i_1 i_2 + i_2^2$$

Therefore,

$$i_1^2 + i_2^2 \neq (i_1 + i_2)^2$$

and the device is nonlinear.

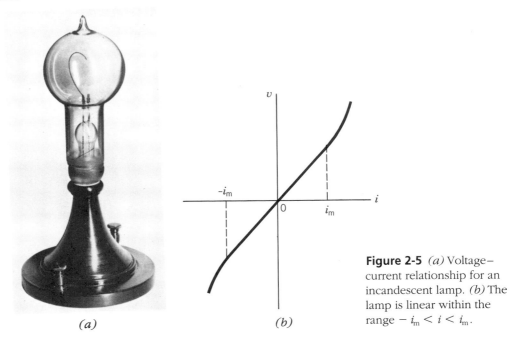

Figure 2-5 (a) Voltage–current relationship for an incandescent lamp. (b) The lamp is linear within the range $-i_m < i < i_m$.

(a)   (b)

The distinguishing feature of an electric circuit component is that its behavior is described in terms of a voltage–current relation. The "v–i" characteristic may be obtained experimentally or from physical principles. Although no device is exactly linear for all values of current, we often assume a range of linear operation. Edison's first commercially available lamp is shown in Figure 2-5a. For example, consider the v–i characteristic of the incandescent lamp, shown in Figure 2-5b. The lamp is essentially linear for the range

$$-i_m < i < i_m$$

Note that i_m indicates the range of current for linearity.

In this book we will be concerned only with linear models of circuit components. In other words, only linear circuits will be considered. Of course, we recognize that these linear models are accurate representations for only some limited range of operation.

EXERCISE 2-1

Determine whether the following element is linear.

$$v = \frac{di}{dt}$$

EXERCISE 2-2

Show that the following element is nonlinear.

$$v = Ri, \quad i \geq 0$$
$$v = 0, \quad i < 0$$

EXERCISE 2-3

The voltage–current relationship for two elements is shown in Figure E 2-3. Determine a linear model for each element and indicate the range of linearity. Assume that the elements of Fig. E 2-3 normally operate around $i = 0$.

Answers: (a) $v = 2.5i$ for $-1 < i < 1$

(b) $v = -1.333i$ for $-1.5 < i < 1.5$

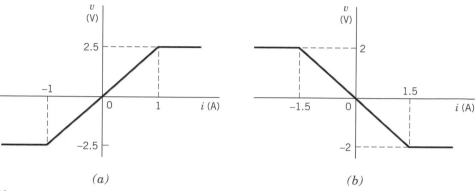

Figure E 2-3

2-4 | ACTIVE AND PASSIVE CIRCUIT ELEMENTS

We may classify circuit elements in two categories, *passive* and *active,* by determining whether they absorb energy or supply energy. An element is said to be passive if the total energy delivered to it from the rest of the circuit is always nonnegative (zero or positive). Then for a passive element, with the current flowing into the + terminal as shown in Figure 2-6*a,* this means that

$$w = \int_{-\infty}^{t} vi \, d\tau \geq 0 \tag{2-3}$$

for all values of *t.*

A **passive element** absorbs energy.

An element is said to be *active* if it is capable of delivering energy. Thus, an active element violates Eq. 2-3 when it is represented by Figure 2-6*a.* In other words, an active element is one that is capable of generating energy. Active elements are potential sources of energy, whereas passive elements are sinks or absorbers of energy. Examples of active elements include batteries and generators. A diagrammatic representation of an active element is shown in Figure 2-6*b.* Note that the current flows into the negative terminal and out of the positive terminal. We then have

$$w = \int_{-\infty}^{t} vi \, d\tau \geq 0 \tag{2-4}$$

for at least one value of *t.*

In this case the element of Figure 2-6*b* is said to be a source of energy.

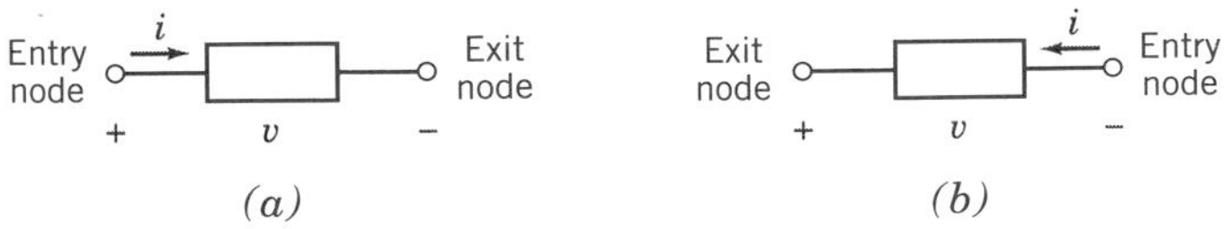

Figure 2-6 *(a)* The entry node of the current *i* is the positive node of the voltage *v*, *(b)* the entry node of the current *i* is the negative node of the voltage *v*. The current flows from the entry node to the exit node.

An **active element** is capable of supplying energy.

Example 2-3

A circuit has an element represented by Figure 2-6*b* where the current is a constant 5 A and the voltage is a constant 6 V. Find the energy supplied over the time interval 0 to *T*.

Solution

Since the current enters the negative terminal, we have an active element, and

$$w = \int_0^T (6)(5) \, d\tau$$

$$w = 30T \quad \text{J}$$

Thus, the device is a generator or an active element, in this case a dc battery.

A collection of circuit elements or components is shown in Figure 2-7. Useful circuits include both active and passive elements that are assembled into a circuit.

EXERCISE 2-4

An active element is shown in Figure 2-6*b*. Find the energy supplied during the period from 0 to 10s when

$$v = 3e^{-t} \text{ V}$$

and

$$i = 2e^{-t} \text{ A}$$

Answer: $3(1 - e^{-20})$ J = 3 J

EXERCISE 2-5

The current entering the positive terminal of an element is $i = -3e^{-2t}$ A and the voltage across the element is $v = 5i$ V for $t \geq 0$. Find whether the energy is delivered to or supplied by the element between $t = 0$ and $t = 2$ s. Determine the total energy between 0 and 2 s. Is the element active or passive? Assume $i = v = 0$ for $t < 0$.

Partial Answer: The element is passive.

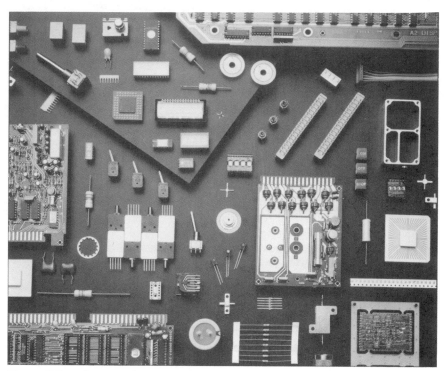

Figure 2-7 Collection of active and passive circuit elements used for an electrical circuit. Courtesy of Hewlett-Packard Co.

2-5 | RESISTORS

The ability of a material to resist the flow of charge is called its *resistivity, ρ*. Materials that are good electrical insulators have a high value of resistivity. Materials that are good conductors of electric current have low values of resistivity. Resistivity values for selected materials are given in Table 2-1. Copper is commonly used for wires since it permits current to flow relatively unimpeded. Silicon is commonly used to provide resistance in semiconductor electric circuits. Polystyrene is used as an insulator.

Table 2-1
Resistivities of Selected Materials

Material	Resistivity ρ (ohm-cm)
Polystyrene	1×10^{18}
Silicon	2.3×10^{5}
Carbon	4×10^{-3}
Aluminum	2.7×10^{-6}
Copper	1.7×10^{-6}

Figure 2-8 Georg Simon Ohm (1787–1854), who determined Ohm's law in 1827. The ohm was chosen as the unit of electrical resistance in his honor.

Resistance is the physical property of an element or device that impedes the flow of current; it is represented by the symbol R.

Georg Simon Ohm was able to show that the current flow in a circuit composed of a battery and a conducting wire of uniform cross section could be expressed as

$$i = \frac{Av}{\rho L} \tag{2-5}$$

where A is the cross-sectional area, ρ the resistivity, L the length, and v the voltage across the wire element. Ohm, who is shown in Figure 2-8, defined the constant resistance R as

$$R = \frac{\rho L}{A} \tag{2-6}$$

Ohm's law, which related the voltage and current, was published in 1827 as

$$v = Ri \tag{2-7}$$

The unit of resistance R was named the ohm in honor of Ohm and is usually abbreviated by the symbol Ω (capital omega), where $1\ \Omega = 1$ V/A. The resistance of a 10-m length of common TV cable is 2 mΩ.

An element that has a resistance R is called a *resistor*. A resistor is represented by the two-terminal symbol shown in Figure 2-9. Ohm's law, Eq. 2-7, requires that the i-versus-v relationship be linear. As shown in Figure 2-10, a resistor may become nonlinear outside its normal rated range of operation. We will assume that a resistor is linear unless stated otherwise. Thus, we will use a linear model of the resistor as represented by Ohm's law.

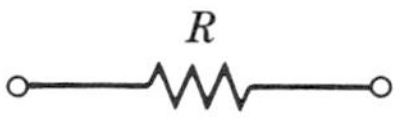

$$R$$
○—◇◇◇—○

Figure 2-9 Symbol for a resistor having a resistance of R ohms.

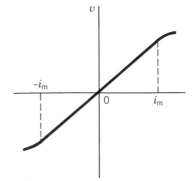

Figure 2-10 A resistor operating within its specified current range, $\pm\, i_{\mathrm{m}}$, can be modeled by Ohm's law.

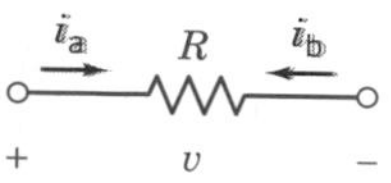

Figure 2-11 A resistor with element current and element voltage.

In Figure 2-11 the element current and element voltage of a resistor are labeled. The relationship between the directions of this current and voltage is important. The voltage direction marks one resistor terminal + and the other −. The current flows from the terminal marked + to the terminal marked −. This relationship between the current and voltage reference directions is a convention called the passive convention. Ohm's law states that when the element voltage and the element current adhere to the passive convention, then

$$v = Ri \qquad (2\text{-}8)$$

Consider Figure 2-11. The element currents i_a and i_b are the same except for the assigned direction, so

$$i_a = -i_b$$

The element current i_a and the element voltage v adhere to the passive convention,

$$v = Ri_a$$

Replacing i_a by $-i_b$ gives

$$v = -Ri_b$$

There is a minus sign in this equation because the element current i_b and the element voltage v do not adhere to the passive convention. We must pay attention to the current direction so that we don't overlook this minus sign.

Ohm's law, equation 2-8, can also be written as

$$i = Gv \qquad (2\text{-}9)$$

where G denotes the *conductance* in siemens (S) and is the reciprocal of R; that is, $G = 1/R$. Many engineers denote the units of conductance as mhos with the symbol ℧, which is an inverted omega (mho is ohm spelled backward). However, we will use SI units and retain siemens as the units for conductance.

Most discrete resistors fall into one of four basic categories: carbon composition, carbon film, metal film, or wirewound. Carbon composition resistors have been in use for nearly 100 years and are still popular. Carbon film resistors have supplanted carbon composition resistors for many general-purpose uses because of their lower cost and better tolerances. Two wirewound resistors are shown in Figure 2-12.

Resistors are sensitive to temperature change from an assumed ambient temperature of 27°C. The temperature sensitivity is defined as

$$k = \frac{1}{R}\frac{dR}{dT}$$

where R is the resistance, T is the temperature in degrees Celsius, and the units of k are parts per million per degree Celsius. A typical carbon film resistor has a coefficient of −400 ppm/°C.

Thick-film resistors, as shown in Figure 2-13, are used in circuits because of their low cost and small size. General-purpose resistors are available in standard values for tolerances of 2, 5, 10, and 20 percent. Carbon composition resistors and some wirewounds have a color code with three to five bands. A color code is a system of standard colors

(a)

(b)

Figure 2-12 *(a)* Wirewound resistor with an adjustable center tap. *(b)* Wirewound resistor with a fixed tap. Courtesy of Dale Electronics.

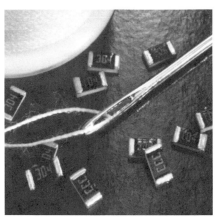

Figure 2-13 Small thick-film resistor chips used for miniaturized circuits. Courtesy of Corning Electronics.

adopted for identification of the resistance of resistors. Figure 2-14 shows a metal film resistor with its color bands. This is a 1/4-watt resistor, implying that it should be operated at or below 1/4 watt of power delivered to it. The normal range of resistors is from less than 1 ohm to 10 megohms. Typical values readily available are $R = N10^n$ ohms, where $n = 1$, 2, . . . , 8.[1]

Figure 2-14 A 1/4-watt metal film resistor. The body of the resistor is 6-mm long. Courtesy of Dale Electronics.

The power delivered to a resistor (when the passive convention is used) is

$$p = vi$$
$$= v\left(\frac{v}{R}\right)$$
$$= \frac{v^2}{R} \tag{2-10a}$$

Alternatively, since $v = iR$, we can write the equation for power as

$$p = vi$$
$$= (iR)i$$
$$= i^2R \tag{2-10b}$$

Thus, the power is expressed as a nonlinear function of the current i through the resistor or of the voltage v across it.

_____

[1] See Appendix F for a description of color code and the standard values for N.

Recall the definition of a passive element as one for which the energy absorbed is always nonnegative. The equation for energy delivered to a resistor is

$$w = \int_{-\infty}^{t} p \, d\tau$$

$$= \int_{-\infty}^{t} i^2 R \, d\tau \tag{2-11}$$

Since i^2 is always positive, the energy is always positive and the resistor is a passive element.

Resistance is a measure of an element's ability to dissipate power irreversibly.

Example 2-4

Let us devise a model for a car battery when the lights are left on and the engine is off. We have all experienced or seen a car parked with its lights on. If we leave the car for a period, the battery will "run down" or "go dead." An auto battery is a 12-V constant-voltage source, and the light bulb can be modeled by a resistor of 6 ohms. The circuit is shown in Figure 2-15. Let us find the current i, the power p, and the energy supplied by the battery for a 4-hour period.

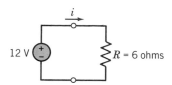

Figure 2-15 Model of a car battery and the headlight lamp.

Solution

According to Ohm's law, Eq. 2-8, we have

$$v = Ri$$

Since $v = 12$ V and $R = 6 \, \Omega$, we have $i = 2$ A.

In order to find the power delivered by the battery, we use

$$p = vi$$
$$= 12(2)$$
$$= 24 \text{ W}$$

Finally, the energy delivered in the 4-hour period is

$$w = \int_{0}^{t} p \, d\tau$$

$$= 24t$$
$$= 24(60 \times 60 \times 4)$$
$$= 3.46 \times 10^5 \text{ J}$$

Since the battery has a finite amount of stored energy, it will deliver this energy and eventually be unable to deliver further energy without recharging. We then say the battery is run down or dead until recharged. A typical auto battery may store 10^6 J in a fully charged condition.

EXERCISE 2-6

Find the power absorbed by a 100-ohm resistor when it is connected directly across a constant 10-V source.
Answer: 1 W

EXERCISE 2-7

A voltage source $v = 10 \cos t$ V is connected across a resistor of 10 ohms. Find the power delivered to the resistor.
Answer: $10 \cos^2 t$ W

EXERCISE 2-8

The current entering a resistor terminal is 1 mA, and the resistance is 1 kΩ. Find (a) the voltage across the resistor, (b) the power absorbed by the resistor, and (c) the conductance of the resistor.
Partial Answer: (a) 1 V

EXERCISE 2-9

Find the minimum required power rating of a resistor when $R = 10$ kΩ and $v = 100$ V.
Answer: 1 W

2-6 INDEPENDENT SOURCES

Some devices are intended to supply energy to a circuit. These devices are called sources. Sources are categorized as being one of two types: voltage sources and current sources. Figure 2-16a shows the symbol that is used to represent a voltage source. The voltage of a voltage source is specified, but the current is determined by the rest of the circuit. A voltage source is described by specifying the function $v(t)$, for example,

$$v(t) = 12 \cos 1000t \quad \text{or} \quad v(t) = 9 \quad \text{or} \quad v(t) = 12 - 2t$$

An active two-terminal element that supplies energy to a circuit is a *source* of energy. An independent source is one for which one variable is independent of the other variable. Thus, an *independent voltage source* provides a specified voltage independent of the current through it and is independent of any other circuit variable.

A **source** is a voltage or current generator capable of supplying energy to a circuit.

An *independent current source* provides a current independent of the voltage across the source element and is independent of any other circuit variable. Thus, when we say a source is independent, we mean it is independent of any other voltage or current in the circuit.

An **independent source** is a voltage or current generator not dependent on other circuit variables.

Suppose the voltage source is a battery and

$$v(t) = 9 \text{ volts}$$

The voltage of this battery is known to be 9 volts regardless of the circuit in which the battery is used. In contrast, the current of the voltage source is not known and depends on the circuit in which the source is used. The current could be 6 amps when the voltage source is connected to one circuit and 6 milliamps when the voltage source is connected to another circuit.

Figure 2-16b shows the symbol that is used to represent a current source. The current of a current source is specified, but the voltage is determined by the rest of the circuit. A current source is described by specifying the function $i(t)$, for example,

$$i(t) = 6 \sin 500t \quad \text{or} \quad i(t) = -0.25 \quad \text{or} \quad i(t) = t + 8$$

A current source specified by $i(t) = -0.25$ milliamps will have a current of -0.25 milliamps in any circuit in which it is used. The voltage across this current source will depend on the particular circuit.

(a)   *(b)*   **Figure 2-16** *(a)* Voltage source. *(b)* Current source.

The preceding paragraphs have ignored some complexities in order to give a simple description of the way sources work. The voltage across a 9-volt battery may not actually be 9 volts. This voltage depends on the age of the battery, the temperature, variations in manufacturing, and the battery current. It is useful to make a distinction between real sources, such as batteries, and the simple voltage and current sources described in the preceding paragraphs. It would be *ideal* if the real sources worked like these simple sources. Indeed, the word *ideal* is used to make this distinction. The simple sources described in the previous paragraph are called the **ideal voltage source** and the **ideal current source**.

The voltage of an **ideal voltage source** is given to be a specified function, say $v(t)$. The current is determined by the rest of the circuit.

The current of an **ideal current source** is given to be a specified function, say $i(t)$. The voltage is determined by the rest of the circuit.

An **ideal source** is a voltage or a current generator independent of the current through the voltage source or the voltage across the current source.

Example 2-5

Consider the plight of the engineer who needs to analyze a circuit containing a 9-volt battery. Is it really necessary for this engineer to include the dependence of battery voltage on the age of the battery, the temperature, variations in manufacturing, and the battery current in this analysis? Hopefully not. We expect the battery to act enough like an ideal 9-volt voltage source that the differences can be ignored. In this case it is said that the battery is **modeled** as an ideal voltage source.

To be specific, consider a battery specified by the plot of voltage versus current shown in Figure 2-17a. This plot indicates that the battery voltage will be $v = 9$ volts when $i \leq 10$

milliamps. As the current increases above 10 milliamps the voltage decreases from 9 volts. When $i \leq 10$ milliamps the dependence of the battery voltage on the battery current can be ignored and the battery can be modeled as an independent voltage source.

Suppose a resistor is connected across the terminals of the battery as shown in Figure 2-17b. The battery current will be

$$i = \frac{v}{R} \tag{2-12}$$

The relationship between v and i shown in Figure 2-17a complicates this equation. This complication can be safely ignored when $i \leq 10$ milliamps. When the battery is modeled as an ideal 9-volt voltage source, the voltage source current is given by

$$i = \frac{9}{R} \tag{2-13}$$

The distinction between these two equations is important. Equation 2-12, involving the v–i relationship shown in Figure 2-17a, is more accurate but also more complicated. Equation 2-13 is simpler but may be inaccurate.

Suppose that $R = 1000$ ohms. Equation 2-13 gives the current of the ideal voltage source:

$$i = \frac{9 \text{ volts}}{1000 \text{ ohms}} = 9 \text{ milliamps} \tag{2-14}$$

Since this current is less than 10 milliamps, the ideal voltage source is a good model for the battery and it is reasonable to expect that the battery current is 9 milliamps.

Suppose instead that $R = 600$ ohms. Once again, Equation 2-13 gives the current of the ideal voltage source.

$$i = \frac{9 \text{ volts}}{600 \text{ ohms}} = 15 \text{ milliamps} \tag{2-15}$$

Since this current is greater than 10 milliamps, the ideal voltage source is not a good model for the battery. In this case it is reasonable to expect that the battery current is different than the current for the ideal voltage source.

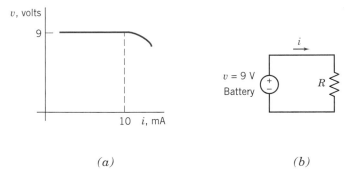

(a)   (b)

Figure 2-17 (a) A plot of battery voltage versus battery current. (b) The battery is modeled as an independent voltage source.

Engineers frequently face a trade-off when selecting a model for a device. Simple models are easy to work with but may not be accurate. Accurate models are usually more complicated and harder to use. The conventional wisdom suggests that simple models be used first. The results obtained using the models must be checked to verify that use of these simple models is appropriate. More accurate models are used when necessary.

The **short circuit** and **open circuit** are special cases of ideal sources. A **short circuit** is an ideal voltage source having $v(t) = 0$. The current in a short circuit is determined by the rest of the circuit. An **open circuit** is an ideal current source having $i(t) = 0$. The voltage across an open circuit is determined by the rest of the circuit. Figure 2-18 shows the symbols used to represent the short circuit and the open circuit. Notice that the power absorbed by each of these devices is zero.

(a) (b) **Figure 2-18** *(a)* Open circuit. *(b)* Short circuit.

Open and short circuits can be added to a circuit without disturbing the branch currents and voltages of all the other devices in the circuit. Figure 2-21 shows how this can be done. Figure 2-21a shows an example circuit. In Figure 2-21b an open circuit and a short circuit have been added to this example circuit. The open circuit was connected between two nodes of the original circuit. In contrast, the short circuit was added by cutting a wire and inserting the short circuit. Adding open circuits and short circuits to a network in this way does not change the network.

Open circuits and short circuits can also be described as special cases of resistors. A resistor with resistance $R = 0$ ($G = \infty$) is a short circuit. A resistor with conductance $G = 0$ ($R = \infty$) is an open circuit.

2.7   VOLTMETERS AND AMMETERS

Measurements of dc current and voltage are made with direct-reading (analog) or digital meters, as shown in Figure 2-19. A direct-reading meter has an indicating pointer whose angular deflection depends on the magnitude of the variable it is measuring. A digital meter displays a set of digits indicating the measured variable value.

To measure a voltage or current, a meter is connected to a circuit using terminals called probes. These probes are color coded to indicate the reference direction of the variable being measured. Frequently, meter probes are colored red and black. In this text, the probes will be shown as red and black. An ideal voltmeter measures the voltage from the red to the black probe. The red terminal is the positive terminal and the black terminal is the negative terminal (see Figure 2-20b).

An ideal ammeter measures the current flowing through its terminals, as shown in Figure 2-20a and has zero voltage, v_m, across its terminals. An ideal voltmeter measures the voltage across its terminals, as shown in Figure 2-20b, and has terminal current, i_m, equal to zero. Practical measuring instruments only approximate the ideal conditions. For a practical ammeter the voltage across its terminals is usually negligibly small. Similarly, the current into a voltmeter is usually negligible.

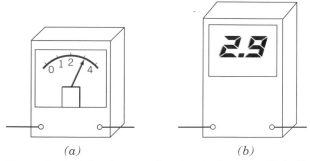

Figure 2-19 *(a)* A direct-reading (analog) meter. *(b)* A digital meter.

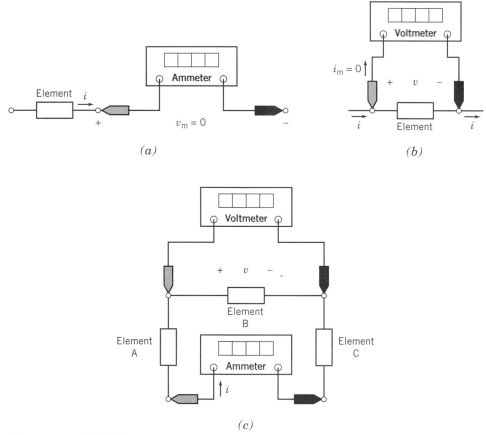

Figure 2-20 *(a)* Ideal ammeter. *(b)* Ideal voltmeter. *(c)* A circuit with ammeter and voltmeter.

Ideal voltmeters act like open circuits, and ideal ammeters act like short circuits. In other words, the model of an ideal voltmeter is an open circuit and the model of an ideal ammeter is a short circuit. Consider the circuit of Figure 2-21*a* and then add an open circuit with a voltage v and a short circuit with a current i as shown in Figure 2-21*b*. In Figure 2-21*c* the open circuit has been replaced by a voltmeter and the short circuit has been replaced by an ammeter. The voltmeter will measure the voltage labeled v in Figure 2-21*b* while the ammeter will measure the current labeled i. Notice that Figure 2-21*c* could be obtained from Figure 2-21*a* by adding a voltmeter and an ammeter. Ideally, adding the voltmeter and ammeter in this way does not disturb the circuit. One more interpretation of Figure 2-21 is useful. Figure 2-21*b* could be formed from Figure 2-21*c* by replacing the voltmeter and the ammeter by their (ideal) models.

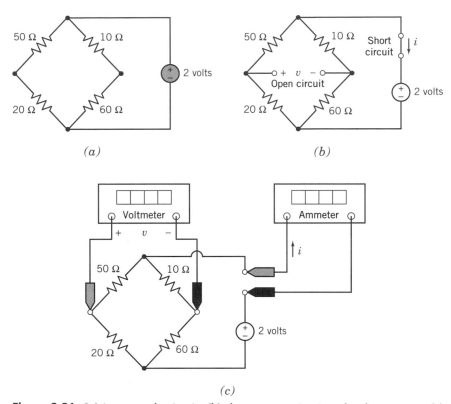

Figure 2-21 *(a)* An example circuit, *(b)* plus an open circuit and a short circuit. *(c)* The open circuit is replaced by a voltmeter and the short circuit is replaced by an ammeter. All resistances are in ohms.

The reference direction is an important part of an element voltage or element current. Figures 2-22 and 2-23 show that attention must be paid to reference directions when measuring an element voltage or element current. Figure 2-22*a* shows a voltmeter. Voltmeters have two color-coded probes. This color coding indicates the reference direction of the voltage being measured. In Figures 2-22*b* and 2-20*c* the voltmeter is used to measure the voltage across the 6-kΩ resistor. When the voltmeter is connected to the circuit as shown in Figure 2-22*b*, the voltmeter measures v_a, with + on the left, at the red probe. When the voltmeter probes are interchanged as shown in Figure 2-22*c*, the voltmeter measures v_b, with + on the right, again at the red probe. Note $v_b = -v_a$.

Figure 2-23*a* shows an ammeter. Ammeters have two color-coded probes. This color coding indicates the reference direction of the current being measured. In Figures 2-23*b,c* the ammeter is used to measure the current in the 6-kΩ resistor. When the ammeter is connected to the circuit as shown in Figure 2-23*b*, the ammeter measures i_a, directed from the red probe toward the black probe. When the ammeter probes are interchanged as shown in Figure 2-23*c*, the ammeter measures i_b, again directed from the red probe toward the black probe. Note $i_b = -i_a$.

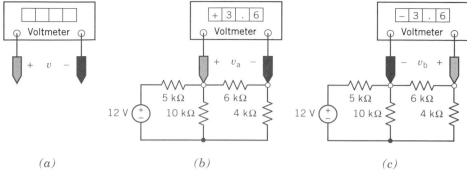

(a)   (b)   (c)

Figure 2-22 *(a)* The correspondence between the color-coded probes of the voltmeter and the reference direction of the measured voltage. In *(b)* the + sign of v_a is on the left, while in *(c)* the + sign of v_b is on the right.

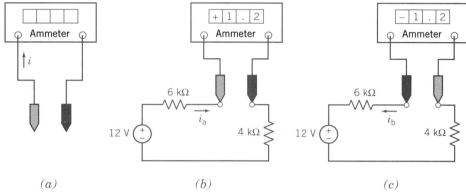

(a)   (b)   (c)

Figure 2-23 *(a)* The correspondence between the color-coded probes of the ammeter and the reference direction of the measured current. In *(b)* the current i_a is directed to the right, while in *(c)* the current i_b is directed to the left.

2-8 ‖ DEPENDENT SOURCES

Some devices, such as transistors and amplifiers, act like controlled sources. For example, the output voltage of an amplifier is controlled by the input voltage of that amplifier. Such devices can be modeled using dependent sources. Dependent sources consist of two elements: the controlling element and the controlled element. The controlling element is either an open circuit or a short circuit. The controlled element is either a voltage source or a current source. There are four types of dependent source that correspond to the four ways of choosing a controlling element and a controlled element. These four dependent sources are called the voltage-controlled voltage source (VCVS), current-controlled voltage source (CCVS), voltage-controlled current source (VCCS), and current-controlled current source (CCCS). The symbols that represent dependent sources are shown in Table 2-2. Notice that the symbol for the controlled element is different from the symbol used to represent independent voltage and current sources. Indeed the terms **independent voltage source** and **independent current source** are used to emphasize the distinction between these sources and dependent sources.

Table 2-2
Dependent Sources

Description	Symbol	
Current-Controlled Voltage Source (CCVS) r is the gain of the CCVS. r has units of volts/ampere.	$v_c = 0$ $+$ $-$ $\downarrow i_c$	$\downarrow i_d$ $v_d = ri_c$
Voltage-Controlled Voltage Source (VCVS) b is the gain of the VCVS. b has units of volts/volt.	$\downarrow i_c = 0$ $+$ v_c $-$	$\downarrow i_d$ $v_d = bv_c$
Voltage-Controlled Current Source (VCCS) g is the gain of the VCCS. g has units of amperes/ volt.	$\downarrow i_c = 0$ $+$ v_c $-$	$+$ v_d $-$ $i_d = gv_c$
Current-Controlled Current Source (CCCS) d is the gain of the CCCS. d has units of amperes/ampere.	$+$ $v_c = 0$ $-$ $\downarrow i_c$	$+$ v_d $-$ $i_d = di_c$

A **dependent source** is a voltage or current generator that depends on another circuit variable.

Consider the CCVS shown in Table 2-2. The controlling element is a short circuit. The element current and voltage of the controlling element are denoted as i_c and v_c. The voltage across a short circuit is zero, so $v_c = 0$. The short circuit current, i_c, is the controlling signal of this dependent source. The controlled element is a voltage source. The element current and voltage of the controlled element are denoted as i_d and v_d. The voltage.v_d is controlled by i_c:

$$v_d = ri_c$$

The constant r is called the gain of the dependent source. The current i_d, like the current in any voltage source, is determined by the rest of the circuit.

Next consider the VCVS shown in Table 2-2. The controlling element is an open circuit. The current in an open circuit is zero, so $i_c = 0$. The open circuit voltage, v_c, is the controlling signal of this dependent source. The controlled element is a voltage source.

The voltage v_d is controlled by v_c:

$$v_d = bv_c$$

The constant b is called the gain of the dependent source. The current i_d is determined by the rest of the circuit.

The controlling element of the VCCS shown in Table 2-2 is an open circuit. The current in this open circuit is $i_c = 0$. The open circuit voltage, v_c, is the controlling signal of this dependent source. The controlled element is a current source. The current i_d is controlled by v_c:

$$i_d = gv_c$$

The constant g is called the gain of the VCCS. The voltage v_d, like the voltage across any current source, is determined by the rest of the circuit.

The controlling element of the CCCS shown in Table 2-2 is a short circuit. The voltage across this open circuit is $v_c = 0$. The short circuit current, i_c, is the controlling signal of this dependent source. The controlled element is a current source. The current i_d is controlled by i_c:

$$i_d = di_c$$

The constant d is called the gain of the CCCS. The voltage v_d, like the voltage across any current source, is determined by the rest of the circuit.

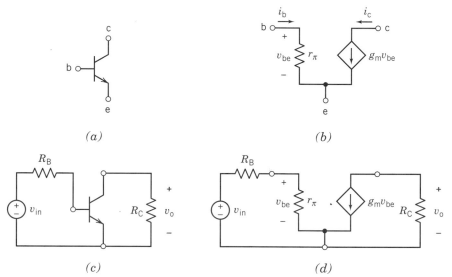

Figure 2-24 (*a*) A transistor. (*b*) A model of the transistor. (*c*) A transistor amplifier. (*d*) A model of the transistor amplifier.

Figure 2-24 illustrates the use of dependent sources to model electronic devices. In certain circumstances, the behavior of the transistor shown in Figure 2-24*a* can be represented using the model shown in Figure 2-24*b*. This model consists of a dependent source and a resistor. The controlling element of the dependent source is an open circuit connected across the resistor. The controlling voltage is v_{be}. The gain of the dependent source is g_m. The dependent source is used in this model to represent a property of the transistor, namely, that the current i_c is proportional to the voltage v_{be}, that is,

$$i_c = g_m v_{be}$$

where g_m has units of amperes/volt. Figures 2-24*c,d* illustrate the utility of this model. Figure 2-24*d* is obtained from Figure 2-24*c* by replacing the transistor by the transistor model. The voltage gain of the transistor amplifier shown in Figure 2-24*c* is defined as

$$A = \frac{v_o}{v_{in}}$$

This gain can be calculated by analyzing Figure 2-24*d* instead of Figure 2-24*c*. Analysis of circuits such as Figure 2-24*d* will be discussed in Chapter 3.

Example 2-6

Determine the power absorbed or supplied by each element in the circuit shown in Figure 2-25. Show that the sum of the power delivered equals the sum of the power absorbed.

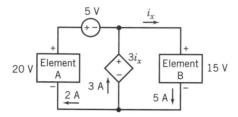

Figure 2-25 Circuit with a dependent source.

Solution

The voltage and current of element B adhere to the passive convention. The power absorbed by element B is

$$P_B = 15 \cdot 5 = 75 \text{ W absorbed}$$

The voltage and current of element A do not adhere to the passive convention. The power delivered by element A is

$$P_A = 20 \cdot 2 = 40 \text{ W delivered}$$

The 5-V source has the 2-A current entering the + terminal and thus absorbing power as

$$P_{5V} = 5 \cdot 2 = 10 \text{ W absorbed}$$

Finally, the dependent source has the 3 A current leave its plus terminal, and it is delivering power as

$$P_x = 3 \cdot 3i_x = 3 \cdot 3(5) = 45 \text{ W delivered}$$

Therefore,

$$P_{delivered} = P_x + P_A = 45 + 40 = 85 \text{ W}$$

The power absorbed is

$$P_{absorbed} = P_B + P_{5V} = 75 + 10 = 85 \text{ W}$$

and $P_{absorbed} = P_{delivered}$.

EXERCISE 2-10

Find the power absorbed by the CCCS in Figure E 2-10. Hint: The controlling element of this dependent source is a short circuit. The voltage across a short circuit is zero. Hence the power absorbed by the controlling element is zero. How much power is absorbed by the controlled element?

Answer: −115.2 watts are absorbed by the CCCS. (The CCCS delivers +115.2 watts to the rest of the circuit.)

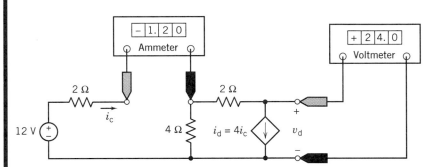

Figure E 2-10 A circuit containing a CCCS. The meters indicate that the current of the controlling branch is $i_c = -1.2$ amperes and that the voltage of the controlled branch is $v_d = 24$ volts.

EXERCISE 2-11

Find the power delivered to the VCVS in Figure E 2-11.
Answer: 6 watts are absorbed by the VCVS.

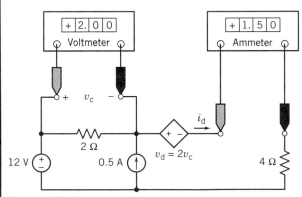

Figure E 2-11 A circuit containing a VCVS. The meters indicate that the voltage of the controlling branch is $v_c = 2.0$ volts and that the current of the controlled branch is $i_d = 1.5$ amperes.

EXERCISE 2-12

Find the power supplied by the VCCS in Figure E 2-12.
Answer: 17.6 watts are supplied by the VCCS. (-17.6 watts are absorbed by the VCCS.)

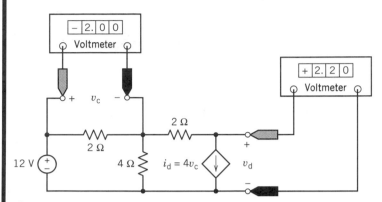

Figure E 2-12 A circuit containing a VCCS. The meters indicate that the voltage of the controlling branch is $v_c = -2.0$ volts and that the voltage of the controlled branch is $v_d = 2.2$ volts.

EXERCISE 2-13

Find the power absorbed by the CCVS in Figure E 2-13.
Answer: 4.375 watts are delivered to the CCVS.

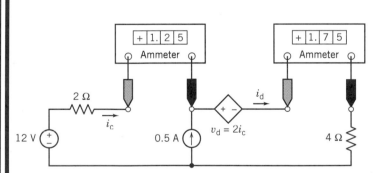

Figure E 2-13 A circuit containing a CCVS. The meters indicate that the current of the controlling branch is $i_c = 1.25$ amperes and that the current of the controlled branch is $i_d = 1.75$ amperes.

EXERCISE 2-14

Electric taxicabs, as shown in Figure E 2-14*a*, were widely used in the first decade of this century. The circuit representing the electric motor and battery is shown in Figure E 2-14*b*. Find the power absorbed by the electric motor when $v = 12$ V and $i_c = 5$ A.
Answer: 120 W

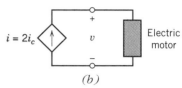

(a)   *(b)*

Figure E 2-14 *(a)* Electric taxicab in New York about 1902. Courtesy of the Smithsonian Institution. *(b)* Circuit representing the electric motor and battery. The current is $i = 2i_c$, where i_c is the controlling current regulated by the throttle.

2-9   TRANSDUCERS

Transducers are devices that convert physical quantities to electrical quantities. This section describes two transducers: potentiometers and temperature sensors. Potentiometers convert position to resistance, and temperature sensors convert temperature to current.

Figure 2-26a shows the symbol for the potentiometer. The potentiometer is a resistor having a third contact, called the wiper, that slides along the resistor. Two parameters, R_p and a, are needed to describe the potentiometer. The parameter R_p specifies the potentiometer resistance ($R_p > 0$). The parameter a represents the wiper position and takes values in the range $0 \le a \le 1$. The values $a = 0$ and $a = 1$ correspond to the extreme positions of the wiper.

Figure 2-26b shows a model for the potentiometer that consists of two resistors. The resistances of these resistors depend on the potentiometer parameters R_p and a.

Frequently, the position of the wiper corresponds to the angular position of a shaft

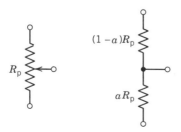

(a)   *(b)*

Figure 2-26 *(a)* The symbol and *(b)* a model for the potentiometer.

connected to the potentiometer. Suppose θ is the angle in degrees and $0 \le \theta \le 360$. Then

$$a = \frac{\theta}{360}$$

Example 2-7

Figure 2-27a shows a circuit in which the voltage measured by the meter gives an indication of the angular position of the shaft. In Figure 2-27b the current source, the potentiometer, and the voltmeter have been replaced by models of these devices. Analysis of Figure 2-27b yields

$$v_m = R_p I a = \frac{R_p I}{360} \theta$$

Solving for the angle gives

$$\theta = \frac{360}{R_p I} v_m$$

Suppose $R_p = 10$ Kohm and $I = 1$ ma. An angle of 163° would cause an output of $v_m = 4.53$ volts. A meter reading of 7.83 volts would indicate that $\theta = 282°$.

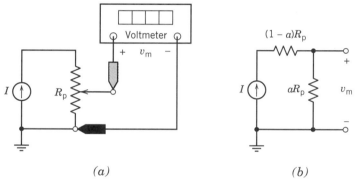

(a) $\qquad\qquad\qquad\qquad\qquad$ (b)

Figure 2-27 (a) A circuit containing a potentiometer. (b) An equivalent circuit containing a model of the potentiometer.

Temperature sensors, such as the AD590 manufactured by Analog Devices, are current sources having current proportional to absolute temperature. Figure 2-28a shows the symbol used to represent the temperature sensor. Figure 2-28b shows the circuit model of the temperature sensor. In order for the temperature sensor to operate properly, the branch voltage v must satisfy the condition

$$4 \text{ volts} \le v \le 30 \text{ volts}$$

When this condition is satisfied the current, i, in microamps, is numerically equal to the temperature T, in degrees Kelvin. The phrase "numerically equal" indicates that the current and temperature have the same value but different units. This relationship can be expressed as

$$i = k \cdot T$$

where $k = 1 \dfrac{\mu A}{°K}$, a constant associated with the sensor.

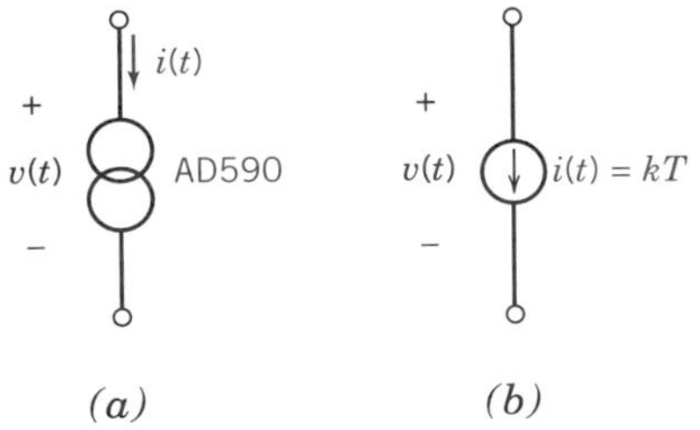

Figure 2-28 (*a*) The symbol and (*b*) a model for the temperature sensor.

(*a*) (*b*)

EXERCISE 2-15

For the potentiometer circuit of Figure 2-27, calculate the meter voltage, v_m, when $\theta = 45°$, $R_p = 20$ kΩ, and $I = 2$ mA.
Answer: $v_m = 5$ V.

EXERCISE 2-16

The voltage and current of an AD590 temperature sensor are 10 V and 280 μA, respectively. Determine the measured temperature.
Answer: $T = 280°$K or approximately 6.8°C.

2-10 **SWITCHES**

Switches have two distinct states: open and closed. Ideally, a switch acts as a short circuit when it is closed and as an open circuit when it is open. Figures 2-29 and 2-30 show several types of switches. In each case, the time when the switch changes state is indicated. Consider first the Single-Pole, Single-Throw (SPST) switches shown in Figure 2-29. The switch in Figure 2-29*a* is initially open. This switch changes state, becoming closed, at time $t = 0$ s. When this switch is modeled as an ideal switch, it is treated like an open circuit when $t < 0$ s and like a short circuit when $t > 0$ s. The ideal switch changes state instantaneously. The switch in Figure 2-29*b* is initially closed. This switch changes state, becoming open, at time $t = 0$ s.

Next, consider the Single-Pole, Double-Throw (SPDT) switch shown in Figure 2-30*a*. This SPDT switch acts like two SPST switches, one between terminals c and a, another between terminals c and b. Before $t = 0$ s, the switch between c and a is closed and the switch between c and b is open. At $t = 0$ s both switches change state, that is, the switch between a and c opens and the switch between c and b closes. Once again, the ideal switches are modeled as open circuits when they are open and as short circuits when they are closed.

In some applications, it makes a difference whether the switch between c and b closes before, or after, the switch between c and a opens. Different symbols are used to represent these two types of double-pole single-throw switch. The break-before-make switch is manufactured so that the switch between c and b closes after the switch between c and a opens. The symbol for the break-before-make switch is shown in Figure 2-30*a*. The make-before-break switch is manufactured so that the switch between c and b closes before the switch between c and a opens. The symbol for the make-before-break switch is shown in Figure 2-30b. Remember, the switch transition from terminal a to terminal b is assumed to take place instantaneously. Thus, the switch wiper makes before breaks, but this transition is very fast compared to the circuit time response.

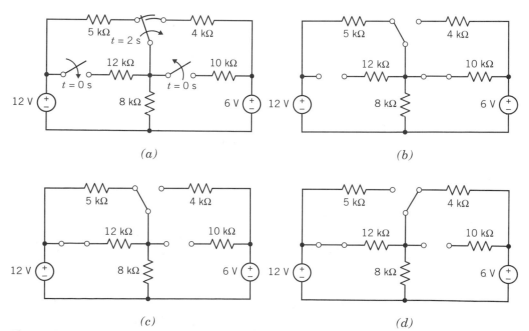

$t = 0$
Initially open

(a)

$t = 0$
Initially closed

(b)

Figure 2-29 SPST switches. (a) Initially open and (b) initially closed.

c

$t = 0$
Break before make

(a)

a

b

c

$t = 0$
Make before break

(b)

a

b

Figure 2-30 SPDT switches. (a) Break before make and (b) make before break.

Example 2-8

Figure 2-31 illustrates the use of open and short circuits for modeling ideal switches. In Figure 2-31a a circuit containing three switches is shown. In Figure 2-31b the circuit is shown as it would be modeled before $t = 0$ s. The two single-pole single-throw switches change state at time $t = 0$ s. Figure 2-31c shows the circuit as it would be modeled when the time is between 0 s and 2 s. The double-pole single-throw switch changes state at time $t = 2$ s. Figure 2-31d shows the circuit as it would be modeled after 2 s.

Figure 2-31 (a) A circuit containing several switches. (b) The equivalent circuit for $t \leq 0$ s. (c) The equivalent circuit for $0 < t < 2$ s. (d) The equivalent circuit for $t > 2$ s.

EXERCISE 2-17

What is the value of the current i in Figure E 2-17 at time $t = 1$ s? At $t = 5$ s?
Answers: $i = 4$ milliamps at $t = 1$ s (the switch is closed), and $i = 0$ amperes at $t = 5$ s (the switch is open).

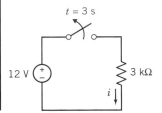

Figure E 2-17 A circuit with an SPST switch that opens at $t = 2$ seconds.

EXERCISE 2-18

What is the value of the current i in Figure E 2-18 at time $t = 4$ s?
Answer: $i = 0$ amperes at $t = 4$ s (both switches are open).

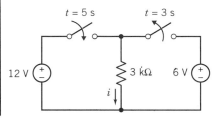

Figure E 2-18 A circuit with two SPST switches.

EXERCISE 2-19

What is the value of the voltage v in Figure E 2-19 at time $t = 4$ s? At $t = 6$ s?
Answers: $v = 6$ volts at $t = 4$ s, and $v = 0$ volts at $t = 6$ s.

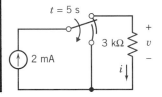

Figure E 2-19 A circuit with a make-before-break SPDT switch.

EXERCISE 2-20

What is the value of the current i in Figure E 2-20 at time $t = 1$ s? At $t = 3$ s?
Answer: $i = 2$ milliamps at $t = 1$ s, and $i = 4$ milliamps at $t = 3$ s.

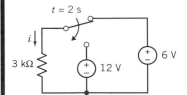

Figure E 2-20 A circuit with a break-before-make SPDT switch.

2-11 VERIFICATION EXAMPLE

A computer analysis of the circuit of Figure V 2-1 has provided the result $v = 48$ V at $t = 3$ s. Verify this result.

Solution

Prior to $t = 1$ s a short circuit draws the 2-A source current and no current flows through the 12-Ω resistor. The switch opens at $t = 1$ s and $v_1 = 12 \times 2 = 24$ V for $t > 1$ s.

The dependent current source provides $i = 0.5v_1 = 12$ A. Then, the current i enters the designated negative terminal for v. Thus, i and v do not adhere to the passive convention. Therefore, $v = -48$ V, not $+48$ V as reported.

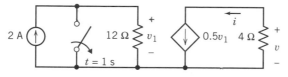

Figure V 2-1 A circuit with a switch and a dependent source.

2-12 DESIGN CHALLENGE SOLUTION

TEMPERATURE SENSOR

Currents can be measured easily using ammeters. A temperature sensor, such as Analog Devices' AD590, can be used to measure temperature by converting temperature to current [Dorf 1993]. Figure 2D-1 shows a symbol used to represent a temperature sensor. In order for this sensor to operate properly, the voltage v must satisfy the condition

$$4 \text{ volts} \leq v \leq 30 \text{ volts}$$

When this condition is satisfied, the current i, in μA, is numerically equal to the temperature T, in °K. The phrase "numerically equal" indicates that the two variables have the same value but different units.

$$i = k \cdot T \quad \text{where} \quad k = 1 \frac{\mu A}{°K}$$

The goal is to design a circuit using the AD590 to measure the temperature of a container of water. In addition to the AD590 and an ammeter, several power supplies and an assortment of standard 2% resistors are available. The power supplies are voltage sources. Power supplies having voltages of 10, 12, 15, 18, or 24 volts are available.

Figure 2D-1 A temperature sensor.

Define the Situation

In order for the temperature transducer to operate properly, its element voltage must be between 4 volts and 30 volts. The power supplies and resistors will be used to establish this voltage. An ammeter will be used to measure the current in the temperature transducer.

The circuit must be able to measure temperatures in the range from 0°C to 100°C since water is a liquid at these temperatures. Recall that the temperature in °C is equal to the temperature in °K minus 273°.

State the Goal

Use the power supplies and resistors to cause the voltage, v, of the temperature transducer to be between 4 volts and 30 volts.

Use an ammeter to measure the current, i, in the temperature transducer.

Generate a Plan

Model the power supply as an ideal voltage source and the temperature transducer as an ideal current source. The circuit shown in Figure 2D-2a causes the voltage across the temperature transducer to be equal to the power supply voltage. Since all of the available power supplies have voltages between 4 volts and 30 volts, any one of the power supplies can be used. Notice that the resistors are not needed.

In Figure 2D-2b a short circuit has been added in a way that does not disturb the network. In Figure 2D-2c this short circuit has been replaced with an (ideal) ammeter. Since the ammeter will measure the current in the temperature transducer, the ammeter reading will be numerically equal to the temperature in °K.

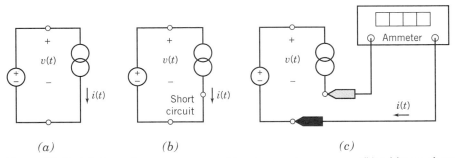

Figure 2D-2 (a) Measuring temperature with a temperature sensor. (b) Adding a short circuit. (c) Replacing the short circuit by an ammeter.

Although any of the available power supplies is adequate to meet the specifications, there may still be an advantage to choosing a particular power supply. For example, it is reasonable to choose the power supply that causes the transducer to absorb as little power as possible.

Take Action on the Plan

The power absorbed by the transducer is

$$p = v \cdot i$$

where v is the power supply voltage. Choosing v as small as possible, 10 volts in this case, makes the power absorbed by the temperature transducer as small as possible. Figure 2D-3a shows the final design. Figure 2D-3b shows a graph that can be used to find the temperature corresponding to any ammeter current.

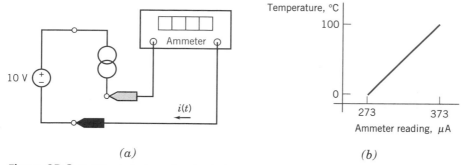

Figure 2D-3 (*a*) Final design of a circuit that measures temperature with a temperature sensor. (*b*) Graph of temperature versus ammeter current.

SUMMARY

Thomas A. Edison, the first electrical engineer, invented and developed many electric circuits and elements. By 1882 Edison had conceived, designed, patented, and tested an electric lighting system to replace the prevalent gas lighting of that period. By 1884 Edison had demonstrated his lighting system in New York and at the Philadelphia Exhibition. In the late 1880s a professional society, the American Institute of Electrical Engineers, was organized. It was also in that period that courses of study in electrical engineering were instituted at several U.S. colleges.

Engineering is a problem-solving activity. It can also be described as the profession that strives to find ways to utilize economically the materials and forces of nature for the benefit of humankind.

The engineer uses models, called circuit elements, to represent the devices that make up a circuit. In this book we consider only linear elements or linear models of devices. A device is linear if it satisfies the properties of both superposition and homogeneity. An example of a physical circuit made up of devices and a circuit model made up of circuit elements is shown in Figure 2-32.

We classify circuit elements as either passive or active. An element is passive if the total energy delivered to it is always nonnegative. Conversely, an active element may supply energy to the rest of the circuit. An active element can be called a source.

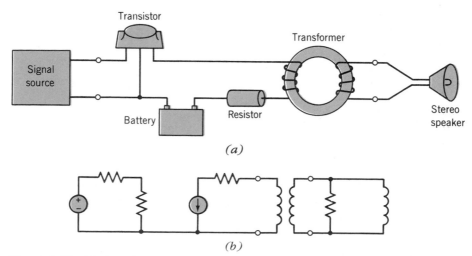

Figure 2-32 (*a*) Physical circuit made up of devices. (*b*) Its circuit model made up of circuit elements.

Sources may be independent of other variables in the circuit, or they may be dependent, having a response that is dependent on a current or voltage that appears elsewhere in the circuit. An independent current source, for example, is an element for which a current flows independently of the voltage across the element.

Resistors are widely used as circuits elements. These passive elements obey Ohm's law and dissipate energy.

Transducers are devices that convert physical quantities, such as rotational position, to an electrical quantity such as voltage. In this chapter we describe two transducers: potentiometers and temperature sensors.

Switches are widely used in circuits to connect and disconnect elements and circuits. They can also be used to create discontinuous voltages or currents.

Electric circuits consist of an interconnection of passive and active elements in a closed path.

TERMS AND CONCEPTS

Active Element   Element that supplies energy to the rest of the circuit. The energy delivered to the element is not always nonnegative. In other words, the energy supplied by the element can be positive.

Dependent Source   Source that provides a current (or a voltage) that is dependent on another variable elsewhere in the circuit.

Engineering   The profession in which knowledge of mathematical, natural, and social sciences gained by study, experience, and practice is applied with judgment to develop ways to utilize economically the materials and forces of nature for the benefit of humankind.

Homogeneity   Property that the response of an element is kv when it is subjected to an excitation ki. The magnitude scale factor is preserved.

Ideal Source   Ideal model of an actual source that assumes that the parameters of the source, such as its magnitude, are independent of other circuit variables.

Independent Source   Source that provides a current (or a voltage) independent of other circuit variables.

Linear Element   Element that satisfies the properties of superposition and homogeneity.

Model   Representation of an element or a circuit.

Ohm's Law   The voltage across the terminals of a resistor is related to the current into the positive terminal as $v = Ri$.

Open Circuit   Condition that exists when the current between two terminals is identically zero, irrespective of the voltage across the terminals.

Passive Element   Element that absorbs energy. The total energy delivered to it is always nonnegative (zero or positive).

Power Delivered to a Resistance $p = i^2R = v^2/R$ watts

Resistance   The physical property of an element to impede current flow. A measure of an element's ability to dissipate power irreversibly.

Resistivity   Ability of a material to resist the flow of charge. The symbol is ρ.

Resistor   Device or element whose primary purpose is to introduce resistance R into a circuit.

Short Circuit   Condition that exists when the voltage across two terminals is identically zero, irrespective of the current between the two terminals.

Source   Voltage or current generator able to supply energy to a circuit.

Superposition   Property that when an excitation i_1 results in a response v_1 and an excitation i_2 results in a response v_2 then an excitation $(i_1 + i_2)$ results in a response $(v_1 + v_2)$.

Switch   Device with two distinct states: open and closed. It acts as a short circuit when it is closed and as an open circuit when open.

Tranducer   Device that converts a physical quantity to an electrical quantity (typically a current or voltage).

REFERENCES   Chapter 2

Clark, W.A. *Edison: The Man Who Made the Future,* Putnam, New York, 1977.
Dorf, Richard C. *Electrical Engineering Handbook,* "Sensors," Chap. 53, CRC Press, Florida, 1993.
Kranzberg, M., and Pursell, C.W. *Technology in Western Civilization,* Oxford Univ. Press, London/ New York, Vol. I, 1967.
McMahon, A.M. *The Making of a Profession: A Century of Electrical Engineering in America,* IEEE Press, New York, 1984.

PROBLEMS

Section 2-3   Linear Models of Circuit Elements

P 2.3-1   An element has a voltage $v = 12$ V and a current $i = 4$ A as shown in Figure P 2.3-1. Determine a linear model of the element and identify whether the energy is absorbed or delivered by the element.
Answer: $p = 48$ W absorbed

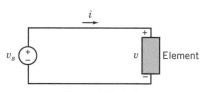

Figure P 2.3-1

P 2.3-2   An element is represented by the relation between current and voltage as

$$v = \sqrt{i}$$

Determine if the element is linear.

P 2.3-3   Determine the range of linear operation for a device with a v–i relationship:

$$v = Ri \qquad i \geq 0$$
$$v = 0 \qquad i < 0$$

P 2.3-4   An element has a v–i characteristic

$$v = 10 - 10^{-11}e^{40i}$$

which applies when $i \geq 0$ and $v \geq 0.3$ V. The element normally has $v = 5$ V.
(a)   Sketch the v–i characteristic curve.
(b)   Determine a linear representation for the circuit ele-

ment and state the range in which the approximation is relatively accurate.

Section 2-4   Active and Passive Circuit Elements

P 2.4-1   Consider the element shown in the circuit in Figure P 2.4-1.
(a)   Determine the power absorbed by the element.
(b)   Determine the energy delivered to the element for the first 10 s.
(c)   Is the element an active or passive element?

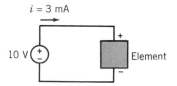

Figure P 2.4-1   Circuit with an independent voltage source $v = 10$ V for $t \geq 0$.

P 2.4-2   (a)   Find the power supplied by the voltage source shown in Figure P 2.4-2 when for $t \geq 0$ we have

$$v = 2 \cos t \text{ V}$$

and

$$i = 10 \cos t \text{ mA}$$

(b)   Determine the energy supplied by this voltage source for the period $0 \leq t \leq 1$ s.

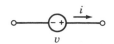

Figure P 2.4-2

P 2.4-3 The current entering the positive terminal of an element is $i = 2 \sin t$ A for $t \geq 0$ and $i = 0$ for $t < 0$. The voltage across the element is

$$v = 2 \frac{di}{dt}$$

Determine whether the element is active or passive by calculating the energy absorbed by the element.

P 2.4-4 The current entering the positive terminal of an element is $i = 2 \sin(t - \pi)$ A, and the voltage across the element is $v = 2 \sin t$ V. Determine whether the element is active or passive.

P 2.4-5 The power delivered to an element is $p = 10e^{-2t}$ mW and the charge entering the positive terminal of the element is $q = 1 - 2e^{-t}$ mC. Find (a) the voltage across the element and (b) the energy absorbed by the element between $t = 0$ and $t = 100$ ms.

P 2.4-6 A battery is being charged. The current i is entering the positive terminal of the battery. Find the energy delivered to a battery when the current is a constant 10 A and the voltage is a constant 12 V for the first hour of charging.
Answer: 432 kJ

P 2.4-7 An electrical element has a current i entering the positive terminal with a voltage v across the element. For $t > 0$ it is known that

$$v(t) = 10 \sin 100t \text{ V}$$
$$i(t) = 2 \cos 100t \text{ mA}$$

(a) Find the power $p(t)$ and plot $p(t)$ versus t.
(b) Determine intervals of time when the element is supplying or absorbing power.
(c) Determine whether the element is active or passive.

P 2.4-8 Repeat Problem 2.4-7 when v remains unchanged but $i = 4 \sin(100t + \pi/3)$ mA.

P 2.4-9 An element represented by Figure 2-4 has the current entering the positive terminal. It is known that $v(t) = 10$ V and $i(t) = 1 - 2 \sin 1000t$ mA. Find the total energy delivered to the element for the interval $0 \leq t \leq \pi/500$ s.

P 2.4-10 The Nelson River direct-current power line in Canada delivers 1.2 GW, at 900 kV, over a distance of 800 km. At the other end, at which the power is generated, the voltage is 950 kV. (a) Draw a circuit representation of this problem. (b) What is the efficiency of the line? That is, what is the power delivered divided by the power generated? (c) Where does the energy lost go? (d) How many joules are transmitted in a day if the conditions do not vary throughout the day?

P 2.4-11 A diagram of an auto battery charger is shown in Figure P 2.4-11a. The linear model of the charger circuit is shown in Figure P 2.4-11b, where element x represents the transformer and diode and element y represents the power losses as heat within the car battery. Determine the power delivered to charge the battery (the 12-V source) and how long it will take to deliver 3360 coulombs of charge. Calculate the power supplied by the charger and verify that it is equal to the power absorbed by the auto battery.

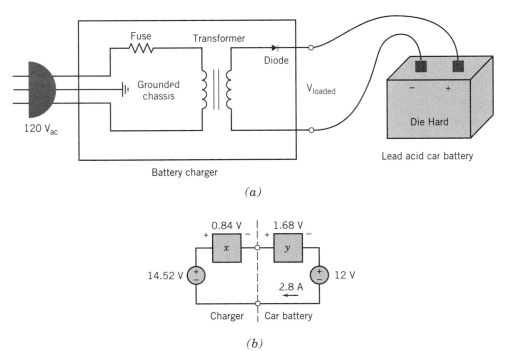

Figure P 2.4-11 *(a)* Auto battery charger. *(b)* Circuit model.

Section 2-5   Resistors

P 2.5-1   The portable lighting equipment for a mine is located 100 meters from its dc supply source. The mine lights use a total of 5 kW and operate at 120 V dc. Determine the required cross-sectional area of the copper wires used to connect the source to the mine lights if we require that the power lost in the copper wires be less than or equal to 5 percent of the power required by the mine lights.

P 2.5-2   An electric heater is connected to a constant 250-V source and absorbs 1000 W. Subsequently, this heater is connected to a constant 210-V source. What power does it absorb from the 210-V source? What is the resistance of the heater?

P 2.5-3   For the circuit of Figure P 2.5-3
(a)   Find the power delivered by the current source.
(b)   Find the power delivered to the 25-Ω resistor.

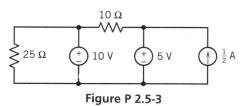

Figure P 2.5-3

P 2.5-4   What is the power delivered to the 2-ohm resistor in Figure P 2.5-4?

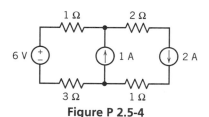

Figure P 2.5-4

P 2.5-5   The 10-V source in Figure P 2.5-5 is known to be delivering 20 W to the circuit. (a) Determine whether the element x is an active or passive element. (b) Determine the power delivered or absorbed by the element x and indicate whether it is delivering or absorbing power.
Answer: $P_x = 4$ W delivered

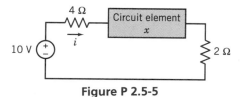

Figure P 2.5-5

P 2.5-6   The 10-V source in Figure P 2.5-5 is known to be delivering 10 W to the circuit. (a) Determine whether the element x is an active or passive element. (b) Determine the power delivered or absorbed by the element x and indicate whether it is delivering or absorbing power.

Section 2-6   Independent Sources

P 2.6-1   A current source and a voltage source are connected in parallel with an element as shown in Figure P 2.6-1. From the vantage point of the element, show that the current source is extraneous.

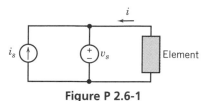

Figure P 2.6-1

P 2.6-2   A current source and a voltage source are connected in series with an element as shown in Figure P 2.6-2. From the vantage point of the element, show that the voltage source is extraneous.

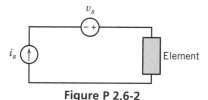

Figure P 2.6-2

P 2.6-3   A current source is connected in a circuit as shown in Figure P 2.6-3 and $i_s = 0.65$ A. Determine i and describe how i changes as different elements are selected.

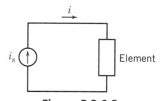

Figure P 2.6-3

Section 2-7   Voltmeters and Ammeters

P 2.7-1   For the circuit of Figure P 2.7-1
(a)   What is the value of the resistance R?
(b)   How much power is delivered by the current source?

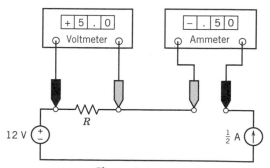

Figure P 2.7-1

P 2.7-2 What values do the meters in Figure P 2.7-2 read?

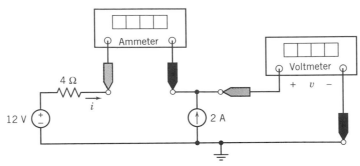

Figure P 2.7-2

Figure P 2.7-3 For the circuit of Figure P 2.7-3, determine the values indicated by the meters.

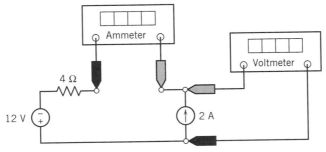

Figure P 2.7-3

Section 2-8 Dependent Sources

P 2.8-1 A modern electric van is shown in Figure P 2.8-1a. The control voltage is a constant voltage, $v_c = 4$ V, and is actuated when the key is turned to start the van's motion, as shown in Figure P 2.8-1b. The dependent voltage source is $v = 3v_c$ and $i = 50$ A.

(a) Find the power p delivered to the electric motor.
(b) Find the energy w expended by the motor during the first 10 minutes of operation.

Answer: (a) $p = 600$ W
(b) $w = 360$ kJ

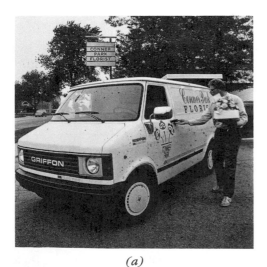

(a)

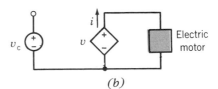

(b)

Figure P 2.8-1 (*a*) Modern electric van using a bank of 12-V batteries that are charged overnight. The van has a top speed of 53 mph and a range of about 60 miles before the battery needs to be charged. Courtesy of the Electric Power Research Institute. (*b*) The electric drive motor control circuit, where $v = bv_c$ and $v_c = 4$ V. When the on–off key is actuated, the motor will start and $b = 3$.

P 2.8-2 Find the power absorbed or supplied by each element in the circuit shown in Figure P 2.8-2. Indicate whether it is absorbed by or supplied by each element.

Answer: $P_{8V} = 24$ mW supplied
$P_{6V} = 18$ mW absorbed
$P_{i_x}/2 = 2$ mW supplied
$P_{2V} = 8$ mW absorbed

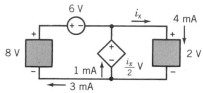

Figure P 2.8-2

P 2.8-3 Find the power supplied or absorbed by each element in the circuit shown in Figure P 2.8-3.

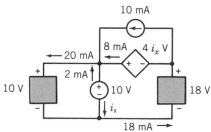

Figure P 2.8-3

Section 2-9 **Transducers**

P 2.9-1 For the potentiometer circuit of Figure 2-27 the current source and potentiometer resistance are 1.1 mA and 100 kΩ, respectively. Calculate the required angle, θ, so that the measured voltage is 23 V.

P 2.9-2 An AD590 sensor has an associated constant $k = 1 \dfrac{\mu A}{°K}$. The sensor has a voltage $v = 20$ V; and the measured current, $i(t)$, as shown in Figure 2-28 is $4\ \mu A < i < 13\ \mu A$ in a laboratory setting. Find the range of measured temperature.

Section 2-10 **Switches**

P 2.10-1 Determine the current, i, at $t = 1$ s and at $t = 4$ s for the circuit of Figure P 2.10-1.

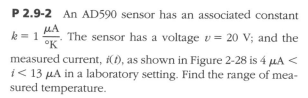

Figure P 2.10-1

P 2.10-2 Determine the voltage, v, at $t = 1$ s and at $t = 4$ s for the circuit shown in Figure P 2.10-2.

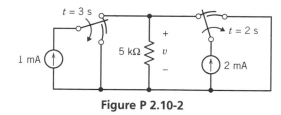

Figure P 2.10-2

VERIFICATION PROBLEMS

VP 2-1 For the circuit of Figure VP 2-1, a computer program has provided the following results:

$$v = 6 \text{ V}$$
$$i = 1.0 \text{ A}$$

Determine if the results are correct.

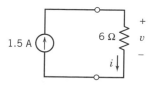

Figure VP 2-1

VP 2-2 For the circuit of Figure VP 2-2, a report states that $v = 20$ V at $t = 2$ s. Determine if this report is correct.

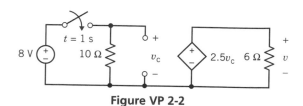

Figure VP 2-2

DESIGN PROBLEMS

DP 2-1 Determine the constant g so that the power absorbed equals the power delivered for the circuit shown in Figure DP 2-1.

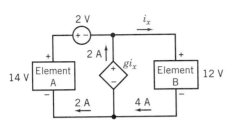

Figure DP 2-1 Circuit with an unspecified constant g.

DP 2-2 Select an appropriate dependent source element for the circuit shown in Figure DP 2-2 so that the sum of the power absorbed is equal to the sum of the power delivered.

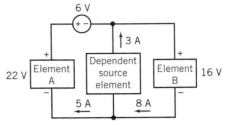

Figure DP 2-2 Circuit with a dependent source to be selected.

RESISTIVE CIRCUITS

PREVIEW

The resistor, with resistance R, is an element commonly used in most electric circuits. In this chapter we consider the analysis of circuits consisting of resistors and sources.

In addition to Ohm's law, we need two laws for relating (1) current flow at connected terminals and (2) the sum of voltages around a closed circuit path. These two laws were developed by Gustav Kirchhoff in 1847.

Using Kirchhoff's laws and Ohm's law, we are able to complete the analysis of resistive circuits and determine the currents and voltages at desired points in the circuit. This analysis may be accomplished for circuits with both independent and dependent sources.

3-1 DESIGN CHALLENGE

ADJUSTABLE VOLTAGE SOURCE

A circuit is required to provide an adjustable voltage. The specifications for this circuit are:

1 It should be possible to adjust the voltage to any value between -5 V and $+5$ V. It should not be possible to accidentally obtain a voltage outside this range.
2 The load current will be negligible.
3 The circuit should use as little power as possible.

The available components are:

1 Potentiometers: resistance values of 10 kΩ, 20 kΩ, and 50 kΩ are in stock;
2 A large assortment of standard 2% resistors having values between 10 Ω and 1 MΩ (see Appendix F);
3 Two power supplies (voltage sources): one 12-V supply and one $-$ 12-V supply; both rated for a maximum of 100 mA (milliamps).

Define the Situation

Figure 3D-1 shows the situation. The voltage v is the adjustable voltage. The circuit that uses the output of the circuit being designed is frequently called the "load." In this case, the load current is negligible, so $i = 0$.

State the Goal

A circuit providing the adjustable voltage

$$-5 \text{ V} \leq v \leq +5 \text{ V}$$

must be designed using the available components.

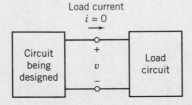

Figure 3D-1 The circuit being designed provides an adjustable voltage, v, to the load circuit.

Generate a Plan

Make the following observations.

1 The adjustability of a potentiometer can be used to obtain an adjustable voltage v.
2 Both power supplies must be used so that the adjustable voltage can have both positive and negative values.
3 The terminals of the potentiometer cannot be connected directly to the power supplies because the voltage v is not allowed to be as large as 12 V or -12 V.

These observations suggest the circuit shown in Figure 3D-2a. The circuit in Figure 3D-2b is obtained by using the simplest model for each component in Figure 3D-2a.

To complete the design, values need to be specified for R_1, R_2, and R_p. Then, several results need to be checked and adjustments made, if necessary.

1 Can the voltage v be adjusted to any value in the range -5 V to 5 V?
2 Are the voltage source currents less than 1 mA? This condition must be satisfied if the power supplies are to be modeled as ideal voltage sources.
3 Is it possible to reduce the power absorbed by R_1, R_2, and R_p?

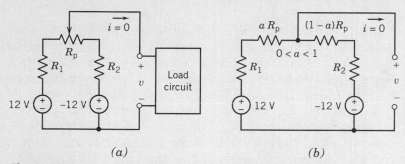

$$(a) \qquad\qquad\qquad (b)$$

Figure 3D-2 (a) A proposed circuit for producing the variable voltage v, and (b) the equivalent circuit after the potentiometer is modeled.

Take Action on the Plan

Taking action on this plan requires analyzing the circuit in Figure 3D-2b. Analysis of this type of circuit is discussed in this chapter. We will return to this problem at the end of this chapter.

3-2 ELECTRIC CIRCUIT APPLICATIONS

Overseas communications have always been of great importance to nations. One of the most brilliant chapters of the history of electrical technology was the development of underwater electric cable circuits. Underwater electric cables were used to carry electric telegraph communications. In late 1852 England and Ireland were connected by cable, and a year later there was a cable between Scotland and Ireland. In June 1853 a cable was strung between England and Holland, a distance of 115 miles.

Figure 3-1 Examples of undersea cable. Courtesy of Bell Laboratories.

Figure 3-2 View of the Sprague electric railway car on the Brookline branch of the Boston system about 1900. This electric railway branch operates as an electric trolley railroad today with modern electric cars. Courtesy of General Electric Company.

It was Cyrus Field and Samuel Morse who saw the potential for a submarine cable across the Atlantic. By 1857 Field had organized a firm to complete the transatlantic telegraph cable and issued a contract for the production of 2500 miles of cable. The cable laying began in June 1858. After several false starts, a cable was laid across the Atlantic by August 5, 1858. However, this cable failed after only a month of operation.

Another series of cable-laying projects commenced, and by September 1865 a successful Atlantic cable was in place. This cable stretched over 3000 miles, from England to eastern Canada. There followed a flurry of cable laying. Approximately 150,000 km (90,000 miles) were in use by 1870, linking all continents and all major islands. An example of modern undersea cable is shown in Figure 3-1.

One of the greatest uses of electricity in the late 1800s was for electric railways. In 1884 the Sprague Electric Railway was incorporated. Sprague built an electric railway for Richmond, Virginia, in 1888. By 1902 the horse-drawn street trolley was obsolete and there were 22,576 miles of electric railway track in the United States. A 1900 electric railway is shown in Figure 3-2.

3-3 | KIRCHHOFF'S LAWS

It is very important to be able to determine the current and voltage relationships when a circuit consists of two or more circuit elements. In this chapter we consider circuits made up of resistors and we show a simple two-resistor circuit in Figure 3-3. This circuit also has a voltage source. Each element is connected at its terminal to another element.

The circuit may conveniently be redrawn as shown in Figure 3-4, where a perfectly conducting wire is used to connect terminals d and c. Thus, this model of the wire has zero resistance. A wire (or two-terminal resistor) with zero voltage across it, irrespective of the current through it, is called a *short circuit*.

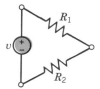

Figure 3-3 Simple two-resistor circuit with a voltage source.

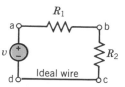

Figure 3-4 Alternative form of the circuit shown in Figure 3-3. The wire connecting terminals d and c is an ideal, perfectly conducting wire.

When we disconnect an element such as R_2 from the circuit of Figure 3-4, we say that we have an *open circuit* at terminals b−c as shown in Figure 3-5. The current between terminals c−b is identically zero, irrespective of the voltage across terminals c−b, when the resistor is removed (or R_2 is infinite). A junction in which two or more elements have a common connection is called a *node*. More correctly, a node is a junction of conductors composed of ideal wires. If we start at terminal a and traverse around the circuit, passing through each node in turn (that is, a to b to c to d, and back to a), then we have traversed a *closed path*.

A *closed path* in a circuit is a traversal through a series of nodes ending at the starting node without encountering a node more than once. A closed path is often called a *loop*.

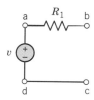

Figure 3-5 An open circuit at terminals b−c obtained by disconnecting R_2 from the circuit of Figure 3-4.

Example 3-1
Identify the closed paths in the circuit of Figure 3-6.

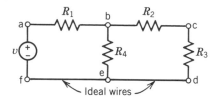

Figure 3-6 Circuit with three closed paths or loops.

Solution
There are three closed paths:

1 a−b−c−d−e−f−a
2 a−b−e−f−a
3 b−c−d−e−b

Note that ideal wires imply that terminals d, e, and f are actually one identical node.

Ohm's law gives the voltage and current relationship for one resistor. However, it remained for Gustav Robert Kirchhoff, a professor at the University of Berlin, to formulate two laws that relate the current and voltage in a circuit with two or more resistors. Kirchhoff, who formulated his laws in 1847, is shown in Figure 3-7.

Figure 3-7 Gustav Robert Kirchhoff (1824–1887). Kirchhoff stated two laws in 1847 regarding the current and voltage in an electrical circuit. Courtesy of the Smithsonian Institution.

Kirchhoff's current law (KCL) states that the algebraic sum of the currents entering any node is identically zero for all instants of time. This statement is a consequence of the fact that charge cannot accumulate at a node.

By **Kirchhoff's current law,** the algebraic sum of the currents into a node at any instant is zero.

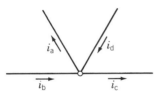

Figure 3-8 Currents at a node. The remaining circuit is not shown.

Let us consider the node shown in Figure 3-8. The sum of the currents entering the node is

$$- i_a + i_b - i_c + i_d = 0$$

Note that we have $- i_a$ since the current i_a is leaving the node. Another way to state KCL is that the sum of currents entering the node is equal to the sum of currents leaving the node.

If the sum of the currents entering a node were not equal to zero, then charge would accumulate at a node. However, a node is a perfect conductor and cannot accumulate or store charge. Thus, the sum of the currents entering a node is equal to zero.

By **Kirchhoff's voltage law** (KVL), the algebraic sum of the voltages around any closed path in a circuit is identically zero for all time.

The term "algebraic" implies the dependency on the voltage polarity encountered as the closed path is traversed.

Let us consider a circuit with two elements and a voltage source connected, as shown in Figure 3-9. The voltage across each element is shown with the sign of the voltage displayed. Starting at node c, the sum of the voltage drops around the loop is

$$- v_1 + v_2 - v_3 = 0$$

A common convention is to use the voltage sign on the first terminal of an element encountered as we traverse a path. Therefore, leaving terminal c, we encounter the minus sign on v_1, then the plus sign on v_2, and finally the minus sign on v_3.

Consider the circuit shown in Figure 3-10, where the voltage for each element is

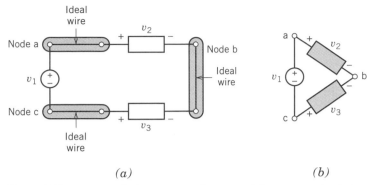

(a)   (b)

Figure 3-9 Circuit with three circuit elements. (a) Circuit with ideal wires and nodes identified. (b) Circuit with ideal wires removed, displaying nodes.

identified with its sign. The ideal wire has zero resistance, and thus the voltage across it is equal to zero. The sum of the voltages around the loop incorporating v_6, v_3, v_4, and v_5 is

$$- v_6 - v_3 + v_4 + v_5 = 0$$

The sum of the voltages around a loop is equal to zero. A circuit loop is a conservative system, which means that the energy required to move a charge around a closed path is zero. Thus, since voltage is work per unit charge, the voltage around the loop is zero.

Nevertheless, it is important to note that not all electrical systems are conservative. An example of a nonconservative system is a radio wave broadcasting system.

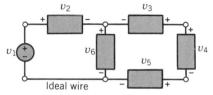

Figure 3-10 Circuit with three closed paths. The ideal wire has zero resistance, and thus the voltage across the wire is zero.

Example 3-2

Consider the circuit shown in Figure 3-11. Notice that the passive convention was used to assign reference directions to the resistor voltages and currents. This anticipates using Ohm's law. Find each current and each voltage when $R_1 = 8\ \Omega$, $v_2 = -10$ V, $i_3 = 2$ A, and $R_3 = 1\ \Omega$. Also, determine the resistance R_2.

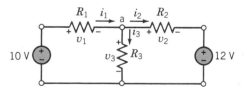

Figure 3-11 Circuit with two constant voltage sources.

Solution

The sum of the currents entering node a is

$$i_1 - i_2 - i_3 = 0$$

Using Ohm's law for R_3, we find that

$$v_3 = R_3 i_3$$
$$= 1(2)$$
$$= 2 \text{ V}$$

Kirchhoff's voltage law for the left-hand loop incorporating v_1, v_3, and the 10-V source is

$$-10 + v_1 + v_3 = 0$$

Therefore,

$$v_1 = 10 - v_3$$
$$= 8 \text{ V}$$

Ohm's law for the resistor R_1 is

$$v_1 = R_1 i_1$$

or

$$i_1 = v_1/R_1$$
$$= 8/8$$
$$= 1 \text{ A}$$

Since we have now found $i_1 = 1$ A and $i_3 = 2$ A as originally stated, then

$$i_2 = i_1 - i_3$$
$$= 1 - 2$$
$$= -1 \text{ A}$$

We can now find the resistance R_2 from

$$v_2 = R_2 i_2$$

or

$$R_2 = v_2/i_2$$
$$= -10/-1$$
$$= 10 \text{ } \Omega$$

Example 3-3

For the circuit shown in Figure 3-12, find i_1 and v_1, given $R_3 = 6 \text{ } \Omega$.

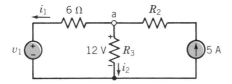

Figure 3-12 Circuit with a voltage source and a current source and where $R_3 = 6 \text{ } \Omega$.

Solution

First, we use KCL to write the equation for the currents at node a as

$$-i_1 - i_2 + 5 = 0$$

However, using Ohm's law for the resistor R_3, we have

$$12 = R_3 i_2$$

or

$$i_2 = 12/6$$
$$= 2 \text{ A}$$

Therefore,

$$i_1 = 5 - i_2$$
$$= 3 \text{ A}$$

We now use KVL clockwise around the left-hand loop to obtain v_1 as

$$-v_1 - 6i_1 + 12 = 0$$

Since $i_1 = 3$ A, we have

$$v_1 = 12 - 6i_1$$
$$= 12 - 18$$
$$= -6 \text{ V}$$

Note that v_1 and i_1 are independent of R_2, since $i_1 + i_2 = 5$ A irrespective of R_2.

EXERCISE 3-1

For the circuit in Figure E 3-1, find i_2 and v_2 when $v_3 = 6$ V.
Answer: $i_2 = -1$ A, $v_2 = 7$ V

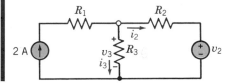

Figure E 3-1 Circuit with a current source and an unknown voltage source. $R_1 = R_2 = 1$ Ω and $R_3 = 2$ Ω. Also $v_3 = 6$ V.

EXERCISE 3-2

Find the voltages v_1 and v_2 in the circuit shown in Figure E 3-2.
Answer: $v_1 = -10/3$ V, $v_2 = 20/3$ V

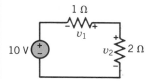

Figure E 3-2

3-4 A SINGLE-LOOP CIRCUIT—THE VOLTAGE DIVIDER

Let us consider a single-loop circuit, as shown in Figure 3-13. In anticipation of using Ohm's law, the passive convention has been used to assign reference directions to resistor voltages and currents. Using KCL at each node, we obtain

a:	$i_s - i_1 = 0$	(3-1)
b:	$i_1 - i_2 = 0$	(3-2)
c:	$i_2 - i_3 = 0$	(3-3)
d:	$i_3 - i_s = 0$	(3-4)

We have four equations, but any one of the four can be derived from the other three equations. In any circuit with n nodes, $n - 1$ independent current equations can be derived from Kirchhoff's current law.

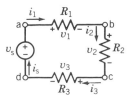

Figure 3-13 Single-loop circuit with a voltage source v_s.

Of course, we also note that

$$i_s = i_1 = i_2 = i_3$$

so that the current i_1 can be said to be the loop current and flows continuously around the loop from a to b to c to d and back to a.

The connection of resistors in Figure 3-13 is said to be a *series* connection since all the elements carry the same current.

In order to determine i_1, we use KVL around the loop to obtain

$$- v_s + v_1 + v_2 + v_3 = 0 \tag{3-5}$$

where v_1 is the voltage across the resistor R_1. Using Ohm's law for each resistor, Eq. 3-5 can be written as

$$- v_s + i_1 R_1 + i_1 R_2 + i_1 R_3 = 0$$

Solving for i_1, we have

$$i_1 = \frac{v_s}{R_1 + R_2 + R_3}$$

Thus, the voltage across the nth resistor R_n is v_n and can be obtained as

$$v_n = i_1 R_n$$
$$= \frac{v_s R_n}{R_1 + R_2 + R_3} \tag{3-6}$$

For example, the voltage across resistor R_2 is

$$v_2 = \frac{R_2}{R_1 + R_2 + R_3} v_s$$

Thus, the voltage appearing across one of a series connection of resistors connected in series with a voltage source will be the ratio of its resistance to the total resistance. This circuit demonstrates the principle of *voltage division*, and the circuit is called a *voltage divider*.

In general, we may represent the voltage divider principle by the equation

$$v_n = \frac{R_n}{R_1 + R_2 + \cdots + R_N} v_s \tag{3-7}$$

where the voltage is across the nth resistor of N resistors connected in series.

Example 3-4

Let us consider the circuit shown in Figure 3-14 and determine the resistance R_2 required so that the voltage across R_2 will be one-fourth of the source voltage when $R_1 = 9 \ \Omega$. Determine the current flowing when $v_s = 12$ V.

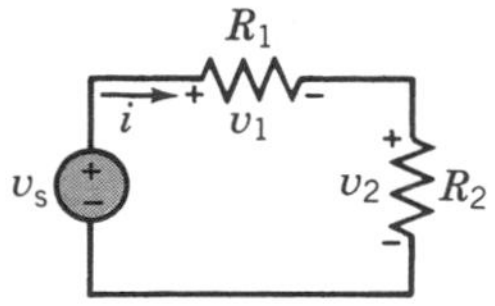

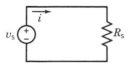

Figure 3-14 Voltage divider circuit with $R_1 = 9\ \Omega$.

Solution

The voltage across resistor R_2 will be

$$v_2 = \frac{R_2}{R_1 + R_2}\, v_s$$

Since we desire $v_2/v_s = 1/4$, we have

$$\frac{R_2}{R_1 + R_2} = \frac{1}{4}$$

or

$$R_1 = 3R_2$$

Since $R_1 = 9\ \Omega$, we require that $R_2 = 3\ \Omega$. Using KVL around the loop, we have

$$-v_s + v_1 + v_2 = 0$$

or

$$v_s = iR_1 + iR_2$$

Therefore,

$$i = \frac{v_s}{R_1 + R_2} \tag{3-8}$$

or

$$i = \frac{12}{12} = 1\ \text{A}$$

Let us consider the simple circuit of the voltage source connected to a resistance R_s as shown in Figure 3-15. For this circuit

$$i = \frac{v_s}{R_s} \tag{3-9}$$

Figure 3-15 Equivalent circuit for a series connection of resistors.

Comparing Eqs. 3-8 and 3-9, we see that the currents are identical when

$$R_s = R_1 + R_2$$

The resistance R_s is said to be an *equivalent resistance* of the series connection of resistors R_1 and R_2. In general, the equivalent resistance of a series of N resistors is

$$R_s = R_1 + R_2 + \cdots + R_N \tag{3-10}$$

We say, then, that the circuit of Figure 3-15 is an equivalent series representation of Figure 3-14. In this specific case

$$R_s = R_1 + R_2 = 9 + 3 = 12 \, \Omega$$

Each resistor in Figure 3-14 absorbs power, so the power absorbed by R_1 is

$$p_1 = \frac{v_1^2}{R_1}$$

and the power absorbed by the second resistor is

$$p_2 = \frac{v_2^2}{R_2}$$

The total power absorbed by the two resistors is

$$\begin{aligned} p &= p_1 + p_2 \\ &= \frac{v_1^2}{R_1} + \frac{v_2^2}{R_2} \end{aligned} \tag{3-11}$$

However, according to the voltage divider principle,

$$v_n = \frac{R_n}{R_1 + R_2} v_s$$

Then we may rewrite Eq. 3-11 as

$$p = \frac{R_1}{(R_1 + R_2)^2} v_s^2 + \frac{R_2}{(R_1 + R_2)^2} v_s^2$$

Since $R_1 + R_2 = R_s$, the equivalent series resistance, we have

$$\begin{aligned} p &= \frac{R_1 + R_2}{R_s^2} v_s^2 \\ &= \frac{v_s^2}{R_s} \end{aligned}$$

Thus, the total power absorbed by the two series resistors is equal to the power absorbed by the equivalent resistance R_s. The power absorbed by the two resistors is equal to that supplied by the source v_s. We show this by noting that the current is

$$i = \frac{v_s}{R_s}$$

and therefore the power supplied is

$$\begin{aligned} p &= v_s i \\ &= \frac{v_s^2}{R_s} \end{aligned}$$

Example 3-5
For the voltage divider of Figure 3-16, find the loop current and the voltage v_2. Then show that the power absorbed by the two resistors is equal to that supplied by the source.

Figure 3-16 Voltage divider for Example 3-5.

Solution

The current i is

$$i = \frac{15}{5 + 10} = 1 \text{ A}$$

Then, using Ohm's law,

$$v_1 = 5 \text{ V and } v_2 = 10 \text{ V}$$

The total power absorbed by the two resistors is

$$\begin{aligned} p &= v_1 i + v_2 i \\ &= (v_1 + v_2)i \\ &= (5 + 10)1 \\ &= 15 \text{ W} \end{aligned}$$

The power supplied by the source is

$$\begin{aligned} p_s &= v_s i \\ &= 15(1) \\ &= 15 \text{ W} \end{aligned}$$

Thus, the power supplied by the source is equal to that absorbed by the series connection of resistors. Of course, since $v_1 + v_2 = v_s$ by Kirchhoff's voltage law, we may readily show that for a series connection of N resistors connected to a source, the power supplied by the source is the total power absorbed by the resistors.

EXERCISE 3-3

For the circuit of Figure E 3-3, find the voltage v_3 and the current i and show that the power delivered to the three resistors is equal to that supplied by the source.
Answer: $v_3 = 3 \text{ V}, i = 1 \text{ A}$

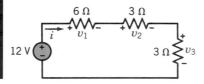

Figure E 3-3 Circuit with three series resistors (for Exercise 3-3).

EXERCISE 3-4

Consider the voltage divider shown in Figure E 3-4 when $R_1 = 6 \text{ }\Omega$. It is desired that the output power absorbed by $R_1 = 6 \text{ }\Omega$ be 6 W. Find the voltage v_o and the required source v_s.
Answer: $v_s = 14 \text{ V}, v_o = 6 \text{ V}$

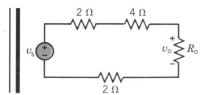

Figure E 3-4 Voltage divider for Exercise 3-4.

EXERCISE 3-5

Again consider the voltage divider shown in Figure E 3-4. Determine the required R_1 when it is desired that $v_o/v_s = 1/6$.

Answer: $R_1 = 8/5 \ \Omega$

EXERCISE 3-6

Consider the voltage divider shown in Figure E 3-4. Determine the required R_1 so that one-half of the power supplied by the source is absorbed by R_1.

Answer: $R_1 = 8 \ \Omega$

EXERCISE 3-7

Determine the equivalent resistance for the series connection of resistors shown in Figure E 3-3.

Answer: $R_s = 12 \ \Omega$

3-5 PARALLEL RESISTORS AND CURRENT DIVISION

Edison reasoned that an electric light system required high-resistance lamps connected in parallel. A parallel combination in which the lamps are strung between lines is shown in Figure 3-17. In this arrangement each lamp is modeled by a resistance R_n and the source

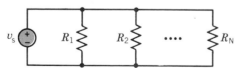

Figure 3-17 Edison's parallel lamp scheme with the nth lamp represented by its resistance R_n. This circuit has a total of N lamps.

voltage v_s appears across each lamp. Each lamp will thus have the same voltage applied, and turning on and off any number of the lamps does not affect the voltage applied to the others. Since all of the lamps have the same voltage across them, the current supplied by the source is the sum of the currents in all of the lamps. (In a series arrangement, on the other hand, the current in the lamps and the lines is the same.) To avoid excessively large and expensive line conductors, the current in each of the lamps in a parallel connection must be kept small. The only answer was to have a lamp whose filament had a high resistance. Edison, therefore, was looking not only for a material that would remain incandescent for a long time when a current was passed through it but also for something that would have a high resistance—a commercial rather than a technical requirement. Edison finally found a thin high-resistance carbonized thread for the filament.

Circuit elements, such as resistors, are connected in *parallel* when the voltage across each element is identical. Elements in parallel are connected at both terminals. The circuit in Figure 3-17 is a parallel circuit since each resistor has v_s across its terminals.

Consider the circuit with two resistors and a current source shown in Figure 3-18.

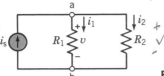

Figure 3-18 Parallel circuit with a current source.

Note that both resistors are connected to terminals a and b and that the voltage v appears across each parallel element. In anticipation of using Ohm's law, the passive convention is used to assign reference directions to the resistor voltages and currents. We may write KCL at node a (or at node b) to obtain

$$i_s - i_1 - i_2 = 0$$

or

$$i_s = i_1 + i_2$$

However, from Ohm's law

$$i_1 = \frac{v}{R_1} \quad \text{and} \quad i_2 = \frac{v}{R_2}$$

Then

$$i_s = \frac{v}{R_1} + \frac{v}{R_2} \tag{3-12}$$

Recall that we defined conductance G as the inverse of resistance R. We may therefore rewrite Eq. 3-12 as

$$i_s = G_1 v + G_2 v$$
$$= (G_1 + G_2)v \tag{3-13}$$

Thus the equivalent circuit for this parallel circuit is a conductance G_P, as shown in Figure 3-19, where

$$G_P = G_1 + G_2$$

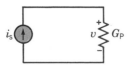

Figure 3-19 Equivalent circuit for a parallel circuit.

The equivalent resistance for the two-resistor circuit is found from

$$G_P = \frac{1}{R_1} + \frac{1}{R_2}$$

Since $G_P = 1/R_P$, we have

$$\frac{1}{R_P} = \frac{1}{R_1} + \frac{1}{R_2}$$

or

$$R_P = \frac{R_1 R_2}{R_1 + R_2} \tag{3-14}$$

Note that the total conductance increases as additional parallel elements are added and that the total resistance declines as each resistor is added.

The circuit shown in Figure 3-18 is called a *current divider* circuit since it divides the source current. Note that

$$i_1 = G_1 v \tag{3-15}$$

Also, since $i_s = (G_1 + G_2)v$, we solve for v, obtaining

$$v = \frac{i_s}{G_1 + G_2} \tag{3-16}$$

Substituting v from Eq. 3-16 into Eq. 3-15, we obtain

$$i_1 = \frac{G_1 i_s}{G_1 + G_2} \tag{3-17}$$

Similarly,

$$i_2 = \frac{G_2 i_s}{G_1 + G_2}$$

Note that we may use $G_2 = 1/R_2$ and $G_1 = 1/R_1$ to obtain the current i_2 in terms of two resistances as follows:

$$i_2 = \frac{R_1 i_s}{R_1 + R_2}$$

The current of the source divides between conductances G_1 and G_2 in proportion to their conductance values.

Figure 3-20 Set of N parallel conductances with a current source i_s.

Let us consider the more general case of current division with a set of N parallel conductors as shown in Figure 3-20. The KCL gives

$$i_s = i_1 + i_2 + i_3 + \cdots + i_N \tag{3-18}$$

for which

$$i_n = G_n v \tag{3-19}$$

for $n = 1, \ldots, N$. We may write Eq. 3-18 as

$$i_s = (G_1 + G_2 + G_3 + \cdots + G_N)v \tag{3-20}$$

Therefore

$$i_s = v \sum_{n=1}^{N} G_n \tag{3-21}$$

Since $i_n = G_n v$, we may obtain v from Eq. 3-21 and substitute it in, obtaining

$$i_n' = \frac{G_n i_s}{\displaystyle\sum_{n=1}^{N} G_n} \tag{3-22}$$

Recall that the equivalent circuit, Figure 3-19, has an equivalent conductance G_P such that

$$G_P = \sum_{n=1}^{N} G_n \tag{3-23}$$

Therefore

$$i_n = \frac{G_n i_s}{G_P} \tag{3-24}$$

which is the basic equation for the current divider with N conductances. Of course, Eq. 3-23 can be rewritten as

$$\frac{1}{R_P} = \sum_{n=1}^{N} \frac{1}{R_n} \tag{3-25}$$

Example 3-6

For the circuit in Figure 3-21 find (a) the current in each branch, (b) the equivalent circuit,

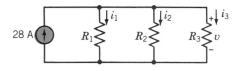

28 A

Figure 3-21 Parallel circuit for Example 3-6.

and (c) the voltage v. The resistors are

$$R_1 = \frac{1}{2}\,\Omega, \qquad R_2 = \frac{1}{4}\,\Omega, \qquad R_3 = \frac{1}{8}\,\Omega$$

Solution

The current divider follows the equation

$$i_n = \frac{G_n i_s}{G_P}$$

so it is wise to find the equivalent circuit, as shown in Figure 3-22, with its equivalent

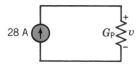

28 A

Figure 3-22 Equivalent circuit for the parallel circuit of Figure 3-21.

conductance G_P. We have

$$\begin{aligned}
G_P &= \sum_{n=1}^{N} G_n \\
&= G_1 + G_2 + G_3 \\
&= 2 + 4 + 8 \\
&= 14 \text{ S}
\end{aligned}$$

Recall that the units for conductance are siemens (S). Then

$$i_1 = \frac{G_1 i_s}{G_P}$$

$$= \frac{2}{14}(28)$$

$$= 4 \text{ A}$$

Similarly,

$$i_2 = \frac{G_2 i_s}{G_P}$$

$$= \frac{4(28)}{14}$$

$$= 8 \text{ A}$$

and

$$i_3 = \frac{G_3 i_s}{G_P}$$

$$= 16 \text{ A}$$

Since $i_n = G_n v$, we have

$$v = \frac{i_1}{G_1}$$

$$= \frac{4}{2}$$

$$= 2 \text{ V}$$

EXERCISE 3-8

A resistor network consisting of parallel resistors is shown in a package used for printed circuit board electronics in Figure E 3-8a. This package is only 2 cm × 0.7 cm and each resistor is 1 kΩ. The circuit is connected to use four resistors as shown in Figure E 3-8b. Find the equivalent circuit for this network. Determine the current in each resistor when i_s = 1 mA.

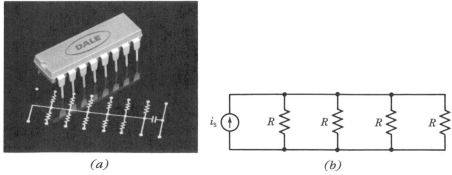

(a)   (b)

Figure E 3-8 (a) A parallel resistor network. Courtesy of Dale Electronics. (b) The connected circuit uses four resistors where R = 1 kΩ.

EXERCISE 3-9

Find the equivalent parallel resistance R_P for the circuit in Figure E 3-9 when $R_1 = 4\ \Omega$ and $R_2 = 2\ \Omega$. Also determine the voltage v.

Answer: $R_P = 4/3\ \Omega$, $v = 4$ V

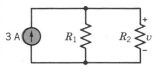

Figure E 3-9 Parallel divider circuit for Exercise 3-9.

3-6 CIRCUIT ANALYSIS

In this section we consider the analysis of a circuit by replacing a set of resistors with an equivalent resistance, thus reducing the network to a form easily analyzed. Following that, we consider the analysis of a circuit that contains a dependent source as well as an independent source.

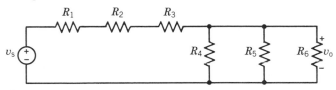

Figure 3-23 Circuit with a set of series resistors and a set of parallel resistors.

Consider the circuit shown in Figure 3-23. Note that it includes a set of resistors that are in series and another set of resistors that are in parallel. It is desired to find the output voltage v_o, so we wish to reduce the circuit to the equivalent circuit shown in Figure 3-24.

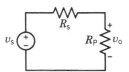

Figure 3-24 Equivalent circuit for the circuit of Figure 3-23.

Two circuits are **equivalent** if they exhibit identical characteristics at the same two terminals.

We note that the equivalent series resistance is

$$R_s = R_1 + R_2 + R_3$$

and the equivalent parallel resistance is

$$R_P = \frac{1}{G_P}$$

where

$$G_P = G_4 + G_5 + G_6$$

Then, using the voltage divider principle, with Figure 3-24 we have

$$v_o = \frac{R_P}{R_s + R_P}\, v_s$$

If a circuit has several combinations of sets of parallel resistors and sets of series resistors, one works through several steps, reducing the network to its simplest form.

Example 3-7

Consider the circuit shown in Figure 3-25a. Find the current i_1 when

$$R_4 = 2 \ \Omega \quad \text{and} \quad R_2 = R_3 = 8 \ \Omega$$

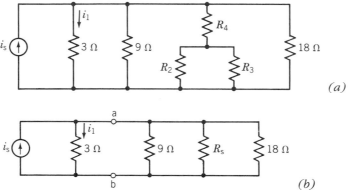

(a)

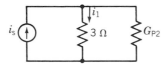

(b)

Figure 3-25 (a) Circuit for Example 3-7. (b) Partially reduced circuit for Example 3-7.

Solution

Since the objective is to find i_1, we will attempt to reduce the circuit so that the 3-Ω resistor is in parallel with one resistor and the current source i_s. Then we can use the current divider principle to obtain i_1. Since R_2 and R_3 are in parallel, we find an equivalent resistance as

$$R_{P1} = \frac{R_2 R_3}{R_2 + R_3}$$
$$= 4 \ \Omega$$

Then adding R_{P1} to R_4 we have a series equivalent resistor

$$R_s = R_4 + R_{P1}$$
$$= 2 + 4$$
$$= 6 \ \Omega$$

Now the R_s resistor is in parallel with three resistors as shown in Figure 3-25b. However, we wish to obtain the equivalent circuit as shown in Figure 3-26 so that we can find i_1.

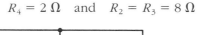

Figure 3-26 Equivalent circuit for Figure 3-25.

Therefore, we combine the 9-Ω resistor, the 18-Ω resistor, and R_s shown to the right of terminals a–b in Figure 3-25b into one parallel equivalent conductance G_{P2}. Thus, we find

$$G_{P2} = \frac{1}{9} + \frac{1}{18} + \frac{1}{R_s}$$
$$= \frac{1}{9} + \frac{1}{18} + \frac{1}{6}$$
$$= \frac{1}{3} \ \text{S}$$

Then, using the current divider principle,

$$i_1 = \frac{G_1 i_s}{G_P}$$

where

$$G_P = G_1 + G_{P2} = \frac{1}{3} + \frac{1}{3} = \frac{2}{3}$$

Therefore,

$$i_1 = \frac{1/3}{2/3} i_s$$

$$= \frac{1}{2} i_s$$

Let us consider a circuit with a dependent source as well as a set of parallel resistors and a set of series resistors. Under these circumstances we reduce the series resistors to their equivalent resistance and the set of parallel conductances to their equivalent conductance. When we reduce the circuit step by step, we must also ensure that we account for the circuit variable that is the excitation for the dependent source.

For example, let us find the current in the circuit shown in Figure 3-27. We can find the equivalent resistance for the two parallel branches on the right of the circuit. First, we note that $R_s = 4 + 8 = 12 \ \Omega$ for the far-right branch. Then, with 12 Ω in parallel with 4 Ω, we have

$$R_P = \frac{4(12)}{4 + 12} = 3 \ \Omega$$

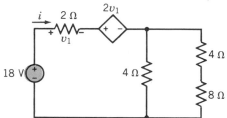

Figure 3-27 Circuit with a VCVS.

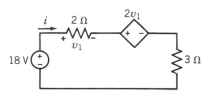

Figure 3-28 Equivalent circuit for Figure 3-27.

In order to preserve the voltage v_1, we will not add R_P to the 2-Ω resistor to reduce the circuit further. Thus we have the equivalent circuit of Figure 3-28. Using KVL around the loop, we obtain

$$-18 + 2i + 2v_1 + 3i = 0$$

However, we know from Ohm's law that

$$v_1 = 2i \qquad\qquad (3\text{-}26)$$

Therefore, we obtain

$$2i + 2(2i) + 3i = 18$$

or

$$i = \frac{18}{9} = 2 \ \text{A}$$

Thus, the introduction of a dependent source required us to obtain only one additional equation, (3-26). This equation was required to solve an otherwise normal KVL equation.

Example 3-8

Find the current i_2 and voltage v for the resistor R in Figure 3-29 when $R = 16\ \Omega$.

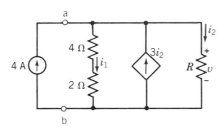

Figure 3-29 Circuit for Example 3-8 with a CCCS.

Solution

We first note that the dependent source is related to the current i_2. Then, writing the KCL equation at node a, we have

$$4 - i_1 + 3i_2 - i_2 = 0 \qquad (3\text{-}27)$$

Note, however from Ohm's law, that

$$i_1 = \frac{v}{4 + 2}$$

$$= \frac{v}{6}$$

and

$$i_2 = \frac{v}{R}$$

$$= \frac{v}{16}$$

Thus, Eq. 3-27 becomes

$$4 - \frac{v}{6} + 2\left(\frac{v}{16}\right) = 0$$

Therefore,

$$v = 96\ \text{V}$$

and

$$i_2 = \frac{v}{16} = 6\ \text{A}$$

Circuit analysis using the reduction of parallel and series subnetworks can often reduce a circuit to a very few equivalent elements. Then, using Kirchhoff's current and voltage laws, the circuit's currents and voltages are readily found.

Example 3-9

Find the current i_2 and the voltage v for the circuit shown in Figure 3-30.

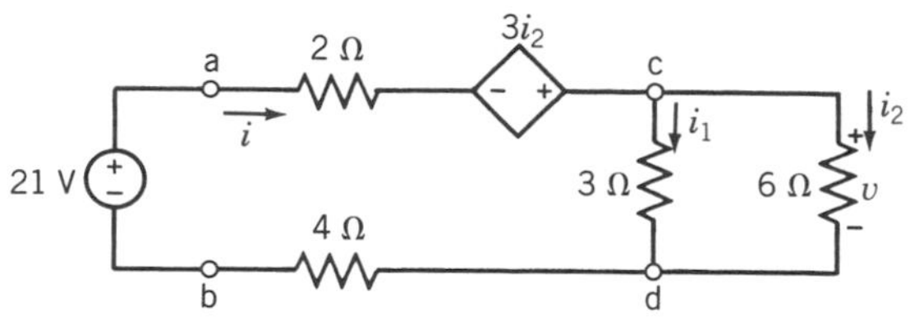

Figure 3-30 Circuit for Example 3-9 with a CCVS.

Solution

First we note that the dependent voltage source depends on the current i_2. Thus, prior to reducing the parallel circuit between terminals c and d, we write the equation for i_2 as

$$i_2 = \frac{v}{6}$$

Then the two parallel resistors can be reduced to

$$R_P = \frac{3(6)}{3 + 6}$$
$$= 2 \ \Omega$$

Hence the total series resistance around the loop is

$$R_s = 2 + R_P + 4$$
$$= 8 \ \Omega$$

Now, using KVL around the loop, we have

$$-21 + 8i - 3i_2 = 0 \qquad (3\text{-}28)$$

Using the current divider principle, we can relate i_2 to i as

$$i_2 = \frac{G_2 \, i}{G_P}$$

where $G_2 = 1/6$ and the conductance resulting from the 3-Ω resistor and the 6-Ω resistor in parallel is $G_P = 1/2$. Therefore, $i_2 = i/3$ or $i = 3i_2$. Substituting this relation for i in KVL Eq. 3-28, we have

$$-21 + 8(3i_2) - 3i_2 = 0$$

Therefore,

$$i_2 = 1 \text{ A} \quad \text{and} \quad v = 6i_2 = 6 \text{ V}$$

In general, we may find the equivalent resistance (or conductance) for a portion of a circuit consisting only of resistors and then replace that portion of the circuit with the equivalent resistance. For example, consider the circuit shown in Figure 3-31. We use

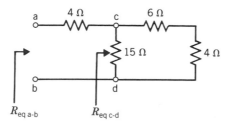

Figure 3-31 The equivalent resistance looking into terminals c–d is denoted as $R_{eq\,c-d}$.

$R_{\text{eq}\,x-y}$ to denote the equivalent resistance seen looking into terminals x−y. We note that the equivalent resistance to the right of terminals c−d is

$$R_{\text{eq}\,c-d} = \frac{15(6 + 4)}{15 + (6 + 4)}$$

$$= \frac{150}{25}$$

$$= 6\ \Omega$$

Then the equivalent resistance of the circuit to the right of terminals a−b is

$$R_{\text{eq}\,a-b} = 4 + R_{\text{eq}\,c-d}$$
$$= 4 + 6$$
$$= 10\ \Omega$$

If the circuit includes a dependent source as in Example 3-9, we must proceed, as noted in that example (see Figure 3-30), to find the current i in order to find the R_{eq} at terminals a−b. For Example 3-9, since $i = 3i_2$, we have $i = 3$ A. Then, using Ohm's law at terminals a−b we have

$$R_{\text{eq}} = \frac{v_s}{i}$$

$$= \frac{21}{3}$$

$$= 7\ \Omega$$

Therefore, in general, when we have a circuit with a dependent source and wish to find the equivalent resistance at two specified terminals x−y, we need only determine the voltage across and current through these terminals and then, using Ohm's law, note that

$$R_{\text{eq}\,x-y} = \frac{v_{xy}}{i_{xy}}$$

where v_{xy} is the voltage across those terminals and i_{xy} is the current entering the terminal x.

Referring back to Example 3-8, which considers the circuit shown in Figure 3-29, we note that the equivalent resistance for the circuit to the right of terminals a−b can readily be found. Since $v = 96$ V and $i_s = 4$ A, we have

$$R_{\text{eq}\,a-b} = \frac{v}{i_s}$$

$$= \frac{96}{4}$$

$$= 24\ \Omega$$

EXERCISE 3-10

Determine i and v_1 for the circuit shown in Figure E 3-10.
Answer: $v_1 = -10$ V, $i = -10/3$ A

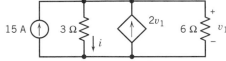

Figure E 3-10

3-7 | VERIFICATION EXAMPLE

The circuit shown in Figure 3V-1a was analyzed by writing and solving a set of simultaneous equations:

$$12 = v_2 + 4i_3, \qquad i_4 = \frac{v_2}{5} + i_3$$

$$v_5 = 4i_3, \qquad \frac{v_5}{2} = i_4 + 5i_4$$

A computer and the program Mathcad (Mathcad User's Guide, 1991) was used to solve the equations as shown in Figure 3 V-1b. It was determined that

$$v_2 = -60 \text{ V}, \qquad i_3 = 18 \text{ A}, \qquad i_4 = 6 \text{ A}, \qquad v_5 = 72 \text{ V}$$

Are these currents and voltages correct?

(a)

```
v2 := 0    i3 := 0    i4 := 0    v5 := 0

Given

      12 ≈ v2 + 4·i3      Apply KVL to loop A.

      i4 ≈ v2 + i3        Apply KCL at node b.
           5

      v5 ≈ 4·i3           Apply KVL to loop B.

      v5 ≈ i4 + 5·i4      Apply KCL at node c.
      2
                          ⎡ -60 ⎤
                          ⎢  18 ⎥
      Find (v2,i3,i4,v5) = ⎢   6 ⎥
                          ⎣  72 ⎦
```

(b)

Figure 3V-1 *(a)* An example circuit and *(b)* computer analysis using Mathcad.

The current i_2 can be calculated from v_2, i_3, i_4, and v_5 in a couple of different ways. First, Ohm's law gives

$$i_2 = \frac{v_2}{5}$$
$$= \frac{-60}{5}$$
$$= -12 \text{ A}$$

Next, applying KCL at node b gives

$$i_2 = i_3 + i_4$$
$$= 18 + 6$$
$$= 24 \text{ A}$$

Clearly i_2 cannot be both -12 and 24 A, so the values calculated for v_2, i_3, i_4, and v_5 cannot be correct. Checking the equations used to calculate v_2, i_3, i_4, and v_5, we find a sign error in the KCL equation corresponding to node b. This equation should be

$$i_4 = \frac{v_2}{5} - i_3$$

After making this correction, v_2, i_3, i_4, and v_5 are calculated to be

$$v_2 = 7.5 \text{ V}, \qquad i_3 = 1.125 \text{ A}, \qquad i_4 = 0.375 \text{ A}, \qquad v_5 = 4.5 \text{ V}$$

Now

$$i_2 = \frac{v_2}{5}$$
$$= \frac{7.5}{5}$$
$$= 1.5$$

and

$$i_2 = i_3 + i_4$$
$$= 1.125 + 0.375$$
$$= 1.5$$

This checks as we expected.

As an additional check, consider v_3. First, Ohm's law gives

$$v_3 = 4i_4$$
$$= 4(1.125)$$
$$= 4.5$$

Next, applying KVL to the loop consisting of the voltage source and the 4-Ω and 5-Ω resistors gives

$$v_3 = 12 - v_2$$
$$= 12 - 7.5$$
$$= 4.5$$

Finally, applying KVL to the loop consisting of the 2-Ω and 4-Ω resistors gives

$$v_3 = v_5$$
$$= 4.5$$

The results of these calculations agree with each other, indicating that

$$v_2 = 7.5 \text{ V}, \qquad i_3 = 1.125 \text{ A}, \qquad i_4 = 0.375 \text{ A}, \qquad v_5 = 4.5 \text{ V}$$

are the correct values.

3-8 DESIGN CHALLENGE SOLUTION

ADJUSTABLE VOLTAGE SOURCE

A circuit is required to provide an adjustable voltage. The specifications for this circuit are:

1 It should be possible to adjust the voltage to any value between -5 V and $+5$ V. It should not be possible to accidentally obtain a voltage outside this range.
2 The load current will be negligible.
3 The circuit should use as little power as possible.

The available components are:

1 Potentiometers: resistance values of 10 kΩ, 20 kΩ, and 50 kΩ are in stock;
2 A large assortment of standard 2% resistors having values between 10 Ω and
3 1 MΩ (see Appendix F);
   Two power supplies (voltage sources): one 12-V supply and one −12-V supply, both are rated at 100 mA (maximum).

Define the Situation

Figure 3D-1 shows the situation. The voltage v is the adjustable voltage. The circuit that uses the output of the circuit being designed is frequently called the "load." In this case, the load current is negligible, so $i = 0$.

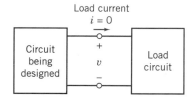

Figure 3D-1 The circuit being designed provides an adjustable voltage, v, to the load circuit.

State the Goal

A circuit providing the adjustable voltage

$$-5 \text{ V} \leq v \leq +5 \text{ V}$$

must be designed using the available components.

Generate a Plan

Make the following observations.

1 The adjustability of a potentiometer can be used to obtain an adjustable voltage v.
2 Both power supplies must be used so that the adjustable voltage can have both positive and negative values.
3 The terminals of the potentiometer cannot be connected directly to the power supplies because the voltage v is not allowed to be as large as 12 V or −12 V.

These observations suggest the circuit shown in Figure 3D-2a. The circuit in Figure 3D-2b is obtained by using the simplest model for each component in Figure 3D-2a.

To complete the design, values need to be specified for R_1, R_2, and R_p. Then, several results need to be checked and adjustments made, if necessary.

1 Can the voltage v be adjusted to any value in the range -5 V to $+5$ V?
2 Are the voltage source currents less than 1 mA? This condition must be satisfied if the power supplies are to be modeled as ideal voltage sources.
3 Is it possible to reduce the power absorbed by R_1, R_2, and R_p?

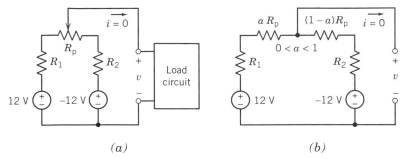

(a)   *(b)*

Figure 3D-2 *(a)* A proposed circuit for producing the variable voltage, v, and *(b)* the equivalent circuit after the potentiometer is modeled.

Take Action on the Plan

It seems likely the R_1 and R_2 will have the same value, so let $R_1 = R_2 = R$. Then it is convenient to redraw Figure 3D-2*b* as shown in Figure 3D-3.

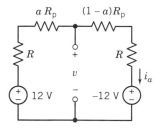

Figure 3D-3 The circuit after setting $R_1 = R_2 = R$.

Applying KVL to the outside loop yields

$$-12 + Ri_a + aR_p i_a + (1 - a)R_p i_a + Ri_a - 12 = 0$$

so

$$i_a = \frac{24}{2R + R_p}$$

Next applying KVL to the left loop gives

$$v = 12 - (R - aR_p)i_a$$

Substituting for i_a gives

$$v = 12 - \frac{24(R + aR_p)}{2R + R_p}$$

When $a = 0$, v must be 5 V, so

$$5 = 12 - \frac{24R}{2R + R_p}$$

Solving for R gives

$$R = 0.7R_p$$

Suppose the potentiometer resistance is selected to be $R_p = 20 \text{ k}\Omega$, the middle of the three available values. Then

$$R = 14 \text{ k}\Omega$$

As a check, notice that when $a = 1$

$$v = 12 - \left(\frac{14 \text{ k} + 20 \text{ k}}{28 \text{ k} + 20 \text{ k}}\right) 24 = 5$$

as required. The specification that

$$-5 \text{ V} \le v \le 5 \text{ V}$$

has been satisfied. The power absorbed by the three resistances is

$$p = i_a^2 (2R + R_p) = \frac{24^2}{2R + R_p}$$

so

$$p = 12 \text{ mW}$$

Notice that this power can be reduced by choosing R_p to be as large as possible, 50 kΩ in this case. Changing R_p to 50 kΩ requires a new value of R:

$$R = 0.7 \, R_p = 35 \text{ k}\Omega$$

Since

$$-5 \text{ V} = 12 - \left(\frac{35 \text{ k} + 50 \text{ k}}{70 \text{ k} + 50 \text{ k}}\right) 24 \le v \le 12 - \left(\frac{35 \text{ k}}{70 \text{ k} + 50 \text{ k}}\right) 24 = 5 \text{ V}$$

the specification that

$$-5 \text{ V} \le v \le 5 \text{ V}$$

has been satisfied. The power absorbed by the three resistances is now

$$p = \frac{24^2}{50 \text{ k} + 70 \text{ k}} = 5 \text{ mW}$$

Finally, the power supply current is

$$i_a = \frac{24}{50 \text{ k} + 70 \text{ k}} = 0.2 \text{ mA}$$

which is well below that 100 mA that the voltage sources are able to supply. The design is complete.

3-9 | DESIGN EXAMPLE—VOLTAGE DIVIDER

Problem

A voltage divider is connected to a source and a voltmeter as shown in Figure 3-32. Ideally, $R_s = 0$ and $R_m = \infty$. However, for one practical circuit, $R_s = 125\ \Omega$ and $R_m = 10\ k\Omega$. Select R_1 and R_2 to minimize the error introduced by R_s and R_m when it is desired that $v/v_s = 0.75$ (Svoboda 1992).

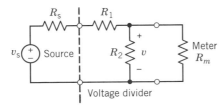

Figure 3-32 A voltage divider with a practical source and a meter.

Define the Situation

1 A voltage divider is used with a practical source resistance R_s.
2 The resistor R_2 is "loaded" by the meter resistance R_m.
3 The desired ratio is $v/v_s = 0.75$.

The Goal

Determine R_1 and R_2 to minimize the difference between the values of the voltage v for the ideal and practical cases.

Generate a Plan

1 Determine the voltage v for the ideal case.
2 Determine the voltage v' for the practical case.
3 Define a measure of the error.
4 Minimize the error and then determine R_2 and R_1.

Take Action Using the Plan

The ideal voltage divider is obtained when $R_s = 0$ and $R_m = \infty$. Then, the divider ratio is

$$v = \frac{R_2}{R_1 + R_2}\, v_s = av_s \tag{3-29}$$

where $a = R_2/(R_1 + R_2)$.

For the practical case, the output voltage v' is

$$v' = \frac{R_p}{R_s + R_1 + R_p}\, v_s$$

Substitute for R_p and $R_1 = R_2(1 - a)/a$ to obtain

$$v' = \frac{aR_2R_m}{(aR_s + R_2)(R_m + R_2) - aR_2^2}\, v_s \tag{3-30}$$

where $R_p = R_2 R_m/(R_2 + R_m)$ and a is as defined in Eq. 3-29. We will define the error as

$$e = \frac{v - v'}{v} = \frac{av_s - v'}{av_s} \tag{3-31}$$

Substituting Eq. 3-30 into Eq. 3-31, we have

$$e = 1 - \frac{R_2 R_m}{(aR_s + R_2)(R_m + R_2) - aR_2^2} \tag{3-32}$$

The objective is to minimize the error e by selecting R_2. We can use calculus and set $de/dR_2 = 0$ to find the best value of R_2. Alternatively, we can use a spreadsheet computer program and determine the best value of R_2. If we use calculus, we find that we require

$$R_2 = (3 \, R_m R_s)^{1/2}$$

when $a = 0.75$. Then, since $R_m = 10 \text{ k}\Omega$ and $R_s = 125 \, \Omega$, we require that

$$R_2 = 1936.5 \, \Omega$$

Also, we have from Eq. 3-29

$$R_2 = a(R_1 + R_2)$$

or

$$R_1 = \frac{R_2(1 - a)}{a}$$

$$= \frac{1936.5 \, (0.25)}{0.75}$$

$$= 645.5 \, \Omega$$

Evaluating Eq. 3-32, we find that the minimum error is 9.3 percent for this case.

SUMMARY

Electric circuits are used in many important applications such as undersea cable systems and electric railways.

One common element in almost all electric circuits is the resistor. A resistor has a resistance R measured in ohms. Ohm's law relates the voltage across the terminals of the resistor to the current into the positive terminal as $v = Ri$.

While the voltage and current of a resistor are linearly related, the power is $p = i^2 R = v^2/R$ watts.

Gustav Robert Kirchhoff formulated the laws that enable us to study a circuit. Kirchhoff's current law (KCL) states that the algebraic sum of the currents entering a node is zero. Kirchhoff's voltage law (KVL) states that the algebraic sum of the voltages around a closed path (loop) is zero.

Using Kirchhoff's laws and Ohm's law, we may analyze a circuit, determining the voltages and currents within the circuit. Two special circuits of interest are the voltage divider and the current divider. The voltage divider uses a series connection of resistors to divide the source voltage v_s in proportion to the ratio of the resistor R_n to the total series resistance $R_s = R_1 + R_2 + \cdots + R_N$.

The current divider, using parallel resistors, divides the source current i_s in proportion to the ratio of G_n to the total parallel conductance $G_P = G_1 + G_2 + \cdots + G_N$.

The equivalent resistance for a series connection of resistors is equal to the sum of the resistances. The equivalent conductance of a set of parallel connected conductances is equal to the sum of the conductances. Thus, an equivalent circuit may be obtained for a more complex circuit by using the equivalence principle for parallel conductors or series resistors.

For a circuit with a dependent source as well as an independent source, in order to find the equivalent circuit we simplify the circuit by replacing a series of resistors by their sum and a parallel set of conductances by their sum. However, we must account for the equation for the dependent source by retaining a relation for the variable on which the controlled source depends.

In general, the equivalent resistance (or conductance) at two terminals may be defined by the ratio of the terminal voltage to the current into the terminal (the inverse of this ratio for a conductance).

TERMS AND CONCEPTS

Closed-path   Traversal through a series of nodes ending at the starting node without encountering a node more than once.

Conductance   The inverse of the resistance; $G = 1/R$ in units of siemens (S). Many engineers use units of mhos ($\mho$).

Current DividerCircuit of parallel resistors that divides the source current i_s so that

$$i_n = \frac{G_n i_s}{G_1 + G_2 + \cdots + G_N}$$

Equivalent Circuit   Arrangement of circuit elements that is equivalent to a more complex arrangement of elements. A circuit equivalent to another circuit exhibits identical characteristics (behavior) between the same terminals.

Kirchhoff's Current Law   The algebraic sum of the currents entering a node is zero.

Kirchhoff's Voltage Law   The algebraic sum of the voltages around a closed path is zero.

Loop   Closed path around a circuit.

Node   Terminal common to two or more branches of a circuit; junction where two or more elements have a common connection.

Ohm's Law   The voltage across the terminals of a resistor is related to the current into the positive terminal as $v = Ri$.

Open Circuit   Condition that exists when the current between two terminals is identically zero, irrespective of the voltage across the terminals.

Parallel Connection   Arrangement of resistors so that each resistor has the same voltage appearing across it.

Resistor   Device or element whose primary purpose is to introduce resistance R into a circuit.

Series Connection   Circuit of a succession of resistors connected so that the same current passes through each resistor.

Short Circuit   Condition that exists when the voltage across two terminals is identically zero, irrespective of the current between the two terminals.

Voltage Divider   Circuit of a series of resistors that divides the input voltage by the ratio of the resistor R_n to the total series resistance, so

$$v_n = \frac{v_s R_n}{R_1 + R_2 + \cdots + R_N}$$

REFERENCES Chapter 3

Feldmann, Peter, and Rohrer, Ronald. "Proof of the Number of Independent Kirchhoff Equations in an Electrical Circuit," *IEEE Trans. Circuits,* July 1991, pp. 681–683.

Mathcad User's Guide, MathSoft Inc., Cambridge, MA, 1991.

Svoboda, James A. "Using Spreadsheets in Introductory Electrical Engineering Courses," *IEEE Trans. Education,* November 1992, pp. 16–21.

PROBLEMS

Section 3-3 Kirchhoff's Laws

P 3.3-1 For the circuit of Figure P 3.3-1, find i, v, and the power absorbed by the unknown circuit element if the power supplied by the 16-V source is 8 W.

Answer: $v = 8$ V, $i = -1/6$ A, $p = -4/3$ W

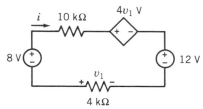

Figure P 3.3-1

P 3.3-2 A solar photovoltaic panel may be represented by the model shown in Figure P 3.3-2, where R_L is the load resistor. The source current is 2 A and $v_{ab} = 40$ V. Find R_1 when $R_2 = 10$ Ω and $R_L = 30$ Ω.

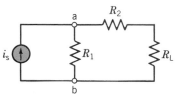

Figure P 3.3-2 Circuit model for solar photovoltaic panel.

P 3.3-3 Find i and the power absorbed by each element shown in Figure P 3.3-3.

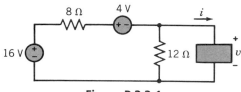

Figure P 3.3-3

P 3.3-4 Find i in the circuit in Figure P 3.3-4.

Answer $i = -2$ mA

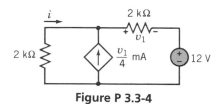

Figure P 3.3-4

P 3.3-5 Find the power absorbed by the 600-Ω resistor in Figure P 3.3-5.

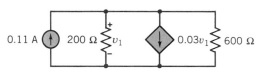

Figure P 3.3-5

P 3.3-6 The current i_2 is 10 A in the circuit shown in Figure P 3.3-6. Determine the resistance R.

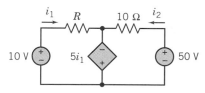

Figure P 3.3-6

P 3.3-7 Determine R_1 when $v/v_s = 0.5$ for the circuit of Figure P 3.3-7.

Answer: $R_1 = 5028$ Ω

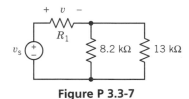

Figure P 3.3-7

P 3.3-8 Consider the circuit shown in Figure P 3.3-8. Some element voltages and element currents are given. What is the power delivered to element a? What is the power delivered to element b?

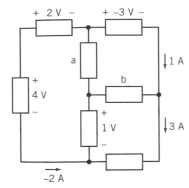

Figure P 3.3-8 Circuit with branch currents in amperes and voltages in volts.

P 3.3-9 For the circuit of Figure P 3.3-9
a. Find the power delivered by each source
b. Find the power delivered to each resistor.
c. Is energy conserved?

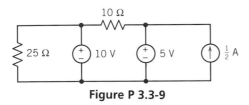

Figure P 3.3-9

P 3.3-10 What is the power delivered to 2-ohm resistor in Figure P 3.3-10? What is the power delivered by the 1-amp current source? What is the power delivered by the 6-volt voltage source?

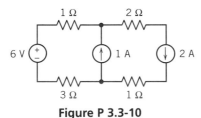

Figure P 3.3-10

Section 3-4   The Voltage Divider and the Equivalent Series Resistor

P 3.4-1 Consider the voltage divider shown in Figure P 3.4-1. It is desired that the power absorbed by R is 8 W when $R = 4\ \Omega$. Determine the required source v_s.

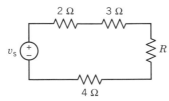

Figure P 3.4-1 Voltage divider circuit.

P 3.4-2 Consider the voltage divider shown in Figure P 3.4-1. It is desired that the power absorbed by R is 6 W when $v_s = 15$ V. Determine the required value of R.

P 3.4-3 The model of a cable and load resistor connected to a source is shown in Figure P 3.4-3. Determine the appropriate cable resistance, R, so that the output voltage, v_o, remains between 9 V and 13 V when the source voltage, v_s, varies between 20 V and 28 V. The cable resistance can only assume integer values when $20 < R < 100\ \Omega$.

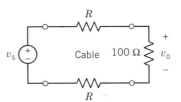

Figure P 3.4-3 Circuit with a cable.

Section 3-5   Parallel Resistors and Current Division

P 3.5-1 For the circuit shown in Figure P 3.5-1, find i using the current divider principle and determine the power absorbed by the 12-Ω resistor.
Answer: 0.3 A, 0.48 W

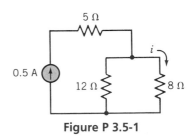

Figure P 3.5-1

P 3.5-2 Find i using current division if $I_0 = 6$ mA in the circuit shown in Figure P 3.5-2.

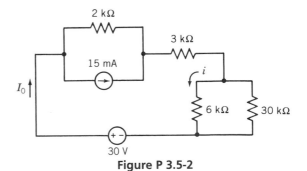

Figure P 3.5-2

P 3.5-3 Consider the circuit shown in Figure P 3.5-3 when $4 \, \Omega \leq R_1 \leq 6 \, \Omega$ and $R_2 = 10 \, \Omega$. Select the source i_s so that v_o remains between 9 V and 13 V.

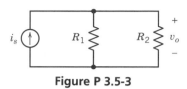

Figure P 3.5-3

P 3.5-4 Determine R and the power delivered to the 6-Ω resistor for the circuit shown in Figure P 3.5-4 when $i = 2$ A.

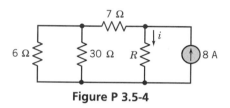

Figure P 3.5-4

P 3.5-5 Determine the voltage v_o using the current division principle for the circuit of Figure P 3.5-5.
Answer: $v_o = 6.09$ V

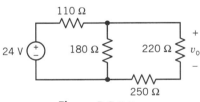

Figure P 3.5-5

Section 3-6   Circuit Analysis (a) Circuits with Independent Sources

P 3.6-1 Find i and i_1 using appropriate circuit reductions and the current divider principle for the circuit of Figure P 3.6-1.
Answer: $i = 2$ A, $i_1 = -3/4$ A

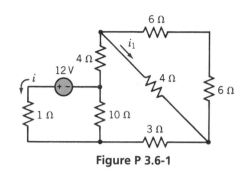

Figure P 3.6-1

P 3.6-2 Find i using appropriate circuit reductions and the current divider principle for the circuit of Figure P 3.6-2.

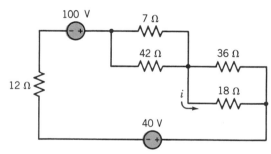

Figure P 3.6-2

P 3.6-3 Find i and the power absorbed by the 12-kΩ resistor for the circuit of Figure P 3.6-3.
Answer: $i = 8$ mA, $p = 85.3$ mW

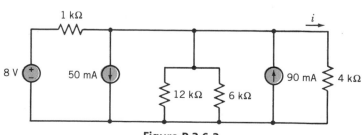

Figure P 3.6-3

P 3.6-4 If $v = 2$ V, find the resistance R of the circuit shown in Figure P 3.6-4.
Answer: $R = 2\ \Omega$

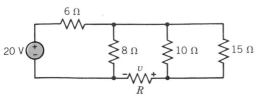

Figure P 3.6-4

P 3.6-5 Most of us are familiar with the effects of a mild electric shock. The effects of a severe shock can be devastating and often fatal. Shock results when current is passed through the body. A person can be modeled as a network of resistances. Consider the model circuit shown in Figure P 3.6-5. Determine the voltage developed across the heart and the current flowing through the heart of the person when he or she firmly grasps one end of a voltage source whose other end is connected to the floor. The heart is represented by R_h. The floor has resistance to current flow equal to R_f, and the person is standing barefoot on the floor. This type of accident might occur at a swimming pool or boat dock. The upper-body resistance R_U and lower-body resistance R_L vary from person to person.

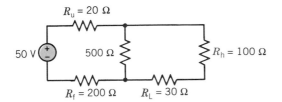

Figure P 3.6-5 The resistance of the heart. $R_h = 10\ \Omega$.

P 3.6-6 Find i for the circuit of Figure P 3.6-6. Note that the units of the conductances are siemens (S).
Answer: $i = 1.6$ A

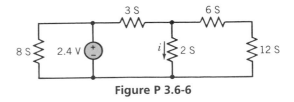

Figure P 3.6-6

P 3.6-7 If $v_1 = 12$ V, find i_s and v for the circuit of Figure P 3.6-7.

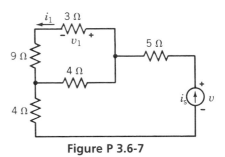

Figure P 3.6-7

P 3.6-8 For the circuit shown in Figure P 3.6-8:
(a) Find the current i.
(b) Find the voltage v.
(c) Find the power absorbed by the 2-A current source.
(d) Find the power delivered by the 8-V voltage source.

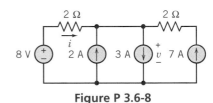

Figure P 3.6-8

P 3.6-9 Find the power supplied by the current source and the power absorbed by the 900-Ω resistor in parallel with the current source of Figure P 3.6-9.

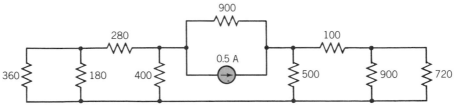

Figure P 3.6-9 All resistances in ohms.

P 3.6-10   Find i_1 in the circuit shown in Figure P 3.6-10.

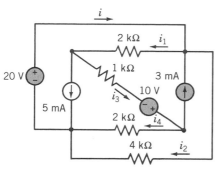

Figure P 3.6-10

P 3.6-11   For the circuit shown in Figure P 3.6-11, find the power absorbed by each element and show that the total circuit neither absorbs nor dissipates energy.

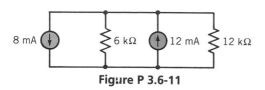

Figure P 3.6-11

P 3.6-12   Electric streetcar railways have been used for about 100 years. An electric railway was established by Werner Siemens from Charlottenbourg to Spandau in Berlin in 1885, as shown in Figure P 3.6-12a. This railway may be represented by the electric circuit shown in Figure P 3.6-12b. Find the power delivered to the motor R_L by the central power source.

(a)

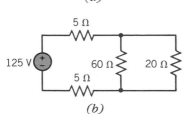

(b)

Figure P 3.6-12   (a) The electric street railway in Berlin in 1885. Courtesy of Burndy Library. (b) Circuit model of the railway, where $R_L = 20\ \Omega$.

Section 3-6   Circuit Analysis (b) Circuits with Independent and Dependent Sources

P 3.6-13   Find the load resistance R_L if $v = 0.06$ V in the circuit of Figure P 3.6-13. This circuit is a model of a transistor amplifier with a load R_L.

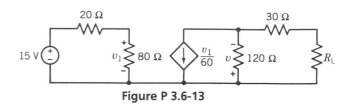

Figure P 3.6-13

P 3.6-14   For the circuit of Figure P 3.6-14, find v_1 and the power absorbed by each element.
Answer: $v_1 = 3$ V

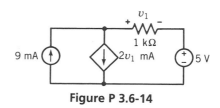

Figure P 3.6-14

P 3.6-15   Find v_a and i in the circuit shown in Figure P 3.6-15.

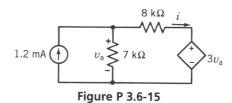

Figure P 3.6-15

P 3.6-16   Find v_x and the power absorbed by the conductance of 5 mS, for the circuit shown in Figure P 3.6-16. The dependent source units are mA/V.

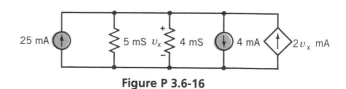

Figure P 3.6-16

P 3.6-17   A model of a common-emitter transistor amplifier is shown in Figure P 3.6-17. Find the voltage v_o when $v_s = 1$ mV.
Answer: $v_o = 4$ V

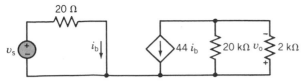

Figure P 3.6-17 Model of a transistor amplifier.

P 3.6-18 Find v_1 and v in the circuit shown in Figure P 3.6-18.

Answer: $v_1 = 4$ V, $v = 15$ V

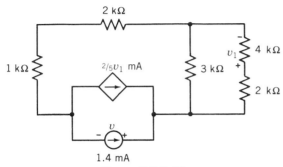

Figure P 3.6-18

P 3.6-19 Find i in the circuit shown in Figure P 3.6-19.

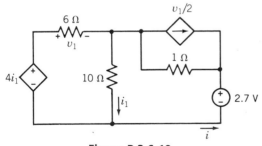

Figure P 3.6-19

Section 3-6 Circuit Analysis (c) Equivalent Resistance of a Circuit

P 3.6-20 Find i and $R_{eq\,a-b}$ if $v_{ab} = 40$ V in the circuit of Figure P 3.6-20.

Answer: $R_{eq\,a-b} = 8\ \Omega$, $i = 5/6$ A

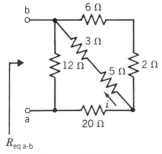

Figure P 3.6-20

P 3.6-21 Find R_{eq}, i, and v if $v_{ab} = 12$ V for the circuit of Figure P 3.6-21.

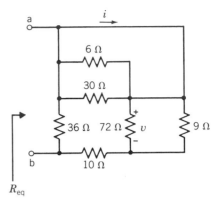

Figure P 3.6-21

P 3.6-22 For the circuit of Figure P 3.6-22, find $R_{eq\,a-b}$. Hint: Note the symmetry of the circuit.

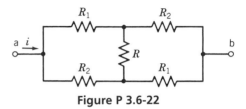

Figure P 3.6-22

P 3.6-23 The source $v_s = 240$ volts is connected to three equal resistors as shown in Figure P 3.6-23. Determine R when the voltage source delivers 1920 W to the resistors.

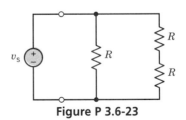

Figure P 3.6-23

P 3.6-24 For the circuit shown in Figure P 3.6-24, find $R_{eq\,a-b}$ and $R_{eq\,cd}$. Also determine the voltage v_{cd} when $v_{ab} = 20$ V.

Answer: $R_{eq\,c-d} = 6\ \Omega$

Figure P 3.6-24

P 3.6-25 Find the R_{eq} at terminals a–b in Figure P 3.6-25. Also determine i, i_1, and i_2.

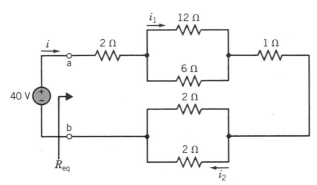

Figure P 3.6-25

P 3.6-26 For the circuit of Figure P 3.6-26, given that $R_{eq} = 9\ \Omega$, find R.
Answer: $R = 15\ \Omega$

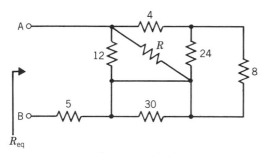

Figure P 3.6-26

P 3.6-27 For the circuit of Figure P 3.6-27, find R_{eq} at terminals a–b and the current i if $v_{ab} = 14$ V.

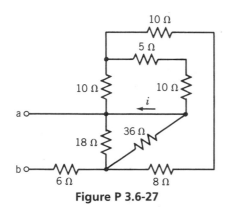

Figure P 3.6-27

P 3.6-28 Determine the equivalent resistance, R_{xy}, of the circuit shown in Figure P 3.6-28.

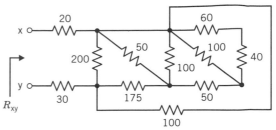

Figure P 3.6-28 All resistances in ohms.

P 3.6-29 The circuit of 1-Ω resistors shown in Figure P 3.6-29 extends to infinity in both directions. Determine the resistance R_{ab} between terminals a and b.

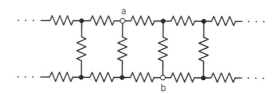

Figure P 3.6-29

P 3.6-30 The model of an amplifier is shown in Figure P 3.6-30. Determine R_i, R_o, and v_o when $R_G = 10\ M\Omega$, $g = 5$ mS, $R_D = 1\ k\Omega$, $R_2 = 100\ \Omega$, and $R_s = 100\ \Omega$.

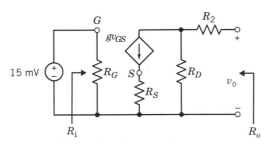

Figure P 3.6-30 Amplifier circuit.

P 3.6-31 Determine R_{eq} for the circuit shown in Figure P 3.6-31 in the form of a continued fraction as

$$R_1 + \cfrac{1}{G_2 + \cdots}$$

Calculate R_{eq} when $R_1 = 1\ \Omega$ and $G_2 = 1/2$ S.

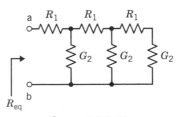

Figure P 3.6-31

P 3.6-32 Each resistance of the infinite network shown in Figure P 3.6-32 is R. Determine the resistance R_{ab}.

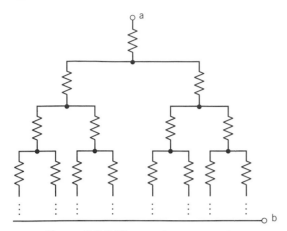

Figure P 3.6-32 An infinite network.

P 3.6-33 The resistances shown in Figure P 3.6-33 form three geometric progressions extending to infinity. Determine the equivalent resistance R_{ab}.

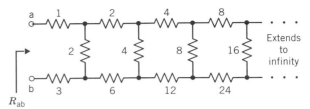

Figure P 3.6-33 All resistances in ohms.

P 3.6-34 Determine R when $R_{eq} = 20 \ \Omega$ for the circuit shown in Figure P 3.6-34.

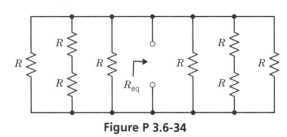

Figure P 3.6-34

VERIFICATION PROBLEMS

VP 3-1 A computer analysis program, used for the circuit of Figure VP 3-1, provides the following branch currents and voltages: $i_1 = -0.833$, $i_2 = -0.333$, $i_3 = -1.167$, and $v = -2.0$. Are these answers correct?

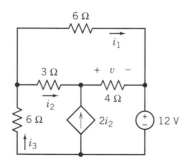

Figure VP 3-1

VP 3-2 For the circuit of Figure VP 3-2, a student notebook reports the current, i, is 1.25 A. Verify this report using current division.

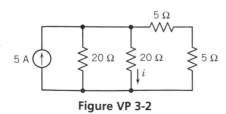

Figure VP 3-2

VP 3-3 For the circuit of Figure VP 3-3, it is reported that v_o is 6.25 V. Verify this report using the voltage divider principle.

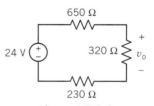

Figure VP 3-3

VP 3-4 The circuit of Figure VP 3-4 represents an auto's electrical system. A report states that $i_H = 9$ A, $i_B = -9$ A, and $i_A = 19.1$ A. Verify if this result is correct.

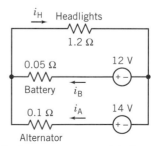

Figure VP 3-4 Electric circuit model of an automobile's electrical system.

DESIGN PROBLEMS

DP 3-1 A circuit is used to supply energy to a heater as shown in Figure DP 3-1. It is desired to supply 3.53 mW to the heater. Determine an appropriate voltage source v_s and a resistance R_1.

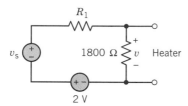

Figure DP 3-1 A heater circuit.

DP 3-2 A circuit is shown in Figure DP 3-2 with an unspecified resistance R. Select R so that the equivalent resistance looking into terminals a–b is 1 Ω.

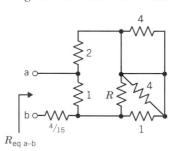

Figure DP 3-2 All resistances in ohms.

DP 3-3 A circuit is shown in Figure DP 3-3 with an unspecified current source constant. Select the constant g so that the voltage $v = 16$ V.

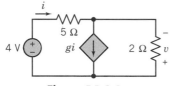

Figure DP 3-3

DP 3-4 A phonograph pickup, stereo amplifier, and speaker are shown in Figure DP 3-4a and redrawn as a circuit model as shown in Figure DP 3-4b. Determine the resistance R so that the voltage v across the load is 16 V. Determine the power delivered to the speaker.

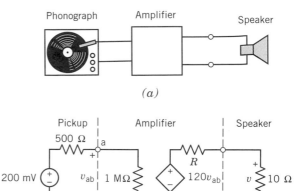

(a)

(b)

Figure DP 3-4 A phonograph stereo system.

DP 3-5 A Christmas tree light set is required that will operate from a 6-V battery on a tree in a city park. The heavy-duty battery can provide 9 A for the 4-hour period of operation each night. Design a parallel set of lights (select the maximum number of lights) when the resistance of each bulb is 12 Ω.

DP 3-6 A circuit with a subcircuit box is shown in Figure DP 3-6. The subcircuit is known to absorb 150 W prior to $t = 5$ s. After the switch is opened at $t = 5$ s, the power absorbed by the box is 65 W. Calculate a suitable value for R and specify your model for the subcircuit.

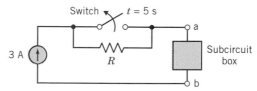

Figure DP 3-6 Circuit with subcircuit box.

DP 3-7 A designer requires the circuit of Figure DP 3-7 to exhibit a resistance R_{eq} at terminals a–b of 24 Ω. Select a suitable value for g and R when $1 \leq g \leq 5$.

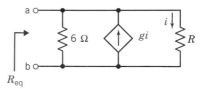

Figure DP 3-7

CHAPTER 4

METHODS OF ANALYSIS OF RESISTIVE CIRCUITS

PREVIEW

Increasingly complex circuits were required to meet the needs of ever-growing communication systems. These circuits were effective in transmitting information over great distances. In Chapter 3 we analyzed circuits consisting of resistors as well as voltage and current sources. As the complexity of circuits increased, analysis techniques were developed that incorporated rigorous systematic methods.

In this chapter we define and utilize two widely used methods of analysis: (1) node voltage and (2) mesh current. These very powerful methods are widely used today for the analysis of large complex circuits in communications and electrical systems.

4-1 DESIGN CHALLENGE

POTENTIOMETER ANGLE DISPLAY

A circuit is needed to measure and display the angular position of a potentiometer shaft. A *potentiometer* is a variable resistor with three terminals. Two terminals are connected to the opposite ends of the resistive element and the third connects to the sliding contact. (Dorf, 1993). The angular position, θ, will vary from -180 degrees to 180 degrees.

Figure 4D-1 illustrates a circuit that could do the job. The $+15$-V and -15-V power supplies, the potentiometer, and resistors R_1 and R_2 are used to obtain a voltage, v_i, that is proportional to θ. The amplifier is used to change the constant of proportionality in order to obtain a simple relationship between θ and the voltage, v_o, displayed by the voltmeter. In this example, the amplifier will be used to obtain the relationship

$$v_o = k \cdot \theta \quad \text{where } k = 0.1 \frac{\text{volt}}{\text{degree}} \tag{4D-1}$$

so that θ can be determined by multiplying the meter reading by 10. For example, a meter reading of -7.32 V indicates that $\theta = -73.2$ degrees.

Define the Situation

The circuit diagram in Figure 4D-2 is obtained by modeling the power supplies as ideal voltage sources, the voltmeter as an open circuit, and the potentiometer by two

118

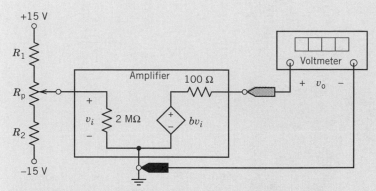

Figure 4D-1 Proposed circuit for measuring and displaying the angular position of the potentiometer shaft.

resistors. The parameter, a, used to model the potentiometer varies from 0 to 1 as θ varies from -180 degrees to 180 degrees. That means

$$a = \frac{\theta}{360°} + \frac{1}{2} \tag{4D-2}$$

Solving for θ gives

$$\theta = \left(a - \frac{1}{2} \right) \cdot 360° \tag{4D-3}$$

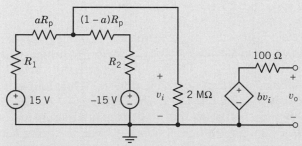

Figure 4D-2 Circuit diagram containing models of the power supplies, voltmeter, and potentiometer.

State the Goal

Specify values of resistors R_1 and R_2, the potentiometer resistance R_p, and the amplifier gain b that will cause the meter voltage, v_o, to be related to the angle θ by Eq. 4D-1.

Generate a Plan

Analyze the circuit shown in Figure 4D-2 to determine the relationship between v_i and θ. Select values of R_1, R_2, and R_p. Use these values to simplify the relationship between v_i and θ. If possible, calculate the value of b that will cause the meter voltage, v_o, to be related to the angle θ by Eq. 4D-1. If this isn't possible, adjust the values of R_1, R_2, and R_p and try again.

Chapter 4 describes two methods, called nodal analysis and mesh analysis, for analyzing circuits like the one shown in Figure 4D-2. We will return to this problem at the end of the chapter after learning about node and mesh analysis.

Figure 4-1 Samuel F. B. Morse, the inventor of the electric telegraph.

4-2 ELECTRIC CIRCUITS FOR COMMUNICATIONS

Samuel F. B. Morse became engaged in the search for an efficient electric telegraph in the 1830s. He worked on the design of a sending device and a code of dots and dashes for letters and numbers. Morse, shown in Figure 4-1, publicly demonstrated his telegraph on January 24, 1838. On February 21, 1838, President Van Buren witnessed the transmission of messages over 10 miles of wires. On April 7, 1838, Morse applied for a patent.

By 1843 the U.S. Senate appropriated funds for the construction of a telegraph line between Baltimore and Washington, D.C. On May 24, 1844, the famous first message was sent: "What hath God wrought!"

By 1851 there were over 50 telegraph companies in the United States. Telegraphs were widely used by railways and the military, in addition to private individuals.

However, the telegraph did not offer the advantages of voice communication and of dialogue, although it did provide an expanding communications network in the United States and Europe. The development of a telephone became a central activity by the 1870s.

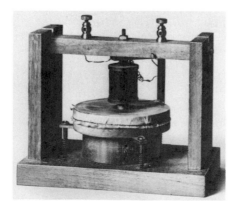

Figure 4-2 The first telephone transmitter with its parchment diaphragm attached to the magnetized metallic reed. This instrument was used to transmit the first speech sounds electrically in 1875. Courtesy of Bell Laboratories.

Figure 4-3 Alexander Graham Bell at the New York end of the circuit to Chicago as this line was opened on October 18, 1892, as part of the ceremonies accompanying the Columbian Exposition. Courtesty of Bell Laboratories.

Alexander Graham Bell (1847–1922) is commonly singled out as the inventor who effectively produced a practical telephone. Bell produced the first telephone transmitter, shown in Figure 4-2, in 1875.

In New Haven, Connecticut, in 1878, twenty-one subscribing parties could use eight lines to the first central switchboard. New York was first connected to Chicago in 1892. Bell is shown in Figure 4-3 at the opening of the Chicago line in 1892.

There were twice as many telephones in the United States—over 155,000—as in Europe in 1885. However, by 1898 the number of lines was greater in Europe. After that time, the United States maintained a widening lead in the use of the telephone. The growth in the number of telephones in the United States from 1925 to 1975 is shown in Figure 4-4. By 1925 there were 20 million telephones in use in the United States. The telephone has become almost universal.

With the advent of the telephone system and the electric power system, complex electric circuits became increasingly common. Thus, it became necessary to develop rigorous useful methods of analysis of complex circuits.

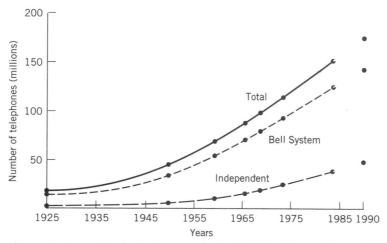

Figure 4-4 Growth of telephone service in the United States. Source: Bell Laboratories.

4-3 NODE VOLTAGE ANALYSIS OF CIRCUITS WITH CURRENT SOURCES

In the previous chapter we analyzed simple circuits that could be reduced to one containing only two nodes. We then obtained a single equation for the one unknown quantity, typically the voltage between the two nodes. We now wish to consider more complicated circuits with two or more nodes and to develop an analysis method for obtaining *node voltages*.

In this chapter we consider only planar circuits, which are defined as circuits that can be drawn on a plane so that no branches cross over each other. A *branch* is a path that connects two nodes. Consider the circuit shown in Figure 4-5*a*. Since an ideal wire connects nodes c and c, these two nodes are both labeled as one, and Figure 4-5*a* may be redrawn as Figure 4-5*b*. Thus, this circuit has three nodes and four branches. We note that this circuit is planar. A *planar* circuit can be drawn on a plane without branches crossing each other. Nonplanar networks are discussed in Chapter 17. A nonplanar circuit is shown in Figure 4-6, where crossover is identified and cannot be eliminated by redrawing the branches.

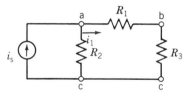
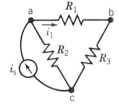

Figure 4-5 Circuit with three nodes.

Analyzing a connected circuit containing n nodes will require $n - 1$ KCL equations. One way to obtain these equations is to apply KCL at each node of the circuit except for one. The node at which KCL is not applied is called the reference node. Any node of the circuit can be selected to be the reference node. We will often choose the node at the bottom of the circuit to be the reference node. When the circuit contains a grounded power supply, the ground node of the power supply is usually selected as the reference node.

The voltage at any node of the circuit, relative to the reference node, is called a node voltage. Considering the circuit shown in Figure 4-5, we may identify node c as the reference node. The voltages v_{ac} and v_{bc} are to be determined. We will consider the voltage at the reference node as being known and typically determine the node voltages v_a and v_b with respect to the reference node c dropping the subscript c on the voltage v_{ac} and v_{bc}. Thus, voltage v_b could be written as v_{bc}, but since c is known as the reference, we

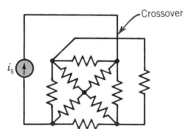

Figure 4-6 Nonplanar circuit with a crossover.

can assume v_b implies the voltage at node b with reference to c. The subscript c can be dropped since the voltage at the reference is assumed to be zero.

In order to determine the voltage at a node, we use Kirchhoff's current law at each of the circuit's nodes, except at the reference node. This set of equations enables us to find the node voltages.

We can arbitrarily choose any node as the reference node. However, it is convenient to choose the node with the most connected branches. If a choice has to be made for the reference node between two nodes having the same number of branches connected, we usually choose the node at the bottom of the circuit.

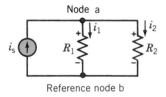

Node a

Reference node b

Figure 4-7 Circuit with two nodes where the lower node is the reference node.

Consider the circuit shown in Figure 4-7. This circuit has two nodes. We assign the lower node as the reference node and then seek to determine v_a. Using Kirchhoff's current law at node a, we have

$$i_s = i_1 + i_2$$
$$= \frac{v_a - v_b}{R_1} + \frac{v_a - v_b}{R_2} \tag{4-1}$$

Therefore, since $v_b = 0$

$$v_a = i_s/(G_1 + G_2)$$

where $G_n = 1/R_n$. Once we have determined v_a with respect to the reference node, we are also able to determine all the currents.

Now let us use Kirchhoff's current law to determine the two node voltages of the circuit shown in Figure 4-5. First we set node c as the reference node. It is usually assumed that the voltage at the reference node is equal to zero. Then at node a we use the KCL equations, where we set the currents entering the node equal to those leaving the node. Also, it is important to note that the current i_1 leaving node a is $i_1 = (v_a - v_b)/R_1$, since $v_{ab} = v_{ac} - v_{bc} = v_a - v_b$ by KVL.

Writing KCL equation at node a, we have

$$i_s = \frac{v_a}{R_2} + \frac{v_a - v_b}{R_1} \tag{4-2}$$

Similarly, the KCL equation at node b is

$$i_1 = \frac{v_b}{R_3} \tag{4-3}$$

However, since

$$i_1 = \frac{v_a - v_b}{R_1}$$

we obtain

$$\frac{v_a - v_b}{R_1} = \frac{v_b}{R_3}$$

or

$$0 = \frac{v_b - v_a}{R_1} + \frac{v_b}{R_3} \tag{4-4}$$

Note that the current *leaving* node b is $(v_b - v_a)/R_1$.

If $R_1 = 1\,\Omega$, $R_2 = R_3 = 0.5\,\Omega$, and $i_s = 4$ A, the two node equations (4-2) and (4-4) may be rewritten as

$$2v_a + \frac{v_a - v_b}{1} = 4 \tag{4-5}$$

$$\frac{v_b - v_a}{1} + 2v_b = 0 \tag{4-6}$$

Rewriting these two equations in terms of the two unknown voltages v_a and v_b, we have

$$3v_a - v_b = 4 \tag{4-7}$$

$$-v_a + 3v_b = 0 \tag{4-8}$$

Adding three times Eq. 4-7 to Eq. 4-8, we have

$$8v_a = 12$$

or

$$v_a = \frac{3}{2} \text{ V} \quad \text{and} \quad v_b = \frac{1}{2} \text{ V}$$

The voltage at node a relative to node b is calculated to be

$$v_a - v_b = 1 \text{ V}$$

Let us consider the three-node circuit shown in Figure 4-8 with two independent current sources and identify node c as the reference node. The lower node includes the ideal wire at node c. We call the reference node the *ground node* since many actual circuits include a connection of one of the nodes to the chassis or the earth (ground).

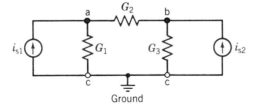

Figure 4-8 Circuit with two unknown node voltages and two independent current sources. The reference node is node c and includes a ground symbol.

Writing KCL equation at node a:

$$i_{s1} = G_1 v_a + G_2(v_a - v_b) \tag{4-9}$$

Writing KCL equation at node b:

$$i_{s2} = G_3 v_b + G_2(v_b - v_a) \tag{4-10}$$

We may rewrite Eqs. 4-9 and 4-10 as

$$(G_1 + G_2)v_a - G_2 v_b = i_{s1} \tag{4-11}$$

and

$$-G_2 v_a + (G_2 + G_3)v_b = i_{s2} \tag{4-12}$$

Note that the coefficient of v_a in Eq. 4-11 (for node a) is the sum of the conductances at node a. The coefficient of v_b is the negative of the conductance between node a and node b. Similarly, in Eq. 4-12 (for node b) the coefficient of v_b is the sum of the conductances at node b and the coefficient of v_a is the negative of the conductance between node b and node a.

In general, for networks containing only conductances and current sources, applying KCL at node k results in the coefficient of v_k being the sum of the conductances at node k. The coefficient of the other terms is the negative of the conductance between those nodes and the kth node. We apply KCL at all the circuits' nodes except the reference node.

TOOLBOX 4-1

MATRICES

The rectangular array of numbers

$$
\mathbf{A} = \begin{bmatrix}
a_{11} & a_{12} & \cdots & a_{1n} \\
a_{21} & a_{22} & \cdots & a_{2n} \\
\cdot & \cdot & & \cdot \\
\cdot & \cdot & & \cdot \\
\cdot & \cdot & & \cdot \\
a_{m1} & a_{m2} & \cdots & a_{mn}
\end{bmatrix} \tag{1}
$$

is known as a *matrix*. The numbers a_{ij} are called *elements* of the matrix, with the subscript i denoting the row and the subscript j denoting the column.

A matrix with m rows and n columns is said to be a matrix of *order* (m, n) or alternatively called an $m \times n$ (m by n) matrix. When the number of the columns equals the number of rows, $m = n$, the matrix is called a *square matrix* of order n. It is common to use boldface capital letters to denote an $m \times n$ matrix.

A matrix consisting of only one column, that is, an $m \times 1$ matrix, is known as a column matrix or, more commonly, a *column vector*. We represent a column vector with boldface lowercase letters as

$$
\mathbf{v} = \begin{bmatrix}
v_1 \\
v_2 \\
\cdot \\
\cdot \\
\cdot \\
v_m
\end{bmatrix} \tag{2}
$$

When the elements of a matrix have a special relationship so that $a_{ij} = a_{ji}$, it is called a *symmetrical* matrix. Thus, for example,

$$
\mathbf{H} = \begin{bmatrix}
3 & -2 & 1 \\
-2 & 6 & 4 \\
1 & 4 & 8
\end{bmatrix} \tag{3}
$$

is a symmetrical matrix of order (3,3).

We can write the node equations more compactly using matrices. For resistive circuits that contain only independent current sources the conductance matrix **G** is given by

$$\mathbf{G} = \begin{bmatrix} \sum_a G & -G_{ab} & -G_{ac} \\ -G_{ab} & \sum_b G & -G_{bc} \\ -G_{ac} & -G_{bc} & \sum_c G \end{bmatrix} \tag{4-13}$$

where $\sum\limits_n G$ is the sum of the conductances at node n and G_{ij} is the sum of the conductances connecting nodes i and j. For a resistive circuit without dependent sources, $\mathbf{G}$ is a
symmetric matrix. Of course, if there are four-node voltages to be solved for, the pattern of
symmetry remains for $\mathbf{G}$ except the matrix is now a 4×4 matrix.

The node voltage matrix equation for a circuit with N unknown node voltages is

$$\mathbf{Gv} = \mathbf{i}_s \tag{4-14}$$

where

$$\mathbf{v} = \begin{bmatrix} v_a \\ v_b \\ \cdot \\ \cdot \\ \cdot \\ v_N \end{bmatrix}$$

which is the vector consisting of the N unknown node voltages. The matrix

$$\mathbf{i}_s = \begin{bmatrix} i_{s1} \\ i_{s2} \\ \cdot \\ \cdot \\ \cdot \\ i_{sN} \end{bmatrix}$$

is the vector consisting of the N current sources where i_{sn} is the sum of all the source
currents *entering* the node n. If the nth current source is not present, then $i_{sn} = 0$.

Example 4-1

Let us consider the circuit shown in Figure 4-9. Find the three node voltages, v_a, v_b, and v_c,
when all the conductances are equal to 1 S.

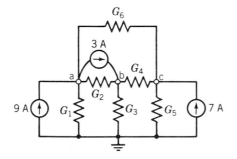

Figure 4-9 Circuit with three unknown node
voltages.

Solution

At node a: $(G_1 + G_2 + G_6)v_a - G_2v_b - G_6v_c = 9 - 3$

At node b: $-G_2v_a + (G_4 + G_2 + G_3)v_b - G_4v_c = 3$

At node c: $-G_6v_a - G_4v_b + (G_4 + G_5 + G_6)v_c = 7$

Substituting the numerical values of the conductances, we see that

$$3v_a - v_b - v_c = 6$$

$$-v_a + 3v_b - v_c = 3$$

$$-v_a - v_b + 3v_c = 7$$

Therefore,

$$\mathbf{G} = \begin{bmatrix} 3 & -1 & -1 \\ -1 & 3 & -1 \\ -1 & 1 & 3 \end{bmatrix}$$

and

$$\mathbf{i}_s = \begin{bmatrix} 6 \\ 3 \\ 7 \end{bmatrix}$$

for Eqn. 4-14.

TOOLBOX 4-2

CRAMER'S RULE AND DETERMINANTS

For a matrix equation

$$\mathbf{Ax} = \mathbf{b} \tag{1}$$

Cramer's rule states that the solution for the unknown, x_k, of the simultaneous equations of Eq. 1 is

$$x_k = \frac{\Delta_k}{\Delta} \tag{2}$$

where Δ is the determinant of $\mathbf{A}$ and Δ_k is Δ with the kth column replaced by the column vector $\mathbf{b}$.

The *determinant* of a matrix is a numerical value. We define the determinant of a square matrix $\mathbf{A}$ as Δ, where

$$\Delta = \begin{vmatrix} a_{11} & a_{12} & \cdots & a_{1n} \\ a_{21} & a_{22} & \cdots & a_{2n} \\ \cdot & \cdot & & \cdot \\ \cdot & \cdot & & \cdot \\ \cdot & \cdot & & \cdot \\ a_{n1} & a_{n2} & \cdots & a_{nn} \end{vmatrix} \tag{3}$$

In general, we are able to determine the determinant Δ in terms of cofactors and minors. The determinant of a submatrix of $\mathbf{A}$ obtained by deleting from $\mathbf{A}$ the ith row and the jth column is called the *minor* of the element a_{ij} and denoted as m_{ij}.

The cofactor c_{ij} is a minor with an associated sign, so that

$$c_{ij} = (-1)^{(i+j)} m_{ij} \tag{4}$$

The rule for evaluating the $n \times n$ determinant Δ is

$$\Delta = \sum_{j=1}^{n} a_{ij} c_{ij} \tag{5}$$

for a selected value of i. Alternatively, we can obtain Δ by using the jth column, and thus

$$\Delta = \sum_{i=1}^{n} a_{ij} c_{ij} \qquad (6)$$

for a selected value of j.

See Appendix B for examples.

The simultaneous equations may be solved by using Cramer's rule. We find

$$v_a = \frac{\begin{vmatrix} 6 & -1 & -1 \\ 3 & 3 & -1 \\ 7 & -1 & 3 \end{vmatrix}}{\Delta}$$

where

$$\Delta = \det \begin{vmatrix} 3 & -1 & -1 \\ -1 & 3 & -1 \\ -1 & -1 & 3 \end{vmatrix}$$

$$= 3(8) - (-1)(-4) - 1(4)$$

$$= 16$$

Then

$$v_a = \frac{1}{\Delta} [6(8) - 3(-4) + 7(4)]$$

$$= \frac{88}{16} \text{ V}$$

Similarly,

$$v_b = \frac{1}{16} \begin{vmatrix} 3 & 6 & -1 \\ -1 & 3 & -1 \\ -1 & 7 & 3 \end{vmatrix}$$

$$= \frac{1}{16} [3(16) + 1(25) - 1(-3)]$$

$$= \frac{76}{16} \text{ V}$$

and

$$v_c = \frac{1}{16} \begin{vmatrix} 3 & -1 & 6 \\ -1 & 3 & 3 \\ -1 & -1 & 7 \end{vmatrix}$$

$$= \frac{1}{16} [3(24) + 1(-1) - 1(-21)]$$

$$= \frac{92}{16} \text{ V}$$

EXERCISE 4-1

Consider the circuit shown in Figure E 4-1. Find the voltages v_a and v_b when $G_1 = 1$ S, $G_2 = 0.5$ S, and $G_3 = 0.25$ S.
Answer: $v_a = 2$ V, $v_b = -4$ V

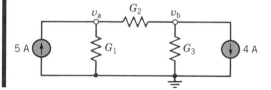

Figure E 4-1

EXERCISE 4-2

(a) Find the sum of the currents at each of the three nodes shown in Figure E 4-2. (b) Show that any one of the equations in (a) can be derived from the remaining two equations.

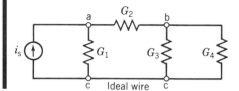

Figure E 4-2

4-4 | NODE VOLTAGE ANALYSIS OF CIRCUITS WITH CURRENT AND VOLTAGE SOURCES

In the preceding section we determined the node voltages of circuits with independent current sources only. In this section we consider circuits with both independent current and voltage sources.

First we consider the circuit with a voltage source between ground and one of the other nodes. Since we are free to select the reference node, this particular arrangement is easily achieved. Such a circuit is shown in Figure 4-10. We immediately note that the source is

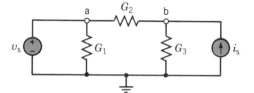

Figure 4-10 Circuit with an independent voltage source and an independent current source.

connected between terminal a and ground and therefore

$$v_a = v_s$$

Thus, v_a is known and only v_b is unknown. We write the KCL equation at node b to obtain

$$i_s = v_b(G_2 + G_3) - v_a G_2 \tag{4-15}$$

However, $v_a = v_s$. Therefore

$$v_b = \frac{i_s + v_s G_2}{G_2 + G_3} \qquad (4\text{-}16)$$

Let us now consider the circuit shown in Figure 4-11. This circuit has an additional

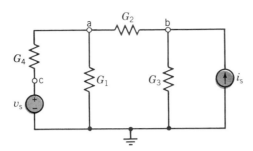

Figure 4-11 Two-node circuit similar to that of Figure 4-10 with an additional conductance G_4 in series with the voltage source.

element G_4 in series with the voltage source. We now have an additional node c, where

$$v_c = v_s$$

Thus, we may write the node equations at nodes a and b, using KCL since v_c is known.

At node b we obtain, as before (see Eq. 4-15),

$$i_s = v_b(G_2 + G_3) - v_a G_2 \qquad (4\text{-}17)$$

At node a we have

$$(v_a - v_b)G_2 + v_a G_1 + (v_a - v_c)G_4 = 0 \qquad (4\text{-}18)$$

or we may rewrite Eq. 4-18, noting that $v_c = v_s$, as

$$v_a(G_2 + G_1 + G_4) - G_2 v_b = v_s G_4 \qquad (4\text{-}19)$$

Note that at node a we have the summation of all the conductances appearing in the coefficient of v_a and the negative coefficient G_2 for v_b, as before. The voltage source appears in a term $v_s G_4$.

We now have the two equations 4-17 and 4-19 and two unknowns v_a and v_b, so we can readily solve for the two unknown voltages.

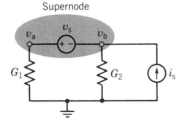

Figure 4-12 Circuit with a supernode that in-corporates v_a and v_b.

Finally, let us consider the circuit of Figure 4-12, which includes a voltage source between two nodes. Since the source voltage is known, use KVL to obtain

$$v_a = v_s + v_b$$

or

$$v_a - v_b = v_s \qquad (4\text{-}20)$$

To account for the fact that the source voltage is known, we consider both node a and node b as part of one larger node represented by the shaded ellipse shown in Figure 4-12.

We require a larger node since v_a and v_b are dependent (see Eq. 4-20). This larger node is often called a *supernode* or a *generalized node*. KCL says that the algebraic sum of the currents entering a supernode is zero. That means that we apply KCL to a supernode in the same way that we apply KCL to a node. We generate an equation obtained by using KCL at the supernode (at both a and b terminals).

A **supernode** consists of two nodes connected by an independent or a dependent voltage source.

We then can write the KCL equation at the supernode as

$$v_a G_1 + v_b G_2 = i_s \tag{4-21}$$

However, since $v_a = v_s + v_b$, we have

$$v_s G_1 + v_b (G_1 + G_2) = i_s \tag{4-22}$$

or

$$v_b = \frac{i_s - v_s G_1}{G_1 + G_2} \tag{4-23}$$

We note that Eq. 4-22 could have been obtained from a circuit equivalent to that of Figure 4-12 but containing a current source connected to node b and in parallel with conductance G_1. The equivalent circuit is shown in Figure 4-13, where $i_1 = v_s G_1$ of Eq. 4-23.

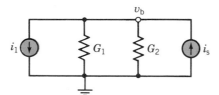

Figure 4-13 Equivalent circuit for that of Figure 4-12, where $i_1 = v_s G_1$.

We can now compile a summary of three methods of dealing with independent voltage sources in a circuit we wish to solve by node voltage methods, as recorded in Table 4-1. These three methods are exemplified in the solution of the circuit shown in Figure 4-14.

Table 4-1
Node Voltage Analysis Methods with a Voltage Source

Case	Method
1. The voltage source connects a node q and the reference node (ground).	Set v_q equal to the source voltage accounting for the polarities and proceed to write the KCL at the remaining nodes.
2. The voltage source lies between two nodes, a and b.	Create a supernode that incorporates a and b and equate the sum of all the currents into the supernode (both nodes a and b) to zero.
3. The voltage source and a series resistor R_1 together lie between two nodes, d and e, with the positive terminal of the source at node d.	Replace the voltage source and series resistor with a parallel combination of a conductance $G_1 = 1/R_1$ and a current source $i_1 = v_s G_1$ entering node d as shown in Figure 4-15.

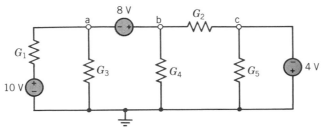

Figure 4-14 Circuit with three independent voltage sources and three nodes. $G_1 = G_2 = G_3 = G_4 = G_5 = 0.5$ S.

The right-most voltage source connected to node c exemplifies method 1. The 8-V source between nodes a and b exemplifies method 2. The left-most voltage source, 10 V, exemplifies method 3.

The third method replaces the voltage source and series resistor with a parallel combination of a conductance $G_1 = 1/R_1$ and a current source $i_1 = G_1 v_s$ entering node d, as shown in Figure 4-15. The two equivalent circuits both exhibit the same conditions at terminals d–e.

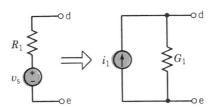

Figure 4-15 An equivalent circuit for a resistor in series with a voltage source is a current source $i_1 = v_s G_1$ and a parallel conductance $G_1 = 1/R_1$. These two circuits are equivalent since they exhibit the same terminal conditions.

Using method 1 for the circuit of Figure 4-14, we note that

$$v_c = -4 \text{ V}$$

With method 2 we have a supernode at nodes a and b. Finally, using method 3 the left-most voltage source and G_1 may be represented by an equivalent current source $i_1 = G_1 v_s = 10 G_1$ and a parallel conductance G_1. Therefore, we obtain the circuit of Figure 4-16 with the current source i_1 and the supernode for nodes a and b.

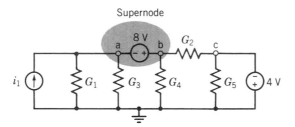

Figure 4-16 The circuit of Figure 4-14 reconfigured with $i_1 = 10 G_1 = 5$ A.

We require two equations: one at the supernode and the other at node c of Figure 4-16. First, at node c we find that $v_c = -4$ V. At the supernode, writing KCL equation at nodes a and b we have

$$v_a(G_1 + G_3) + v_b(G_4 + G_2) - G_2 v_c = i_1 \tag{4-24}$$

and the relation between v_a and v_b is

$$v_b - v_a = 8 \tag{4-25}$$

Substituting $v_c = -4$ V and $v_b = 8 + v_a$ from Eq. 4-25 into Eq. 4-24, we obtain

$$v_a(G_1 + G_3) + (8 + v_a)(G_4 + G_2) = i_1 + G_2 v_c$$

or

$$v_a(G_1 + G_3 + G_4 + G_2) = i_1 + G_2(-4) - 8(G_4 + G_2)$$

Since $i_1 = G_1(10) = 5$ A, we obtain

$$v_a = \frac{5 - 4(0.5) - 8(1)}{G_1 + G_2 + G_3 + G_4}$$

$$= \frac{-5}{2} \text{ V}$$

Then

$$v_b = 8 + v_a$$

$$= \frac{11}{2} \text{ V}$$

Note that G_5, being in parallel with a voltage source, did not affect the solution. However, we can still find the current, i_5, flowing downward in G_5 as $i_5 = (-4)(1/2) = -2$ A.

EXERCISE 4-3

Find the voltage v_a for the circuit of Figure E 4-3. Assume that $G_1 = G_3 = 1$ S and $G_2 = 0.5$ S.

Answer: 6 V

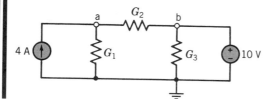

Figure E 4-3

EXERCISE 4-4

Find the voltages v_a and v_b for the circuit of Figure E 4-4. Let $G_1 = G_3 = 1$ S and $G_2 = G_4 = 0.5$ S.

Answer: $v_a = 1$ V

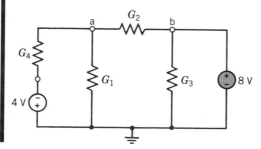

Figure E 4-4

EXERCISE 4-5

Find v_a, v_b, and v_c when $G_1 = G_2 = G_3 = G_4 = 1$ S for the circuit of Figure E 4-5.
Answer: $v_a = -5$ V, $v_b = -5/3$ V, $v_c = 4/3$ V

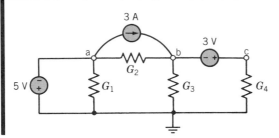

Figure E 4-5

4-5 NODE VOLTAGE ANALYSIS WITH DEPENDENT SOURCES

If the circuit contains a dependent source, the node voltage equations must be supplemented with an additional equation resulting from each dependent source. If the dependent source is a voltage source and connects two nodes other than the reference node, we use a supernode at the source. If the dependent source is a current source, we may include the current in the summation at a node. For a dependent voltage source, it is recommended that you use the methods summarized in Table 4-1.

Let us consider a circuit with a dependent voltage source as shown in Figure 4-17. We sum the currents at node a, obtaining

$$v_a(G_1 + G_2) - G_2 v_b = 5 \tag{4-26}$$

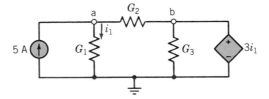

Figure 4-17 Circuit with a dependent voltage source. $G_1 = G_2 = G_3 = 1$ S.

The dependent voltage source sets $v_b = 3i_1$. Since $i_1 = G_1 v_a$ and $G_1 = 1$, we can write

$$v_b = 3i_1$$
$$= 3v_a$$

Therefore, Eq. 4-26 becomes

$$2v_a - 3v_a = 5$$

or

$$v_a = -5 \text{ V}$$

Let us consider a circuit with a dependent voltage source and a dependent current source, as shown in Figure 4-18. Because the voltage source between terminals a and b forms a supernode, we sum the currents at the supernode to obtain

$$v_a G_1 + v_b G_2 = 5 - 3i_1 \tag{4-27}$$

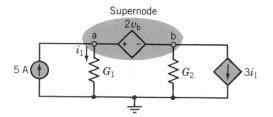

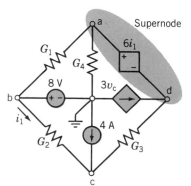

Figure 4-18 Circuit with a dependent voltage source and a dependent current source.

The additional constraining equation due to the dependent voltage source is

$$v_a - v_b = 2v_b$$

or

$$v_a = 3v_b$$

The dependent current source depends on i_1, which is $i_1 = v_a G_1$. Therefore, Eq. 4-27, in terms of v_a, becomes

$$v_a G_1 + \frac{1}{3} v_a G_2 = 5 - 3G_1 v_a$$

Therefore,

$$\left(G_1 + \frac{G_2}{3} + 3G_1 \right) v_a = 5$$

or

$$v_a = \frac{5}{G_1 + G_2/3 + 3G_1} \qquad (4\text{-}28)$$

It is clear that the analysis of a circuit with dependent voltage sources follows the procedures we summarized earlier in Table 4-1. We must, however, be careful to record the proper relationships that constrain the dependent sources.

Let us now consider a complex circuit illustrating all the node voltage methods.

Example 4-2

Consider the complex circuit shown in Figure 4-19. Find the node voltages v_c and v_d when $G_1 = G_2 = G_3 = G_4 = 0.5$ S.

Figure 4-19 The circuit of Example 4-2. This circuit has two dependent sources. A supernode incorporating nodes a and d is shown with shading.

Solution

The reference is conveniently identified as the center node because it has the most connected branches. The independent voltage source requires that $v_b = 8$ V. The dependent voltage source depends on i_1, which is

$$i_1 = G_2(v_b - v_c) \tag{4-29}$$

Finally, we note that there is a supernode at nodes a and d because of the dependent voltage source. Therefore, we need to write two KCL node equations: one at node c and the other at the supernode. At node c we have

$$v_c(G_2 + G_3) - G_2v_b - G_3v_d = 4 \tag{4-30}$$

At the supernode, we note that $v_a = v_d + 6i_1$. Writing the current equation at the supernode, we obtain

$$v_a(G_1 + G_4) - G_1v_b + G_3(v_d - v_c) = 3v_c \tag{4-31}$$

Since we already know $v_b = 8$ and $v_a = v_d + 6i_1$, let's eliminate v_b and v_a from Eqs. 4-30 and 4-31, obtaining the two equations

$$v_c(G_2 + G_3) - 8G_2 - G_3v_d = 4 \tag{4-32}$$

and

$$(v_d + 6i_1)(G_1 + G_4) - 8G_1 + G_3(v_d - v_c) = 3v_c \tag{4-33}$$

Recalling that $G_1 = G_2 = G_3 = G_4 = 0.5$ and that $i_1 = G_2(8 - v_c)$, we have

$$v_c - 0.5v_d = 8 \tag{4-34}$$

and

$$-6.5v_c + 1.5v_d = -20 \tag{4-35}$$

Therefore, we obtain the two node voltages

$$v_d = -18.3 \text{ V} \quad \text{and} \quad v_c = -1.14 \text{ V}$$

EXERCISE 4-6

Find the node voltage v_b for the circuit shown in Figure E 4-6 when the element is (a) conductance $G = 0.25$ S, (b) a dependent current source flowing to the right where $i_s = 3v_a$.
Answer: (a) 30 V, (b) 15/4 V

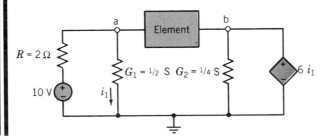

Figure E 4-6

EXERCISE 4-7

A model of a transistor amplifier is shown in Figure E 4-7. Find the voltage v_c in terms of the source voltage v_s.

Answer: $v_c = \dfrac{12G_1 - (G_1 + G_2)v_s}{(G_1 + G_2 + G_3/11)}$

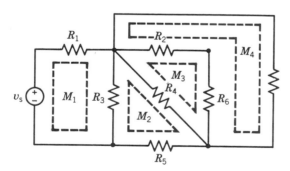

Figure E 4-7

4-6 MESH CURRENT ANALYSIS WITH INDEPENDENT VOLTAGE SOURCES

In this and succeeding sections, we consider the analysis of circuits using Kirchhoff's voltage law (KVL) around a closed path. A *closed path* or a *loop* is drawn by starting at a node and tracing a path such that we return to the original node without passing an intermediate node more than once.

A mesh is a special case of a loop. A *mesh* is a loop that does not contain any other loops within it. Mesh current analysis is applicable only to planar networks. For planar networks, the meshes in the network look like "windows." Analysis of nonplanar networks is discussed in Chapter 17. There are four meshes in the circuit shown in Figure 4-20. They are identified as M_i. Mesh 2 contains the elements R_3, R_4, and R_5. Note that the resistor R_3 is common to both mesh 1 and mesh 2.

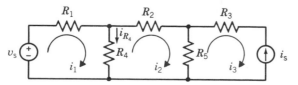

Figure 4-20 Circuit with four meshes. Each mesh is identified by dashed lines.

We define a *mesh current* as the current that flows through the elements constituting the mesh. We will use the convention of a mesh current flowing clockwise, as shown in Figure 4-21. This circuit has three mesh currents. Note that the current in an element common to two meshes is the algebraic sum of the mesh currents. Therefore, for Figure

Figure 4-21 Circuit with three mesh currents.

4-21 the current in R_4 flowing downward is

$$i_{R_4} = i_1 - i_2$$

Let us consider the two-mesh circuit of Figure 4-22. A mesh cannot have other loops within it. Thus we cannot choose the outer loop $v_s \rightarrow R_1 \rightarrow R_2 \rightarrow v_s$ as one mesh, since it would contain the loop $v_s \rightarrow R_1 \rightarrow R_3 \rightarrow v_s$ within it. We are required to choose the two mesh currents as shown in Figure 4-23.

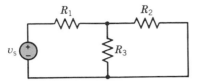

Figure 4-22 Circuit with two meshes.

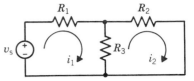

Figure 4-23 Mesh currents for the circuit of Figure 4-22.

We may use Kirchhoff's voltage law around each mesh. We will use the convention of summing the voltage drops around the mesh traveling clockwise. It is, however, equally appropriate to adopt a convention of summing the voltage rises around the mesh clockwise. Thus, for the circuit of Figure 4-23, we have

$$\text{mesh 1:} \qquad -v_s + R_1 i_1 + R_3(i_1 - i_2) = 0 \qquad (4\text{-}36)$$

$$\text{mesh 2:} \qquad R_3(i_2 - i_1) + R_2 i_2 = 0 \qquad (4\text{-}37)$$

Note that the voltage across R_3 in mesh 1 is determined from Ohm's law, where

$$\begin{aligned} v &= R_3 i_a \\ &= R_3(i_1 - i_2) \end{aligned}$$

where i_a is the actual element current flowing downward through R_3.

The two equations 4-36 and 4-37 will enable us to determine the two mesh currents i_1 and i_2. Rewriting the two equations, we have

$$i_1(R_1 + R_3) - i_2 R_3 = v_s$$

and

$$-i_1 R_3 + i_2(R_3 + R_2) = 0$$

If $R_1 = R_2 = R_3 = 1\ \Omega$, we have

$$2i_1 - i_2 = v_s$$

and

$$-i_1 + 2i_2 = 0$$

Add twice the first equation to the second equation obtaining $3i_1 = 2v_s$. Then we have

$$i_1 = \frac{2v_s}{3} \quad \text{and} \quad i_2 = \frac{v_s}{3}$$

Thus, we have obtained two independent mesh current equations that are readily solved for the two unknowns. If we have N meshes and write N mesh equations in terms of N mesh currents, we can obtain N independent mesh equations. This set of N equations are independent, and thus guarantees a solution for the N mesh currents.

A circuit that contains only independent voltage sources and resistors results in a specific format of equations that can readily be obtained. Consider a circuit with three meshes, as shown in Figure 4-24. Assign the clockwise direction to all of the mesh currents. Using KVL, the three mesh equations are

mesh 1: $\quad -v_s + R_1 i_1 + R_4(i_1 - i_2) = 0$

mesh 2: $\quad R_2 i_2 + R_5(i_2 - i_3) + R_4(i_2 - i_1) = 0$

mesh 3: $\quad R_5(i_3 - i_2) + R_3 i_3 + v_g = 0$

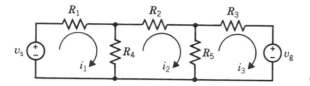

Figure 4-24 Circuit with three mesh currents and two voltage sources.

These three mesh equations can be rewritten by collecting coefficients for each mesh current as

mesh 1: $\quad (R_1 + R_4)i_1 - R_4 i_2 = v_s$

mesh 2: $\quad -R_4 i_1 + (R_4 + R_2 + R_5)i_2 - R_5 i_3 = 0$

mesh 3: $\quad -R_5 i_2 + (R_3 + R_5)i_3 = -v_g$

Hence, we note that the coefficient of the mesh current i_1 for the first mesh is the sum of resistances in the loop and that the coefficient of the second mesh current is the negative of the resistance common to meshes 1 and 2. In general, we state that for mesh current i_n, the equation for the nth mesh with independent voltage sources only is obtained as follows:

$$ -\sum_{q=1}^{Q} R_k i_q + \sum_{j=1}^{P} R_j i_n = \sum_{n=1}^{N} v_{sn} \tag{4-38} $$

In words, for mesh n we multiply i_n by the sum of all resistances R_j around the mesh. Then we add the terms due to the resistances in common with another mesh as the negative of the connecting resistance R_k, multiplied by the mesh current in the adjacent mesh i_q for all Q adjacent meshes. Finally, the independent voltage sources around the loop appear on the right side of the equation as the negative of the voltage sources encountered as we traverse the loop in the direction of the mesh current. Remember that the above result is obtained assuming all mesh currents flow clockwise.

The general matrix equation for the mesh current analysis for independent voltage sources present in a circuit is

$$ \mathbf{R i} = \mathbf{v}_s \tag{4-39} $$

where $\mathbf{R}$ is a symmetric matrix with a diagonal consisting of the sum of resistances in each mesh and the off-diagonal elements are the negative of the resistances connecting two meshes. The matrix $\mathbf{i}$ consists of the mesh currents as

$$ \mathbf{i} = \begin{bmatrix} i_1 \\ i_2 \\ \cdot \\ \cdot \\ \cdot \\ i_N \end{bmatrix} $$

For N mesh currents. The source matrix $\mathbf{v}_s$ is

$$\mathbf{v}_s = \begin{bmatrix} v_{s1} \\ v_{s2} \\ \cdot \\ \cdot \\ \cdot \\ v_{sN} \end{bmatrix}$$

where v_{sj} is the sum of the sources in the jth mesh with the appropriate sign assigned to each source.

For the circuit of Figure 4-24 and the matrix Eq. 4-39, we have

$$\mathbf{R} = \begin{bmatrix} (R_1 + R_4) & -R_4 & 0 \\ -R_4 & (R_2 + R_4 + R_5) & -R_5 \\ 0 & -R_5 & (R_3 + R_5) \end{bmatrix}$$

Note that $\mathbf{R}$ is a symmetric matrix, as we expected.

EXERCISE 4-8

Use mesh analysis to determine the current i for the circuit shown in Figure E4-8.
Answer: $i = 1$ A

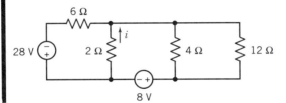

Figure E 4-8

4-7 ❙ MESH CURRENT ANALYSIS WITH CURRENT SOURCES

Heretofore, we have considered only circuits with independent voltage sources for analysis by the mesh current method. If the circuit has an independent current source, as shown in Figure 4-25, we recognize that the second mesh current is equal to the negative of the current source. We can then write

$$i_2 = -i_s$$

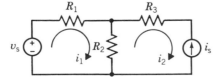

Figure 4-25 Circuit with an independent voltage source and an independent current source.

and we need only determine the first mesh current i_1. Writing KVL for the first mesh, we obtain

$$(R_1 + R_2)i_1 - R_2 i_2 = v_s$$

Since $i_2 = -i_s$, we have

$$i_1 = \frac{v_s - R_2 i_s}{R_1 + R_2} \tag{4-40}$$

where i_s and v_s are sources of known magnitude.

If we encounter a circuit as shown in Figure 4-26, we have a current source i_s that has an unknown voltage v_{ab} across its terminals. We can readily note that

$$i_2 - i_1 = i_s \tag{4-41}$$

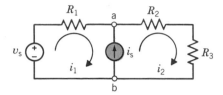

Figure 4-26 Circuit with an independent current source common to both meshes.

by writing KCL at node a. The two mesh equations are

mesh 1: $R_1 i_1 + v_{ab} = v_s$ $\tag{4-42}$

mesh 2: $(R_2 + R_3)i_2 - v_{ab} = 0$ $\tag{4-43}$

We note that if we add Eqs. 4-42 and 4-43 we eliminate v_{ab}, obtaining

$$R_1 i_1 + (R_2 + R_3)i_2 = v_s$$

However, since $i_2 = i_s + i_1$, we obtain

$$R_1 i_1 + (R_2 + R_3)(i_s + i_1) = v_s$$

or

$$i_1 = \frac{v_s - (R_2 + R_3)i_s}{R_1 + R_2 + R_3} \tag{4-44}$$

Thus, we account for independent current sources by recording the relationship between the mesh currents and the current source. If the current source influences *only one* mesh current, we record that constraining equation and write the KVL equations for the remaining meshes. If the current source influences two mesh currents, we write the KVL equation for both meshes, assuming a voltage v_{ab} across the terminals of the current source. Then, adding these two mesh equations, we obtain an equation independent of v_{ab}.

Example 4-3

Consider the circuit of Figure 4-27 where $R_1 = R_2 = 1\ \Omega$ and $R_3 = 2\ \Omega$. Find the three mesh currents.

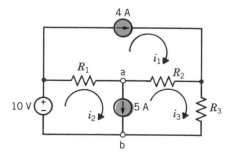

Figure 4-27 Circuit with two independent current sources.

Solution

Since the 4-A source flows only in mesh 1, we note that

$$i_1 = 4$$

For the 5-A source, we have

$$i_2 - i_3 = 5 \tag{4-45}$$

Writing KVL for mesh 2 and mesh 3, we obtain

$$\text{mesh 2:} \qquad R_1(i_2 - i_1) + v_{ab} = 10 \tag{4-46}$$

$$\text{mesh 3:} \qquad R_2(i_3 - i_1) + R_3 i_3 - v_{ab} = 0 \tag{4-47}$$

We substitute $i_1 = 4$ and add Eqs. 4-46 and 4-47 to obtain

$$R_1(i_2 - 4) + R_2(i_3 - 4) + R_3 i_3 = 10 \tag{4-48}$$

From Eq. 4-45, $i_2 = 5 + i_3$. Substituting into Eq. 4-48, we have

$$R_1(5 + i_3 - 4) + R_2(i_3 - 4) + R_3 i_3 = 10$$

Using the values for the resistors, we obtain

$$i_2 = \frac{33}{4} \text{ A} \quad \text{and} \quad i_3 = \frac{13}{4} \text{ A}$$

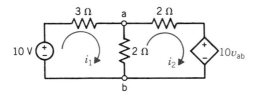

Figure 4-28 Circuit with one dependent voltage source.

If the circuit includes dependent sources, we must add the constraining equation imposed by each dependent source. For example, the circuit shown in Figure 4-28 contains one dependent voltage source. Writing the KVL equations for the two meshes, we have

$$\text{mesh 1:} \qquad 5i_1 - 2i_2 = 10 \tag{4-49}$$

$$\text{mesh 2:} \qquad -2i_1 + 4i_2 = -10v_{ab} \tag{4-50}$$

However, $v_{ab} = 2(i_1 - i_2)$. Therefore, we obtain for mesh 2 (Eq. 4-50)

$$-2i_1 + 4i_2 = -20(i_1 - i_2)$$

or

$$18i_1 - 16i_2 = 0 \tag{4-51}$$

Subtracting 8 times Eq. 4-49 from Eq. 4-51, we obtain $22i_1 = 80$. Therefore, we have

$$i_1 = \frac{80}{22} \text{ A} \quad \text{and} \quad i_2 = \frac{90}{22} \text{ A}$$

A more general technique for the mesh analysis method when a current source is common to two meshes involves the concept of a supermesh. A *supermesh* is one mesh created from two meshes that have a current source in common, as shown in Figure 4-29. We then reduce the number of meshes by one when we have a common current source between two meshes. This current source is an element common to the two meshes and thus reduces the number of independent mesh equations by one.

A **supermesh** is one larger mesh created from two meshes that have an independent or dependent current source in common.

For example, consider the circuit of Figure 4-29. The 5-A current source is common to

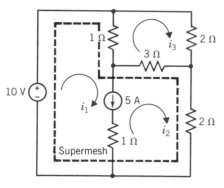

Figure 4-29 Circuit with a supermesh that incorporates mesh 1 and mesh 2. The supermesh is indicated by the dashed line.

mesh 1 and mesh 2. The supermesh consists of the interior of mesh 1 and mesh 2. Writing KVL around the periphery of the supermesh shown by the heavy dashed lines, we obtain

$$- 10 + 1(i_1 - i_3) + 3(i_2 - i_3) + 2i_2 = 0$$

For mesh 3, we have

$$1(i_3 - i_1) + 2i_3 + 3(i_3 - i_2) = 0$$

Finally, the constraint equation required by the current source common to meshes 1 and 2 is

$$i_1 - i_2 = 5$$

Then the three equations may be reduced to

supermesh: $\qquad 1i_1 + 5i_2 - 4i_3 = 10$

mesh 3: $\qquad - 1i_1 - 3i_2 + 6i_3 = 0$

current source: $\qquad 1i_1 - 1i_2 \qquad = 5$

Therefore, solving the three equations simultaneously we find that $i_2 = 2.5$, $i_1 = 7.5$, and $i_3 = 2.5$.

The methods of mesh current analysis utilized when a current source is present are summarized in Table 4-2.

As a final example, let us consider the circuit shown in Figure 4-30. This circuit includes a current source common to two meshes and a dependent voltage source. We select a

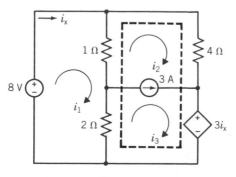

Figure 4-30 Circuit with a supermesh and a dependent voltage source. The supermesh is identified by the heavy dashed line.

Table 4-2
Mesh Current Analysis Methods with a Current Source

Case	Method
1. A current source appears on the periphery of only one mesh, n.	Equate the mesh current i_n to the current source, accounting for the direction of the current source.
2. A current source is common to two meshes	A. Assume a voltage v_{ab} across the terminals of the current source, write the KVL equations for the two meshes, and add them to eliminate v_{ab}. or B. Create a supermesh as the periphery of the two meshes and write one KVL equation around the periphery of the supermesh. In addition write the constraining equation for the two mesh currents in terms of the current source.

supermesh since mesh 2 and mesh 3 have a current source in common. We obtain a KVL equation for mesh 1 and the supermesh as follows:

$$\text{mesh 1:} \qquad 3i_1 - i_2 - 2i_3 = 8 \qquad (4\text{-}52)$$

$$\text{supermesh:} \qquad -3i_1 + 5i_2 + 2i_3 + 3i_x = 0 \qquad (4\text{-}53)$$

We note that

$$i_x = i_1$$

Also, the constraint equation for the current source is

$$i_3 - i_2 = 3$$

Substituting $i_x = i_1$ and $i_3 = 3 + i_2$ into Eqs. 4-52 and 4-53, we obtain

$$3i_1 - i_2 - 2(3 + i_2) = 8$$

and

$$-3i_1 + 5i_2 + 2(3 + i_2) + 3i_1 = 0$$

Then, we rearrange these equations obtaining

$$3i_1 - 3i_2 = 14 \qquad (4\text{-}54)$$

and

$$7i_2 = -6 \qquad (4\text{-}55)$$

From Eq. 4-55, we have

$$i_2 = \frac{-6}{7} \text{ A}$$

Then, from Eq. 4-54, we obtain

$$i_1 = \frac{80}{21} \text{ A}$$

EXERCISE 4-9

Modern household electric appliances are commonplace today. However, imagine the pleasure of those who first purchased a vacuum cleaner, as shown in Figure E 4-9a. During the first half of the twentieth century, the average American household was transformed by the introduction of electric appliances. The circuit model of the cleaner and its power source is shown in Figure E 4-9b. Find the voltage source v_s required to deliver 150 W to the motor connected between terminals a and b.
Answer: 60 V

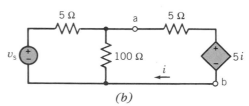

(a)   *(b)*

Figure E 4-9 *(a)* The modern convenience of the vacuum cleaner is serenely demonstrated in about 1910. Courtesy of Brown Brothers. *(b)* Circuit model of the cleaner and its power source.

EXERCISE 4-10

Determine the current i in the circuit shown in Figure E 4-10.
Answer: $i = 3$ A

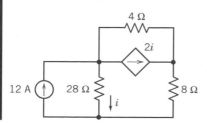

Figure E 4-10

4-8 | THE NODE VOLTAGE METHOD AND MESH CURRENT METHOD COMPARED

The analysis of a complex circuit can usually be accomplished by either the node voltage or the mesh current method. The advantage of using these methods is the systematic procedures provided for obtaining the simultaneous equations.

There are some cases where one method is clearly preferred over another. For example, when the circuit contains only voltage sources, it is probably easier to use the mesh current method. When the circuit contains only current sources, it will usually be easiest to use the node voltage method.

If a circuit has both current sources and voltage sources, it can be analyzed by either method. One approach is to compare the number of equations required for each method. If the circuit has fewer nodes than meshes, it may be wise to select the node voltage method. If the circuit has fewer meshes than nodes, it may be easier to use the mesh current method.

Another point to consider when choosing between the two methods is what information is required. If you need to know several currents, it may be wise to proceed directly with mesh current analysis. Remember, mesh current analysis only works for planar networks. Nonplanar circuits are discussed in Chapter 17.

It is often helpful to determine which method is more appropriate for the problem requirements and to consider both methods.

Example 4-4

Determine the best analysis method for the circuits of Figure 4-31 when it is required to determine
(a) The voltage v_{ab} in Figure 4-31a.
(b) The current through resistor R_2 in Figure 4-31b.
(c) The current i in Figure 4-31c.

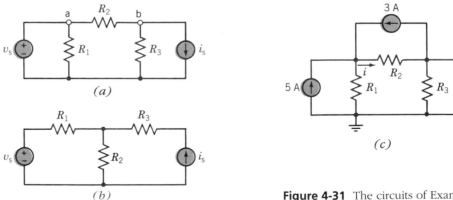

Figure 4-31 The circuits of Example 4-4.

Solution
(a) The circuit of Figure 4-31a is most suitable for the node voltage method. Since $v_a = v_s$, we need only write KCL at node b.
(b) The circuit of Figure 4-31b is most suitable for mesh current analysis. Since the current source defines the right-hand mesh current, we need only write the KVL equation for the left-hand mesh.

(c) The circuit has two nodes in addition to the ground node, so two node equations would be required. The circuit has four meshes; however, the currents in three meshes are defined by the three current sources. Therefore, only one mesh current, that of the mesh including R_1, R_2, and R_3, is unknown. It would be easiest, therefore, to find i by using the mesh current method.

EXERCISE 4-11

Determine the best analysis method for the circuits of Figures E 4-11a and E 4-11b. For circuit (a), it is required to find the voltage v_a. For circuit (b), we wish to find the current i.

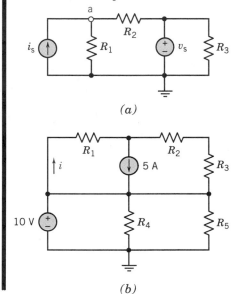

(a)

(b)

Figure E 4-11

4-9 | DC ANALYSIS USING PSpice[1]

The calculation of currents and voltages in a complicated circuit can be facilitated by using PSpice. In this section we will illustrate the utility of PSpice while also using algebraic methods for insight.

First, let us consider a simple two-mesh circuit as shown in Figure 4-32 and determine i_1 and i_2. The two mesh equations are

$$\text{mesh 1:} \quad 5i_1 - 4i_2 = 9 \tag{4-56}$$

$$\text{mesh 2:} \quad -4i_1 + 8i_2 = 0 \tag{4-57}$$

Multiplying Eq. 4-57 by one-half and adding it to Eq. 4-56, we obtain

$$3i_1 = 9$$

---

[1] This section may be omitted or considered with a later chapter. See Appendix G for an introduction to PSpice.

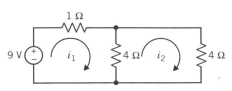

Figure 4-32 A simple two-mesh circuit.

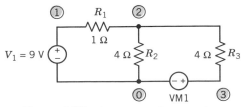

Figure 4-33 The circuit of Figure 4-37 redrawn with node numbers indicated. The reference node is denoted as zero.

or

$$i_1 = 3 \text{ A}$$

Then, using Eq. 4-57, we find that $i_2 = 1.5$ A. This simple circuit is easily analyzed using algebraic methods. We now proceed to use PSpice to determine i_1 and i_2.

First, we redraw the circuit numbering the nodes with the reference node chosen as zero, as shown in Figure 4-33. The dummy source VM1 (VM1 = 0) is inserted to measure the current i_2. Recall that the source will measure the current with the convention of current flowing from + to − through the source. The PSpice program is provided in Figure 4-34. Note that the title statement is the first line. Also note that the second line beginning with an asterisk is a comment line and allows us to set up column headings. The output is provided in Figure 4-35. It is verified that $i_1 = 3$ A and $i_2 = 1.5$ A. Note that the printout gives a current through the source of − 3 A because PSpice convention has the current

```
     ***     TWO-MESH ANALYSIS   (Title Statement)

  *  ELEM    NODE     NODE      VALUE
     R1       1        2          1
     R2       2        0          4
     R3       2        3          4
     V1       1        0         DC    9
     VM1      3        0         DC    0
     .END
```

Figure 4-34 PSpice program for the circuit of Figure 4-33.

```
*TWO-MESH ANALYSIS

****     SMALL SIGNAL BIAS SOLUTION        TEMPERATURE =   27.000 DEG C

NODE   VOLTAGE   NODE   VOLTAGE   NODE   VOLTAGE   NODE   VOLTAGE
(  1)    9.0000 (  2)    6.0000 (  3)    0.0000

VOLTAGE SOURCE CURRENTS
NAME           CURRENT
V1            -3.000E+00
VM1            1.500E+00
TOTAL POWER DISSIPATION  2.70E+01   WATTS

   JOB CONCLUDED
   TOTAL JOB TIME           .50
```

Figure 4-35 Output of PSpice calculation.

flow from + to − through the source. The printout also provides the voltage at each numbered node with reference to the ground (0) node.

In this section we are using the single-point dc analysis, which is called the "small signal bias solution," in order to calculate the dc currents and node voltages.

Clearly, the utility of PSpice increases as the circuit becomes more complex. Consider the circuit shown in Figure 4-36, where we wish to determine i_1 and i_3. The mesh equations are

$$\text{mesh 1:} \quad 11i_1 - 8i_2 \qquad = 42$$

$$\text{mesh 2:} \quad -8i_1 + 18i_2 - 6i_3 = 0$$

$$\text{mesh 3:} \qquad \quad -6i_2 + 18i_3 = 0$$

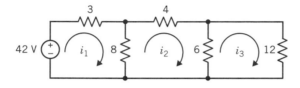

Figure 4-36 A three-mesh circuit. All resistances in ohms.

Rewriting these equations in matrix form, we obtain

$$\mathbf{R}\mathbf{i} = \mathbf{v}_s \qquad (4\text{-}58)$$

where

$$\mathbf{R} = \begin{bmatrix} 11 & -8 & 0 \\ -8 & 18 & -6 \\ 0 & -6 & 18 \end{bmatrix}$$

and

$$\mathbf{v}_s = \begin{bmatrix} 42 \\ 0 \\ 0 \end{bmatrix}$$

We evaluate the determinant of $\mathbf{R}$ as $\Delta = 2016$. Then, using Cramer's rule, we find that $i_1 = 6$ A and $i_3 = 1$ A. This relatively simple circuit would require more complicated calculations as controlled and other independent sources are added so that all the elements of $\mathbf{R}$ and $\mathbf{v}_s$ become nonzero. This circuit with three meshes is about as far as we can go without introducing calculation tedium. A four-mesh circuit with many sources would be difficult to solve without causing calculation errors. Thus, PSpice is a real aid when the circuit becomes complicated.

In this section we use the simple single-point dc analysis referred to as an operating point analysis and discussed in appendix section G-4. The computer printout will call this dc analysis a "small signal bias solution" (see Figure 4-39).

The three-mesh circuit of Figure 4-36 is redrawn for PSpice analysis as shown in Figure

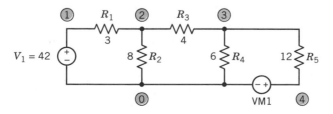

Figure 4-37 The three-mesh circuit redrawn for PSpice analysis. All resistances in ohms.

4-37. The PSpice program is provided in Figure 4-38. Note that it is verified that $i_1 = 6$ A and $i_3 = 1$ A as shown in the output of Figure 4-39.

```
 ***     THREE-MESH ANALYSIS

 *  ELEM   NODE    NODE    VALUE
    R1      1       2        3
    R2      2       0        8
    R3      2       3        4
    R4      3       0        6
    R5      3       4       12
    V1      1       0       DC    42
    VM1     4       0       DC     0
    .END
```

Figure 4-38 The PSpice program for the three-mesh circuit.

```
*THREE-MESH ANALYSIS

****     SMALL SIGNAL BIAS SOLUTION      TEMPERATURE =    27.000 DEG C

NODE   VOLTAGE   NODE   VOLTAGE   NODE    VOLTAGE   NODE   VOLTAGE
(  1)    42.0000 (  2)    24.0000 (  3)     12.0000 (  4)     0.0000

VOLTAGE SOURCE CURRENTS
NAME            CURRENT
V1              -6.000E+00
VM1              1.000E+00
TOTAL POWER DISSIPATION  2.52E+02  WATTS
```

Figure 4-39 The output of the PSpice program for the three-mesh circuit.

At the end of Chapter 5 we will demonstrate the analysis of circuits with dependent sources and the use of the **.DC** command.

4-10 VERIFICATION EXAMPLES

Problem

The circuit shown in Figure 4V-1*a* was analyzed using PSpice. The PSpice output file, Figure 4V-1*b*, includes the node voltages of the circuit. Are these node voltages correct?

Solution

The node equation corresponding to node 2 is

$$\frac{V(2) - V(1)}{100} + \frac{V(2)}{200} + \frac{V(2) - V(3)}{100} = 0$$

where, for example, $V(2)$ is the node voltage at node 2. When the node voltages from Figure 4V-1*b* are substituted into the left-hand side of this equation, the result is

$$\frac{7.2727 - 12}{100} + \frac{7.2727}{200} + \frac{7.2727 - 5.0909}{100} = 0.011$$

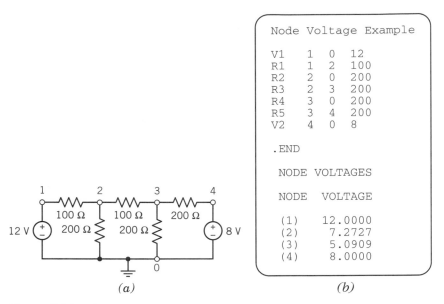

Figure 4V-1 *(a)* A circuit and *(b)* the node voltages calculated using PSpice. The bottom node has been chosen as the reference node, which is indicated by the ground symbol and the node number 0. The voltages and resistors have units of volts and ohms, respectively.

The right-hand side of this equation should be 0 instead of 0.011. It looks like something is wrong. Is a current of only 0.011 negligible? Probably not in this case. If the node voltages were correct, then the currents of the 100-Ω resistors would be 0.047 A and 0.022 A, respectively. The current of 0.011 A does not seem negligible when compared to currents of 0.047 A and 0.022 A.

Is it possible that PSpice would calculate the node voltages incorrectly? Probably not, but the PSpice input file could easily contain errors. In this case, the value of the resistance connected between nodes 2 and 3 has been mistakenly specified to be 200 Ω. After changing this resistance to 100 Ω, PSpice calculates the node voltages to be

$$V(1) = 12.0 \qquad V(2) = 7.0 \qquad V(3) = 5.5 \qquad V(4) = 8.0$$

Substituting these voltages into the node equation gives

$$\frac{7.0 - 12.0}{100} + \frac{7.0}{200} + \frac{7.0 - 5.5}{100} = 0.0$$

so these node voltages do satisfy the node equation corresponding to node 2.

Problem
The circuit shown in Figure 4V-2*a* was analyzed using PSpice. The PSpice output file, Figure 4V-1*b*, includes the mesh currents of the circuit. Are these mesh currents correct?

The PSpice output file will include the currents through the voltage sources. Recall that PSpice uses the passive convention so that the current in the 8-V source will be $-i_1$ instead of i_1. The two 0-V sources have been added to include mesh currents i_2 and i_3 in the PSpice output file.

Solution
The mesh equation corresponding to mesh 2 is

$$200(i_2 - i_1) + 500i_2 + 250(i_2 - i_3) = 0$$

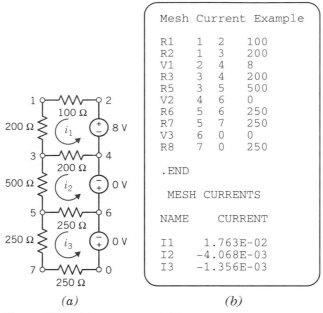

```
Mesh Current Example

R1   1   2   100
R2   1   3   200
V1   2   4   8
R3   3   4   200
R5   3   5   500
V2   4   6   0
R6   5   6   250
R7   5   7   250
V3   6   0   0
R8   7   0   250

.END

   MESH  CURRENTS

NAME      CURRENT

I1    1.763E-02
I2   -4.068E-03
I3   -1.356E-03
```

(a)   (b)

Figure 4V-2 (a) A circuit and (b) the mesh currents calculated using PSpice. The voltages and resistances are given in volts and ohms, respectively.

When the mesh currents from Figure 4V-2b are substituted into the left-hand side of this equation, the result is

$$200(-0.004068 - 0.01763) + 500(-0.004068)$$
$$+ 250(-0.004068 - (-0.001356)) = 1.629$$

The right-hand side of this equation should be 0 instead of 1.629. It looks like something is wrong. Most likely, the PSpice input file contains an error. This is indeed the case. The nodes of both 0-V voltage sources have been entered in the wrong order. Recall that the first node should be the positive node of the voltage source. After correcting this error, Pspice gives

$$i_1 = 0.01763, \qquad i_2 = 0.004068, \qquad i_3 = 0.001356$$

Using these values in the mesh equation gives

$$200(0.004068 - 0.01763) + 500(0.004068) + 250(0.004068 - 0.001356) = 0.0$$

These mesh currents do indeed satisfy the mesh equation corresponding to mesh 2.

4-11 DESIGN CHALLENGE SOLUTION

POTENTIOMETER ANGLE DISPLAY

A circuit is needed to measure and display the angular position of a potentiometer shaft. The angular position, θ, will vary from -180 degrees to 180 degrees.

Figure 4D-1 illustrates a circuit that could do the job. The $+15$-V and -15-V power supplies, the potentiometer, and resistors R_1 and R_2 are used to obtain a voltage, v_i, that is proportional to θ. The amplifier is used to change the constant of proportionality to obtain a simple relationship between θ and the voltage, v_o, displayed by the voltmeter. In this example, the amplifier will be used to obtain the relationship

$$v_o = k \cdot \theta \quad \text{where } k = 0.1 \frac{\text{volt}}{\text{degree}} \qquad \text{(4D-1)}$$

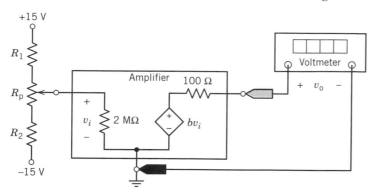

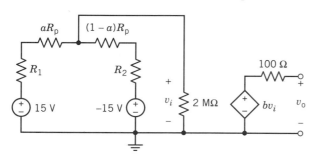

Figure 4D-1 Proposed circuit for measuring and displaying the angular position of the potentiometer shaft.

so that θ can be determined by multiplying the meter reading by 10. For example, a meter reading of -7.32 V indicates that $\theta = -73.2$ degrees.

Define the Situation

The circuit diagram in Figure 4D-2 is obtained by modeling the power supplies as ideal voltage sources, the voltmeter as an open circuit, and the potentiometer by two

Figure 4D-2 Circuit diagram containing models of the power supplies, voltmeter and potentiometer.

resistors. The parameter, a, used to model the potentiometer varies from 0 to 1 as θ varies from -180 degrees to 180 degrees. That means

$$a = \frac{\theta}{360°} + \frac{1}{2} \qquad \text{(4D-2)}$$

Solving for θ gives

$$\theta = \left(a - \frac{1}{2} \right) \cdot 360° \qquad \text{(4D-3)}$$

State the Goal

Specify values of resistors R_1 and R_2, the potentiometer resistance R_p, and the amplifier gain b that will cause the meter voltage, v_o, to be related to the angle θ by Eq. 4D-1.

Generate a Plan

Analyze the circuit shown in Figure 4D-2 to determine the relationship between v_i and θ. Select values of R_1, R_2, and R_p. Use these values to simplify the relationship between

v_i and θ. If possible, calculate the value of b that will cause the meter voltage, v_o, to be related to the angle θ by Eq. 4D-1. If this isn't possible, adjust the values of R_1, R_2, and R_p and try again.

Take Action on the Plan

The circuit has been redrawn in Figure 4D-3. A single node equation will provide the relationship between v_i and θ:

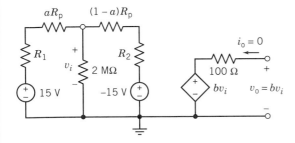

Figure 4D-3 The redrawn circuit showing the node v_i.

$$\frac{v_i}{2 \text{ M}\Omega} + \frac{v_i - 15}{R_1 + aR_p} + \frac{v_i - (-15)}{R_2 + (1 - a)R_p} = 0$$

Solving for v_i gives

$$v_i = \frac{2 \text{ M}\Omega (R_p(2a - 1) + R_1 - R_2)15}{(R_1 + aR_p)(R_2 + (1 - a)R_p) + 2 \text{ M}\Omega (R_1 + R_2 + R_p)} \tag{4D-4}$$

This equation is quite complicated. Let's put some restrictions on R_1, R_2, and R_p that will make it possible to simplify this equation. First, let $R_1 = R_2 = R$. Second, require that both R and R_p be much smaller than 2 MΩ (for example R < 20 kΩ). Then

$$(R + aR_p)(R + (1 - a)R_p) \ll 2 \text{ M}\Omega (2R + R_p)$$

In words, the first term in the denominator of the left side of Eq. 4D-4 is negligible compared to the second term. Eq. 4D-4 can be simplified to

$$v_i = \frac{R_p(2a - 1) \, 15}{2R + R_p}$$

Next, using Eq. 4D-2

$$v_i = \left(\frac{R_p}{2R + R_p}\right)\left(\frac{15 \text{ V}}{180°}\right) \theta$$

It is time to pick values for R and R_p. Let $R = 5$ kΩ and $R_p = 10$ kΩ; then

$$v_i = \left(\frac{7.5 \text{ V}}{180°}\right) \theta$$

Referring to Figure 4D-2, the amplifier output is given by

$$v_o = bv_i \tag{4D-5}$$

so

$$v_o = b\left(\frac{7.5 \text{ V}}{180°}\right) \theta$$

Comparing this equation to Eq. 4D-1 gives

$$b\left(\frac{7.5\ \text{V}}{180°}\right) = 0.1\ \frac{\text{volt}}{\text{degree}}$$

or

$$b = \frac{180}{7.5}(0.1) = 2.4$$

The final circuit is shown in Figure 4D-4. As a check, suppose $\theta = 150°$. From Eq. 4D-2 we see that

$$a = \frac{150°}{360°} + \frac{1}{2} = 0.9167$$

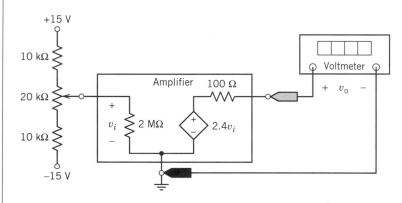

Figure 4D-4 The final designed circuit.

Using Eq. 4D-4, we calculate

$$v_i = \frac{2\ \text{M}\Omega\ (10\ \text{k}\Omega(2 \times 0.9167 - 1))15}{(5\ \text{k}\Omega + 0.9167 \times 10\ \text{k}\Omega)(5\ \text{k}\Omega + (1 - 0.9167)10\ \text{k}\Omega) + 2\ \text{M}\Omega(2 \times 5\ \text{k}\Omega + 10\ \text{k}\Omega)} = 6.24$$

Finally, Eq. 4D-5 indicates that the meter voltage will be

$$v_o \times 2.4 \cdot 6.24 = 14.98$$

This voltage will be interpreted to mean that the angle was

$$\theta = 10 \cdot v_o = 149.8°$$

which is correct to three significant digits.

4-12 DESIGN EXAMPLE—MAXIMUM POWER TO A LOAD

Problem

A model of a complex electronic circuit is shown in Figure 4-40. The goal is to select R in order to deliver maximum power to a load resistor R. The resistor is constrained so that $0 \le R \le 5\ \Omega$. Find R that will result in the maximum power being absorbed by R and the magnitude of that power.

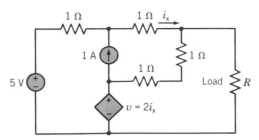

Figure 4-40 The model of an electronic circuit.

Define the Situation, State the Assumptions, and Develop a Model

1 The current source of 1 A is common to two meshes.
2 The dependent voltage source depends upon the current i_x.
3 $0 \leq R \leq 5\ \Omega$.

The Goal

Find R that will maximize the power p absorbed by the resistor, and determine the magnitude of the maximum power.

Generate a Plan

1 Select a method of analysis: node analysis or mesh analysis.
2 Determine the current through R.
3 Find the power absorbed by R.
4 Determine the magnitude of R that will result in the maximum power being absorbed by R.
5 Find the maximum power absorbed by R.

Prepare the Plan

Goal	Equation	Need	Information
Select either node or mesh analysis.		Determine whether number of mesh equations is less than number of node equations	
Obtain the equations to complete the analysis method required.		Mesh or node equations	
Find the current, i, through R.		Solution of the mesh or node equations	
Find the power absorbed by R.	$p = i^2 R$	i	
Determine the magnitude of R that results in the maximum p.			$0 \leq R \leq 5\ \Omega$

Take Action Using the Plan

1 Select either node or mesh analysis. We select mesh analysis since we seek the current through R, and mesh analysis will require only two mesh equations plus the constraining equation resulting from the 1-A source.

2   Select the three mesh currents as shown in Figure 4-41. Note that $i_x = i_2$ and that the 1-A source is common to mesh 1 and mesh 2.

3   Create a supermesh as shown by the dotted lines in Figure 4-41.

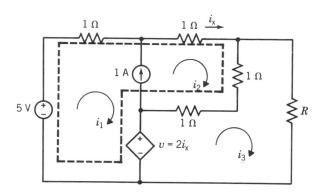

Figure 4-41   The circuit of Figure 4-40 with the three mesh currents identified.

4   Write the two mesh equations and the constraining equation.

supermesh:	$-5 + 1i_1 + 3i_2 + 2i_x - 2i_3 = 0$
mesh 3:	$(2 + R)i_3 - 2i_x - 2i_2 = 0$
constraining equation:	$i_2 - i_1 = 1$

5   Rearrange the two mesh equations and substitute $i_2 = 1 + i_1$ and $i_x = i_2$ to obtain the two equations

$$6i_1 - 2i_3 = 0$$
$$-4i_1 + (2 + R)i_3 = 4$$

6   Solve for i_3, obtaining

$$i_3 = \frac{24}{4 + 6R}$$
$$= \frac{12}{2 + 3R}$$

7   Find the power absorbed by R:

$$p = i_3^2 R$$
$$= \left(\frac{12}{2 + 3R}\right)^2 R$$

8   Determine the magnitude of R that maximizes p. Since it appears that p will be maximum when R is near 1, we will evaluate p when $R = 1/4, 1/2, 2/3, 3/4$, and 1. The results of the calculations are recorded in the following table. Clearly, the maximum power is attained when $R = 2/3\ \Omega$.

The Power Absorbed by Resistance R

R (ohms)	1/4	1/2	2/3	3/4	1
p (watts)	4.76	5.88	6.75	5.98	5.76

An alternative approach is to obtain dp/dR and set the derivative to zero. Solving for R, we find $R = 2/3\ \Omega$ when maximum power is absorbed by R.

SUMMARY

Electric circuits are widely used for communication systems such as the telegraph and telephone. These circuits are increasingly complex and require a disciplined method of analysis.

The node voltage method of circuit analysis identifies the nodes of a circuit where two or more elements are connected. Once we set one node equal to the reference node, we can obtain the KCL equations at each node. Solution of the simultaneous equations results in knowledge of the node voltages.

For circuits that contain only independent current sources we obtain the node voltage matrix equation $\mathbf{Gv} = \mathbf{i}_s$, where $\mathbf{G}$ is a symmetric matrix, $\mathbf{v}$ is the vector of unknown node voltages, and $\mathbf{i}_s$ is the vector of independent current sources.

When a circuit has voltage sources as well as current sources, we can still use the node voltage method by utilizing the concept of a supernode. A supernode is a "large node" that includes two nodes connected by a known voltage source. If the voltage source is directly connected between a node q and the reference node, we may set $v_q = v_s$ and write the KCL equations at the remaining nodes. If the circuit contains dependent sources, the node voltage equations must be supplemented with the constraint equation resulting from each dependent source.

Mesh current analysis with independent voltage sources uses the KVL around the periphery of each mesh or "window." A mesh current flows around the periphery of the window. The matrix equation for mesh current analysis with independent voltage sources is $\mathbf{Ri} = \mathbf{v}_s$, where $\mathbf{R}$ is a symmetric matrix, $\mathbf{i}$ is the vector of unknown mesh currents, and $\mathbf{v}_s$ is the vector of known voltage sources.

If a current source is common to two adjoining meshes, we define the interior of the two meshes as a supermesh. We then write the mesh current equation around the periphery of the supermesh. If a current source appears at the periphery of only one mesh, we may define that mesh current as equal to the current of the source, accounting for the direction of the current source.

In general, either node voltage or mesh current analysis can be used to obtain the currents or voltages in a circuit. However, a circuit with fewer node equations than mesh current equations may require that we select the node voltage method. Conversely, mesh current analysis is readily applicable for a circuit with fewer mesh current equations than node voltage equations.

TERMS AND CONCEPTS

Branch   Path that connects two nodes.

Loop   Closed path progressing from node to node and returning to the starting node without passing an intermediate node more than once.

Mesh   Loop that does not contain any other loops within it.

Mesh Current   The current that flows around the periphery of a mesh; the current that flows through the elements constituting the mesh.

Node   Terminal common to two or more branches of a circuit.

Node Voltage   Voltage from a selected node to the reference node.

Planar Circuit   Circuit that can be drawn on a plane without branches crossing each other.

Reference Node   Node selected as the reference for all other nodes.

Supermesh   One larger mesh created from two meshes that have an independent or dependent current source in common.

Supernode A large node that includes two nodes connected by an independent or dependent voltage source.

REFERENCES Chapter 4

Dorf, Richard, The Electrical Engineering Handbook, CRC Press, Chap. 1, 1993.
Frank, Peter H. "New Use of Electricity Seen in Medicine," *New York Times,* July 29, 1987, page 28.

PROBLEMS

Section 4-3   Node Voltage Analysis of Circuits with Independent Current Sources Only

P 4.3-1   Find the node voltages v_A and v_B for the circuit shown in Figure P 4.3-1.
Answer: $v_A = 10$ V, $v_B = 28$ V

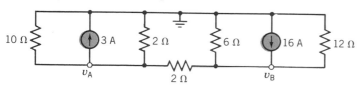

Figure P 4.3-1

P 4.3-2   Consider the circuit shown in Figure P 4.3-2. Find the voltages v_a and v_b when $R_1 = 100$ Ω, $R_2 = 200$ Ω, and $R_3 = 400$ Ω.

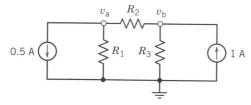

Figure P 4.3-2

P 4.3-3   Find v_1 in the circuit shown in Figure P 4.3-3.
Answer: $v_1 = 1$ V

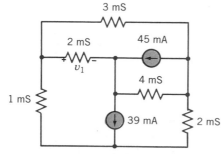

Figure P 4.3-3

P 4.3-4   A circuit with three node voltages and a reference node is shown in Figure P 4.3-4. (a) Find the three node voltages. (b) Find the power absorbed by the 1-mA current source.

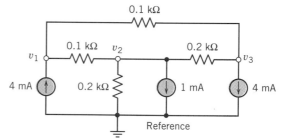

Figure P 4.3-4

P 4.3-5   Find the voltage between each of the lettered nodes (a, b, c, and d) and ground in Figure P 4.3-5.

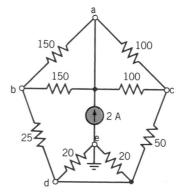

Figure P 4.3-5 All resistances in ohms.

Section 4-4   Node Voltage Analysis of Circuits with Independent Voltage and Current Sources

P 4.4-1   Determine the node voltage v_a for the circuit of Figure P 4.4-1.
Answer: 2.18 V

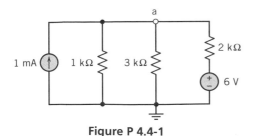

Figure P 4.4-1

P 4.4-2 Find v in Figure P 4.4-2 using nodal analysis.

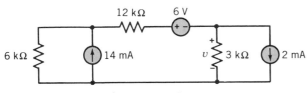

Figure P 4.4-2

P 4.4-3 Find v_a for the circuit of Figure P 4.4-3 when $i_s = 2.7$ mA and $v_s = 35$ V.
Answer: $v_a = -43.5$ V

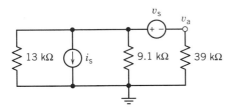

Figure P 4.4-3

P 4.4-4 Find v_x using nodal analysis for the circuit of Figure P 4.4-4.
Answer: $v_x = -1.125$ V

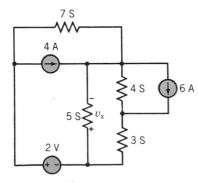

Figure P 4.4-4

P 4.4-5 Two voltage sources are labeled v_1 and v_2. Devise a resistor network that will yield

$$v_o = \frac{2}{3} v_1 + \frac{1}{3} v_2$$

Use a minimum number of resistors.

P 4.4-6 An electronic instrument incorporates a 15-V power supply. A digital display is added that requires a 5-V power supply. Unfortunately, the project is over budget and you are instructed to use the existing power supply. Using a voltage divider, as shown in Figure P 4.4-6, you are able to obtain 5 V. The specification sheet for the digital display shows that the display will operate properly over a supply voltage range of 4.8 V to 5.4 V. Further, the display will draw 300 mA (I) when the display is active and 100 mA when quiescent (no activity).
(a) Select values of R_1 and R_2 so that the display will be supplied with 4.8 V to 5.4 V under all conditions of current I.
(b) Calculate the maximum power dissipated by each resistor, R_1 and R_2, and the maximum current drawn from the 15-V supply.
(c) Is the use of the voltage divider a good engineering solution? If not, why? What problems might arise?

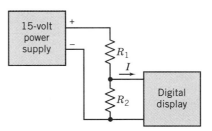

Figure P 4.4-6 A digital display circuit.

P 4.4-7 Find v_a for the circuit shown in Figure P 4.4-7.
Answer: $v_a = 19.37$ V

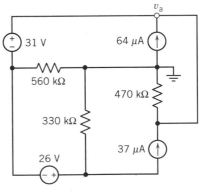

Figure P 4.4-7

P 4.4-8 Find the voltage v for the circuit of Figure P 4.4-8.
Answer: $v = 3.33$ V

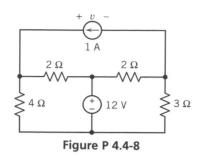

Figure P 4.4-8

P 4.4-9 Determine v and i for the circuit of Figure P 4.4-9.
Answer: $v = 1.286$ V, $i = 0.4$ A

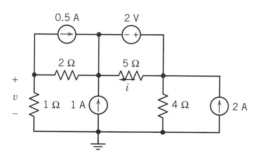

Figure P 4.4-9

Section 4-5 Node Voltage Analysis with Dependent Sources

P 4.5-1 Using nodal analysis, find v_1 in Figure P 4.5-1.
Answer: $v_1 = 24$ V

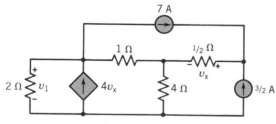

Figure P 4.5-1

P 4.5-2 Using node voltage analysis, find v_x and the power absorbed by the 2-Ω resistor in Figure P 4.5-2.
Answer: $v_x = 104$ V, $p = 32$ W

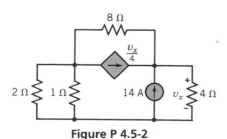

Figure P 4.5-2

P 4.5-3 Find v_1 and i_2 for the circuit shown in Figure P 4.5-3.

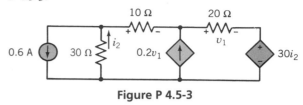

Figure P 4.5-3

P 4.5-4 (a) Find i_x and v_x in Figure P 4.5-4 using nodal analysis methods. (b) Find power absorbed by the 1-Ω resistor and the 10-A source.
Answer: $i_x = 6$ A, $v_x = 10$ V

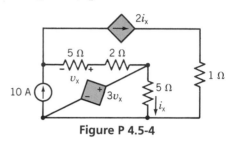

Figure P 4.5-4

P 4.5-5 Determine i_1 in the circuit shown in Figure P 4.5-5 using nodal analysis.

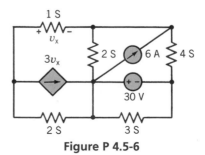

Figure P 4.5-5

P 4.5-6 Using nodal analysis methods, find v_x in the circuit of Figure P 4.5-6.
Answer: $v_x = 6$ V

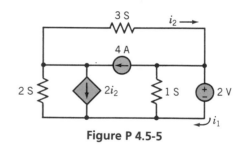

Figure P 4.5-6

P 4.5-7 Determine the current i of the circuit shown in Figure P 4.5-7 using node voltage analysis.

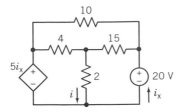

Figure P 4.5-7 Resistances in ohms.

P 4.5-8 Determine v for the circuit of Figure P 4.5-8.

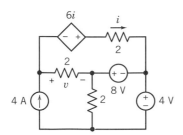

Figure P 4.5-8 Resistances in ohms.

P 4.5-9 Determine v_x for the circuit shown in Figure P 4.5-9.

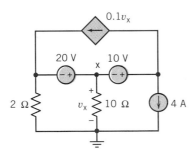

Figure P 4.5-9

Section 4-6 Mesh Current Analysis with Independent Voltage Sources Only

P 4.6-1 For the circuit shown in Figure P 4.6-1, (a) find i_1, i_2, and i_3 when $R_1 = R_2 = 3\ \Omega$ and $R_3 = 6\ \Omega$ and (b) show that energy is conserved in the circuit.
Answer: (a) $i_1 = 0$, $i_2 = \frac{1}{3}$ A, $i_3 = \frac{1}{3}$ A

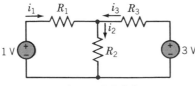

Figure P 4.6-1

P 4.6-2 Determine the three mesh currents of the circuit of Figure P 4.6-2.

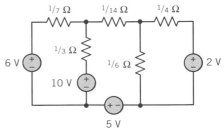

Figure P 4.6-2

P 4.6-3 Find the voltage between node a and ground in Figure P 4.6-3.

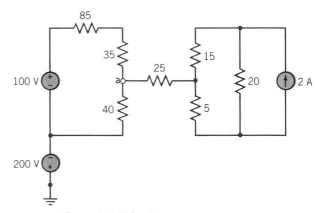

Figure P 4.6-3 All resistances in ohms.

P 4.6-4 Find i using mesh current analysis for the circuit shown in Figure P 4.6-4.
Answer: $i = -211.4\ \mu$A

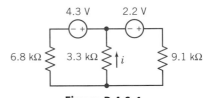

Figure P 4.6-4

P 4.6-5 Determine the mesh currents of the circuit shown in Figure P 4.6-5.

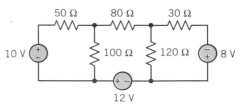

Figure P 4.6-5

P 4.6-6 Find the current i for the circuit of Figure P 4.6-6. Hint: A short circuit can be treated as a 0-V voltage source.

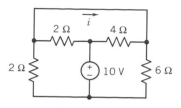

Figure P 4.6-6

P 4.6-7 Find the currents i_1, i_2, and i_3 for the circuit shown in Figure P 4.6-7.

Answer: $i_1 = \dfrac{21}{64}$ A, $i_2 = \dfrac{10}{64}$ A, $i_3 = \dfrac{-2}{64}$ A

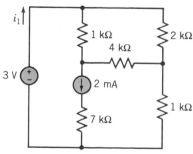

Figure P 4.6-7

Section 4-7 Mesh Current Analysis with Voltage and Current Sources (a) Independent Sources Only

P 4.7-1 Using mesh analysis, find i_1 for the circuit shown in Figure P 4.7-1.

Answer: $i_1 = 3$ mA

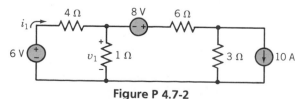

Figure P 4.7-1

P 4.7-2 Find i_1 and v_1 using mesh analysis methods for the circuit of Figure P 4.7-2.

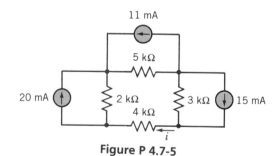

Figure P 4.7-2

P 4.7-3 Use mesh analysis to find i_1 for the circuit of Figure P 4.7-3.

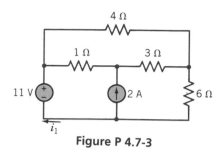

Figure P 4.7-3

P 4.7-4 Find the power absorbed by each element in the circuit shown in Figure P 4.7-4.

Partial Answer: $P_{6V} = 24$ W, $P_{4A} = -24$ W

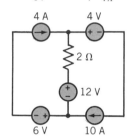

Figure P 4.7-4

P 4.7-5 Use mesh analysis to find i for the circuit of Figure P 4.7-5.

Answer: $i = 2.143$ mA

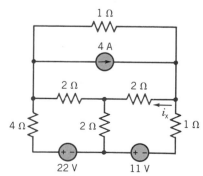

Figure P 4.7-5

P 4.7-6 Find i_x in the circuit of Figure P 4.7-6.

Figure P 4.7-6

P 4.7-7 Determine the current i for the circuit shown in Figure P 4.7-7 using the mesh current method.

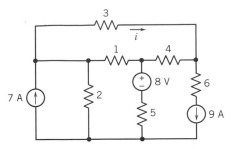

Figure P 4.7-7 All resistances in ohms.

P 4.7-8 Find v_1 and i_x in the circuit shown in Figure P 4.7-8 using the mesh currents shown.
Answer: $i_x = -3/4$ A, $v_1 = 7/4$ V

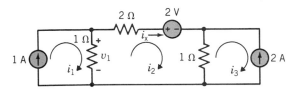

Figure P 4.7-8

P 4.7-9 Use mesh analysis to find i_x in the circuit of Figure P 4.7-9. Use the concept of a supermesh.
Answer: $i_x = 5$ A

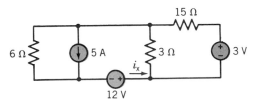

Figure P 4.7-9

P 4.7-10 Find v_{ab} for the circuit of Figure P 4.7-10.

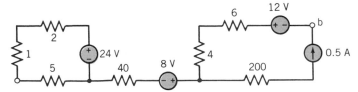

Figure P 4.7-10 All resistances in ohms.

P 4.7-11 Obtain the four mesh currents for the circuit shown in Figure P 4.7-11.

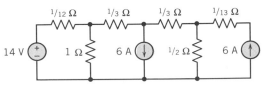

Figure P 4.7-11

P 4.7-12 Determine v and i for the circuit of Figure P 4.7-12.
Answer: $v = 1.286$ V, $i = 0.4$ A

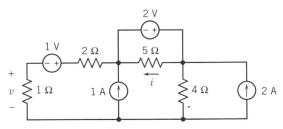

Figure P 4.7-12

Section 4-7 Mesh Current Analysis with Voltage and Current Sources (b) Independent and Dependent Sources

P 4.7-13 Determine the current i in the circuit shown in Figure P 4.7-13.
Answer: $i = -\frac{16}{3}$ A

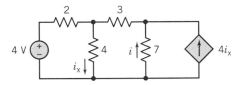

Figure P 4.7-13 All resistances in ohms.

P 4.7-14 Although scientists continue to debate exactly why and how it works, the process of utilizing electricity to aid in the repair and growth of bones—which has been used mainly with fractures—soon may be extended to an array of other problems, ranging from osteoporosis and osteoarthritis to spinal fusions and skin ulcers.

An electric current is applied to bone fractures that have not healed in the normal period of time. The process seeks to imitate natural electrical forces within the body. It takes only a small amount of electric stimulation to accelerate bone recovery. The direct current method uses an electrode that is implanted at the bone. This method has a success rate approaching 80 percent.

The implant is shown in Figure P 4.7-14a and the circuit model is shown in Figure P 4.7-14b. Find the energy delivered to the cathode during a 24-hour period. The bone fissure and cathode model is represented by the dependent voltage source and the 100-kΩ resistor.

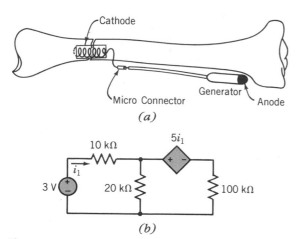

(a)

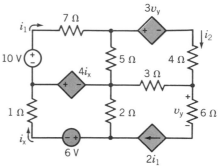

Figure P 4.7-17

(b)

Figure P 4.7-14 *(a)* Electric aid to bone repair. *(b)* Circuit model.

P 4.7-15 For the circuit shown in Figure P 4.7-15, find i_x and the power supplied by the 8-V source by mesh analysis methods. The units for i_x are milliamps.

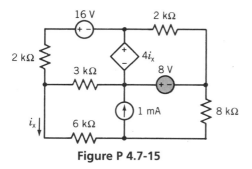

Figure P 4.7-15

P 4.7-16 Use mesh analysis to determine four equations expressed in terms of the four currents i_1, i_2, i_3, and i_4 for the circuit of Figure P 4.7-16. Use the concept of a supermesh.

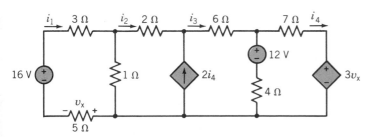

Figure P 4.7-16

P 4.7-17 Use mesh analysis methods to find three equations in terms of i_1, i_2, and i_x for the circuit of Figure P 4.7-17.
Partial Answer: $12i_1 - 5i_2 - 4i_x = 10$

P 4.7-18 Find i in the circuit of Figure P 4.7-18.

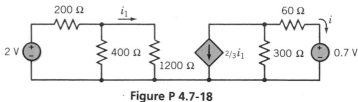

Figure P 4.7-18

P 4.7-19 Obtain the four mesh currents for the circuit shown in Figure P 4.7-19.

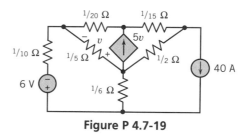

Figure P 4.7-19

P 4.7-20 Determine the current i in the 1-Ω resistor for the circuit of Figure P 4.7-20.

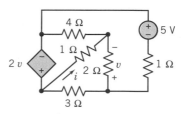

Figure P 4.7-20

P 4.7-21 The model of a bipolar junction transistor (BJT) amplifier is shown in Figure P 4.7-21.
(a) Determine the gain ratio v_o/v_i.
(b) Calculate the required value of g in order to obtain a gain $v_o/v_i = -170$ when $R_L = 5$ kΩ, $R_1 = 100$ Ω, and $R_2 = 1$ kΩ.

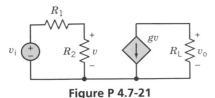

Figure P 4.7-21

P 4.7-22 An amplifier provides a gain in voltage between the input voltage v_{in} and the output voltage v_o. A model of a transistor amplifier is shown in Figure P 4.7-22, where $d = 10$. Find the ratio v_o/v_{in}.

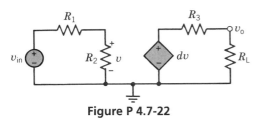

Figure P 4.7-22

Section 4-8 Node Voltage Method and Mesh Current Method Compared

P 4.8-1 Determine v_a using (a) mesh analysis and (b) node analysis for the circuit shown in Figure P 4.8-1.
Answer: $v_a = 6$ V.

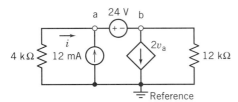

Figure P 4.8-1

P 4.8-4 A two-stage transistor amplifier is shown in Figure P 4.8-4a. For midband frequencies and small signals, the circuit can be simplified to the circuit shown in Figure P 4.8-4b. Find $v_o/v_s = A =$ gain if $R_1 = R_3 = 265$ kΩ, $R_2 = R_4 = 2$ kΩ, $h_{ie} = 1.3$ kΩ, and $\beta = 50$.
Answer: $v_o/v_s = 2320$

P 4.8-2 Determine the voltage v_c and the current i_x of the circuit of Figure P 4.8-2 using (a) mesh analysis and (b) node analysis.

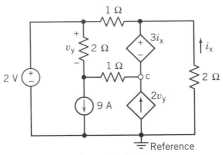

Figure P 4.8-2

P 4.8-3 A transistor is connected as shown in Figure P 4.8-3. It is assumed that $v_{be} = 0.7$ V and $i_c = 0.9i_e$. Develop a model of the circuit and determine i_b and v_{ce}.

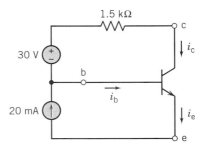

Figure P 4.8-3 A transistor circuit.

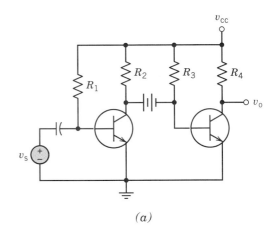

(a)

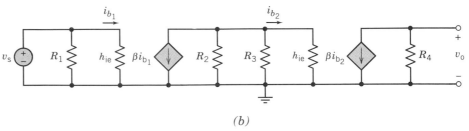

(b)

Figure P 4.8-4 Two-stage transistor amplifier. (*a*) Complete circuit. (*b*) Linear model.

P 4.8-5 For the circuit of Figure P 4.8-5, determine i_2 using mesh current analysis and repeat using node voltage analysis when $g = 900$ nmhos and $r = 1$ MΩ. Which method is preferable in this case?

Answer: $i_2 = 110.63$ nA

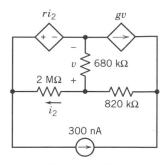

Figure P 4.8-5

P 4.8-6 For the circuit shown in Figure P 4.8-6, determine the values displayed by the meters.

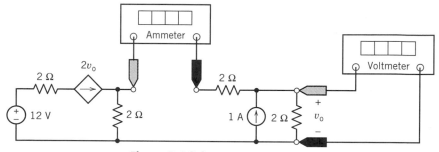

Figure P 4.8-6 Circuit with meters.

Section 4-9   DC Analysis Using PSpice

PSpice PROBLEMS

SP 4-1 Use PSpice to solve Problem 4.5-1.

SP 4-2 Use PSpice to solve Problem 4.5-2.

SP 4-3 Use PSpice to solve Problem 4.5-4.

SP 4-4 Use PSpice to determine the three node voltages of the circuit shown in Figure SP 4-4.

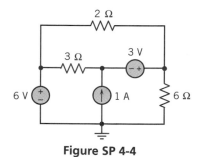

Figure SP 4-4

SP 4-5 Determine v_a and v_b of Example 4-1 using PSpice.

SP 4-6 Determine v_d of Example 4-2 using PSpice.

SP 4-7 Determine the node voltage v_a for the circuit of Figure P 4.4-1.

SP 4-8 Determine the current i for the circuit of Figure P 4.7-13.

SP 4-9 Determine the voltage v across the output resistor R_L when $R_L = 75$ Ω in the circuit of Figure SP 4-9.

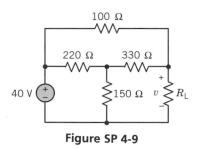

Figure SP 4-9

SP 4-10 Determine the voltage v and the current i for the circuit shown in Figure SP 4-10 when $R = 3$ kΩ.

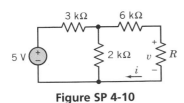

Figure SP 4-10

SP 4-11 Find i_2 of Example 3-9 using PSpice.

SP 4-12 Find $R_{\text{eq a-b}}$ of Example 3-10 using PSpice.

SP 4-13 Using PSpice, find the voltage v when $R = 10\ \Omega$ for the circuit of DP 4-4.

SP 4-14 Find $v_a - v_c$ for the circuit of Figure SP 4-14.

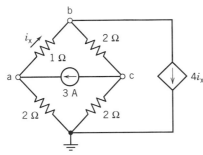

Figure SP 4-14

SP 4-15 Determine the current i, shown in Figure SP 4-15.

Answer: $i = 0.56$ A

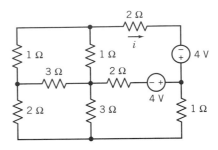

Figure SP 4-15

SP 4-16 Determine the node voltages v_a, v_b, and v_c for the circuit shown in Figure SP 4-16.

Answer: $v_a = \dfrac{18}{41}$ V, $v_b = \dfrac{13}{41}$ V, $v_c = \dfrac{34}{41}$ V

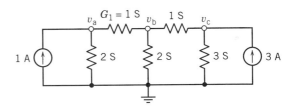

Figure SP 4-16

VERIFICATION PROBLEMS

VP 4-1 For the circuit shown in Figure VP 4-1, your lab partner reports her measurements for the node voltages as $v_a = 5.2$ V, $v_b = -4.8$ V, and $v_c = 3.0$ V. Is she correct?

VP 4-2 A report asserts that the node voltages of the circuit of Figure VP 4-2 are $v_a = 4$ V, $v_b = 20$ V, and $v_c = 12$ V. Are these correct?

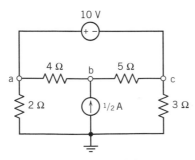

Figure VP 4-1

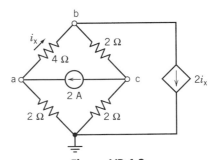

Figure VP 4-2

DESIGN PROBLEMS

DP 4-1 For the circuit shown in Figure DP 4-1, it is desired that $v_{ba} = 3$ V by adjusting the current source i_s. Select i_s to achieve the desired voltage v_{ba}.

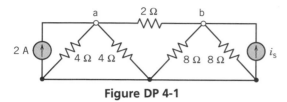

Figure DP 4-1

DP 4-2 For the circuit shown in Figure DP 4-2, determine the required voltage source v_s so that the current i equals 3.0 A.

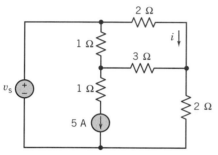

Figure DP 4-2

DP 4-3 For the circuit shown in Figure DP 4-3, it is desired to set the voltage at node a equal to 0 V in order to control an electric motor. Select voltages v_1 and v_2 in order to achieve $v_a = 0$ V when v_1 and v_2 are less than 20 V and greater than zero and $R = 2$ Ω.

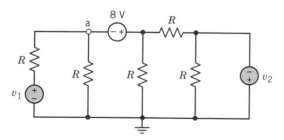

Figure DP 4-3

DP 4-4 A wiring circuit for a special lamp in a home is shown in Figure DP 4-4. The lamp has a resistance of 2 Ω, and the designer selects $R = 100$ Ω. The lamp will light when $I \geq 50$ mA but will burn out when $I > 75$ mA.
(a) Determine the current in the lamp and determine if it will light for $R = 100$ Ω.
(b) Select R so that the lamp will light but will not burn out if R changes by ±10% because of temperature changes in the home.

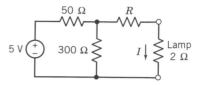

Figure DP 4-4 A lamp circuit.

DP 4-5 In order to control a device using the circuit shown in Figure DP 4-5, it is necessary that $v_{ab} = 10$ V. Select the resistors when it is required that all resistors be greater than 1 Ω and $R_3 + R_4 = 20$ Ω.

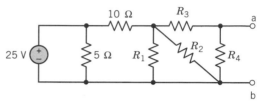

Figure DP 4-5 A circuit for earth fault line stress measurement.

DP 4-6 The current i shown in the circuit of Figure DP 4-6 is used to measure the stress between two sides of an earth fault line. Voltage v_1 is obtained from one side of the fault, and v_2 is obtained from the other side of the fault. Select the resistances R_1, R_2, and R_3 so that the magnitude of the current i will remain in the range between 0.5 mA and 2 mA when v_1 and v_2 may each vary independently between +1 V and +2 V ($1 V \leq v_n \leq 2$ V).

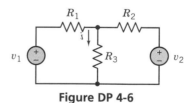

Figure DP 4-6

DP 4-7 For the circuit of Figure DP 4-7, determine the required value of g so that $v_a - v_c = \frac{20}{3}$ V.
Answer: $g = 4$

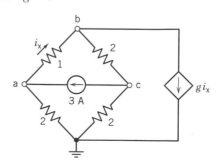

Figure DP 4-7 All resistances in ohms.

CHAPTER 5

CIRCUIT THEOREMS

PREVIEW

Many electric circuits are complex, but it is the engineer's goal to reduce their complexity in order to analyze them readily. In this chapter we consider the reduction of complex circuits to a simpler form. We also show how some simple sources can be transformed from a current source form to a voltage source form, and vice versa.

Furthermore, using the property of superposition for linear circuits, we develop a relatively simple method of analyzing a complex circuit when we wish to determine only one voltage or current within the circuit.

The function of many circuits is to deliver maximum power to a load such as an audio speaker in a stereo system. Here we develop the required relationship between a load resistor and a fixed series resistor that can represent the remaining portion of the circuit.

5-1 DESIGN CHALLENGE

STRAIN GAUGE BRIDGE

Strain gauges are transducers that measure mechanical strain, which is a deformation caused by force. Electrically, the strain gauges are resistors. The strain causes a change in resistance that is proportional to the strain.

Figure 5D-1 shows four strain gauges connected in a configuration called a bridge. Strain gauge bridges are used to measure force or pressure (Doeblin, 1966).

The bridge output is usually a small voltage. In Figure D 5-1 an amplifier multiplies the bridge output, v_i, by a gain to obtain a larger voltage, v_o, which is displayed by the voltmeter.

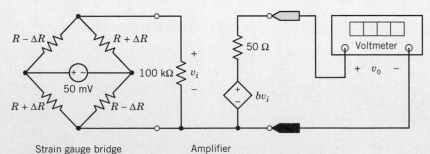

Figure 5D-1 Design problem involving a strain gauge bridge.

Define the Situation

A strain gauge bridge is used to measure force. The strain gauges have been positioned so that the force will increase the resistance of two of the strain gauges while, at the same time, decreasing the resistance of the other two strain gauges.

The strain gauges used in the bridge have nominal resistances of $R = 120\ \Omega$. (The nominal resistance is the resistance when the strain is zero.) This resistance is expected to increase or decrease by no more than $2\ \Omega$ due to strain. This means that

$$-2\ \Omega \le \Delta R \le 2\ \Omega \tag{5D-1}$$

The output voltage, v_o, is required to vary from -10 V to $+10$ V as ΔR varies from $-2\ \Omega$ to $2\ \Omega$.

State the Goal

Determine the amplifier gain, b, needed to cause v_o to be related to ΔR by the equation

$$v_o = 5\ \frac{\text{volt}}{\text{ohm}} \cdot \Delta R \tag{5D-2}$$

Generate a Plan

Analyze the circuit shown in Figure 5D-1 to determine the relationship between v_i and ΔR. Calculate the amplifier gain needed to satisfy Eq. 5D-2.

The required analysis could be accomplished using node or mesh equations. Chapter 5 presents Thévenin's Theorem. This theorem will significantly simplify the required analysis. We will postpone the analysis until the end of the chapter so that we can make use of Thévenin's Theorem.

5-2 ELECTRIC POWER FOR CITIES

The outstanding advantage of electricity compared to other power sources is its transmittability and flexibility. Electrical energy can be moved to any point along a pair of wires and, depending on the user's requirements, converted to light, heat, motion, or other forms.

Although the first demand for large-scale generation of electrical power was for lighting, that use was soon matched by other applications such as street railways and industrial processes. In all cases, the basic problems are the same: (1) to generate the electric power from another energy source such as coal or oil, (2) to transmit it to the place of use, and (3) to convert it to the form in which it is to be used. The first two decades of this century

Figure 5-1 Dynamo Room at the Pearl Street Station. This was Edison's first central station for incandescent electric lighting. It began operation in New York City in 1882. Courtesy of General Electric Company.

brought solutions to these three problems and thus the ascendancy of electric power as the predominant choice.

Perhaps one of the most interesting steps in the beginning of power generation was the opening of the Pearl Street Station, as shown in Figure 5-1. With the transmission of power, cities were able to use this power for lighting and electric street railways and could develop the infrastructure necessary for their modern operation.

Although lighting and vehicle traction were the first uses of electric power, they were soon eclipsed by industrial use. By 1920 industry's use of electricity was predominant. Factories used electric power for motors, heating, machines, and electrochemistry, among others. One of the first products to be made on a large scale by electric heating was carborundum (silicon carbide). Later, in 1906, the electrolytic process for producing aluminum was introduced.

By 1910 electric automobiles, as shown in Figure 5-2, were commonplace. Nevertheless, they were replaced by gasoline-fueled automobiles by 1920 because electric cars operated at lower top speeds and over shorter ranges without recharging than gasoline cars could achieve. However, the availability of electric motive power remained a critical factor in the development of cities. Electrically powered elevators permitted the construction of high-rise, multistory office and apartment complexes. Also, the modern electric railway and subway permitted cities to spread out into suburbs while accommodating the needs for mass transit. A modern electric railway is shown in Figure 5-3. With the construction of dense, modern cities, the demand for electric power became intensive. The

Figure 5-2 Baker electric car, 1910. Courtesy of Motor Vehicle Manufacturers Assoc.

Figure 5-3 Bay Area Rapid Transit (BART) railway. Photograph copyright © by Ron May.

skyline of nighttime New York City, shown in Figure 5-4, illustrates the dependence on electric power for lighting and the elevator.

Consumption of electric energy in the home increased as electric appliances became widely available. By 1940 about one-half of American homes were using electric vacuum cleaners and clothes washers. By 1980 over 95 percent of American homes had vacuum cleaners and over 40 percent had electric dishwashers.

The early development of the electric phonograph can be traced to Edison's tinfoil phonograph, shown in Figure 5-5. With the addition of a battery-driven electric motor in 1893, Edison's phonograph became one of the most widely sought-after entertainment devices. By 1893 Edison had 65 patents on the phonograph and its improvements.

Figure 5-4 Skyline of New York City at night. Photograph copyright © by Ron May.

Figure 5-5 Original tinfoil phonograph patented by Edison in 1877. Courtesy of Science Museum, London.

5-3 | SOURCE TRANSFORMATIONS

We found in Chapter 4 that it is generally easier to use mesh current analysis when all the sources are voltage sources. Similarly, it is usually easier to use node current analysis when all the sources are current sources. If we have both voltage and current sources in a circuit, it is valuable to make a set of adjustments to the circuit so that all the sources are of one type.

It is possible to transform an independent voltage source in series with a resistor into a current source in parallel with a resistance, or vice versa. Consider the pair of circuits shown in Figures 5-6a and 5-6b. We will find the appropriate relationship required between these two circuits so that they are interchangeable, both providing the same response at terminals a–b.

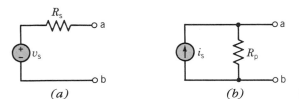

(a)   *(b)*   **Figure 5-6** Two equivalent circuits.

A **source transformation** is a procedure for transforming one source into another while retaining the terminal characteristics of the original source.

A source transformation rests on the concept of equivalence. An *equivalent circuit* is one whose terminal characteristics remain identical to those of the original circuit. It is important to note that equivalence implies an identical effect at the terminals but not *within* the equivalent circuits themselves.

We want the circuit of Figure 5-6a to transform into that of Figure 5-6b. We then require that both circuits have the same characteristic for all values of an external resistor R connected between terminals a–b (Figures 5-7a and 5-7b). We will try the two extreme values $R = 0$ and $R = \infty$.

When the external resistance $R = 0$, we have a short circuit across terminals a–b. First, we require the short-circuit current to be the same for each circuit. The short-circuit

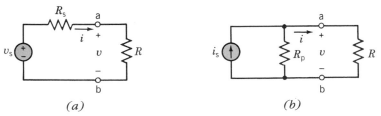

(a)   *(b)*

Figure 5-7 *(a)* Voltage source with an external resistor R. *(b)* Current source with an external resistance R.

current for Figure 5-7a is

$$i_s = \frac{v_s}{R_s} \tag{5-1}$$

The short-circuit current for Figure 5-7b is i_s. Therefore, we require that

$$i_s = \frac{v_s}{R_s} \tag{5-2}$$

For the open-circuit condition R is infinite, and from Figure 5-7a we have the voltage $v_{ab} = v_s$. For the open-circuit voltage of Figure 5-7b we have

$$v_{ab} = i_s R_p \tag{5-3}$$

Since v_{ab} must be equal for both circuits to be equivalent, we require that

$$v_s = i_s R_p$$

Also, from Eq. 5-2 we require $i_s = v_s/R_s$. Therefore, we must have

$$v_s = \left(\frac{v_s}{R_s}\right) R_p$$

and, therefore, we require that

$$R_s = R_p \tag{5-4}$$

Equations 5-2 and 5-4 must be true simultaneously for both circuits for the two sources to be equivalent. Of course, we have proved that the two sources are equivalent at two values ($R = 0$ and $R = \infty$). We have not proved that the circuits are equal for all R, but we assert that the equality relationship holds for all R for these two circuits as we show below.

For the circuit of Figure 5-7a we use KVL to obtain

$$v_s = iR_s + v$$

Dividing by R_s gives

$$\frac{v_s}{R_s} = i + \frac{v}{R_s}$$

If we use KCL for the circuit of Figure 5-7b we have

$$i_s = i + \frac{v}{R_p}$$

Thus the two circuits are equal when $i_s = v_s/R_s$ and $R_s = R_p$.

Any voltage source v_s with its associated series resistance R_s can be replaced by an equivalent current source i_s with an associated parallel conductance G_p where $i_s = v_s/R_s$ and $G_p = 1/R_s$. Any parallel combination of i_s and G_p can be replaced by an equivalent series combination of v_s and R_s.

Source transformations are useful for circuit simplification and also may be useful in node or mesh analysis. The method of transforming one form of source into the other form is summarized in Figure 5-8.

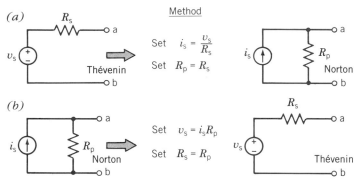

Figure 5-8 Method of source transformations.

Circuits are said to be *dual* when the characterizing equations of one network can be obtained from the other by simply interchanging v and i and interchanging G and R. There is a duality between resistance and conductance, current and voltage, and a series circuit and a parallel circuit. The two equivalent sources of Figure 5-8 are said to be dual circuits. Note that $R_s = 1/G_p$ is one element of duality. The equation for the voltage source with a short circuit is

$$i_s = G_p v_s$$

and the equation for the current source with an open circuit is

$$v_s = R_s i_s$$

Note that the voltage and the current are dual, the resistance R_s is the dual of the conductance G_p, and the short circuit is the dual of the open-circuit condition.

Therefore, *dual circuits* are defined by the same characterizing equations with v and i interchanged and G and R interchanged. When the set of transforms that converts one circuit into another also converts the second into the first, the circuits are said to be *duals*. Note that the transform of Figure 5-8a converts the first circuit into the second, while the transform of Figure 5-8b converts the second circuit into the first.

Dual circuits are two circuits such that the equations describing the first circuit, with v and i interchanged and R and G interchanged, describe the second circuit.

Example 5-1
Find the source transformation for the circuits shown in Figures 5-9a and 5-9b.

Solution
Using the method summarized in Figure 5-8, we note that the voltage source of Figure 5-9a can be transformed to a current source with $R_p = R_s = 14\,\Omega$. The current source is

$$i_s = \frac{v_s}{R_s} = \frac{28}{14} = 2\text{ A}$$

The resulting transformed source is shown on the right side of Figure 5-9a.

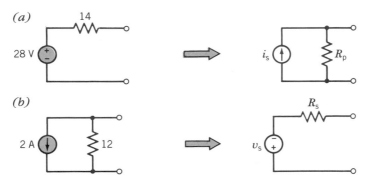

Figure 5-9 The circuits of Example 5-1. All resistances in ohms.

Starting with the current source of Figure 5-9b, we have $R_s = R_p = 12 \ \Omega$. The voltage source is

$$v_s = i_s R_p$$
$$= 2(12)$$
$$= 24 \ \text{V}$$

The resulting transformed source is shown on the right side of Figure 5-9b. Note that the positive sign of the voltage source v_s appears on the lower terminal since the current source flows downward.

Example 5-2

A circuit is shown in Figure 5-10. Find the current i by reducing the circuit to the right of terminals a−b to its simplest form using source transformations.

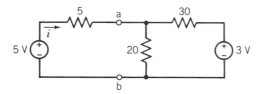

Figure 5-10 The circuit of Example 5-2. Resistances in ohms.

Solution

The first step is to transform the 30-Ω series resistor and the 3-V source to a current source with a parallel resistance. First, we note that $R_p = R_s = 30 \ \Omega$. The current source is

$$i_s = \frac{v_s}{R_p} = \frac{3}{30} = 0.1 \ \text{A}$$

as shown in Figure 5-11a. Combining the two parallel resistances in Figure 5-11a, we have $R_{p2} = 12 \ \Omega$, as shown in Figure 5-11b.

The parallel resistance of 12 Ω and the current source of 0.1 A can be transformed to a voltage source in series with $R_s = 12 \ \Omega$, as shown in Figure 5-11c. The voltage source v_s is found using Eq. 5-2:

$$v_s = i_s R_s$$
$$= 0.1(12)$$
$$= 1.2 \ \text{V}$$

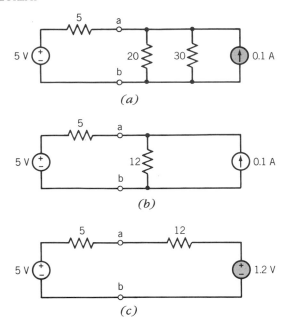

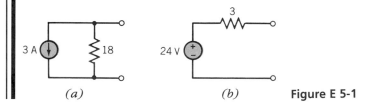

Figure 5-11 Source transformation steps for Example 5-2. All resistances in ohms.

Then the current i is found by using KVL around the loop of Figure 5-11c, yielding $i = 3.8/17 = 0.224$ A.

EXERCISE 5-1

Find the source transformation for the circuits of Figures E 5-1a and E 5-1b. All resistances are in ohms.

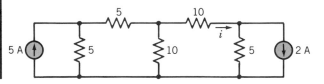

Figure E 5-1

EXERCISE 5-2

Using source transformations, find the current i for the circuit of Figure E 5-2.
Answer: 1.125 A

Figure E 5-2 All resistances in ohms.

EXERCISE 5-3

Using the method of source transformations, find the current i for the circuit shown in Figure E 5-3 when $v_1 = -e^{-t}$ V and $v_2 = e^{-2t}$ V.
Answer: $i = \frac{1}{6}(e^{-2t} - e^{-t})$ A

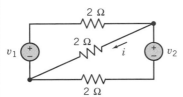

Figure E 5-3

EXERCISE 5-4

Determine the voltage at the terminals, v_{ab}, for the circuit shown in Figure E 5-4.
Answer: $v_{ab} = v_1 + v_2$

Figure E 5-4

5-4   SUPERPOSITION

We discussed linear elements in Section 2-2 and noted that one of the properties of a linear element is that the principle of superposition holds. A linear element satisfies superposition when it satisfies the following response and excitation relationship:

$$i_1 \longrightarrow v_1$$
$$i_2 \longrightarrow v_2 \tag{5-5}$$

then

$$i_1 + i_2 \longrightarrow v_1 + v_2$$

where the arrow implies the excitation causation and the resulting response. Thus, we can state that a device, if excited by current i_1, will exhibit response v_1. Similarly, an excitation i_2 will cause response v_2. Then if we use an excitation $i_1 + i_2$ we will find a response $v_1 + v_2$.

The principle of *superposition* states that for a linear circuit consisting of linear elements and independent sources, we can determine the total response by finding the response to each *independent* source with all other *independent* sources set to zero and then summing the individual responses. In this case, the response we seek may be a current or a voltage.

The **superposition principle** requires that the total effect of several causes acting simultaneously is equal to the sum of the effects of the individual causes acting one at a time.

The *superposition principle* may be restated as follows. In a linear circuit containing independent sources, the voltage across (or the current through) any element may be obtained by adding algebraically all the individual voltages (or currents) caused by each independent source acting alone, with all other independent voltage sources replaced by short circuits and all other independent current sources replaced by open circuits.

The principle of superposition requires that we deactivate (disable) all but one independent source and find the response due to that source. We then repeat the process by disabling all but a second source and determining the response. We find the response to each separate source, and then the total response is the sum of all the responses.

First, we note that when considering one independent source, we set the other independent sources to zero. Thus an independent voltage source appears as a short circuit with zero voltage across it. Similarly, if an independent current source is set to zero, no current flows and it appears as an open circuit. Also, it is important to note that if a dependent source is present, it must remain active (unaltered) during the superposition process.

Example 5-3

Find the current i in the 6-Ω resistor using the principle of superposition for the circuit of Figure 5-12.

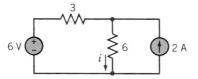

Figure 5-12 Circuit of Example 5-3. Resistances in ohms.

Solution

First set the current source to zero; and then the current source appears as an open circuit as shown in Figure 5-13a. The portion of the current i due to the first source we call i_1 and, therefore,

$$i_1 = \frac{6}{9} \text{ A}$$

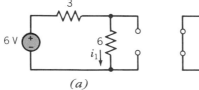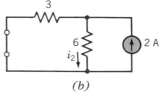

| (a) | (b) |

Figure 5-13 (a) Circuit of Example 5-3 with the current source set equal to zero. (b) Circuit of Example 5-3 with the voltage source set equal to zero. All resistances in ohms.

For the second step, set the voltage source to zero, replacing it with a short circuit. Then we have the circuit of Figure 5-13b. The portion of the current i due to the second source is called i_2, and we obtain i_2 by the current divider principle as

$$i_2 = \frac{3}{3 + 6} \, 2$$
$$= \frac{6}{9} \, \text{A}$$

The total current is then the sum of i_1 and i_2:

$$i = i_1 + i_2$$
$$= \frac{12}{9} \, \text{A}$$

In the following, let us use the superposition principle when the circuit contains both independent and dependent sources.

Example 5-4

Find the current i for the circuit of Figure 5-14. All resistances are in ohms.

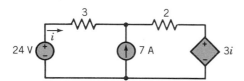

Figure 5-14 Circuit for Example 5-4.

Solution

We need to find the current i due to two independent sources. First, we will find the current resulting from the independent voltage source. The current source is replaced with an open circuit, and we have the circuit as shown in Figure 5-15a. The current i_1 represents the portion of current i resulting from the first source.

Kirchhoff's voltage law around the loop gives

$$-24 + (3 + 2)i_1 + 3i_1 = 0$$

or

$$i_1 = 3$$

When we set the independent voltage source to zero and determine the current i_2 due to the current source, we obtain the circuit of Figure 5-15b. Writing the KCL equation at node a, we obtain

$$-i_2 - 7 + \frac{v_a - 3i_2}{2} = 0 \qquad (5\text{-}6)$$

Noting that $-i_2 = v_a/3$ in the left-hand branch, we can state that $v_a = -3i_2$. Substituting for v_a in Eq. 5-6, we obtain

$$-i_2 - 7 + \frac{-3i_2 - 3i_2}{2} = 0$$

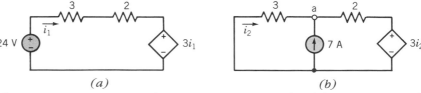

Figure 5-15 Circuit with (a) the voltage source activated and the current source deactivated and (b) the current source activated and the voltage source deactivated. All resistances are in ohms.

Therefore

$$i_2 = -\frac{7}{4}$$

Thus, the total current is

$$i = i_1 + i_2$$

$$= 3 - \frac{7}{4}$$

$$= \frac{5}{4} \text{ A}$$

EXERCISE 5-5

Using the superposition principle, find the current i in the circuit of Figure E 5-5.
Answer: $i = 0.224$ A

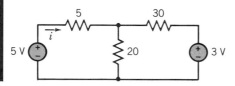

Figure E 5-5 Resistances in ohms.

EXERCISE 5-6

Using the superposition principle, find the voltage v in the circuit of Figure E 5-6.
Answer: 4 V

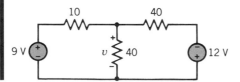

Figure E 5-6 Resistances in ohms.

EXERCISE 5-7

Using the superposition principle, find the current i in the circuit of Figure E 5-7.
Answer: $i = -1$ A

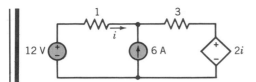

Figure E 5-7 Resistances in ohms.

5-5 | THÉVENIN'S THEOREM

In the preceding section we saw that the analysis of a circuit may be significantly simplified by use of the superposition principle. A theorem was developed by M. L. Thévenin, a French engineer, who first published the principle in 1883. Thévenin, who is credited with the theorem, probably based his work on earlier work by Hermann von Helmholtz (see Figure 5-16).

Figure 5-16 Hermann von Helmholtz (1821–1894), who is often credited with the basic work leading to Thévenin's theorem. Courtesy of the New York Public Library.

The goal of Thévenin's theorem is to reduce some portion of a circuit to an equivalent source and a single element. This reduced equivalent circuit connected to the remaining part of the circuit will enable us to find the current or voltage of interest. Thévenin's theorem rests on the concept of equivalence. A circuit equivalent to another circuit exhibits identical characteristics at identical terminals.

This circuit simplification process can be illustrated by the circuit of Figure 5-17a. If we wish to determine the current or power delivered to R_L, the rest of the circuit can be reduced to an equivalent circuit, shown in Figure 5-17b. The Thévenin equivalent circuit is shown to the left of terminals a–b and consists of a voltage v_t and a resistance R_t. Thévenin's principle is particularly useful when we wish to find the resulting current, voltage, or power delivered to a single element, especially when the element is variable. We reduce the rest of the circuit to R_t in series with a voltage source v_t, as shown in Figure 5-17b, and then later reconnect it to the element.

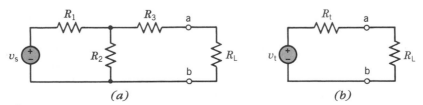

(a)   *(b)*

Figure 5-17 *(a)* Circuit and *(b)* a Thévenin circuit connected to a load resistor.

Thévenin's theorem may be stated as follows: Given any linear circuit, divide it into two circuits, A and B, each connected at the same pair of terminals. If either circuit contains a dependent source, it and its control variable must be in the same circuit. Then find the Thévenin equivalent for circuit A. Define v_{oc} as the open-circuit voltage of circuit A when circuit B is disconnected from the two terminals. Then the equivalent circuit of A is a voltage source v_{oc} in series with R_t, where R_t is the resistance looking into the terminals of circuit A with all its independent sources deactivated (set equal to zero).

Thévenin's theorem requires that, for any circuit of resistance elements and energy sources with an identified terminal pair, the circuit can be replaced by a series combination of an ideal voltage source v_t and a resistance R_t, where v_t is the open-circuit voltage at the two terminals and R_t is the ratio of the open-circuit voltage to the short-circuit current at the terminal pair.

Thévenin's theorem is summarized in Figure 5-18. If we examine the circuits of Figures 5-18*b* and 5-18*c*, they must exhibit the same circuit characteristics at the terminals. Since the circuit of Figure 5-18*b* must have the same open-circuit voltage at the two terminals as that of Figure 5-18*c*, we may state that $v_t = v_{oc}$, where v_{oc} is the open-circuit voltage of circuit A.

Step	Action	
(a)	Identify circuit A and circuit B.	
(b)	Separate circuit A from circuit B.	
(c)	Replace circuit A with its Thévenin equivalent.	
(d)	Reconnect circuit B and determine the variable of interest (e.g., current i).	

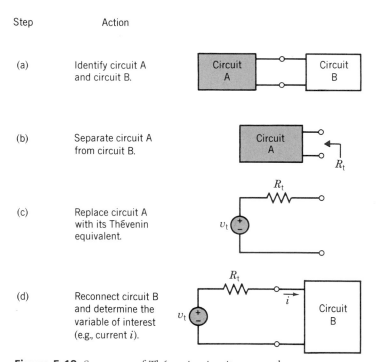

Figure 5-18 Summary of Thévenin circuit approach.

The resistance calculated looking into circuit A must be the same for the circuits of Figures 5-18*b* and 5-18*c*. If we choose to set all the independent sources of circuit A equal to zero, we may determine the equivalent resistance between the terminals, which is then required to be equal to R_t. Thévenin's theorem is often used to separate a circuit into a simple part and a second part where the components are nonlinear, unknown, or not yet specified. The simple part is selected to be as large as possible and is replaced by its Thévenin equivalent. Another use of Thévenin's theorem is to determine R_t of a circuit so that the load resistance can be selected for maximum power transfer.

In summary, Thévenin's equivalent circuit for circuit A is a voltage source $v_t = v_{oc}$ and a resistor R_t, where R_t is the resistance of circuit A seen at its terminals. (Many engineers use R_{th} instead of R_t as notation for the Thévenin resistance.) The Thévenin equivalent circuit is R_t in series with v_{oc}.

Example 5-5

Using Thévenin's theorem, find the current i through the resistor R in the circuit of Figure 5-19. All resistances are in ohms.

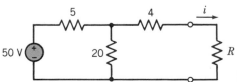

Figure 5-19 Circuit for Example 5-5.

Solution

Since we are interested in the current i, we identify the resistor R as circuit B. Then circuit A is as shown in Figure 5-20a. The Thévenin resistance R_t is found from Figure 5-20b, where we have deactivated the voltage source. We calculate the equivalent resistance looking into the terminals, obtaining $R_t = 8\ \Omega$.

(a)

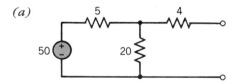

(b)

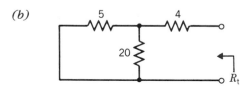

(c)

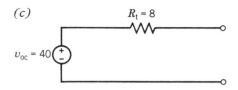

(d)

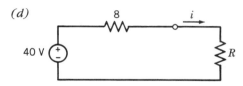

Figure 5-20 Steps for determining the Thévenin equivalent circuit for the circuit left of the terminals of Figure 5-19.

The Thévenin voltage v_t is equal to v_{oc}, the open-circuit voltage. Using the voltage divider principle with the circuit of Figure 5-20a, we find $v_{oc} = 40$ V.

Reconnecting circuit B to the Thévenin equivalent circuit as shown in Figure 5-20*d*, we obtain

$$i = \frac{40}{R + 8} \quad \text{A}$$

Example 5-6

Find the Thévenin equivalent circuit for the circuit shown in Figure 5-21.

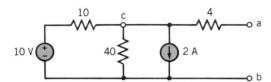

Figure 5-21 Circuit for Example 5-6. Resistances in ohms.

Solution

One approach is to find the open-circuit voltage and the circuit's Thévenin equivalent resistance R_t. First, let us find the resistance R_t. Deactivating the sources results in a short circuit for the voltage source and an open circuit for the current source, as shown in Figure 5-22. Look into the circuit at terminals a–b to find R_t. The 10-Ω resistor in parallel with the 40-Ω resistor results in an equivalent resistance of 8 Ω. Adding 8 Ω to 4 Ω in series, we obtain

$$R_t = 12 \ \Omega$$

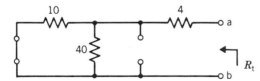

Figure 5-22 Circuit of Figure 5-21 with all the sources deactivated. Resistances in ohms.

Next, we wish to determine the open-circuit voltage at terminals a–b. Since no current flows through the 4-Ω resistor, the open-circuit voltage is identical to the voltage across the 40-Ω resistor, v_c. Using the bottom node as the reference, we write KCL at node c of Figure 5-21 to obtain

$$\frac{v_c - 10}{10} + \frac{v_c}{40} + 2 = 0$$

Solving for v_c yields

$$v_c = -8 \ \text{V}$$

Therefore, the Thévenin equivalent circuit is as shown in Figure 5-23.

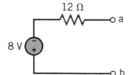

Figure 5-23 Thévenin equivalent circuit for the circuit of Figure 5-21.

Some circuits contain one or more dependent sources as well as independent sources. The presence of the dependent source prevents us from directly obtaining R_t from simple circuit reduction using the rules for parallel and series resistors.

A procedure for determining R_t is: (1) determine the open-circuit voltage v_{oc}, and (2) determine the short-circuit current i_{sc} when terminals a−b are connected by a short circuit, as shown in Figure 5-24; then

$$R_t = \frac{v_{oc}}{i_{sc}}$$

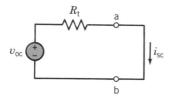

Figure 5-24 Thévenin circuit with a short circuit at terminals a−b.

This method is attractive since we already need the open-circuit voltage for the Thévenin equivalent circuit. We can show that $R_t = v_{oc}/i_{sc}$ by writing the KVL equation for the loop of Figure 5-24, obtaining

$$-v_{oc} + R_t i_{sc} = 0$$

Clearly, $R_t = v_{oc}/i_{sc}$.

Example 5-7

Find the Thévenin equivalent circuit for the circuit shown in Figure 5-25, which includes a dependent source. All resistances are in ohms.

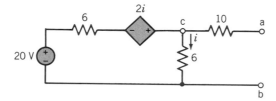

Figure 5-25 Circuit of Example 5-7.

Solution

First, we find the open-circuit voltage $v_{oc} = v_{ab}$. Writing KVL around the mesh of Figure 5-25 (using i as the mesh current), we obtain

$$-20 + 6i - 2i + 6i = 0$$

Therefore,

$$i = 2 \text{ A}$$

Since there is no current flowing through the 10-Ω resistor, the open-circuit voltage is identical to the voltage across the resistor between terminals c and b. Therefore,

$$v_{oc} = 6i$$
$$= 12 \text{ V}$$

The next step is to determine the short-circuit current for the circuit of Figure 5-26. Using the two mesh currents indicated, we have

$$-20 + 6i_1 - 2i + 6(i_1 - i_2) = 0$$

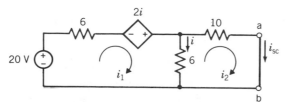

Figure 5-26 Circuit of Figure 5-25 with output terminals a−b short circuited.

and

$$6(i_2 - i_1) + 10i_2 = 0$$

Substitute $i = i_1 - i_2$ and rearrange the two equations to obtain

$$10i_1 - 4i_2 = 20$$

and

$$-6i_1 + 16i_2 = 0$$

Therefore, we find that $i_2 = i_{sc} = 120/136$ A. The Thévenin resistance is

$$R_t = \frac{v_{oc}}{i_{sc}}$$

$$= \frac{12}{120/136}$$

$$= 13.6 \ \Omega$$

Example 5-8

Find the Thévenin equivalent circuit to the left of terminals a–b for the circuit of Figure 5-27. All resistances in ohms.

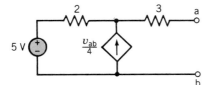

Figure 5-27 Circuit for Example 5-8.

Solution

First, we need to determine the open-circuit voltage $v_{ab} = v_{oc}$. Noting that the current source $v_{ab}/4$ provides a current source common to two meshes, we create a supermesh that includes v_{ab} and write one KVL equation around the periphery of the supermesh (see Table 4-2 to review the concept of supermesh). Also note that the current in the 3-Ω resistor is zero. The KVL equation around the periphery is

$$-5 + 2 \left(\frac{-v_{ab}}{4} \right) + 3(0) + v_{ab} = 0$$

Therefore

$$v_{ab} = 10 \text{ V}$$

To find the short-circuit current, we establish a short circuit across a–b, as shown in Figure 5-28. Since $v_{ab} = 0$, the current source is set to zero (i.e., replaced by an open-circuit). Then, using KVL around the periphery of the supermesh, we have

$$-5 + (2 + 3)i_{sc} = 0$$

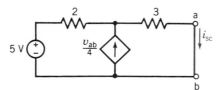

Figure 5-28 Circuit of Figure 5-27 with terminals a–b short-circuited. Resistances in ohms.

or

$$i_{sc} = 1 \text{ A}$$

Therefore, the Thévenin resistance is

$$R_t = \frac{v_{oc}}{i_{sc}}$$
$$= 10 \text{ } \Omega$$

Another arrangement we may confront is a circuit containing no independent sources but one or more dependent sources. Consider the circuit without independent sources shown in Figure 5-29. We wish to determine the Thévenin equivalent circuit. Thus, we will determine v_{oc} and R_t at the terminals a–b.

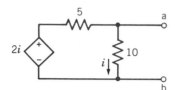

Figure 5-29 A circuit for which we seek its Thévenin equivalent. Resistances in ohms.

Since the circuit has no independent sources, $i = 0$ when terminals a–b are open. Therefore, $v_{oc} = 0$. Similarly, we find that $i_{sc} = 0$.

It remains to determine R_t. Since $v_{oc} = 0$ and $i_{sc} = 0$, R_t cannot be determined from $R_t = v_{oc}/i_{sc}$. Therefore, we choose to connect a current source of 1 A at terminals a–b, as shown in Figure 5-30. Then, determining v_{ab}, the Thévenin resistance is

$$R_t = \frac{v_{ab}}{1}$$

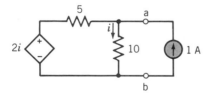

Figure 5-30 Circuit of Figure 5-29 with a 1-A source connected at terminals a–b. Resistances in ohms.

Writing KCL at a and setting b as the reference, we obtain

$$\frac{v_a - 2i}{5} + \frac{v_a}{10} - 1 = 0$$

For the 10-Ω resistor,

$$i = \frac{v_a}{10}$$

and therefore we have

$$\frac{v_a - 2(v_a/10)}{5} + \frac{v_a}{10} = 1$$

or

$$v_a = \frac{50}{13} \text{ V}$$

The Thévenin resistance is $R_t = v_a/1$ or

$$R_t = \frac{50}{13} \, \Omega$$

The Thévenin equivalent is shown in Figure 5-31. It is worth noting that one could alternatively find R_t by connecting a 1-V voltage source at terminals a–b and then find the current from b to a. The concept of finding R_t by connecting a 1-A source between the terminals a–b may also be used for circuits containing independent sources. In that case, set all of the independent sources to zero and use the 1-A source at terminals a–b to find v_{ab} and thus $R_t = v_{ab}/1$.

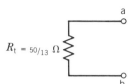

a

$R_t = 50/13 \, \Omega$

b

Figure 5-31 Thévenin equivalent circuit for the circuit of Figure 5-29.

The three common methods of finding a Thévenin equivalent circuit are summarized in Table 5-1. When separating a circuit containing dependent sources into two subcircuits to make a Thévenin equivalent, remember that the dependent source and its control voltage or current must be in the same subcircuit.

A laboratory procedure for determining the Thévenin equivalent of a black box circuit (see Figure 5-32a) is to measure i and v for two or more values of v_s and a fixed value

Table 5-1

Methods of Finding a Thévenin Equivalent Circuit

Number of Method	If the Circuit Contains:	Method
1	Resistors and independent sources	(a) Deactivate the sources and find R_t by circuit resistance reduction. (b) Find v_{oc} with the independent sources activated.
2	Resistors and independent and dependent sources **or** Resistors and independent sources	(a) Determine the open-circuit voltage v_{oc} with the sources activated. (b) Find the short-circuit current i_{sc} when a short circuit is applied to terminals a–b. (c) $R_t = v_{oc}/i_{sc}$.
3	Resistors and dependent sources (no independent sources)	(a) Note that $v_{oc} = 0$. (b) Connect a 1-A current source to terminals a–b and determine v_{ab}. (c) $R_t = v_{ab}/1$.

R_t

a

v_{oc}

b

Thévenin equivalent circuit

of R. For the circuit of Figure 5-32b we replace the test circuit with its Thévenin equivalent obtaining

$$v = v_{oc} + iR_t \tag{5-7}$$

The procedure is to measure v and i for a fixed R and several values of v_s. For example, let $R = 10\ \Omega$ and consider the two measurement results

1) $v_s = 49\text{ V} : i = 0.5\text{ A}, v = 44\text{ V}$

and

2) $v_s = 76\text{ V} : i = 2\text{ A}, \quad v = 56\text{ V}$

Then we have two simultaneous equations (using Eq. 5-7):

$$44 = v_{oc} + 0.5\ R_t$$

$$56 = v_{oc} + 2R_t$$

Solving these simultaneous equations we get $R_t = 8\ \Omega$ and $v_{oc} = 40\text{ V}$, thus obtaining the Thévenin equivalent of the black box circuit.

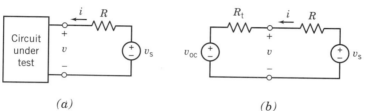

(a)   *(b)*

Figure 5-32 (a) Circuit under test with laboratory source v_s and resistor R, (b) Circuit of (a) with Thévenin equivalent circuit replacing test circuit.

EXERCISE 5-8

Find the Thévenin equivalent for the circuit of Figure E 5-8.
Answer: $R_t = 19\ \Omega$
               $v_{oc} = 16\text{ V}$ with plus terminal at b

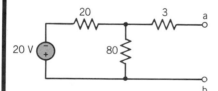

Figure E 5-8 Resistances in ohms.

EXERCISE 5-9

Find the Thévenin equivalent circuit to the left of terminals a−b for the circuit of Figure E 5-9.
Answer: $v_{oc} = 16\text{ V}, R_t = \frac{16}{3}\ \Omega$

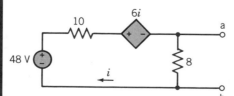

Figure E 5-9 Resistances in ohms.

EXERCISE 5-10

Find the Thévenin equivalent for the circuit shown in Figure E 5-10.
Answer: $v_{oc} = 0$, $R_t = \frac{2}{5}\,\Omega$

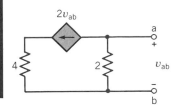

Figure E 5-10 Resistances in ohms.

5-6 ‖ NORTON'S EQUIVALENT CIRCUIT

An American engineer, E. L. Norton at Bell Telephone Laboratories, proposed an equivalent circuit for circuit A of Figure 5-18 using a current source and an equivalent resistance. The Norton equivalent circuit is a dual of the Thévenin equivalent circuit. Norton published his method in 1926, 43 years after Thévenin. The Norton equivalent is the source transformation of the Thévenin equivalent.

Norton's theorem may be stated as follows: Given any linear circuit, divide it into two circuits, A and B. If either A or B contains a dependent source, its controlling variable must be in the same circuit. Consider circuit A and determine its short-circuit current i_{sc} at its terminals. Then the equivalent circuit of A is a current source i_{sc} in parallel with a resistance R_n, where R_n is the resistance looking into circuit A with all its independent sources deactivated.

Norton's theorem requires that, for any circuit of resistance elements and energy sources with an identified terminal pair, the circuit can be replaced by a parallel combination of an ideal current source i_{sc} and a conductance G_n, where i_{sc} is the short-circuit current at the two terminals and G_n is the ratio of the short-circuit current to the open-circuit voltage at the terminal pair.

We therefore have the Norton circuit for circuit A as shown in Figure 5-33. Since this is the dual of the Thévenin circuit, it is clear that $R_n = R_t$ and $v_{oc} = R_t i_{sc}$. It is clear that the Norton equivalent is simply the source transformation of the Thévenin equivalent.

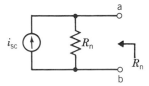

Figure 5-33 Norton equivalent circuit for a linear circuit A.

Example 5-9
Find the Norton equivalent circuit for the circuit of Figure 5-34.

Solution
Since the circuit contains only an independent source, we can deactivate the source and find R_n by circuit reduction. Replacing the voltage source by a short circuit we have a 6-k Ω

resistor in parallel with $(8 \text{ k}\Omega + 4 \text{ k}\Omega) = 12 \text{ k}\Omega$. Therefore,

$$R_n = \frac{6 \times 12}{6 + 12} = 4 \text{ k}\Omega$$

To determine i_{sc} we short circuit the output terminals with the voltage source activated as shown in Figure 5-35. Writing KCL at node a we have

$$-\frac{15 \text{ V}}{12 \text{ k}\Omega} + i_{sc} = 0$$

or

$$i_{sc} = 1.25 \text{ mA}$$

Thus, the Norton equivalent (Figure 5.33) has $R_n = 4 \text{ k}\Omega$ and $i_{sc} = 1.25 \text{ mA}$.

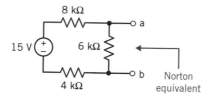

Figure 5-34 Circuit of Example 5-9.

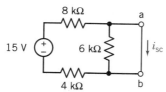

Figure 5-35 Short circuit connected to output terminals.

Example 5-10

Find the Norton equivalent circuit for the circuit of Figure 5-36.

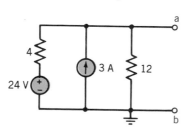

Figure 5-36 Circuit of Example 5-10. Resistances in ohms.

Solution

First, determine the current i_{sc} for the short-circuit condition shown in Figure 5-37. Writing KCL at a, we obtain

$$-\frac{24}{4} - 3 + i_{sc} = 0$$

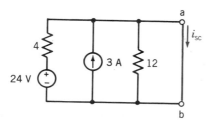

Figure 5-37 Short circuit connected to terminals a–b of the circuit of Figure 5-36. Resistances in ohms.

Note that no current flows in the 12-Ω resistor since it is in parallel with a short circuit. Also, because of the short circuit, the 24-V source causes 24 V to appear across the 4-Ω resistor. Therefore,

$$i_{sc} = \frac{24}{4} + 3$$

$$= 9 \text{ A}$$

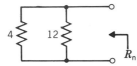

Figure 5-38 Circuit of Figure 5-36 with its sources deactivated. The voltage source becomes a short circuit, and the current source is re-placed by an open circuit. Resistances in ohms.

Now determine the equivalent resistance $R_n = R_t$ by deactivating the sources in the circuit as shown in Figure 5-38. Clearly, $R_n = 3 \Omega$. Thus, we obtain the Norton equivalent circuit as shown in Figure 5-39.

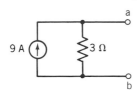

Figure 5-39 Norton equivalent of the circuit of Figure 5-36.

Table 5-2

Methods of Finding a Norton Equivalent Circuit

Number of Method	If the Circuit Contains:	Method
1	Resistors and independent sources	(a) Deactivate the sources and find R_t by circuit reduction. (b) Find i_{sc} with the sources activated.
2	Resistors and independent and dependent sources **or** Resistors and independent sources	(a) Determine the short-circuit current i_{sc} with all sources activated. (b) Find the open-circuit voltage v_{oc}. (c) $R_t = v_{oc}/i_{sc}$.
3	Resistors and dependent sources (no independent sources)	(a) Note that $i_{sc} = 0$. (b) Connect a 1-A current source to terminals a–b and determine v_{ab}. (c) $R_t = v_{ab}/1$.

The Norton equivalent circuit

Note: $R_n = R_t$

We may encounter circuits with dependent sources as well as independent sources and wish to obtain the Norton equivalent. Since the Norton equivalent is the dual of the Thévenin equivalent, we may develop a table of methods parallel to that summarizing the Thévenin methods (Table 5-1). Table 5-2 summarizes the three methods of determining the Norton equivalent.

Example 5-11

Find the Norton equivalent to the left of terminals a−b for the circuit of Figure 5-40.

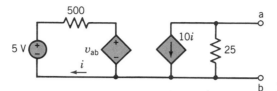

Figure 5-40 The circuit of Example 5-11. Resistances in ohms.

Solution

First, we need to determine the short-circuit current i_{sc} using Figure 5-41. Note that $v_{ab} = 0$ when the terminals are short-circuited. Then,

$$i = 5/500$$
$$= 10 \text{ mA}$$

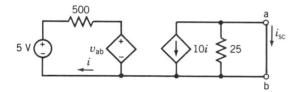

Figure 5-41 Circuit of Figure 5-40 with a short circuit at the terminals a−b. Resistances in ohms.

Therefore, for the right-hand portion of the circuit

$$i_{sc} = -10i$$
$$= -100 \text{ mA}$$

Now, to obtain R_t, we need $v_{oc} = v_{ab}$ from Figure 5-40, where i is the current in the first (left-hand) mesh. Writing the mesh current equation, we have

$$-5 + 500i + v_{ab} = 0$$

Also, for the right-hand mesh of Figure 5-40 we note that

$$v_{ab} = -25(10i)$$
$$= -250i$$

Therefore,

$$i = \frac{-v_{ab}}{250}$$

Substituting i into the first mesh equation, we obtain

$$500\left(\frac{-v_{ab}}{250}\right) + v_{ab} = 5$$

Therefore,

$$v_{ab} = -5 \text{ V}$$

and

$$R_t = \frac{v_{ab}}{i_{sc}}$$

$$= \frac{-5}{-0.1}$$

$$= 50 \ \Omega$$

The Norton equivalent circuit is shown in Figure 5-42.

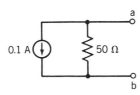

Figure 5-42 The Norton equivalent circuit for Example 5-11.

EXERCISE 5-11

Find the Norton equivalent of the circuit shown in Figure E 5-11.
Answer: $i_{sc} = 6$ A, $R_t = 6 \ \Omega$

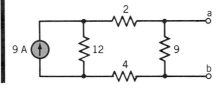

Figure E 5-11 Resistances in ohms.

EXERCISE 5-12

Find the Norton equivalent of the circuit shown in Figure E 5-12.
Answer: $R_t = 6.44$ kΩ, $i_{sc} = -0.15$ A

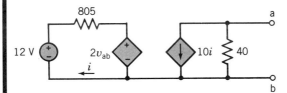

Figure E 5-12 Resistances in ohms.

EXERCISE 5-13

Use Thévenin's theorem to formulate a general expression for the current i in terms of the variable resistance R shown in Figure E 5-13.
Answer: $i = 20/(8 + R)$ A

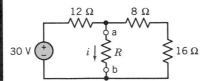

Figure E 5-13

5-7 ║ MAXIMUM POWER TRANSFER

Many applications of circuits require that the maximum power available from a source be transferred to a load resistor R_L. Consider the circuit A shown in Figure 5-43, terminated

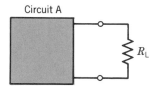

Figure 5-43 Circuit A contains resistors and independent and dependent sources. The load is the resistor R_L.

with a load R_L. As demonstrated in Section 5-5, circuit A can be reduced to its Thévenin equivalent, as shown in Figure 5-44.

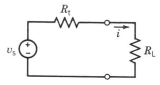

Figure 5-44 The Thévenin equivalent is substituted for circuit A. Here we use v_s for the Thévenin source voltage.

The general problem of power transfer can be discussed in terms of efficiency and effectiveness. Power utility systems are designed to transport the power to the load with the greatest efficiency by reducing the losses on the power lines. Thus, the effort is concentrated on reducing R_t, which would represent the resistance of the source plus the line resistance. Clearly, the idea of using superconducting lines that would exhibit no line resistance is exciting to power engineers.

In the case of signal transmission, as in the electronics and communications industries, the problem is to attain the maximum signal strength at the load. Consider the signal received at the antenna of an FM radio receiver from a distant station. It is the engineer's goal to design a receiver circuit so that the maximum power ultimately ends up at the output of the amplifier circuit connected to the antenna of your FM radio. Thus, we may represent the FM antenna and amplifier by the Thévenin equivalent shown in Figure 5-44.

Let us consider the general circuit of Figure 5-44. We wish to find the value of the load R_L such that the maximum power is delivered to it. First, we need to find the power from

$$p = i^2 R_L$$

Since the current i is

$$i = \frac{v_s}{R_L + R_t}$$

we find that the power is

$$p = \left(\frac{v_s}{R_L + R_t}\right)^2 R_L \tag{5-8}$$

Assuming that v_s and R_t are fixed for a given source, the maximum power is a function of R_L. To find the value of R_L that maximizes the power, we use the differential calculus to find where the derivative dp/dR_L equals zero. Taking the derivative, we obtain

$$\frac{dp}{dR_L} = v_s^2 \frac{(R_t + R_L)^2 - 2(R_t + R_L)R_L}{(R_L + R_t)^4}$$

The derivative is zero when

$$(R_t + R_L)^2 - 2(R_t + R_L)R_L = 0 \tag{5-9}$$

or

$$(R_t + R_L)(R_t + R_L - 2R_L) = 0 \tag{5-10}$$

Solving Eq. 5-10, we obtain

$$R_L = R_t \tag{5-11}$$

To confirm that Eq. 5-11 is a maximum, it should be shown that $d^2p/dR_L^2 < 0$. Therefore, the maximum power is transferred to the load when R_L is equal to the Thévenin equivalent resistance R_t.

The maximum power, when $R_L = R_t$, is then obtained by substituting $R_L = R_t$ in Eq. 5-8 to yield

$$p_{max} = \frac{v_s^2 R_L}{(2R_L)^2}$$

$$= \frac{v_s^2}{4R_L}$$

The power delivered to the load will differ from the maximum attainable as the load resistance R_L departs from $R_L = R_t$. The power attained as R_L varies from R_t is portrayed in Figure 5-45.

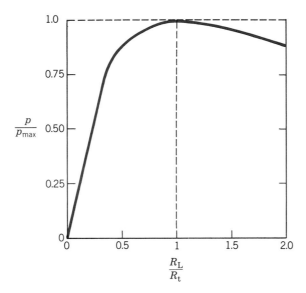

Figure 5-45 Power actually attained as R_L varies in relation to R_t.

The **maximum power transfer** theorem states that the maximum power delivered by a source represented by its Thévenin equivalent circuit is attained when the load R_L is equal to the Thévenin resistance R_t.

The *efficiency of power transfer* is defined as the ratio of the power delivered to the load, p_{out}, to the power supplied by the source, p_{in}. Therefore, we have the efficiency η as

$$\eta = p_{out}/p_{in}$$

For maximum power transfer, when $R_s = R_L$, we have

$$p_{in} = v_s i$$

$$= v_s \left(\frac{v_s}{R_s + R_L} \right)$$

$$= \frac{v_s^2}{2R_L}$$

The output power delivered to the load was found above to be

$$p_{out} = p_{max} = \frac{v_s^2}{4R_L}$$

Therefore, $p_{out}/p_{in} = 1/2$, and only 50% efficiency can be achieved at maximum power transfer conditions. Circuits may have efficiencies less than 50% at the maximum power condition.

We may also use Norton's equivalent circuit to represent the circuit A. We then have a circuit with a load resistor R_L as shown in Figure 5-46. The current i may be obtained from the current divider principle to yield

$$i = \frac{R_t}{R_t + R_L} i_s$$

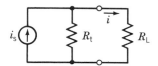

Figure 5-46 Norton's equivalent circuit representing the source circuit and a load resistor R_L. Here we use i_s as the Norton source current.

Therefore, the power p is

$$p = i^2 R_L$$

$$= \frac{i_s^2 R_t^2 R_L}{(R_t + R_L)^2} \qquad (5\text{-}12)$$

Using calculus, it is possible to show that the maximum power occurs when

$$R_L = R_t \qquad (5\text{-}13)$$

Then the maximum power delivered to the load is

$$p_{max} = \frac{R_L i_s^2}{4} \qquad (5\text{-}14)$$

Example 5-12

Find the load R_L that will result in maximum power delivered to the load for the circuit of Figure 5-47. Also determine the p_{max}.

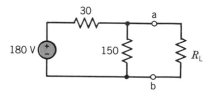

Figure 5-47 Circuit for Example 5-12. Resistances in ohms.

Solution

First, we determine the Thévenin equivalent circuit for the circuit to the left of terminals a–b. Disconnect the load resistor. The Thévenin voltage source v_t is

$$v_t = \frac{150}{180} \times 180$$

$$= 150 \text{ V}$$

The Thévenin resistance R_t is

$$R_t = \frac{30 \times 150}{30 + 150}$$

$$= 25 \ \Omega$$

The Thévenin circuit connected to the load resistor is shown in Figure 5-48. Maximum power transfer is obtained when $R_L = R_t = 25 \ \Omega$.

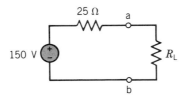

Figure 5-48 Thévenin equivalent circuit connected to R_L for Example 5-12.

Then the maximum power is

$$p_{max} = \frac{v_s^2}{4 \ R_L}$$

$$= \frac{(150)^2}{4 \times 25}$$

$$= 225 \text{ W}$$

The Thévenin source v_t actually provides a total power of

$$p_s = v_t \ i$$
$$= 150 \times 3$$
$$= 450 \text{ W}$$

Thus, we note that one-half the power is dissipated by the resistance R_t.

The actual source of the circuit of Figure 5-47 is 180 V. This source delivers a power $p = 180 i_1$ where i_1 is the current through the source when $R_L = 25 \ \Omega$. We readily calculate that $i_1 = 3.5$ A. Therefore, the actual source delivers 630 W to the total circuit, resulting in an efficiency of 35.7%.

If you prefer to use the Norton equivalent circuit, it is shown in Figure 5-46.

Example 5-13

Find the load R_L that will result in maximum power delivered to the load of the circuit of Figure 5-49. Also determine p_{max} delivered.

Solution

We will use Method 2 of Table 5-1 to obtain the Thévenin equivalent circuit for the circuit of Figure 5-49a. First we find v_{oc} with the sources activated and the load resistor disconnected as shown in Figure 5-49b. The KVL gives

$$-6 + 10i - 2v_{ab} = 0$$

Also, we note that $v_{ab} = v_{oc} = 4i$. Therefore

$$10i - 8i = 6$$

or $i = 3$ A. Therefore, $v_{oc} = 4i = 12$ V.

To determine the short-circuit current, we add a short circuit as shown in Figure 5-49c. Writing KVL we have

$$-6 + 6i_{sc} = 0$$

Hence $i_{sc} = 1$ A.

Therefore, $R_t = v_{oc}/i_{sc} = 12 \ \Omega$. The Thévenin equivalent circuit is shown in Figure 5-49d with the load resistor reconnected.

Maximum power is achieved when $R_L = R_t = 12 \ \Omega$. Then

$$p_{max} = \frac{v_{oc}^2}{4R_L} = \frac{12^2}{4(12)} = 3 \text{ W}$$

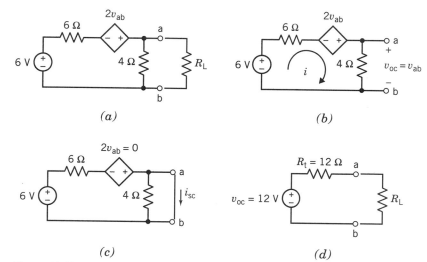

(a)

(b)

(c)

(d)

Figure 5-49 Determination of maximum power transfer to a load R_L.

EXERCISE 5-14

Find the maximum power delivered to R_L for the circuit of Figure E 5-14 using a Thévenin equivalent circuit. All resistances in ohms.

Answer: 9 W

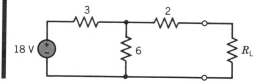

Figure E 5-14

EXERCISE 5-15

Find the maximum power delivered to R_L for the circuit of Figure E 5-15 using a Norton equivalent circuit. All resistances in ohms.

Answer: 175 W

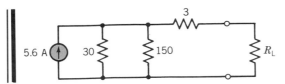

Figure E 5-15

EXERCISE 5-16

For the circuit of Figure E 5-16, find the power delivered to the load when R_L is fixed and R_t may be varied between 1 Ω and 5 Ω. Select R_t so that maximum power is delivered to R_L.
Answer: 13.9 W

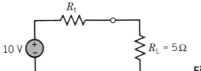

Figure E 5-16

EXERCISE 5-17

A resistive circuit was connected to a variable resistor, and the power delivered to the resistor was measured as shown in Figure E 5-17. Determine the Thévenin equivalent circuit.
Partial Answer: $R_t = 20\ \Omega$

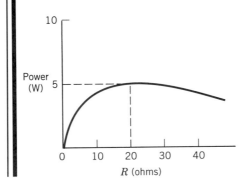

Figure E 5-17

5-8 FURTHER USE OF PSpice FOR DC ANALYSIS[1]

In this section we will illustrate the utility of PSpice analysis when the circuit contains dependent sources and demonstrate the usefulness of the **.DC** command and the **.PRINT** command. We also will demonstrate the utility of the **.TF** statement.

First, let us demonstrate the utility of the **.DC** command, which permits the analysis of a circuit over a range of values of a source voltage or current. In this example we will change a voltage source over a specified range (often called a sweep analysis). Let us

----

[1] This section may be omitted or considered with a later chapter. See Appendix G for an introduction to PSpice.

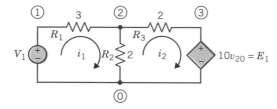

Figure 5-50 A two-mesh circuit with a dependent source and a variable input source V_1. Resistances in ohms.

consider a circuit that contains a dependent source shown in Figure 5-50, which was discussed earlier (see Figure 4-28). The nodes are labeled with numerals within the circles, and the goal is to obtain the two mesh currents when V_1 equals two values: 10 V and 12 V. This type of analysis is useful; for example, the voltage source might change when it switches from standby to full on. See Appendix G-5 for a review of the .DC statement. The PSpice program for the circuit is provided in Figure 5-51, where the VCVS is represented by E1. The controlling voltage is designated as the voltage v_{20} and the multiplier is 10.

```
  **      TWO MESH WITH DEPENDENT SOURCE

 *ELEM     NODE  NODE    VALUE
  R1        1     2        3
  R2        2     0        2
  R3        2     3        2
  V1        1     0       DC    10
 *ELEM     NOD   NOD    CNOD   CNOD   MULT
  E1        3     0       2      0     10
  .DC       V1    10      12     2
  .PRINT    DC    I(R1)  I(R3)
  .END
```

Figure 5-51 The PSpice program for the circuit of Figure 5-50.

The .DC statement sweeps the voltage V_1 between 10 V and 12 V in one increment of 2 V. The .PRINT statement (discussed in Appendix G-5) provides the output of the currents in R_1 and R_3 since these two currents are i_1 and i_2. The output is shown in Figure 5-52.

```
 ****         DC TRANSFER CURVES

 ********************************************
   V1            I(R1)          I(R3)
  1.000E+01     3.636E+00      4.091E+00
  1.200E+01     4.364E+00      4.909E+00
```

Figure 5-52 Output for the PSpice program of Figure 5-51. The currents in R_1 and R_3 are provided for V_1 equal to 10 V and 12 V.

The .TF statement can be used to determine the dc transfer (gain ratio) function and the Thévenin equivalent of a circuit. The format is

$$.\text{TF} < \text{Output Variable} > < \text{Input Source} >$$

If v_o is the output variable and v_s is the input source, then the computer printout provides v_o/v_s, R_{in} seen by the input source, and the output resistance $R_o = R_t$ seen at v_o as shown in Figure 5-53. The .TF statement provides these three calculated values as the printout and a .PRINT statement is not necessary. The .TF calculation provides $v_o = v_t$ and R_t, the Thévenin equivalent circuit elements.

Figure 5-53 A circuit with an input source v_s.

Let us use .TF analysis to determine the Thévenin equivalent circuit for the circuit of Example 5-7 as redrawn in Figure 5-54. The circuit is redrawn for PSpice analysis in Figure 5-55, where the CCVS is represented by H1 and the dummy source VM1 is used to provide i that controls H1. The 10-Ω resistor is omitted and will be added later to R_o of the circuit of Figure 5-55. The PSpice program for this circuit is shown in Figure 5-56. The output variable is $v_{30} = v_3$ referenced to the ground node.

The output of the calculation is shown in Figure 5-57, and $v_o = 0.6v_1 = 12$ V. The output resistance is $R_o = 3.6$ Ω and thus $R_t = 3.6 + 10 = 13.6$ Ω. These calculations verify the results obtained in Example 5-7.

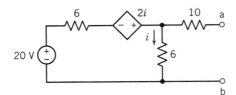

Figure 5-54 The circuit of Example 5-7. Resistances in ohms.

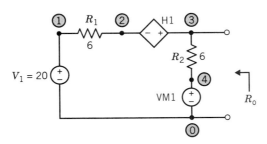

Figure 5-55 The circuit of Figure 5-54 redrawn for PSpice. Resistances in ohms.

```
**      THEVENIN EQUIVALENT

V1      1     0      DC    20
R1      1     2      6
R2      3     4      6
VM1     4     0      DC    0
H1      3     2      VM1   2
.TF     V(3)  V1
.END
```

Figure 5-56 The PSpice program for Example 5-7. The dummy source VM1 generates the control current.

```
****      SMALL-SIGNAL CHARACTERISTICS

      V(3)/V1 = 6.000E-01
      INPUT RESISTANCE AT V1 = 1.000E+01
      OUTPUT RESISTANCE AT V(3) = 3.600E+00
```

Figure 5-57 Output of PSpice program for Example 5-7.

5-9 ‖ VERIFICATION EXAMPLE

Using the **.TF** statement, PSpice reports that $R_t = -1\ \Omega$ for the circuit of Figure 5-58a. Verify this result.

Solution

At first, it looks odd that the Thévenin resistance is negative. Since we have a CCVS in the circuit, let us set the controlling current, I_x, equal to 1 A as shown in Figure 5-58b. Then writing the KCL at node a we obtain

$$\frac{v_a + 2I_x}{2} + \frac{v_a}{4} - 1 = 0$$

Then $v_a = -\frac{4}{3}\ \text{V}$ when $I_x = 1$. Since $R_t = v_a/I_x$, we have $R_t = -\frac{4}{3}\ \Omega$, not $-1\ \Omega$ as stated. Clearly, the PSpice program contains an error although the resistance is negative as reported.

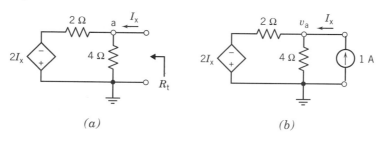

(a)   *(b)*

Figure 5-58 Circuit with dependent voltage source.

5-10 DESIGN CHALLENGE SOLUTION

STRAIN GAUGE BRIDGE

Strain gauges are transducers that measure mechanical strain. Electrically, the strain gauges are resistors. The strain causes a change in resistance that is proportional to the strain.

Figure 5D-1 shows four strain gauges connected in a configuration called a bridge. Strain gauge bridges are used to measure force or pressure (E.O. Doebelin, 1966).

The bridge output is usually a small voltage. In Figure 5D-1 an amplifier multiplies the bridge output, v_i, by a gain to obtain a larger voltage, v_o, which is displayed by the voltmeter.

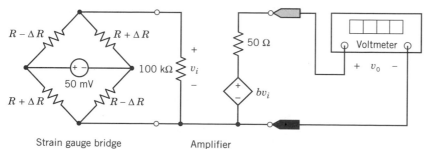

Strain gauge bridge   Amplifier

Figure 5D-1 Design problem involving a strain gauge bridge.

Define the Situation

A strain gauge bridge is used to measure force. The strain gauges have been positioned so that the force will increase the resistance of two of the strain gauges while, at the same time, decreasing the resistance of the other two strain gauges.

The strain gauges used in the bridge have nominal resistances of $R = 120\ \Omega$. (The nominal resistance is the resistance when the strain is zero.) This resistance is expected to increase or decrease by no more than $2\ \Omega$ due to strain. This means that

$$-2\ \Omega \le \Delta R \le 2\ \Omega \tag{5D-1}$$

The output voltage, v_o, is required to vary from -10 V to $+10$ V as ΔR varies from $-2\ \Omega$ to $2\ \Omega$.

State the Goal

Determine the amplifier gain, b, needed to cause v_o to be related to ΔR by

$$v_o = 5\,\frac{\text{volt}}{\text{ohm}} \cdot \Delta R \tag{5D-2}$$

Generate a Plan

Use Thévenin's Theorem to analyze the circuit shown in Figure 5D-1 to determine the relationship between v_i and ΔR. Calculate the amplifier gain needed to satisfy Eq. 5D-2.

Take Action on the Plan

We begin by finding the Thévenin equivalent of the strain gauge bridge. This requires two calculations: one to find the open circuit voltage V_t and the other to find the Thévenin resistance R_t. Figure 5D-2a shows the circuit used to calculate V_t. Begin by finding the currents i_1 and i_2.

$$i_1 = \frac{50\ \text{mV}}{(R - \Delta R) + (R + \Delta R)} = \frac{50\ \text{mV}}{2R}$$

Similarly

$$i_2 = \frac{50\ \text{mV}}{(R + \Delta R) + (R - \Delta R)} = \frac{50\ \text{mV}}{2R}$$

Then

$$V_t = (R + \Delta R)i_1 - (R - \Delta R)i_2$$

$$= (2\Delta R)\frac{50\ \text{mV}}{2R}$$

$$= \frac{\Delta R}{R}\,50\ \text{mV}$$

$$= \frac{50\ \text{mV}}{120\ \Omega}\,\Delta R$$

$$= (0.4167 \times 10^{-3})\Delta R \tag{5D-3}$$

Figure 5D-2b shows the circuit used to calculate R_t. This figure shows that R_t is comprised of a series connection of two resistances, each of which is a parallel connection of two strain gauge resistances

$$R_t = \frac{(R - \Delta R)(R + \Delta R)}{(R - \Delta R) + (R + \Delta R)} + \frac{(R + \Delta R)(R - \Delta R)}{(R + \Delta R) + (R - \Delta R)}$$

$$= 2\,\frac{R^2 - \Delta R^2}{2R}$$

Since R is much larger than ΔR, this equation can be simplified to

$$R_t = R$$

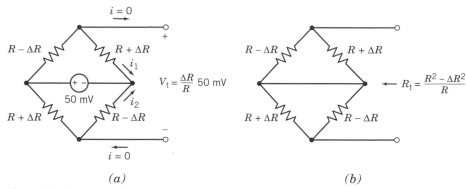

Figure 5D-2 Calculating (a) the open circuit voltage and (b) the Thévenin resistance of the strain gauge bridge.

In Figure 5D-3 the strain gauge bridge has been replaced by its Thévenin equivalent circuit. This simplification allows us to calculate v_i using voltage division

$$v_i = \frac{100\ \text{k}\Omega}{100\ \text{k}\Omega\ +\ R_t}\,V_t$$

$$= 0.9988\ V_t$$

$$= (0.4162 \times 10^{-3})\Delta R \qquad (5D\text{-}4)$$

Model the voltmeter as an ideal voltmeter. Then the voltmeter current is $i = 0$ as shown in Figure 5D-3. Applying KVL to the right-hand mesh gives

$$v_o + 50(0) - bv_i = 0$$

or

$$v_o = bv_i$$

$$= b(0.4162 \times 10^{-3})\Delta R \qquad (5D\text{-}5)$$

Comparing Eq. 5D-5 to Eq. 5D-2 shows that the amplifier gain, b, must satisfy

$$b(0.4162 \times 10^{-3}) = 5$$

Hence, the amplifier gain is

$$b = 12{,}013$$

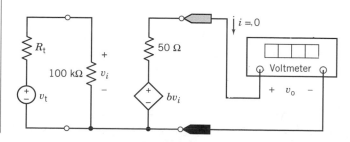

Figure 5D-3 Solution to the design problem.

5-11 DESIGN EXAMPLE—MAXIMUM POWER TRANSFER

Problem

It is desired to deliver maximum power to the load resistor R of Figure 5-59 when $v_s = V_0$ volts. The resistor is constrained so that $1\,\Omega \le R \le 10\,\Omega$ and the constant of the dependent source is $1 \le b \le 2.5$. Select b and R when b can be changed in increments of 0.1 and R can be changed in increments of $1\,\Omega$.

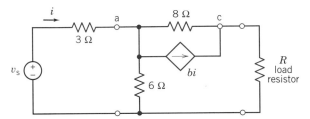

Figure 5-59 A circuit with a dependent current source.

Define the Situation

1 The parameters are constrained.
2 The source is a constant with a value of V_0.
3 The parameters can be changed by prescribed increments.

The Goal

Maximize the power delivered to R.

Generate a Plan

1 Use a source transformation to change the parallel combination of the 8-Ω resistor and the dependent current source to a dependent voltage source and a resistor in series.
2 Use mesh analysis to determine i_2, the current through the load resistor, as shown in Figure 5-60.
3 Determine the power delivered to R.
4 Determine the best values of R and b in order to maximize the power using a spreadsheet computer program.

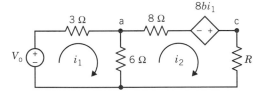

Figure 5-60 A circuit equivalent to the circuit of Figure 5-59.

Take Action Using the Plan

Using a source transformation for the current source and the parallel resistor of Figure 5-59, we obtain the series resistance and the voltage source as shown in Figure 5-60.

The mesh equations for Figure 5-60 are

$$\text{mesh 1:} \qquad 9i_1 - 6i_2 = V_0$$

$$\text{mesh 2:} \qquad -(8b + 6)i_1 + (14 + R)i_2 = 0$$

						R					
b		1	2	3	4	5	6	7	8	9	10
	1	37.91	75.24	112.21	148.97	185.56	222.04	258.44	294.78	331.06	367.31
	1.1	38.32	75.88	112.99	149.83	186.48	223.00	259.43	295.79	332.09	368.35
	1.2	38.84	76.64	113.90	150.82	187.52	224.07	260.52	296.90	333.21	369.47
	1.3	39.49	77.58	114.97	151.96	188.70	225.27	261.73	298.11	334.43	370.69
	1.4	40.35	78.74	116.25	153.29	190.05	226.63	263.08	299.45	335.76	372.01
	1.5	41.53	80.22	117.81	154.86	191.61	228.17	264.60	300.94	337.22	373.46
	1.6	43.25	82.18	119.77	156.76	193.44	229.93	266.31	302.60	338.84	375.04
	1.7	45.99	84.89	122.27	159.08	195.61	231.98	268.25	304.47	340.64	376.78
	1.8	51.02	88.89	125.61	161.98	198.22	234.38	270.49	306.58	342.64	378.70
	1.9	63.18	95.35	130.25	165.73	201.42	237.22	273.08	308.98	344.90	380.83
	2	128.44	107.56	137.17	170.74	205.46	240.67	276.13	311.74	347.45	383.22
	2.1	11.11	138.89	148.56	177.78	210.68	244.91	279.76	314.96	350.36	385.90
	2.2	11.11	355.56	170.73	188.39	217.71	250.26	284.15	318.74	353.72	388.95
	2.3	19.24	5.56	231.68	206.18	227.68	257.23	289.59	323.25	357.62	392.42
	2.4	23.18	22.22	833.33	241.98	242.88	266.67	296.47	328.73	362.23	396.43
	2.5	25.48	37.56	1.33	348.44	268.89	280.17	305.48	335.53	367.74	401.11

Figure 5-61 Spreadsheet for power as a function of *b* and *R*.

Using Cramer's rule, we obtain i_2 as

$$i_2 = \frac{(6 + 8b)V_0}{9(14 + R) - 6(6 + 8b)}$$

Therefore, the power delivered to the resistor R is $p = i_2^2 R$. Since V_0 is unspecified, we obtain p/V_0^2 as

$$\frac{p}{V_0^2} = \frac{(6 + 8b)^2 R}{(90 + 9R - 48b)^2}$$

We then establish a spreadsheet program for the prescribed range and increment the values for R and b. The resulting output grid is shown in Figure 5-61.

The maximum power is delivered to R when $R = 3\ \Omega$ and $b = 2.4$. Then

$$p = 833.3 V_0^2\ \text{W}$$

SUMMARY

Electrical power is used for many applications such as transportation vehicles, heating and lighting, and communications. It is important that a series of analytical methods be available for reducing the complexity of the electrical circuits used for these myriad applications.

The first analytical method we considered was the source transformation, where we transform one source into another source while retaining the terminal characteristics. We are interested in transforming a voltage source into a current source and vice versa. We found that the voltage source v_s in series with a resistor R_s can be transformed into a current source consisting of $i_s = v_s/R_s$ and a parallel resistor R_p that is equal to R_s.

A source consisting of a current source i_s in parallel with a resistor R_p can be transformed into a voltage source $v_s = R_p i_s$ in series with a resistor R_s that is equal to R_p.

The superposition theorem permits us to determine the total response of a linear circuit by finding the response to each independent source and then adding the responses algebraically. The superposition theorem often permits us to determine a voltage or current by repeatedly using the voltage divider and current divider principle.

Since we are often interested in the current, voltage, or power of a single element such as a load resistor R_L, we reduce the rest of the circuit to its Thévenin or Norton equivalent. The Thévenin equivalent consists of a voltage source v_t in series with the resistance R_t. The Norton equivalent consists of a current source i_n in parallel with a resistor R_t.

The Thévenin equivalent circuit is found by noting that $v_t = v_{oc}$, the open-circuit voltage. The resistance R_t is calculated for the circuit with all its independent sources deactivated.

The Norton equivalent circuit is found by determining i_{sc}, the short-circuit current. Also, R_t is calculated for the circuit with all its independent sources deactivated.

The goal of many electronic and communications circuits is to deliver maximum power to a load resistor R_L. Maximum power is attained when R_L is set equal to the resistance R_t of the Thévenin equivalent circuit. This results in maximum power at the load when the series resistance R_t cannot be reduced.

For the case of a power transmission line, our goal is to deliver as much as possible of the source power to the load R_L. In this case, we strive to reduce the Thévenin series resistance R_t to a minimum. It is the goal of power engineers to reduce R_t to zero by using superconducting power lines.

TERMS AND CONCEPTS

Dual Circuits   Two circuits such that the equations describing the first circuit, with v and i interchanged and R and G interchanged, decribe the second circuit.

Efficiency of Power Transfer   Ratio of the power delivered to the load to the power supplied by the source.

Equivalent Circuit   Arrangement of circuit elements that is equivalent to a more complex arrangement of elements. A circuit that exhibits identical characteristics (behavior) to another circuit at identical terminals.

Maximum Power Transfer Theorem   The maximum power delivered by a circuit represented by its Thévenin equivalent is attained when the load resistor R_L is equal to the Thévenin resistance R_t.

Norton's Theorem   For a linear circuit, divide it into two parts, A and B. For circuit A, determine its short-circuit current at its terminals. The equivalent circuit of A is a current source i_{sc} in parallel with a resistance R_n, where R_n is the resistance calculated with all its independent sources deactivated.

Source Transformation   Transformation of one source into another while retaining the terminal characteristics. A voltage source may be transformed to a current source and vice versa.

Superposition Theorem   For a linear circuit containing independent sources, the voltage across (or the current through) any element may be obtained by adding algebraically all the individual voltages (or currents) caused by each independent source acting alone with all other sources set to zero.

Thévenin's Theorem   Divide a circuit into two parts, A and B, connected at a pair of terminals. Determine v_{oc} as the open-circuit voltage of A with B disconnected. Then the equivalent circuit of A is a source voltage v_{oc} in series with R_t, where R_t is the resistance seen at the terminals of circuit A when all the independent sources are deactivated.

REFERENCE   Chapter 5

Doebelin, E. O. *Measurement Systems*, McGraw-Hill, New York, 1966.
Edelson, Edward "Solar Cell Update," *Popular Science*, June 1992, pp. 95–99.

PROBLEMS

Section 5-3   Source Transformations

P 5.3-1   For the circuit of Figure P 5.3-1, find the current i and the power absorbed by the resistor R_L when $R_L = 2\,\Omega$ by using successive source transformations.
Answer: $i = 2\,\text{A}, p = 8\,\text{W}$

P 5.3-2   Consider the circuit of Figure P 5.3-2. Find i_a by simplifying the circuit (using source transformations) to a single-loop circuit so that you need to write only one KVL equation to find i_a.

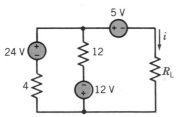

Figure P 5.3-1   Resistances in ohms.

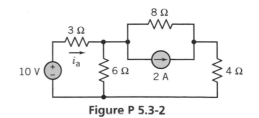

Figure P 5.3-2

P 5.3-3 Use source transformations to find the voltage v across the 1-mA current source for the circuit shown in Figure P 5.3-3.

Answer: $v = 3$ V

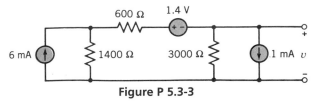

Figure P 5.3-3

P 5.3-4 Find v_0 using source transformations if $i = 5/2$ A in the circuit shown in Figure P 5.3-4.

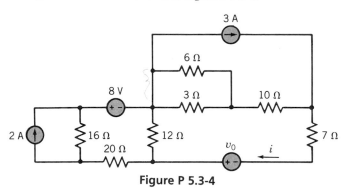

Figure P 5.3-4

P 5.3-5 Determine if the current i has the same value in the circuits *(a)* and *(b)* of Figure P 5.3-5.

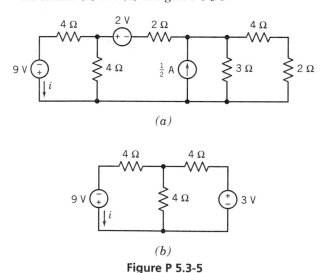

(a)

(b)

Figure P 5.3-5

Section 5-4   Superposition

P 5.4-1 Use superposition to find v for the circuit shown in Figure P 5.4-1.

Answer: $v = -6$ V

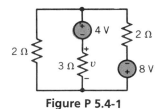

Figure P 5.4-1

P 5.4-2 Use superposition to find v for the circuit of Figure P 5.4-2.

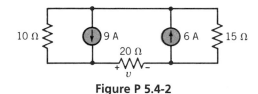

Figure P 5.4-2

P 5.4-3 Use superposition to find i for the circuit of Figure P 5.4-3.

Answer: $i = -2$ mA

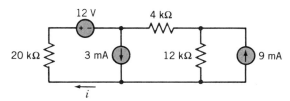

Figure P 5.4-3

P 5.4-4 Use superposition to find v_x for the circuit of Figure P 5.4-4.

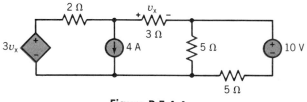

Figure P 5.4-4

P 5.4-5 Use superposition to find i for the circuit of Figure P 5.4-5.

Answer: $i = 3.5$ mA

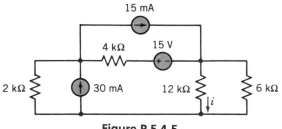

Figure P 5.4-5

P 5.4-6 Determine the current i of the circuit shown in Figure P 5.4-6.

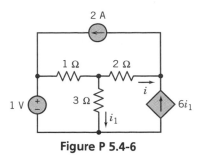

Figure P 5.4-6

P 5.4-7 Determine the voltage v using the principle of superposition for the circuit of Figure P 5.4-7.

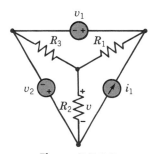

Figure P 5.4-7

P 5.4-8 For the circuit shown in Figure P 5.4-8, use superposition to find v in terms of the R's and source values.

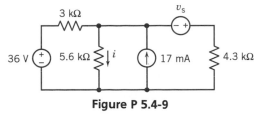

Figure P 5.4-8

P 5.4-9 The current i of the circuit shown in Figure P 5.4-9 is required to be 5.62 mA. Determine the source voltage v_s to satisfy the current requirement.
Answer: $v_s = 24$ V

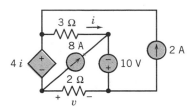

Figure P 5.4-9

Section 5-5 Thévenin's Theorem

P 5.5-1 Obtain the Thévenin equivalent for the circuit shown in Figure P 5.5-1. Resistances in ohms.

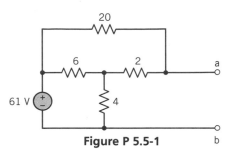

Figure P 5.5-1

P 5.5-2 Find the Thévenin equivalent circuit for the circuit of Figure P 5.5-2.

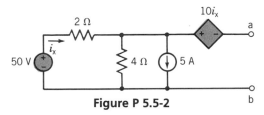

Figure P 5.5-2

P 5.5-3 Find R_t for the circuit of Figure P 5.5-3.
Answer: $R_t = 3 \ \Omega$

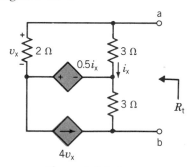

Figure P 5.5-3

P 5.5-4 Find the Thévenin resistance for the circuit shown in Figure P 5.5-4.

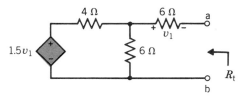

Figure P 5.5-4

P 5.5-5 Find the Thévenin equivalent circuit for the circuit shown in Figure P 5.5-5.
Answer: $R_t = 3 \ \Omega$, $v_{oc} = 3$ V

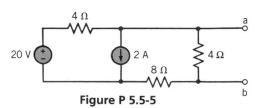

Figure P 5.5-5

P 5.5-6   Find the Thévenin equivalent circuit shown in Figure P 5.5-6 for $R = 1\ \Omega$.
Answer: $R_t = -3\ \Omega$, $v_{oc} = 25$ V

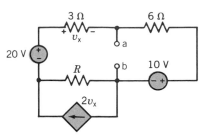

Figure P 5.5-6

P 5.5-7   Find the Thévenin equivalent circuit for the circuit of Figure P 5.5-7.
Answer: $R_t = 10\ \Omega$, $v_{oc} = -24$ V

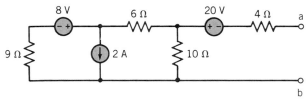

Figure P 5.5-7

P 5.5-8   Find the Thévenin equivalent for terminals a–b for the circuit of Figure P 5.5-8.

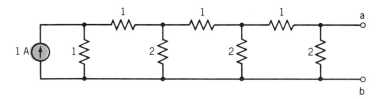

Figure P 5.5-8   All resistances in ohms.

P 5.5-9   Determine the Thévenin equivalent circuit for the circuit of Figure P 5.5-9 at the output terminals a–b.

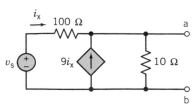

Figure P 5.5-9

P 5.5-10   Determine the Thévenin equivalent circuit for the circuit shown in Figure P 5.5-10.

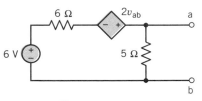

Figure P 5.5-10

P 5.5-11   Determine the Thévenin equivalent circuit for the circuit shown in Figure P 5.5-11.

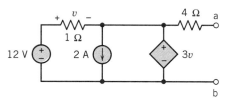

Figure P 5.5-11

P 5.5-12   Measurements made on terminals a–b of a linear circuit, Figure P 5.5-12*a*, that is known to be made up only of independent and dependent voltage sources and current sources and resistors yield the current–voltage characteristics shown in Figure P 5.5-12*b*. Find the Thévenin equivalent circuit.

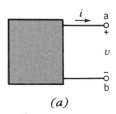

(a)
Figure P 5.5-12

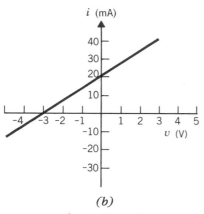

(b)

Figure P 5.5-12

P 5.5-13 Find the Thévenin equivalent of the circuit of Figure P 5.5-13 at terminals a–b.

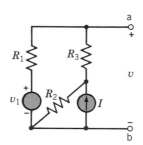

Figure P 5.5-13

P 5.5-14 For the circuit of Figure P 5.5-14, $R_{bc} = 10^6 \ \Omega$, $R_{ce} = 100 \ \text{k}\Omega$, $R_{bb} = 100 \ \Omega$, $R_{be} = 2 \ \text{k}\Omega$, and $g_m = 50 \ \text{mA/V}$. (a) Find R_{in} (open C–E). (b) Find R_{out} (short B–E).
Answer: $R_{in} = 299 \ \Omega$

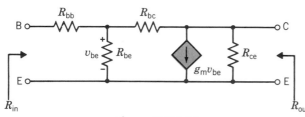

Figure P 5.5-14

P 5.5-15 A resistor, R, is connected to a circuit box as shown in Figure P 5.5-15. The voltage, v, is measured as 6 V and 2 V when R is 2 kΩ and 4 kΩ, respectively. Determine the Thévenin equivalent of the circuit within the box and predict the voltage, v, when $R = 8$ kΩ.

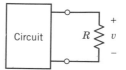

Figure P 5.5-15 Circuit box connected to resistor, R.

P 5.5-16 Determine the Thévenin equivalent circuit at terminals a–b for the circuit shown in Figure P 5.5-16.

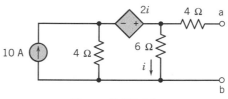

Figure P 5.5-16

P 5.5-17 The model of a transistor amplifier is shown in Figure P 5.5-17. Determine the output resistance at terminals a–b.

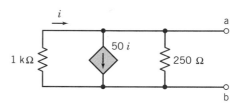

Figure P 5.5-17 Model of transistor amplifier.

P 5.5-18 Determine the Thévenin equivalent circuit at terminals a–b of the circuit shown in Figure P 5.5-18.

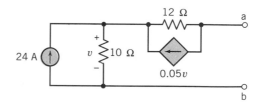

Figure P 5.5-18

P 5.5-19 Determine the current i_L through the load resistor R_L by first determining the Thévenin equivalent circuit to the left of terminals a–b in Figure P 5.5-19 and then connecting $R_L = 2.2 \ \Omega$.

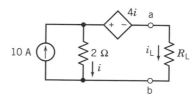

Figure P 5.5-19

P 5.5-20 An engineer has an unknown circuit A, as shown in Figure P 5.5-20. She wishes to determine whether the network is linear and, if it is, to determine its Thévenin equivalent. The only equipment available to the engineer is a voltmeter (assumed ideal) and a 10-kΩ and a 100-kΩ test resistor that can be placed across the terminals during a measurement. The following data were recorded:

Test Resistor	Meter Reading
Absent	1.5 V
10 kΩ	0.25 V
100 kΩ	1.0 V

What should the engineer conclude about the network from these results? Support your conclusion with plots of the network's $v-i$ characteristics.

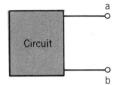

Figure P 5.5-20 Box with unknown circuit.

P 5.5-21 For the circuit shown in Figure P 5.5-21, find the Thévenin equivalent circuit between points A and B. (R_d and diode are the load.) The diode is a commonly used electronic device.
Answer: $R_t = 4.8$ kΩ, $v_{oc} = 18$ V

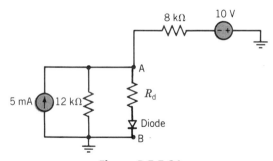

Figure P 5.5-21

P 5.5-22 Using Thévenin's theorem, show that the circuit in Figure P 5.5-22a is equivalent to Figure P 5.5-22b where

$$R_b = \frac{R_1 R_2}{R_1 + R_2}$$

and

$$v = V_{cc}\frac{R_2}{R_1 + R_2}$$

The transistor Q is an electronic device.

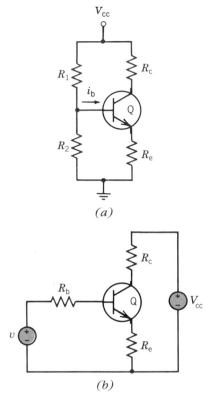

(a)

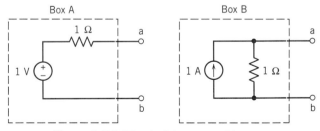

(b)

Figure P 5.5-22

P 5.5-23 Two black boxes are shown in Figure P 5.5-23. Box A contains the Thévenin equivalent of some linear circuit, and box B contains the Norton equivalent of the same circuit. With access to just the outsides of the boxes and their terminals, how can you determine which is which, using only one shorting wire?

Figure P 5.5-23 Black boxes problem.

P 5.5-24 A student uses a voltmeter with an internal resistance of 5 kΩ to investigate a network with a constant Thévenin voltage. With the voltmeter across the open-circuited network terminals, the meter reads 91.25 V. When the student places a 2-kΩ resistor in parallel with the meter, across the network terminals, the meter reads 37.5 V. What are the Thévenin voltage and resistance for the network?

P 5.5-25 The "tunnel diode" is a high-impurity-density *p-n* junction device. It can exhibit a negative resistance characteristic for certain current areas. Figure P 5.5-25 shows the *i-v* characteristic of a tunnel diode. In three ranges, it exhibits a linear *i-v* characteristic. Find the Thévenin equivalent circuit of the tunnel diode in these three areas.

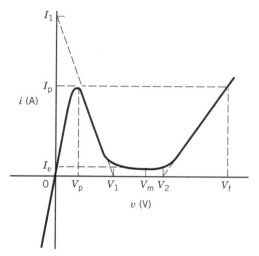

Figure P 5.5-25 Tunnel diode characteristic.

Section 5-6   Norton's Theorem

P 5.6-1 Determine the Norton equivalent of the circuit shown in Figure P 5.6-1.

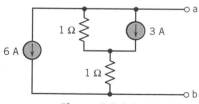

Figure P 5.6-1

P 5.6-2 Find the Norton equivalent circuit for the circuit shown in Figure P 5.6-2.
Answer: $R_t = 5\ \Omega$, $i_{sc} = -6.6$ A

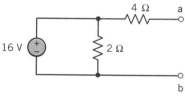

Figure P 5.6-2

P 5.6-3 (a) Find the Norton equivalent circuit for the circuit shown in Figure P 5.6-3.
  (b) If a resistor R is connected between termi-

nals a–b, what is the maximum power that it could absorb?

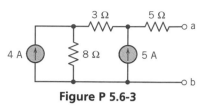

Figure P 5.6-3

P 5.6-4 Consider the circuit shown in Figure P 5.6-4. Find and draw the Thévenin and Norton equivalent circuits. Express everything in terms of α, R, and I_o.
Answer: $R_t = R/(1 - \alpha R)$
   $i_{sc} = I_o$
   $v_{oc} = RI_o/(1 - \alpha R)$

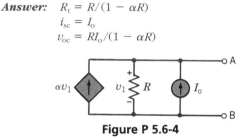

Figure P 5.6-4

P 5.6-5 Find the Norton equivalent circuit for the circuit shown in Figure P 5.6-5.

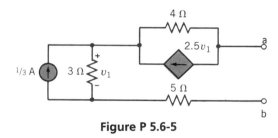

Figure P 5.6-5

P 5.6-6 Find R_t between terminals a–b and draw the Norton equivalent for the circuit shown in Figure P 5.6-6. Current i_x has units of mA.
Answer: $R_t = 3$ kΩ, $i_{sc} = 1$ mA

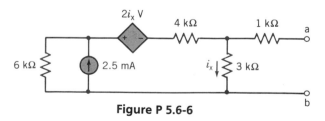

Figure P 5.6-6

P 5.6-7 Find the Norton equivalent circuit between terminals a–b for $R = 0$ for the circuit shown in Figure P 5.5-6.

P 5.6-8 Determine the Norton equivalent circuit for the circuit shown in Figure P 5.6-8.

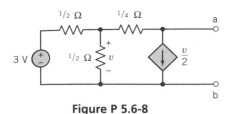

Figure P 5.6-8

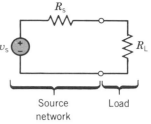

Figure P 5.7-3

Section 5-7   Maximum Power Transfer

P 5.7-1   The circuit model for a photovoltaic cell is given in Figure P 5.7-1. (Edelson, 1992). The current i_s is proportional to the solar insolation (kW/m²). (a) Find the load resistance for maximum power transfer. (b) Find the maximum power transferred when $i_s = 1$ A.

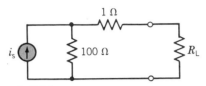

Figure P 5.7-1   Circuit model of photovoltaic cell.

P 5.7-2   For the circuit in Figure P 5.7-2 (a) find R such that maximum power is dissipated in R and (b) calculate the value of maximum power.
Answer: $R = 60\ \Omega$, $P_{max} = 54$ mW

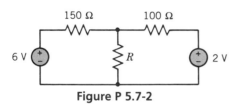

Figure P 5.7-2

P 5.7-3   For the circuit in Figure P 5.7-3, prove that for R_s variable and R_L fixed, the power dissipated in R_L is maximum when $R_s = 0$.

P 5.7-4   Find the maximum power to the load R_L if the maximum power transfer condition is met for the circuit of Figure P 5.7-4.
Answer: max $p_L = 0.75$ W

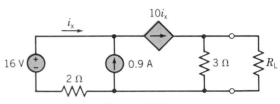

Figure P 5.7-4

P 5.7-5   Consider the circuit of Figure P 5.7-5. (a) Find R_L such that R_L absorbs maximum power. (b) If maximum $p_L = 54$ W, find I_0.
Answer: $R_L = 1.5\ \Omega$, $I_0 = 18$ A

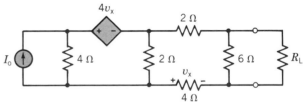

Figure P 5.7-5

P 5.7-6   For the circuit of Figure P 5.7-6 (a) find the Thévenin equivalent of the network to the left of terminals a–b and (b) find the value of α such that maximum power is delivered to the circuit to the right of terminals a–b.
Answer: (a) $v_{oc} = 30$ V, $R_t = 6\ \Omega$; (b) $\alpha = 6$

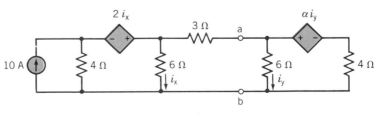

Figure P 5.7-6

P 5.7-7 Determine the maximum power that can be absorbed by a resistor, R, connected to terminals a–b of the circuit shown in Figure P 5.7-7. Specify the value of R.

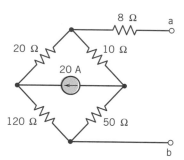

Figure P 5.7-7 Bridge circuit.

P 5.7-8 Many communication circuits are designed to transfer maximum power to an electronic receiver as shown in Figure P 5.7-8. Select R in order to maximize the power to the receiver.

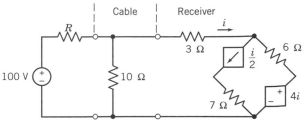

Figure P 5.7-8 Communication circuit.

P 5.7-9 For the circuit of Figure P 5.7-9, determine the resistance R so that the power to R is maximum. For that value of R, determine the power delivered to R.

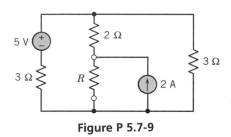

Figure P 5.7-9

PSpice PROBLEMS

SP 5-1 Determine the current i for the circuit of Problem P 5.3-1 using PSpice.

SP 5-2 Determine the voltage v for the circuit shown in Figure P 5.4-7 using PSpice.

SP 5-3 Determine the voltage v for the circuit of Problem P 5.4-1 using PSpice.

SP 5-4 Use PSpice to aid in the determination of the Thévenin equivalent of the circuit of Problem P 5.7-6.

SP 5-5 Obtain the Thévenin equivalent circuit for Problem P 5.5-1 using PSpice.

SP 5-6 Determine the output voltage v for the circuit shown in Figure SP 5-6.

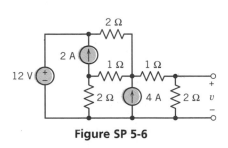

Figure SP 5-6

SP 5-7 Determine the current i for the circuit shown in Figure SP 5-7.

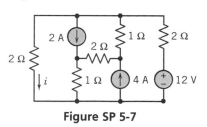

Figure SP 5-7

SP 5-8 Determine v for the circuit shown in Figure SP 5-8.

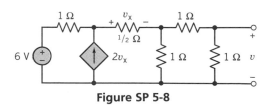

Figure SP 5-8

SP 5-9 Determine the current i of Problem P 5.4-6.

SP 5-10 Determine the Thévenin equivalent circuit for Problem P 5.5-11.

SP 5-11 A circuit with a constant voltage source and a variable current source is shown in Figure SP 5-11. Use PSpice to obtain a graphic plot of v_2 versus i_s when i_s is a constant I_o, which varies between 0 and 2 mA.

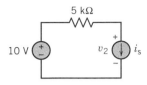

Figure SP 5-11

SP 5-12 A transistor amplifier circuit is shown in Figure SP 5-12. Use PSpice to calculate i.

Answer: $i = 9.52$ mA

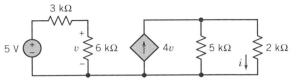

Figure SP 5-12 Transistor amplifier circuit.

SP 5-13 Use PSpice to determine the Thévenin equivalent of the circuit of Figure SP 5-13.

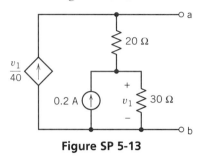

Figure SP 5-13

VERIFICATION PROBLEMS

VP 5-1 For the circuit of Figure VP 5-1 the current has been measured for three different values of R and is listed in the table. Are the data consistent?

R (Ω)	i (mA)
5000	16.5
500	43.8
0	97.2

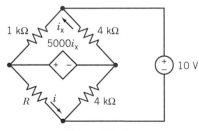

Figure VP 5-1

VP 5-2 For the circuit of Figure VP 5-2, a computer program states that maximum power is delivered to a resist-

ance, R_L, connected to a–b when $R_L = 72$ Ω. Check this result, and calculate the power absorbed by R_L.

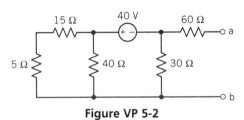

Figure VP 5-2

VP 5-3 For the circuit of Figure VP 5-3, a laboratory report states that maximum power is delivered to the load R_L when $R_L = 60$ Ω. Verify this result when $R = 110$ Ω. Calculate the power delivered to the load resistance R_L.

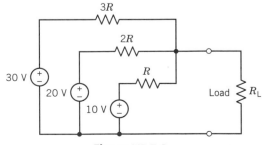

Figure VP 5-3

DESIGN PROBLEMS

DP 5-1   It is desired to deliver 150 W to the load resistor R_L of the circuit shown in Figure DP 5-1. Determine a suitable value for R_L.

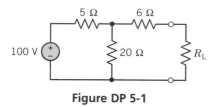

Figure DP 5-1

DP 5-2   It is desired to deliver maximum power to a load resistor R_L as shown in Figure DP 5-2. Determine the required R_L and find the power delivered to the load.

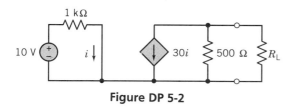

Figure DP 5-2

DP 5-3   The model of a transistor amplifier is shown in Figure DP 5-3. The output resistance at terminals a–b is desired to be 50 kΩ. Determine the required coefficient g when $d = 4 \times 10^{-4}$.

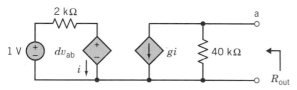

Figure DP 5-3   Transistor amplifier.

DP 5-4   A transistor amplifier can be represented by the circuit shown in Figure DP 5-4 with $R_L = 30\ \Omega$. The input terminals are to be connected to a source. Determine the required constant, b, so that the input resistance is 2705 Ω.

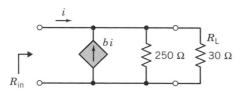

Figure DP 5-4   Transistor amplifier.

DP 5-5   For the circuit shown in Figure DP 5-5, choose the constant d so that the Thévenin equivalent resistance is 64 Ω. Calculate the resulting v_t at terminals a–b.

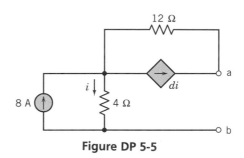

Figure DP 5-5

DP 5-6   Select v_s so that $v = 0$ in the circuit of Figure DP 5-6.

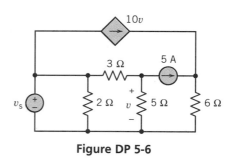

Figure DP 5-6

DP 5-7   A model of a transistor amplifier is shown in Figure DP 5-7. It is specified that the output resistance, R_{ab}, is greater than 60 Ω and less than 70 Ω. Select an appropriate value for the constant b. Determine the resulting magnitude of the output voltage v_{ab}.

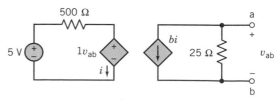

Figure DP 5-7   Transistor amplifier.

DP 5-8 A transmitter and antenna circuit is shown in Figure DP 5-8.

(a) Find the Thévenin equivalent circuit for the transmitter circuit (left of terminals a–b).

(b) Determine the value of β required to yield R_t of the transmitter so that it matches the load resistance, R_L.

(c) What value of β results in maximum power delivered to R_L?

(d) What practical engineering constraints could influence the value of β selected for the circuit?

DP 5-10 If the resistive power divider of Design Problem 5-9 is changed to obtain an N-way power divider by adding additional series sections of R and a 50-ohm load resistor in shunt at a–b, derive an expression for the value of resistance, R, that is required as a function of N. Note that power is lost in the resistive divider circuits and that not all of the total power is delivered to the N loads.

DP 5-11 Design a four-way resistive power divider for the source and load resistances of Design Problems 5-9 and 5-10, and verify that the output powers are equal. Compute the relative voltage loss or attenuation by comparing the voltage across one of the loads to the input voltage (V_{load}/V_s).

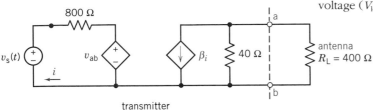

Figure DP 5-8 Transmitter and antenna circuit.

DP 5-9 There are many applications in dc, ac, and radio frequency (RF) circuits where power division is required. Consider the two-way resistive power divider circuit shown in Figure DP 5-9. If maximum power transfer to each TV set and equal power division are required at the interface a–b, what is the value of the resistor, R, that is necessary?

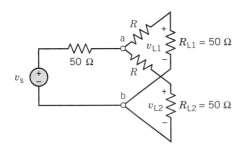

Figure DP 5-9 Power divider circuit.

CHAPTER 6

THE OPERATIONAL AMPLIFIER

PREVIEW

With the invention of the electron tube in 1907, the field of electronics was born. In this chapter we consider an active element called the operational amplifier, which has many useful applications in electronics.

When the operational amplifier is connected to resistors and capacitors, we obtain various configurations of amplifier circuits. Several useful forms of amplifier circuits are described and analyzed. We look at the time domain response of these circuits as well as their response when the input source is a steady-state sinusoid. The benefits of an amplifier circuit include isolation between input and output terminals and a power gain between the input and output terminals.

6-1 DESIGN CHALLENGE

TRANSDUCER INTERFACE CIRCUIT

A customer wants to automate a pressure measurement, which requires converting the output of the pressure transducer to a computer input. This conversion can be done using a standard integrated circuit called an *Analog-to-Digital Converter* (ADC) (Dorf, 1993). The ADC requires an input voltage between 0 V and 10 V, while the pressure transducer output varies between − 250 mV and 250 mV. Design a circuit to interface the pressure transducer with the ADC. That is, design a circuit that translates the range − 250 mV to 250 mV to the range 0 V to 10 V.

Define the Situation

The situation is shown in Figure 6D-1.

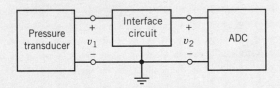

Figure 6D-1 Interfacing a pressure transducer with an Analog-to-Digital Converter (ADC).

The specifications state that

$$- 250 \text{ mV} \leq v_1 \leq 250 \text{ mV}$$

$$0 \text{ V} \leq v_2 \leq 10 \text{ V}$$

A simple relationship between v_2 and v_1 is needed so that information about the pressure is not obscured. Consider

$$v_2 = a \cdot v_1 + b$$

The coefficients, a and b, can be calculated by requiring that $v_2 = 0$ when $v_1 = - 250$

226

mV and that $v_2 = 10$ V when $v_1 = 250$ mV, that is,

$$0 \text{ V} = a \cdot -250 \text{ mV} + b$$

$$10 \text{ V} = a \cdot 250 \text{ mV} + b$$

Solving these simultaneous equations gives $a = 20$ V and $b = 5$ V.

State the Goal

Design a circuit having input voltage v_1 and output voltage v_2. These voltages should be related by

$$v_2 = 20v_1 + 5 \text{ V}$$

In this chapter, we will see that such a circuit can be designed using operational amplifiers.

6-2 ELECTRONICS

As the use of radio began to expand in the 1920s, the industry associated with radio and electron tubes became important to electrical engineering. As we will see in Chapter 9, it was the invention of the triode vacuum tube by Lee De Forest in 1907 that enabled radio to grow. The use of the triode as an amplifier and detector demonstrated that radio circuits could readily be designed.

By 1929 the vacuum tube became known as the electron tube. With the publication of *Electronics* magazine in 1930, a new word was born and an industry identified. *Electronics* is the engineering field and industry that uses electron devices in circuits and systems. During the first half of the century, electronics was dominated by the vacuum tube.

Engineers were interested in miniaturization of electronics and turned to experimentation with semiconductors, which led to the invention of the transistor.

The fundamental properties of semiconductors differentiate them from metals and insulators. A *semiconductor* is an electronic conductor with a resistivity in the range between the resistivity of metals and that of insulators. The first radio detectors used galena crystals, silicon, or silicon carbide crystals. With the development of the theory of quantum mechanics in the period 1926–1936, the understanding of semiconductors increased.

By the 1930s several early semiconductor devices had been built in the laboratory. The analogy between the semiconductor diode and the vacuum diode was obvious, and several people attempted to build a three-terminal semiconductor device.

In July 1945 a semiconductor research program was set up at Bell Laboratories. John Bardeen, William Shockley, and Walter Brattain, shown in Figure 6-1, were the inventors of the first transistor. One of the first transistors using a germanium crystal was assembled as shown in Figure 6-2. A *transistor* is an active semiconductor device with three or more terminals. Transistors became commercially available in 1954 as the manufacturing methods were perfected.

Texas Instruments (TI) was one of the early manufacturers of the transistor. One of the leading engineers at TI was Patrick E. Haggerty, who saw the opportunity for a miniature

Figure 6-1 Nobel Prize winners John Bardeen, William Shockley, and Walter H. Brattain (left to right), shown at Bell Telephone Laboratories in 1948 with apparatus used in the first investigations that led to the invention of the transistor. The trio received the 1956 Nobel Prize in physics for their invention of the transistor, which was announced by Bell Laboratories in 1948. Courtesy of Bell Telephone Laboratories.

Figure 6-2 The first transistors assembled by their inventors at Bell Laboratories (in 1947) were primitive by today's standards. Yet they revolutionized the electronics industry and changed our way of life. The first transistor, a "point-contact" type, amplified electrical signals by passing them through a solid semiconductor material, basically the same operation as performed by present "junction" transistors. The three terminal wires can be seen on the top of the transistor. The actual record of the first transistor operation was December 23, 1947. Courtesy of Bell Telephone Laboratories.

radio using transistors and small components. The goal was to build a miniature radio by late 1954 and sell it for $50. The first transistor radio built by TI is shown in Figure 6-3, and an inside view of the radio is shown in Figure 6-4. The radio was so popular that 100,000 were sold during 1955.

With the advent of the transistor, several electronic circuits became available. The motivation to miniaturize these electronic circuits was based on factors such as the complexity of circuits, the need for reliability, and the desire to reduce power consumption, weight, and of course cost. By the late 1950s engineers were discussing the possibility that complete circuit functions could be formed within a single block of semiconductor. The first patent for an *integrated circuit* was filed on May 21, 1953, by Harwick Johnson of RCA.

Figure 6-3 The first commercial transistor radio. The molded plastic case of the Regency TR-1 radio was designed to fit into the pocket of a man's dress shirt. It was 5 by 3 by $1\frac{1}{4}$ inches and sold for $49.95. This radio was introduced on October 18, 1954. Courtesy of Texas Instruments.

Figure 6-4 Inside view of the radio in Figure 6-3. The specially designed miniature components barely fit inside the 5 by 3 by $1\frac{1}{4}$ inch case. The four transistors, discrete resistors, capacitors, and other miniaturized components were mounted, along with the $2\frac{3}{4}$-inch speaker, in the front half of the plastic case. Courtesy of Texas Instruments.

By the 1960s integrated circuits were available for many electronic functions. An *integrated circuit* is defined as a combination of interconnected circuit elements inseparably associated on or within a continuous semiconductor (often called a chip).

It is the purpose of this chapter to discuss the use and operation of one integrated electronic device, the operational amplifier.

6-3 ‖ THE OPERATIONAL AMPLIFIER

The first operational amplifiers, using vacuum tubes, were available in the 1940s. They were used to perform the mathematical operations such as addition and integration in an electrical system called the analog computer. Some engineers call the operational amplifier by the shortened name *op amp*.

The *operational amplifier* is an active element with a high gain ratio designed to be used with other circuit elements to perform a specified signal-processing operation. The μA741 operational amplifier is shown in Figure 6-5a. It has eight pin connections, whose functions are indicated in Figure 6-5b.

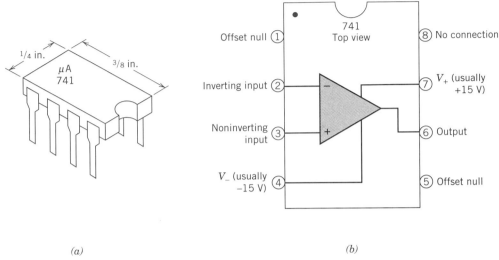

(a)   (b)

Figure 6-5 (*a*) A μA741 integrated circuit has eight connecting pins. (*b*) The correspondence between the circled pin numbers of the integrated circuit and the nodes of the operational amplifier.

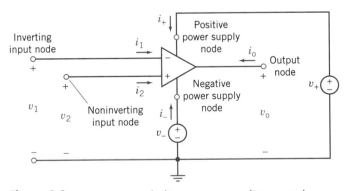

Figure 6-6 An op amp, including power supplies v_+ and v_-.

The operational amplifier shown in Figure 6-6 has five terminals. The names of these terminals are shown in both Figure 6-5b and Figure 6-6. Notice the plus and minus signs on the triangular part of the symbol of the operational amplifier. The plus sign identifies the noninverting input, and the minus sign identifies the inverting input.

The power supplies are used to bias the operational amplifier. In other words, the power supplies cause certain conditions that are required for the operational amplifier to function properly. Once the operational amplifiers have been biased, the power supplies have little effect on the way that the operational amplifiers work. It is inconvenient to include the power supplies in drawings of operational amplifier circuits. These power supplies tend to clutter drawings of operational amplifier circuits, making them harder to read. Consequently, the power supplies are frequently omitted from drawings that accompany explanations of the function of operational amplifier circuits, such as the drawings found in textbooks. It is understood that power supplies are part of the circuit even though they are not shown. (Schematics, the drawings used to describe how to assemble a circuit, are a different matter.) The power supplies are shown in Figure 6-6, denoted as v_+ and v_-.

Since the power supplies are frequently omitted from the drawing of an operational amplifier circuit, it is easy to overlook the power supply currents. This mistake is avoided by careful application of *Kirchhoff's Current Law* (KCL). As a general rule, it is not helpful to apply KCL in a way that involves any power supply current. Two specific cases are of particular importance. First, the ground node in Figure 6-6 is a terminal of both power supplies. Both power supply currents would be involved if KCL was applied to the ground node. These currents must not be overlooked. It is best simply to refrain from applying KCL at the ground node of an operational amplifier circuit. Second, KCL requires that the sum of all currents into the operational amplifier be zero

$$i_1 + i_2 + i_o + i_+ + i_- = 0$$

Both power supply currents are involved in this equation. Once again, these currents must not be overlooked. It is best simply to refrain from applying KCL to sum the currents into an operational amplifier when the power supplies are omitted from the diagram.

6-4 ‖ THE IDEAL OPERATIONAL AMPLIFIER

Operational amplifiers are complicated devices that exhibit both linear and nonlinear behavior. The operational amplifier output voltage and current, v_o and i_o, must satisfy three conditions in order for an operational amplifier to be linear, that is:

$$|v_o| \leq v_{sat}$$

$$|i_o| \leq i_{sat} \tag{6-1}$$

$$\left| \frac{dv_o(t)}{dt} \right| \leq SR$$

The saturation voltage, v_{sat}, the saturation current, i_{sat}, and the slew rate limit, SR, are all parameters of an operational amplifier. For example, if a μA741 operational amplifier is biased using $+15$-V and -15-V power supplies, then

$$v_{sat} = 14 \text{ V} \qquad i_{sat} = 2 \text{ mA} \qquad SR = 500{,}000 \, \frac{\text{V}}{\text{s}} \tag{6-2}$$

These restrictions reflect the fact that operational amplifiers cannot produce arbitrarily large voltages or arbitrarily large currents or change output voltage arbitrarily quickly.

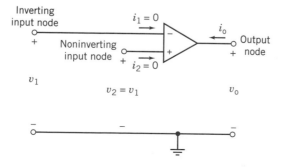

Figure 6-7 The ideal operational amplifier.

Figure 6-7 describes the *ideal operational amplifier.* The ideal operational amplifier is a simple model of an operational amplifier that is linear. The ideal operational amplifier is characterized by restrictions on its input currents and voltages. The currents into the input terminals of an ideal operational amplifier are zero. Consequently, in Figure 6-7

$$i_1 = 0 \quad \text{and} \quad i_2 = 0$$

The node voltages at the input nodes of an ideal operational amplifier are equal. Consequently, in Figure 6-7

$$v_2 = v_1$$

The ideal operational amplifier is a model of a linear operational amplifier, so the operational amplifier output current and voltage must satisfy the restrictions in Eq. 6-1. If they do not, then the ideal operational amplifier is not an appropriate model of the real operational amplifier. The output current and voltage depend on the circuit in which the operational amplifier is used. The ideal op amp conditions are summarized in Table 6-1.

Table 6-1
Operating Conditions for an Ideal Operational Amplifier

Variable	Ideal Condition
Inverting node input current	$i_1 = 0$
Noninverting node input current	$i_2 = 0$
Voltage difference between inverting node voltage v_1 and noninverting node voltage v_2	$v_2 - v_1 = 0$

Example 6-1
Consider the circuit shown in Figure 6-8*a.* Suppose the operational amplifier is a μA741 operational amplifier. Model the operational amplifier as an ideal operational amplifier. Determine how the output voltage, v_o, is related to the input voltage, v_s.

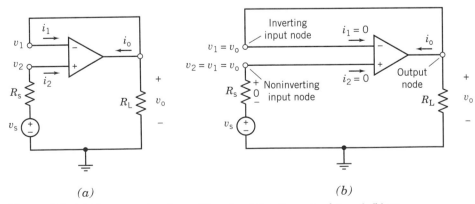

(a)   *(b)*

Figure 6-8 *(a)* The operational amplifier circuit for Example 6-1 and *(b)* an equivalent circuit showing the consequences of modeling the operational amplifier as an ideal operational amplifier. The voltages v_1, v_2, and v_o are node voltages and hence are the voltages at the node with respect to ground.

Solution

Figure 6-8*b* shows the circuit when the operational amplifier of Figure 6-8a is modeled as an ideal operational amplifier.

1 The inverting input node and output node of the operational amplifier are connected by a short circuit, so the node voltages at these nodes are equal

$$v_1 = v_0$$

2 The voltages at the inverting and noninverting nodes of an ideal op amp are equal

$$v_2 = v_1 = v_o$$

3 The currents into the inverting and noninverting nodes of an operational amplifier are zero, so

$$i_1 = 0 \quad \text{and} \quad i_2 = 0$$

4 The current in resistor R_s is $i_2 = 0$, so the voltage across R_s is 0 V. The voltage across R_s is $v_s - v_2 = v_s - v_o$, hence

$$v_s - v_o = 0$$

or

$$v_s = v_o$$

Does this solution satisfy the requirements of Eq. 6.1 and 6.2? The output current of the operational amplifier must be calculated. Apply KCL at the output node of the operational amplifier to get

$$i_1 + i_o + \frac{v_o}{R_L} = 0$$

Since $i_1 = 0$,

$$i_o = -\frac{v_o}{R_L}$$

Now Eqs. 6.1 and 6.2 require

$$|v_s| \leq 14 \text{ V}$$

$$\left| \frac{v_s}{R_L} \right| \leq 2 \text{ mA}$$

$$\left| \frac{d}{dt} v_s \right| \leq 500,000 \frac{\text{V}}{\text{s}}$$

For example, when $v_s = 10$ V and $R_L = 20$ kΩ, then

$$|v_s| = 10 \text{ V} < 14 \text{ V}$$

$$\left| \frac{v_s}{R_L} \right| = \frac{10 \text{ V}}{20 \text{ k}\Omega} = \frac{1}{2} \text{ mA} < 2 \text{ mA}$$

$$\left| \frac{d}{dt} v_s \right| = 0 < 500,000 \frac{\text{V}}{\text{s}}$$

This is consistent with the use of the ideal operational amplifier. On the other hand, when

$v_s = 10$ V and $R_L = 2$ kΩ, then

$$\frac{v_s}{R_L} = 5 \text{ mA} > 2 \text{ mA}$$

so it is not appropriate to model the μA741 as an ideal operational amplifier when $v_s = 10$ V and $R_L = 2$ kΩ. When $v_s = 10$ V we require $R_L > 5$ kΩ in order to satisfy Eq. 6.1.

EXERCISE 6-1

Find the ratio v_o/v_s of the circuit shown in Figure E 6-1 where an ideal op amp is assumed.

Answer: $\dfrac{v_o}{v_s} = 1 + \dfrac{R_2}{R_1}$

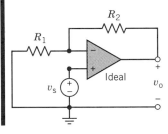

Figure E 6-1

EXERCISE 6-2

A noninverting op amp circuit is shown in Figure E 6-2.
(a)   Find the voltage ratio v_o/v_s.
(b)   Determine v_o/v_s when $R_2 \gg R_1$.

Answer: (a) $\dfrac{v_o}{v_s} = \dfrac{R_2}{R_1 + R_2}\left(1 + \dfrac{R_4}{R_3}\right)$

(b) $\dfrac{v_o}{v_s} = 1 + \dfrac{R_4}{R_3}$

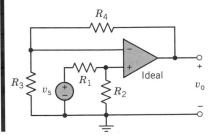

Figure E 6-2

6-5 | NODAL ANALYSIS OF CIRCUITS CONTAINING IDEAL OPERATIONAL AMPLIFIERS

It is convenient to use node equations to analyze circuits containing ideal operational amplifiers. There are three things to remember.

1 The node voltages at the input nodes of ideal operational amplifiers are equal. Thus, one of these two node voltages can be eliminated from the node equations. For example, in Figure 6-9, the voltages at the input nodes of the ideal operational amplifier are v_1 and v_2. Since

$$v_1 = v_2$$

v_2 can be eliminated from the node equations.

2 The currents in the input leads of an ideal operational amplifier are zero. These currents are involved in the KCL equations at the input nodes of the operational amplifier.

3 The output current of the operational amplifier is not zero. This current is involved in the KCL equations at the output node of the operational amplifier. Applying KCL at this node adds another unknown to the node equations. If the output current of the operational amplifier is not to be determined, then it is not necessary to apply KCL at the output node of the operational amplifier.

Example 6-2

The circuit shown in Figure 6-9 is called a difference amplifier. The operational amplifier has been modeled as an ideal operational amplifier. Use node equations to analyze this circuit and determine v_o in terms of the two source voltages, v_a and v_b.

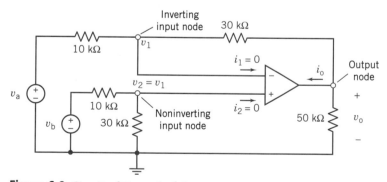

Figure 6-9 Circuit of Example 6-2.

Solution

The node equation at the noninverting node of the ideal operational amplifier is

$$\frac{v_2}{30\ \text{k}\Omega} + \frac{v_2 - v_b}{10\ \text{k}\Omega} + i_2 = 0$$

Since $v_2 = v_1$ and $i_2 = 0$, this equation becomes

$$\frac{v_1}{30 \text{ k}\Omega} + \frac{v_1 - v_b}{10 \text{ k}\Omega} = 0$$

Solving for v_1 we have

$$v_1 = 0.75 \cdot v_b$$

The node equation at the noninverting node of the ideal operational amplifier is

$$\frac{v_1 - v_a}{10 \text{ k}\Omega} + \frac{v_1 - v_o}{30 \text{ k}\Omega} + i_1 = 0$$

Since $v_1 = 0.75 v_b$ and $i_1 = 0$, this equation becomes

$$\frac{0.75 \cdot v_b - v_a}{10 \text{ k}\Omega} + \frac{0.75 \cdot v_b - v_o}{30 \text{ k}\Omega} = 0$$

Solving for v_o we have

$$v_o = 3(v_b - v_a)$$

The difference amplifier takes its name from the fact that the output voltage, v_o, is a function of the difference, $v_b - v_a$, of the input voltages.

Example 6-3

Next, consider the circuit shown in Figure 6-10a. This circuit is called a bridge amplifier. The part of the circuit that is called a bridge is shown in Figure 6-10b. The operational amplifier and resistors R_5 and R_6 are used to amplify the output of the bridge. The operational amplifier in Figure 6-10a has been modeled as an ideal operational amplifier. As a consequence, $v_1 = 0$ and $i_1 = 0$ as shown. Determine the output voltage, v_o in terms of the source voltage, v_s.

Solution

Here is an opportunity to use Thévenin's Theorem. Figure 6-10c shows the Thévenin equivalent of the bridge circuit. Figure 6-10d shows the bridge amplifier after the bridge has been replaced by its Thévenin equivalent. Figure 6-10a is simpler than Figure 6-10a. It is easier to write and solve the node equations representing Figure 6-10d than it is to write and solve the node equations representing Figure 6-10a. Thévenin's Theorem assures us that the voltage v_o in Figure 6-10d is the same as the voltage v_o in Figure 6-10a.

Let us write node equations representing the circuit in Figure 6-10d. First notice that the node voltage v_a is given by (using KVL)

$$v_a = v_1 + v_{oc} + R_t i_1$$

Since $v_1 = 0$ and $i_1 = 0$,

$$v_a = v_{oc}$$

Now writing the node equation at node a

$$i_1 + \frac{v_a - v_o}{R_5} + \frac{v_a}{R_6} = 0$$

Since $v_a = v_{oc}$ and $i_1 = 0$,

$$\frac{v_{oc} - v_o}{R_5} + \frac{v_{oc}}{R_6} = 0$$

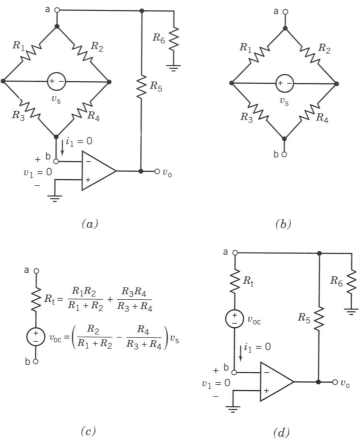

Figure 6-10 *(a)* A bridge amplifier, including the bridge circuit. *(b)* The bridge circuit and, *(c)* its Thévenin equivalent circuit. *(d)* The bridge amplifier, including the Thévenin equivalent of the bridge.

Solving for v_o we have

$$v_o = \left(1 + \frac{R_6}{R_5}\right) v_{oc}$$

$$= \left(1 + \frac{R_6}{R_5}\right)\left(\frac{R_2}{R_1 + R_2} - \frac{R_4}{R_3 + R_4}\right) v_s$$

EXERCISE 6-3

Find the relationship v_o/v_s for the circuit shown in Figure E 6-3.

Answer: $\dfrac{v_o}{v_s} = \dfrac{R_2}{R_1 + R_2}$

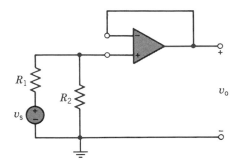

Figure E 6-3

6-6 DESIGN USING OPERATIONAL AMPLIFIERS

One of the early applications of operational amplifiers was to build circuits that performed mathematical operations. Indeed, the operational amplifier takes its name from this important application. Many of the operational amplifier circuits that perform mathematical operations are used so often that they have been given names. These names are part of an electrical engineers vocabulary. Figure 6-11 shows several standard operational amplifier circuits. The next several examples show how to use Figure 6-11 to design simple operational amplifier circuits.

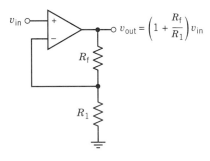

(b) Noninverting amplifier

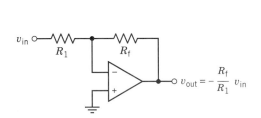

(a) Inverting amplifier

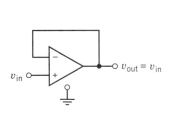

(c) Voltage follower (buffer amplifier)

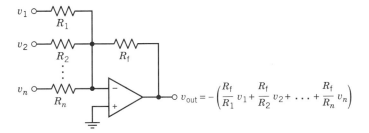

(d) Summing amplifier

Figure 6-11 A brief catalog of operational amplifier circuits. Note that all node voltages are referenced to the ground node.

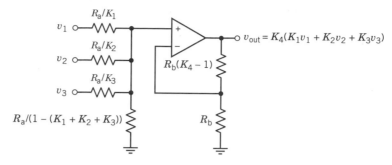

$$v_{out} = K_4(K_1v_1 + K_2v_2 + K_3v_3)$$

(e) Noninverting summing amplifier

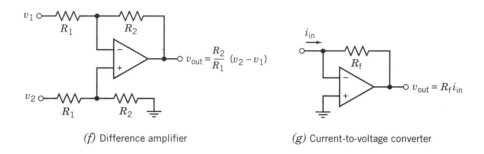

$$v_{out} = \frac{R_2}{R_1}(v_2 - v_1)$$

(f) Difference amplifier

$$v_{out} = R_f i_{in}$$

(g) Current-to-voltage converter

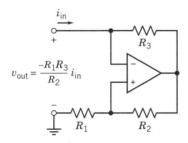

$$v_{out} = \frac{-R_1R_3}{R_2} i_{in}$$

(h) Negative resistance convertor

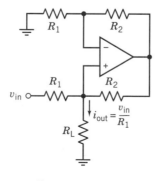

$$i_{out} = \frac{v_{in}}{R_1}$$

(i) Voltage-controlled
current source (VCCS)

Figure 6-11 A brief catalog of operational amplifier circuits (continued).

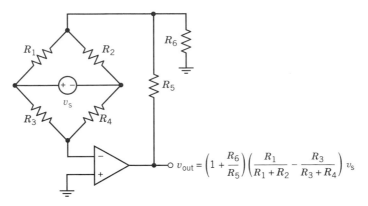

$$v_{\text{out}} = \left(1 + \frac{R_6}{R_5}\right)\left(\frac{R_1}{R_1 + R_2} - \frac{R_3}{R_3 + R_4}\right)v_s$$

(j) Bridge amplifier

Figure 6-11 A brief catalog of operational amplifier circuits (continued).

Example 6-4

This example illustrates the use of a voltage follower to prevent loading. The voltage follower is shown in Figure 6-11c. Loading can occur when two circuits are connected. Consider Figure 6-12. In Figure 6-12a the output of Circuit #1 is the voltage v_a. In Figure 6-12b, Circuit #2 is connected to Circuit #1. The output of Circuit #1 is used as the input to Circuit #2. Unfortunately, connecting Circuit #2 to Circuit #1 can change the output of Circuit #1. This is called *loading*. Referring again to Figure 6-12, Circuit #2 is said to load Circuit #1 if $v_b \neq v_a$. The current i_b is called the load current. Circuit #1 is required to provide this current in Figure 6-12b but not in Figure 6-12a. This is the cause of the loading. The load current can be eliminated using a voltage follower as shown in Figure 6-12c. The voltage follower copies voltage v_a from the output of Circuit #1 to the input of Circuit #2 without disturbing Circuit #1.

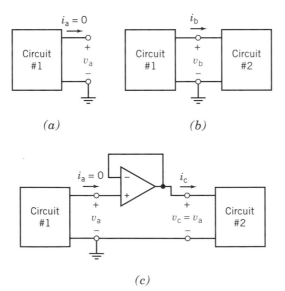

Figure 6-12 Circuit #1 *(a)* before and *(b)* after Circuit #2 is connected. *(c)* Preventing loading using a voltage follower.

Solution

As a specific example, consider Figure 6-13. The voltage divider shown in Figure 6-13*a* can be analyzed by writing a node equation at node a:

$$\frac{v_a - v_{in}}{20 \text{ k}\Omega} + \frac{v_a}{60 \text{ k}\Omega} = 0$$

Solving for v_a we have

$$v_a = \frac{3}{4} v_{in}$$

In Figure 6-13*b*, a resistor is connected across the output of the voltage divider. This circuit can be analyzed by writing a node equation at node a

$$\frac{v_b - v_{in}}{20 \text{ k}\Omega} + \frac{v_b}{60 \text{ k}\Omega} + \frac{v_b}{30 \text{ k}\Omega} = 0$$

Solving for v_b we have

$$v_b = \frac{1}{2} v_{in}$$

Since $v_b \neq v_a$, connecting the resistor directly to the voltage divider loads the voltage divider. This loading is caused by the current required by the 30-kΩ resistor. Without the voltage follower, the voltage divider must provide this current.

In Figure 6-13*c*, a voltage follower is used to connect the 30-kΩ resistor to the output of the voltage divider. Once again, the circuit can be analyzed by writing a node equation at node a

$$\frac{v_c - v_{in}}{20 \text{ k}\Omega} + \frac{v_c}{60 \text{ k}\Omega} = 0$$

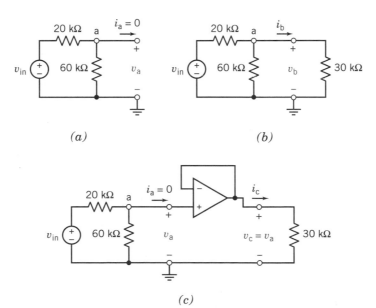

Figure 6-13 A voltage divider *(a)* before and *(b)* after a 30-kΩ resistor is added. *(c)* A voltage follower is added to prevent loading.

Solving for v_c we have

$$v_c = \frac{3}{4} v_{in}$$

Since $v_c = v_a$, loading is avoided when the voltage follower is used to connect the resistor to the voltage divider. The voltage follower, not the voltage divider, provides the current required by the 30-kΩ resistor.

Example 6-5

A common application of operational amplifiers is to scale a voltage, that is, to multiply a voltage by a constant, K, so that

$$v_o = K v_{in}$$

This situation is illustrated in Figure 6-14a. The input voltage, v_{in}, is provided by an ideal voltage source. The output voltage, v_o, is the branch voltage of a 100-kΩ resistor. This

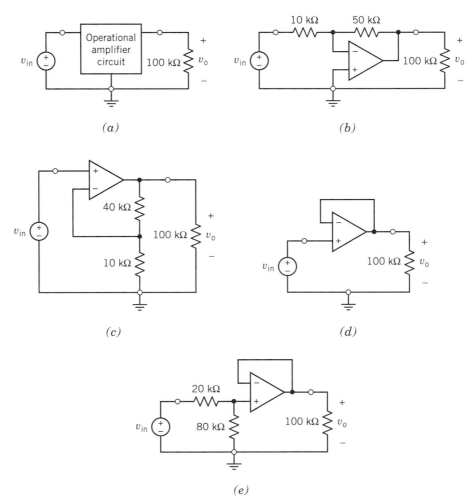

Figure 6-14 (a) An amplifier is required to make $v_o = K v_{in}$. The choice of amplifier circuit depends on the value of the gain K. Four cases are shown: (b) $K = -5$, (c) $K = 5$, (d) $K = 1$, and (e) $K = 0.8$.

resistor, sometimes called a load resistor, represents the circuit that will use the voltage v_o as its input. Circuits that perform this operation are usually called amplifiers. The constant K is called the gain of the amplifier.

The required value of the constant K will determine which of the circuits is selected from Figure 6-14. There are four cases to consider: $K < 0$, $K > 1$, $K = 1$, and $0 < K < 1$.

Solution

Since resistor values are positive, the gain of the inverting amplifier, shown in Figure 6-11a, is negative. Accordingly, when $K < 0$ is required, an inverting amplifier is used. For example, suppose we require $K = -5$. From Figure 6-11a

$$-5 = -\frac{R_f}{R_1}$$

so

$$R_f = 5\,R_1$$

As a rule of thumb, it is a good idea to choose resistors in operational amplifier circuits that have values between 5 kΩ and 500 kΩ when possible. Choosing

$$R_1 = 10\ \text{k}\Omega$$

gives

$$R_f = 50\ \text{k}\Omega$$

The resulting circuit is shown in Figure 6-14b.

Next suppose we require $K = 5$. The noninverting amplifier, shown in Figure 6-14b, is used to obtain gains greater than 1. From Figure 6-11b

$$5 = 1 + \frac{R_f}{R_1}$$

so

$$R_f = 4\,R_1$$

Choosing $R_1 = 10$ kΩ gives $R_f = 40$ kΩ. The resulting circuit is shown in Figure 6-14c.

Consider using the noninverting amplifier of Figure 6-11b to obtain a gain $K = 1$. From Figure 6-11b

$$1 = 1 + \frac{R_f}{R_1}$$

so

$$\frac{R_f}{R_1} = 0$$

This can be accomplished by replacing R_f by a short circuit ($R_f = 0$) or by replacing R_1 by an open circuit ($R_1 = \infty$) or both. Doing both converts a noninverting amplifier into a voltage follower. The gain of the voltage follower is 1. In Figure 6-14d a voltage follower is used for the case $K = 1$.

There is no amplifier in Figure 6-11 that has a gain between 0 and 1. Such a circuit can be obtained using a voltage divider together with a voltage follower. Suppose we require $K = 0.8$. First, design a voltage divider to have an attenuation equal to K,

$$0.8 = \frac{R_2}{R_1 + R_2}$$

so

$$R_2 = 4 \cdot R_1$$

Choosing $R_1 = 20\,\text{k}\Omega$ gives $R_2 = 80\,\text{k}\Omega$. Adding a voltage follower gives the circuit shown in Figure 6-14e.

Example 6-6

Design a circuit having one output, v_o, and three inputs v_1, v_2, and v_3. The output must be related to the inputs by

$$v_o = 2\,v_1 + 3\,v_2 + 4\,v_3$$

In addition, the inputs are restricted to have values between -1 V and 1 V, that is,

$$|v_i| \le 1\,\text{V} \qquad i = 1,\,2,\,3$$

Consider using an operational amplifier having $i_{\text{sat}} = 2$ mA and $v_{\text{sat}} = 15$ V, and design a circuit to implement the circuit.

Solution

The required circuit must multiply each input by a separate positive number and add the results. The noninverting summer shown in Figure 6-11e can do these operations. This circuit is represented by six parameters: K_1, K_2, K_3, K_4, R_a, and R_b. Designing the noninverting summer amounts to choosing values for these six parameters. Notice that $K_1 + K_2 + K_3 < 1$ is required to ensure that all of the resistors have positive values. Pick $K_4 = 10$ (a convenient value that is just a little larger than $2 + 3 + 4 = 9$). Then

$$v_o = 2\,v_1 + 3\,v_2 = 4\,v_3 = 10\,(0.2\,v_1 + 0.3\,v_2 + 0.4\,v_3)$$

That is $K_4 = 10$, $K_1 = 0.2$, $K_2 = 0.3$, and $K_3 = 0.4$. Figure 6-11e does not provide much guidance in picking values of R_a and R_b. Try $R_b = 100\,\Omega$. Then

$$R_a = (K_4 - 1)\,R_b = (10 - 1)100 = 900\,\Omega$$

Figure 6-15 shows the resulting circuit. It is necessary to check this circuit to ensure that it satisfies the specifications. Writing node equations

$$\text{node a:} \quad \frac{v_a - v_1}{500} + \frac{v_a - v_2}{333} + \frac{v_a - v_3}{250} + \frac{v_a}{1000} = 0$$

$$\text{output node:} \quad -\frac{v_o - v_a}{900} + \frac{v_a}{100} = 0$$

and solving these equations yield

$$v_o = 2\,v_1 + 3\,v_2 = 4\,v_3 \quad \text{and} \quad v_a = \frac{v_o}{10}$$

The output current of the operational amplifier is given by

$$i_{\text{oa}} = \frac{v_a - v_o}{900} = -\frac{v_o}{1000} \tag{6-3}$$

How large can the output voltage be? We know that

$$|v_o| = |2\,v_1 + 3\,v_2 + 4\,v_3|$$

so

$$|v_o| \le 2|v_1| + 3|v_2| + 4|v_3| = 9\,\text{V}$$

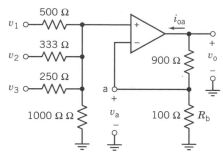

Figure 6-15 The proposed noninverting summing amplifier.

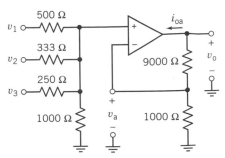

Figure 6-16 The final design of the noninverting summing amplifier.

The operational amplifier output voltage will always be less than v_{sat}. That's good. Now what about the output current? Notice that $|v_o| \leq 9$ V. From Eq. 6-3

$$| i_{oa} | = \left| \frac{-v_o}{1000 \ \Omega} \right| \leq \left| \frac{-9 \ \text{V}}{1000 \ \Omega} \right| = 9 \ \text{mA}$$

The operational amplifier output current exceeds $i_{sat} = 2\text{mA}$. This is not allowed. Increasing R_b will reduce i_o. Try $R_b = 1000 \ \Omega$. Then

$$R_a = (K_4 - 1) \ R_b = (10 - 1) \ 1000 = 9000 \ \Omega$$

This produces the circuit shown in Figure 6-16. Increasing R_a and R_b does not change the operational amplifier output voltage. As before

$$| v_o | \leq 2| v_1 | + 3| v_2 | + 4| v_3 | = 9 \ \text{V}$$

Increasing R_a and R_b does reduce the operational amplifier output current. Now

$$| i_{oa} | = \left| \frac{-v_o}{R_a + R_b} \right| \leq \left| \frac{-9 \ \text{V}}{10000 \ \Omega} \right| = 0.9 \ \text{mA}$$

so $|i_o| < 2$ mA and $|v_a| < 15$ V, as required.

Example 6-7

Two things will be done in this example. First the operational amplifier circuit that is labeled a negative resistance converter in Figure 6-11h will be analyzed to confirm that it does indeed act like a negative resistor. Next, the operational amplifier negative resistance converter will be used to design an amplifier with a gain greater than 1.

Solution

The operational amplifier negative resistance converter is redrawn in Figure 6-17a. The voltages v_a, v_b, and v_c are the node voltages at nodes a, b, and c, respectively. Writing a node equation at node a gives

$$i_1 + \frac{v_a - v_b}{R_3} = i_a$$

Writing a node equation at node c gives

$$i_2 + \frac{v_c - v_b}{R_2} + \frac{v_c}{R_1} = 0$$

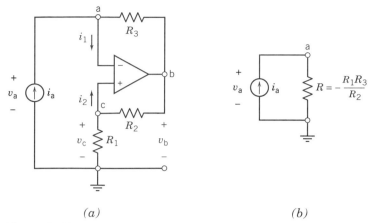

(a)   *(b)*

Figure 6-17 *(a)* An operational amplifier circuit that implements a negative resistance converter and *(b)* its equivalent circuit. The negative resistance appears between terminal *a* and ground.

We assume that the operational amplifier is modeled as an ideal operational amplifier. Then $i_1 = 0$, $i_2 = 0$, and $v_c = v_a$. The node equations become

$$0 + \frac{v_a - v_b}{R_3} = i_a$$

and

$$0 + \frac{v_a - v_b}{R_2} + \frac{v_a}{R_1} = 0$$

Eliminating v_b and solving for v_a gives

$$v_a = -\frac{R_1 R_3}{R_2} i_a$$

Let

$$R = -\frac{R_1 R_3}{R_2}$$

Then

$$v_a = R i_a$$

which is Ohm's Law. In other words, the operational amplifier circuit shown in Figure 6-17*a* acts like the negative resistor shown in Figure 6-17*b*. Notice that one node of the negative resistor must be the ground node and the other node is the *a* terminal.

Next consider the voltage divider shown in Figure 6-18*a*. This circuit can be analyzed using voltage division

$$v_{out} = \frac{-10 \text{ k}\Omega}{-10 \text{ k}\Omega + 5 \text{ k}\Omega} v_{in} = 2 v_{in}$$

In other words, this voltage divider acts like an amplifier with a gain equal to 2.

In Figure 6-18*b* the negative resistor is replaced by the operational amplifier negative resistance converter. To complete the design, the resistances R_1, R_2, and R_3 must be

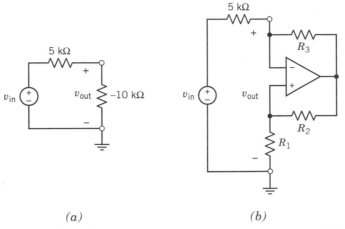

Figure 6-18 *(a)* A voltage divider using a negative resistor. *(b)* The same voltage divider using the negative resistance converter.

chosen so that

$$-10 \text{ k}\Omega = -\frac{R_1 R_3}{R_2}$$

This equation does not have a unique solution. One solution is $R_1 = 10 \text{ k}\Omega$, $R_2 = 10 \text{ k}\Omega$, and $R_3 = 10 \text{ k}\Omega$.

Example 6-8

It is desired to design an inverting amplifier with an ideal operational amplifier and standard resistor values. The inverting amplifier, shown in Figure 6-11*a*, must have a gain of -23. The resistors R_1 and R_f must each be a single standard value resistor. The standard values are

$$R = r \, 10^n$$

where

$$r \in \{ \ 1.0, \ 1.1, \ 1.2, \ 1.3, \ 1.5, \ 1.6, \ 1.8, \ 2.0, \ 2.2, \ 2.4, \ 2.7, \ 3.0, \ 3.3,$$
$$3.6, \ 3.9, \ 4.3, \ 4.7, \ 5.1, \ 5.6, \ 6.2, \ 6.8, \ 7.5, \ 8.2, \ 9.1 \ \}$$

The operation of this inverting amplifier is characterized by

$$\frac{v_{out}}{v_s} = -\frac{R_f}{R_1} = -G$$

We need to select standard resistor values for R_1 and R_f so that $G = R_f / R_1$ is equal to 23. First, express the gain G as

$$G = g10^m$$

where $0 \leq g < 1$. Next, let

$$R_1 = r_1 10^n \quad \text{and} \quad R_f = r_f 10^{n+m}$$

Define the error as

$$e_{ij} = \left| g - \frac{r_f}{r_1} \right|$$

rj

ri	1.0	1.1	1.2	1.3	1.5	1.6	1.8	2.0	2.2	2.4	2.7	3.0	3.3	3.6	3.9	4.3	4.7	5.1	5.6	6.2	6.8	7.5	8.2	9.1
1.1	1.30	1.20	1.10	1.00	0.80	0.70	0.50	0.30	0.10	0.10	0.40	0.70	1.00	1.30	1.60	2.00	2.40	2.80	3.30	3.90	4.50	5.20	5.90	6.80
1.2	1.39	1.30	1.21	1.12	0.94	0.85	0.66	0.48	0.30	0.12	0.15	0.43	0.70	0.97	1.25	1.61	1.97	2.34	2.79	3.34	3.88	4.52	5.15	5.97
1.2	1.47	1.38	1.30	1.22	1.05	0.97	0.80	0.63	0.47	0.30	0.05	0.20	0.45	0.70	0.95	1.28	1.62	1.95	2.37	2.87	3.37	3.95	4.53	5.28
1.5	1.53	1.45	1.38	1.30	1.15	1.07	0.92	0.76	0.61	0.45	0.22	0.01	0.24	0.47	0.70	1.01	1.32	1.62	2.01	2.47	2.93	3.47	4.01	4.70
1.6	1.63	1.57	1.50	1.43	1.30	1.23	1.10	0.97	0.83	0.70	0.50	0.30	0.10	0.10	0.30	0.57	0.83	1.10	1.43	1.83	2.23	2.70	3.17	3.77
1.8	1.68	1.61	1.55	1.49	1.36	1.30	1.18	1.05	0.93	0.80	0.61	0.43	0.24	0.05	0.14	0.39	0.64	0.89	1.20	1.58	1.95	2.39	2.82	3.39
2.0	1.74	1.69	1.63	1.58	1.47	1.41	1.30	1.19	1.08	0.97	0.80	0.63	0.47	0.30	0.13	0.09	0.31	0.53	0.81	1.14	1.48	1.87	2.26	2.76
2.2	1.80	1.75	1.70	1.65	1.55	1.50	1.40	1.30	1.20	1.10	0.95	0.80	0.65	0.50	0.35	0.15	0.16	0.25	0.50	0.80	1.10	1.45	1.80	2.25
2.4	1.85	1.80	1.75	1.71	1.62	1.57	1.48	1.39	1.30	1.21	1.07	0.94	0.80	0.66	0.53	0.35	0.34	0.18	0.28	0.52	0.79	1.11	1.43	1.84
2.7	1.88	1.84	1.80	1.76	1.68	1.63	1.55	1.47	1.38	1.30	1.18	1.05	0.93	0.80	0.68	0.51	0.56	0.41	0.03	0.28	0.53	0.83	1.12	1.49
3.0	1.93	1.89	1.86	1.82	1.74	1.71	1.63	1.56	1.49	1.41	1.30	1.19	1.08	0.97	0.86	0.71	0.73	0.60	0.23	0.00	0.22	0.48	0.74	1.07
3.3	1.97	1.93	1.90	1.87	1.80	1.77	1.70	1.63	1.57	1.50	1.40	1.30	1.20	1.10	1.00	0.87	0.88	0.75	0.43	0.23	0.03	0.20	0.43	0.73
3.6	2.00	1.97	1.94	1.91	1.85	1.82	1.75	1.69	1.63	1.57	1.48	1.39	1.30	1.21	1.12	1.00	0.99	0.88	0.60	0.42	0.24	0.03	0.18	0.46
3.9	2.02	1.99	1.97	1.94	1.88	1.86	1.80	1.74	1.69	1.63	1.55	1.47	1.38	1.30	1.22	1.11	1.09	0.99	0.74	0.58	0.41	0.22	0.02	0.23
4.3	2.04	2.02	1.99	1.97	1.92	1.89	1.84	1.79	1.74	1.68	1.61	1.53	1.45	1.38	1.30	1.20	1.21	1.11	0.86	0.71	0.56	0.38	0.20	0.03
4.7	2.07	2.04	2.02	2.00	1.95	1.93	1.88	1.83	1.79	1.74	1.67	1.60	1.53	1.46	1.39	1.30	1.30	1.21	1.00	0.86	0.72	0.56	0.39	0.18
5.1	2.09	2.07	2.04	2.02	1.98	1.96	1.92	1.87	1.83	1.79	1.73	1.66	1.60	1.53	1.47	1.39	1.38	1.30	1.11	0.98	0.85	0.70	0.56	0.36
5.6	2.10	2.08	2.06	2.05	2.01	1.99	1.95	1.91	1.87	1.83	1.77	1.71	1.65	1.59	1.54	1.46	1.46	1.39	1.20	1.08	0.97	0.83	0.69	0.52
6.2	2.12	2.10	2.09	2.07	2.03	2.01	1.98	1.94	1.91	1.87	1.82	1.76	1.71	1.66	1.60	1.53	1.54	1.48	1.30	1.19	1.09	0.96	0.84	0.68
6.8	2.14	2.12	2.11	2.09	2.06	2.04	2.01	1.98	1.95	1.91	1.86	1.82	1.77	1.72	1.67	1.61	1.61	1.55	1.40	1.30	1.20	1.09	0.98	0.83
7.5	2.15	2.14	2.12	2.11	2.08	2.06	2.04	2.01	1.98	1.95	1.90	1.86	1.81	1.77	1.73	1.67	1.67	1.62	1.48	1.39	1.30	1.20	1.09	0.96
8.2	2.17	2.15	2.14	2.13	2.10	2.09	2.06	2.03	2.01	1.98	1.94	1.90	1.86	1.82	1.78	1.73	1.73	1.68	1.55	1.47	1.39	1.30	1.21	1.09
9.1	2.18	2.17	2.15	2.14	2.12	2.10	2.08	2.06	2.03	2.01	1.97	1.93	1.90	1.86	1.82	1.78	1.78	1.74	1.62	1.54	1.47	1.39	1.30	1.19
	2.19	2.18	2.17	2.16	2.14	2.12	2.10	2.08	2.04	2.08	2.00	1.97	1.94	1.90	1.87	1.83	1.78	1.74	1.68	1.62	1.55	1.48	1.40	1.30

ri

THIS DISPLAYS THE ERROR eij FOR THE COMBINATIONS OF ri AND rj

Figure 6-19 The spreadsheet for all standard values of r_i and r_j.

Use a spreadsheet computer program to set up a table that displays the error for each of the $24 \times 24 = 576$ possible combinations of r_1 and r_f.

The desired value of the gain is 23, and therefore $g = 2.3$ and $m = 1$. We then seek the minimum error where

$$e_{ij} = \left| 2.3 - \frac{r_f}{r_1} \right|$$

and use a spreadsheet to generate all combinations of r_1 and r_f as shown in Figure 6-19.

The spreadsheet identifies $r_f = 6.2$ and $r_1 = 2.7$ as providing minimum error, and then the actual gain, G_a, is

$$G_a = \frac{r_f \times 10^m}{r_1} = \frac{6.2 \times 10^1}{2.7} = 22.96$$

EXERCISE 6-4

Determine the ratio, v_o/v_{in}, of the circuit of Figure 6-11b when $R_f = 100$ kΩ and $R_1 = 25$ kΩ.

Answer: $v_o/v_{in} = 5$

6-7 CHARACTERISTICS OF PRACTICAL OPERATIONAL AMPLIFIERS

The ideal operational amplifier is the simplest model of an operational amplifier. This simplicity is obtained by ignoring some imperfections of practical operational amplifiers. This section considers some of these imperfections and provides alternate operational amplifier models to account for these imperfections.

Consider the operational amplifier shown in Figure 6-20a. If this operational amplifier is ideal, then

$$i_1 = 0, \; i_2 = 0 \quad \text{and} \quad v_1 - v_2 = 0 \tag{6-4}$$

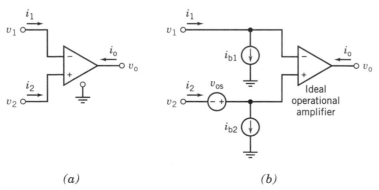

(a)   (b)

Figure 6-20 (a) An operational amplifier and (b) the offsets model of an operational amplifier.

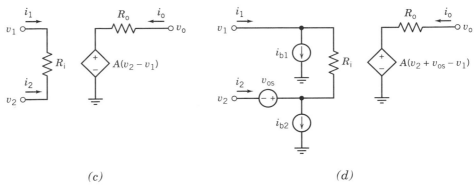

(c)   (d)

Figure 6-20 (continued) (c) The finite gain model of an operational amplifier. (d) The offsets and finite gain model of an operational amplifier.

In contrast, the operational amplifier model shown in Figure 6-20d accounts for several nonideal parameters of practical operational amplifiers, namely:

nonzero bias currents,
nonzero input offset voltage,
finite input resistance,
nonzero output resistance,
finite voltage gain.

This model more accurately describes practical operational amplifiers than does the ideal operational amplifier. Unfortunately the more accurate model of Figure 6-20d is much more complicated and much more difficult to use than the ideal operational amplifier. The models in Figures 6-20b and 6-20c provide a compromise. These models are more accurate than the ideal operational amplifier but easier to use than the model in Figure 6-20d. It will be convenient to have names for these models. The model in Figure 6-20b will be called "the offsets model" of the operational amplifier. Similarly, the model in Figure 6-20c will be called "the finite gain model" of the operational amplifier and the model in Figure 6-20d will be called "the offsets and finite gain model" of the operational amplifier.

The operational amplifier model shown in Figure 6-20b accounts for the nonzero bias current and nonzero input offset voltage of practical operational amplifiers but not the finite input resistance, the nonzero output resistance, or the finite voltage gain. This model consists of three independent sources and an ideal operational amplifier. In contrast to the ideal operational amplifier, the operational amplifier model that accounts for offsets is represented by the equations

$$i_1 = i_{b1}, \quad i_2 = i_{b2} \quad \text{and} \quad v_1 - v_2 = v_{os} \tag{6-5}$$

The voltage v_{os} is a small, constant voltage called the input offset voltage. The currents i_{b1} and i_{b2} are called the bias currents of the operational amplifier. They are small, constant currents. The difference between the bias currents is called the input offset current, i_{os}, of the amplifier:

$$i_{os} = i_{b1} - i_{b2}$$

Notice that when the bias currents and input offset voltage are all zero, Eq. 6-5 is the same as Eq. 6-4. In other words, the offsets model reverts to the ideal operational amplifier when the bias currents and input offset voltage are zero.

Frequently, the bias currents and input offset voltage can be ignored because they are very small. However, when the input signal to a circuit is itself small, the bias currents and input voltage can become important.

Manufacturers specify a maximum value for the bias currents, the input offset current, and the input offset voltage. For the μA741 the maximum bias current is specified to be 500 nA, the maximum input offset current is specified to be 200 nA, and the maximum input offset voltage is specified to be 5 mV. These specifications guarantee that

$$|i_{b1}| \leq 500 \text{ nA} \quad \text{and} \quad |i_{b2}| \leq 500 \text{ nA}$$
$$|i_{b1} - i_{b2}| \leq 200 \text{ nA}$$
$$|v_{os}| \leq 5 \text{ mV}$$

Table 6-2 shows the bias currents, offset current, and input offset voltage *typical* of several types of operational amplifier.

Table 6-2

Selected Parameters of Typical Operational Amplifiers

Parameter	Units	μA741	LF351	TL051C	OPA101 AM	OP-07E
Saturation voltage, v_{sat}	V	13	13.5	13.2	13	13
Saturation current, i_{sat}	mA	2	15	6	30	6
Slew rate, SR	V/μs	0.5	13	23.7	6.5	0.17
Bias current, i_b	nA	80	0.05	0.03	0.012	1.2
Offset current, i_{os}	nA	20	0.025	0.025	0.003	0.5
Input offset voltage, v_{os}	mV	1	5	0.59	0.1	0.03
Input resistance, R_i	MΩ	2	10^6	10^6	10^6	50
Output resistance, R_o	Ω	75	1000	250	500	60
Differential gain, A	V/mV	200	100	105	178	5000
Common mode rejection ratio, $CMRR$	V/mv	31.6	100	44	178	1413
Gain bandwidth product, B	MHz	1	4	3.1	20	0.6

Example 6-9

The inverting amplifier shown in Figure 6-21a contains a μA741 operational amplifier. This inverting amplifier designed in Example 6-5 has a gain of -5, that is,

$$v_o = -5 \cdot v_{in}$$

The design of the inverting amplifier is based on the ideal model of an operational amplifier and so did not account for the bias currents and input offset voltage of the μA741 operational amplifier. In this example, the offsets model of an operational amplifier will be used to analyze the circuit. This analysis will tell us what effect the bias currents and input offset voltage have on the performance of this circuit.

Solution

In Figure 6-21b the operational amplifier has been replaced by the offsets model of an operational amplifier. Notice that the operational amplifier in Figure 6-21b is the ideal operational amplifier that is part of the model of the operational amplifier used to account for the offsets. The circuit in Figure 6-21b contains four inputs that correspond to the four independent sources v_{in}, i_{b1}, i_{b2}, and v_{os}. (The input v_{in} is obtained by connecting a voltage source to the circuit. In contrast, the "inputs" i_{b1}, i_{b2}, and v_{os} are the results of

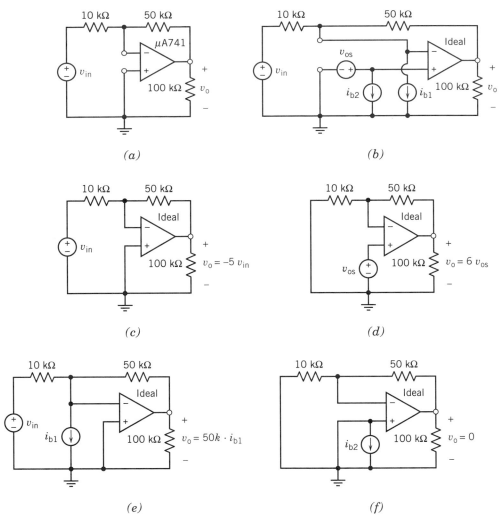

Figure 6-21 *(a)* An inverting amplifier and *(b)* an equivalent circuit that accounts for the input offset voltage and bias currents of the operational amplifier. *(c)–(f)* Analysis using superposition.

imperfections of the operational amplifier. These inputs are part of the operational amplifier model and do not need to be added to the circuit.) Superposition can be used to good advantage in analyzing this circuit. Figures 6-21*c* through 6-20*f* illustrate this process. In each of these figures, all but one input has been set to zero and the output due to that one input has been calculated.

Figure 6-21*c* shows the circuit used to calculate the response to v_{in} alone. The other inputs, i_{b1}, i_{b2}, and v_{os}, have all been set to zero. Recall that zero current sources act like open circuits and zero voltage sources act like short circuits. Figure 6-21*c* is obtained from Figure 6-20*b* by replacing the current sources i_{b1}, i_{b2} by open circuits and by replacing the voltage source v_{os} by a short circuit. The operational amplifier in Figure 6-21*c* is the ideal operational amplifier that is part of the offset model. Analysis of the inverting amplifier in Figure 6-21*c* gives

$$v_o = -5 \cdot v_{in}$$

Next consider Figure 6-21*d*. This circuit is used to calculate the response to v_{os} alone.

The other inputs, v_{in}, i_{b1}, and i_{b2} have all been set to zero. Figure 6-21d is obtained from Figure 6-20b by replacing the current sources i_{b1}, i_{b2} by open circuits and by replacing the voltage source v_{os} by a short circuit. Again, the operational amplifier is the ideal operational amplifier from the offsets model. The circuit in Figure 6-21d is one we have seen before; it is the noninverting amplifier (Figure 6-11b). Analysis of this noninverting amplifier gives

$$v_o = \left(1 + \frac{50 \text{ k}\Omega}{10 \text{ k}\Omega}\right) \cdot v_{os}$$
$$= 6 \, v_{os}$$

Next consider Figure 6-21e. This circuit is used to calculate the response to i_{b1} alone. The other inputs, v_{in}, v_{os}, and i_{b2} have all been set to zero. Figure 6-21e is obtained from Figure 6-20b by replacing the current source i_{b2} by an open circuit and by replacing the voltage sources v_{in} and v_{os} by short circuits. Notice that the voltage across the 10-kΩ resistor is zero because this resistor is connected between the input nodes of the ideal operational amplifier. Ohm's Law says that the current in the 10-kΩ resistor must be zero. The current in the 50-kΩ resistor is i_{b1}. Finally, paying attention to the reference directions,

$$v_o = 50 \text{ k}\Omega \cdot i_{b1}$$

Figure 6-21f is used to calculate the response to i_{b2} alone. The other inputs, v_{in}, v_{os}, and i_{b1}, have all been set to zero. Figure 6-21f is obtained from Figure 6-20b by replacing the current source i_{b1} by an open circuit and by replacing the voltage sources v_{in} and v_{os} by short circuits. Replacing v_{os} by a short circuit inserts a short circuit across the current source i_{b2}. Again, the voltage across the 10-kΩ resistor is zero, so the current in the 10-kΩ resistor must be zero. Kirchoff's Current Law shows that the current in the 50-kΩ resistor is also zero. Finally,

$$v_o = 0$$

The output caused by all four inputs working together is the sum of the outputs caused by each input working alone. Therefore

$$v_o = -5 \cdot v_{in} + 6 \cdot v_{os} + (50 \text{ k}\Omega) i_{b1}$$

When the input of the inverting amplifier, v_{in}, is zero, the output v_o also should be zero. However, v_o is nonzero when we have a finite v_{os} or i_{b1}. Let

$$\text{output offset voltage} = 6 \cdot v_{os} + (50 \text{ k}\Omega) i_{b1}$$

Then

$$v_o = -5 \cdot v_{in} + \text{output offset voltage}$$

Recall that when the operational amplifier is modeled as an ideal operational amplifier, analysis of this inverting amplifier gives

$$v_o = -5 \cdot v_{in}$$

Comparing these last two equations shows that bias currents and input offset voltage cause the output offset voltage. Modeling the operational amplifier as an ideal operational amplifier amounts to assuming that the output voltage is not important and thus ignoring it. Using the operational amplifier model that accounts for offsets is more accurate, but also more complicated.

How large is the output offset voltage of this inverting amplifier? The input offset voltage of a μA741 operational amplifier will be at most 5 mV, and the bias current will be at most 500 nA, so

$$\text{output offset voltage} \leq 6 \cdot 5 \text{ mV} + (50 \text{ k}\Omega)\,500 \text{ nA} = 55 \text{ mV}$$

We note that we can only ignore the effect of the offset voltage when $5v_{\text{in}} > 500$ mV or $v_{\text{in}} > 100$ mV. The output offset error can be reduced by using a better operational amplifier, that is, one that guarantees smaller bias currents and input offset voltage.

Now, let us turn our attention to different parameters of practical operational amplifiers. The operational amplifier model shown in Figure 6-20c accounts for the finite input resistance, the nonzero output resistance, and the finite voltage gain of practical operational amplifiers but not the nonzero bias current and nonzero input offset voltage. This model consists of two resistors and a VCVS.

The finite gain model reverts to an ideal operational amplifier when the gain, A, becomes infinite. To see that this is so notice that in Figure 6-20c

$$v_{\text{o}} = A(v_2 - v_1) + R_{\text{o}}i_{\text{o}}$$

so

$$v_2 - v_1 = \frac{v_{\text{o}} - R_{\text{o}}i_{\text{o}}}{A}$$

The models in Figure 6-20, as well as the model of the ideal operational amplifier, are valid only when v_{o} and i_{o} satisfy Eq. 6-2. Therefore,

$$|v_{\text{o}}| \leq v_{\text{sat}} \quad \text{and} \quad |i_{\text{o}}| \leq i_{\text{sat}}$$

Then

$$|v_2 - v_1| \leq \frac{v_{\text{sat}} + R_{\text{o}}i_{\text{sat}}}{A}$$

Therefore,

$$\lim_{A \to \infty} (v_2 - v_1) = 0$$

Next, since

$$i_1 = -\frac{v_2 - v_1}{R_{\text{i}}} \quad \text{and} \quad i_2 = \frac{v_2 - v_1}{R_{\text{i}}}$$

we conclude that

$$\lim_{A \to \infty} i_1 = 0 \quad \text{and} \quad \lim_{A \to \infty} i_2 = 0$$

Thus, i_1, i_2, and $v_2 - v_1$ satisfy Eq. 6-4. In other words, the finite gain model of the operational amplifier reverts to the ideal operational amplifier as the gain becomes infinite. The gain for practical op amps ranges from 100,000 to 10^7.

Example 6-10

In Figure 6-22a a voltage follower is used as a buffer amplifier. Analysis based on the ideal operational amplifier shows that the gain of the buffer amplifier is

$$\frac{v_{\text{o}}}{v_{\text{s}}} = 1$$

What effects will the input resistance, output resistance, and finite voltage gain of a practical operational amplifier have on the performance of this circuit? To answer this question, replace the operational amplifier by the operational amplifier model that accounts for finite voltage gain. This gives the circuit shown in Figure 6-22b.

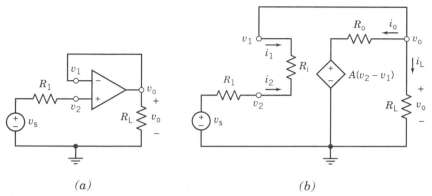

(a)   (b)

Figure 6-22 (a) A voltage follower used as a buffer amplifier and (b) an equivalent circuit with the operational amplifier model that accounts for finite voltage gain.

Solution

To be specific, suppose $R_1 = 1$ kΩ; $R_L = 10$ kΩ; and the parameters of the practical operational amplifier are $R_i = 100$ kΩ, $R_o = 100 \, \Omega$, and $A = 10^5$ V/V.

Suppose that $v_o = 10$ V. We can find the current, i_L, in the output resistor as

$$i_L = \frac{v_o}{R_L}$$

$$= \frac{10 \text{ V}}{10^4 \, \Omega}$$

$$= 10^{-3} \text{ A}$$

Apply KCL at the top node of R_L to get

$$i_1 + i_o + i_L = 0$$

It will turn out that i_1 will be much smaller than both i_o and i_L. It is useful to make the approximation that $i_1 = 0$. (We will check this assumption later in this example.) Then

$$i_o = -i_L$$

Next, apply KVL to the mesh consisting of the VCVS, R_o, and R_L to get

$$-A(v_2 - v_1) - i_o R_o + i_L R_L = 0$$

Combining the last two equations and solving for $(v_2 - v_1)$ gives

$$v_2 - v_1 = \frac{i_L(R_o + R_L)}{A}$$

$$= \frac{10^{-3}(100 + 10{,}000)}{10^5}$$

$$= 1.01 \times 10^{-4} \text{ V}$$

Now i_1 can be calculated using Ohm's Law,

$$i_1 = \frac{v_1 - v_2}{R_i}$$

$$= \frac{-1.01 \times 10^{-4} \text{ V}}{100 \text{ k}\Omega}$$

$$= -1.01 \times 10^{-9} \text{ A}$$

This justifies our earlier assumption that i_1 is negligible compared with i_o and i_L.

Applying KVL to the outside loop gives

$$-v_s - i_1 R_s - i_1 R_i + v_o = 0$$

Now let us do some algebra to determine v_s.

$$v_s = v_o - i_1(R_1 + R_i)$$

$$= v_o + \frac{v_2 - v_1}{R_i} \times (R_1 + R_i)$$

$$= v_o + \frac{i_L(R_o + R_L)}{A} \times \frac{R_1 + R_i}{R_i}$$

$$= v_o + \frac{v_o}{R_L} \times \frac{R_o + R_L}{A} \times \frac{R_1 + R_i}{R_i}$$

The gain of this circuit is

$$\frac{v_o}{v_s} = \frac{1}{1 + \frac{1}{A} \times \frac{R_o + R_L}{R_L} \times \frac{R_i + R_1}{R_i}}$$

This equation shows that the gain will be approximately 1 when A is very large, $R_o \ll R_L$, and $R_1 \ll R_i$. In this example, for the specified A, R_o and R_i we have

$$\frac{v_o}{v_s} = \frac{1}{1 + \frac{1}{10^5} \times \frac{100 + 10{,}000}{10{,}000} \times \frac{10^5 + 1000}{1000}}$$

$$= \frac{1}{1.00001}$$

$$= 0.99999$$

Thus, the input resistance, output resistance, and voltage gain of the practical operational amplifier have only a small, essentially negligible, combined effect on the performance of the buffer amplifier.

Example 6-11

A sensor is used to measure a variable such as temperature or fluid flow rate. The sensor is normally connected to a meter with a resistance R_L equal to 100 Ω. The sensor is represented by a model with a voltage source v_s in series with a resistor R_1 equal to 2000 Ω. The sensor may be directly connected to the meter, or it may be connected through a buffer amplifier as shown in Figure 6-21a. Find the voltage ratio v_o/v_s for the direct connection and for the amplifier connection, where v_o is the voltage across the load resistor. The input resistance R_i is 10 MΩ for the operational amplifier, R_o is 1 Ω, and A is 10^6. Also find the ratio of the power delivered to the meter with and without the operational amplifier.

Solution

Since A is 10^6 and R_L is much larger than the output resistance of the amplifier, we will assume that the voltage gain with the operational amplifier in place is $v_o/v_s = 1$.

If the source is connected directly to the meter, without the amplifier, the output voltage is

$$v_o = \frac{R_L}{R_1 + R_L}\, v_s$$

$$= \frac{100}{2100}\, v_s$$

Therefore,

$$\frac{v_o}{v_s} = 0.048$$

With a unity gain amplifier in place, the power to the meter is

$$P_a = \frac{v_o^2}{R_L}$$

$$= \frac{v_s^2}{R_L}$$

The power at the meter when the source is directly connected (without the amplifier) to the load is

$$P_d = \frac{v_o^2}{R_L} = \frac{(0.048\, v_s)^2}{R_L} = \frac{0.0023\, v_s^2}{R_L}$$

The responses of the circuit with and without the operational amplifier are shown in Table 6-3.

Table 6-3

	Voltage Gain v_o/v_s	Power to Meter
Direct connection	0.048	$\dfrac{0.0023 v_s^2}{R_L}$
Connection to meter through buffer amplifier	1.0	v_s^2/R_L

The ratio of the power delivered at the output terminals with the amplifier to the power delivered without the amplifier connected is called the *power gain* and is written as P_a/P_d. In this case, we have

$$\frac{P_a}{P_d} = \frac{v_s^2/R_L}{0.0023\, v_s^2/R_L}$$

$$= 435$$

The power gain of an operational amplifier circuit is a great advantage indeed.

Table 6-2 lists two other parameters of practical operational amplifiers that have not yet been mentioned. They are the *Common Mode Rejection Ratio (CMRR)* and the gain bandwidth product. Consider first the common mode rejection ratio. In the finite gain model, the voltage of the dependent source is

$$A(v_2 - v_1)$$

In practice, we find that dependent source voltage is more accurately expressed as

$$A(v_2 - v_1) + A_{cm}\left(\frac{v_1 + v_2}{2}\right)$$

where

$$v_2 - v_1 \text{ is called the differential input voltage}$$

$$\frac{v_1 + v_2}{2} \text{ is called the common mode input voltage}$$

and

$$A_{cm} \text{ is called the common mode gain}$$

The gain A is sometimes called the differential gain to distinguish it from A_{cm}. The common mode rejection ratio is defined to be the ratio of A to A_{cm}

$$CMRR = \frac{A}{A_{cm}}$$

The dependent source voltage can be expressed using A and *CMRR* as

$$A(v_2 - v_1) + A_{cm}\frac{v_1 + v_2}{2} = A(v_2 - v_1) + \frac{A}{CMRR}\frac{v_1 + v_2}{2}$$

$$= A\left[\left(1 + \frac{1}{2\ CMRR}\right)v_2 - \left(1 - \frac{1}{2\ CMRR}\right)v_1\right]$$

CMRR can be added to the finite gain model by changing the voltage of the dependent source. The appropriate change is

$$\text{Replace } A(v_2 - v_1) \text{ by } A\left[\left(1 + \frac{1}{2\ CMRR}\right)v_2 - \left(1 - \frac{1}{2\ CMRR}\right)v_1\right]$$

This change will make the model more accurate but also more complicated. Table 6-2 shows that *CMRR* is typically very large. For example, a typical LF351 operational amplifier has $A = 100$ V/mV and $CMRR = 100$ V/mV. This means that

$$A\left[\left(1 + \frac{1}{2\ CMRR}\right)v_2 - \left(1 - \frac{1}{2\ CMRR}\right)v_1\right] = 100,000.5\ v_2 - 99,999.5\ v_1$$

compared to

$$A(v_2 - v_1) = 100,000\ v_2 - 100,000\ v_1$$

In most cases, negligible error is caused by ignoring the *CMRR* of the operational amplifier. The *CMRR* does not need to be considered unless accurate measurements of very small differential voltages must be made in the presence of very large common mode voltages.

Next we consider the gain bandwidth product of the operational amplifier. The finite gain model indicates that the gain, A, of the operational amplifier in a constant. Suppose

$$v_1 = 0 \quad \text{and} \quad v_2 = M \sin \omega t$$

so that

$$v_2 - v_1 = M \sin \omega t$$

The voltage of the dependent source in the finite gain model will be

$$A(v_2 - v_1) = A \cdot M \sin \omega t$$

The amplitude, $A \cdot M$, of this sinusoidal voltage does not depend on the frequency, ω. Practical operational amplifiers do not work this way. The gain of a practical amplifier is a function of frequency, say $A(\omega)$. For many practical amplifiers, $A(\omega)$ can be adequately represented as

$$A(\omega) = \frac{B}{j\omega}$$

It is not necessary to know now how this function behaves. Functions of this sort will be discussed in Chapter 13. For now it is enough to realize that the parameter B is used to describe the dependence of the operational amplifier gain on frequency. The parameter B is called the gain bandwidth product of the operational amplifier.

EXERCISE 6-5

The input offset voltage of a *typical* μA741 amplifier is 1 mV and the bias current is 80 nA. Suppose the operational amplifier in Figure 6-21a is a typical μA741. Show that the output offset voltage of the inverting amplifier will be 10 mV.

EXERCISE 6-6

Suppose the 10-kΩ resistor in Figure 6-21a is changed to 2 kΩ and the 50-kΩ resistor is changed to 10 kΩ. (These changes will not change the gain of the inverting amplifier. It will still be -5.) Show that the *maximum* output offset voltage is reduced to 35 mV. (Use $i_b = 500$ nA and $v_{os} = 5$ mV to calculate the *maximum* output offset voltage that could be caused by the μA741 amplifier.)

EXERCISE 6-7

Suppose the μA741 operational amplifier in Figure 6-21a is replaced with a *typical* OPA101AM operational amplifier. Show that the output offset voltage of the inverting amplifier will be 0.6 mV.

EXERCISE 6-8

(a) Determine the voltage ratio v_o/v_s, the input resistance R_{in}, and the output resistance R_{out} for the op amp circuit shown in Figure E 6-8.
(b) Calculate v_o/v_s, R_{in}, and R_{out} for a practical op amp with $A = 10^5$, $R_o = 100\ \Omega$, and $R_i = 500$ kΩ. The circuit resistors are $R_s = 10$ kΩ, $R_f = 50$ kΩ, and $R_a = 25$ kΩ.
Answer: (b) $v_o/v_s = -2$, $R_{out} = 3$ mΩ

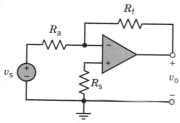

Figure E 6-8

EXERCISE 6-9

A noninverting buffer amplifier (Figure 6-11c) is used to connect a sensor to a meter. The sensor provides 10 V in series with a resistor equal to 1000 Ω. The amplifier has a unity gain and an output resistance R_o equal to 10 Ω. The meter has a resistance R_L equal to 500 Ω. Find the ratio of the power delivered by the amplifier to the load, P_2, to the power delivered to the load if the sensor is connected directly, P_1.

Answer: $P_2/P_1 = 8.65$

6-8 | THE OPERATIONAL AMPLIFIER AND PSpice

Frequently, engineers design circuits based on a simple model of an operational amplifier, such as the ideal operational amplifier. These circuits are then checked by analyzing them using a more accurate, more complicated operational amplifier model. Fortunately, PSpice takes the drudgery out of this analysis.

PSpice does not recognize operational amplifiers as devices. Instead, the PSpice "subcircuit" feature includes operational amplifiers in a PSpice input file. The syntax for a subcircuit is

```
.subckt<subcircuit name><interface node list>
circuit element statements defining subcircuit
.ENDS<subcircuit name>
```

Table 6-4 provides subcircuits that implements the models of operational amplifiers that have been discussed in this chapter. The statement

```
Xname<interface node list><subcircuit name>
```

inserts a subcircuit into the circuit. (Xname is a PSpice device name that begins with the letter X. We will use XOAk to name the *k*th operational amplifier, for example, XOA1, XOA2, XOA3, . . .)

Example 6-12

The circuit shown in Figure 6-23a consists of two summing amplifiers. Assuming ideal operational amplifiers, analysis of this circuit is used to determine v_o for the ideal and a practical op amp.

Solution

We have v_4 as:

$$v_4 = -\left(\frac{100 \text{ k}\Omega}{50 \text{ k}\Omega} v_1 + \frac{100 \text{ k}\Omega}{25 \text{ k}\Omega} v_2\right)$$
$$= -(2v_1 + 4v_2)$$

where v_4 is the node voltage at node 4. Similarly

$$v_o = -\left(\frac{50 \text{ k}\Omega}{10 \text{ k}\Omega} v_3 + \frac{50 \text{ k}\Omega}{25 \text{ k}\Omega} v_4\right)$$
$$= -(5 v_3 + 2 v_4)$$

Table 6-4

PSpice models and subcircuits for operational amplifier (parameter values correspond to a typical μA741)

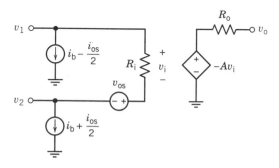

Ideal operational amplifier

```
.subckt ideal_op_amp 1 2 3
* op amp nodes listed in order: − + o
E   3 0-1   1 2-1   − 1G
.ends ideal_op_amp
```

The offsets model of an operational amplifier

```
.subckt offsets_op_amp 1 2 4
* op amp nodes listed in order: − + o
Ib1   1   0   70nA
Ib2   2   0   90nA
Vos   3   2   1mV
Ri    1   3   1G
E     4   0   1   3   − 1G
.ends offsets_op_amp
```

The finite gain model of an operational amplifier

```
.subckt finite_gain_op_amp 1 2 4
* op amp nodes listed in order: − + o
Ri   1   2   2MEG
E    3   0   1   2   − 200000
Ro   4   3   75
.ends finite_gain_op_amp
```

The offsets and finite gain model of an operational amplifier

```
.subckt op_amp 1   2   5
* op amp nodes listed in order: − + o
Ib1   1   0   70nA
Ib2   2   0   90 nA
Vos   3   2   1mV
Ri    1   3   2MEG
E     4   0   1   3   − 200000
Ro    4   5   75
.ends op_amp
```

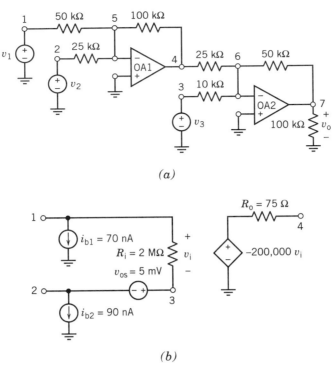

(a)

(b)

Figure 6-23 (a) A circuit using operational amplifiers and (b) a PSpice model of a typical μA741 operational amplifier.

Combining these equations yields

$$v_o = -(5\,v_3 - 2(2\,v_1 + 4\,v_2))$$
$$= 4\,v_1 + 8\,v_2 - 5\,v_3$$

How will this circuit perform when the operational amplifiers are typical μA741s? This question can be answered by replacing each operational amplifier by the model that accounts for both offsets and finite gain and then analyzing the resulting circuit. Fortunately, PSpice can be used to do this analysis. Figure 6-24 shows a PSpice input file that corresponds to the circuit in Figure 6-23a. The lines

```
XOA1  5  0  4  uA741_op_amp
```

and

```
XOA2  6  0  7  uA741_op_amp
```

indicate that XOA1 and XOA2 are the device names of the operational amplifiers and that both operational amplifiers are to be modeled using the subcircuit named "uA741_op_amp." The operational amplifier nodes are listed in the same order as given in the .subckt statement. That means that node 5 is the inverting input node, node 0 is the noninverting input node, and node 4 is the output node of OA1. Similarly, node 6 is the inverting input node, node 0 is the noninverting input node, and node 7 is the output node of OA2.

PSpice will replace each operational amplifier with a copy of the subcircuit "uA741_op_amp." In doing so, the devices and nodes of the subcircuit will be renamed by appending the device names of the operational amplifiers. For example, the device

```
Ib1  1  0  70nA
```

in subcircuit uA741_op_amp becomes

```
XOA1.Ib1  XOA1.1  0  70nA
```

when the subcircuit uA741_op_amp is substituted for XOA1 and

```
XOA2.Ib1  XOA2.1  0  70nA
```

when the subcircuit uA741_op_amp is substituted for XOA2. Let us notice some things about node numbers. First, node 0 always represents the ground node. Node 0 is the same node throughout the PSpice file, including inside subcircuits. For this reason, it was not necessary to include node 0 in the list of interface nodes in the .subckt statement.

Second, it is not necessary to keep the node numbers of the subcircuit separate from the node numbers of the main circuit. For example, consider node 4 of the subcircuit UA741_op_amp. This node becomes XOA1.4 when the subcircuit is substituted for XOA1 and XOA2.4 when the subcircuit is substituted for XOA2. Neither of these node numbers will be confused with node 4 of the main circuit.

Third, some nodes have two numbers—one as part of the main circuit and one as part of a subcircuit. In particular, the nodes in the statements

```
XOA1  5  0  4  uA741_op_amp
```

and

```
.subckt  uA741_op_amp  1  2  5
```

are listed in the same order. So nodes 5, 0, and 4 will correspond to nodes OA1.1, OA1.2, and OA1.5 when uA741_op_amp is substituted for XOA1. PSpice will use the node numbers from the main circuit rather than the node numbers from the subcircuit. This means that

```
XOA1.Ib1  XOA1.1  0  70nA
```

becomes

```
XOA1.Ib1  5  0  70nA
```

```
Two Summing Amplifiers

V1    1   0   200mV
V2    2   0   125mV
V3    3   0   250mV
R1    1   5   50k
R2    2   5   25k
R3    5   4   100k
XOA1  5   0   4  uA741_op_amp
R4    4   6   25k
R5    3   6   10k
R6    6   7   50k
R7    7   0   100k
XOA2  6   0   7  uA741_op_amp

.subckt uA741_op_amp  1   2   5
*                     -   +   o
Ib1   1   0   70nA
Ib2   2   0   90nA
Vos   3   2   1mV
Ri    1   3   2Meg
E     4   0   1  3   -200000
Ro    4   5   75
.ends uA741_op_amp

.end
```

Figure 6-24 PSpice input file for the summing amplifier.

and

$$\text{XOA2.Ib1} \quad \text{XOA2.1} \quad 0 \quad 70\text{nA}$$

becomes

$$\text{XOA2.Ib1} \quad 6 \quad 0 \quad 70\text{nA}$$

Figure 6-25 shows the PSpice device list after the subcircuits have been substituted for the operational amplifiers. PSpice does not show this list to the user but does use the device names and node names from the list in output files and error messages.

```
Two Summing Amplifiers

V1    1   0   200mV
V2    2   0   125mV
V3    3   0   250mV
R1    1   5   50k
R2    2   5   25k
R3    5   4   100k

XOA1.Ib1   5   0    70nA
XOA1.Ib2   0   0    90nA
XOA1.Vos   XOA1.3   0    1mV
XOA1.Ri    5   XOA1.3   2Meg
XOA1.E     XOA1.4   0    5   XOA1.3    -200000
XOA1.Ro    XOA1.4   4    75

R4    4   6   25k
R5    3   6   10k
R6    6   7   50k
R7    7   0   100k

XOA2.Ib1   6   0    70nA
XOA2.Ib2   0   0    90nA
XOA2.Vos   XOA2.3   0    1mV
XOA2.Ri    6   XOA2.3   2Meg
XOA2.E     XOA2.4   0    6   XOA2.3    -200000
XOA2.Ro    XOA2.4   7    75

.end
```

Figure 6-25 PSpice list showing substitution of subcircuits.

Figure 6-26 shows PSpice output corresponding to the input file in Figure 6-24. Notice that $v_1 = 200$ mV, $v_2 = 125$ mV, and $v_3 = 250$ mV, so it is expected that $v_4 = -900$ mV and $v_o = 550$ mV if the operational amplifiers are ideal. PSpice shows that $v_4 = -886$ mV and $v_o = 534.6$ mV when the operational amplifiers are typical μA741s.

NODE	VOLTAGE	NODE	VOLTAGE
(1)	.2000	(2)	.1250
(3)	.2500	(4)	-.8860
(5)	.0010	(6)	997.3E-06
(7)	.5334	(XOA1.3)	.0010
(XOA1.4)	-.8893	(XOA2.3)	-.0010
(XOA2.4)	.5346		

Figure 6-26 PSpice output for the summing amplifier.

EXERCISE 6-10

Replace the subcircuit uA741_op_amp in Figure 6-24 with the subcircuit ideal_op_amp from Table 6-3. Make corresponding changes to the lines describing XOA1 and XOA2. Use PSpice to confirm that $v_4 = -900$ mV and $v_o = 550$ mV when the operational amplifiers are ideal.

EXERCISE 6-11

Modify the subcircuit uA741_op_amp to obtain a subcircuit TL051_op_amp that represents a typical TL051 operational amplifier. (See Table 6-1.) Use PSpice to find the node voltages v_o and v_4 of the summing amplifier in Figure 6-23a when $v_1 = 200$ mV, $v_2 = 125$ mV, $v_3 = 250$ mV, and the operational amplifiers are all a typical TL051.

EXERCISE 6-12

Redesign the summing amplifier of Figure 6-23a by dividing all the resistor values by 10. Notice that the performance of the summing amplifier depends on ratios of resistor values. Dividing all resistances by 10 does not change these ratios. The performance of the summing amplifier should be unchanged. Use PSpice to confirm that $v_4 = -900$ mV and $v_o = 55$ mV when the operational amplifiers are ideal and $v_1 = 200$ mV, $v_2 = 125$ mV, and $v_3 = 250$ mV. Use PSpice to find the node voltages v_o and v_4 when the operational amplifiers are all a typical uA741.

6-9 VERIFICATION EXAMPLE

The circuit in Figure 6V-1a was analyzed by writing and solving the following set of simultaneous equations

$$\frac{v_6}{10} + i_5 = 0$$

$$10\,i_5 = v_4$$

$$\frac{v_4}{10} + i_3 = i_2$$

$$3 = 5\,i_2 + 10\,i_3$$

$$20\,i_3 = v_6$$

(These equations use units of volts, milliamps, and k ohms.) A computer and the program Mathcad was used to solve these equations as shown in Figure 6V-1b. The solution of these equations indicates that

$$i_2 = -0.6 \text{ mA} \qquad i_3 = 0.6 \text{ mA} \qquad v_4 = -12 \text{ V}$$
$$i_5 = 1.2 \text{ mA} \quad \text{and} \quad v_6 = 12 \text{ V}$$

Are these voltages and currents correct?

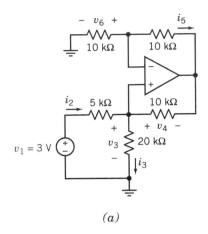

(a)

i2 := 0   i3 := 0   v4 := 0   i5 := 0   v6 := 0

Given

$$\frac{v6}{10} + i5 \approx 0 \qquad 10 \cdot i5 \approx v4 \qquad \frac{v4}{10} + i3 \approx i2$$

$$3 \approx 5 \cdot i2 + 10 \cdot i3 \qquad\qquad 20 \cdot i3 \approx v6$$

$$\text{Find } (i2, i3, v4, i5, v6) = \begin{bmatrix} -0.6 \\ 0.6 \\ -12 \\ -1.2 \\ 12 \end{bmatrix}$$

(b)

Figure 6V-1 *(a)* An example circuit and *(b)* computer analysis using MathCad.

Solution

Consider the voltage v_3. Using Ohm's Law

$$v_3 = 20\, i_3$$
$$= 20\,(0.6)$$
$$= 12\text{ V}$$

Remember resistances are in k ohms and currents in milliamps. Applying KVL to the mesh consisting of the voltage source and the 5-kΩ and 20-kΩ resistors gives

$$v_3 = 3 - 5\, i_2$$
$$= 3 - 5\,(-0.6)$$
$$= 6\text{ V}$$

Clearly, v_3 can not be both 12 and 6, so the values obtained for i_2, i_3, v_4, i_5, and v_6 cannot all be correct. Checking the simultaneous equations, we find that a resistor value has been entered incorrectly. The KVL equation corresponding to the mesh consisting of the voltage source and the 5-kΩ and 20-kΩ resistors should be

$$3 = 5\, i_2 + 20\, i_3$$

Note that $10\, i_3$ was incorrectly used in the fourth line of the Mathcad program of Figure

6V-1. After making this correction, i_2, i_3, v_4, i_5, and v_6 are calculated to be

$$i_2 = -0.2 \text{ mA} \qquad i_3 = 0.2 \text{ mA} \qquad v_4 = -4 \text{ V}$$
$$i_5 = 0.4 \text{ mA} \quad \text{and} \quad v_6 = 4 \text{ V}$$

Now

$$v_3 = 20\ i_3$$
$$= 20\ (0.2)$$
$$= 4$$

and

$$v_3 = 3 - 5\ i_2$$
$$= 3 - 5\ (-0.2)$$
$$= 4$$

This agreement suggests that the new values of i_2, i_3, v_4, i_5, and v_6 are correct. As an additional check, consider v_5. First, Ohm's Law gives

$$v_5 = 10\ i_5$$
$$= 10\ (-0.4)$$
$$= -4$$

Next, applying KVL to the loop consisting of the two 10-kΩ resistors and the input of the operational amplifier gives

$$v_5 = 0 + v_4$$
$$= 0 + (-4)$$
$$= -4$$

This increases our confidence that the new values of i_2, i_3, v_4, i_5, and v_6 are correct.

6-10 DESIGN CHALLENGE SOLUTION

TRANSDUCER INTERFACE CIRCUIT

A customer wants to automate a pressure measurement, which requires converting the output of the pressure transducer to a computer input. This conversion can be done using a standard integrated circuit called an Analog-to-Digital Converter (ADC) (Dorf, 1993). The ADC requires an input voltage between 0 V and 10 V, while the pressure transducer output varies between −250 mV and 250 mV. Design a circuit to interface the pressure transducer with the ADC. That is, design a circuit that translates the range −250 mV to 250 mV to the range 0 V to 10 V.

Define the Situation

The situation is shown in Figure 6D-1.

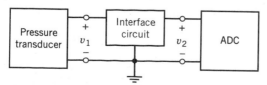

Figure 6D-1 Interfacing a pressure transducer with an Analog-to-Digital Converter (ADC).

The specifications state that

$$- 250 \text{ mV} \leq v_1 \leq 250 \text{ mV}$$

$$0 \text{ V} \leq v_2 \leq 10 \text{ V}$$

A simple relationship between v_2 and v_1 is needed so that information about the pressure is not obscured. Consider

$$v_2 = a \cdot v_1 + b$$

The coefficients, a and b, can be calculated by requiring that $v_2 = 0$ when $v_1 = -250$ mV and that $v_2 = 10$ V when $v_1 = 250$ mV, that is,

$$0 \text{ V} = a \cdot -250 \text{ mV} + b$$

$$10 \text{ V} = a \cdot 250 \text{ mV} + b$$

Solving these simultaneous equations gives $a = 20$ V and $b = 5$ V.

State the Goal

Design a circuit having input voltage v_1 and output voltage v_2. These voltages should be related by

$$v_2 = 20\, v_1 + 5 \text{ V}$$

Generate a Plan

Figure 6D-2 shows a plan (or a structure) for designing the interface circuit. The operational amplifiers are biased using $+15$-V and -15-V power supplies. The constant 5-V input is generated from the 15-V power supply by multiplying by a gain of 1/3. The input voltage, v_1, is multiplied by a gain of 20. The summer (adder) adds the outputs of the two amplifiers in order to obtain v_2.

Each block in Figure 6D-2 will be implemented using an operational amplifier circuit.

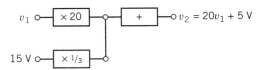

Figure 6D-2 A structure (or plan) for the interface circuit.

Take Action on the Plan

Figure 6D-3 shows one proposed interface circuit. Some adjustments have been made to the plan. The summer is implemented using the inverting summing amplifier from Figure 6-11d. The inputs to this inverting summing amplifier must be $-20v_i$ and -5 V instead of 20 v_i and 5 V. Consequently, an inverting amplifier is used to multiply v_1 by -20. A voltage follower is used to prevent the summing amplifier from loading the voltage divider. To make the signs work out correctly, the -15-V power supply provides the input to the voltage divider.

The circuit shown in Figure 6D-3 is not the only circuit that solves this design challenge. There are several circuits that implement

$$v_2 = 20v_1 + 5 \text{ V}$$

We will be satisfied with having found one circuit that does the job.

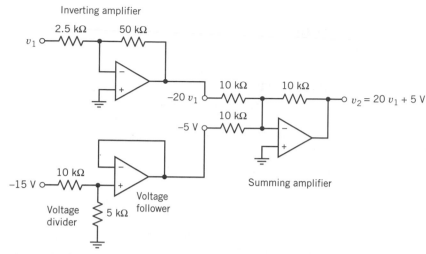

Figure 6D-3 One implementation of the interface circuit.

SUMMARY

With the invention of the electron tube in 1907, the field of electronics was born. Electronics became a very important field with myriad applications. In 1947 the transistor was invented, which revolutionized the field of electronics. With the small transistor available, miniature radios and calculators were made possible. By the 1960s integrated electronic circuits were available for use in applications requiring many circuits and functions.

Several models are available for operational amplifiers. Simple models are easy to use. Accurate models are more complicated. The simplest model of the operational amplifier is the ideal operational amplifier. The currents into the input terminals of an ideal operational amplifier are zero and the voltages at the input nodes of an ideal operational amplifier are equal. It is convenient to use node equations to analyze circuits that contain ideal operational amplifiers.

Operational amplifiers are used to build circuits that perform mathematical operations. Many of these circuits have been used so often that they have been given names. The inverting amplifier gives a response of the form $v_o = -Kv_i$ where K is a positive constant. The noninverting amplifier gives a response of the form $v_o = Kv_i$ where K is a positive constant. Another useful operational amplifier circuit is the noninverting amplifier with a gain of $K = 1$, often called a voltage follower or buffer. The output of the voltage follower faithfully follows the input voltage. The voltage follower reduces loading by isolating its output terminal from its input terminal. Figure 6-11 is a catalog of some of frequently used operational amplifier circuits.

Practical operational amplifiers have properties that are not included in the ideal operational amplifier. These include the input offset voltage, bias current, dc gain, input resistance and output resistance. More complicated models are needed to account for these properties. PSpice can be used to reduce the drudgery of analyzing operational amplifier circuits with complicated models.

TERMS AND CONCEPTS

Buffer Amplifier   Isolating amplifier used to separate the output from the input. Usually $v_o = 1\, v_s$.

Common Mode Rejection Ratio (CMRR)   Ratio of the gain A that multiplies the differential input $(v_1 - v_2)$ to the common mode gain A_{CM} that multiplies the common mode component of the input.

Electronics   The engineering field and industry that uses electron devices in circuits and systems.

Ideal Operational Amplifier   The simplest model of an operational amplifier. This model is characterized by $i_1 = 0$, $i_2 = 0$, and $v_1 = v_2$.

Integrated Circuit   Combination of interconnected circuit elements inseparably associated on or within a continuous semiconductor.

Inverting Amplifier   Amplifier with a relationship of the form $v_o = -Kv_s$, where K is a positive constant, v_o is the output voltage, and v_s is the source voltage.

Noninverting Amplifier   Amplifier with a relationship $v_o = Kv_s$, where K is a positive constant, v_o is the output voltage, and v_s is the source voltage.

Operational Amplifier   Amplifier with a high gain designed to be used with other circuit elements to perform a specified signal processing operation (often called an op amp).

Output Offset Voltage   A small, nonzero output voltage that occurs for an operational amplifier when the input to the circuit is zero.

Semiconductor   Electronic conductor with resistivity in the range between metals and insulators.

Slew Rate (SR)   Maximum rate at which the output voltage of an operational amplifier can change. It is normally expressed in $V/\mu s$.

Transistor   Active semiconductor device with three or more terminals.

Vacuum Tube   Electron tube evacuated so that its electrical characteristics are unaffected by the presence of residual gas or vapor.

Virtual Ground   Terminal in a circuit that appears to the observer to be essentially (virtually) connected to ground.

Voltage Follower   Amplifier with a voltage gain of one so that the output voltage follows the input voltage.

REFERENCES   Chapter 6

Dorf, Richard C. *Electrical Engineering Handbook,* A/D Converters, Chap. 31, CRC Press, Boca Raton, FL, 1993.

Graeme, Jerald. "Feedback Linearizes Current Source," *Electronic Design,* January 23, 1992, pp. 69–70.

Svoboda, James A. "Using Spreadsheets in Introductory Electrical Engineering Courses," *IEEE Transactions on Education,* November 1992, pp. 16–21.

PROBLEMS

Section 6-4 The Ideal Operational Amplifier

P 6.4-1 An operational amplifier circuit is shown in Figure P 6.4-1. Assume that the operational amplifier is ideal. Show that

(a) $v_2 = 0$

(b) $\dfrac{v_o}{R_b} + \dfrac{v_s}{R_a} = 0$

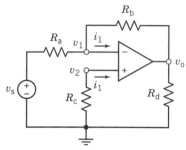

Figure P 6.4-1 v_1, v_2, and v_o are node voltages.

P 6.4-2 An operational amplifier circuit is shown in Figure 6-8. It is proposed to use a μA741 biased with $+15$-V and -15-V power supplies. The source voltage, v_s, has a maximum value of 12 V as shown in Figure P 6.4-2. Determine the minimum value of R_L for which it is appropriate to model the operational amplifier as an ideal operational amplifier.

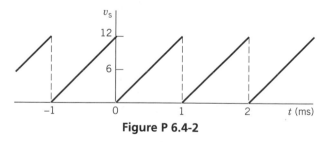

Figure P 6.4-2

P 6.4-3 Determine the permissible range of R for the circuit shown in Figure P 6.4-3 when the op amp saturates at $V_o = \pm 14$ V. Assume an ideal op amp.

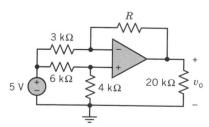

Figure P 6.4-3

Section 6-5 Nodal Analysis of Circuits Containing Ideal Operational Amplifiers

P 6.5-1 Determine the gain ratio v_2/v_1 for the circuit shown in Figure P 6.5-1. Assume an ideal op amp.

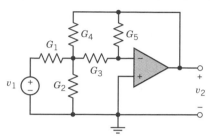

Figure P 6.5-1

P 6.5-2 For the circuit shown in Figure P 6.5-2, calculate the voltage v_o and the current i_o. Assume an ideal op amp.

Answer: $v_o = 12$ V, $i_o = 5.6$ mA

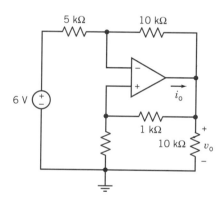

Figure P 6.5-2

P 6.5-3 One common application uses an operational amplifier to drive a low-resistance load, R_L, with a voltage source with a high resistance R_s as shown in Figure P 6.5-3. This circuit is called a voltage-to-current converter. Assume an ideal op amp and find i_L/v_s.

Answer: $i_L/v_s = 1/R$

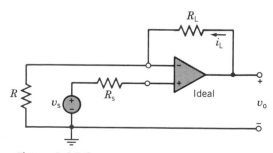

Figure P 6.5-3 A voltage-to-current converter.

P 6.5-4 If $R_1 = 4.8 \text{ k}\Omega$ and $R_2 = R_4 = 30 \text{ k}\Omega$, find v_o/v_s for the circuit shown in Figure P 6.5-4 when $R_3 = 1 \text{ k}\Omega$.
Answer: $v_o/v_s = -200$

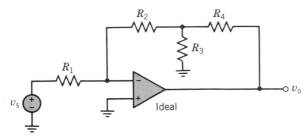

Figure P 6.5-4

P 6.5-5 For the circuit shown in Figure P 6.5-5, show that the magnitude of the load current i_L is 100 times the source current for any load that satisfies $R_L \ll R_2$. Use the ideal operational amplifier and let $R_1 = 100 \text{ k}\Omega$ and $R_2 = 1 \text{ k}\Omega$.

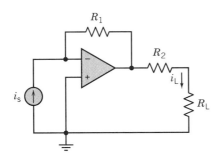

Figure P 6.5-5 A current multiplier circuit.

P 6.5-6 The circuit shown in Figure P 6.5-6 includes a simple strain gauge. The resistor R changes its value by ΔR when it is twisted or bent. Derive a relation for the voltage gain v_o/v_s and show that it is proportional to the fractional change in R, namely $\Delta R/R_o$.

Answer: $v_o = -v_s \dfrac{R_o}{R_o + R_1} \dfrac{\Delta R}{R_o}$

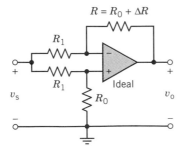

Figure P 6.5-6 A strain gauge circuit.

P 6.5-7 Find $v_o(t)$ for the circuit of Figure P 6.5-7 when $v_s = 175 \text{ mV}$. Assume an ideal op amp.
Answer: $v_o = 4.375 \text{ V}$

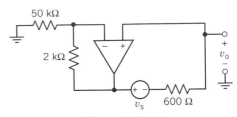

Figure P 6.5-7

P 6.5-8 Find $v_o(t)$ for the circuit of Figure P 6.5-8 when $v_s = 50 \text{ mV}$. Assume an ideal op amp.
Answer: $v_o = 11.05 \text{ V}$

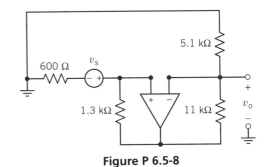

Figure P 6.5-8

P 6.5-9 Show that the circuit shown in Figure P 6.5-9 behaves as a current amplifier with a gain of

$$\frac{i_L}{i_s} = -\frac{R_1 + R_2}{R_1}$$

when the op amp is ideal.

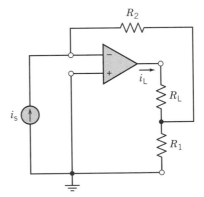

Figure P 6.5-9 A current amplifier circuit.

P 6.5-10 An operational amplifier can be used to convert a current source to a voltage, v_o, as shown in Figure P 6.5-10. Find the ratio v_o/i_s, using an ideal operational amplifier.

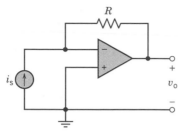

Figure P 6.5-10 A current-to-voltage converter.

P 6.5-11 For the circuit of Figure P 6.5-11, find the ratio v_o/v_s as a function of the four resistor values R_1, R_2, R_3, and R_4. Use an ideal op amp.

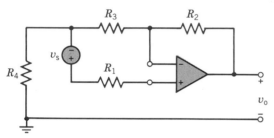

Figure P 6.5-11

P 6.5-12 A circuit with two sources is shown in Figure P 6.5-12. Find v_o when the operational amplifier is assumed to be ideal.
Answer: $v_o = -2$ V

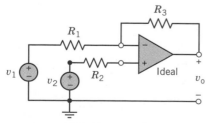

Figure P 6.5-12

P 6.5-13 Assuming an ideal op amp, find the output voltage v_o in terms of the two input voltages v_1 and v_2 of the circuit of Figure P 6.5-13.

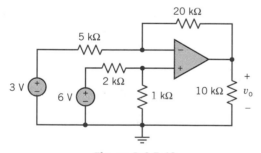

Figure P 6.5-13

P 6.5-14 For the circuit of Figure P 6.5-14 determine the required relationship between R and R_L so that the load current i_L is independent of R_L. Find an expression for i_L when $v_1 = V_o$, a constant voltage, for an ideal op amp.

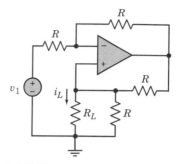

Figure P 6.5-14 Constant current source circuit.

P 6.5-15 Find v and i for the circuit of Figure P 6.5-15. Assume an ideal op amp.

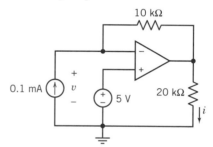

Figure P 6.5-15

P 6.5-16 Find v_o and i_o for the circuit of Figure P 6.5-16. Assume an ideal op amp.

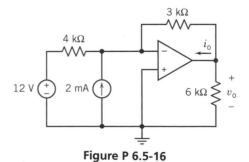

Figure P 6.5-16

P 6.5-17 Determine the Thévenin equivalent circuit for the circuit of Figure P 6.5-17 at the output terminals.

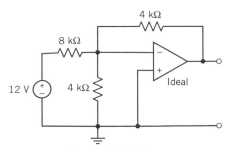

Figure P 6.5-17

P 6.5-18 Find v_o and i_o for the circuit of Figure P 6.5-18. Assume ideal op amps.

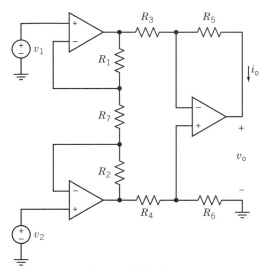

Figure P 6.5-18

P 6.5-19 Find v_o for the circuit shown in Figure P 6.5-19. Assume an ideal operational amplifier.

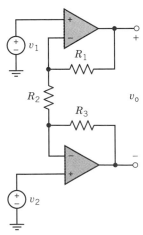

Figure P 6.5-19

P 6.5-20 A linear current source circuit provides a load current $i(t)$ as shown in Figure P 6.5-20. A voltage reference V_R appears across a potentiometer R and then $v = xV_R$ (Graeme, 1992). The resistance xR adjusts the op amp circuit gain. Show that the current is $i = xV_R/R_s$. Determine the range of current that can be achieved. Assume an ideal op amp.

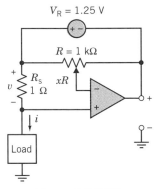

Figure P 6.5-20

P 6.5-21 An inverting amplifier is shown in Figure P 6.5-21. Determine the value of R required so that $v_o/v_s = -1000$. Assume an ideal op amp.

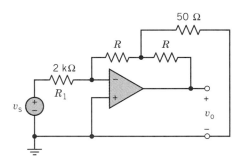

Figure P 6.5-21

Section 6-6   Design Using Operational Amplifiers

P 6.6-1 Design the operational amplifier circuit in Figure P 6.6-1 so that

$$v_{out} = r \cdot i_{in}$$

where

$$r = 20 \frac{V}{mA}$$

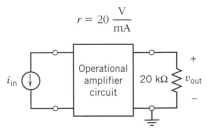

Figure P 6.6-1

P 6.6-2 Design the operational amplifier circuit in Figure P 6.6-2, so that

$$i_{out} = g \cdot v_{in}$$

where

$$g = 2 \frac{mA}{V}$$

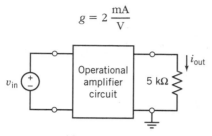

Figure P 6.6-2

P 6.6-3 Design the operational amplifier circuit in Figure P 6.6-3 so that

$$v_{out} = 5 \cdot v_1 + 2 \cdot v_2$$

Figure P 6.6-3

P 6.6-4 Design the operational amplifier circuit in Figure P 6.6-3 so that

$$v_{out} = 5 \cdot (v_1 - v_2)$$

P 6.6-5 Design the operational amplifier circuit in Figure P 6.6-3 so that

$$v_{out} = 5 \cdot v_1 - 2 \cdot v_2$$

P 6.6-6 The voltage divider shown in Figure P 6.6-6 has a gain of

$$\frac{v_{out}}{v_{in}} = \frac{-10 \ k\Omega}{5 \ k\Omega + (-10 \ k\Omega)}$$

$$= 2$$

Design an operational amplifier circuit to implement the -10-kΩ resistor.

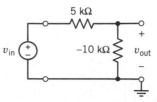

Figure P 6.6-6 A circuit with a negative resistor.

P 6.6-7 Design the operational amplifier circuit in Figure P 6.6-7 so that

$$i_{in} = 0 \quad \text{and} \quad v_{out} = 3 \cdot v_{in}$$

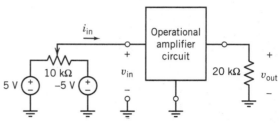

Figure P 6.6-7

P 6.6-8 A voltage-subtracting circuit is shown in Figure P 6.6-8. (a) Show that v_o can be expressed as

$$v_o = \frac{1 + R_2/R_1}{1 + R_3/R_4} v_2 - \frac{R_2}{R_1} v_1$$

(b) Design a circuit with an output $v_o = 4v_2 - 11v_1$.
Answer: $R_1 = 10 \ k\Omega$ (one solution)
$R_2 = 110 \ k\Omega$
$R_3 = 20 \ k\Omega$
$R_4 = 10 \ k\Omega$

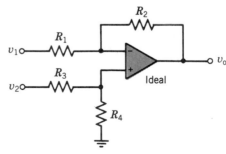

Figure P 6.6-8 A voltage subtracting circuit.

P 6.6-9 Design an operational amplifier circuit with output $v_o = 6v_1 + 2v_2$, where v_1 and v_2 are input voltages. Assume an ideal op amp.

P 6.6-10 For the circuit shown in Figure P 6.6-10, find v_L if $R_L = 12 \ k\Omega$. Assume ideal operational amplifiers.
Answer: $v_L = 1 \ V$

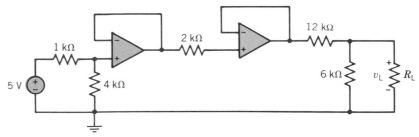

Figure P 6.6-10

P 6.6-11 The input voltage is $v_s = 62.4$ mV for the circuit of Figure P 6.6-11. It is desired that $v_o = 13.0$ V. Determine R_1 to achieve the desired output. Assume ideal op amps.
Answer: $R_1 = 220\ \Omega$

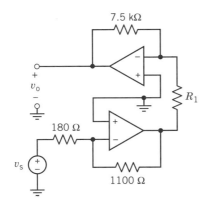

Figure P 6.6-11

P 6.6-12 Find the required value of R for the circuit of Figure P 6.6-12 so that $v_o = 12$ V when $v_s = -4$ V.

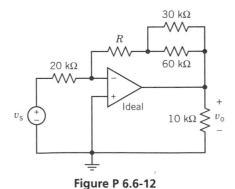

Figure P 6.6-12

P 6.6-13 For the op-amp circuit shown in Figure P 6.6-13, find and list all the possible voltage gains that can be achieved by connecting the resistor terminals to either the input or the output voltage terminals. Assume an ideal op amp.

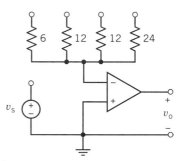

Figure P 6.6-13 Resistances in kΩ.

P 6.6-14 A source voltage delivers a voltage of 1 cos ωt V and a maximum current of 10 mA. Design an amplifier circuit that will develop an output voltage of -10 cos ωt V when driven by the source.

P 6.6-15 Determine v_o for the circuit of Figure P 6.6-15 when all the op amps are ideal.

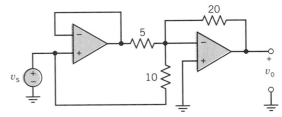

Figure P 6.6-15 Resistances in kΩ.

P 6.6-16 (a) Determine an expression for v_o in terms of v_a, v_b, and the resistors in Figure P 6.6-16. (b) What function does this circuit perform? (c) Design an equivalent circuit that requires only two operational amplifiers and few resistors. The voltages shown are referenced to ground. Assume ideal operational amplifiers.

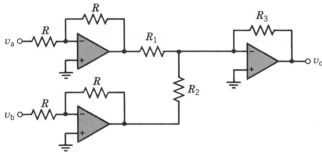

Figure P 6.6-16

P 6.6-17 Assuming that all the op amps are ideal, determine v_o in terms of v_1 and v_s for the circuit shown in Figure P 6.6-17.

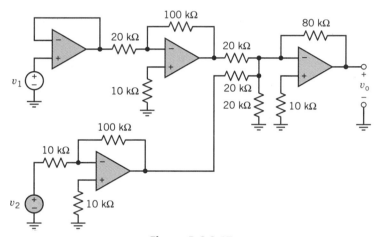

Figure P 6.6-17

P 6.6-18 Find v_o/v_s for the circuit of Figure P 6.6-18. Assume that the operational amplifiers are ideal.
Answer: $v_o/v_s = 22$

P 6.6-19 The two-op amp circuit shown in Figure P 6.6-19 is used to obtain an output voltage in terms of two input voltages. Obtain an expression for v_o. Select R_1, R_2, and R_G so that v_o is 10 times $(v_1 - v_2)$. Assume ideal op amps.

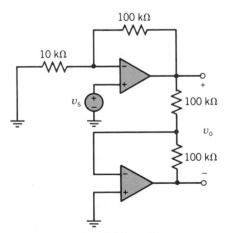

Figure P 6.6-18

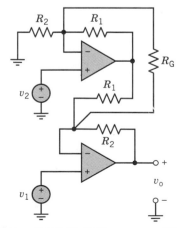

Figure P 6.6-19 Difference amplifier.

P 6.6-20 Determine v_o for the circuit shown in Figure P 6.6-20. Assume ideal op amps.

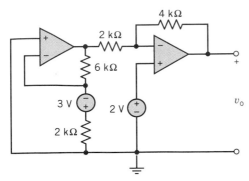

Figure P 6.6-20

P 6.6-21 For the circuit of Figure P 6.6-21, find the value of R such that $v_o = -100v_s$ when $R_1 = 10$ kΩ. Assume an ideal op amp.

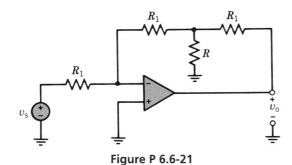

Figure P 6.6-21

P 6.6-22 Determine v_o and i_o for the op amp circuit shown in Figure P 6.6-22. Assume an ideal op amp.

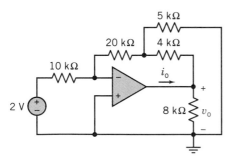

Figure P 6.6-22

P 6.6-23 An op-amp circuit is shown in Figure P 6.6-23. Assume ideal op amps and find v_x so that $v_1 - v_2 = 13$ V.

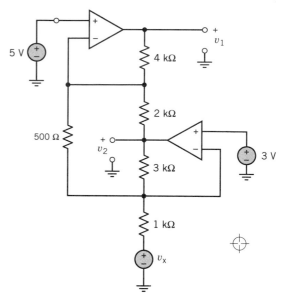

Figure P 6.6-23

P 6.6-24 Determine the required value of R so that $v_o = -1.95$ for the circuit of Figure P 6.6-24. Assume an ideal op amp.

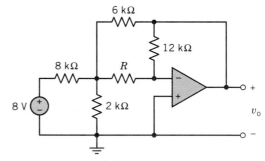

Figure P 6.6-24

P 6.6-25 Find v_o and i_o for the circuit of Figure P 6.6-25 when $R_1 = 7$ kΩ, $R_2 = 98$ kΩ, $R_3 = 10$ kΩ, $R_4 = 20$ kΩ, $R_L = 2$ kΩ, and $v_s = 0.1$ V.

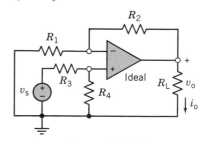

Figure P 6.6-25

P 6.6-26   The circuit shown in Figure P 6.6-26 is called the inverted R-2R ladder digital-to-analog converter (DAC). The input to this circuit is a binary code represented by $b_1\, b_2\, \ldots\, b_n$, where b_i is either 1 or 0. Each switch shown in the figure is controlled by only one of the digits of the binary code. If $b_i = 1$, the switch will be at the left position, whereas if $b_i = 0$, the switch will be at right position. Depending on the position of the switch, each current I_i is diverted either to true ground bus (adding to I^+) or the virtual ground bus (adding to I^-).

(a)   Show that $I = V_R/R$ regardless of the digital input code.

(b)   Show that the output voltage can be expressed as

$$V_o = -\frac{R_f}{R}\, V_R\, (b_1\, 2^{-1} + b_2\, 2^{-2} + \cdots$$
$$+ b_{n-1}\, 2^{-n+1} + b_n\, 2^{-n})$$

(c)   Show that $I^+ + I^- = (1 - 2^{-n})\, V_R/R$ regardless of the binary code.

(d)   Given $R_f = R = 10\ \text{k}\Omega$, $V_R = -16$ V and assuming a four digit binary input code, find the output voltage V_o for each combination of the input code, ranging from 0000 to 1111. Explain the relationship between the output and the input code.

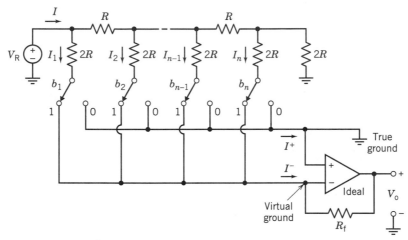

Figure P 6.6-26   Digital to Analog converter circuit.

Section 6-7   Characteristics of the Practical Operational Amplifier

P 6.7-1   Consider the inverting amplifier shown in Figure P 6.7-1. The operational amplifier is a typical OP-07E (Table 6-2). Use the offsets model of the operational amplifier to calculate the output offset voltage. (Recall that the input, v_{in}, is set to zero when calculating the output offset voltage.)
Answer: 0.45 mV

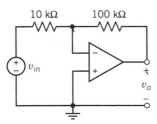

Figure P 6.7-1

P 6.7-2   Consider the noninverting amplifier shown in Figure P 6.7-2. The operational amplifier is a typical LF351 (Table 6-2). Use the offsets model of the operational am-

plifier to calculate the output offset voltage. (Recall that the input, v_{in}, is set to zero when calculating the output offset voltage.)

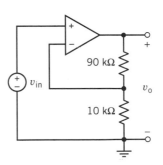

Figure P 6.7-2

P 6.7-3   Consider the inverting amplifier shown in Figure P 6.7-3. Use the finite gain model of the operational amplifier (Figure 6-20c) to calculate the gain of the inverting amplifier. Show that

$$\frac{v_o}{v_{\text{in}}} = \frac{R_{\text{in}}(R_o - AR_2)}{(R_1 + R_{\text{in}})(R_o + R_2) + R_1 R_{\text{in}}(1 + A)}$$

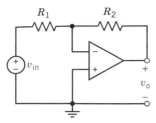

Figure P 6.7-3

P 6.7-4 Consider the inverting amplifier shown in Figure P 6.7-3. Suppose the operational amplifier is ideal, $R_1 = 5$ kΩ, and $R_2 = 50$ kΩ. The gain of the inverting amplifier will be

$$\frac{v_o}{v_{in}} = -10$$

Use the results of Problem P 6.7-3 to find the gain of the inverting amplifier in each of the following cases:

(a) The operational amplifier is ideal, but 2% resistors are used and $R_1 = 5.1$ kΩ and $R_2 = 49$ kΩ.
(b) The operational amplifier is represented using the finite gain model with $A = 200{,}000$, $R_i = 2$ MΩ, and $R_o = 75\Omega$; $R_1 = 5$ kΩ and $R_2 = 50$ kΩ.
(c) The operational amplifier is represented using the finite gain model with $A = 200{,}000$, $R_i = 2$ MΩ, and $R_o = 75\Omega$; $R_1 = 5.1$ kΩ and $R_2 = 49$ kΩ.

P 6.7-5 The circuit in Figure P 6.7-5 is called a direct coupled difference amplifier and is used for instrumentation circuits. The output of a measuring element is represented by the common mode signal v_{cm} and the differential signal $(v_n - v_p)$. Using an ideal operational amplifier, show that

$$v_o = -\frac{R_4}{R_1}(v_n - v_p)$$

when

$$\frac{R_4}{R_1} = \frac{R_3}{R_2}$$

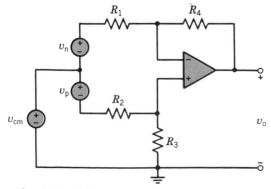

Figure P 6.7-5 A direct coupled difference amplifier.

P 6.7-6 A noninverting amplifier circuit is shown in Figure P 6.7-6. Using the finite gain model of the operational amplifier when $A > 10^5$, prove that the input resistance seen by the source voltage is

$$R_{in} \cong \left(\frac{AR_1}{R_1 + R_2 + R_o}\right) R_i$$

and the output resistance at terminals a–b is

$$R_{out} \cong \left(\frac{R_1 + R_2}{AR_1}\right) R_o$$

Hint: Assume that $R_i \gg R_2$ and use the finite gain model of Figure 6-20a.

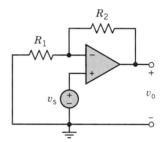

Figure P 6.7-6

P 6.7-7 A circuit has three inputs v_1, v_2, and v_3 with $|v_i| \leq 1$ V for $i = 1, 2, 3$. The op amps have $i_{sat} = 2$ mA and $v_{sat} = 15$ V. Design an op amp circuit to produce

(a) $v_o = -(4v_1 + 2 v_2)$
(b) $v_o = 4v_1 + 2v_2 - 3v_3$

P 6.7-8 An op amp circuit is shown in Figure P 6.7-8. Determine the maximum value of v_1 to retain linear operation when the output voltage saturates at $v_o = \pm 12$ V. Assume an ideal op amp in the linear region.

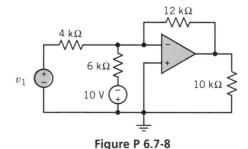

Figure P 6.7-8

P 6.7-9 An op amp circuit is shown in Figure P 6.7-9 with a load resistor. It is desired to determine the output voltage for (a) an ideal op amp and (b) an op amp with $A = 10^4$, $R_i = 200$ kΩ, and $R_o = 5$ kΩ. Compare the results.

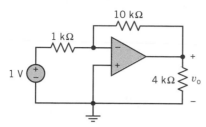

Figure P 6.7-9

P 6.7-10 Consider the inverting amplifier circuit shown in Figure P 6.7-10. Determine the input resistance seen by the source, v_s, using the finite gain model of Figure 6-20a. Assume that $A \geq 10^5$ and R_o is negligible.
Answer: $R_{in} \cong R_1'$

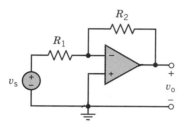

Figure P 6.7-10

P 6.7-11 An inverting operational amplifier is shown in Figure P 6.7-11. (a) Using the finite gain model of Figure 6.20a, show that the output resistance seen at the output terminal is

$$R_{out} \cong \frac{R_o}{Ak}$$

where

$$k = \frac{R_1}{R_1 + R_2}$$

Assume that R_i is much greater than R_1 and R_2. (b) If $R_o = 1000 \ \Omega$, $A = 10^5$, and $k = 0.1$, find R_{out} at the output terminals.
Answer: (b) $R_{out} = 0.1 \ \Omega$

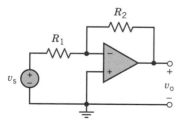

Figure P 6.7-11

P 6.7-12
(a) Determine the voltage ratio v_o/v_1 and the input resistance for the circuit shown in Figure P 6.7-12. Use the finite gain model.
(b) Evaluate the voltage ratio and the input resistance when $R_1 = R_2 = 10 \ k\Omega$ and the op amp has $A = 10^4$ and $R_i = 100 \ k\Omega$. Assume R_o is negligible.

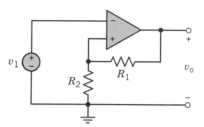

Figure P 6.7-12

P 6.7-13 A sensor is normally connected to a meter with a resistance equal to 500 Ω. The sensor is represented by a voltage source in series with R_s, a 500-Ω resistance. An engineer designs a circuit with a buffer amplifier with an input resistance of 1 MΩ and the gain A of the operational amplifier is 10^5. Find (a) the ratio v_o/v_s with and without the amplifier and (b) the power gain using the buffer amplifier. (c) A power gain of 16 is desired. Determine the value of the meter resistance that will be required if the sensor resistance remains unchanged.

P 6.7-14 An operational amplifier circuit which appears to the load as a current source so that $i_L \cong v_s/R_s$ is shown in Figure P 6.7-14. (a) Using a finite gain model, show that the output resistance seen at the terminals of R_L is $R_{out} \cong AR_s + R_o \approx AR_s$. (b) Find R_{out} when $A = 10^5$, $R_s = 10 \ k\Omega$, and $R_o = 1 \ k\Omega$.
Answer: (b) $R_{out} = 10^9 \ \Omega$

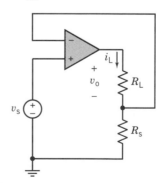

Figure P 6.7-14 The output is the voltage across R_L.

PSpice PROBLEMS

SP 6-1 Consider the op amp circuit of Figure 6-22.

(a) Determine v_o/v_s for an ideal op amp.
(b) Determine R_{in} and v_o/v_s when $R_i = 80$ kΩ, $R_o = 100$ Ω, $A = 10^4$, $R_1 = 1$ kΩ, and $R_L = 5$ kΩ.

SP 6-2 Determine the output voltage for Problem 6.7-9 using PSpice for both the ideal and the finite gain case.

SP 6-3 Determine the voltage ratio v_o/v_s and the input resistance R_{in} for the circuit of Figure 6-22 when $R_i = 100$ kΩ, $A = 10^4$, $R_1 = 1$ kΩ, $R_o = 1$ kΩ and $R_L = 10$ kΩ.

SP 6-4 Determine the output resistance R_{out} for the op amp circuit of P 6.7-14.

SP 6-5 Determine v_o/v_s and R_{in} for the noninverting circuit of Figure 6-22 when the op amp has $R_i = 100$ kΩ, $R_o = 100$ Ω, and $A = 10^4$. The circuit resistors are $R_1 = 10$ kΩ and $R_L = 10$ kΩ.

SP 6-6 Determine the voltage ratio v_o/v_s, R_{in}, and R_{out} of the circuit of Exercise 6-8 for the practical values given. Compare the results with those of an ideal op amp.

SP 6-7 Use PSpice to obtain the input resistance, output resistance, and voltage ratio for the op amp circuit of Figure SP 6-7. Assume an ideal op amp.

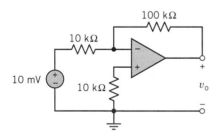

Figure SP 6-7

SP 6-8 A circuit with its nodes identified is shown in Figure SP 6-8. Determine v_{34}, v_{23}, v_{50}, and i_o.

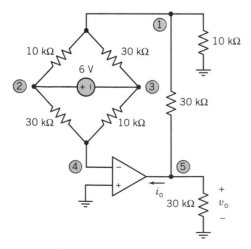

Figure SP 6-8 Bridge circuit.

SP 6-9 Use PSpice to analyze the VCCS shown in Figure SP 6-9. Consider two cases:

(a) The operational amplifier is ideal.
(b) The operational amplifier is a typical μA741 represented by the "offsets and finite gain" model.

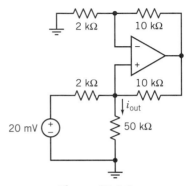

Figure SP 6-9

SP 6-10 Use PSpice to analyze the noninverting summing amplifier shown in Figure SP 6-10. Consider two cases:

(a) The operational amplifier is ideal.
(b) The operational amplifier is a typical μA741 represented by the "offsets and finite gain" model.

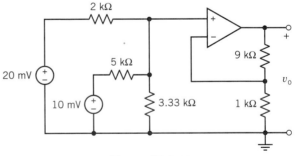

Figure SP 6-10

VERIFICATION PROBLEMS

VP 6-1 A computer analysis program for the circuit of Figure VP 6-1 states that $i_o = 312.6 \ \mu A$ when $i_s = 24 \ \mu A$. Verify this result.

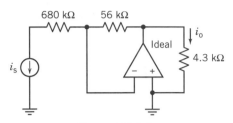

Figure VP 6-1

VP 6-2 A laboratory report states that the output voltage is $v_o = +1.5$ V for the circuit of Figure VP 6-2. Determine if this report is correct. Assume ideal op amps.

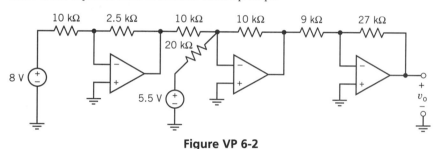

Figure VP 6-2

VP 6-3 A student uses a computer to determine $v_o = -4$ V for the circuit of Figure VP 6-3. Verify this result.

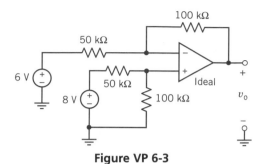

Figure VP 6-3

DESIGN PROBLEMS

DP 6-1 Design the operational amplifier circuit in Figure DP 6-1 so that

$$i_{out} = \frac{1}{4} \cdot i_{in}$$

Figure DP 6-1

DP 6-2 Design an inverting amplifier with a gain of -100 and an input resistance seen by the voltage source of 1 kΩ. Use a μA741 op amp for the circuit.

DP 6-3 An operational amplifier has a very high input resistance, and therefore an operational amplifier circuit is a good choice for building a voltmeter as shown in Figure DP 6-3. The voltage to be measured is v_s. The voltmeter is represented by resistance R_m. If the meter requires a current of 0.1 mA for full-scale deflection, find the appropriate resistor R for the meter to record in the range 0 to 1 V when $R_m = 10$ kΩ. Assume an ideal operational amplifier.

Answer: $R = 10$ kΩ.

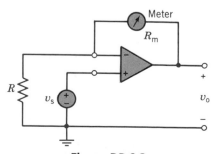

Figure DP 6-3

DP 6-4 A circuit used to measure temperature is shown in Figure DP 6-4. The temperature sensor has a resistance R_T that varies with temperature according to $R_T = 1000e^{-T/25°C} \, \Omega$ where $T =$ temperature in °C. The output voltage, v_o, is 0 V at $-55°C$.

(a) Determine R_x required.
(b) A full-scale deflection of the meter is desired at 125°C and is achieved when the meter current is 1 mA. Determine the required R_y.

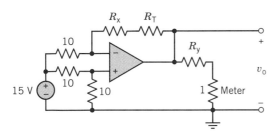

Figure DP 6-4 Temperature measurement circuit. Resistances in kΩ.

DP 6-5 A microphone has an unloaded voltage $v_s = 20$ mV, as shown in Figure DP 6-5a. An op amp is available as shown in Figure DP 6-5b. It is desired to provide an output voltage of 4 V. Design an inverting circuit and a noninverting circuit and contrast the input resistance at terminals x–y seen by the microphone. Which configuration would you recommend in order to achieve good performance in spite of changes in the microphone resistance R_s?

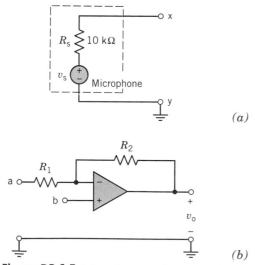

(a)

(b)
Figure DP 6-5 Microphone and op amp circuit.

DP 6-6 An amplifier circuit is shown in Figure DP 6-6. Assume an ideal op amp and determine the required resistance R so that the magnitude of the input resistance is 1 MΩ. Discuss the sign of the input resistance. Note carefully the positive and negative terminals of the op amp.

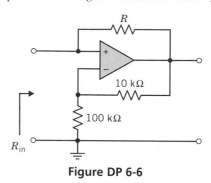

Figure DP 6-6

DP 6-7 Determine the resistance R in Figure DP 6-7 so that the output resistance is 570 Ω. Assume an ideal op amp. Note the positive and negative terminals of the op amp.

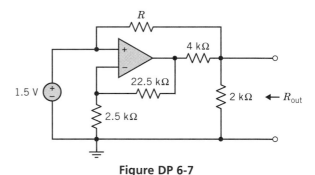

Figure DP 6-7

CHAPTER 7

ENERGY STORAGE ELEMENTS

PREVIEW

The energy storage in electric circuit elements is another aspect in the development of flexible and useful circuits. In this chapter we describe the characteristics of two energy storage elements: the inductor and the capacitor. These two terminal elements have been widely used by electrical engineers for over 100 years.

We may describe the storage of energy in electrical elements as analogous to the storage of information on a tablet or in a file drawer. As will be seen, the stored energy can later be retrieved and used for complex purposes. With the addition of inductors and capacitors to the familiar resistor, we will be ready to construct important and useful electric circuits. Since these circuits often contain one or more switches that open or close at a specified time, we also consider the effects of these switch changes on the circuit's behavior.

7-1 DESIGN CHALLENGE

INTEGRATOR AND SWITCH

This design challenge involves an integrator and a voltage-controlled switch.

An integrator is a circuit that performs the mathematical operation of integration. The output of an integrator, say $v_o(t)$, is related to the input of the integrator, say $v_s(t)$, by the equation

$$v_o(t_2) = K \cdot \int_{t_1}^{t_2} v_s(t) \, dt + v_o(t_1) \tag{7D-1}$$

The constant K is called the gain of the integrator.

Integrators have many applications. One application of an integrator is to measure an interval of time. Suppose $v_s(t)$ is a constant voltage. Then

$$v_o(t_2) = K \cdot (t_2 - t_1) \cdot v_s + v_o(t_1) \tag{7D-2}$$

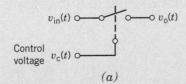

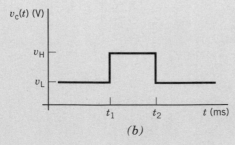

Figure 7D-1 The voltage-controlled switch. (a) Switch symbol. (b) Typical control voltage.

286

This equation indicates that the output of the integrator at time t_2 is a measure of the time interval $t_2 - t_1$.

Switches can be controlled electronically. Figure 7D-1 illustrates an electronically controlled SPST switch. The symbol shown in Figure 7D-1a is sometimes used to emphasize that a switch is controlled electronically. The node voltage $v_c(t)$ is called the control voltage. Figure 7D-1b shows a typical control voltage. This voltage-controlled switch is closed when $v_c(t) = v_H$ and open when $v_c(t) = v_L$. The switch shown in Figure 7D-1 is open before time t_1. It closes at time t_1 and stays closed until time t_2. The switch opens at time t_2 and remains open.

Consider Figure 7D-2. The voltage $v_c(t)$ controls the switch. The integrator converts the time interval $t_2 - t_1$ to a voltage that is displayed using the voltmeter. The time

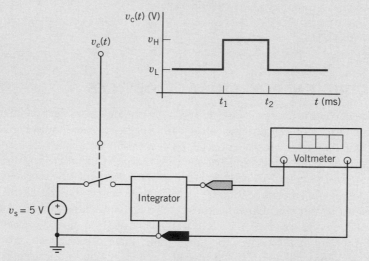

Figure 7D-2 Using an integrator to measure an interval of time.

interval to be measured could be as small as 5 ms or as large as 200 ms. The challenge is to design the integrator. The available components include:

standard 2% resistors (see Appendix F);
1 μF, 0.2 μF and 0.1 μF capacitors;
operational amplifiers;
+ 15-V and − 15-V power supplies;
1-kΩ, 10-kΩ, and 100-kΩ potentiometers;
voltage-controlled SPST switches.

Define the Situation

It is convenient to set the integrator output to zero at time t_1. The relationship between the integrator output voltage and the time interval should be simple. Accordingly, let

$$v_o(t_2) = \frac{10 \text{ V}}{200 \text{ ms}} \cdot (t_2 - t_1) \tag{7D-3}$$

Figure 7D-2 indicates that $v_s = 5$ V. Comparing Eqs. 7D-2 and 7D-3 yields

$$K \cdot v_s = \frac{10 \text{ V}}{200 \text{ ms}} \quad \text{and therefore} \quad K = 10 \frac{\text{V}}{\text{s}} \tag{7D-4}$$

State the Goal

Design an integrator satisfying both

$$K = 10 \; \frac{V}{s} \quad \text{and} \quad v_o(t_1) = 0 \qquad \qquad (7D\text{-}5)$$

An energy storage element—either a capacitor on an inductor—is needed to make an integrator. Energy storage elements are described in this chapter. We will return to this design challenge at the end of this chapter, after learning about capacitors and inductors.

7-2 ELECTRIC ENERGY STORAGE DEVICES

Storage of electric energy in devices has been pursued from the time of the Leyden jar. Part of the energy stored in these devices can later be released and supplied to a load.

In 1746 Pieter van Musschenbrock, professor of physics at Leyden, in Holland, stored a charge in a bottle containing water. This charge could then be released to provide a discharge or shock. The Leyden jar, the first manmade capacitor, had arrived, and it provided the first means of storing electric charge. It was shown that the charge stored was inversely proportional to the thickness of the glass and directly proportional to the surface area of the conductors. For a while the glass was thought to be essential, but that was proved wrong in 1762 when the first parallel-plate capacitor was made from two large boards covered with metal foil. It was the Leyden jar, used singly or grouped into banks of capacitors, that became the standard laboratory equipment.

With the development of the early capacitor, the concept of charge storage was explored further by Charles-Augustin de Coulomb and others as they developed the early theory of electricity. The study of electricity moved to quantitative description with the work of Coulomb, who succeeded in describing the ideas of electrostatics. Henry Cavendish proved the inverse-square law for electrostatics in 1772–1773 and performed de-

Figure 7-1 Hans C. Oersted (1777–1851), the first person to observe the magnetic effects of an electric current. Courtesy of Burndy Library.

Figure 7-2 Michael Faraday's electrical discoveries were not his only legacy; his published account of them inspired much of the scientific work of the later nineteenth century. His *Experimental Researches in Electricity* remains today as one of the greatest accounts of scientific work ever written. Courtesy of Burndy Library.

tailed measurements on capacitance and conductivity. However, he did not publish his findings, and they came to full light only when Maxwell published them in 1879.

Many scientists were also interested in the theory of magnetic force. Hans Christian Oersted, a professor at the University of Copenhagen, discovered the magnetic field associated with an electric current. Oersted established the fundamental fact that a compass needle was affected only when a current was flowing nearby and not by the mere presence of a voltage or charge. He concluded that the magnetic field was circular and was dispersed in the space around the wire. Oersted, who first published his discovery in Latin in a pamphlet dated July 21, 1820, is shown in Figure 7-1.

The results of Oersted's discovery spread quickly, and at a public demonstration in September 1820 one member of the audience was André-Marie Ampère. Within weeks, Ampère published a paper on the mutual action of an electric current on a magnet and showed that two coils carrying currents behaved like magnets.

Michael Faraday and Sir Humphrey Davy repeated Oersted's experiments in 1821 at the British Royal Institution. Faraday's experiments with magnetism and electricity continued for decades. Faraday constructed an iron ring with two coils wound on opposite sides. On August 29, 1831, Faraday connected one coil to a battery and the other coil to a galvanometer. He noted the transitory nature of the current induced in the second coil, which occurred only when the current in the first coil was started or stopped by connecting or disconnecting the battery. Faraday read an account of his work to the Royal Society on November 24, 1831, and published it in 1832. Faraday is shown in Figure 7-2. To honor Faraday, the unit of capacitance is named the farad.

During the same period, Joseph Henry, an American, was exploring the concepts of electromagnetism. During the years 1827 to 1831, Henry studied the effects of magnetism with electromagnets such as the one shown in Figure 7-3. Although Faraday deserves the credit for the discovery of electromagnetic induction between two coils, it was Henry who first discovered self-induction with a single coil. Henry noted the principle of self-induction when a vivid spark was produced as a long coil of wire was disconnected from a

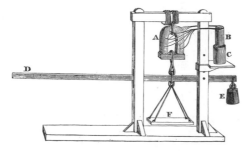

Figure 7-3 Joseph Henry's electromagnet. Direct current from the voltaic pile (B–C) was applied to a coil wound around an iron horseshoe core (A) to produce a powerful electromagnet. From Joseph Henry, *Galvanic Multiplier,* 1831. Courtesy of Burndy Library.

battery. Henry was honored by the use of his name for the unit of self-induction. In 1847 Henry became the first director of the Smithsonian Institution, where he served for 32 years.

7-3 | CAPACITORS

A capacitor is a two-terminal element that is a model of a device consisting of two conducting plates separated by a nonconducting material. Electric charge is stored on the plates, as shown in Figure 7-4, and the area between the plates is filled with a dielectric material. The capacitance value is proportional to the dielectric constant and surface area of the dielectric and is inversely proportional to its thickness. To obtain greater capacitance, a very thin structure of large area is required. For this configuration we may state the capacitance C as

$$C = \frac{\epsilon A}{d}$$

where ϵ is the dielectric constant, A the area of plates, and d the space between plates. Dielectric constants of some materials are given in Table 7-1. (The dielectric constant is that property that determines the energy stored per unit volume for unit voltage difference across a capacitor.) Another common term for dielectric constant is permittivity (Dorf, 1993).

We describe the charge $+q$ on one plate as identical to $-q$ on the other plate. The energy to move the charge q to the positive plate from the other plate is provided by the battery voltage v. The capacitor has been charged to the voltage v, which will be proportional to the charge q. Thus, we write

$$q = Cv \tag{7-1}$$

where C is the constant of proportionality, which is called the capacitance. The unit of capacitance is coulomb per volt and is called farad (F) in honor of Faraday. A capacitor is a linear element as long as it retains the relationship represented by Eq. 7-1.

Capacitance is a measure of the ability of a device to store energy in the form of separated charge or an electric field.

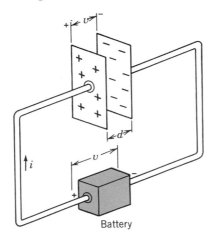

Figure 7-4 Capacitor connected to a battery.

Battery

Table 7-1

Relative Dielectric Constant, ϵ_r

Material	$\epsilon_r = \epsilon/\epsilon_0$
Glass	7
Nylon	2
Bakelite	5

ϵ_0 = permittivity of free space = 8.85×10^{-12} farad/meter. Dielectric constant is also called relative permittivity.

A capacitor can be used to model charge storage in any device. Most capacitors used in electronic equipment are linear. However, some are nonlinear. The charge associated with some reverse-biased semiconductor diodes, for example, varies as the 2/3 power of voltage because the distance across the junction, d, is a function of voltage. Unless otherwise stated, we will assume for our purposes that capacitors are linear.

When we first connect a battery to the capacitor shown in Figure 7-4, a current flows while the charges flow from one plate to the other. Since we use the conventional current i of positive charge flow, we can represent i as shown in Figure 7-4. Since the current is

$$i = \frac{dq}{dt}$$

we differentiate Eq. 7-1 to obtain

$$i = C\frac{dv}{dt} \tag{7-2}$$

Equation 7-2 is the current–voltage relationship for the model of a capacitor and can readily be shown to be a linear relation. Use Eq. 7-2 and the properties of superposition and homogeneity as described in Section 2-2 to show the capacitor relation is that of a linear element.

As the current flows toward the left-hand plate, it causes that plate (or its terminal) to have a positive voltage relative to the right-hand plate. The circuit symbol for a capacitor is shown in Figure 7-5. Again, the passive sign convention assumes that the current flows into the positive terminal of a capacitor as shown in Figure 7-5.

Figure 7-5 Circuit symbol for a capacitor.

In summary, we may define a *capacitor* as a two-terminal element whose primary purpose is to introduce capacitance into an electric circuit. Capacitance is defined as the ratio of the charge stored to the voltage difference between the two conducting plates or wires, $C = q/v$.

Capacitors use various dielectrics and are built in several forms (Trotter, 1988). Some common capacitors use impregnated paper for a dielectric; whereas others use mica sheets, ceramics, and organic and metal films. Miniature metal film capacitors are shown in Figure 7-6. Miniature hermetically sealed polycarbonate capacitors are shown in Figure 7-7.

Figure 7-6 Miniature metal film capacitors ranging from 1 mF to 50 mF. Courtesy of Electronic Concepts Inc.

Figure 7-7 Miniature hermetically sealed polycarbonate capacitors ranging from 1 μF to 50 μF. Courtesy of Electronic Concepts Inc.

Capacitors come in a wide range of values. Two pieces of insulated wire about an inch long when twisted together will have a capacitance of about 1 pF. On the other hand, a power supply capacitor about an inch in diameter and a few inches long could have a capacitance of 0.01 F.

Reflecting on Eq. 7-2, we note that a circuit will have a current that depends on the voltage across the capacitor, v. Remember that v implies the voltage is a function of time and could be written as $v(t)$. If the voltage is constant, then $i = 0$. If the voltage in Figure 7-5 is

$$v = Kt \ \text{V}$$

then, since K is a constant

$$i = C\frac{dv}{dt}$$
$$= CK \ \text{A}$$

Let us find the current through a capacitor when $v = 5 \sin t$ V in Figure 7-4:

$$i = C\frac{dv}{dt}$$
$$= 5C \cos t \ \text{A}$$

Example 7-1

Find the current for a capacitor $C = 1$ mF when the voltage across the capacitor is represented by the signal shown in Figure 7-8.

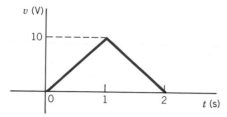

Figure 7-8 Waveform of the voltage across a capacitor for Example 7-1. The units are volts and seconds.

Solution

The voltage (with units of volts) is given by

$$
\begin{aligned}
v &= 0 & t \le 0 \\
&= 10t & 0 < t \le 1 \\
&= 20 - 10t & 1 < t \le 2 \\
&= 0 & t > 2
\end{aligned}
$$

Then, since $i = C\, dv/dt$, where $C = 10^{-3}$ F, we obtain

$$
\begin{aligned}
i &= 0 & t < 0 \\
&= 10^{-2} & 0 < t < 1 \\
&= -10^{-2} & 1 < t < 2 \\
&= 0 & t > 2
\end{aligned}
$$

Therefore, the resulting current is a series of two pulses of magnitudes 10^{-2} A and -10^{-2} A, respectively, as shown in Figure 7-9.

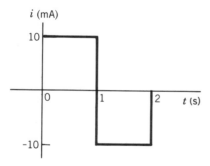

Figure 7-9 Current for Example 7-1.

Now consider the waveform shown in Figure 7-10, where the voltage changes from a constant voltage of zero to another constant voltage of 1 over an increment of time Δt. Since $i = C\, dv/dt$, we obtain

$$
\begin{aligned}
i &= 0 & t < 0 \\
&= C(1/\Delta t) & 0 < t < \Delta t \\
&= 0 & t > \Delta t
\end{aligned}
$$

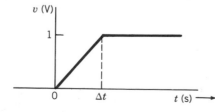

Figure 7-10 Voltage waveform where the change in voltage occurs over an increment of time Δt.

Thus, we obtain a pulse of height equal to $C/\Delta t$. As Δt decreases, the current will increase. Clearly, Δt cannot decline to zero or we would experience an infinite current. An infinite current is an impossibility, since it would require infinite power and an instantaneous movement of charge to occur at the capacitor terminals. According to the requirements of conservation of charge, the amount of the charge cannot change instantaneously. Thus, an instantaneous ($\Delta t = 0$) change of voltage across the capacitor is not possible.

The principle of conservation of charge states that the amount of electric charge cannot change instantaneously. Thus, $q(t)$ must be continuous over time. Recall that $q(t) = Cv(t)$. Thus, the voltage across the capacitor cannot change instantaneously; that is, we cannot have a discontinuity in $v(t)$.

The voltage across a capacitor cannot change instantaneously.

Now let us find the voltage $v(t)$ in terms of $i(t)$ by integrating both sides of Eq. 7-2. We obtain

$$v = \frac{1}{C} \int_{t_0}^{t} i \, d\tau + v(t_0) \tag{7-3}$$

where $v(t_0) = q(t_0)/C$ is the voltage across the capacitor at time t_0. We typically use $t_0 = 0$, since we can arbitrarily select t_0.

Example 7-2

Find the voltage v for a capacitor $C = 1/2$ F when the current is as shown in Figure 7-11 and $v(t_0) = v(0) = 0$.

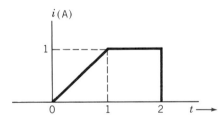

Figure 7-11 Current waveform for Example 7-2. The units are amperes and seconds.

Solution

First, we write the equation for $i(t)$ as

$$\begin{aligned}
i &= 0 & t \leq 0 \\
&= t & 0 < t \leq 1 \\
&= 1 & 1 < t \leq 2 \\
&= 0 & 2 < t
\end{aligned}$$

Then since

$$v = \frac{1}{C} \int_{0}^{t} i \, d\tau$$

and $C = 1/2$, we have

$$\begin{aligned}
v &= 0 & t \leq 0 \\
&= 2 \int_{0}^{t} \tau \, d\tau & 0 < t \leq 1 \\
&= 2 \int_{1}^{t} (1) \, d\tau + v(1) & 1 < t \leq 2 \\
&= v(2) & 2 < t
\end{aligned}$$

with units of volts. Therefore, for $0 < t \le 1$, we have

$$v = t^2$$

For the period $1 < t \le 2$, we note that $v(1) = 1$ and therefore we have

$$v = 2(t - 1) + 1 = (2t - 1) \text{ V}$$

The resulting voltage waveform is shown in Figure 7-12. The voltage changes as t^2 during the first 1 s, changes linearly with t during the period from 1 to 2 s, and stays constant equal to 3 V after $t = 2$ s.

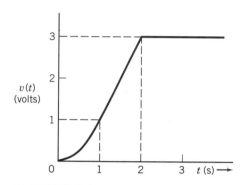

Figure 7-12 Voltage waveform for Example 7-2.

Actual capacitors have some resistance associated with them. Fortunately, it is easy to include approximate resistive effects in the circuit models. In capacitors the dielectric material between the plates is not a perfect insulator and has some small conductivity. This can be represented by a very high resistance in parallel with the capacitor. Ordinary capacitors can hold a charge for hours, and the parallel resistance is then hundreds of megaohms. For this reason, the resistance associated with a capacitor is usually ignored. It is important to note that the voltage waveform cannot change instantaneously, but the current waveform may do so.

EXERCISE 7-1

Consider the circuit of Figure E 7-1. A constant current of 10 mA is flowing into a 10-μF capacitor at the positive terminal. Find the charge and voltage on the capacitor after 100 ms if the initial voltage on the capacitor is zero.
Answer: 100 V, 1 mC

Figure E 7-1

EXERCISE 7-2

A 1-μF capacitor has a voltage as shown in Figure E 7-2. Find the current for $t > 0$.

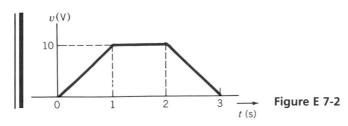

Figure E 7-2

EXERCISE 7-3

A 100-μF capacitor has a current waveform as shown in Figure E 7-3, where the units are amperes and seconds. Find the voltage for $t > 0$ when $v(0) = 0$.

Figure E 7-3

EXERCISE 7-4

The current $i(t)$ of the circuit shown in Figure E 7-4 is zero for $t < 0$ and $i(t) = 2 + 2 \cos 5t$ mA, $t \geq 0$. The initial capacitor voltage is 1/5 V. Determine $v(t)$ for $t \geq 0$.

Answer: $\quad v(t) = \frac{1}{5} + \frac{2}{5}t + \frac{2}{25} \sin 5t$ V

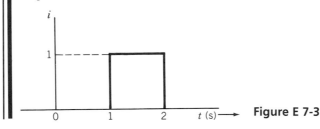

Figure E 7-4

7-4 | ENERGY STORAGE IN A CAPACITOR

Consider a capacitor that has been connected to a battery of voltage v. A current flows and a charge is stored on the plates of the capacitor, as shown in Figure 7-4. Eventually, the voltage across the capacitor is a constant, and the current through the capacitor is zero. The capacitor has stored energy by virtue of the separation of charges between the capacitor plates. These charges have an electrical force acting on them.

The forces acting on the charges stored in a capacitor are said to result from an electric field. An *electric field* is defined as the force acting on a unit positive charge in a specified region. Since the charges have a force acting on them along a direction x, we recognize that the energy required originally to separate the charges is now stored by the capacitor in the electric field.

The energy stored in a capacitor is

$$w_c(t) = \int_{-\infty}^{t} vi \, d\tau$$

Remember that v and i are both functions of time and could be written as $v(t)$ and $i(t)$. Since

$$i = C \frac{dv}{dt}$$

we have

$$w_c = \int_{-\infty}^{t} v \, C \frac{dv}{d\tau} \, d\tau$$

$$= C \int_{v(-\infty)}^{v(t)} v \, dv$$

$$= \frac{1}{2} C v^2 \Big|_{v(-\infty)}^{v(t)}$$

Since the capacitor was uncharged at $t = -\infty$, set $v(-\infty) = 0$. Therefore,

$$w_c(t) = \frac{1}{2} C v^2(t) \ \text{J} \tag{7-4a}$$

Therefore, as a capacitor is being charged and $v(t)$ is changing, the energy stored, w_c, is changing. Note that $w_c(t) \geq 0$ for all $v(t)$, so the element is said to be passive (see Section 2-4).

Since $q = Cv$, we may rewrite Eq. 7-4a as

$$w_c = \frac{1}{2C} q^2(t) \ \text{J} \tag{7-4b}$$

The capacitor is a storage element that stores but does not dissipate energy. For example, consider a 100-mF capacitor that has a voltage of 100 V across it. The energy stored is

$$w_c = \frac{1}{2} C v^2$$

$$= \frac{1}{2} (0.1)(100)^2$$

$$= 500 \ \text{J}$$

As long as the capacitor remains unconnected to any other element, the energy of 500 J remains stored. Now if we connect the capacitor to the terminals of a resistor, we expect a current to flow until all the energy is dissipated as heat by the resistor. After all the energy dissipates, the current is zero and the voltage across the capacitor is zero.

As noted in the previous section, the requirement of conservation of charge implies that the voltage on a capacitor is continuous. Thus, the *voltage and charge on a capacitor cannot change instantaneously.* This statement is summarized by the equation

$$v(0^+) = v(0^-)$$

where the time just prior to $t = 0$ is called $t = 0^-$ and the time immediately after $t = 0$ is called $t = 0^+$. The time between $t = 0^-$ and $t = 0^+$ is infinitely small. Nevertheless, the voltage will not change abruptly.

To illustrate the continuity of voltage for a capacitor, consider the circuit shown in Figure 7-13. For the circuit shown in Figure 7-13a the switch has been closed for a long time and the capacitor voltage has become $v_c = 10$ V. At time $t = 0$ we open the switch as

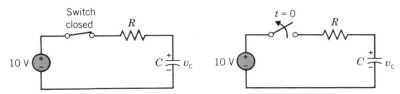

Figure 7-13 A circuit *(a)* where the capacitor is charged and $v_c = 10$ V and *(b)* the switch is opened at $t = 0$.

shown in Figure 7-13*b*. Since the voltage on the capacitor is continuous,

$$v_c(0^+) = v_c(0^-) = 10 \text{ V}$$

Example 7-3

A 10-mF capacitor is charged to 100 V, as shown in the circuit of Figure 7-14. Find the energy stored by the capacitor and the voltage of the capacitor at $t = 0^+$ after the switch is opened.

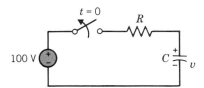

Figure 7-14 Circuit of Example 7-3 with $C = 10$ mF.

Solution

The voltage of the capacitor is $v = 100$ at $t = 0^-$. Since the voltage at $t = 0^+$ cannot change from the voltage at $t = 0^-$, we have

$$v(0^+) = v(0^-) = 100 \text{ V}$$

The energy stored by the capacitor at $t = 0^+$ is

$$w_c = \frac{1}{2} Cv^2$$

$$= \frac{1}{2} (10^{-2})(100)^2$$

$$= 50 \text{ J}$$

Example 7-4

The voltage across a 5-mF capacitor varies as shown in Figure 7-15. Determine and plot the capacitor current, power, and energy.

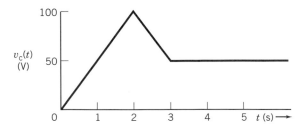

Figure 7-15 The voltage across a capacitor.

Solution

The current is determined from $i_c = C \, dv/dt$ and is shown in Figure 7-16*a*. The power is $v(t)i(t)$—the product of the current curve (Figure 7-16*a*) and the voltage curve (Figure

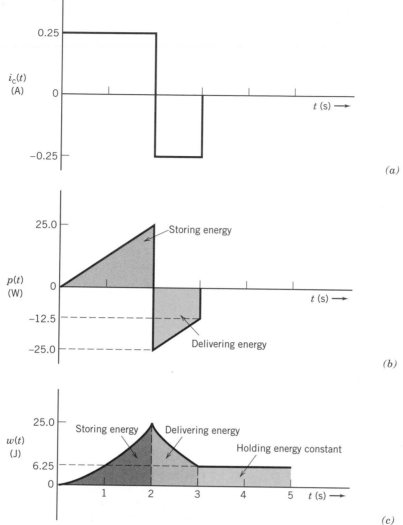

Figure 7-16 The current, power, and energy of the capacitor of Example 7-4.

7-15)—and is shown in Figure 7-16*b*. The capacitor receives energy during the first two seconds and then delivers energy for the period $2 < t < 3$.

The energy is $\omega = \int p\, dt$ and can be found as the area under the $p(t)$ curve. The curve for the energy is shown in Figure 7-16*c*. Note that the capacitor increasingly stores energy from $t = 0$ s to $t = 2$ s, reaching a maximum energy of 25 J. Then the capacitor delivers a total energy of 18.75 J to the external circuit from $t = 2$ s to $t = 3$ s. Finally, the capacitor holds a constant energy of 6.25 J after $t = 3$ s.

EXERCISE 7-5

A 200-μF capacitor has been charged to 100 V. Find the energy stored by the capacitor. Find the capacitor voltage at $t = 0^+$ if $v(0^-) = 100$ V.
Answer: $w = 1$ J; $v(0^+) = 100$ V

EXERCISE 7-6

A constant current $i = 2$ A flows into a capacitor of 100 μF after a switch is closed at $t = 0$. The voltage of the capacitor was equal to zero at $t = 0^-$. Find the energy stored at (a) $t = 1$ s and (b) $t = 100$ s.
Answer: $w(1) = 20$ kJ; $w(100) = 200$ MJ

EXERCISE 7-7

The initial capacitor voltage of the circuit shown in Figure E 7-7 is $v_c(0^-) = 3$ V. Determine (a) the voltage $v(t)$ and (b) the energy stored in the capacitor at $t = 0.2$ s and $t = 0.8$ s when

$$i(t) = \begin{cases} 3e^{5t} \text{ A} & 0 < t < 1 \\ 0 & t \geq 1 \text{ s} \end{cases}$$

Answers: (a) $3e^{5t}$ V, $0 \leq t < 1$
(b) $w(0.2) = 6.65$ J, $w(0.8) = 2.68$ kJ

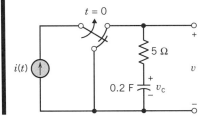

Figure E 7-7

7-5 SERIES AND PARALLEL CAPACITORS

First, let us consider the parallel connection of N capacitors as shown in Figure 7-17. We wish to determine the equivalent circuit for the N parallel capacitors as shown in Figure 7-18.

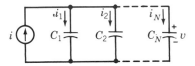

Figure 7-17 Parallel connection of N capacitors.

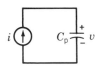

Figure 7-18 Equivalent circuit for N parallel capacitors.

Using KCL, we have

$$i = i_1 + i_2 + i_3 + \cdots + i_N$$

Since

$$i_n = C_n \frac{dv}{dt}$$

and v appears across each capacitor, we obtain

$$i = C_1 \frac{dv}{dt} + C_2 \frac{dv}{dt} + C_3 \frac{dv}{dt} + \cdots + C_N \frac{dv}{dt}$$

$$= (C_1 + C_2 + C_3 + \cdots + C_N) \frac{dv}{dt}$$

$$= \left(\sum_{n=1}^{N} C_n \right) \frac{dv}{dt} \tag{7-5}$$

For the equivalent circuit shown in Figure 7-16,

$$i = C_p \frac{dv}{dt} \tag{7-6}$$

Comparing Eqs. 7-5 and 7-6, it is clear that

$$C_p = C_1 + C_2 + C_3 + \cdots + C_N$$

$$= \sum_{n=1}^{N} C_n$$

Thus the equivalent capacitance of a set of N parallel capacitors is simply the sum of the individual capacitances. It must be noted that all the parallel capacitors will have the same initial condition, $v(0)$. Note the similarity with the results we found in Section 3-5 for parallel conductances, where $G_P = \Sigma\, G_n$.

Now let us determine the equivalent capacitance C_s of a set of N series-connected capacitances, as shown in Figure 7-19. The equivalent circuit for the series of capacitors is shown in Figure 7-20.

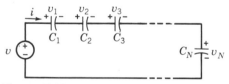

Figure 7-19 Series connection of N capacitors.   **Figure 7-20** Equivalent circuit for N series capacitors.

Using KVL for the loop of Figure 7-19, we have

$$v = v_1 + v_2 + v_3 + \cdots + v_N \tag{7-7}$$

Since, in general,

$$v_n = \frac{1}{C_n} \int_{t_0}^{t} i \, d\tau + v_n(t_0)$$

where i is common to all capacitors, we obtain

$$v = \frac{1}{C_1} \int_{t_0}^{t} i \, d\tau + v_1(t_0) + \cdots + \frac{1}{C_N} \int_{t_0}^{t} i \, d\tau + v_N(t_0)$$

$$= \left(\frac{1}{C_1} + \frac{1}{C_2} + \cdots + \frac{1}{C_N} \right) \int_{t_0}^{t} i \, d\tau + \sum_{n=1}^{N} v_n(t_0)$$

$$= \sum_{n=1}^{N} \frac{1}{C_n} \int_{t_0}^{t} i \, d\tau + \sum_{n=1}^{N} v_n(t_0) \tag{7-8}$$

From Eq. 7-7 we note that at $t = t_0$

$$v(t_0) = v_1(t_0) + v_2(t_0) + \cdots + v_N(t_0)$$

$$= \sum_{n=1}^{N} v_n(t_0) \tag{7-9}$$

Substituting Eq. 7-9 into Eq. 7-8, we obtain

$$v = \left(\sum_{n=1}^{N} \frac{1}{C_n} \int_{t_0}^{t} i \, d\tau \right) + v(t_0) \tag{7-10}$$

The KVL for the loop of the equivalent circuit of Figure 7-20 yields

$$v = \frac{1}{C_s} \int_{t_0}^{t} i \, d\tau + v(t_0) \tag{7-11}$$

Comparing Eqs. 7-10 and 7-11, we find that

$$\frac{1}{C_s} = \sum_{n=1}^{N} \frac{1}{C_n} \tag{7-12}$$

For the case of two series capacitors, Eq. 7-12 becomes

$$\frac{1}{C_s} = \frac{1}{C_1} + \frac{1}{C_2}$$

or

$$C_s = \frac{C_1 C_2}{C_1 + C_2} \tag{7-13}$$

Example 7-5

Find the equivalent capacitance for the circuit of Figure 7-21 when $C_1 = C_2 = C_3 = 2$ mF, $v_1(0) = 10$ V, and $v_2(0) = v_3(0) = 20$ V.

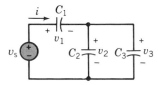

Figure 7-21 Circuit for Example 7-5.

Solution

Since C_2 and C_3 are in parallel, we replace them with C_p, where

$$C_p = C_2 + C_3$$
$$= 4 \text{ mF}$$

The voltage at $t = 0$ across the equivalent capacitance C_p is equal to the voltage across C_2 or C_3, which is $v_2(0) = v_3(0) = 20$ V. As a result of replacing C_2 and C_3 with C_p, we obtain the circuit shown in Figure 7-22.

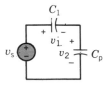

Figure 7-22 Circuit resulting from Figure 7-21 by replacing C_2 and C_3 with C_p.

We now want to replace the series of two capacitors C_1 and C_p with one equivalent capacitor. Using the relationship of Eq. 7-13, we obtain

$$C_s = \frac{C_1 C_p}{C_1 + C_p}$$

$$= \frac{(2 \times 10^{-3})(4 \times 10^{-3})}{(2 \times 10^{-3}) + (4 \times 10^{-3})}$$

$$= \frac{8}{6} \text{ mF}$$

The voltage at $t = 0$ across C_s is

$$v(0) = v_1(0) + v_p(0)$$

where $v_p(0) = 20$ V, the voltage across the capacitance C_p at $t = 0$. Therefore, we obtain

$$v(0) = 10 + 20$$
$$= 30 \text{ V}$$

Thus, we obtain the equivalent circuit shown in Figure 7-23.

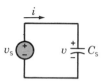

Figure 7-23 Equivalent circuit for the circuit of Example 7-5.

EXERCISE 7-8

Find the equivalent capacitance for the circuit of Figure E 7-8.
Answer: $C_{eq} = 4$ mF

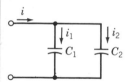

Figure E 7-8

EXERCISE 7-9

Find the relationship for the division of current between two parallel capacitors as shown in Figure E 7-9.
Answer: $i_n = iC_n/(C_1 + C_2)$, $n = 1, 2$

Figure E 7-9

EXERCISE 7-10

Determine the equivalent capacitance C_{eq} for the circuit shown in Figure E 7-10.

Answer: 10/19 mF

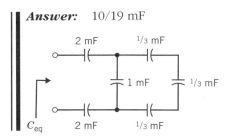

Figure E 7-10

7-6 ‖ INDUCTORS

A wire may be shaped as a multiturn coil, as shown in Figure 7-24. If we use a current source i_s, we find that the voltage across the coil is proportional to the rate of change of the current $i = i_s$, the current through the inductor. This proportional relationship may be expressed by the equation

$$v = L\frac{di}{dt} \tag{7-14}$$

where L is the constant of proportionality called *inductance* and is measured in henrys (H). Remember v and i are both functions of time (Dorf, 1993).

We define an *inductor* as a two-terminal element consisting of a winding of N turns for introducing inductance into an electric circuit. Inductance is defined as the property of an electric device by which a time-varying current through the device produces a voltage across it.

An ideal inductor is a coil wound with resistanceless wire. When current exists in the wire, energy is stored in the magnetic field around the coil. A constant current i in the coil results in zero voltage across the coil. A current that varies with time produces a self-induced voltage. Note that an abrupt (or instantaneous) change in current is impossible since an infinite voltage would be required.

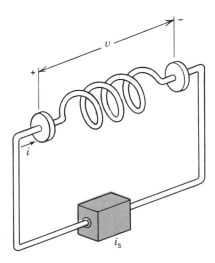

Figure 7-24 Coil of wire connected to a current source.

Single-layer coils wound in a helix are often called solenoids. An example is shown in Figure 7-25. If the length of the coil is greater than one-half the diameter and the core is of nonferromagnetic material, the inductance of the coil is given by

$$L = \frac{\mu_0 N^2 A}{l + 0.45d} \text{ H}$$

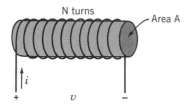

N turns

Area A

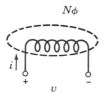

i

v

Figure 7-25 Coil wound as a tight helix on a core of area A.

where N is the number of turns; A the cross-sectional area in square meters; l the length in meters; d the diameter in meters; and $\mu_0 = 4\pi \times 10^{-7}$ H/m, a constant known as the permeability of free space.

Iron cores have higher permeabilities than air, thereby concentrating the magnetic flux. This effect increases the inductance of the coil.

The force experienced by two neighboring current-carrying wires can be described in terms of the existence of a magnetic field, which can be described in terms of magnetic flux that forms a loop around the coil, as shown in Figure 7-26. A *magnetic flux* $\phi(t)$ is associated with a current i in a coil. In this case we have an N-turn coil, and each flux line passes through all turns. Then the total flux is said to be $N\phi$.

$N\phi$

i

v

Figure 7-26 Model of the inductor.

Let us assume that the coil of wire has N turns and the core material has a relatively high permeability so that the magnetic flux ϕ is concentrated within the area A. According to Faraday, the changing flux creates an induced voltage in each turn equal to the derivative of the flux ϕ, so the total voltage v across N turns is

$$v = N\frac{d\phi}{dt} \tag{7-15}$$

However, since the total flux $N\phi$ is proportional to the current i in the coil, we have

$$N\phi = Li \tag{7-16}$$

where L, inductance, is the constant of proportionality. Substituting Eq. 7-16 into 7-15, we obtain

$$v = L\frac{di}{dt} \tag{7-17}$$

The circuit symbol for an inductor is shown in Figure 7-27. The passive sign convention for an inductor requires the current to flow into the positive terminal as shown in Figure 7-25. From the perspective of modeling electrical devices, the capacitor is a circuit element often used to model the effect of electric fields. Correspondingly, the inductor models the effects of magnetic fields. A magnetic field is a state, produced either by current flow in a

Figure 7-27 Circuit symbol for an inductor.

wire or by a permanent magnet, that can induce voltage in a conductor when the magnetic field changes.

Inductance is a measure of the ability of a device to store energy in the form of a magnetic field.

Inductors include the actual resistance of the copper wire used in the coil. For this reason, inductors are far from ideal elements and are typically modeled by an ideal inductance in series with a small resistance.

An example of a coil with a large inductance is shown in Figure 7-28. Inductors are

Figure 7-28 Coil with a large inductance. Courtesy of MuRata Company.

wound in various forms, as shown in Figure 7-29. Practical inductors have inductances ranging from 1 μH to 10 H.

Figure 7-29 Elements with inductances arranged in various forms of coils. Courtesy of Dale Electronic Inc.

Examining Eq. 7-17, we note that if the current i is constant, then the voltage across the inductor is zero. As the current changes more rapidly, the voltage will increase. Let us consider the voltage of an inductor when the current changes at $t = 0$ from zero to a constantly increasing current and eventually levels off as shown in Figure 7-30. Let us

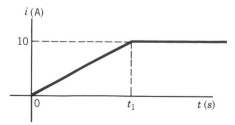

Figure 7-30 A current waveform. The current is in amperes.

determine the voltage of the inductor. We may describe the current (in amperes) by

$$i = 0 \qquad\qquad t \le 0$$
$$= \frac{10t}{t_1} \qquad 0 < t < t_1$$
$$= 10 \qquad\qquad t \ge t_1$$

Let us consider a 0.1-H inductor and find the voltage waveform. Since $v = L\,di/dt$, we have (in volts)

$$v = 0 \qquad\qquad t \le 0$$
$$= \frac{1}{t_1} \qquad 0 < t < t_1$$
$$= 0 \qquad\qquad t \ge t_1$$

The resulting voltage pulse waveform is shown in Figure 7-31. Note that as t_1 decreases, the magnitude of the voltage increases. Clearly, we cannot let $t_1 = 0$, since the voltage required would then become infinite and we would require infinite power at the terminals of the inductor. Thus, instantaneous changes in the current through an inductor are not possible.

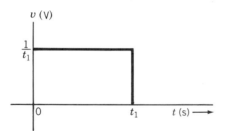

Figure 7-31 Voltage response for the current waveform of Figure 7-30 when $L = 0.1$ H.

Faraday, Ampére, and Oersted developed the concept of magnetic flux $\phi(t)$ associated with the current in the inductor. For a linear inductor $\phi(t) = Mi(t)$, where M is a constant. Just as the principle of conservation of charge applies to a capacitor, the principle of conservation of flux applies to the inductor. Thus, the flux $\phi(t)$ cannot have discontinuities, and therefore $i(t)$ through the inductor cannot have discontinuities.

The current in an inductance cannot change instantaneously.

The limiting requirements on the current through the inductor and the voltage across a capacitor are summarized in Table 7-2.

The current in an inductor in terms of the voltage across it may be determined by integrating the relationship

Table 7-2
Limiting Requirements for Inductors and Capacitors

	Limiting Requirements
Capacitor	The voltage across a capacitor may not change instantaneously (change discontinuously).
Inductor	The current in an inductor may not change instantaneously (change discontinuously).

$$v = L\frac{di}{dt} \tag{7-18}$$

from t_0 to t. We obtain from Eq. 7-18 that

$$di = \frac{v}{L}\,dt$$

Integrating, we have

$$i = \frac{1}{L}\int_{t_0}^{t} v\,d\tau + i(t_0) \tag{7-19}$$

where $i(t_0)$ is the current that accumulates from $t = -\infty$ to t_0. Normally, we select $t_0 = 0$.

For example, let us consider the voltage waveform shown in Figure 7-32 for an inductor when $L = 0.1$ H and $i(t_0) = 2$ A. Since $v(t) = 2$ V between $t = 0$ and $t = 2$, we have

$$i = 10\int_{0}^{t}(2)\,d\tau + i(t_0)$$

$$= 2(10t + 1)\text{ A}$$

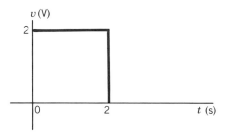

Figure 7-32 Voltage waveform for an inductor (in volts).

This current waveform is shown in Figure 7-33.

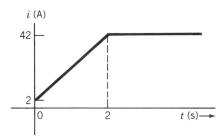

Figure 7-33 Current waveform for an inductor $L = 0.1$ H corresponding to the voltage waveform of Figure 7-32.

Example 7-6

Find the voltage across an inductor, $L = 0.1$ H, when the current in the inductor is

$$i = 20\,te^{-2t}\text{ A}$$

for $t > 0$ and $i(0) = 0$.

Solution

The voltage for $t > 0$ is

$$v = L\frac{di}{dt}$$

$$= (0.1)\frac{d}{dt}(20te^{-2t})$$

$$= 2(-2te^{-2t} + e^{-2t})$$

$$= 2e^{-2t}(1 - 2t)\text{ V}$$

The voltage is equal to 2 V when $t = 0$, as shown in Figure 7-34b. The current waveform is shown in Figure 7-34a. Note that the current reaches a maximum value and the voltage is zero at $t = 0.5$ s.

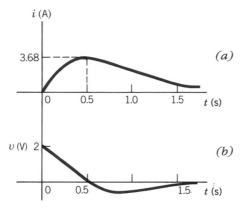

Figure 7-34 Voltage and current waveforms for Example 7-6.

EXERCISE 7-11

Find the voltage across an inductor, $L = 0.05$ H, if the current is $i = 20(1 - e^{-100t})$ A for $t \geq 0$.
Answer: $v = 100e^{-100t}$ V

EXERCISE 7-12

A current flowing in a 2-mH inductor is

$$i = 5e^{-200t} - 5e^{-800t} \text{ A} \qquad t \geq 0$$
$$= 0 \qquad t < 0$$

Find the voltage across the inductor. Sketch the current and the voltage.
Answer: $v = 8e^{-800t} - 2e^{-200t}$ V, $t \geq 0$

EXERCISE 7-13

Find the current in an inductor, $L = 1$ H, when $v = 9t^2$ V for $t \geq 0$ and $i = 0$ for $t < 0$.
Answer: $i = 3t^3$ A, $t \geq 0$

7-7 ENERGY STORAGE IN AN INDUCTOR

The power in an inductor is

$$p = vi = \left(L\frac{di}{dt} \right) i \qquad (7\text{-}20)$$

The energy stored in the inductor is stored in the magnetic field and is

$$w = \int_{t_0}^{t} p \, d\tau$$

$$= L \int_{i(t_0)}^{i(t)} i \, di$$

Integrating the current between $i(t_0)$ and $i(t)$, we obtain

$$w = \frac{L}{2} [i^2(t)]_{i(t_0)}^{i(t)}$$

$$= \frac{L}{2} i^2(t) - \frac{L}{2} i^2(t_0) \tag{7-21}$$

Usually, we select $t_0 = -\infty$ for the inductor and then the current $i(-\infty) = 0$. Then we have

$$w = \frac{1}{2} L i^2 \tag{7-22}$$

Note that $w(t) \geq 0$ for all $i(t)$, so the inductor is a passive element. The inductor does not generate or dissipate energy but only stores energy. It is important to note that inductors and capacitors are fundamentally different from other devices considered in earlier chapters in that they have memory.

Example 7-7
Find the current in an inductor, $L = 0.1$ H, when the voltage across the inductor is

$$v = 10te^{-5t} \text{ V}$$

Assume that the current is zero for $t \leq 0$.

Solution
The voltage as a function of time is shown in Figure 7-35a. Note that the voltage reaches a maximum at $t = 0.2$ s.
The current is

$$i = \frac{1}{L} \int_{0}^{t} v \, d\tau + i(t_0)$$

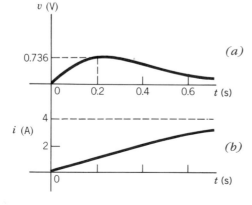

Figure 7-35 Voltage and current for Example 7-7.

Since the voltage is zero for $t < 0$, the current in the inductor at $t = 0$ is $i(0) = 0$. Then we have

$$i = 10 \int_0^t 10\tau e^{-5\tau} \, d\tau$$

$$= 100 \left[\frac{-e^{-5\tau}}{25} (1 + 5\tau) \right]_0^t$$

$$= 4(1 - e^{-5t}(1 + 5t)) \text{ A}$$

The current as a function of time is shown in Figure 7-35b.

Example 7-8

Find the power and energy for an inductor of 0.1 H when the current and voltage are as shown in Figures 7-36a and 7-36b.

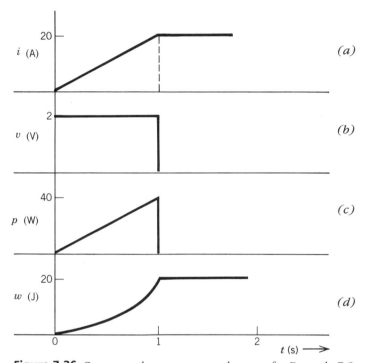

Figure 7-36 Current, voltage, power, and energy for Example 7-8.

Solution

First, we write the expression for the current and the voltage. The current is

$$\begin{aligned} i &= 0 & t &< 0 \\ &= 20t & 0 &\le t \le 1 \\ &= 20 & 1 &< t \end{aligned}$$

The voltage is expressed as

$$\begin{aligned} v &= 0 & t &< 0 \\ &= 2 & 0 &< t < 1 \\ &= 0 & 1 &< t \end{aligned}$$

You may verify the voltage by using $v = L\,di/dt$. Then the power is

$$p = vi$$
$$= 40t \ \text{W}$$

for $0 \le t < 1$ and zero for all other time.

The energy, in joules, is then

$$w = \frac{1}{2} Li^2$$

$$= 0.05(20t)^2 \quad 0 \le t \le 1$$
$$= 0.05(20)^2 \quad 1 < t$$

and zero for all $t < 0$.

The power and energy are shown in Figures 7-36c and 7-36d.

Example 7-9

Find the power and the energy stored in a 0.1-H inductor when $i = 20te^{-2t}$ A and $v = 2e^{-2t}(1 - 2t)$ V for $t \ge 0$ and $i = 0$ for $t < 0$. (See Example 7-6.)

Solution

The power is

$$p = iv$$
$$= (20te^{-2t})[2e^{-2t}(1 - 2t)]$$
$$= 40te^{-4t}(1 - 2t) \ \text{W} \qquad t > 0$$

The energy is then

$$w = \frac{1}{2} Li^2$$

$$= 0.05(20te^{-2t})^2$$
$$= 20t^2e^{-4t} \ \text{J} \qquad t > 0$$

Note that w is positive for all values of $t > 0$. The energy stored in the inductor is shown in Figure 7-37.

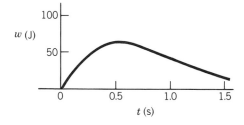

Figure 7-37 Energy stored in the inductor of Example 7-9.

EXERCISE 7-14

The current in an inductor, $L = 1/4$ H, is $i = 4te^{-t}$ A for $t \ge 0$ and $i = 0$ for $t < 0$. Find the voltage, power, and energy in this inductor.

Partial Answer: $w = 2t^2e^{-2t}$ J

EXERCISE 7-15

The current through the inductor of a television tube deflection circuit is shown in Figure E 7-15 when $L = 1/2$ H. Find the voltage, power, and energy in the inductor.
Partial Answer: $p = 2t$ for $0 \le t < 1$
$\qquad\qquad\qquad = 2(t - 2)$ for $1 < t < 2$
$\qquad\qquad\qquad = 0$ for other t

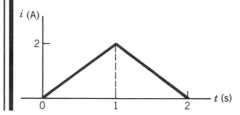

Figure E 7-15

7-8 SERIES AND PARALLEL INDUCTORS

A series and parallel connection of inductors can be reduced to an equivalent simple inductor. Consider a series connection of N inductors as shown in Figure 7-38. The voltage across the series connection is

$$v = v_1 + v_2 + \cdots + v_N$$

$$= L_1 \frac{di}{dt} + L_2 \frac{di}{dt} + \cdots + L_N \frac{di}{dt}$$

$$= \sum_{n=1}^{N} L_n \frac{di}{dt}$$

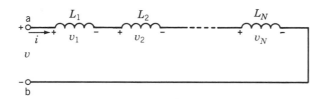

Figure 7-38 Series of N inductors.

Since the equivalent series inductor L_s, as shown in Figure 7-39, is represented by

$$v = L_s \frac{di}{dt}$$

Figure 7-39 Equivalent inductor L_s for N series inductors.

we require that

$$L_s = \sum_{n=1}^{N} L_n \qquad (7\text{-}23)$$

Thus, an equivalent inductor for a series of inductors is the sum of the N inductors.

Now, consider the set of N inductors in parallel, as shown in Figure 7-40. The current i is

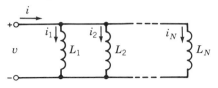

Figure 7-40 Connection of N parallel inductors.

equal to the sum of the currents in the N inductors:

$$i = \sum_{n=1}^{N} i_n$$

However, since

$$i_n = \frac{1}{L_n} \int_{t_0}^{t} v \, d\tau + i_n(t_0)$$

we may obtain the expression

$$i = \sum_{n=1}^{N} \frac{1}{L_n} \int_{t_0}^{t} v \, d\tau + \sum_{n=1}^{N} i_n(t_0) \qquad (7\text{-}24)$$

The equivalent inductor L_p, as shown in Figure 7-41, is represented by the equation

$$i = \frac{1}{L_p} \int_{t_0}^{t} v \, d\tau + i(t_0) \qquad (7\text{-}25)$$

Figure 7-41 Equivalent inductor L_p for the connection of N parallel inductors.

When Eqs. 7-24 and 7-25 are set equal to each other, we have

$$\frac{1}{L_p} = \sum_{n=1}^{N} \frac{1}{L_n} \qquad (7\text{-}26)$$

and

$$i(t_0) = \sum_{n=1}^{N} i_n(t_0) \qquad (7\text{-}27)$$

Example 7-10

Find the equivalent inductance for the circuit of Figure 7-42. All the inductor currents are zero at t_0.

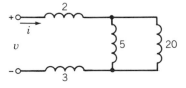

Figure 7-42 The circuit of Example 7-10. All inductances in millihenries.

Solution

First, we find the equivalent inductance for the 5-mH and 20-mH inductors in parallel. From Eq. 7-26 we obtain

$$\frac{1}{L_p} = \frac{1}{L_1} + \frac{1}{L_2}$$

or

$$L_p = \frac{L_1 L_2}{L_1 + L_2}$$

$$= \frac{5 \times 20}{5 + 20}$$

$$= 4 \text{ mH}$$

This equivalent inductor is in series with the 2-mH and 3-mH inductors. Therefore, using Eq. 7-23, we obtain

$$L_{eq} = \sum_{n=1}^{N} L_n$$

$$= 2 + 3 + 4$$

$$= 9 \text{ mH}$$

EXERCISE 7-16

Find the equivalent inductance of the circuit of Figure E 7-16.
Answer: $L_{eq} = 14 \text{ mH}$

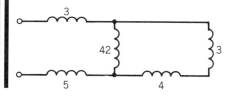

Figure E 7-16 All inductances in mH.

EXERCISE 7-17

Find the equivalent inductance of the circuit of Figure E 7-17.
Answer: $L_{eq} = 4 \text{ mH}$

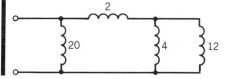

Figure E 7-17 All inductances in mH.

EXERCISE 7-18

Determine the current ratio i_1/i for the circuit shown in Figure E 7-18. Assume that the initial currents are zero at t_0.

Answer: $\dfrac{i_1}{i} = \dfrac{L_2}{L_1 + L_2}$

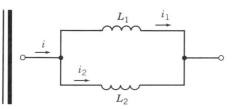

Figure E 7-18

7-9 ‖ INITIAL CONDITIONS OF SWITCHED CIRCUITS

In this section we concentrate on finding the change in selected variables in a circuit when a switch is thrown from open to closed or vice versa. The time of throwing the switch is considered to be $t = 0$, and we want to determine the value of the variable at $t = 0^-$ and $t = 0^+$, immediately before and after throwing the switch. Thus, a *switched circuit* is an electric circuit with one or more switches that open or close at time $t_0 = 0$.

We are particularly interested in the change in the current and voltage of energy storage elements after the switch is thrown since these variables along with the sources will dictate the behavior of the circuit for $t > 0$. In Table 7-3 we summarize the important characteristics of the behavior of an inductor and a capacitor. Note that we assume $t_0 = 0$. Recall that an instantaneous change in the inductor current and the capacitor voltage is not permitted. However, it is possible to change instantaneously an inductor's voltage and a capacitor's current.

Table 7-3
Characteristics of Energy Storage Elements

Variable	Inductors	Capacitors
Passive sign convention	$\overset{i}{\to}\ \ \overset{L}{\underset{+\ \ v\ \ -}{}}$	$\overset{i}{\to}\ \ \overset{C}{\underset{+\ \ v\ \ -}{}}$
Voltage	$v = L\dfrac{di}{dt}$	$v = \dfrac{1}{C}\displaystyle\int_0^t i\, d\tau + v(0)$
Current	$i = \dfrac{1}{L}\displaystyle\int_0^t v\, d\tau + i(0)$	$i = C\dfrac{dv}{dt}$
Power	$p = Li\dfrac{di}{dt}$	$p = Cv\dfrac{dv}{dt}$
Energy	$w = \dfrac{1}{2}Li^2$	$w = \dfrac{1}{2}Cv^2$
An instantaneous change is not permitted for the element's:	Current	Voltage
Will permit an instantaneous change in the element's:	Voltage	Current
This element acts as a: (see note below)	Short circuit to a constant current into its terminals	Open circuit to a constant voltage across its terminals

Note: Assumes that the element is in a circuit with a steady-state condition.

We will assume that the switches in a circuit have been in position for a long time at $t = 0$, the switching time. Thus, we say the circuit conditions are in *steady state* at the time of switching. The steady-state response of a circuit is that which exists after a long time following any switching operation. Furthermore, when a circuit is excited by only dc sources, and the circuit attains steady state, all the branch currents and voltages are constant.

When a constant current flows into an inductor, the voltage, $v = L\,di/dt$, is zero across the element and appears as a short circuit.

An inductance behaves as a short circuit to a dc current.

Similarly, if a constant voltage is applied across a capacitor, it appears as an open circuit since $i = C\,dv/dt$ is equal to zero.

A capacitance behaves as an open circuit to a dc voltage.

First, let us consider a circuit with an inductor as shown in Figure 7-43. The symbol for the switch implies it is open at $t = 0^-$ and then closes at $t = 0$. Before the switch is thrown we assume that the circuit has attained steady-state conditions. The source current at $t = 0^-$ divides between i_1 and i_L so that

$$i_1 + i_L = i_s$$

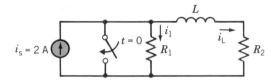

Figure 7-43 An *RL* circuit. $R_1 = R_2 = 1\ \Omega$. The switch is open for $t < 0$ and is closed at $t = 0$.

Note that i_L is a constant current, so the inductor voltage will be zero. Then using the current divider principle,

$$i_L = \frac{R_1}{R_1 + R_2}\, i_s$$

Since $i_s = 2$ A, we have the current at $t = 0^-$. When $R_1 = R_2 = 1$, we obtain

$$i_L(0^-) = \left(\frac{1}{2}\right) 2$$
$$= 1\text{ A}$$

Since the current cannot change instantaneously for the inductor, we have

$$i_L(0^+) = i_L(0^-)$$
$$= 1\text{ A}$$

However, note that the current in the resistor can change instantaneously. Prior to $t = 0$ we have

$$i_1(0^-) = 1\text{ A}$$

After the switch is thrown, we require that the voltage across R_1 be equal to zero because of the switched short circuit, and therefore

$$i_1(0^+) = 0$$

Thus, the current in the resistor changes abruptly from 1 to 0.

Now let us consider the circuit with a capacitor shown in Figure 7-44. Prior to $t = 0$, the switch has been closed for a long time. Since the source is a constant, the current in the capacitor is zero for $t < 0$ because the capacitor appears as an open circuit in a steady-

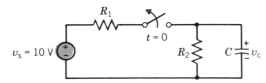

Figure 7-44 An *RC* circuit. $R_1 = R_2 = 1\ \Omega$. The switch is closed for $t < 0$ and opens at $t = 0$.

state condition. Therefore, the voltage across the capacitor for $t < 0$ can be obtained from the voltage divider principle with R_1 and R_2 so that

$$v_c = \frac{R_2}{R_1 + R_2}\, v_s$$

Therefore, at $t = 0^-$, when $v_s = 10$ and $R_1 = R_2 = 1\ \Omega$, we obtain

$$v_c(0^-) = \left(\frac{1}{2}\right) 10$$
$$= 5\ \text{V}$$

However, since the voltage across a capacitor cannot change instantaneously, we have

$$v_c(0^-) = v_c(0^+) = 5\ \text{V}$$

When the switch is opened at $t = 0$, the source is removed from the circuit, but $v_c(0^+)$ remains equal to 5 V.

If a circuit has both a capacitor and an inductor, we consider the current and voltage prior to $t = 0$ and find $v_c(0^-)$ and $i_L(0^-)$. As an example, consider the circuit shown in Figure 7-45. We assume that the switch has been closed a long time and steady-state conditions exist. In order to find $v_c(0^-)$ and $i_L(0^-)$ we replace the capacitor by an open circuit and the inductor by a short circuit, as shown in Figure 7-46. Immediately, we note that

$$i_L(0^-) = \frac{10}{5}$$
$$= 2\ \text{A}$$

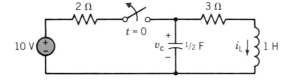

Figure 7-45 Circuit with an inductor and a capacitor. The switch is closed for a long time prior to opening at $t = 0$.

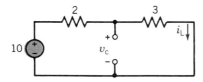

Figure 7-46 Circuit of Figure 7-45 for $t < 0$.

The capacitor is an open circuit as shown but does have a voltage v_c across its terminals. Using the voltage divider principle, we note that

$$v_c(0^-) = \left(\frac{3}{5}\right) 10$$
$$= 6\ \text{V}$$

Then we note that

$$v_c(0^+) = v_c(0^-) = 6 \text{ V}$$

and

$$i_L(0^+) = i_L(0^-) = 2 \text{ A}$$

Example 7-11

Find $i_L(0^+)$, $v_c(0^+)$, $dv_c(0^+)/dt$, and $di_L(0^+)/dt$ for the circuit of Figure 7-47. We will use $dv_c(0^+)/dt$ to denote $dv_c(t)/dt \mid_{t=0^+}$.

Assume that switch 1 has been open and switch 2 has been closed for a long time and steady-state conditions prevail at $t = 0^-$.

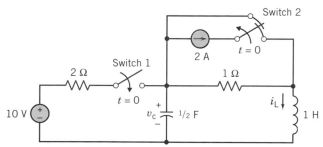

Figure 7-47 Circuit for Example 7-11. Switch 1 closes at $t = 0$ and switch 2 opens at $t = 0$.

Solution

First, we redraw the circuit for $t = 0^-$ by replacing the inductor with a short circuit and the capacitor with an open circuit, as shown in Figure 7-48. Then we note that

$$i_L(0^-) = 0$$

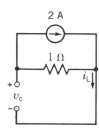

Figure 7-48 Circuit of Figure 7-47 at $t = 0^-$.

and

$$v_c(0^-) = -2 \text{ V}$$

Therefore, we have

$$i_L(0^+) = i_L(0^-) = 0$$

and

$$v_c(0^+) = v_c(0^-) = -2 \text{ V}$$

In order to find $dv_c(0^+)/dt$ and $di_L(0^+)/dt$ we throw the switch at $t = 0$ and redraw the circuit of Figure 7-47, as shown in Figure 7-49 (we disconnected the current source since its switch is open).

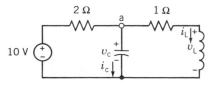

Figure 7-49 Circuit of Figure 7-47 at $t = 0^+$ with the switch closed and the current source disconnected.

Since we wish to find $dv_c(0^+)/dt$, we recall that

$$i_c = C \frac{dv_c}{dt}$$

so

$$\frac{dv_c(0^+)}{dt} = \frac{i_c(0^+)}{C}$$

Similarly, since for the inductor

$$v_L = L \frac{di_L}{dt}$$

we may obtain $di_L(0^+)/dt$ as

$$\frac{di_L(0^+)}{dt} = \frac{v_L(0^+)}{L}$$

Using KVL for the right-hand mesh of Figure 7-49, we obtain

$$v_L - v_c + 1 i_L = 0$$

Therefore, at $t = 0^+$

$$v_L(0^+) = v_c(0^+) - i_L(0^+)$$
$$= -2 - 0$$
$$= -2 \text{ V}$$

Hence, we obtain

$$\frac{di_L(0^+)}{dt} = -2 \text{ A/s}$$

Similarly, to find i_c we write KCL at node a to obtain

$$i_c + i_L + \frac{v_c - 10}{2} = 0$$

Consequently, at $t = 0^+$

$$i_c(0^+) = \frac{10 - v_c(0^+)}{2} - i_L(0^+)$$
$$= 6 - 0$$
$$= 6 \text{ A}$$

Accordingly,

$$\frac{dv_c(0^+)}{dt} = \frac{i_c(0^+)}{C}$$
$$= \frac{6}{1/2}$$
$$= 12 \text{ V/s}$$

Thus, we found that at the switching time $t = 0$, the current in the inductor and the voltage of the capacitor remained constant. However, the inductor voltage did change instantaneously from $v_L(0^-) = 0$ to $v_L(0^+) = -2$ V and we determined that $di_L(0^+)/dt = -2$ A/s.

Also, the capacitor current changed instantaneously from $i_c(0^-) = 0$ to $i_c(0^+) = 6$ A and we found that $dv_c(0^+)/dt = 12$ V/s.

7-10 ‖ THE OPERATIONAL AMPLIFIER AND *RC* CIRCUITS

The circuit elements connected to the operational amplifier are not limited to resistors. It is useful to consider the connection of a combination of resistors and capacitors to an operational amplifier. First, let us consider the *RC* circuit and operational amplifier as shown in Figure 7-50.

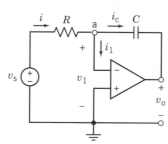

Figure 7-50 An integrator implemented using an operational amplifier.

Applying KVL to the mesh containing the independent voltage source gives

$$v_s = iR + v_1$$

Applying KCL at node a gives

$$i_c = i - i_1$$

$$= \frac{v_s - v_1}{R} - i_1$$

Applying KVL to the loop consisting of the operational amplifier input, capacitor, and operational amplifier output gives

$$v_o = v_1 - \frac{1}{C}\int_{-\infty}^{t} i_c \, dt$$

$$= v_1 - \frac{1}{C}\int_{-\infty}^{t} \left(\frac{v_s - v_1}{R} - i_1\right) dt$$

When the operational amplifier is modeled as an ideal operational amplifier, then $v_1 = 0$ and $i_1 = 0$, so

$$v_o = -\frac{1}{C}\int_{-\infty}^{t} \frac{v_s}{R} \, dt$$

$$= -\left(\frac{1}{RC}\int_{0}^{t} v_s \, dt + v_o(0)\right)$$

Therefore, the output voltage of this circuit is proportional to the integral of the input voltage. This circuit is commonly called an *integrator*. If we choose $RC = 1$, then

$$v_o = -\int_{-\infty}^{t} v_s \, dt$$

$$= -\left(\int_{0}^{t} v_s \, dt + v_C(0)\right)$$

Next, consider the circuit shown in Figure 7-51. Applying KVL to the mesh containing the independent voltage source gives

$$v_s = v_c + v_1$$

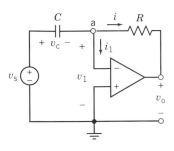

Figure 7-51 A differentiator implemented using an operational amplifier.

Applying KCL at node a gives

$$i - i_1 = C \frac{dv_c}{dt}$$

$$= C \frac{d(v_s - v_1)}{dt}$$

Applying KVL to the loop consisting of the operational amplifier input, resistor, and operational amplifier output gives

$$v_o = v_1 - Ri$$

$$= v_1 - R\left(C \frac{d(v_s - v_1)}{dt} + i_1\right)$$

When the operational amplifier is modeled as an ideal operational amplifier, then $v_1 = 0$ and $i_1 = 0$, so

$$v_o = -RC \frac{dv_s}{dt}$$

Therefore, the output voltage of this circuit is proportional to the derivative of the input voltage. This circuit is commonly called a *differentiator*. Caution is advised in the use of this circuit, since it can accentuate the noise present within the circuit.

EXERCISE 7-19

For the integrator circuit of Figure 7-50, we have $R = 1\ k\Omega$, $C = 500\ \mu F$ and the initial capacitor voltage is $v_c(0) = 2$ V. When $v_s = 5$ V for $t > 0$ and $v_s = 0$ for $t < 0$, determine and sketch the output voltage $v_o(t)$. Assume the circuit is in steady state at $t = 0^-$.

Answer: $v_o = 2 - 10t$ V, $t \geq 0$

EXERCISE 7-20

For the differentiator circuit of Figure 7-51 we have $R = 20\ \text{k}\Omega$, $C = 50\ \mu\text{F}$ and the initial capacitor voltage is $v_c(0) = 0$. Assume the circuit is in steady state at $t = 0^-$. When $v_s(t) = 0.2 \sin 10t$ V, for $t \geq 0$ determine $v_o(t)$.

Answer: $v_o(t) = -2 \cos 10t$ V, $t \geq 0$

7-11 VERIFICATION EXAMPLE

A laboratory report regarding the circuit of Figure 7V -1a states that $v_c(0^+) = 0.5$ V and $i_L(0^+) = 2$ mA. Verify these results.

Solution

First, we redraw the circuit at $t = 0^-$ when the switch is closed. Replacing the capacitor with an open circuit and the inductor with a short circuit, we obtain the steady-state circuit model of Figure 7V-1b at $t = 0^-$. Then $i_L = 5/500 = 10$ mA at $t = 0^-$. Using the voltage divider principle, we obtain $v_c(0^-) = \frac{4}{5}v_s = 4$ V. Therefore, $v_c(0^+) = 4$ V and $i_L(0^+) = 10$ mA. The reported results are incorrect.

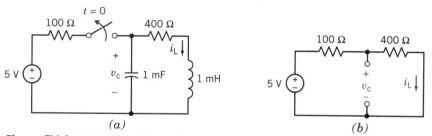

Figure 7V-1 (*a*) Circuit, (*b*) steady-state model of circuit.

7-12 DESIGN CHALLENGE SOLUTION

INTEGRATOR AND SWITCH

This design challenge involves an integrator and a voltage-controlled switch.

An integrator is a circuit that performs the mathematical operation of integration. The output of an integrator, say $v_o(t)$, is related to the input of the integrator, say $v_s(t)$, by the equation

$$v_o(t_2) = K \cdot \int_{t_1}^{t_2} v_s(t)\ dt + v_o(t_1) \tag{7D-1}$$

The constant K is called the gain of the integrator.

Integrators have many applications. One application of an integrator is to measure an interval of time. Suppose $v_s(t)$ is a constant voltage. Then

$$v_o(t_2) = K \cdot (t_2 - t_1) \cdot v_s + v_o(t_1) \tag{7D-2}$$

This equation indicates that the output of the integrator at time t_2 is a measure of the time interval $t_2 - t_1$.

Switches can be controlled electronically. Figure 7D-1 illustrates an electronically controlled SPST switch. The symbol shown in Figure 7D-1*a* is sometimes used to

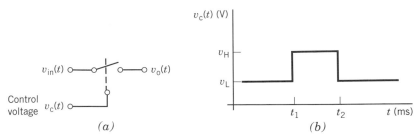

Figure 7D-1 The voltage controlled switch. (*a*) Switch symbol. (*b*) Typical control voltage.

emphasize that a switch is controlled electronically. The node voltage $v_c(t)$ is called the control voltage. Figure 7D-1*b* shows a typical control voltage. This voltage-controlled switch is closed when $v_c(t) = v_H$ and open when $v_c(t) = v_L$. The switch shown in Figure 7D-1 is open before time t_1. It closes at time t_1 and stays closed until time t_2. The switch opens at time t_2 and remains open.

Consider Figure 7D-2. The voltage $v_c(t)$ controls the switch. The integrator converts the time interval $t_2 - t_1$ to a voltage that is displayed using the voltmeter. The time

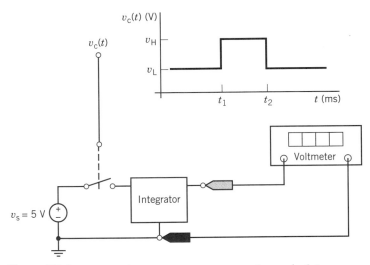

Figure 7D-2 Using an integrator to measure an interval of time.

interval to be measured could be as small as 5 ms or as large as 200 ms. The challenge is to design the integrator. The available components include:

> standard 2% resistors (see Appendix F);
> 1 μF, 0.2 μF, and 0.1 μF capacitors;
> operational amplifiers;
> + 15-V and − 15-V power supplies;
> 1-kΩ, 10-kΩ, and 100-kΩ potentiometers;
> voltage-controlled SPST switches.

Define the Situation

It is convenient to set the integrator output to zero at time t_1. The relationship between the integrator output voltage and the time interval should be simple. Accordingly, let

$$v_o(t_2) = \frac{10 \text{ V}}{200 \text{ ms}} \cdot (t_2 - t_1) \tag{7D-3}$$

Figure 7D-2 indicates that $v_s = 5$ V. Comparing Eqs. 7D-2 and 7D-3 yields

$$K \cdot v_s = \frac{10 \text{ V}}{200 \text{ ms}} \quad \text{and therefore} \quad K = 10 \frac{\text{V}}{\text{s}} \tag{7D-4}$$

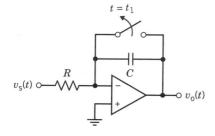

Figure 7D-3 An integrator using an operational amplifier.

State the Goal

Design an integrator satisfying both

$$K = 10 \frac{\text{V}}{\text{s}} \quad \text{and} \quad v_o(t_1) = 0 \tag{7D-5}$$

Generate a Plan

Let us use the integrator described in Section 7-9. Adding a switch as shown in Figure 7D-3 satisfies the condition $v_o(t_1) = 0$. The analysis performed in Section 7-9 showed that

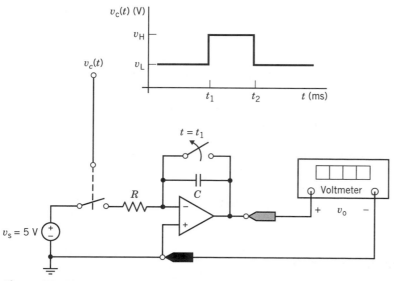

Figure 7D-4 Using an operational amplifier integrator to measure an interval of time.

$$v_o(t_2) = \frac{1}{RC} \cdot \int_{t_1}^{t_2} v_s(t) \, dt \qquad \text{(7D-6)}$$

so R and C must be selected to satisfy

$$\frac{1}{RC} = K = 10 \, \frac{V}{s} \qquad \text{(7D-7)}$$

Take Action on the Plan

Any of the available capacitors would work. Select $C = 1 \, \mu F$. Then

$$R = \frac{1}{10 \, \dfrac{V}{s} \cdot 1 \, \mu F} = 100 \, k\Omega \qquad \text{(7D-8)}$$

The final design is shown in Figure 7D-4.

SUMMARY

The storage of electric energy in electric circuit elements provides one more important step in the development of flexible and useful circuits. Electric energy may be stored in capacitors and inductors.

The work performed on moving a charge results in energy storage in a capacitor. The current flowing through a capacitor is equal to the derivative of the voltage across the capacitor multiplied by the constant C, where C is the capacitance in farads. A capacitor is a linear element, and the forces acting on the charge stored in a capacitor are said to result from an electric field. To satisfy the law of conservation of charge, the voltage and charge of a capacitor cannot change instantaneously. A set of series or parallel capacitors can readily be reduced to an equivalent capacitance.

An inductor is an energy storage element consisting of a coil of N turns for which a time-varying current induces a voltage across the device. The energy in an inductor is said to be stored in a magnetic field. Since the stored energy may not change instantaneously, the inductor current may not change instantaneously. A set of series or parallel inductors can readily be reduced to an equivalent inductor.

The energy stored in an inductor or a capacitor is always equal to or greater than zero, and they are classified as passive elements since they do not generate or dissipate energy. They only store energy.

Since we often encounter circuits that contain one or more switches that open or close at a time denoted as t_0, we must determine the effect of these changes on the inductor's or capacitor's voltage and current. Using the facts that an inductor's current and a capacitor's voltage cannot change instantaneously, we may determine the effect of switch changes in a circuit. We will assume, unless stated otherwise, that a circuit prior to switching at $t = 0$ has attained steady-state conditions at $t = 0^-$.

The characteristics of energy storage elements are summarized in Table 7-3.

TERMS AND CONCEPTS

Capacitance Ratio of the charge stored to the voltage difference between the two conducting plates or wires; $C = q/v$.

Capacitor   Two-terminal element whose primary purpose is to introduce capacitance into an electric circuit.

Dielectric Constant   Property that determines the energy stored per unit volume for unit voltage difference in a capacitor.

Electric Field   State of a region in which charged bodies are subject to forces by virtue of their charge; the force acting on a unit positive charge.

Energy Storage   Work performed in moving a charge resulting in energy storage in a capacitor, or work performed to establish a magnetic field resulting in energy storage in an inductor.

Inductance   Property of an electric device by virtue of which a time-varying current produces a voltage across the device.

Inductor   Two-terminal element consisting of a winding of N turns for introducing inductance into an electric circuit.

Initial Time   The time, t_o, when a new action is initiated in a circuit; usually, $t_o = 0$.

Magnetic Field   State produced either by current flow or by a permanent magnet that can induce voltage in a conductor when the magnetic field changes.

Switched Circuit   Electric circuit with one or more switches that open or close at time t_o.

REFERENCES   Chapter 7

Ashley, S., "Surging Ahead with Ultracapacitors," *Mechanical Engineering,* February 1995, pp. 76–80.

Cantor, G., *Michael Faraday,* St. Martin's Press, New York, 1991.

Dorf, Richard, *Electrical Engineering Handbook, Capacitors and Inductors,* Chapter 1.2, CRC Press, Boca Raton, FL, 1993.

Thomas, J. M., *Michael Faraday and the Royal Institution,* American Institute of Physics, New York, 1991.

Trotter, D. M., *Capacitors, Scientific American,* Vol. 259, No. 1, 1988. pp. 86–90.

PROBLEMS

Section 7-3   Capacitors and Their *v-i* Equation

P 7.3-1   The voltage across a 40-μF capacitor is 25 V at $t = 0$. If the current through the capacitor as a function of time is given by $i(t) = 6e^{-6t}$ mA for $t > 0$, find $v(t)$ for $t > 0$.
Answer: $v(t) = 50 - 25e^{-6t}$ V

P 7.3-2   A 15-μF capacitor has a voltage of 5 V across it at $t = 0$. If a constant current of 25 mA flows through the capacitor, how long will it take for the capacitor to hold a charge of 150 μC?
Answer: $t = 3$ ms

P 7.3-3   A current source, i_s, as shown in Figure P 7.3-3, is connected to an uncharged capacitor at $t = 0$. Determine the voltage waveform from $t = 0$ s to $t = 2.5$, and sketch the waveform when $C = 1$ mF.

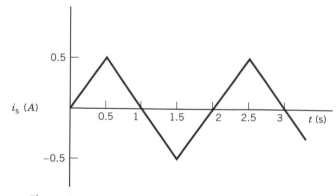

i_s (A)

Figure P 7.3-3 Waveform of current source i_s.

P 7.3-4   Determine $v(t)$ for the circuit shown in Figure P 7.3-4a when the $i_s(t)$ is as shown in Figure P 7.3-4b and $v_o(0^-) = -1$ mV.

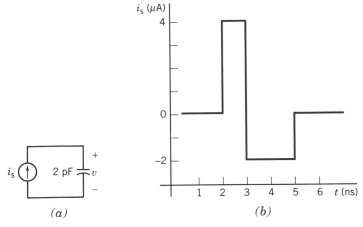

(a)

(b)

Figure P 7.3-4 (a) Circuit and (b) waveform of current source.

P 7.3-5 The capacitance of a two-plate capacitor with air between the plates is 1.2 μF. The capacitor is connected to a 12 V battery and charged. Then, a slab of dielectric material is inserted between the plates. As a result, 2.6×10^{-5} C of additional charge flows from one plate through the battery and on to the other plate. What is the relative dielectric constant of the material?

Answer: 1.8

P 7.3-6 The current, i, through a capacitor is shown in Figure P 7.3-6. When $v(0) = 0$ and $C = 0.5$ F, determine and plot $v(t)$, $p(t)$, and $w(t)$ for 0 s $< t <$ 6 s.

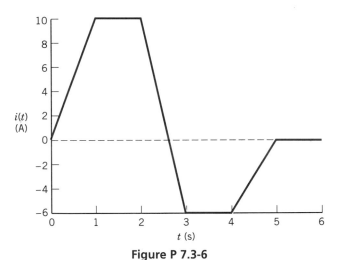

Figure P 7.3-6

P 7.3-7 Find i for the circuit of Figure P 7.3-7 if $v = 5(1 - 2e^{-2t})$ V.

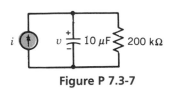

Figure P 7.3-7

P 7.3-8 Determine $v(t)$ for the circuit of Figure P 7.3-8 when $v(0) = -4$ mV.

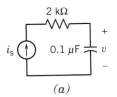

(a)

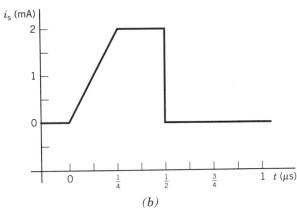

(b)

Figure P 7.3-8 (a) Circuit and (b) waveform of current source.

Section 7-4 Energy Storage in a Capacitor

P 7.4-1 A 0.1-mF capacitor is initially charged to 10 V. Between $t = 0$ ms and $t = 10$ ms, the capacitor is discharged with the current shown in Figure P 7.4-1.

(a) What is the capacitor voltage at $t = 6$ ms and $t = 10$ ms?

(b) How much energy was drained from the capacitor between $t = 0$ ms and $t = 10$ ms?

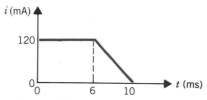

Figure P 7.4-1 Capacitor current.

P 7.4-2 In a pulse power circuit the voltage of a 10-μF capacitor is zero for $t < 0$ and

$$v = 5(1 - e^{-4000t}) \text{ V} \quad t \geq 0$$

Determine i_c and the energy stored in the capacitor at $t = 0$ ms and $t = 10$ ms.

P 7.4-3 If $v_c(t)$ is given by the waveform shown in Figure P 7.4-3, sketch the capacitor current for $-1\text{ s} \leq t \leq 2$ s. Sketch the power and the energy for the capacitor over the same time interval when $C = 1$ mF.

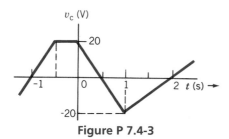

Figure P 7.4-3

P 7.4-4 The current through a 2-μF capacitor is $50\cos(10t + \pi/6)$ μA for all time. The average voltage across the capacitor is zero. What is the maximum value of the energy stored in the capacitor? What is the first nonnegative value of t at which the maximum energy is stored?

P 7.4-5 The energy stored by a 1-mF capacitor used in a laser power supply is given as $w = 4e^{-10t}$ J for $t \geq 0$. Find the capacitor voltage and current at $t = 0.1$ s.

P 7.4-6 If a capacitor can store energy, as does a battery, could not a capacitor be used to power an electric train? Such a capacitor (electrolytic) would be about 1 cm³ per 10 μF, for the voltage level required. Suppose we start with a charge at 100 V and can extract all the energy from the capacitor to drive the train for 1 hour at 30 kW. How big would the capacitor have to be? Is this practical?

P 7.4-7 A capacitor is used in the electronic flash unit of a camera. A small battery with a constant voltage of 6 V is used to charge a capacitor with a constant current of 10 μA. How long does it take to charge the capacitor when $C = 10$ μF? What is the stored energy?

P 7.4-8 If a capacitor can store energy, as does a battery, could not a capacitor be used to power an electric car? Such a capacitor (electrolytic) would be about 1 cm³ per 100 μF, for the voltage level required. Suppose we start with an initial voltage of 500 V and can extract all the

energy from the capacitor to drive the car for 1 hour. The auto drive train requires 1.5 kW. How big would the capacitor have to be? Is this practical?

Section 7-5 Series and Parallel Capacitors

P 7.5-1 Find the equivalent capacitance C_{eq} at terminals a–b for the circuit of Figure P 7.5-1.
Answer: $C_{eq} = 16$ mF

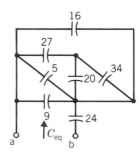

Figure P 7.5-1 All capacitances are given in millifarads.

P 7.5-2 Find C_{eq} looking into terminals a–b for the circuit of Figure P 7.5-2 when all the capacitances are in microfarads.
Answer: $C_{eq} = 4$ μF

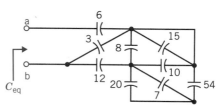

Figure P 7.5-2

P 7.5-3 Find C_{eq1}, C_{eq2}, v_1, v_2, and v for the circuit of Figure P 7.5-3.
Answer: $C_{eq1} = 24$ μF
$C_{eq2} = 16$ μF
$v_2 = 45$ V
$v_1 = 15$ V
$v = 15/2$ V

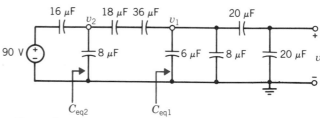

Figure P 7.5-3 The symbol C_{eq} represents the equivalent capacitance looking into the designated terminals.

P 7.5-4 Determine each capacitor voltage in the circuit of Figure P 7.5-4. Assume the circuit is at steady state.

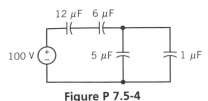

Figure P 7.5-4

P 7.5-5 For the series connection of N capacitors, as shown in Figure 7.17, determine the voltage divider relationship v_N/v where N is the capacitor for which we seek the output voltage. When we have three equal capacitors, determine the ratio v_N/v. Assume the initial voltage of each capacitor is zero.

Section 7-6 Inductors

P 7.6-1 Nikola Tesla (1857–1943) was an American electrical engineer who experimented with electric induction. Tesla built a large coil with a very large inductance, shown in Figure P 7.6-1. The coil was connected to a source current

$$i_s = 100 \sin 400t \text{ A}$$

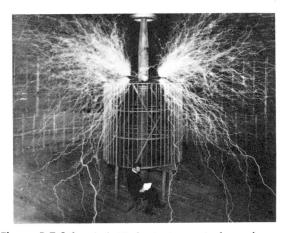

Figure P 7.6-1 Nikola Tesla sits impassively as alternating current induction coils discharge millions of volts with a roar audible 10 miles away (about 1910). Courtesy of Burndy Library.

so that the inductor current $i_L = i_s$. Find the voltage across the inductor and explain the discharge in the air shown in the figure. Assume that L = 200 H, and the average discharge distance is 2 m. Note the dielectric strength of air is 3×10^6 V/m.

P 7.6-2 The model of an electric motor consists of a series combination of a resistor and inductor. A current $i(t) = 4te^{-t}$ A flows through the series combination of a 10-Ω resistor and 0.1-henry inductor. Find the voltage across the combination.

Answer: $v(t) = 0.4e^{-t} + 39.6te^{-t}$ V

P 7.6-3 The current through a 20-mH inductor is shown in Figure P 7.6-3. Find the inductor voltage at $t = 1$ ms and $t = 6$ ms.

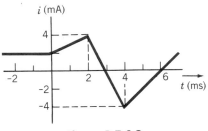

Figure P 7.6-3

P 7.6-4 Find R of the circuit shown in Figure P 7.6-4 if $v_1 = 1e^{-200t}$ V for $t \geq 0$.

Answer: $R = 80 \, \Omega$

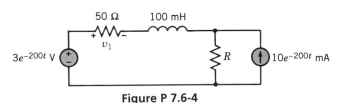

Figure P 7.6-4

P 7.6-5 Determine $i_L(t)$ for $t > 0$ when $i_L(0) = -2 \, \mu$A for the circuit of Figure P 7.6-5a when $v_s(t)$ is as shown in Figure P 7.6-5b.

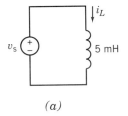

(a)

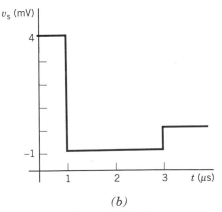

(b)

Figure P 7.6-5

P 7.6-6 Determine $v(t)$ for $t > 0$ for the circuit of Figure P 7.6-6*a* when $i_L(0) = 0$ and i_s is as shown in Figure P 7.6-6*b*.

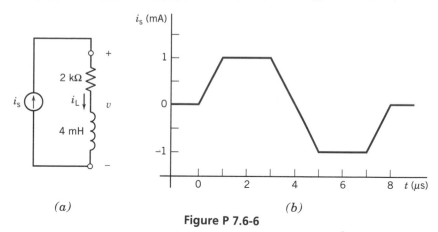

(a)

(b)

Figure P 7.6-6

P 7.6-7 A *gyrator* consists of the two voltage-controlled sources shown in Figure P 7.6-7. It is shown connected to a capacitor at the terminals c–d. Show that the *i–v* relationship at terminals a–b is that of an inductor. Assume that the capacitor is initially uncharged.

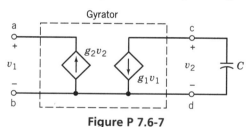

Figure P 7.6-7

Section 7-7 Energy Storage In An Inductor

P 7.7-1 The current through a 10-mH inductor used in a fusion power experiment is given by $i(t) = 2 \sin^2 t$ mA for $t > 0$. Find the instantaneous power in the inductor and the energy stored in the inductor at $t = 3\pi/4$ s.
Answer: $P(t) = 8 \times 10^{-8}(\sin^3 t) \cos t$ W
$\qquad w = 5$ nJ

P 7.7-2 A current of $5e^{-4t}$ mA flows through a series combination of a 10-Ω resistor and a 0.1-H inductor. Find the total power absorbed by this series combination as a function of time.

P 7.7-3 The current in a 10-mH inductor is zero for $t < 0$ and $5te^{-t}$ A for $t \geq 0$. Find the time when the maximum power is delivered to the inductor. At what time is the energy stored in the inductor a maximum?

P 7.7-4 A 1.0-mH inductor connected in a telephone circuit is found to have a voltage across it as shown in Figure P 7.7-4. The initial inductor current is given by $i(0) = 20$ mA. Find (a) $i(t)$ for $t > 0$ with t in milliseconds and i in milliamps, (b) the initial energy in the inductor at

$t = 0$, and (c) the net amount of energy that was transferred to the inductor between $t = 0$ ms and $t = 10$ ms. Hint: Use units of ms, mA and mH.
Answer: (a) $\quad i(t) = 20 + 6t^2$ mA $(0$ ms $< t < 2$ ms$)$
$\qquad i(t) = -4 + 24t$ mA $(2$ ms $< t < 4$ ms$)$
$\qquad i(t) = -36 + 40t - 2t^2$ mA
$\qquad (4$ ms $< t < 10$ ms$)$
$\qquad i(t) = 164$ mA $(t > 10$ ms$)$
$\qquad$(b) $\quad w(0) = 0.2\ \mu$J

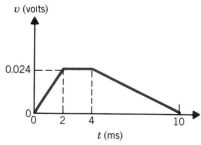

Figure P 7.7-4

P 7.7-5 For the circuit shown in Figure P 7.7-5, determine $v(t)$ and the energy stored in L_2 when $i_s = 3e^{-6t}$ mA, $t \geq 0$.
Partial Answer: $w(t) = 1.125e^{-12t}\ \mu$J

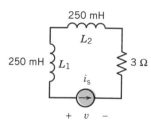

Figure P 7.7-5

Section 7-8   Series and Parallel Inductors

P 7.8-1   Find the equivalent inductance L_{eq} for terminals a–b for the circuit of Figure P 7.8-1 when all the inductances are in microhenrys.
Answer: $L_{eq} = 27\ \mu H$

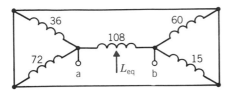

Figure P 7.8-1 The symbol L_{eq} represents the equivalent inductance looking into the designated terminals.

P 7.8-2   Find the equivalent inductance L_{eq} at terminals a–b of the circuit of Figure P 7.8-2 when the switch S is at (a) position A and (b) position B.

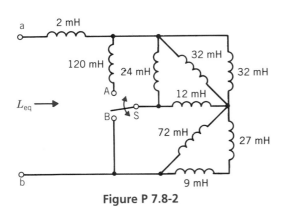

Figure P 7.8-2

P 7.8-3   Given that $i_1 = 0.5$ A and $i_3 = 0.2$ A at $t = 0$, find $i(t)$ for $t \geq 0$ for the circuit of Figure P 7.8-3.
Answer: $i(t) = 1.06 - 0.48e^{-5000t}$ A

Section 7-9   Initial Conditions of Switched Circuits

P 7.9-1   Find $v(0^+)$ and $dv_c(0^+)/dt$ for the circuit of Figure P 7.9-1.

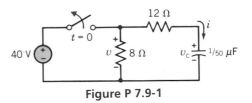

Figure P 7.9-1

P 7.9-2   Find $dv_c(0^+)/dt$ and $di_L(0^+)/dt$ for the circuit of Figure P 7.9-2 if $v_c(0^-) = 8$ V and $i_L(0^-) = 4$ A.
Answer: $dv_c(0^+)/dt = +60$ V/μs
$di_L(0^+)/dt = -29$ mA/μs

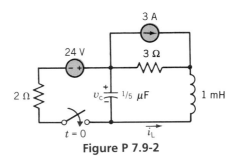

Figure P 7.9-2

P 7.9-3   Find $di_L(0^-)/dt$ for the circuit of Figure P 7.9-3 if $i_L(0^-) = 5$ mA.

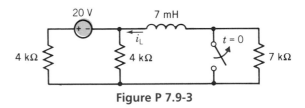

Figure P 7.9-3

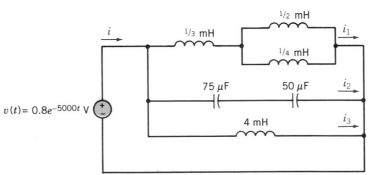

Figure P 7.8-3

P 7.9-4 Find $v_c(0^+)$ and $dv_c(0^+)/dt$ if $v(0^-) = 15$ V for the circuit of Figure P 7.9-4.

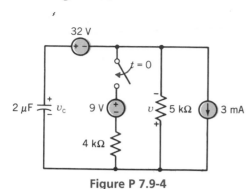

Figure P 7.9-4

P 7.9-5 For the circuit shown in Figure P 7.9-5, find $dv_c(0^+)/dt$, $di_L(0^+)/dt$, and $i(0^+)$ if $v(0^-) = 16$ V. Assume that the switch was closed for a long time prior to $t = 0$.

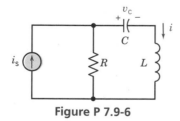

Figure P 7.9-5

P 7.9-6 Determine

$$\left.\frac{dv_c}{dt}\right|_{t=0^+} \quad \text{and} \quad \left.\frac{di}{dt}\right|_{t=0^+}$$

for the *RLC* circuit shown in Figure P 7.9-6 when $v_c(0^-) = -1$ V and $i_s = 2e^{-t}$ for $t \geq 0$ and $i_s = 0$ for $t < 0$.

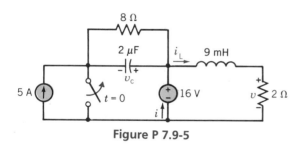

Figure P 7.9-6

P 7.9-7 For the circuit of Figure P 7.9-7, determine the current and voltage of each passive element at $t = 0^-$ and $t = 0^+$. The current source is $i_s = 0$ for $t < 0$ and $i_s = 4$ A for $t \geq 0$.

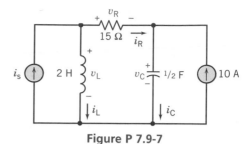

Figure P 7.9-7

P 7.9-8 Determine $i(0^+)$, $v(0^+)$, $dv(0^+)/dt$, and $di(0^+)/dt$ for the circuit of Figure P 7.9-8 when $C = 1/5$ F and $L = 1/2$ H.

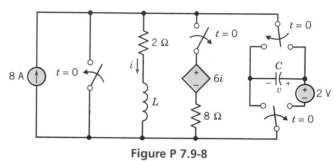

Figure P 7.9-8

P 7.9-9 For the circuit of Figure P 7.9-9, determine the value of *R* such that the energy stored in both the capacitor and the inductor is the same at steady state (when $t \to \infty$).

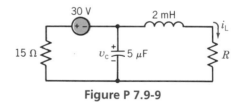

Figure P 7.9-9

Section 7-10 The Operational Amplifier and *RC* Circuits

P 7.10-1 For the integrator circuit of Figure 7-50, we have $R = 10$ kΩ and the initial capacitor voltage is $v_c(0) = 1$ V. Select v_s and C so that $v_o = (1 + t)$ V, $t \geq 0$. Assume the circuit is in steady state at $t = 0^-$.

P 7.10-2 For the differentiator circuit of Figure 7-51, we have $R = 100$ kΩ and $C = 10$ μF. Determine and sketch v_o if

$$v_s = 2 - 3t \qquad t \geq 0$$
$$= 2 \qquad t < 0$$

P 7.10-3 Consider the circuit shown in Figure 7-51. (a) Find $v_o(t)$ and show that the circuit performs as a differentiator. (b) Find values for R and C so that $v_o = -\left(\frac{1}{10}\right) dv_s/dt$. Assume an ideal op amp.

VERIFICATION PROBLEMS

VP 7-1 A student plans to build the circuit shown in Figure VP 7-1*a* and use a voltage source as shown in Figure VP 7-1*b*. The resistance of the wires is considered negligible in the model shown in Figure VP 7-1*a*. A computer model estimates infinite current, $i(t)$, at $t = 1$ s and $t = 3$ s. Verify this result and provide advice to the student experimenter.

VP 7-2 A student report for the circuit of Figure VP 7-2 states that $v_c(0^-) = 8$ V and $i_L(0^-) = 2$ A. Verify these results.

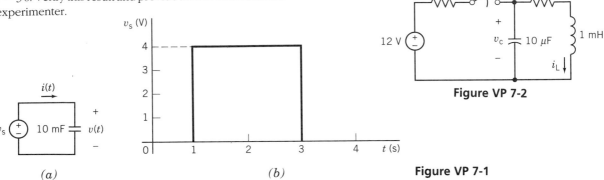

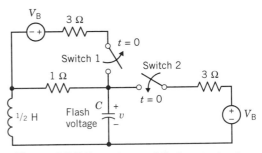

Figure VP 7-2

(a)   (b)   **Figure VP 7-1**

DESIGN PROBLEMS

DP 7-1 Select the resistance R for the circuit shown in Figure DP 7-1 so that $v(0) = 20$ V and $i(0) = 5$ A. Assume that the switch has been closed for a long time.

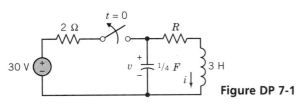

Figure DP 7-1

DP 7-2 A laser pulse power circuit is shown in Figure DP 7-2. It is required that $v(0) = 7.4$ V and $i(0) = 3.7$ A. Determine the required resistance R. Assume that the switch has been closed for a long time before it is opened at $t = 0$.

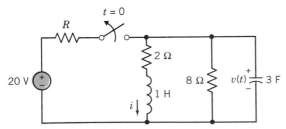

Figure DP 7-2 Laser pulse power circuit.

DP 7-3 A high-speed flash unit for sports photography requires a flash voltage $v(0^+) = 3$ V and

$$\left. \frac{dv(t)}{dt} \right|_{t=0^+} = 24 \text{ V/s}$$

The flash unit uses the circuit shown in Figure DP 7-3. Switch 1 has been closed a long time, and switch 2 has been open a long time at $t = 0$. Actually, the long time in this case is 3 s. Determine the required battery voltage, V_B, when $C = 1/8$ F.

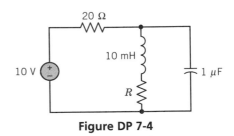

Figure DP 7-3 High-speed flash unit circuit.

DP 7-4 For the circuit shown in Figure DP 7-4, select a value of R so that the energy stored in the inductor is equal to the energy stored in the capacitor at steady state.

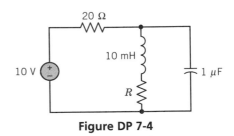

Figure DP 7-4

CHAPTER 8

THE NATURAL RESPONSE OF A FIRST-ORDER *RL* OR *RC* CIRCUIT

PREVIEW

In the last chapter, we examined the characteristics of two electrical energy storage elements: the inductor and the capacitor. In this chapter, we consider the response of a simple circuit consisting of a resistor connected to a capacitor or an inductor. These simple circuits can be described by one first-order differential equation. We determine the response of the circuit with all independent sources disconnected and obtain the circuit's natural response.

8-1 DESIGN CHALLENGE

SECURITY ALARM CIRCUIT

Problem

A switching circuit is used in a security alarm circuit for a building. The two switches in the circuit of Figure 8D-1 are thrown simultaneously at $t = 0$ by an intruder interrupting a sonar sensor that then activates the switches. The current i_2 activates alarms at the security office and the police station when it changes from a positive quantity to a negative quantity. The goal is to achieve this change in current i_2 within 0.4 s. Select L_2 so that the alarm system meets the requirements.

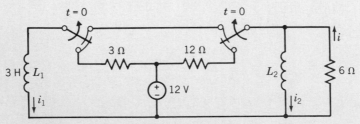

Figure 8D-1 Switching circuit for security alarm.

Define the Situations, State the Assumptions, and Develop a Model

The two switches have been at the lower setting for a long time prior to switching at $t = 0$.

Figure 8D-2 shows the circuit after time $t = 0$. The inductors are in parallel, so they can be replaced by a single equivalent inductor. The circuit will then consist of a single inductor and a single resistor. This chapter shows how to analyze this type of circuit to express the inductor current as a function of time. We will need to do this analysis to solve the design problem, so we will set the problem aside for now and return to it at the end of the chapter.

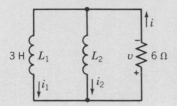

Figure 8D-2 The security alarm circuit of Figure 8D-1 for $t \geq 0$.

8-2 TELECOMMUNICATIONS AND SWITCHED CIRCUITS

Professor Charles Wheatstone of King's College in London teamed with William Cooke to experiment with telegraph circuits in 1837. Their first patent, the world's first for electrical communication, was sealed on June 12, 1837.

Wheatstone, shown in Figure 8-1, was also the inventor of the bridge circuit now associated with his name.

Figure 8-1 Charles Wheatstone (1802–1875), who was one of the first to experiment with telegraph circuits. Courtesy of the Institution of Electrical Engineers.

As we saw earlier, S. F. B. Morse independently invented and developed the telegraph in America. He opened the telegraph to the public on May 24, 1844. Meanwhile, Wheatstone introduced his automatic Morse transmitter, a machine for which he was knighted and that was widely used for over half a century. In many cases, punched paper tape was used to store the message for transmission and to feed the data to the telegraph key. This automatic system was used almost exclusively by the British.

The principle of the telegraph required an electric circuit to be closed and opened in a meaningful, patterned way. In its simplest form, the concept can be represented by the circuit shown in Figure 8-2. The sender of the telegraph message operated the sender key, which caused a current to flow, activating the receiving key. The source voltage v_s was a battery, but the current varied as a function of time. This type of switched circuit became important, and the analysis of its behavior became critical.

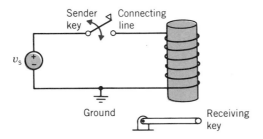

Figure 8-2 Telegraph with a sending and receiving key or switch.

A **telegraph** is a device for communication of messages using the activation of a switch or key to send a code.

On December 12, 1901, Guglielmo Marconi, shown in Figure 8-3, received a wireless signal from Cornwall, England, 2800 km across the Atlantic Ocean. Marconi started in Italy

Figure 8-3 Guglielmo Marconi (1874–1937). Courtesy of the Institution of Electrical Engineers.

but went to England to work on his spark transmitter. He received a patent in 1896, and Marconi and his supporters formed a company called the Wireless Telegraph Signal Company. Marconi and others built a transmitter consisting of a spark gap of 3 mm fed by a voltage source and an inductor, as shown in Figure 8-4. Again, the switching action was

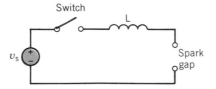

Figure 8-4 Circuit model of Marconi's transmitter for radio waves.

important, as it initiated a series of sparks across the gap. The spark caused a series of radiated waveforms to be transmitted through the air and received by a similar circuit some distance away.

It is important to note that the wireless transreceiver, or radio, was in part invented by Nikola Tesla with his patent filed on September 8, 1897. Nevertheless, it was Marconi who turned his and Tesla's inventions into a practical radio system.

Radio transmission is the sending of communication messages by means of radiated electromagnetic waves other than heat or light waves.

It is clear, then, that a method of analyzing a switched circuit became very critical to the design and development of electrical telecommunications. By the turn of the century, electrical engineering students, such as those shown in Figure 8-5, needed robust analytical tools when studying switched circuits. Today we use switched circuits widely. For example, a circuit board that controls a numerical indicator, as shown in Figure 8-6, uses resistors, capacitors, and switches to direct the numerals on the indicator.

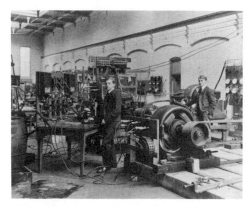

Figure 8-5 Circuits laboratory of Electrical Engineering Department, Massachusetts Institute of Technology, about 1900. Courtesy of the MIT Museum.

Figure 8-6 Circuit board that controls a numerical indicator. This circuit contains resistors, capacitors, and switches. Courtesy of DAC Inc.

In this chapter we consider the behavior of a circuit containing one or more resistors, a switch, and one energy storage element. The energy storage element may be a capacitor *or* an inductor.

8-3 | THE SOURCE-FREE RESPONSE OF A CIRCUIT

As we noted in the preceding chapter, the key attribute of an energy storage element is its ability to store energy. Consider the circuit shown in Figure 8-7 with a resistor and an inductor connected to a source when $t < 0$. We then open the switch at $t = 0$ and wish to calculate the current i_L for $t \geq 0$. Note that the original source is used to establish $i_L(0^-)$, and then the switch is opened at $t = 0$ and the source is disconnected. We assume, unless otherwise stated, that the switch has been closed for a long time prior to $t = 0$ and thus steady-state conditions prevail at $t = 0^-$.

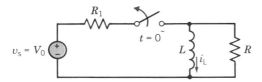

Figure 8-7 Circuit with an inductor and resistor switched at $t = 0$.

A similar situation exists for the capacitor in the circuit shown in Figure 8-8. The constant-voltage source will establish a capacitor voltage $v_c(0^-)$ before the switch is thrown to terminal b at $t = 0$. We assume that the switch connecting the capacitor to the voltage source has been connected to terminal a for a long time before $t = 0$ so that we can readily find $v_c(0^-)$ assuming steady-state conditions. Then the switch is thrown to position b and the resulting *RC* circuit is disconnected from the source.

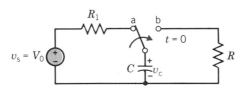

Figure 8-8 Circuit with a capacitor switched at $t = 0$.

Notice in both Figures 8.7 and 8.8 the source voltage is $v_s = V_0$, indicating a constant voltage source. The role of the voltage source in these circuits is to establish the initial condition, $v_C(0^-)$ or $i_L(0^-)$. A constant voltage source is used for convenience and without loss of generality.

Figures 8-7 and 8-8 incorporate models of physical devices that we call switches. Figure 8-7 shows a switch that can be either open or closed. Figure 8-8 shows a switch that can be connected to either of two terminals. In both cases we assume that the switch is actuated at the time indicated (usually $t = t_0 = 0$) and thrown in the direction indicated by the arrow. We also assume that the switch action occurs in a negligible period of time. Thus, the switch shown in Figure 8-8 moves from terminal a to terminal b in a negligible period of time.

In this chapter we are interested in the source-free response of a resistor–capacitor (*RC*) circuit or of a resistor–inductor (*RL*) circuit. That is, we wish to determine the response of a circuit free of independent sources. The source-free response will depend only on the initial energy stored in the energy storage element. Since the response depends only on the nature of the *RL* or *RC* circuit and not on external sources, it is called the *natural response*.

The **natural response** of a circuit depends only on the internal energy storage of the circuit and not on external sources.

Let us find the initial value of the inductor current at $t = 0^+$ for the circuit of Figure 8-7. If we consider the circuit prior to $t = 0$, we assume the switch has been closed for a long time. Then the inductor carries a constant current, and it appears as a short circuit.

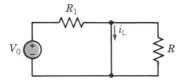

Figure 8-9 The *RL* circuit of Figure 8-7 immediately prior to $t = 0$.

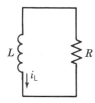

Figure 8-10 The *RL* circuit of Figure 8-7 disconnected from the independent source at $t = 0$.

Therefore, we have the equivalent circuit as shown in Figure 8-9. Clearly, $i_L(0^-) = V_0/R_1$. Since $i(0^+) = i(0^-)$, we now know the initial current in the inductor at $t = 0$ and can proceed to determine the response of the circuit shown in Figure 8-10 for $t \geq 0$.

Let us proceed to determine the initial conditions of the *RC* circuit shown in Figure 8-8. Prior to $t = 0$ when the switch is thrown, the capacitor appears as an open circuit and $v_c(0^-) = V_0$. After the switch is thrown, the initial voltage across the capacitor is

$$v_c(0^+) = v_c(0^-) = V_0$$

Then we may proceed to find the response of the circuit for $t \geq 0$ by using the circuit shown in Figure 8-11 with the initial voltage $v_c(0^+)$ known.

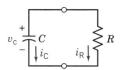

Figure 8-11 The *RC* circuit of Figure 8-8 disconnected from the independent source at $t = 0$.

Now let us find the voltage $v_c(t)$ for time greater than zero. Writing the KCL at the top node of the circuit of Figure 8-11, we have

$$i_c + i_R = 0$$

Since the reference directions of i_c and v_c adhere to the passive convention

$$i_c = C \frac{dv_c}{dt}$$

Notice that the voltage across the resistor is v_c. Since v_c and i_R adhere to the passive convention

$$i_R = \frac{v_c}{R}$$

Combining these equations we obtain

$$C \frac{dv_c}{dt} + \frac{v_c}{R} = 0 \tag{8-1}$$

This equation may be rewritten as

$$\frac{dv_c}{dt} + \frac{1}{RC} v_c = 0 \tag{8-2}$$

For the *RL* circuit shown in Figure 8-10 we write the KVL around the mesh, obtaining

$$L \frac{di_L}{dt} + Ri_L = 0$$

or

$$\frac{di_L}{dt} + \frac{R}{L}i_L = 0 \tag{8-3}$$

Examining Eqs. 8-2 and 8-3, we note that in both cases we have obtained a first-order differential equation. It is called a first-order constant-coefficient differential equation because the highest order derivative term is one. It is also called a first-order linear differential equation because there are no square or higher order power derivative terms. Therefore, we have the general form

$$\frac{dx}{dt} + ax = 0 \tag{8-4}$$

where $a = 1/RC$ for the *RC* circuit and $a = R/L$ for the *RL* circuit. Of course, $x = v_c$ for the *RC* circuit and $x = i_L$ for the *RL* circuit.

Equation 8-4 is called the source-free equation in electrical engineering, the homogeneous equation in mathematics, and the zero-input equation by others.

In the next three sections we will examine methods for obtaining the solution, $x(t)$, to the first-order differential equation (8-4).

8-4 | SOLUTION OF THE FIRST-ORDER DIFFERENTIAL EQUATION

We need to solve the differential equation that describes a circuit in order to determine the dependent variable (either v or i) as a function of time. We will consider three methods for solving the first-order differential equation (8-4): (1) separation of the variables, (2) use of an assumed exponential solution, and (3) use of differential operators.

A **first-order differential equation** is an equation in which the highest order derivative term is the first derivative, that is, dx/dt.

The first method is to separate the variables x and t of Eq. 8-4. We repeat Eq. 8-4 as

$$\frac{dx}{dt} + ax = 0 \tag{8-5}$$

and rearrange the equation to obtain

$$\frac{dx}{x} = a\,dt$$

Taking the integral of both sides, we have

$$\int \frac{dx}{x} = -a \int dt$$

Consequently we obtain

$$\ln x = -at + K \tag{8-6}$$

where K is a constant of integration that will be selected to satisfy the initial condition $x(0^+)$. Henceforth, we use the notation $x(0)$ instead of $x(0^+)$ for convenience. Thus, we have $x(0^-)$ prior to the switch throw and $x(0)$ immediately after.

For the RC circuit, recall that $x = v_c$ and $v_c(0) = v_c(0^-) = V_0$. Therefore, at $t = 0$ from Eq. 8-6, we obtain

$$\ln v_c(0) = K$$

or

$$\ln V_0 = K$$

Substituting this value into Eq. 8-6 and using $x = v_c$, we have

$$\ln v_c = -at + \ln V_0$$

or

$$\ln v_c - \ln V_0 = -at$$

Therefore,

$$\ln \frac{v_c}{V_0} = -at$$

Since $e^{\ln y} = y$, we have

$$\frac{v_c}{V_0} = e^{-at}$$

or

$$v_c(t) = V_0 e^{-at}$$
$$= V_0 e^{-t/RC}$$

for the RC circuit where V_0 is the initial capacitor voltage.

We leave it as an exercise to prove that

$$i_L = I_0 e^{-Rt/L}$$

is the response of the RL circuit where I_0 is the initial inductor current $i_L(0)$. Since the response is characterized by the circuit elements and the initial voltage, it is called the natural response.

EXERCISE 8-1

Find the natural response for the inductor current i_L for the circuit of Figure 8-7.
Answer: $i_L = (V_0/R_1)e^{-Rt/L} \qquad t \geq 0$

EXERCISE 8-2

A practical capacitor will ultimately lose its charge over time and can be modeled by the RC circuit of Figure E 8-2 where R is the leakage resistance. Determine the voltage ratio $v_C(t)/V_0$ at $t = 100$ s for $C = 1$ mF and $R = 1$ MΩ .
Answer: 0.905

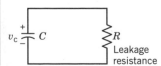

Figure E 8-2 Capacitor with leakage resistance.

8-5 THE EXPONENTIAL SOLUTION TO THE FIRST-ORDER DIFFERENTIAL EQUATION

Again, we consider the general first-order differential equation, repeating Eq. 8-5, as

$$\frac{dx}{dt} + ax = 0 \tag{8-7}$$

We will solve the equation by using an exponential solution of the form

$$x = Ae^{st} \tag{8-8}$$

where A and s are constants to be determined. The exponential form is an appropriate form for the solution since x and the derivative of x will both be exponentials.

An **exponential solution** is an assumed solution of the form $x = Ae^{st}$ where A and s are constants to be determined.

Let us substitute Eq. 8-8 into Eq. 8-7 to obtain

$$\frac{d(Ae^{st})}{dt} + aAe^{st} = 0$$

or

$$sAe^{st} + aAe^{st} = 0$$

Therefore, we have

$$(s + a)Ae^{st} = 0$$

and to be true either $(s + a)$ or Ae^{st} must be zero. We note that Ae^{st} cannot equal zero or the trivial solution, $x = 0$, results for all time. Consequently, we find that

$$(s + a) = 0$$

or $s = -a$. Substituting $s = -a$ into Eq. 8-8, we have

$$x = Ae^{-at} \tag{8-9}$$

where A will be determined by using the initial condition of x.

As an example of this method, let us consider the *RL* circuit of Figure 8-10. Recall this circuit evolved from Figure 8-7. The differential equation for the circuit is

$$\frac{di}{dt} + \frac{R}{L} i = 0 \tag{8-10}$$

If we assume the exponential form of the solution, we write the solution as

$$i = Ae^{st}$$

Since

$$\frac{di}{dt} = sAe^{st}$$

we have from Eq. 8-9

$$sAe^{st} + \frac{R}{L} Ae^{st} = 0$$

Table 8-1

Source-Free Response of RL and RC Circuits

Circuit	Source-Free Response
RL	$i_L(t) = i_L(0) e^{-Rt/L}$
RC	$v_c(t) = v_c(0) e^{-t/RC}$

and therefore $s = -R/L$ is required. Then we know that

$$i = Ae^{-Rt/L} \tag{8-11}$$

In order to find A, we recall from Figure 8-7 that $i(0) = i_L(0^-) = V_0/R_1$. Hence, substituting $t = 0$ in Eq. 8-11 yields

$$i(0) = A$$

and so

$$A = \frac{V_0}{R_1}$$

Therefore, we have the final solution for the RL circuit of Figure 8-10 as

$$i = \frac{V_0}{R_1} e^{-Rt/L}$$

In Table 8-1 we summarize the general source-free solution of the RL and RC circuits.

8-6 ∥ DIFFERENTIAL OPERATORS

In this section we introduce the differential operator s.

An **operator** is a symbol that represents a mathematical operation.

We can define a differential *operator* s such that

$$sx = \frac{dx}{dt} \quad \text{and} \quad s^2x = \frac{d^2x}{dt^2}$$

Thus, the operator s denotes differentiation of the variable with respect to time. The utility of the operator s is that it can be treated as an algebraic quantity. This permits the replacement of differential equations with algebraic equations, which are easily handled.

Again, let us consider the first-order differential equation

$$\frac{dx}{dt} + ax = 0 \tag{8-12}$$

Using the operator s as the differential operator, we have

$$sx + ax = 0$$

which may be rewritten as

$$(s + a)x = 0 \tag{8-13}$$

Since we do not permit x to be equal to zero (a trivial solution), we require $s = -a$. Furthermore, the solution can be postulated to be an exponential of the form

$$x = Ae^{st} \tag{8-14}$$

and the value of x can be found by substituting $s = -a$ into Eq. 8-14 to obtain

$$x = Ae^{-at}$$

Then, using the initial condition of x as before, we determine the value of A.

Use of the s operator is particularly attractive when higher order differential equations are involved. Then we use the s operator so that

$$s^n x = \frac{d^n x}{dt^n} \quad \text{for} \quad n \geq 0$$

We assume that $n = 0$ represents no differentiation so that

$$s^0 = 1$$

which implies $s^0 x = x$.

Because integration is the inverse of differentiation, we define

$$\frac{1}{s} x = \int_{-\infty}^{t} x \, d\tau \tag{8-15}$$

The operator $1/s$ must be shown to satisfy the usual rules of algebraic manipulations. Of these rules, the commutative multiplication property presents the only difficulty. Thus, we require

$$s \cdot \frac{1}{s} = \frac{1}{s} \cdot s = 1 \tag{8-16}$$

Is this true for the operator s? First, we examine Eq. 8-15. Multiplying Eq. 8-15 by s yields

$$s \cdot \frac{1}{s} x = \frac{d}{dt} \int_{-\infty}^{t} x \, d\tau$$

or

$$x = x$$

as required. Now we try the reverse order by multiplying sx by the integration operator to obtain

$$\frac{1}{s} sx = \int_{-\infty}^{t} \frac{dx}{d\tau} \, d\tau$$
$$= x(t) - x(-\infty)$$

Therefore

$$\frac{1}{s} sx = x$$

only when $x(-\infty) = 0$. From a physical point of view, we require that all capacitor voltages and inductor currents be zero at $t = -\infty$. Then the operator $1/s$ can be said to satisfy Eq. 8-16 and can be manipulated as an ordinary algebraic quantity.

We will use operators in this and succeeding chapters since they permit relatively easy algebraic manipulation of the equations resulting from the use of the operator.

8-7 | THE RESPONSE OF A FIRST-ORDER CIRCUIT

In the preceding sections we found that an RL circuit and an RC circuit can both be described by the general first-order differential equation

$$\frac{dx}{dt} + ax = 0$$

The solution of this differential equation was shown to be

$$x = Ae^{-at}$$

where A is determined by the initial condition $x(0)$. The response may also be written as

$$x = Ae^{-t/\tau}$$

where $\tau = 1/a$. The constant τ is called the *time constant* of the circuit and is denoted by the Greek letter tau. Note that the units of the time constant are seconds.

For an RL circuit it is easy to show that $\tau = L/R$. For an RC circuit, we find that $\tau = RC$. Since the circuit will exhibit an exponentially decaying waveform, we note the values of $e^{-t/\tau}$ in Table 8-2 for $t = n\tau$. We show the behavior of the response $x(t)$ in Figure 8-12. It is clear that the response of the RL or RC circuit is a function of the magnitude of L/R or RC,

Table 8-2
Values of $e^{-t/\tau}$ for $t = n\tau$

$t = n\tau$	τ	2τ	3τ	4τ	5τ
$e^{-t/\tau}$	0.368	0.135	0.050	0.018	0.007

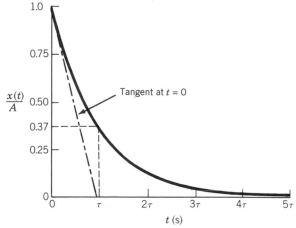

Figure 8-12 Response of $x(t)/A$ for the first five time constants. The dashed line is the tangent at $t = 0$.

respectively. If an *RC* circuit is to respond rapidly, a small *RC* is needed. On the other hand, to hold the capacitor voltage near its original value, a large *RC* is needed. Thus, the design engineer can use the selection of the circuit elements to obtain the desired circuit response.

The capacitor voltage of the *RC* circuit shown in Figure 8-11 is described by the equation

$$v_C = V_0 e^{-t/\tau}$$

where $\tau = RC$. The response v_C has declined to 36.8 percent of its initial value at $t = \tau$ as shown in Figure 8-12. Also, note that the slope of the tangent of v_C at $t = 0$ is equal to $-V_0/T$.

If we wish to hold the capacitor voltage nearly equal to V_0 for 10 s, we will need a large *RC* product. For example, if we wished the voltage to decay only 10 percent, we would require

$$0.9\ V_0 = V_0 e^{-t/\tau}$$

and, therefore we require $t/\tau = 0.105$. Since $t = 10$ s, we need $\tau = 95$. This can be obtained by selecting $R = 10\ k\Omega$ and $C = 9.5\ mF$.

Note that after $t = 5\tau$ the response $x(t)$ is less than 1 percent of its initial value (see Table 8-2). Thus, we often say that the response of the circuit has reached its final value in approximately five time constants. Since the response is brief and transitory, it is often called the *transient response, $x(t)$*, of the circuit. The response that exists a long time ($t > 5\tau$) after the switching operation is performed at $t = 0$ is called the *steady-state response*.

The **transient response** is the temporary response that disappears with time.

It is clear that the response of a first-order source-free circuit may be determined by (1) the circuit time constant and (2) the initial condition.

Example 8-1

Find the response $i(t)$ for $t > 0$ for the circuit of Figure 8-13. The resistances are $R_1 = 3\ \Omega$ and $R_2 = 2\ \Omega$.

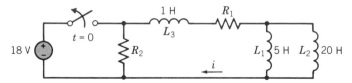

Figure 8-13 Circuit for Example 8-1.

Solution

First, after the switch is opened we note that the parallel inductors can be reduced to one equivalent inductance L_p. Then we have

$$L_p = \frac{L_1 L_2}{L_1 + L_2}$$
$$= \frac{5 \times 20}{5 + 20}$$
$$= 4\ H$$

The total inductance L_T in the loop is

$$L_T = L_3 + L_p$$
$$= 1 + 4$$
$$= 5 \text{ H}$$

The total resistance in the loop, R_T, is

$$R_T = R_1 + R_2$$
$$= 5 \ \Omega$$

Thus, we may represent the circuit of Figure 8-13 by its equivalent, for $t \geq 0$, as shown in Figure 8-14. Notice that the current i in Figure 8-14 is equal to the current i in Figure 8-13. Then the response for $t \geq 0$ will be

$$i = i(0)e^{-t/\tau}$$

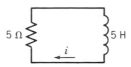

5 Ω   5 H   **Figure 8-14** Equivalent circuit for Example
                8-1 when $t > 0$.

where

$$\tau = \frac{L_T}{R_T}$$
$$= \frac{5}{5}$$
$$= 1 \text{ s}$$

We need to find $i(0)$ by examining the circuit of Figure 8-13 with the switch closed. Assuming the switch has been closed for a long time, we obtain the equivalent circuit at $t = 0^-$ as shown in Figure 8-15, where each inductor has been replaced by a short circuit. Then we have $i(0) = 6$ A.

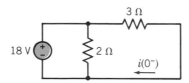

Figure 8-15 Equivalent circuit for Example
8-1 at $t = 0^-$.

Therefore, the current i is

$$i = 6e^{-t} \qquad t > 0$$

After a long time, as $t \to \infty$, the current is

$$i = 0$$

Thus, the steady-state current is zero.

The **steady-state** response is that which exists after a long time following any switching operation.

Example 8-2

Consider the circuit shown in Figure 8-16. Find the voltage across the capacitor, v, for $t \geq 0$. The switch is moved to position b at $t = 0$. Assume that the switch was in position a for a long time prior to switching at $t = 0$.

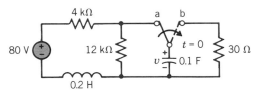

Figure 8-16 Circuit of Example 8-2.

Solution

The first step is to find the initial voltage across the capacitor, v, at $t = 0^-$. When the switch is in position a for a long time, the capacitor appears as an open circuit and the inductor appears as a short circuit to the constant voltage source. Therefore, the equivalent circuit for $t = 0^-$ is shown in Figure 8-17. Using the voltage divider principle,

$$v(0^-) = \frac{12}{16} 80$$
$$= 60 \text{ V}$$

At $t = 0$, $v(0) = v(0^-)$. When the circuit is switched to position b, the remaining circuit involving the capacitor is represented by Figure 8-18.

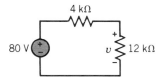

Figure 8-17 Equivalent circuit for Example 8-2 for $t = 0^-$.

Figure 8-18 Equivalent circuit for Example 8-2 for $t > 0$.

The time constant of the circuit is $\tau = RC = 3$. Recall that the voltage can then be described by

$$v = v(0)e^{-t/\tau}$$
$$= 60e^{-t/3} \text{ V} \qquad t \geq 0$$

EXERCISE 8-3

Find the current i for $t \geq 0$ for the circuit shown in Figure E 8-3 when $i(0^-) = 8$ A.
Answer: $i = 8e^{-1800t}, t \geq 0$

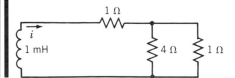

Figure E 8-3

EXERCISE 8-4

Find the voltage v for $t \geq 0$ for the circuit shown in Figure E 8-4 when $v(0^-) = 16$ V.
Answer: $v = 16e^{-1.875t}$ V

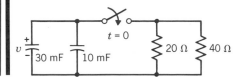

Figure E 8-4

EXERCISE 8-5

Determine the current $i(t)$ for $t \geq 0$ for the circuit shown in Figure E 8-5. Assume that the switch was closed a long time.

Answer: $i = \dfrac{2}{7} e^{-100t}$ A

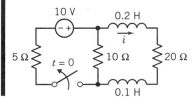

Figure E 8-5

8-8 ⏐⏐ ENERGY STORAGE AND DISSIPATION IN *RL* AND *RC* CIRCUITS

So far we have found the transient response of the *RL* and *RC* circuits. Next, we turn to a consideration of the energy storage and dissipation in these circuits. Again, let us consider the *RL* or *RC* circuit as redrawn in Figures 8-19 and 8-20, respectively. The voltage across

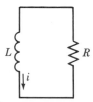

Figure 8-19 An *RL* circuit.

Figure 8-20 An *RC* circuit.

the capacitor for the circuit of Figure 8-20 is

$$v = V_0 e^{-t/\tau} \qquad t \geq 0$$

where $V_0 = v(0^+)$ and $\tau = RC$. Similarly, we found that the inductor current for the circuit of Figure 8-19 is

$$i = i(0)e^{-t/\tau} \qquad t \geq 0$$

where $\tau = L/R$. The instantaneous power absorbed by the resistor for the *RL* circuit is

$$p = i^2R$$
$$= i^2(0) \, Re^{-2t/\tau} \qquad t \geq 0$$

The instantaneous power absorbed by the resistor in the *RC* circuit is

$$p = \frac{v^2}{R}$$
$$= \frac{V_0^2}{R} e^{-2t/\tau} \qquad t \geq 0$$

The power for both the *RL* and *RC* circuits is of the form $p = p_0 e^{-2t/\tau}$. Therefore, the energy absorbed by the resistor after $t = 0$ is

$$w = \int_0^t p \, dt$$
$$= \int_0^t p_0 e^{-2t/\tau} \, dt$$
$$= p_0 \frac{\tau}{2} (1 - e^{-2t/\tau})$$

For the *RL* circuit we have $\tau = L/R$ so that

$$w = i^2(0)R \left(\frac{L}{2R}\right)(1 - e^{-2Rt/L})$$

or

$$w = \frac{1}{2} LI_0^2(1 - e^{-(2R/L)t}) \qquad t \geq 0 \tag{8-17}$$

where $I_0 = i(0)$.

For the *RC* circuit we have $\tau = RC$ and

$$w = \left(\frac{V_0^2}{R}\right)\frac{RC}{2}(1 - e^{-2t/RC})$$
$$= \frac{1}{2} CV_0^2(1 - e^{-2t/RC}) \qquad t \geq 0 \tag{8-18}$$

The initial energy stored in the capacitor at $t = 0$ is $\frac{1}{2}CV_0^2$, and this energy is transferred to the resistor as time progresses. For the *RC* circuit (using Eq. 8-18), we have $w = 0$ at $t = 0$, and no energy has yet been dissipated by the resistor. At $t = \infty$, we have from Eq. 8-18

$$w = \frac{1}{2} CV_0^2$$

and so all the energy originally stored by the capacitor has been dissipated by the resistor. Similarly, the inductor has an initially stored energy of

$$w_{\text{L}}(0) = \frac{1}{2} LI_0^2$$

at $t = 0$. However, at $t = \infty$ the inductor has transferred all its energy to the resistor and using Eq. 8-17 the total energy dissipated by the resistor is given by

$$w = \frac{1}{2} LI_0^2$$

Example 8-3

For the circuit of Figure 8-21, find the current i and the energy dissipated in the 10-Ω resistor for $t > 0$. At what time is 90 percent of the energy stored in the inductor dissipated by the resistor?

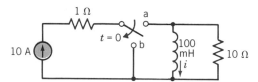

Figure 8-21 Circuit of Example 8-3. The switch moves from a to b instantaneously at $t = 0$.

Solution

First, we wish to find the current i for $t > 0$. We assume that the switch has been in position a for a long time and steady state prevails at $t = 0^-$. Since we require the initial current in the inductor, we note that the inductor appears as a short circuit to the constant current source at $t = 0^-$. Therefore, $i(0^-) = 10$ A. Also, $i(0^-) = i(0)$.

After the switch is moved to position b, we have $i(0) = I_0 = 10$ A and $\tau = L/R = 0.1/10 = 0.01$ s. The energy stored by the inductor at $t = 0$ is

$$w_L(0) = \frac{1}{2} LI_0^2$$

$$= \left(\frac{0.1}{2}\right)(10)^2$$

$$= 5 \text{ J}$$

The current i is

$$i = I_0 e^{-t/\tau}$$

where $\tau = L/R = 0.01$. Therefore, we obtain the current

$$i(t) = 10 \; e^{-100t}$$

The energy dissipated by the resistor is

$$w = \frac{1}{2} LI_0^2(1 - e^{-2t/\tau})$$

$$= 5(1 - e^{-200t})$$
$$= w(0)(1 - e^{-200t}) \text{ J}$$

We wish to find t_1 such that $w(t_1) = 0.1w(0)$ and the resistor has dissipated 90 percent of the original energy stored in the inductor. This occurs at $w(t_1)/w(0) = 0.1$ or

$$0.1 = 1 - e^{-200t_1}$$

Therefore, $t_1 = 0.53$ ms. Thus, in 0.53 ms this circuit dissipates 90 percent of the originally stored energy.

EXERCISE 8-6

For the circuit of Figure E 8-3, find the time when 50 percent of the energy originally stored in the inductor has been dissipated by the resistors.

EXERCISE 8-7

Consider the circuit of Figure E 8-4. Find the energy $w(t)$ dissipated by the two parallel resistors for $t \geq 0$.

EXERCISE 8-8

A capacitor is short-circuited by a switch and wire, as shown in Figure E 8-8, where R represents the resistance of the switch and wire and $R = 150$ mΩ. The original energy stored in the capacitor is 5 J. Determine the percent of the original energy remaining in the capacitor 0.6 ms after the switch closes. Also determine the current at $t = 0.8$ ms.
Answer: 1.8%, 32.75 A

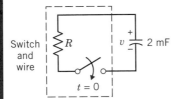

Figure E 8-8 Capacitor short-circuited by a switch and wire.

8-9 | *RL* AND *RC* CIRCUITS WITH DEPENDENT SOURCES

We have found that the response of an *RC* or *RL* circuit is of the form

$$x = Ae^{-t/\tau}$$

where the time constant is RC or L/R for the RC or RL circuit, respectively.

If a circuit is free of independent sources but includes a dependent source, the time constant of the circuit incorporates the effect of the dependent source.

Consider the *RL* circuit shown in Figure 8-22, which includes a current-controlled

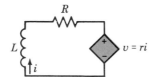

Figure 8-22 An *RL* circuit with a dependent voltage source.

voltage source. We will assume the initial current is $i(0) = I_0$. The KVL equation for the loop is

$$L \frac{di}{dt} + Ri + ri = 0$$

This loop equation may be rewritten as

$$\frac{di}{dt} + \frac{R+r}{L} i = 0$$

Clearly, the effect of the controlled voltage source is to change the time constant to

$$\tau = \frac{L}{R+r}$$

Then the solution for the current is

$$i = I_0 e^{-t/\tau}$$

Example 8-4

Find the capacitor voltage v for $t > 0$ when $v(0) = V_0$ for the circuit of Figure 8-23.

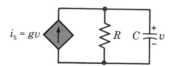

Figure 8-23 An *RC* circuit with a dependent current source.

Solution

First, we use KCL to find the sum of the currents at the upper node as

$$\frac{v}{R} + C\frac{dv}{dt} - gv = 0$$

This equation can be rewritten as

$$C\frac{dv}{dt} + \left(\frac{1}{R} - g\right)v = 0$$

or

$$\frac{dv}{dt} + \frac{k}{C}v = 0 \tag{8-19}$$

where

$$k = \left(\frac{1}{R} - g\right)$$

Since the solution of Eq. 8-19 is known to be

$$v = V_0 e^{-t/\tau}$$

the new time constant is given by

$$\tau = C/k$$

Of course, when $g = 0$, the dependent source is removed, and we have the familiar time constant $\tau = RC$. Note that for the special case $g = 1/R$ we have $k = 0$ and $v(t) = V_0$ for $t \geq 0$.

Example 8-5

Find the inductor current i and the capacitor voltage v for $t \geq 0$ for the circuit of Figure 8-24*a*. Assume the circuit is in steady state at $t = 0^-$.

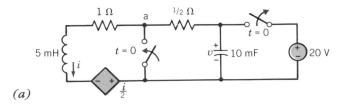

(a)

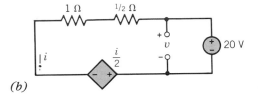

(b)

Figure 8-24 *(a)* Circuit of Example 8-5.
(b) Circuit of Example 8-5 at $t = 0^-$.

Solution

First, we redraw the circuit, as shown in Figure 8-24*b*, in order to determine the voltage $v(0^-)$ and the current $i(0^-)$. The inductor is replaced with a short circuit and the capacitor with an open circuit, as shown. Examining Figure 8-24*b*, we find that

$$v(0^-) = 20 \text{ V}$$

In order to find $i(0^-)$, we write KVL around the loop to obtain

$$\frac{3}{2} i - \frac{i}{2} = 20$$

and, hence,

$$i(0^-) = 20 \text{ A}$$

Therefore, the initial conditions are $i(0) = 20$ A and $v(0) = 20$ V.

After the switches are thrown at $t = 0$, we have the circuit shown in Figure 8-25. The short-circuit wire has the effect of bringing the circuit together at node a, as shown. Since *i* must enter a and leave a, it is clear that no current flows from the right-hand mesh into the left-hand mesh and vice versa. Thus, the two meshes are independent. We may then solve the response for the *RL* circuit and the *RC* circuit independently.

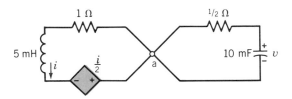

Figure 8-25 Circuit of Figure 8-24*a* at
$t = 0^+$.

The *RC* circuit has a time constant $\tau = RC = 5 \times 10^{-3}$ s and $v(0) = 20$ V. Therefore,

$$v = V(0)e^{-t/\tau}$$
$$= 20e^{-200t}$$

The sum of the voltages in the *RL* circuit is

$$L \frac{di}{dt} + Ri - \frac{i}{2} = 0$$

Consequently, we have

$$\frac{di}{dt} + \frac{R - \frac{1}{2}}{L} i = 0$$

and for $R = 1$ and $L = 5 \times 10^{-3}$ we have

$$\frac{di}{dt} + \frac{1}{\tau} i = 0$$

where

$$\tau = \frac{L}{R - \frac{1}{2}} = 10^{-2} \text{ s}$$

The solution for i is then

$$i = i(0)e^{-t/\tau}$$
$$= 20e^{-100t} \text{ A} \qquad t \geq 0$$

Example 8-6

Determine the inductor current $i_L(t)$ for $t \geq 0$ for the circuit of Figure 8-26. Assume the switch was at position 1 for a long time.

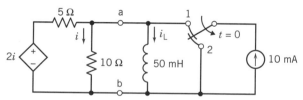

Figure 8-26 An inductive circuit with a dependent source.

Solution

Examining the circuit at $t = 0^-$, we readily determine that $i_L(0^-) = i_L(0^+) = 10$ mA. Note that the circuit to the left of the terminals a–b can be represented by a Thévenin resistance R_t if we disconnect the inductance L. Disconnect the inductor and connect a 1-A source to the terminals a–b as was accomplished on page 191 for the circuit of Figure 5-30. As shown on page 192, we determine that $R_t = \frac{50}{13}$ Ω. Then, reconnecting L to R_t, we obtain the simple RL circuit of Figure 8-10 with $i_L(0) = 10$ mA. Therefore

$$i_L = i_L(0)e^{-t/\tau}$$

Since

$$\tau = \frac{L}{R} = \frac{50 \times 10^{-3}}{(50/13)} = 13 \text{ ms}$$

Then

$$i_L(t) = 10e^{-76.9t} \text{ mA}$$

EXERCISE 8-9

For the circuit shown in Figure E 8-9, find i for $t > 0$ if $i(0) = 2$ A.
Answer: $i = 2e^{-300t}$ A

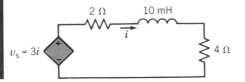

Figure E 8-9

EXERCISE 8-10

For the circuit shown in Figure E 8-10, find v when $v(0) = 10$ V.
Answer: $v = 10e^{-5000t}$ V

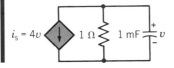

Figure E 8-10

EXERCISE 8-11

Find $i(t)$ for the circuit shown in Figure E 8-11 when $v(0^-) = 3$ V and $C = 4$ mF.

Answer: $i = \dfrac{3}{7} e^{-1321t}$ A

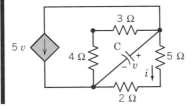

Figure E 8-11

8-10 SEQUENTIAL SWITCHING

Often circuits have several switches that are not activated at the same time t_0. For example, a circuit may have two switches where the first switch activates at $t = 0$ and the second switch activates at $t = t_1$. The voltages and currents for the circuit can be determined by the methods described in the preceding sections, using the fact that the inductor current and the capacitor voltage do not change instantaneously at the switching instant.

Sequential switching is the action of two or more switches activated at different instants of time in a circuit.

As an example of sequential switching, consider the circuit shown in Figure 8-27. This circuit is a model of the switching occurring in a modern computer. The first switch moves from point a to point b at $t = 0$ and the second switch closes at $t = 1$ ms. We wish to find the current i for $t \geq 0$.

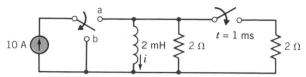

Figure 8-27 Circuit with sequential switching.

First, we find the current at $t = 0^-$, assuming the first switch has been at point a and the second switch has been open for a long time. Then we note that $i(0^-) = 10$ A. Immediately after $t = 0$ we have an RL circuit with $\tau = L/R = 1$ ms. Therefore, the current for $0 \le t \le 1$ ms is

$$i = 10e^{-1000t} \text{ A}$$

At $t = t_1 = 1$ ms we find

$$i(t_1) = 10e^{-1}$$
$$= 3.68 \text{ A}$$

As the second switch is thrown at $t = 1$ ms, we have an equivalent resistance of 1Ω resulting from the two resistors in parallel. Therefore, $\tau_1 = L/R = \frac{2}{1} = 2$ ms.

The expression for i for $t \ge 1$ ms is

$$i = i(t_1)e^{-(t-t_1)/\tau_1}$$
$$= 3.68e^{-(t-t_1)/\tau_1} \text{ A}$$

Note that we include $(t - t_1)$ to shift the expression to the new switch time $t_1 = 1$ ms. Here i starts at $t = t_1$. The current waveform for $t \ge 0$ is shown in Figure 8-28. The slope at t_1^- is -3.68 V/ms and changes to $-3.68/2$ V/ms at t_1^+.

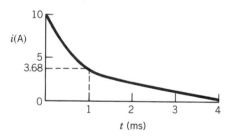

Figure 8-28 Current waveform for $t \ge 0$. The exponential has a different time constant for $0 \le t < t_1$ and for $t \ge t_1$ where $t_1 = 1$ ms.

EXERCISE 8-12

In the circuit shown in Figure E 8-12, the switch a is a "make before break" switch so that the inductor current is never interrupted. The interval of time to switch is assumed to be negligible. Find the inductor current for $t \ge 0$.

Answer: $i = 4e^{-200t} \text{ A}$ $0 \text{ ms} \le t \le 10 \text{ ms}$
          $= 0.54e^{-250(t-0.01)} \text{ A}$ $t > 10 \text{ ms}$

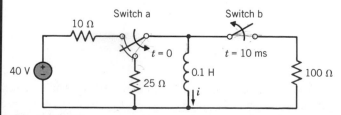

Figure E 8-12

EXERCISE 8-13

Find the voltage v for the circuit shown in Figure E 8-13 for $t \geq 0$.

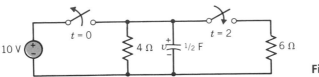

Figure E 8-13

8-11 NATURAL RESPONSE USING PSpice

In this section we illustrate the use of PSpice to determine the transient analysis of an unforced *RL* or *RC* circuit. See Appendix G-6 for a general discussion of the transient analysis of source-free circuits. Let us reconsider Example 8-2, which is concerned with a capacitor charged initially to $v(0) = 60$ V, as shown in Figure 8-29. The goal is to determine when $v(t)$ has decayed to 52 V.

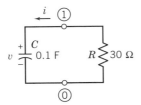

Figure 8-29 An *RC* circuit with $v(0) = 60$ V.

We will use the transient analysis statement .TRAN, which takes the form

$$.\text{TRAN} \ \langle \text{tstep} \rangle \ \langle \text{tstop} \rangle \ \langle \text{tstart} \rangle \ \langle \text{step max} \rangle \ \langle \text{UIC} \rangle$$

where tstep is the time step interval. The optional step max allows the user to set a limit on the step size. Usually, tstart = 0. Since 201 points can be calculated and plotted, we require tstop ≤ 201 tstep. The phrase UIC means "use initial conditions." We also have indicated that $i(t)$ flows from plus to the minus terminal through C as required by the PSpice convention.

The initial condition is attached to the capacitor statement, and the PSpice program is written as shown in Figure 8-30. The time constant is $RC = 3$ s, and we choose a step size

```
**       RC  TRANSIENT

R        0    1      30
C        1    0       0.1    IC = 60
.TRAN    20M         500 M   UIC
.PLOT    TRAN        V(1)
.PROBE
.END
```

Figure 8-30 The PSpice program for the *RC* circuit of Figure 8-29.

equal to 20 ms. We wish to print out 25 points over the first half second. Thus, we obtain a .TRAN statement as shown in Figure 8-30. Then, tstop = 500×10^{-3}, and we let tstart = 0 by default as shown in Figure 8-30. The .PLOT statement will give a printout as shown in Figure 8-31. We use the .PROBE statement to obtain the superior form of the plot as shown in Figure 8-32. Remember that PROBE is only available in PSpice. Examining Figure 8-31, we find that $v(t) = 52$ V at $t = 0.43$ s.

```
  TIME          V(1)
(*)----------   4.5000E+01   5.0000E+01   5.5000E+01   6.0000E+01   6.5000E+01
                - - - - - - - - - - - - - - - - - - - - - - - - - - -
 0.000E+00   6.000E+01 ·            ·            ·            *          ·
 2.000E-02   5.960E+01 ·            ·            ·          * .          ·
 4.000E-02   5.921E+01 ·            ·            ·         *  .          ·
 6.000E-02   5.881E+01 ·            ·            ·        *   .          ·
 8.000E-02   5.842E+01 ·            ·            ·       *    .          ·
 1.000E-01   5.803E+01 ·            ·            ·      *     .          ·
 1.200E-01   5.765E+01 ·            ·            ·     *      .          ·
 1.400E-01   5.726E+01 ·            ·            ·    *       .          ·
 1.600E-01   5.688E+01 ·            ·            ·   *        .          ·
 1.800E-01   5.651E+01 ·            ·            ·  *         .          ·
 2.000E-01   5.613E+01 ·            ·            . *          .          ·
 2.200E-01   5.576E+01 ·            ·            .*           .          ·
 2.400E-01   5.539E+01 ·            ·           *             .          ·
 2.600E-01   5.502E+01 ·            ·         *.              .          ·
 2.800E-01   5.465E+01 ·            ·        * .              .          ·
 3.000E-01   5.429E+01 ·            ·       *  .              .          ·
 3.200E-01   5.393E+01 ·            ·      *   .              .          ·
 3.400E-01   5.357E+01 ·            ·     *    .              .          ·
 3.600E-01   5.322E+01 ·            ·    *     .              .          ·
 3.800E-01   5.286E+01 ·            ·  *       .              .          ·
 4.000E-01   5.251E+01 ·            ·  *       .              .          ·
 4.200E-01   5.216E+01 ·            · *        .              .          ·
 4.400E-01   5.182E+01 ·            ·*         .              .          ·
 4.600E-01   5.147E+01 ·           . *        .              .          ·
 4.800E-01   5.113E+01 ·           .*         .              .          ·
 5.000E-01   5.079E+01 ·           *          .              .          ·
                - - - - - - - - - - - - - - - - - - - - - - - - - - -
```

Figure 8-31 PRINT output for PSpice program of Figure 8-30.

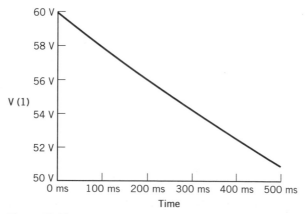

Figure 8-32 PROBE output for PSpice program of Figure 8-30.

8-12 | VERIFICATION EXAMPLES

Example 8V-1

Consider the circuit and corresponding transient response shown in Figure 8V-1. How can we tell if the transient response is correct? There are two things that need to be verified: the initial current $i_L(t_0)$ and the time constant τ when the switch is open.

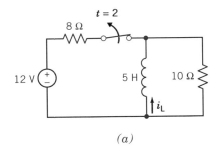

(a)

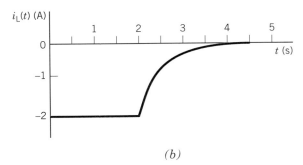

(b)

Figure 8V-1 (a) An example RL circuit and (b) its natural response.

Solution

Consider first the initial current $i_L(t_0)$. (In this example, $t_0 = 2$ s.) Before time $t_0 = 2$ s, the switch is closed and has been closed long enough for the circuit to reach steady state, that is, for any transients to have died out. To calculate $i_L(t_0)$ we simplify the circuit in two ways. First, replace the switch with a short circuit because the switch is closed. Second, replace the inductor with a short circuit because inductors act like short circuits when all the inputs are constants and the circuit is at a steady state. The resulting circuit is shown in Figure 8V-2a. The voltage across the 10-Ω resistor is zero, so the current in the 10-Ω resistor will also be zero. The current $i_L(t_0)$ is given by

$$i_L(t_0) = \frac{0 - 12}{8}$$

$$= -1.5 \text{ A}$$

In contrast, the transient response shown in Figure 8V-1b indicates that the initial current is $i_L(t_0) = -2$ A. This transient response is not correct.

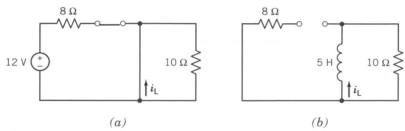

(a)   *(b)*

Figure 8V-2 *(a)* The example circuit before $t = 2$ s and *(b)* after $t = 2$ s.

Let us see if the time constant is also wrong. The time constant is calculated from the circuit shown in Figure 8V-2b. This circuit has been simplified by setting the input to zero (a zero voltage source acts like a short circuit) and replacing the switch by an open circuit. The time constant is

$$\tau = \frac{5}{10}$$

$$= \frac{1}{2} \text{ s}$$

Figure 8V-3 shows how the time constant can be determined from the plot of the transient response. The dashed line is tangent to $i_L(t)$ at $t = t_0$. The value of τ obtained from the transient response is the same as the calculated value. The transient response has the correct time constant.

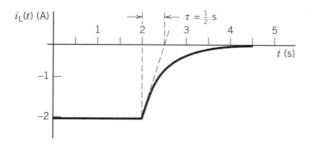

Figure 8V-3 Determining the time constant from the natural response.

The transient response shown in Figure 8V-1b has the correct time constant but the wrong initial current. Notice that the time constant depends on the value of the inductor and of the 10-Ω resistor while the initial current, $i_L(t_0)$, depends on the values of the two resistors. It seems likely that the error in determining the transient response involved the 8-Ω resistor.

Example 8V-2

Consider the circuit and corresponding transient response shown in Figure 8V-4. How can we tell if the transient response is correct? There are two things that need to be verified: the initial capacitor voltage $v_C(t_0)$ and the time constant when the switch is open.

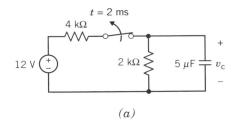

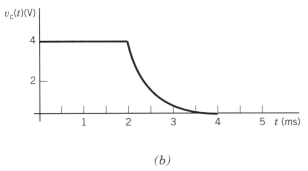

Figure 8V-4 (*a*) An example RC circuit and (*b*) its natural response.

Solution

Before time $t_0 = 20$ ms, the switch is closed and has been closed long enough for the circuit to reach steady state, that is, for any transients to have died out. Figure 9 V-5*a* shows the circuit used to calculate the initial capacitor voltage. This circuit has been simplified in two ways. First, the switch has been replaced with a short circuit. Second, the capacitor has been replaced with an open circuit. The steady-state capacitor voltage is calculated using voltage division as

$$v_c(0) = \frac{2\,\text{k}\Omega}{2\,\text{k}\Omega + 4\,\text{k}\Omega}\,12$$

$$= 4\,\text{V}$$

The transient response shown in Figure 8V-4*b* indicates that the initial voltage is $v_C(t_0) = +4$ V. This aspect of the transient response is correct.

Figure 8V-5*b* shows the circuit used to calculate the time constant when the switch is open. This circuit has been simplified in two ways. First, the switch has been replaced with

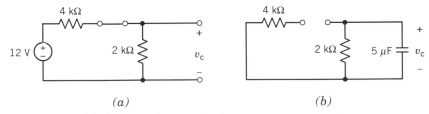

Figure 8V-5 (*a*) The example circuit before $t = 2$ ms and (*b*) after $t = 2$ ms.

an open circuit. Second, the input has been set to zero (a zero voltage source acts like a short circuit). The time constant is

$$\tau = 2\,\text{k}\Omega \cdot 5\,\mu\text{F}$$

$$= 10\,\text{ms}$$

Figure 8V-6 shows that the time constant of the transient response is $\frac{1}{2}$ ms. The transient response is not correct.

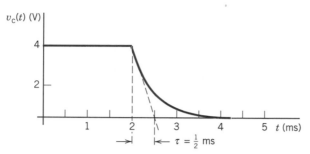

Figure 8V-6 Determining the time constant from the natural response.

The transient response shown in Figure 8V-4*b* has the correct initial condition but the wrong time constant. Notice that the time constant depends on the values of the capacitor and of the 2-Ω resistor while the initial voltage, $v_C(t_0)$, depends on the values of the two resistors. It seems likely that the error in determining the transient response involved the capacitor.

8-13 DESIGN CHALLENGE SOLUTION

SECURITY ALARM CIRCUIT

Problem

A switching circuit is used in a security alarm circuit for a building. The two switches in the circuit of Figure 8D-1 are thrown simultaneously at $t = 0$ by an intruder interrupt-

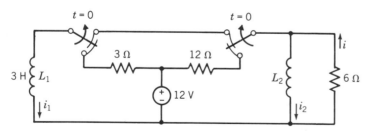

Figure 8D-1 Switching circuit for security alarm.

ing a sonar sensor that then activates the switches. The current i_2 activates alarms at the security office and the police station when it changes from a positive quantity to a negative quantity. The goal is to achieve this change in current i_2 within 0.4 s. Select L_2 so that the alarm system meets the requirements.

Define the Situations, State the Assumptions, and Develop a Model

The two switches have been at the lower setting for a long time prior to switching at $t = 0$.

The Goal

Determine $i_2(t)$ for $t > 0$ and select the inductance L_2.

Generate a Plan

1 Find $i_1(0^-)$ and $i_2(0^-)$ prior to switching.
2 Set $i_1(0) = i_1(0^-)$ and $i_2(0) = i_2(0^-)$ since the inductor currents cannot change at $t = 0$.
3 Find the time constant of the switched circuit and try a value for L_2.
4 Note that after the switch is thrown, $i = i_1 + i_2$ and $i(t) = i(0)e^{-t/\tau}$.
5 Find $v(t)$ across the 6-Ω resistor and thus the voltage across each inductor.
6 Find $i_2(t)$.

Prepare the Plan

Goal	Equation	Need	Information
Find $i_1(0^-)$ and $i_2(0^-)$.		Steady-state currents at $t = 0^-$	Draw circuit at $t = 0^-$.
Find $i_1(0)$ and $i_2(0)$.	$i_1(0) = i_1(0^-)$ $i_2(0) = i_2(0^-)$		
Find the time constant of the switched circuit.	$\tau = L_p/R$	L_p for the two parallel inductors	Draw circuit for $t \geq 0$.
Select a trial value for L_2.		Try $L_2 \approx L_1$	
Find $i(t)$ for $t \geq 0$. Find $v(t)$ across the resistor.	$i = i(0) \, e^{-t/\tau}$ $v = Ri(t)$	$i(0) = i_1(0) + i_2(0)$ $i(t)$	$R = 6\Omega$
Find $i_1(t)$ and $i_2(t)$.	$i_n(t) = \dfrac{1}{L_n}\displaystyle\int_0^t - v(\tau)\,d\tau + i_n(0)$		

Take Action Using the Plan

First determine $i_1(0^-)$ and $i_2(0^-)$ using the circuit of Figure 8D-2 at $t = 0^-$. Accordingly,

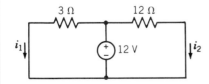

Figure 8D-2 The circuit of Figure 8D-1 at $t = 0^-$. It is assumed that this circuit is in steady state at $t = 0^-$.

we obtain

$$i_1(0^-) = \frac{12}{3} = 4 \text{ A}$$

and

$$i_2(0^-) = \frac{12}{12} = 1 \text{ A}$$

Next, determine $i_1(0)$ and $i_2(0)$. Since the inductor currents do not change instantaneously, we have

$$i_1(0) = i_1(0^-) = 4 \text{ A}$$

and

$$i_2(0) = i_2(0^-) = 1 \text{ A}$$

Third, find the time constant of the circuit for $t \geq 0$. We redraw the switched circuit as shown in Figure 8D-3. The time constant is $\tau = L_p/R$ where $R = 6$ and L_p is the

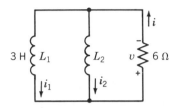

Figure 8D-3 The circuit of Figure 8D-1 for $t \geq 0$.

parallel equivalent inductance for L_1 and L_2. Therefore,

$$L_p = \frac{3 \times L_2}{3 + L_2}$$

and the time constant is $\tau = L_p/R$. We will try a time constant less than 0.4 s and see if the current changes from positive to negative in less than 0.4 s. Since $R = 6\ \Omega$, we need L_p less than 1.8 H. If we select $L_2 = 2$ H, we have $L_p = 1.2$ H. Then

$$\tau = \frac{L_p}{R} = \frac{1.2}{6} = 0.2 \text{ s}$$

which should meet our needs. Fourth, find $i(t)$ for $t \geq 0$, recalling that

$$i(t) = i(0)\ e^{-t/\tau}$$

Examining Figure 8D-3, we note that

$$\begin{aligned} i(0) &= i_1(0) + i_2(0) \\ &= 4 + 1 \\ &= 5 \text{ A} \end{aligned}$$

Therefore,

$$\begin{aligned} i(t) &= i(0)\ e^{-t/\tau} \\ &= 5\ e^{-5t} \text{ A} \qquad t \geq 0 \end{aligned}$$

For the fifth step, find $v(t)$ across the resistor as

$$\begin{aligned} v(t) &= Ri(t) \\ &= 6 \times 5\ e^{-5t} \text{ V} \qquad t \geq 0 \end{aligned}$$

Consequently, find $i_1(t)$ and $i_2(t)$ by noting that the voltage across each inductor is $- v(t)$. Then, for $i_2(t)$, we have

$$\begin{aligned} i_2(t) &= \frac{1}{L_2} \int_0^t - v(\tau)\, d\tau + i_2(0) \\[6pt] &= \frac{1}{2} \int_0^t - 30\ e^{-5\tau}\, d\tau + 1 \\[6pt] &= - 2 + 3\ e^{-5t} \text{ A} \end{aligned}$$

Thus, as expected, $i_2(t)$ changes from $+1$ A at $t = 0$ to -2 A in approximately 1 s. The current i_2 passes through zero when $3 e^{-5t} = 2$ or $t = 0.08$ s. Thus, $L_2 = 2$ H results in a circuit meeting the goal. Of course, other values of L could also satisfy the requirements.

SUMMARY

Circuits with one or more switches activated at different times can be used as important parts of telecommunications and control systems. In this chapter we considered the unforced (natural) response of an *RL* or *RC* circuit.

We formulated the first-order differential equation representing the *RL* or *RC* circuit and went on to determine the natural response $x(t)$ of the circuit. We then found that the natural response of an *RL* or *RC* circuit depends on the characteristics of the circuit and not on any external sources.

To find the solution of the first-order differential equation, we can use the method of separation of variables or an assumed exponential solution. Also, a differential operator s can be used in tandem with the exponential solution to determine the solution of the differential equation. The response is then dictated by the initial condition of the energy storage element and the circuit time constant.

The response $x(t)$ is called the transient response, and the response that occurs after a long time ($t \rightarrow \infty$) is called the steady-state response. The initial energy stored by the energy storage element is dissipated over time in the circuit resistance.

If a circuit is free of independent sources but includes a dependent source, the time constant will incorporate the effect of the dependent source. Often, circuits have sequential switches that activate at a sequence of times. We use our method of solution of the first-order differential equation, recognizing the changes in the circuit as the switches open and close sequentially.

We have, then, explored a set of methods for determining the unforced, natural response of a circuit. In the next chapter we will explore methods for determining the response of an *RC* or *RL* circuit when it is subjected to an independent source.

TERMS AND CONCEPTS

Exponential Solution   Assumed solution of the form $x = Ae^{st}$ where A and s are constants to be determined.

First-Order Differential Equation   Differential equation where the highest order derivative term is the first derivative, i.e., dx/dt.

Natural Response   Response of an *RL* or *RC* circuit that depends only on the nature of the circuit and not on external sources.

Operator   A symbol that represents a mathematical operation. Differential operator s such that $s^n x = d^n x / dt^n$.

Radio Transmission   Transmission of communication messages by means of radiated electromagnetic waves other than heat or light waves.

Sequential Switching   Action of two or more switches activated at different instants of time in a circuit.

Steady-State Response   Response that exists after a long time following any switch activation.

Telegraph   Device for communication of messages using the activation of a switch or key to send a code.

Time Constant   Value in seconds, τ, in an exponential response $Ae^{-t/\tau}$. The constant τ is the time to complete 63 percent of the decay.

Transient Response   Time response $x(t)$ of a circuit to a stimulus or to a response resulting from a switch activation.

REFERENCES   Chapter 8

Orlando, T. *Foundations of Applied Superconductivity,* Addison-Wesley, Reading, MA, 1991.

Quint, David W. "Measurement of R, L and C Parameters in VLSI Packages," *Hewlett-Packard Journal,* October 1990, pp. 73–77.

Scott, David, "Electronic Pistol," *Popular Science,* August 1991, p. 70.

PROBLEMS

Section 8-4   Solution of the First-Order Differential Equation (Source-Free Response)

P 8.4-1   Determine the inductor current $i_L(t)$ for the circuit of Figure P 8.4-1.

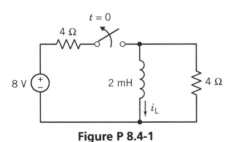

Figure P 8.4-1

P 8.4-2   Determine the capacitor voltage $v_C(t)$ for the circuit of Figure P 8.4-2.

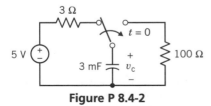

Figure P 8.4-2

P 8.4-3   A circuit is shown in Figure P 8.4-3. It is desired to select the capacitance C so that $v(t) = 189$ V at $t = 0.25$ s. Assume that the switch was connected to terminal 1 for a long time.

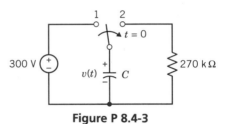

Figure P 8.4-3

P 8.4-4   From a practical viewpoint, the crucial difference between copper wire and superconductors is their respective current-carrying capabilities. Standard copper wire used in homes, for example, will carry about 10^7 A/m². Practical superconductors used in wire can carry current densities of 10^{10} A/m² or higher. The magnetic field produced by an electromagnet is proportional to the current in the windings. We can increase the strength of the field created by using the higher current-carrying capabilities of superconductors. Thus, we can generate the same field strength with a *smaller* superconducting electromagnet as compared to the standard copper-wound equivalent. This is important in electric generator design; it is possible to scale down the size of the generator while still producing the same power. A power plant could therefore be shrunk in size, becoming more efficient (Orlando, 1991).   Consider the loop of metallic wire shown in Figure P 8.4-4*a*. The loop is initially placed in a static magnetic field such that flux threads the hole. After the system reaches the steady state, the field is suddenly turned off (at time $t = 0$). A current $i(t)$ will therefore be induced in the wire. One model of the loop is the circuit shown in Figure P 8.4-4*b* with $i(0) = 1$ A and $L = 0.1$ mH. Since the loop is a superconducting wire, we can estimate $R = 10^{-10}$ Ω. Determine the percentage reduction in the current $i(t)$ after one day.

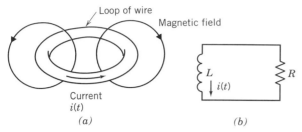

(a)   *(b)*

Figure P 8.4-4 A superconducting loop of wire.

P 8.4-5 Figure P 8.4-5*a* shows astronaut Dale Gardner using the manned maneuvering unit to dock with the spinning Westar VI satellite on November 14, 1984. Gardner used a large tool called the apogee capture device (ACD) to stabilize the satellite and capture it for recovery, as shown in Figure P 8.4-5*a*. The ACD can be modeled by the circuit of Figure P 8.4-5*b*. Find the inductor current i_L for $t > 0$.
Answer: $i_L(t) = 6e^{-20t}$ A

(a)

(b)

Figure P 8.4-5 *(a)* Astronaut Dale Gardner using the manned maneuvering unit to dock with the Westar VI satellite. Courtesy of NASA. *(b)* Model of the apogee capture device. Assume that the switch has been in position a for a long time at $t = 0^-$.

P 8.4-6 A laser fusion experiment uses the circuit of Figure P 8.4-6. Find $v(t)$ for $t > 0$ for the circuit. Assume that the switch has been closed a long time at $t = 0^-$.

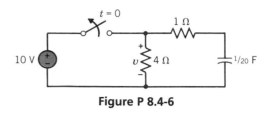

Figure P 8.4-6

P 8.4-7 The switch of the circuit shown in Figure P 8.4-7 is moved from A to B at $t = 0$, after being at A for a long time. Find $v(t)$ for $t > 0$.
Answer: $v(t) = -0.18e^{-100t}$ V

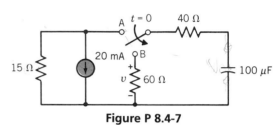

Figure P 8.4-7

Section 8-5 The Exponential Solution to the First-Order Differential Equation

P 8.5-1 Find $i(t)$, $t \geq 0$, for the circuit shown in Figure P 8.5-1.
Answer: $i(t) = 1e^{-3 \times 10^4 t}$ mA

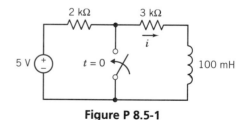

Figure P 8.5-1

P 8.5-2 The goal is to measure a capacitance in a laboratory. A student has a stopwatch, a voltmeter with an internal resistance of $R_m = 30$ kΩ, and a 100-V battery. The battery was first connected to the capacitor, which was then charged to steady state. The capacitor was then switched to the voltmeter, which read 36.7 V at $t = 13.5$ s. Draw the circuit used and find the value of the capacitance.

P 8.5-3 The switch in the circuit of Figure P 8.5-3 is opened at $t = 0$ after having been closed for a long time. Determine and plot $i(t)$ for 0 s $\leq t \leq 0.5$ s.

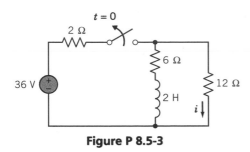

Figure P 8.5-3

P 8.5-4 The switch shown in Figure P 8.5-4 is opened at $t = 0$ after having been closed for a long time. Determine $v_o(t)$ and plot $v_o(t)$ for $0 \le t \le 3$ s.

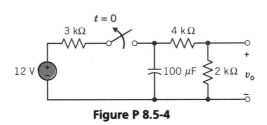

Figure P 8.5-4

P 8.5-5 For the circuit of Figure P 8.5-5, find $i(t)$ for $t > 0$. Assume that the switches have been in their respective positions for a long time at $t = 0^-$.

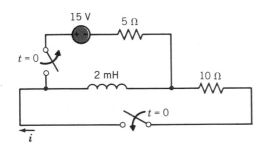

Figure P 8.5-5

P 8.5-6 A security alarm for an office building door is modeled by the circuit of Figure P 8.5-6. The switch represents the door interlock and v is the alarm indicator voltage. Find $v(t)$ for $t > 0$ for the circuit of Figure P 8.5-6. The switch has been closed for a long time at $t = 0^-$.

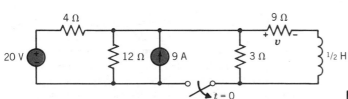

Section 8-7 The Response of a First-Order Circuit

P 8.7-1 A complex electronic circuit for a communication satellite is represented by Figure P 8.7-1. Find $i(t)$ for $t > 0$ for the circuit shown in Figure P 8.7-1. The circuit is in steady state at $t = 0^-$.
Answer: $i = (1/6)e^{-3000t}$ mA

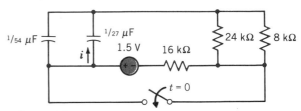

Figure P 8.7-1 A circuit for a communication satellite.

P 8.7-2 In the circuit shown in Figure P 8.7-2, the switch is flipped from A to B at $t = 0$, after being at A for a long time. Find $v_c(t)$ and $i_L(t)$ for $t > 0$.

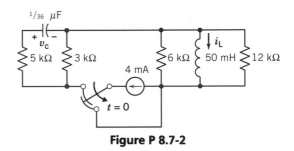

Figure P 8.7-2

P 8.7-3 A complex circuit used in a telegraph transmitter is shown in Figure P 8.7-3. For the circuit of Figure P 8.7-3, find $i(t)$ for $t > 0$. The switch has been in position a for a long time at $t = 0^-$.

Figure P 8.5-6 A security alarm circuit.

Answer: $i(t) = -3e^{-4000t}$ mA

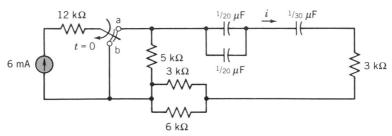

Figure P 8.7-3 A telegraph transmitter circuit.

P 8.7-4 The switch of the circuit of Figure P 8.7-4*a* is in position 1 for a long time and is moved to position 2 at $t = 0$. The current $i(t)$ is measured and plotted as shown in Figure P 8.7-4*b*. Determine the value of *C*.

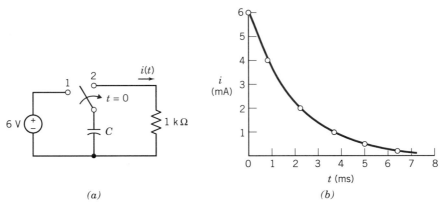

(*a*)

(*b*)

Figure P 8.7-4

P 8.7-5 For the circuit of Figure P 8.7-5, find $i(t)$ for $t > 0$. Assume that the switch has been closed a long time at $t = 0^-$.

Answer: $i(t) = \frac{2}{5} e^{-10t}$ A

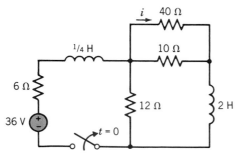

Figure P 8.7-5

P 8.7-6 Determine $i(t)$ for the circuit shown in Figure P 8.7-6. Find the initial rate of change of current and plot $i(t)$ for four time constants.

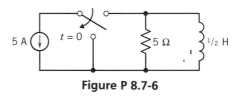

Figure P 8.7-6

P 8.7-7 An electronic flash on a camera uses the circuit shown in Figure P 8.7-7. Harold E. Edgerton invented the electronic flash in 1930. A capacitor builds a steady-state voltage and then discharges it as the shutter switch is pressed. The discharge produces a very brief light discharge. Determine the elapsed time t_1 to reduce the capacitor voltage to one-half of its initial voltage. Find the current at $t = t_1$.

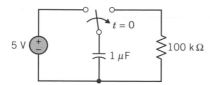

Figure P 8.7-7 Electronic flash circuit.

P 8.7-8 Inductance is measured by applying a ramp current $i_s = At$, $t \geq 0$, as shown in Figure P 8.7-8 (Quint, 1990). Determine $v_o(t)$ and discuss how R_x and L_x can be estimated when an inductor is represented by R_x and L_x in series.

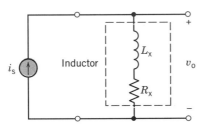

Figure P 8.7-8 Circuit for measuring inductance.

Section 8-8 Energy Storage and Dissipation in *RL* and *RC* Circuits

P 8.8-1 Efficient use of energy can be achieved by load management programs of electric utilities. The utility hooks up electronically activated controllers to appliances such as water heaters, air conditioners, and swimming-pool pumps at the homes of volunteer customers (they receive a rebate for participating). During peak hours, signals go out over the power lines ordering controllers at selected houses to turn off for periods of typically 15–30 minutes; the shutdown call then rotates to another group of houses. The effect is barely noticeable to the consumer (water stays fairly hot in the tank, for example), but the cumulative effect should be significant.

A model of a local switch and the energy load from a home is shown in Figure P 8.8-1. Find the energy w dissipated in the two load resistors for the first minute after the switch is opened at $t = 0$ after having been closed for a long time.

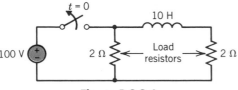

Figure P 8.8-1

P 8.8-2 A circuit is shown in Figure P 8.8-2 with two capacitors where $C_1 = C_2 = 1$ F. At $t = 0$ the switch is thrown from position a to position b. (a) Find the initial voltage and energy stored in the left-hand capacitor, C_1, if the switch has been in position a for a long time. The voltage on the right-hand capacitor, C_2, at $t = 0^-$ is zero. (b) Find the energy stored by the two capacitors and the energy dissipated by the resistor for $t \geq 0$.

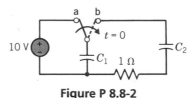

Figure P 8.8-2

P 8.8-3 The most important safety feature of new automobiles is the airbag. The airbag deploys when a pendulum sensor detects a sudden deceleration of greater than 10 g and closes a switch. An equivalent circuit of the airbag deployment device is shown in Figure P 8.8-3. Determine the time required before the energy absorbed by the resistor reaches 25 mJ and triggers the explosive deployment of the bag. The capacitor is precharged to 24 V, $R = 100$ Ω, and $C = 0.1$ mF.

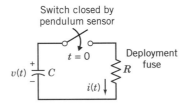

Figure P 8.8-3 Airbag deployment circuit.

P 8.8-4 A home handyman decides to fix his own TV set, despite the warning on the back. He switches it off, pulls the plug, removes the back cover, reaches inside, and receives a large electrical shock. Assume he touches the inside within 2 s after switching off the set. The dc source that drives the set's picture tube is 20 kV. The capacitance and resistance that form its load are 1 μF and 1 MΩ, as shown in Figure P 8.8-4. Why did he get a shock, and what was the voltage? If his body resistance was 2 MΩ, how much energy did his body absorb if he held on for 10 s?

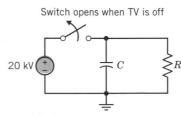

Figure P 8.8-4 TV set picture tube circuit.

Section 8-9 *RL* and *RC* Circuits with Dependent Sources

P 8.9-1 For the circuit shown in Figure P 8.9-1, find the voltage $v(t)$ and the current $i_L(t)$ for $t > 0$. Assume that the circuit is in steady state at $t = 0^-$.

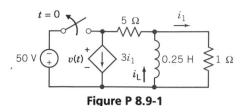

Figure P 8.9-1

P 8.9-2 Find $v(t)$ for $t > 0$ for the circuit shown in Figure P 8.9-2. The circuit is in steady state at $t = 0^-$.
Answer: $v(t) = -216e^{-96t}$ V

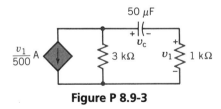

Figure P 8.9-2

P 8.9-3 For the circuit of Figure P 8.9-3, find $v_c(t)$ for $t > 0$ if $v_c(0) = 6$ V.

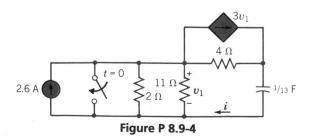

Figure P 8.9-3

P 8.9-4 Find $i(t)$ for $t > 0$ for the circuit of Figure P 8.9-4. The circuit is in steady state at $t = 0^-$.
Answer: $i(t) = -14.3e^{-3.25t}$ A

Figure P 8.9-4

P 8.9-5 Find $i(t)$ for $t > 0$ for the circuit shown in Figure P 8.9-5. The circuit is in steady state at $t = 0^-$.
Answer: $i(t) = \frac{2}{3}e^{-6t}$ A   for $0 \le t \le 51$ ms
$i(t) = 1.47e^{-14(t-0.051)}$ A   for $t > 51$ ms

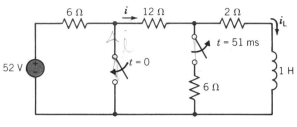

Figure P 8.9-5

Section 8-10 Sequential Switching

P 8.10-1 Sequential switching is used repetitively to generate communication signals. For the circuit shown in Figure P 8.10-1, switch a has been in position 1 and switch b has been open for a long time. At $t = 0$, switch a moves to position 2. Then, 100 ms after switch a moves, switch b closes. Find the capacitor voltage v for $t \ge 0$.

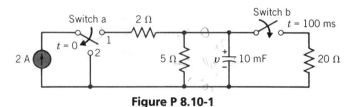

Figure P 8.10-1

P 8.10-2 For the circuit of Figure P 8.10-2, determine $v_c(t)$ and $v_x(t)$ for $t > 0$ when $C = 0.2$ F. Assume that the circuit is in steady state when $t = 0^-$.

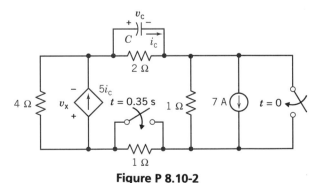

Figure P 8.10-2

P 8.10-3 For the circuit shown in Figure P 8.10-3, switch 1 has been closed for a long time and switch 2 has been in position a for a long time. At $t = 0$ switch 1 is opened, and at $t = 4$ ms switch 2 is changed to position b. Find i for $t \ge 0$ when (a) $k = 10$ and (b) $k = -11$.

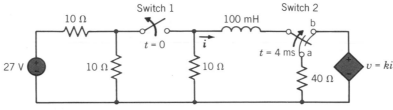

Figure P 8.10-3

P 8.10-4 Cardiac pacemakers are used by people to maintain regular heart rhythm when they have a damaged heart. The circuit of a pacemaker can be represented as shown in Figure P 8.10-4. The resistance of the wires, R, can be neglected since $R < 1$ mΩ. The heart's load resistance, R_L, is 1 kΩ. The first switch is activated at $t = t_0$ and the second switch is activated at $t_1 = t_0 + 10$ ms. This cycle is repeated every second. Find $v(t)$ for $t_0 \le t \le 1$. Note that it is easiest to consider $t_0 = 0$ for this calculation. The cycle repeats by switch 1 returning to position a and switch 2 returning to its open position. Hint: Use $q = Cv$ to determine $v(0^-)$ for the 100-μF capacitor.

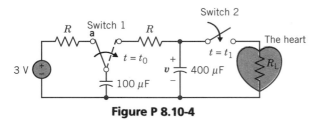

Figure P 8.10-4

P 8.10-5 Determine the current $i(t)$ and the voltage $v(t)$ for the circuit shown in Figure P 8.10-5 when $L = 0.2$ H and $C = 0.2$ F. Sketch the current and voltage versus time. Assume that the circuit is in steady state at $t = 0^-$.

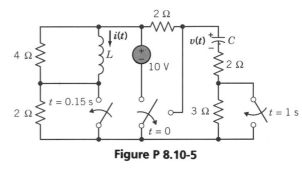

Figure P 8.10-5

PSpice PROBLEMS

SP 8-1 Determine and plot $i(t)$ for the circuit of Figure SP 8-1.

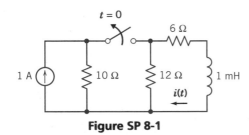

Figure SP 8-1

SP 8-2 Determine and plot $i_L(t)$ for Problem P 8.4-5.

SP 8-3 Determine and plot $i(t)$ for Problem P 8.9-4.

SP 8-4 Find $v_c(t)$ and $i_L(t)$ for Problem P 8.7-2.

SP 8-5 Find $v_o(t)$ for Problem P 8.5-4 for $0 \le t \le 3$ s.

SP 8-6 An RC circuit is shown in Figure SP 8-6 with $R = 1$ kΩ and $C = 20$ μF. Determine the time, t_1, when $v(t_1) = 4.00$ V, given $v(0^-) = 12$ V.

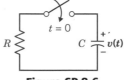

Figure SP 8-6

VERIFICATION PROBLEMS

VP 8-1 A computer program for the circuit of Figure VP 8-1 states that $v(t) = 1.31$ V at $t = 2$ s. Assume the switch was in position A for a long time prior to switching at $t = 0$. Verify the resulting $v(t)$ at $t = 2$ s.

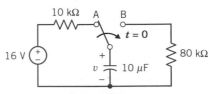

Figure VP 8-1

VP 8-2 A student report states that the current, $i_L = 120e^{-80,000t}$ μA for the circuit shown in Figure VP 8-2. Assume the switch was closed for a long time. Verify the reported result.

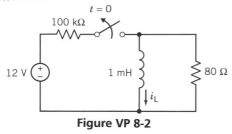

Figure VP 8-2

VP 8-3 A student solution states that $i(t)$ is equal to 100 mV at $t = 2$ s for the circuit of Figure VP 8-3. Verify this solution.

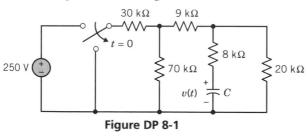

Figure VP 8-3 Assume that switch 1 was closed for a long time.

VP 8-4 For the circuit of Figure VP 8-4, it is reported that $v_o = -3.68$ V at $t = 2$ s. Verify this result.

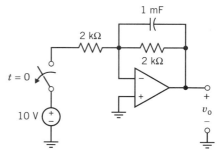

Figure VP 8-4 Assume the switch was closed for a long time.

DESIGN PROBLEMS

DP 8-1 The response of the circuit shown in Figure DP 8-1 is represented by the capacitor voltage $v(t)$. It is specified that the voltage $v(t)$ decays to less than 1 percent of its initial value, $v(0)$, within 600 ms (5 time constants). Determine the required capacitance C to achieve the specified response. Assume that the switch has been in the initial position for a long time.

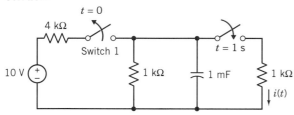

Figure DP 8-1

DP 8-2 The switch in Figure DP 8-2 has been closed for a long time and is opened at $t = 0$. Determine the inductance L required for $i(t) = 1.14$ A at $t = 0.1$ s.

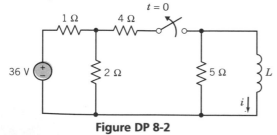

Figure DP 8-2

DP 8-3 In order to obtain energy conservation in offices, a designer has developed a sensor that turns on the lights after determining that a person has entered a room. However, if the person does not push the continue button, the lights will again switch off in a predetermined time equal to 120 s. Design a precharged *RC* circuit that will hold the lights on for 120 s and then turn them off again unless a continue switch is pressed.

DP 8-4 An electronically controlled target pistol has been designed (Scott, 1991). The firing pin is tripped by a battery-powered solenoid. The control circuit is shown in Figure DP 8-4. Select the inductance *L* so that the solenoid voltage exceeds 20 V and the magnitude of the inductor current exceeds 0.5 A during the first 25 ms following $t = 0$. Assume that the circuit is in steady state at $t = 0^-$. The trigger is pulled at $t = 0$.

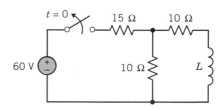

Figure DP 8-4 Target pistol trigger circuit.

DP 8-5 For the design challenge of Section 8-13, find the maximum value of L_2 at which the current i_2 will change from positive to negative at exactly 0.4 s.

CHAPTER 9

THE RESPONSE OF *RL* AND *RC* CIRCUITS TO A FORCING FUNCTION

Problems

PREVIEW

Many electric devices, such as radio and television receivers, are built to process electric signals sent to the receiver from broadcasting stations. In this chapter we consider the response of electric circuits to input signals, often called independent sources.

The total response of an electric circuit to an input signal or forcing function is made up of the natural response (considered in Chapter 8) and the forced response. We will consider forcing functions that are constants, exponentials, and sinusoids. Also, we will consider circuits that have more than one forcing function present and use the principle of superposition to determine the total response.

9-1 DESIGN CHALLENGE

A DRIVER AND CABLE

It is frequently necessary to connect two pieces of electronic equipment together so that the output from one device can be used as the input to another device. For example, this situation occurs when a printer is connected to a computer. Consider the circuit shown in Figure 9D-1a. Here a cable is used to connect two circuits. The driver

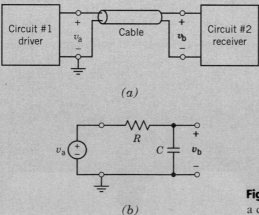

(a)

(b)

Figure 9 D-1 (a) Two circuits connected by a cable and (b) an equivalent circuit.

(source) is used to send a signal through the cable to the receiver. Let us replace the driver, cable, and receiver with simple models. Model the driver as a voltage source, the cable as an RC circuit, and the receiver as an open circuit. The values of resistance and capacitance used to model the cable will depend on the length of the cable. For example, when RG58 coaxial cable is used,

$$R = r \cdot L \quad \text{where } r = 0.54 \frac{\Omega}{\text{m}}$$

and

$$C = c \cdot L \quad \text{where } c = 88 \frac{\text{pF}}{\text{m}}$$

and L is the length of the cable in meters. Figure 9D-1b shows the equivalent circuit.

Suppose that the circuits connected by the cable are digital circuits. The driver will send 1's and 0's to the receiver. These 1's and 0's will be represented by voltages. The output of the driver will be one voltage, called V_{OH} (voltage out high), to represent logic 1 and another voltage, V_{OL} (voltage out low), to represent a logic 0. For example, one popular type of logic, called TTL (transistor-transistor logic) logic, uses $V_{OH} = 2.4$ V and $V_{OL} = 0.4$ V. The receiver uses two different voltages, V_{IH} and V_{IL}, to represent 1's and 0's. (This is done to provide noise immunity, but that is another story.) The receiver will interpret its input, v_b, to be a logic 1 whenever $v_b > V_{IH}$ and to be a logic 0 whenever $v_b < V_{IL}$. (Voltages between V_{IH} and V_{IL} will only occur during transitions between logic 1 and logic 0. These voltages will sometimes be interpreted as logic 1 and other times as logic 0.) TTL logic uses $V_{IH} = 2.0$ V and $V_{IL} = 0.8$ V.

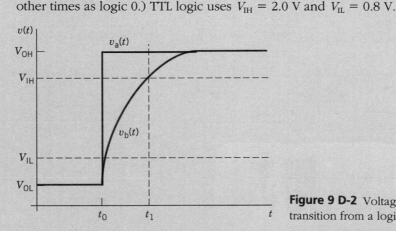

Figure 9 D-2 Voltages that occur during a transition from a logic 0 to a logic 1.

Figure 9D-2 shows what happens when the driver output changes from logic 0 to logic 1. Before time t_0,

$$v_a = V_{OL} \quad \text{and} \quad v_b < V_{IL} \quad \text{for } t \le t_0$$

In words, a logic 0 is sent and received. The driver output switches to V_{OH} at time t_0. The receiver input, v_b, makes this transition more slowly. Not until time t_1 does the receiver input become large enough to be interpreted as a logic 1. That is,

$$v_b > V_{IH} \quad \text{for } t \ge t_1$$

The time that it takes for the receiver to recognize the transition from logic 0 to logic 1

$$\Delta t = t_1 - t_0$$

is called the delay. This delay is important because it puts a limit on how fast 1's and 0's can be sent from the driver to the receiver. To ensure that the 1's and 0's are received reliably, each 1 and each 0 must last for at least Δt. The rate at which 1's and 0's are sent from the driver to the receiver is inversely proportional to the delay.

Suppose two TTL circuits are connected using RG58 coaxial cable. What restriction must be placed on the length of the cable to ensure that the delay, Δt, is less than 2 ns?

This chapter describes procedures for solving this type of problem. We will return to this problem at the end of the chapter.

9-2 ‖ RADIO AND MESSAGES

Attempts to communicate effectively have been pursued since society organized. A principal application of electric circuits is the communication of messages. Typically, the message is available in the form of *electric signals,* which may be a voltage or current varying in time in a manner that conveys information. The varying current in a telephone receiver is an example of an electric signal.

Perhaps the form of signal best known to schoolchildren is that used by Paul Revere to inform the patriot leaders of the advance of the British army. Revere, in 1775, used the visual signal method: "one lantern if by land, two if by sea."

Electric signals or waveforms are processed by electric circuits to yield the desired output form. Examples of electrical communication devices include underwater cable, radio, and television.

A *signal* is defined as a real-valued function of time. By real valued, we mean that for any fixed value of time, the value of the signal at that time is a real number. Examples of common forms of signals are shown in Figure 9-1.

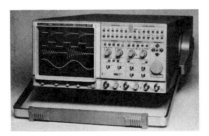

Figure 9-1 Common forms of electrical signals shown on the screen of an oscilloscope. Courtesy of Panasonic Industrial Co.

Although others performed important work, Guglielmo Marconi is the person to whom we owe the greatest debt for the development of the radio. Marconi developed the spark gap transmitter and realized that it needed to be connected to an elevated metal object, which acted as an antenna. This early radio transmitter and antenna can be modeled by the circuit shown in Figure 9-2. The signal to be transmitted is represented by the voltage source, v_s. Several signal waveforms are shown in Table 9-1.

An **electric signal** is a voltage or current varying with time in a manner that conveys information.

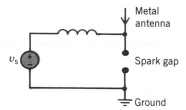

Figure 9-2 Model of Marconi's radio transmitter and antenna, about 1900.

Table 9-1

Common Signal Waveforms

Name	Equation	Waveform
1 Continuous or dc	$v(t) = V_0$	
2 Step	$v(t) = 0 \quad t < 0$ $\quad\quad = V_0 \quad t > 0$	
3 Decaying exponential	$v(t) = 0 \quad\quad t < 0$ $\quad\quad = V_0 e^{-at} \quad t > 0$	
4 Ramp	$v(t) = 0 \quad t < 0$ $\quad\quad = Kt \quad t \geq 0$	
5 Pulse	$v(t) = V_0 \quad 0 \leq t \leq t_1$ $\quad\quad = 0 \quad\quad \text{elsewhere}$	
6 Sinusoid	$v(t) = V_0 \sin(\omega t + \theta)$	

Note: Voltage is used to designate the waveform equation; current could be used equally as well.

The electromagnetic wave transmitted from the antenna was then received at another antenna. To achieve detection and reproduction of the received signal, a detector was required. By 1906 it was found that crystals were excellent materials for detectors. After World War I, and especially after commercial broadcasting commenced, amateurs built crystal-receiving sets in great numbers.

Realizing the need for a more efficient detector, many set about to adapt the Edison bulb to that purpose. In 1897 J. J. Thomson identified the electron as a negatively charged particle in a cathode ray tube beam. The Edison effect is the liberation of electrons from a solid as a result of its thermal energy. The discovery of the electron made it possible to explain the Edison effect, and in 1904 J. A. Fleming of England utilized Edison's effect in a rectifier to detect radio waves. Fleming made an evacuated bulb in which part of the heated filament was surrounded by a metal cylinder connected to a separate terminal. This device, patented in 1904 and called the Fleming valve, served as a detector in a radio receiver. The Fleming valve was a two-terminal device and did not provide amplification of the received signal. The output of Fleming's diode, or two-terminal tube, produced an audible signal from the radio signal. The diode was stable but not very sensitive as a detector, and therefore carborundum, silicon, and galena crystals continued to be used as detectors.

Radio transmits signals between a transmitter and a receiver without connecting wires.

Lee De Forest of the United States, at a meeting of the American Institute of Electrical Engineers held in New York in 1907, made his first announcement of the invention of a three-terminal device. He called his triode tube the "audion," and with this three-terminal tube, amplification of low-power radio signals became possible.

The development of vacuum tube devices was an extension of Edison's early work, followed by work of Fleming, Langmuir, Thomson, and others. Perhaps the scientific key to all these developments was the identification of the electron as a negatively charged particle by J. J. Thomson in 1897. Thomson, Langmuir, and Coolidge are shown in Figure 9-3. The chronology of early work on electrical communications for the period 1838–1920 is summarized in Table 9-2.

Figure 9-3 Photograph taken in the General Electric Research Laboratories in 1923, showing, from the left, Irving Langmuir, Joseph J. Thomson, and William D. Coolidge. They are inspecting an early pliotron, a high-vacuum tube that was used both as a powerful amplifier and as an oscillator. Thomson was the discoverer of the electron, Langmuir was one of the pioneers in the development of thermionic-emission electron tubes, and Coolidge was a pioneer in the development of high-intensity x-ray tubes. Courtesy of General Electric Co.

We transmit information in electrical form with the use of signals. Using electrical circuits to process these signals resulted in improved communication systems. It is the purpose of this chapter to consider the response of *RC* and *RL* circuits subjected to voltage and/or current sources.

Table 9-2

Chronology of Early Electrical Communication

Year	Event
1838–1866	The birth of telegraphy. Morse perfects his telegraph. Transatlantic cables installed by Cyrus Field.
1864	"A Dynamical Theory of the Electromagnetic Field," by James Clerk Maxwell, establishes the concept of electromagnetic radiation.
1876–1899	The birth of telephony. Acoustic transducer perfected by Alexander Graham Bell. First telephone exchange, in New Haven, Connecticut, with eight lines (1878).
1887–1907	Wireless telegraphy. Heinrich Hertz verifies Maxwell's theory. Demonstrations by Marconi and Popov. Marconi patents a complete wireless telegraph system (1897).
1904–1920	Radio. Lee De Forest invents the audion (triode based on Fleming's valve). First commercial broadcasting station, KDKA, opens in Pittsburgh in 1920.

9-3 THE RESPONSE OF AN *RC* OR *RL* CIRCUIT EXCITED BY A CONSTANT SOURCE

In the preceding chapter we examined the behavior of an *RC* or *RL* circuit when there were no external sources within the circuit. All independent sources were switched out of the circuit prior to or at $t = t_0$ and the natural response of the circuit was determined.

In this section we consider circuits that, in addition to having initial stored energy, include independent sources. Since the response of the circuit results from its being subjected to a source, the response is called the *forced response*.

The **forced response** of a circuit is the behavior exhibited in reaction to one or more independent signal sources.

Consider the *RC* circuit shown in Figure 9-4. The sources we consider in this and the next three sections are limited to constant sources and, therefore, $i_s = I_0$. We assume that the initial voltage on the capacitor is $v(0) = V_0$. Again, for convenience we use the notation $v(0)$ for the voltage at $t = 0^+$. However, we always use $v(0^-)$ for the voltage at $t = 0^-$.

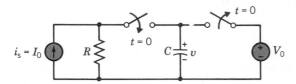

Figure 9-4 An *RC* circuit with a constant current source when $t \geq 0$. The initial voltage $v(0) = V_0$ due to the voltage source V_0, which is disconnected at $t = 0$.

Writing the KCL equations at the upper node for $t > 0$, we obtain

$$C\frac{dv}{dt} + \frac{v}{R} - I_0 = 0$$

or

$$\frac{dv}{dt} + \frac{v}{RC} = \frac{I_0}{C} \qquad (9\text{-}1)$$

Before we proceed toward the solution of Eq. 9-1, let us consider the *RL* circuit subjected to a constant voltage source as shown in Figure 9-5. Using KVL around the loop after $t = 0$, we have

$$L\frac{di}{dt} + Ri - V_0 = 0$$

or

$$\frac{di}{dt} + \frac{R}{L}i = \frac{V_0}{L} \qquad (9\text{-}2)$$

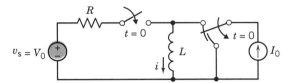

Figure 9-5 An *RL* circuit with a constant voltage source after $t \geq 0$. The initial current is $i(0) = I_0$ due to the current source I_0, which is disconnected at $t = 0$.

Examining Eqs. 9-1 and 9-2, we see that we have an equation of the form

$$\frac{dx}{dt} + \frac{x}{\tau} = K \qquad (9\text{-}3)$$

where τ is the time constant. We may rewrite Eq. 9-3 as

$$\frac{dx}{dt} = \frac{K\tau - x}{\tau}$$

or, separating the variables, we have

$$\frac{dx}{x - K\tau} = \frac{-dt}{\tau}$$

Forming the indefinite integral, we obtain

$$\int \frac{dx}{x - K\tau} = \frac{-1}{\tau} \int dt + D$$

where D is a constant of integration.

Performing the integration, we have

$$\ln(x - K\tau) = \frac{-t}{\tau} + D$$

This result can be written as

$$x - K\tau = e^{-t/\tau}e^D$$

or

$$x = K\tau + Ae^{-t/\tau} \qquad (9\text{-}4)$$

where $A = e^D$, which is determined by the initial condition $x(0)$.

Now let us refer back to the *RC* circuit represented by Eq. 9-1, where $\tau = RC$, $K = I_0/C$, and $v(0) = V_0$. Then, using Eq. 9-4, we have

$$v = \frac{I_0}{C}\tau + Ae^{-t/RC}$$

In order to find A, we see from Figure 9-4 that $v(0) = V_0$ and therefore at $t = 0$ for the preceding equation

$$v(0) = \frac{I_0}{C}\tau + A$$

or

$$V_0 = I_0R + A$$

Since $A = V_0 - I_0R$, the equation for the voltage becomes

$$v = I_0R + (V_0 - I_0R)e^{-t/RC}$$

or

$$v(t) = v(\infty) + [v(0) - v(\infty)]e^{-t/\tau} \tag{9-5}$$

where $v(\infty)$ is the steady-state value of $v(t)$ at $t = \infty$.

Equation 9-5 is the complete response, v, of the *RC* circuit of Figure 9-4 to a sudden application of a constant current source. This response is shown in Figure 9-6 with the assumption that $V_0 > RI_0$.

Examining Eq. 9-5 and Figure 9-6, we note that the transient natural response decays to zero after five time constants and the steady-state response is a constant voltage. Note that $e^{-5} = 0.0067 \approx 0$ compared to $e^{-0} = 1$ at $t = 0$.

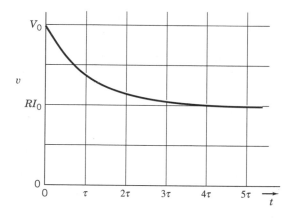

Figure 9-6 Complete response v of the *RC* circuit of Figure 9-4. The initial voltage on the capacitor is V_0 at $t = 0$ (actually $t = 0^+$, but we have dropped the + for convenience).

The steady-state response to a dc (constant) source is a constant.

Let us proceed to find the complete current response of the *RL* circuit. Reviewing Eq. 9-2, we note that $\tau = L/R$, $K = V_0/L$, and the initial current is $i(0) = I_0$. Substituting into Eq. 9-4, we obtain

$$i = \frac{V_0}{L}\tau + Ae^{-t/\tau}$$

or

$$i = \frac{V_0}{R} + Ae^{-Rt/L}$$

We describe the inductor current as consisting of two parts: the forced response, i_f, and the natural response, i_n.

$$i = i_f + i_n \tag{9-6}$$

where

$$i_f = \frac{V_o}{R}$$

and

$$i_n = Ae^{-Rt/L}$$

Sometimes these are also called the steady-state and transient responses. The sum of the forced response and the natural response is called the complete response.

The next step is to evaluate A by letting $t = 0$, obtaining

$$i(0) = \frac{V_0}{R} + A$$

Substituting $i(0) = I_0$, we obtain

$$I_0 = \frac{V_0}{R} + A$$

Then $A = I_0 - V_0/R$, and we obtain the solution for the inductor current as

$$i = \frac{V_0}{R} + \left(I_0 - \frac{V_0}{R} \right) e^{-Rt/L}$$

$$= i(\infty) + (i(0) - i(\infty))e^{-t/\tau} \tag{9-7}$$

Equation 9-7 is the total response of the *RL* circuit of Figure 9-5. The response of the *RL* circuit is shown in Figure 9-7 with the assumption that $I_0 > V_0/R$.

The value of di/dt of Eq. 9-6 at $t = 0$ is

$$\left. \frac{di}{dt} \right|_{t=0} = \frac{-1}{\tau} \left(I_0 - \frac{V_0}{R} \right)$$

This slope at $t = 0$ is shown for the interval $0 < t < \tau$ in Figure 9-7.

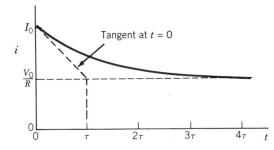

Figure 9-7 Complete response of the *RL* circuit of Figure 9-5. The initial current in the inductor is I_0 at $t = 0$. The tangent at $t = 0$ is also shown.

Any circuit that can be reduced to an equivalent circuit consisting of a resistor and a capacitor (or inductor) is characterized by a first-order linear differential equation.

It is worth noting that the response of an *RC* or *RL* circuit to a switched dc source or a step function can be represented by Eq. 9-5 or Eq. 9-6 in terms of the initial and final values of the variable and the time constant. In general, the response is

$$x(t) = x(\infty) + [x(t_0) - x(\infty)]e^{-(t-t_0)/\tau} \tag{9-8}$$

valid for $t > t_0$, t_0 being the time at which a switch(es) changes position.

Example 9-1

Find the capacitor voltage for $t \geq 0$ for the circuit shown in Figure 9-8. At what time does the voltage v pass through zero? The switch was in position a for a long time prior to $t = 0^-$.

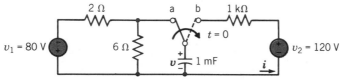

Figure 9-8 The *RC* circuit of Example 9-1.

Solution

We will use the general form of the response of a switched dc source given in Eq. 9-8. Therefore,

$$v(t) = v(\infty) + [v(0) - v(\infty)]e^{-t/\tau}$$

First, we need to find the initial capacitor voltage $v(0^-)$. Since the source v_1 is a constant and the switch has been in position a for a long time, the capacitor appears as an open circuit. Then the voltage across the capacitor is equal to the voltage across the 6-Ω resistor, or

$$v(0^-) = v(0) = -60 \text{ V}$$

Note the negative sign due to the reference sign for v.

At time $t = 0$, the switch advances to position b. The time constant of the circuit is then

$$\tau = RC = 1 \text{ s}$$

The source $v_2 = 120$ V is the constant source voltage or forcing function. Therefore, the value of $v(t)$ at $t = \infty$ is $v(\infty) = 120$ V. Then, $v(0) - v(\infty) = -60 - 120 = -180$ V. Therefore, we obtain the complete response as

$$v = v(\infty) + [v(0) - v(\infty)]e^{-t/\tau}$$
$$= 120 - 180e^{-t}$$

The response of this *RC* circuit is shown in Figure 9-9.

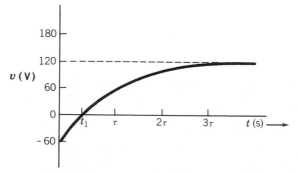

Figure 9-9 Response of the *RC* circuit for Example 9-1.

The time t_1 when the voltage passes through zero is obtained from

$$0 = 120 - 180e^{-t_1}$$

or

$$e^{-t_1} = \frac{120}{180}$$

Therefore,

$$t_1 = 0.41 \text{ s}$$

EXERCISE 9-1

Find the response of the inductor current i for the circuit of Figure E 9-1. The switch is activated at $t = 0$ after being in the a position for a long time.

Answer: $i = -4 + 12e^{-1000t}$ A, $t \geq 0$

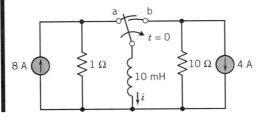

Figure E 9-1

EXERCISE 9-2

Determine the capacitor voltage $v(t)$ for the circuit shown in Figure E 9-2. Assume that the circuit is in steady state at $t = 0^-$.

Answer: $v(t) = 50(1 + e^{-100t})$ V

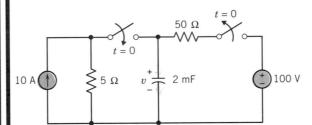

Figure E 9-2

9-4 A SIMPLE PROCEDURE FOR FINDING THE RESPONSE OF AN *RC* OR *RL* CIRCUIT TO A CONSTANT SOURCE

We have seen that the response of an *RC* or *RL* circuit is simply

$$x = x_n + x_f$$

where x_n is the natural response, $Ae^{-t/\tau}$, of the circuit and x_f is the response to the constant forcing function (a constant source). It would be wise, then, to proceed directly to find x_n and x_f separately and add them to find x. Then, we use $x(0)$ to find A and we have the complete response, $x(t)$.

Let us find the response i of the circuit shown in Figure 9-10 when it is given that $i(0) = 3$ A. We know that

$$i = i_n + i_f$$

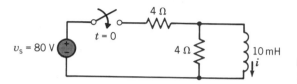

Figure 9-10 An *RL* circuit with $i(0) = 3$ A.

First, we find the unforced (natural) response for $t \geq 0$ when $v_s = 0$. Then we have the circuit shown in Figure 9-11, where we have replaced the source with a short circuit since $v_s = 0$. Therefore,

$$i_n = Ae^{-t/\tau}$$

where

$$\tau = \frac{L}{R_p} = \frac{0.01}{2} = 5 \text{ ms}$$

since the parallel resistors reduce to $R_p = 2\ \Omega$.

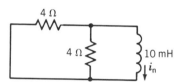

Figure 9-11 The *RL* circuit of Figure 9-10 with $v_s = 0$ and $t \geq 0$.

At this point, it is worth noting that we can determine τ of a first-order circuit by using Thévenin's principle to evaluate the resistance seen at the terminals of a given capacitor or inductor.

The forced response is a constant, so it does not matter what time we determine i_f once the steady-state condition is reached. For ease, we use $t = \infty$ when the transient has died out and the circuit responds only to the constant source. At $t = \infty$, or steady state, the inductor behaves as a short circuit and the current through it is a constant. We then represent the circuit as shown in Figure 9-12. Therefore,

$$i_f = \frac{80}{4} = 20 \text{ A}$$

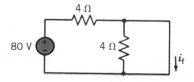

Figure 9-12 The *RL* circuit in the steady-state condition.

Thus, we have the total current as

$$i = i_n + i_f$$
$$= Ae^{-200t} + 20$$

Recall that

$$i(0) = 3 \text{ A}$$

Hence we have

$$i(0) = A + 20$$

or

$$3 = A + 20$$

Therefore, $A = -17$, and we have

$$i = -17e^{-200t} + 20 \text{ A}$$

Remember to apply the initial condition to the complete response, not just the natural response, because the initial condition is given for the total current or voltage, not just part of it.

Example 9-2
Find the response v of the *RC* circuit shown in Figure 9-13 when it is known that $v(0) = 10$ V.

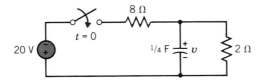

Figure 9-13 Circuit of Example 9-2.

Solution
After the switch closes at $t = 0$, we have

$$v = v_n + v_f$$

First, we find v_n, the natural response, by setting the voltage source to zero (replace it with a short circuit). We may then determine the circuit time constant $\tau = R_p C$, where R_p is the parallel equivalent resistance for the 8-Ω and 2-Ω parallel resistors. Therefore,

$$R_p = \frac{8 \times 2}{8 + 2} = 1.6 \text{ } \Omega$$

Then

$$\tau = R_p C = 1.6 \left(\frac{1}{4}\right) = 0.4 \text{ s}$$

Therefore, the natural response is

$$v_n = Ae^{-t/\tau}$$
$$= Ae^{-2.5t}$$

The forced response to the constant voltage source can be determined for the steady-state condition as t approaches infinity. Then the capacitor becomes an open circuit and it remains to find v_f, the voltage across the open circuit. Since v_f is the voltage across the 2-Ω resistor, we use the voltage divider principle to determine that

$$v_f = \frac{2}{10}(-20)$$

$$= -4\,\text{V}$$

Therefore, we obtain

$$v = v_n + v_f$$
$$= Ae^{-2.5t} - 4$$

It is given that $v(0) = 10$ V. We may determine A from v at $t = 0$ as

$$v(0) = A - 4$$

or

$$10 = A - 4$$

Therefore, $A = 14$ and we have the complete solution

$$v = 14e^{-2.5t} - 4 \ \ \text{V}$$

Example 9-3

Find the current i for $t \geq 0$ for the circuit of Figure 9-14a when it can be shown that $v(0) = 5$ V.

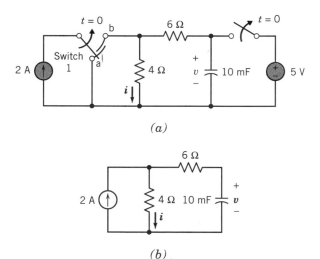

(a)

(b)

Figure 9-14 *(a)* Circuit of Example 9-3. Switch 1 moves from position a to position b at $t = 0$. *(b)* The circuit for $t > 0$.

Solution

The circuit for $t \geq 0$ is shown in Figure 9-14b. The current i is given by the equation

$$i = i_n + i_f$$

Since the circuit is linear, the natural response of every current or voltage in the circuit has the same form as v_n. This is true since a linear operation on an exponential

$$v_n = Be^{-t/\tau}$$

will yield an exponential of the same form with only a change in the constant. Therefore, the natural response for the current is

$$i_n = Ae^{-t/\tau}$$

where $\tau = RC$, which is determined for the source-free circuit. Setting the current source to zero is equivalent to an open circuit for the source. Then $\tau = RC = 10(10^{-2}) = 0.1$ s. Therefore,

$$i_n = Ae^{-10t}$$

In the steady-state condition after switch 1 is moved to position b and stays there a long time, the capacitor behaves as an open circuit in Figure 9-14b. Then $i_f = 2$ A.
    Therefore, the current is

$$i = i_n + i_f$$
$$= Ae^{-10t} + 2$$

It remains to find A by using $i(0)$. Again, we use the notation $i(0)$ for $i(0^+)$. At $t = 0$ we have the KVL equation for the right-hand mesh of Figure 9-14b as

$$-4i(0) + v(0) + 6[2 - i(0)] = 0$$

Since we can show that for the capacitor

$$v(0) = v(0^-) = 5$$

we have

$$10i(0) = v(0) + 12$$

or

$$i(0) = 1.7 \text{ A}$$

Using $i(0)$, we obtain

$$i(0) = A + 2$$

or

$$1.7 = A + 2$$

Therefore, $A = -0.3$ and the complete solution is

$$i = (-0.3e^{-10t} + 2) \text{ A}$$

EXERCISE 9-3

Find v for $t > 0$ for the circuit shown in Figure E 9-3, when $v(0) = 10$ V.
Answer: $v = (8e^{-10t} + 2)$ V

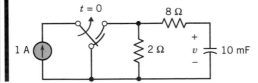

Figure E 9-3

EXERCISE 9-4

Find the voltage v for $t \geq 0$ for the circuit of Figure E 9-4 when $i(0) = 1/2$ A.
Answer: $v = (-97.5e^{-6400t} + 37.5)$ V

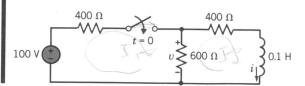

Figure E 9-4

9-5 ∥ THE UNIT STEP SOURCE

In the preceding section we consider circuits where the sources are switched in and out of the circuit at $t = t_0$. At the instant t_0 when these sources are applied to or disconnected from the circuit, many of the currents and voltages within the circuit change abruptly.

The application of a battery (constant source) by means of two switches, as shown in Figure 9-15, may be considered equivalent to a source that is zero up to t_0 and equal to the voltage V_0 thereafter.

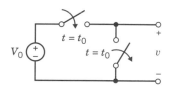

Figure 9-15 Application of a constant voltage source at $t = t_0$ using two switches both acting at $t = t_0$.

We define the *unit step forcing function* as a function of time that is zero for $t < t_0$ and unity for $t > t_0$. At $t = t_0$ the magnitude changes from zero to one. We represent the unit step function by $u(t - t_0)$ where

$$u(t - t_0) = \begin{cases} 0 & t < t_0 \\ 1 & t > t_0 \end{cases} \tag{9-9}$$

The value of $u(t - t_0)$ is not defined at $t = t_0$, where it switches instantaneously from a magnitude of zero to one. The unit step function is shown in Figure 9-16. We will often consider $t_0 = 0$.

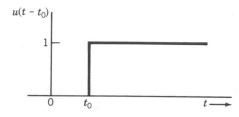

Figure 9-16 Unit step forcing function, $u(t - t_0)$.

The unit step function is dimensionless since it is a mathematical function. If we wish to represent the voltage of Figure 9-15, we use the representation

$$v(t) = V_0 u(t - t_0)$$

which implies the application of a voltage V_0 at $t = t_0$.

An exact one-switch equivalent of the source $v = V_0 u(t - t_0)$ requires that v be zero for $t < t_0$. As shown in Figure 9-17, v is defined as zero for $t < t_0$ and jumps to V_0 at $t = t_0$.

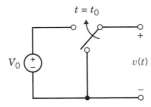

Figure 9-17 Single-switch equivalent circuit for the step voltage source.

Henceforth, we will use the symbol as shown in Figure 9-18 to represent the step voltage source. The *step response* of a circuit is the response of the circuit to the sudden application of a constant source $v = V_0 u(t - t_0)$ when all the initial conditions of the circuit are equal to zero.

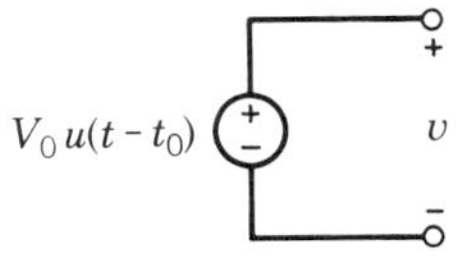

Figure 9-18 Symbol for the step voltage source of magnitude V_0 applied at $t = t_0$.

It is worth noting that $u(-t)$ simply implies that we have a value of 1 for $t < 0$, so that

$$u(-t) = \begin{cases} 1 & t < 0 \\ 0 & t > 0 \end{cases}$$

For the case where the change occurs at t_0, we have

$$u(t_0 - t) = \begin{cases} 1 & t < t_0 \\ 0 & t > t_0 \end{cases}$$

Let us consider the *pulse* source:

$$v(t) = \begin{cases} 0 & t < t_0 \\ V_0 & t_0 < t < t_1 \\ 0 & t_1 < t \end{cases}$$

which is shown in Figure 9-19*a*. As shown in Figure 9-19*b*, the pulse can be obtained from two step voltage sources, the first of magnitude V_0 occurring at $t = t_0$ and the second equal to $-V_0$ occurring at $t = t_1$. Thus, the two step sources of magnitude V_0 shown in Figure 9-20 will yield the desired pulse. We have $v(t) = V_0 u(t - t_0) - V_0 u(t - t_1)$ to provide the pulse. Notice how easy it is to use two step function symbols to represent this pulse source. The pulse is said to have a duration of $(t_1 - t_0)$ s.

We recognize that the unit step function is an ideal model. No real element can switch instantaneously at $t = t_0$. However, if it switches in a very short time (say, 1 ns) we can

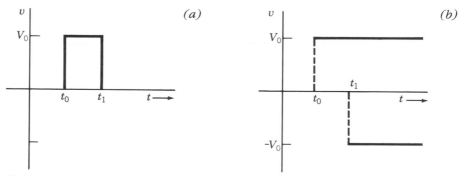

Figure 9-19 *(a)* Rectangular voltage pulse. *(b)* Two step voltage waveforms that yield the voltage pulse.

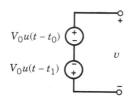

Figure 9-20 Two step voltage sources that yield a rectangular voltage pulse, $v(t)$ with a magnitude of V_0 and a duration of $(t_1 - t_0)$ where $t_0 < t_1$.

consider the switching as instantaneous for most of the medium speed circuits. As long as the switching time is small compared to the time constant of the circuit, it can be ignored.

Let us consider the application of a pulse to an *RL* circuit as shown in Figure 9-21. Here we let $t_0 = 0$ for convenience. The duration of the pulse is t_1 s.

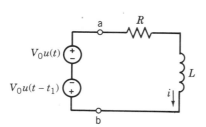

Figure 9-21 Pulse source connected to an *RL* circuit. The pulse is generated by the combination of two step voltage sources each with units in volts.

A **pulse signal** has a constant nonzero value for a time duration of $\Delta t = t_1 - t_0$.

The pulse is applied to the *RL* circuit when $i(0) = 0$.

We wish to determine i for the pulse source. Since the circuit is linear, we may use the principle of superposition so that

$$i = i_1 + i_2$$

where i_1 is the response to $V_0 u(t)$ and i_2 is the response to $V_0 u(t - t_1)$.

We recall that the response of an *RL* circuit to a constant forcing function applied at $t = t_n$ is (see Eq. 9-6 with $I_0 = 0$)

$$i = \frac{V_0}{R}(1 - e^{-a(t - t_n)}) \text{ A} \qquad t > t_n$$

where $a = 1/\tau = R/L$. Consequently, we may add the two solutions to the two step sources, carefully noting $t_0 = 0$ and t_1 as the start of each response. Therefore,

$$i_1 = \frac{V_0}{R}(1 - e^{-at}) \text{ A} \qquad t \geq 0$$

and

$$i_2 = \frac{-V_0}{R}(1 - e^{-a(t-t_1)}) \text{ A} \qquad t > t_1$$

Adding the responses, we have

$$i = \begin{cases} \dfrac{V_0}{R}(1 - e^{-at}) & 0 < t \leq t_1 \\ \dfrac{V_0}{R}e^{-at}(e^{at_1} - 1) & t > t_1 \end{cases}$$

Of course, the current i is zero for $t < 0$. The response i is shown in Figure 9-22. The magnitude of the response at $t = t_1$ is

$$i(t_1) = \frac{V_0}{R}(1 - e^{-t_1/\tau}) \text{ A}$$

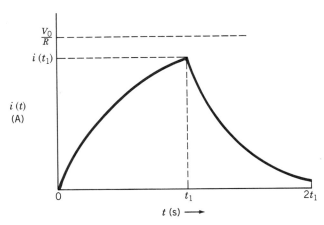

Figure 9-22 Response of the *RL* circuit shown in Figure 9-21.

If t_1 is greater than τ, the response will approach V_0/R before starting its decline as shown in Figure 9-22. The response at $t = 2t_1$ is

$$i(2t_1) = \frac{V_0}{R}e^{-2at_1}(e^{at_1} - 1)$$

$$= \frac{V_0}{R}(e^{-at_1} - e^{-2at_1}) \text{ A}$$

where $a = 1/\tau$.

EXERCISE 9-5

Find v for the circuit shown in Figure E 9-5 when $v(0^-) = 0$, $I_0 = 1$ A, and $t_1 = 0.5$ s. Plot the response for $0 < t < 1$ s.

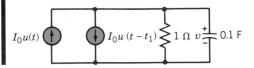

Figure E 9-5

EXERCISE 9-6

A pulse is transmitted along a telephone line represented by the circuit shown in Figure E 9-6. Find the received voltage $v(t)$. The length of the pulse is 100 ms and, therefore, $t_1 = 100$ ms. The line is represented by $C = 0.1 \ \mu\text{F}$ and $R = 200 \ \text{k}\Omega$.

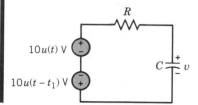

Figure E 9-6

EXERCISE 9-7

The circuit shown in Figure E 9-7*a* has a current source as shown in Figure E 9-7*b*. Determine the current $i(t)$ in the inductor.

Answer: $\quad i(t) = \begin{cases} 5(1 - e^{-10t}) \ \text{A} & t \leq 0.2 \ \text{s} \\ 4.32e^{-10(t-0.2)} \ \text{A} & t \geq 0.2 \ \text{s} \end{cases}$

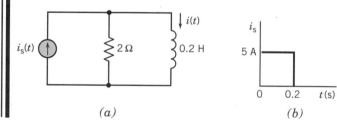

(a) $\qquad\qquad\qquad$ (b) $\qquad$ **Figure E 9-7**

USING THE SUPERPOSITION PRINCIPLE TO SOLVE CIRCUITS
9-6 ‖ **WITH MULTIPLE SOURCES**

Many circuits have multiple sources switched in or out of the circuit (or activated or deactivated) at $t = t_0$. Since the circuits are linear, we can use the principle of superposition to determine the resulting response to each source and then add the responses to determine the total response.

As an example of the use of the principle of superposition, consider the circuit shown in Figure 9-23. We will assume that $i(0^-)$ is known to be equal to I_0 and will determine $i(t)$ for $t > 0$. Since the circuit has two sources activated at $t = 0$, we find the response to each source and then add them.

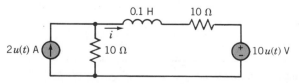

Figure 9-23 Circuit with two constant sources.

First, consider the response to the current source while the voltage source is set to zero. Replacing the voltage source with a short circuit, we have the circuit of 9-24*a*, where i_1 is the inductor current

where _____ equal to 1 A using the curr___

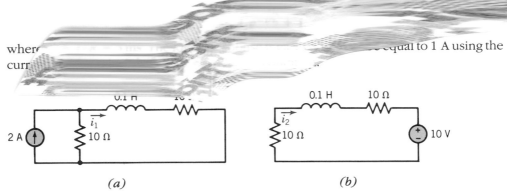

(a)   *(b)*

Figure 9-24 *(a)* Circuit of Figure 9-23 with the voltage source deactivated for $t > 0$. *(b)* Circuit of Figure 9-23 with the current source deactivated for $t > 0$.

The response i_2 to the second source (the voltage source) is obtained from the circuit of Figure 9-24*b*, where the current source is deactivated and replaced by an open circuit. The forced current is easily found by replacing the inductor with a short circuit and determining that $i_{f_2} = -0.5$. Then we have

$$i_2 = i_{n_2} + i_{f_2}$$
$$= A_2 e^{-t/\tau_2} + (-0.5)$$

and $\tau_2 = 5$ ms. Then we can obtain the total response as

$$i = i_1 + i_2$$
$$= (A_1 + A_2)e^{-t/\tau} + 0.5$$

since $\tau_1 = \tau_2 = \tau$. Using $i(0) = I_0$, we have

$$I_0 = (A_1 + A_2) + 0.5$$

or

$$A_1 + A_2 = I_0 - 0.5$$

Therefore, we obtain

$$i = (I_0 - 0.5)e^{-200t} + 0.5 \text{ A}$$

The use of the superposition principle can be particularly valuable in the solution of complex circuits with many sources.

Example 9-4

Find the current $i(t)$ for $t > 0$ for the circuit in Figure 9-25. Assume that the circuit has reached steady state at $t = 0^-$.

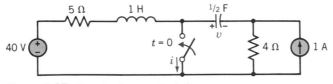

Figure 9-25 Circuit of Example 9-4.

Solution

When the switch is closed at $t = 0$, the circuit becomes two separate circuits with the current in the short circuit, i, equal to the sum of the currents due to the separate circuits. Therefore, $i = i_1 + i_2$.

First, let us determine the initial conditions at $t = 0^-$ for the inductor current and the capacitor voltage. The inductor is replaced by a short circuit and the capacitor by an open circuit as shown in Figure 9-26. Since no current can flow through the open circuit, the inductor current is $i_L(0^-) = 0$. The capacitor voltage can be obtained by summing the voltages around the left-hand loop to obtain

$$-40 + v + 4 = 0$$

or

$$v(0^-) = 36 \text{ V}$$

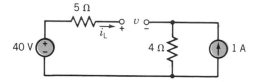

Figure 9-26 Circuit of Example 9-4 for $t = 0^-$.

After the switch is activated at $t = 0$, we have the two subcircuits shown in Figures 9-27a and 9-27b. The total current in the short circuit is $i = i_1 + i_2$. Notice that $i_1(0)$ does not depend on the current source and $i_2(0)$ does not depend on the voltage source.

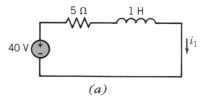
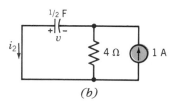

(a)	(b)

Figure 9-27 The circuit of Example 9-4 for $t > 0$ becomes two distinct circuits with $i = i_1 + i_2$.

Considering the circuit shown in Figure 9-27a, we have

$$i_1 = Ae^{-t/\tau} + 8$$

where $\tau = L/R = 1/5$ s. Since $i_1(0) = 0$, we obtain

$$i_1 = -8e^{-5t} + 8 \text{ A}$$

For the circuit with the capacitor shown in Figure 9-27b, we have

$$i_2 = -C\frac{dv}{dt}$$

where it remains to find $v(t)$. We obtain from the circuit of Figure 9-27b

$$v(t) = Be^{-t/\tau} - 4$$

where $\tau = RC = 2$ s. Since $v(0) = 36$, we have

$$36 = B - 4$$

or $B = 40$. Therefore,

$$v(t) = 40e^{-t/2} - 4$$

Th⋯ ⋯⋯⋯ *i* i⋯ th⋯⋯

$$= -\frac{1}{2}(-20e^{-t/2})$$

Hence, we obtain the current in the short-circuit wire, *i*, as

$$i = i_1 + i_2$$
$$= (-8e^{-5t} + 8 + 10e^{-t/2}) \text{ A}$$

EXERCISE 9-8

Find the current *i* for $t > 0$ when $i(0^-) = 0$ for the circuit of Figure E 9-8.
Answer: $i = 2.067(1 - e^{-30,000t}) A$

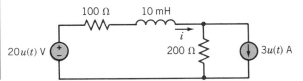

Figure E 9-8

EXERCISE 9-9

Find *i* for $t > 0$ for the circuit of Figure E 9-9. Assume the circuit was in steady state at $t = 0^-$.
Answer: $i = -3 + 3e^{-1000t} - 3e^{-10t} A$

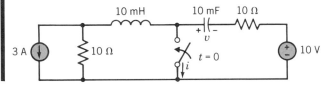

Figure E 9-9

EXERCISE 9-10

Determine and sketch the voltage $v(t)$ and the current $i(t)$ for the circuit shown in Figure E 9-10. Assume that the circuit is in steady state at $t = 0^-$.
Partial Answer: $v(t) = 13 - 4e^{-416.7t} \text{ V}$

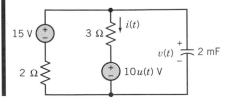

Figure E 9-10

9-7 THE RESPONSE OF AN *RL* OR *RC* CIRCUIT TO A NONCONSTANT SOURCE

In the previous sections we have wisely used the fact that the forced response to a constant source will be a constant itself. It now remains to determine what the response will be when the forcing function is not a constant.

The differential equation described by an *RL* or *RC* circuit is represented by the general form (see Eq. 9-3)

$$\frac{dx(t)}{dt} + ax(t) = y(t) \tag{9-10}$$

where $y(t)$ is a constant only when we have a constant current or constant voltage source and where $a = 1/\tau$ is a constant.

In this section we introduce the *integrating factor method,* which consists of multiplying Eq. 9-10 by a factor that makes the left-hand side a perfect derivative, and then integrating both sides.

Consider the derivative of a product of two terms such that

$$\frac{d}{dt}(xe^{at}) = \frac{dx}{dt}e^{at} + axe^{at}$$

$$= \left(\frac{dx}{dt} + ax\right)e^{at} \tag{9-11}$$

The term within the parentheses on the right-hand side of Eq. 9-11 is exactly the form on the left-hand side of Eq. 9-10.

Therefore, if we multiply both sides of Eq. 9-10 by e^{at}, the left-hand side of the equation can be represented by the perfect derivative $d(xe^{at})/dt$. Carrying out these steps, we show that

$$\left(\frac{dx}{dt} + ax\right)e^{at} = ye^{at}$$

or

$$\frac{d}{dt}(xe^{at}) = ye^{at}$$

Integrating both sides of the second equation, we have

$$xe^{at} = \int ye^{at}\,dt + K$$

where K is a constant of integration. Therefore, solving for $x(t)$, we multiply by e^{-at} to obtain

$$x = e^{-at}\int ye^{at}\,dt + Ke^{-at} \tag{9-12}$$

For the case where the source is a constant so that $y(t) = M$, we have

$$x = e^{-at}M \int e^{at}\, dt + Ke^{-at}$$

$$= \frac{M}{a} + Ke^{-at}$$

$$= x_f + x_n$$

where the natural response is $x_n = Ke^{-at}$ and the forced response is $x_f = M/a$, a constant.

Now consider the case where $y(t)$, the forcing function, is not a constant. Considering Eq. 9-12, we see that the natural response remains, $x_n = Ke^{-at}$. However, the forced response is

$$x_f = e^{-at} \int y(t)e^{at}\, dt$$

Thus, the forced response will be dictated by the form of $y(t)$. Let us consider the case where $y(t)$ is an exponential function so that $y(t) = e^{bt}$. We assume that $(a + b)$ is not equal to zero. Then we have

$$x_f = e^{-at} \int e^{bt}e^{at}\, dt$$

$$= e^{-at} \int e^{(a+b)t}\, dt$$

$$= \frac{1}{a+b} e^{-at}e^{(a+b)t}$$

$$= \frac{e^{bt}}{a+b} \tag{9-13}$$

Therefore, the forced response of an *RL* or *RC* circuit to an exponential forcing function is of the same form as the forcing function itself. When $a + b$ is not equal to zero, we assume that the forced response will be of the same form as the forcing function itself, and we try to obtain the relationship that will be satisfied under those conditions.

Example 9-5
Find the current i for the circuit of Figure 9-28 for $t > 0$ when

$$v_s = 10e^{-2t}u(t) \text{ V}$$

Assume the circuit is in steady state at $t = 0^-$.

Solution
Since the forcing function is an exponential, we expect an exponential for the forced response i_f. Therefore, we expect i_f to be

$$i_f = Be^{-2t}$$

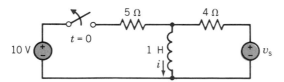

Figure 9-28 Circuit for Example 9-5.

for $t > 0$. Writing KVL around the loop, we have

$$L\frac{di}{dt} + Ri = v_s$$

or

$$\frac{di}{dt} + 4i = 10e^{-2t}$$

for $t > 0$. Substituting $i_f = Be^{-2t}$, we have

$$-2Be^{-2t} + 4Be^{-2t} = 10e^{-2t}$$

or

$$(-2B + 4B)e^{-2t} = 10e^{-2t}$$

Hence, $B = 5$ and

$$i_f = 5e^{-2t}$$

The natural response can be obtained by considering the equation

$$L\frac{di}{dt} + Ri = 0$$

Using the operator $s = d/dt$, we have

$$(Ls + R)i = 0$$

so that

$$s = -\frac{R}{L}$$

and, therefore,

$$i_n = Ae^{-Rt/L}$$
$$= Ae^{-4t}$$

The complete response is

$$i = i_n + i_f$$
$$= Ae^{-4t} + 5e^{-2t}$$

The initial condition $i(0) = i(0^-)$ is found from consideration of the 10-V source connected prior to $t = 0$. Recall that v_s is zero for $t < 0$. Therefore, $i(0^-) = 10/5 = 2$ A. Using the initial condition with Eq. 9-13, we have

$$2 = A + 5$$

or $A = -3$. Therefore,

$$i = (-3e^{-4t} + 5e^{-2t})\ \text{A} \qquad t > 0$$

The voltage source of Example 9-5 is a decaying exponential of the form

$$v_s = 10e^{-2t}u(t)\ \text{V}$$

This source is said to be *aperiodic* (nonperiodic). A periodic source is one that repeats itself exactly after a fixed length of time. Thus the signal $f(t)$ is *periodic* if there is a number

T such that for all *t*

$$f(t + T) = f(t) \tag{9-14}$$

The smallest positive number *T* that satisfies Eq. 9-14 is called the *period*. The period defines the duration of one complete cycle of $f(t)$. Thus, any source for which there is no value of *T* satisfying Eq. 9-14 is said to be aperiodic. An example of a periodic source is $10 \sin 2t\, u(t)$, which we consider in Example 9-6. The period of this sinusoidal source is π s.

A **sinusoidal signal** varies in accordance with a sine or cosine function of time.

Example 9-6

Find the response $v(t)$ for $t > 0$ for the circuit of Figure 9-29. The initial voltage $v(0) = 0$ and the current source is $i_s = (10 \sin 2t)u(t)$ A.

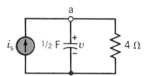

Figure 9-29 Circuit of Example 9-6.

Solution

Since the forcing function is a sinusoidal function, we expect that v_f is of the same form. Writing KCL at node a, we obtain

$$C\frac{dv}{dt} + \frac{v}{R} = i_s$$

or

$$0.5\frac{dv}{dt} + \frac{v}{4} = 10 \sin 2t \tag{9-15}$$

for $t > 0$. We assume that v_f will consist of the sinusoidal function $\sin 2t$ and its derivatives.

Examining Eq. 9-15, $v_f/4$ plus $0.5\, dv_f/dt$ must equal $10 \sin 2t$. However, $d(\sin 2t)/dt = 2 \cos 2t$. Therefore, the trial v_f needs to contain both $\sin 2t$ and $\cos 2t$ terms. Thus, we try the proposed solution

$$v_f = A \sin 2t + B \cos 2t$$

The derivative of v_f is then

$$\frac{dv_f}{dt} = 2A \cos 2t - 2B \sin 2t$$

Substituting v_f and dv_f/dt into Eq. 9-15, we obtain

$$(A \cos 2t - B \sin 2t) + \frac{1}{4}(A \sin 2t + B \cos 2t) = 10 \sin 2t$$

Therefore, equating $\sin 2t$ terms and $\cos 2t$ terms we obtain

$$\left(\frac{A}{4} - B\right) = 10 \quad \text{and} \quad \left(A + \frac{B}{4}\right) = 0$$

Solving for A and B, we obtain

$$A = \frac{40}{17} \quad \text{and} \quad B = \frac{-160}{17}$$

Consequently,

$$v_f = \frac{40}{17} \sin 2t - \frac{160}{17} \cos 2t$$

It is necessary that v_f be made up of $\sin 2t$ and $\cos 2t$ since the solution has to satisfy the differential equation. Of course, the derivative of $\sin \omega t$ is $\omega \cos \omega t$.

We need to obtain the natural response from Eq. 9-15 by considering the unforced differential equation

$$C\frac{dv_n}{dt} + \frac{v_n}{R} = 0$$

If $v_n = De^{-st}$, we can use the operator s to obtain

$$\left(Cs + \frac{1}{R}\right) v_n = 0$$

Therefore, $s = -1/RC$ and the natural response is

$$v_n = De^{-t/RC}$$

The total response is then

$$v = v_n + v_f$$

$$= De^{-t/2} + \frac{40}{17} \sin 2t - \frac{160}{17} \cos 2t$$

Since $v(0) = 0$, we obtain at $t = 0$

$$0 = D - \frac{160}{17}$$

or

$$D = \frac{160}{17}$$

Then the complete response is

$$v = \left(\frac{160}{17} e^{-t/2} + \frac{40}{17} \sin 2t - \frac{160}{17} \cos 2t\right) V$$

A special case for the forced response of a circuit may occur when the forcing function is a damped exponential where we have $y(t) = e^{-bt}$. Referring back to Eq. 9-13, we can show that

$$x_f = \frac{e^{-bt}}{a - b}$$

when $y(t) = e^{-bt}$. Note that here we have e^{-bt} while we used e^{bt} for Eq. 9-13. For the special case where $a = b$, we have $a - b = 0$, and this form of the response is indeterminate. For this special case, we must use $x_f = te^{-bt}$ as the forced response. The solution x_f, for the forced response, when $a = b$, will satisfy the original differential equation (9-10). Thus, when the natural response already contains a term of the same form as the forcing function, we need to multiply the assumed form of the forced response by t.

Table 9-3
Steady-State Response to a Forcing Function

Forcing Function, $y(t)$	Steady-State Response, $x_f(t)$
1. Constant $y(t) = M$	$x_f = N$, a constant
2. Exponential $y(t) = Me^{-bt}$	$x_f = Ne^{-bt}$
3. Sinusoid $y(t) = M \sin(\omega t + \theta)$	$x_f = A \sin \omega t + B \cos \omega t$

The steady-state response to selected forcing functions is summarized in Table 9-3. We note that if a circuit is linear, at steady state and excited by a single sinusoidal source having frequency ω, then all the branch currents and voltages are sinusoids having frequency ω.

EXERCISE 9-11

The electrical power plant for the orbiting space station shown in Figure E 9-11*a* uses photovoltaic cells to store energy in batteries. The charging circuit is modeled by the circuit shown in Figure E 9-11*b*, where $v_s = 10 \sin 20t$ V. If $v(0^-) = 0$, find $v(t)$ for $t > 0$.
Answer: $v = 40e^{-10t} - 40 \cos 20t + 20 \sin 20t$ V

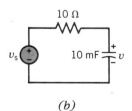

(*a*)   (*b*)

Figure E 9-11 (*a*) The NASA space station design shows in use the longer habitable modules that would house an orbiting scientific laboratory. (*b*) The circuit for energy storage for the laboratories. Photograph courtesy of the National Aeronautics and Space Administration.

EXERCISE 9-12

Find the current i in the circuit of Figure E 9-12 for $t > 0$ when $i_s = 10e^{-5t}u(t)$ A and $i(0^-) = 0$.

Answer: $i = 10.53(e^{-5t} - e^{-100t})$ A

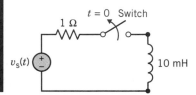

Figure E 9-12

EXERCISE 9-13

An experimenter is working in her laboratory with an electromagnet as shown in Figure E 9-13. She notices that whenever she turns off the electromagnet, a big spark appears at the switch contacts. Explain the occurrence of the spark. Suggest a way to suppress the spark by adding one element.

Figure E 9-13

9-8 ‖ THE FORCED RESPONSE USING PSpice

In this section we consider the use of the transient analysis statement (.TRAN) in order to obtain the response of an RC or RL circuit to a forcing function with an initial condition. Let us consider the RC circuit shown in Figure 9-30a. The initial capacitor voltage is 5 V, and the magnitude of the step voltage input is 25 V. The circuit redrawn for PSpice is shown in Figure 9-30b, where $v(0) = 5$ V. We use the piecewise linear input source V1 to generate a

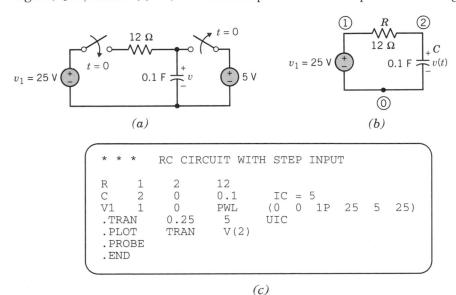

Figure 9-30 (*a*) RC circuit. (*b*) Circuit redrawn for PSpice. (*c*) PSpice program.

step as described in Appendix G-6. The piecewise linear waveform starts with a value of zero at $t = 0$ and jumps to 25 V after 1 ps and remains at 25 V. The PSpice program is shown in Figure 9-30c. We choose a 250-ms step size and plot 5 s of response (20 points). The response $v(t)$ is printed in Figure 9-31a, and the .PROBE statement is used to obtain a higher resolution plot as shown in Figure 9-31b. Examining the printout of Figure 9-31a, we find that the capacitor voltage attains 98.8 percent of its final value at $t = 5$ s.

```
  TIME           V(2)
(*)-----------     5.0000E+00    1.0000E+01    1.5000E+01    2.0000E+01    2.5000E+01
               - - - - - - - - - - - - - - - - - - - - - - - - - - - -
0.000E+00  5.000E+00 *           .             .             .             .
2.500E-01  8.749E+00 .       *   .             .             .             .
5.000E-01  1.182E+01 .           .     *       .             .             .
7.500E-01  1.429E+01 .           .             . *           .             .
1.000E+00  1.631E+01 .           .             .       *     .             .
1.250E+00  1.794E+01 .           .             .             .   *         .
1.500E+00  1.927E+01 .           .             .             .       *     .
1.750E+00  2.035E+01 .           .             .             .         .*  .
2.000E+00  2.123E+01 .           .             .             .           .*
2.250E+00  2.193E+01 .           .             .             .             . *
2.500E+00  2.251E+01 .           .             .             .             .   *
2.750E+00  2.298E+01 .           .             .             .             .     *
3.000E+00  2.336E+01 .           .             .             .             .       *
3.250E+00  2.367E+01 .           .             .             .             .         *
3.500E+00  2.392E+01 .           .             .             .             .         *
3.750E+00  2.412E+01 .           .             .             .             .           *
4.000E+00  2.429E+01 .           .             .             .             .           *
4.250E+00  2.442E+01 .           .             .             .             .           *
4.500E+00  2.453E+01 .           .             .             .             .            *.
4.750E+00  2.462E+01 .           .             .             .             .            *.
5.000E+00  2.469E+01 .           .             .             .             .            *.
               - - - - - - - - - - - - - - - - - - - - - - - - - - - -
```

(a)

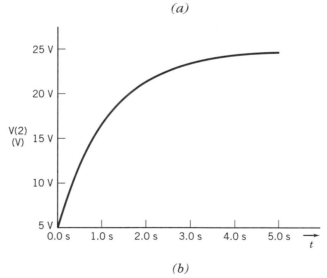

(b)

Figure 9-31 PRINT and PROBE output for the RC circuit of Figure 9-30.

9-9 | VERIFICATION EXAMPLES

Example 9 V-1

Consider the circuit and corresponding transient response shown in Figure V 9-1. How can we tell if the transient response is correct? There are three things that need to be verified: the initial voltage, $v_o(t_0)$; the final voltage, $v_o(\infty)$; and the time constant, τ.

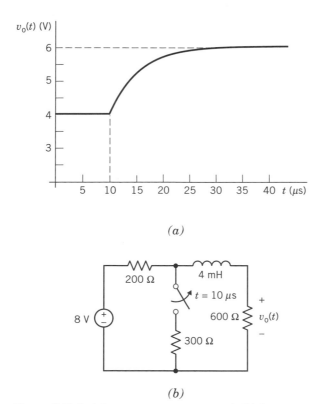

(a)

(b)

Figure 9 V-1 (a) A transient response and (b) the corresponding circuit.

Solution

Consider first the initial voltage, $v_o(t_0)$. (In this example, $t_0 = 10$ ms.) Before time $t_0 = 10$ ms, the switch is closed and has been closed long enough for the circuit to reach steady state, that is, for any transients to have died out. To calculate $v_o(t_0)$ we simplify the circuit in two ways. First, replace the switch with a short circuit because the switch is closed. Second, replace the inductor with a short circuit because inductors act like short circuits when all the inputs are constants and the circuit is at steady state. The resulting circuit is shown in Figure 9 V-2a. After replacing the parallel 300-Ω and 600-Ω resistors with the equivalent 200-Ω resistor, the initial voltage is calculated using voltage division as

$$v_o(t_0) = \frac{200}{200 + 200} 8$$

$$= 4 \text{ V}$$

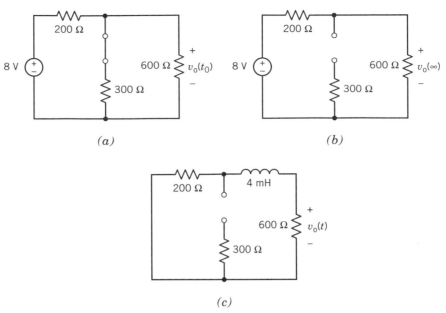

(a)   (b)

(c)

Figure 9 V-2 Circuits used to calculate the *(a)* initial voltage, *(b)* final voltage, and *(c)* time constant.

Next consider the final voltage, $v_o(\infty)$. In this case, the switch is open and the circuit has reached steady state. Again, the circuit is simplified in two ways. The switch is replaced with an open circuit because the switch is open. The inductor is replaced by a short circuit because inductors act like short circuits when all the inputs are constants and the circuit is at steady state. The simplified circuit is shown in Figure 9 V-2*b*. The final voltage is calculated using voltage division as

$$v_o(\infty) = \frac{200}{200 + 600} 8$$

$$= 6 \text{ V}$$

The time constant is calculated from the circuit shown in Figure 9 V-2*c*. This circuit has been simplified by setting the input to zero (a zero voltage source acts like a short circuit)

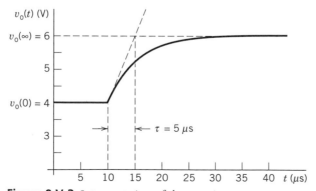

Figure 9 V-3 Interpretation of the transient response.

and replacing the switch by an open circuit. The time constant is

$$\tau = \frac{4 \text{ mH}}{200 \ \Omega + 800 \ \Omega}$$

$$= 5 \ \mu s$$

Figure 9 V-3 shows how the initial voltage, final voltage, and time constant can be determined from the plot of the transient response. (Recall that a procedure for determining the time constant graphically was illustrated in Figure 9-7.) Since the values of $v_o(t_0)$, $v_o(\infty)$, and τ obtained from the transient response are the same as the values obtained by analyzing the circuit, we conclude that the transient response is indeed correct.

Example 9 V-2

Consider the circuit and corresponding transient response shown in Figure 9 V-4. How can we tell if the transient response is correct? There are four things that need to be verified: the steady-state capacitor voltage when the switch is open, the steady-state capacitor voltage when the switch is closed, the time constant when the switch is open, and the time constant when the switch is closed.

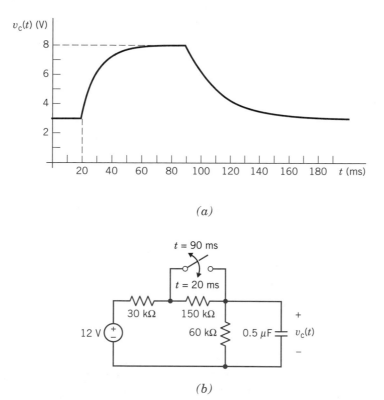

(a)

(b)

Figure 9 V-4 (a) A transient response and (b) the corresponding circuit.

Solution

Figure 9 V-5a shows the circuit used to calculate the steady-state capacitor voltage when the switch is open. The circuit has been simplified in two ways. First, the switch has been replaced with an open circuit. Second, the capacitor has been replaced with an open circuit because capacitors act like open circuits when all the inputs are constants and the

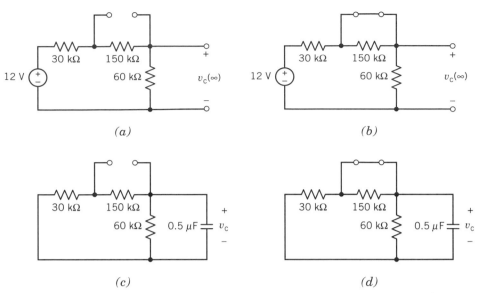

Figure 9 V-5 Circuits used to calculate (*a*) the steady-state voltage when the switch is open, (*b*) the steady-state voltage when the switch is closed, (*c*) the time constant when the switch is open, and (*d*) the time constant when the switch is closed.

circuit is at steady state. The steady-state capacitor voltage is calculated using voltage division as

$$v_c(\infty) = \frac{60}{60 + 30 + 150} 12$$
$$= 3 \text{ V}$$

Figure 9 V-5*b* shows the circuit used to calculate the steady-state capacitor voltage when the switch is closed. Again, this circuit has been simplified in two ways. First, the switch has been replaced with a short circuit. Second, the capacitor has been replaced with an open circuit. The steady-state capacitor voltage is calculated using voltage division as

$$v_c(\infty) = \frac{60}{60 + 30} 12$$
$$= 8 \text{ V}$$

Figure 9 V-5*c* shows the circuit used to calculate the time constant when the switch is open. This circuit has been simplified in two ways. First, the switch has been replaced with an open circuit. Second, the input has been set to zero (a zero voltage source acts like a short circuit). Notice that 180 kΩ in parallel with 60 kΩ is equivalent to 45 kΩ. The time constant is

$$\tau = 45 \text{ k}\Omega \cdot 0.5 \ \mu\text{F}$$
$$= 22.5 \text{ ms}$$

Figure 9 V-5*d* shows the circuit used to calculate the steady-state capacitor voltage when the switch is closed. The switch has been replaced with a short circuit and the input has been set to zero. Notice that 30 kΩ in parallel with 60 kΩ is equivalent to 20 kΩ. The time constant is

$$\tau = 20 \text{ k}\Omega \cdot 0.5 \ \mu\text{F}$$
$$= 10 \text{ ms}$$

Having done these calculations, we expect the capacitor voltage to be 3 V until the switch closes at $t = 20$ ms. The capacitor voltage will then increase exponentially to 8 V, with a time constant equal to 10 ms. The capacitor voltage will remain 8 V until the switch opens at $t = 90$ ms. The capacitor voltage will then decrease exponentially to 3 V, with a time constant equal to 22.5 ms. Figure 9 V-6 shows that the transient response satisfies this description. We conclude that the transient response is correct.

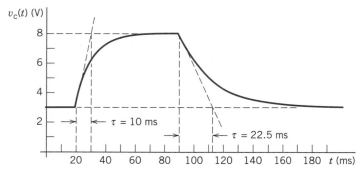

Figure 9 V-6 Interpretation of the transient response.

9-10 DESIGN CHALLENGE SOLUTION

A DRIVER AND CABLE

It is frequently necessary to connect two pieces of electronic equipment together so that the output from one device can be used as the input to another device. This situation is illustrated in Figure 9 D-1a. Here a cable is used to connect two circuits. The

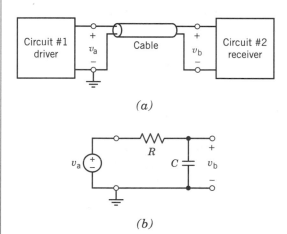

Figure 9 D-1 (a) Two circuits connected by a cable and (b) an equivalent circuit.

driver is used to send a signal through the cable to the receiver. Let us replace the driver, cable, and receiver with simple models. Model the driver as a voltage source, the cable as an *RC* circuit, and the receiver as an open circuit. The values of resistance and capacitance used to model the cable will depend on the length of the cable. For example, when RG58 coaxial cable is used,

$$R = r \cdot L \quad \text{where } r = 0.54 \, \frac{\Omega}{m}$$

and

$$C = c \cdot L \quad \text{where } c = 88 \, \frac{pF}{m}$$

and *L* is the length of the cable in meters. Figure 9 D-1*b* shows the equivalent circuit.

Suppose that the circuits connected by the cable are digital circuits. The driver will send 1's and 0's to the receiver. These 1's and 0's will be represented by voltages. The output of the driver will be one voltage, V_{OH}, to represent logic 1 and another voltage, V_{OL}, to represent a logic 0. For example, one popular type of logic, called TTL logic, uses $V_{OH} = 2.4$ V and $V_{OL} = 0.4$ V. The receiver uses two different voltages, V_{IH} and V_{IL}, to represent 1's and 0's. (This is done to provide noise immunity, but that is another story.) The receiver will interpret its input, v_b, to be a logic 1 whenever $v_b > V_{IH}$ and to be a logic 0 whenever $v_b < V_{IL}$. (Voltages between V_{IH} and V_{IL} will only occur during transitions between logic 1 and logic 0. These voltages will sometimes be interpreted as logic 1 and other times as logic 0.) TTL logic uses $V_{IH} = 2.0$ V and $V_{IL} = 0.8$ V.

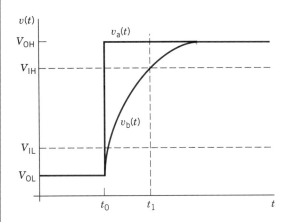

Figure 9 D-2 Voltages that occur during a transition from a logic 0 to a logic 1.

Figure 9 D-2 shows what happens when the driver output changes from logic 0 to logic 1. Before time t_0,

$$v_a = V_{OL} \quad \text{and} \quad v_b < V_{IL} \quad \text{for } t \leq t_0$$

In words, a logic 0 is sent and received. The driver output switches to V_{OH} at time t_0. The receiver input, v_b, makes this transition more slowly. Not until time t_1 does the receiver input become large enough to be interpreted as a logic 1. That is,

$$v_b > V_{IH} \quad \text{for } t \geq t_1$$

The time that it takes for the receiver to recognize the transition from logic 0 to logic 1

$$\Delta t = t_1 - t_0$$

is called the delay. This delay is important because it puts a limit on how fast 1's and 0's can be sent from the driver to the receiver. To ensure that the 1's and 0's are received reliably, each 1 and each 0 must last for at least Δt. The rate at which 1's and 0's are sent from the driver to the receiver is inversely proportional to the delay.

Suppose two TTL circuits are connected using RG58 coaxial cable. What restriction must be placed on the length of the cable to ensure that the delay, Δt, is less than 2 ns?

Define the Situation

The voltage $v_b(t)$ is the capacitor voltage of an RC circuit. The RC circuit is at steady state just before time t_0.

The input to the RC circuit is $v_a(t)$. Before time t_0, $v_a(t) = V_{OL} = 0.4$ V. At time t_0, $v_a(t)$ changes abruptly. After time t_0, $v_a(t) = V_{OH} = 2.4$ V.

Before time t_0, $v_b(t) = V_{OL} = 0.4$ V. After time t_0, $v_b(t)$ increases exponentially. Eventually $v_b(t) = V_{OH} = 2.4$ V.

The time constant of the RC circuit is

$$\tau = R \cdot C$$
$$= rcL^2$$
$$= 47.52 \times 10^{-12} \cdot L^2$$

where L is the cable length in meters.

State the Goal

Calculate the maximum value of the cable length, L, for which $v_b > V_{IH} = 2.0$ V by time $t = t_0 + \Delta t$, where $\Delta t = 2$ ns.

Generate a Plan

Calculate the voltage $v_b(t)$ in Figure 9 D-1 b. The voltage $v_b(t)$ will depend on the length of the cable, L, because the time constant of the RC circuit is a function of L. Set $v_b = V_{IH}$ at time $t = t_0 + \Delta t$. Solve the resulting equation for the length of the cable.

Take Action on the Plan

Using the notation introduced in this chapter,

$$v_b(0) = V_{OL} = 0.4 \text{ V}$$

$$v_b(\infty) = V_{OH} = 2.4 \text{ V}$$

and

$$\tau = 47.52 \times 10^{-6} \cdot L^2$$

Using Eq. 9-8, the voltage $v_b(t)$ is expressed as

$$v_b(t) = V_{OH} - (V_{OL} - V_{OH})e^{-(t - t_0)/\tau}$$

The capacitor voltage, v_b, will be equal to V_{IH} at time $t_1 = t_0 + \Delta t$, so

$$V_{IH} = V_{OH} - (V_{OL} - V_{OH})e^{-\Delta t/\tau}$$

Solving for the delay, Δt, gives

$$\Delta t = -\tau \ln\left[\frac{V_{IH} - V_{OH}}{V_{OL} - V_{OH}}\right]$$

$$= -47.52 \times 10^{-12} \cdot L^2 \cdot \ln\left[\frac{V_{IH} - V_{OH}}{V_{OL} - V_{OH}}\right]$$

In this case

$$L = \sqrt{\frac{-\Delta t}{47.52 \times 10^{-12} \cdot \ln\left[\dfrac{V_{IH} - V_{OH}}{V_{OL} - V_{OH}}\right]}}$$

and therefore

$$L = \sqrt{\frac{-2 \cdot 10^{-9}}{47.52 \times 10^{-12} \cdot \ln\left[\dfrac{2.0 - 2.4}{0.4 - 2.4}\right]}}$$

$$= 5.11 \text{ m}$$
$$= 16.8 \text{ ft}$$

SUMMARY

Electrical signals are processed by electric circuits to yield desired output waveforms. It is important to determine the response of *RL* or *RC* circuits to a source voltage or current acting as an input signal after a specified time t_0. Since the response of the circuit results from its being subjected to an independent source, the response is called the forced response.

The total response of a circuit consists of the natural response plus the forced response. The forced response to an independent source that is a constant, an exponential, or a sinusoid will be of the same form as the forcing function. Thus, if the independent source is a constant, the forced response will be a constant. The response of a circuit to the sudden application of a constant source is called the step response.

The unit step forcing function is a function of time that is zero for $t < t_0$ and unity for $t > t_0$. At $t = t_0$ the magnitude changes from zero to one. A unit pulse is also a useful signal that is zero for $t < t_0$, unity for $t_0 < t < t_1$, and equal to zero for $t > t_1$.

We may use the superposition principle to solve circuits with multiple sources. Since the circuit is linear, we may determine the response to each source and then add the responses to obtain the total response.

Finally, we found that we can determine the response to a circuit with a nonconstant source as well as a constant source.

TERMS AND CONCEPTS

Complete Response   The sum of the forced response and the transient response.

Electric Signal   Voltage or current varying in time in a manner that conveys information.

Forced Response   Steady State Response of a circuit to an independent source.

Integrating Factor Method   Method for obtaining the solution of a differential equation by multiplying the equation by an exponential factor that makes one side of the equation a perfect derivative and then integrating both sides of the equation.

Pulse   Function of time that is zero for $t < t_0$, has magnitude M for $t_0 < t < t_1$, and is equal to zero for $t > t_1$.

Signal   Real-valued function of time; waveform that conveys information.

Sinusoidal Signal   A waveform that varies in accordance with a sine or cosine function of time.

Step Response   Response of a circuit to the sudden application of a constant source when all the initial conditions of the circuit are equal to zero.

Step Voltage Source   Voltage source, v, represented by $v = Vu(t - t_0)$.

Unit Step Forcing Function   Function of time that is zero for $t < t_0$ and unity for $t > t_0$. At $t = t_0$ the magnitude changes from zero to one. The unit step is dimensionless.

REFERENCES   Chapter 9

Gould, Larry A. "Electronic Valve Timing," *Automotive Engineering*, April 1991, pp. 19–24.

Jurgen, Ronald. "Electronic Handgun Trigger Proposed," *IEEE Institute*, February 1989, p. 5.

Pierce, John R., and A. M. Noll. "Signals: The Science of Telecommunications," *Scientific American Library*, W. H. Freeman, San Francisco, 1990.

Wright, A. "Construction and Application of Electric Fuses," *Power Engineering Journal*, Vol. 4, No. 3, 1990, pp. 141–148.

PROBLEMS

Section 9-3   The Response of an *RC* or *RL* Circuit Excited by a Constant Source

P 9.3-1   Find v and i for $t > 0$ if the circuit in Figure P 9.3-1 is in steady state at $t = 0^-$. All resistances in ohms.

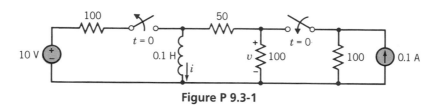

Figure P 9.3-1

P 9.3-2   The uncharged capacitor of the circuit shown in Figure P 9.3-2 is switched from position a to position b at $t = 0$ and remains there for 200 ms before being switched to position c, where it remains indefinitely. Find v and plot v for $0 < t < 500$ ms.

Answer: $v = \begin{cases} 200(1 - e^{-10t}), & 0 < t < 200 \text{ ms} \\ 173e^{-20(t - 0.2)}, & t > 200 \text{ ms} \end{cases}$

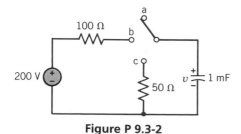

Figure P 9.3-2

P 9.3-3   For the circuit of Figure P 9.3-3, find $v_L(t)$.
Answer: $v_L(t) = 10e^{-20,000t}$ V, $t > 0$

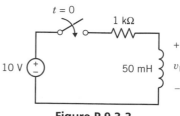

Figure P 9.3-3

P 9.3-4   Consider the circuit shown in Figure P 9.3-4. The switch is moved from A to B at $t = 0$ after being at A for a long time. Let $v_2(0^-) = 0$ V. Find $v_1(0^-)$, $v_1(0)$, $v_2(0)$, $v_R(0)$, $v_R(t)$, $i(t)$, $v_1(t)$, and $v_2(t)$ for $t > 0$.

Answer: $v_1(0^-) = 40$ V
$$v_1(0) = 40 \text{ V}$$
$$v_2(0) = 0 \text{ V}$$
$$v_R(0) = 40 \text{ V}$$
$$v_R(t) = 40e^{-18t} \text{ V}$$
$$i(t) = 8e^{-18t} \text{ A}$$
$$v_1(t) = \tfrac{80}{3} + \tfrac{40}{3}e^{-18t} \text{ V}$$
$$v_2(t) = \tfrac{80}{3} - \tfrac{80}{3}e^{-18t} \text{ V}$$

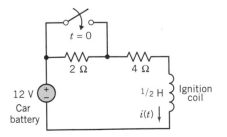

Figure P 9.3-4

P 9.3-7 A capacitor is connected to a voltage source at position 1 as shown in Figure P 9.3-7. The switch is moved instantaneously to position 2 at $t = 0$. Determine the voltage appearing across each capacitor, $v_1(t)$ and $v_2(t)$, for $t > 0$.

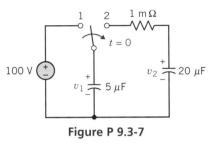

Figure P 9.3-7

P 9.3-5 A neon bulb flashes on and off depending on the current passing through it. The neon bulb shown in the circuit of Figure P 9.3-5 has the following behavior. The bulb remains off and acts as an open circuit until the bulb voltage v reaches a threshold value of $V_T = 65$ V. Once v reaches V_T, a discharge occurs and the bulb acts like a simple resistor of value $R_N = 1$ kΩ; the discharge is maintained as long as the bulb current i remains above the value $I_s = 10$ mA needed to sustain the discharge (even if the voltage v drops below V_T). As soon as i drops below 10 mA, the bulb again becomes an open circuit.

(a) Find $v(t)$ and $i(t)$ for one period of operation.
(b) Estimate the flashing rate for the bulb.

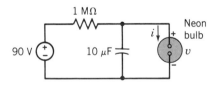

Figure P 9.3-5 Neon bulb flasher circuit.

P 9.3-8 A photoflash for a camera has a circuit shown in Figure P 9.3-8. The photoflash tube will conduct when the voltage across it reaches 240 V and the conduction continues until the voltage across it drops to 30 V. Assume that the resistance of the tube is negligibly small when it is conducting and infinite when it is not. Select a value of R so that the minimum time between flashes will be 10 s. For the same value of R, what percentage increase in energy delivered to the tube is achieved by waiting 20 s between flashes instead of 10 s?

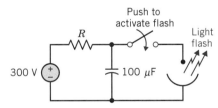

Figure P 9.3-8 Photoflash circuit.

P 9.3-6 A car battery is connected to an ignition coil as shown in Figure P 9.3-6. A short circuit as represented by the switch occurs at $t = 0$. Determine $i(t)$ for $t > 0$.
Answer: $i = 3 - 1e^{-8t}$ A, $t \geq 0$

P 9.3-9 Find $i(t)$ for $t > 0$ for the circuit shown in Figure P 9.3-9. Assume the circuit reached steady state at $t = 0^-$.
Answer: $i(t) = \tfrac{3}{2}e^{-6000t} + \tfrac{1}{2}$ mA

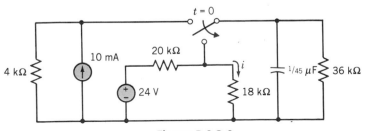

Figure P 9.3-6 Automobile ignition circuit.

Figure P 9.3-9

P 9.3-10 Switch 1 has been open and switch 2 has been closed for a long time at $t = 0^-$ in the circuit of Figure P 9.3-10. At $t = 0$, switch 1 is closed and then switch 2 is opened at $t = 3$ s. Determine $i(t)$ and plot $i(t)$ for $0 \leq t \leq 8$ s.

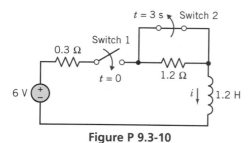

Figure P 9.3-10

P 9.3-11 The switch of the circuit of Figure P 9.3-11 has been connected to terminal 1 for a long time and is switched to terminal 2 at $t = 0$ for a duration of 30 s. At $t = 30$ s, the switch is returned to terminal 1. Determine the voltage $v(t)$ and plot $v(t)$ for $0 \leq t \leq 80$ s.

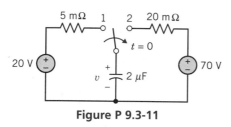

Figure P 9.3-11

P 9.3-12 Determine $v(t)$ and $i(t)$ for the circuit shown in Figure P 9.3-12. Assume that the switch has been connected to position 1 for a long time before it is moved to position 2 at $t = 0$. Sketch $i(t)$ and $v(t)$ for $-1 < t < 6$ s.

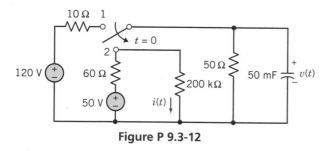

Figure P 9.3-12

P 9.3-13 An electronic flash of a camera uses a small battery to charge a capacitor. Then, when the flash is activated, the capacitor is switched across the flashbulb. Assume that the battery is a 6-V battery that should not be operated with a current above 100 μA. The capacitor is to be selected. (a) Draw a circuit model that will represent the charging and discharging action. (b) It is desired to

charge the capacitor within 5 s and to discharge it within $\frac{1}{2}$ s. Select the appropriate values for the elements in the circuit. Assume the value of the bulb resistance is 10 kΩ. Assume that the capacitor is charged or discharged in five time constants.

Section 9-4 A Simple Procedure for Finding the Complete Response of an *RC* or *RL* Circuit to a Constant Source

P 9.4-1 Find the response $v(t)$ for the circuit of Figure P 9.4-1 when it is known that $v(0) = 12$ V.

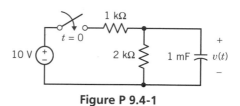

Figure P 9.4-1

P 9.4-2 Find $i(t)$ for $t > 0$ if $v_c(0) = 12$ V for the circuit of Figure P 9.4-2.

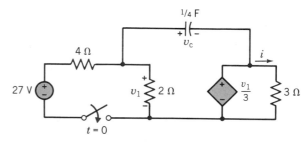

Figure P 9.4-2

P 9.4-3 Find $v_c(t)$ for $t > 0$ for the circuit shown in Figure P 9.4-3. Assume steady state at $t = 0^-$.
Answer: $v_c(t) = -3e^{-4t} - 9$ V

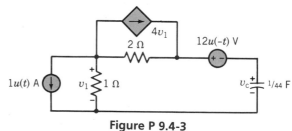

Figure P 9.4-3

P 9.4-4 Determine $v(t)$ and $i(t)$ for the circuit shown in Figure P 9.4-4. Assume a steady-state condition at $t = 0^-$, $v(0^-) = 2.14$ V, and let $C = 2$ mF.

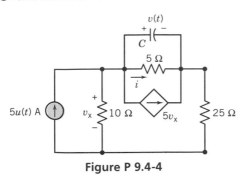

Figure P 9.4-4

Answer: $v = -10 + 10(1 - e^{-4t})u(t)$ V

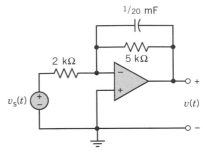

Figure P 9.4-7

P 9.4-5 Determine $v(t)$ for $t > 0$ for the circuit shown in Figure P 9.4-5. Assume that the circuit is in steady state at $t = 0^-$.

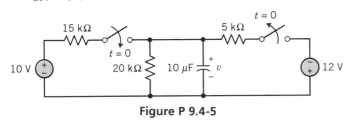

Figure P 9.4-5

P 9.4-6 Determine and sketch $i(t)$ for the circuit shown in Figure P 9.4-6. Calculate the time required for $i(t)$ to reach 99 percent of its final value.

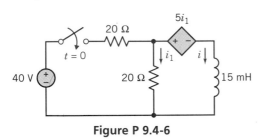

Figure P 9.4-6

P 9.4-7 A voltage signal is the input of the circuit of Figure P 9.4-7 and it is desired to determine $v(t)$. The pulse is

$$v_s(t) = 4 - 4u(t)$$

Find $v(t)$, assuming an ideal op amp.

Section 9-5 The Response to Step and Pulse Sources

P 9.5-1 Find $i(t)$ for $t > 0$ for the circuit of Figure P 9.5-1. Assume the circuit is in steady state at $t = 0^-$.

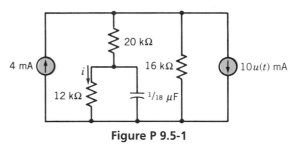

Figure P 9.5-1

P 9.5-2 Find the step response $v_c(t)$ of the circuit shown in Figure P 9.5-2 when $v_s = 20u(t)$ V. The initial voltage $v_c(0)$ is zero.

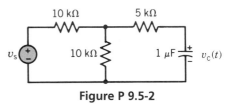

Figure P 9.5-2

P 9.5-3 Find $v(t)$ for $t > 0$ for the circuit shown in Figure P 9.5-3 when $v_s = 15e^{-t}[u(t) - u(t - 1)]$ V.

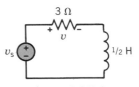

Figure P 9.5-3

P 9.5-4 Use step functions to represent the signal of Figure P 9.5-4.

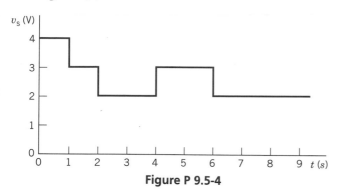

Figure P 9.5-4

P 9.5-5 The initial voltage of the capacitor of the circuit shown in Figure P 9.5-5 is zero. Determine the voltage $v(t)$ when the source is a pulse, so that

$$v_s = \begin{cases} 0 & t < 1 \text{ s} \\ 4 \text{ V} & 1 < t < 2 \text{ s} \\ 0 & t > 2 \text{ s} \end{cases}$$

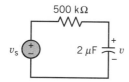

Figure P 9.5-5

P 9.5-6 For the circuit shown in Figure P 9.5-6, find $v(t)$ for $t > 0$. Assume steady state at $t = 0^-$.

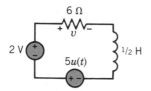

Figure P 9.5-6

P 9.5-7 For the circuit of Figure P 9.5-7, find $v_c(t)$ for $t > 0$. Note that the current source on the left is $3u(-t)$ A and will be zero for $t > 0$. Assume steady state at $t = 0^-$.
Answer: $v_c(t) = 18e^{-2t} + 6$ V

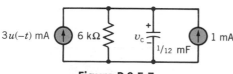

Figure P 9.5-7

P 9.5-8 For the circuit shown in Figure P 9.5-8, (a) find $i(t)$ for $t > 0$ when $i(0) = I_0$; (b) for $\alpha > 0$, draw the equivalent circuit with only passive elements; (c) for $\alpha < 0$, determine whether it is possible to obtain an equivalent passive circuit and state your reasons.

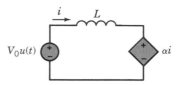

Figure P 9.5-8

P 9.5-9 Studies of an artificial insect are being used to understand the nervous system of animals (Beer, 1991). A model neuron in the nervous system of the artificial insect is shown in Figure P 9.5-9. A series of pulses, called synapses, is the input signal, v_s. The switch generates a pulse by opening at $t = 0$ and closing at $t = 0.5$ s. Assume that the circuit is in steady state and that $v(0^-) = 10$ V. Determine the voltage $v(t)$ for $0 < t < 2$ s.

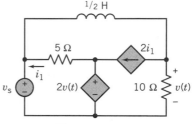

Figure P 9.5-9 Neuron circuit model.

P 9.5-10 Determine and sketch $v(t)$ for the circuit shown in Figure P 9.5-10 when $v_s = 4[u(t) - u(t - 1.8)]$ V. Assume that the circuit is in steady state at $t = 0^-$.

Answer: $v = \begin{cases} 4 - \dfrac{20}{9} e^{-t/0.45} \text{ V} & 0 \le t < 1.8 \text{ s} \\ 4e^{-(t - 1.8)/0.45} \text{ V} & t \ge 1.8 \text{ s} \end{cases}$

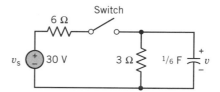

Figure P 9.5-10

P 9.5-11 Human cell membranes resemble capacitors with a capacitance of 1 μF. The pulse stimulus $i_s(t)$ has a magnitude of I_m and a duration d, so that

$$i_s = I_m \qquad 0 < t < d$$
$$= 0 \qquad \text{elsewhere}$$

where I_m = 1 mA and d = 5 ms. Determine and plot $v(t)$. The resistance R is equal to 1 kΩ for the heart membrane, as shown in Figure P 9.5-11.

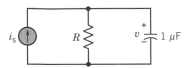

Figure P 9.5-11 Model for cell membrane circuit.

P 9.5-12 A circuit used in a mobile telephone is shown in Figure P 9.5-12*a* where the input source v_s is a pulse defined as shown in Figure P 9.5-12*b*. Find the time response $v_0(t)$ for this circuit when the initial voltage of the capacitor is zero.

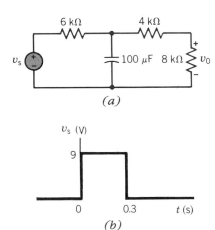

(a)

(b)

Figure P 9.5-12 A circuit used in a mobile telephone.

P 9.5-13 An electronic circuit can be used to replace the springs and levers normally used to detonate a shell in a handgun (Jurgen, 1989). The electric trigger would eliminate the clicking sensation, which may cause a person to misaim. The proposed trigger uses a magnet and a solenoid with a trigger switch. The circuit of Figure P 9.5-13 represents the trigger circuit with $i_s(t) = 40[u(t) - u(t - t_0)]$ A where t_0 = 1 ms. Determine and plot $v(t)$ for $0 < t < 0.3$ s.

Answer:

$$v = \begin{cases} 480\,(1 - e^{-1000t}) & 0 < t < 1\text{ ms} \\ 480\,(1 - e^{-1})e^{-1000(t - t_0)} & t > 1\text{ ms}, \ t_0 = 1\text{ ms} \end{cases}$$

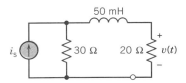

Figure P 9.5-13 Electric trigger circuit for handgun.

P 9.5-14 An experimental electronic valve timing system for an automobile engine uses pneumatic (compressed air) actuators to lift the valves, actuated by electromagnetically operated pneumatic valves (Gould, 1991). The engine valve must open within 2 ms after being commanded to open by the engine-control computer.

The pneumatic valve will begin to open when the current in the valve solenoid coil exceeds 300 mA, its guaranteed "on" current. The valve will begin to close when the solenoid coil current drops below 225 mA, its guaranteed "off" current. To ensure that the engine valve opens on time, only 0.1 ms has been allowed for the current in the solenoid to reach its on current after the applied voltage changes from 0 V to 12 V and 0.2 ms is allowed for the current to decay below its off current after the supply voltage is removed and the coil is shorted (the applied voltage changes from 12 V to 0 V). Either response may be faster. The solenoid coil can be modeled as an inductance of 3 mH with a series resistance that is determined by the diameter (gauge) of wire chosen for the coil. Smaller gauge wire will cause higher resistance and lower power dissipation and allow a smaller solenoid package. What is the maximum value of solenoid resistance that will meet both timing specifications? With this resistance, what is the maximum current in the solenoid coil? Assume that the solenoid is to be energized 25 percent of the time. What is the *approximate* average power dissipation in this solenoid coil? What simplifying assumptions did you make?

In practice, a solenoid is not directly shorted to turn it off, but is shorted through a device called a diode with a voltage drop of 0.7 V, as this is easier to implement. Why would you not simply open-circuit the coil to achieve zero current?

P 9.5-15 Determine $v(t)$ for $t > 0$ for the circuit of Figure P 9.5-15.

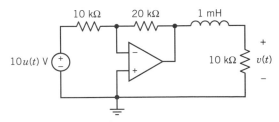

Figure P 9.5-15

Section 9-6   Using the Superposition Principle to Solve Circuits with Multiple Sources

P 9.6-1 Using the superposition principle, find the current i of the circuit of Figure P 9.6-1.

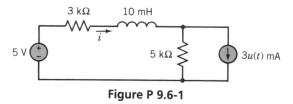

Figure P 9.6-1

P 9.6-2 Find $i(t)$ for $t > 0$ for the circuit shown in Figure P 9.6-2. Assume that the circuit is in steady state at $t = 0^-$.

Answer: $i(t) = -\frac{1}{4}e^{-25t} + 2$ A

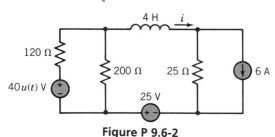

Figure P 9.6-2

P 9.6-3 Determine and sketch $v(t)$ for the circuit shown in Figure P 9.6-3 when $v_s = -30 + 60u(t)$ V and $i_s = 6 - 3u(t)$ A.

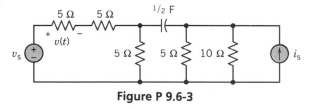

Figure P 9.6-3

P 9.6-4 Determine $v_c(t)$ for $t > 0$ for the circuit of Figure P 9.6-4.

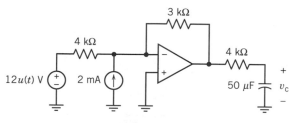

Figure P 9.6-4

P 9.6-5 Determine $v_c(t)$ for $t > 0$ for the circuit of Figure P 9.6-5 when $v_s = 2u(-t) + 12u(t)$ V.

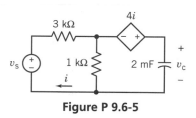

Figure P 9.6-5

Section 9-7   The Response of an *RL* or *RC* Circuit to a Nonconstant Source

P 9.7-1 Find $v_c(t)$ for $t > 0$ for the circuit shown in Figure P 9.7-1 when $v_1 = 8e^{-5t}u(t)$ V. Assume the circuit is in steady state at $t = 0^-$.

Answer: $v_c(t) = 4e^{-9t} + 18e^{-5t}$ V

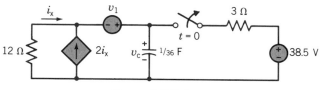

Figure P 9.7-1

P 9.7-2 Find $v(t)$ for $t > 0$ for the circuit shown in Figure P 9.7-2. Assume steady state at $t = 0^-$.

Answer: $v(t) = 20e^{-10t/3} - 12e^{-2t}$ V

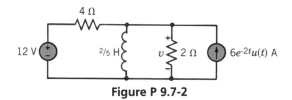

Figure P 9.7-2

P 9.7-3 Find $v(t)$ for $t > 0$ for the circuit shown in Figure P 9.7-3 when $v_1 = (25 \sin 4000t)u(t)$ V. Assume steady state at $t = 0^-$.

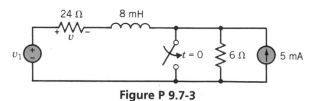

Figure P 9.7-3

P 9.7-4 Find $v_c(t)$ for $t > 0$ for the circuit shown in Figure P 9.7-4 when $i_s = [2 \cos 2t]u(t)$ mA.

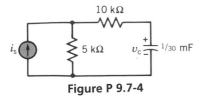

Figure P 9.7-4

P 9.7-5 Many have witnessed the use of an electrical megaphone for amplification of speech to a crowd. A model of a microphone and speaker is shown in Figure P 9.7-5*a*, and the circuit model is shown in Figure P 9.7-5*b*. Find $v(t)$ for $v_s = 10(\sin 100t)u(t)$, which could represent a person whistling or singing a pure tone.

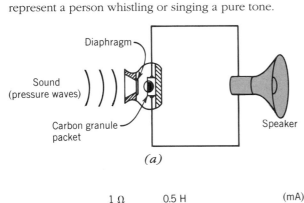

Figure P 9.7-5 Megaphone circuit.

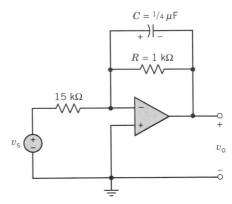

Figure P 9.7-6 Integrator circuit.

P 9.7-7 Most television sets use magnetic deflection in the cathode ray tube. To move the electron beam across the screen it is necessary to have a ramp of current as shown in Figure P 9.7-7*a*, to flow through the deflection coil. The deflection coil circuit is shown in Figure P 9.7-7*b*. Find the waveform v_1 that will generate the current ramp.

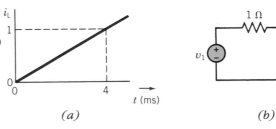

Figure P 9.7-7 Television deflection circuit.

P 9.7-8 Determine $v(t)$ for the circuit shown in Figure P 9.7-8.

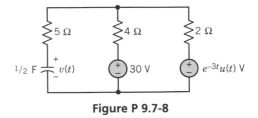

Figure P 9.7-8

P 9.7-6 A lossy integrator is shown in Figure P 9.7-6. It can be seen that the lossless capacitor of the ideal integrator circuit has been replaced with a model for the lossy capacitor, namely, a lossless capacitor in parallel with a 1-Ω resistor. If $v_s = 15e^{-2t}u(t)$ V and $v_o(0) = 10$ V, find $v_o(t)$ for $t > 0$. Assume an ideal op amp.

P 9.7-9 (a) In the circuit of Figure P 9.7-9, given $v_{c1}(0) = 10$ V, $v_{c2}(0) = 20$ V, $C_1 = 4$ F, find $v_o(t)$ in terms of $v_1(t)$ and $v_2(t)$ for $t > 0$. (b) If $v_1(t) = 10e^{-2t}$ V and $v_2(t) = 20e^{-t}$ V, find $v_o(t)$ for $t > 0$.

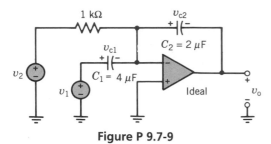

Figure P 9.7-9

P 9.7-10 For the circuit shown in Figure P 9.7-10 find $v_c(t)$, $t \geq 0$, when $v_c(0^-) = 3$ V.
Answer: $v_c(t) = 4 - e^{-250t}$ V, $t \geq 0$

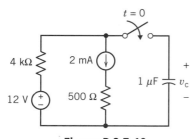

Figure P 9.7-10

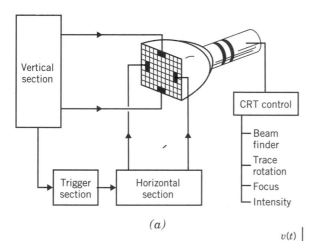

(a)

P 9.7-11 Determine $v(t)$ for the circuit shown in Figure P 9.7-11a when v_s varies as shown in Figure P 9.7-11b. The initial capacitor voltage is $v_c(0) = 0$.

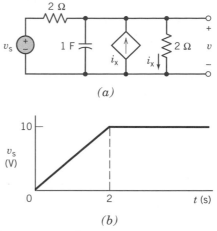

Figure P 9.7-11

P 9.7-12 The electron beam, which is used to "draw" signals on an oscilloscope, is moved across the face of a cathode-ray tube (CRT) by a force exerted on electrons in the beam. The basic system is shown in Figure P 9.7-12a. The force is created from a time-varying, ramp-type voltage applied across the vertical or the horizontal plates. As an example, consider the simple circuit of Figure P 9.7-12b for horizontal deflection where the capacitance between the plates is C.

(a) Derive an expression for the voltage across the capacitance. If $v(t) = kt$ and $R_s = 625$ kΩ, $k = 1000$, and $C = 2000$ pF, compute v_c as a function of time. Sketch $v(t)$ and $v_c(t)$ on the same graph for time less than 10 ms. Does the voltage across the plates track the input voltage?

(b) Describe the deflection of the electron beam if the force is $F = qE$, where q is the charge and E is the electric field; E is the ratio of the voltage across the plates to the spacing of the plates ($E = v/S$). Assume a zero initial condition for the capacitor.

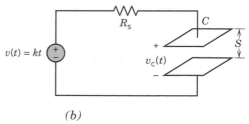

(b)

Figure P 9.7-12 Cathode-ray tube beam circuit.

PSpice PROBLEMS

SP 9-1   Determine the current *i* of the circuit shown in Figure 9-10 using PSpice. Plot *i*(*t*) for five time constants and find *i*(*t*) at *t* = 10 ms assuming *i*(0) = 3 A.

SP 9-2   Determine *i*(*t*) for Problem P 9.4-6.

SP 9-3   Determine and plot $v_c(t)$ for P 9.7-12.

SP 9-4   For the circuit shown in Figure SP 9-4 find $v_c(t)$, *t* ≥ 0, when $v_c(0^-)$ = 5 V. Plot $v_c(t)$ for five time constants.

SP 9-5   Determine and plot *v*(*t*) for P 9.7-11.

SP 9-6   Determine and plot $v_c(t)$ for P 9.4-3.

SP 9-7   Determine and plot *i*(*t*) of the circuit of Figure 9-14 with *v*(0) = 5 V.

SP 9-8   Determine and plot *v*(*t*) for P 9.4-7.

SP 9-9   An *RC* circuit as shown in Figure SP 9-9 has *v*(0) = 0. It is desired to plot the response of the circuit for four time constants: 2, 4, 8, and 16 ms. Select the appropriate value of *R* and use a PSpice program to plot the step response for the four time constants on one graphical plot.

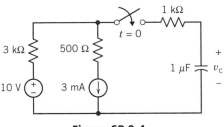

Figure SP 9-4

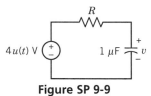

Figure SP 9-9

VERIFICATION PROBLEMS

VP 9-1   The circuit of Figure VP 9-1*a* has a pulse input as shown in Figure VP 9-1*b*. A computer simulation states that $v_c(t)$ is 0.478 V at *t* = 2 s. Assume that $v_c(0)$ = 0. Verify the result.

VP 9-2   The circuit of Figure VP 9-2 is studied by a student who reports that *i*(*t*) = 4.39 A when *t* = 0.5 s. Verify this result.

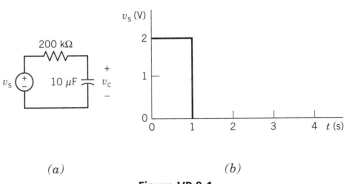

(a)   (b)

Figure VP 9-1

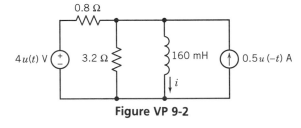

Figure VP 9-2

DESIGN PROBLEMS

DP 9-1   For the circuit shown in Figure DP 9-1, it is desired that *i*(*t*) = 2.5 A at *t* = 47 ms. Determine the resistance *R* that meets this specification when v_s = 300 V. The initial voltage on the capacitor is $v_c(0)$ = 100 V.

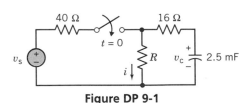

Figure DP 9-1

DP 9-2 For the circuit shown in Figure DP 9-2, specify the inductance L so that the current in the inductor reaches its steady-state value (five time constants) in 3.1 μs after the switch is closed at $t = 0$. Determine the energy stored in the inductor after $t > 3.1$ μs. Assume that the circuit was unexcited prior to $t = 0$.

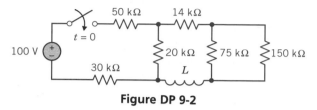

Figure DP 9-2

DP 9-3 For the circuit shown in Figure DP 9-3, select the inductance L so that the inductor current $i(t) = 3$ A at $t = 14$ ms. What is $i(t)$ when $t = 1$ s?

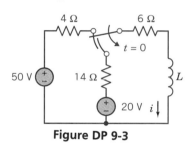

Figure DP 9-3

DP 9-4 Solenoids are electromechanical components that convert electrical energy into mechanical work and move a plunger. Current passing through a helical coil winding of closely spaced turns of copper magnet wire produces a magnetic field that surrounds the coil. All solenoids develop magnetizing force, which has a relationship to the current and number of turns in the coil. Solenoid size determines the amount of work a solenoid can perform. A large unit will develop more force at a given stroke than a small solenoid (with the same coil current) because its greater physical volume will accommodate more turns of wire on the coil. It is important that a solenoid selected for a particular application have a rated force as close to the load requirements as possible. Too much force reduces solenoid life because the unit must absorb the excess energy. If the force is too low, the result will be unsatisfactory performance because the plunger will not pull in or seat properly.

The circuit model of a simple solenoid is shown in Figure DP 9-4. The source is a 12-V dc battery with a resistance $R_s = 1$ Ω. Select the required R and L of the solenoid and $v_s = V_0 u(t)$ so that the solenoid will close in 100 ms while restricting the force to less than 0.5 N. The force-developed relationship is measured as $f = i(t)$ N.

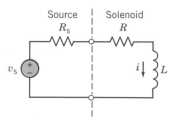

Figure DP 9-4 Model of solenoid circuit.

DP 9-5 The switch in the circuit shown in Figure DP 9-5 is at position 1 for a long time and is switched to position 2 at $t = 0$. The output voltage, $v(t)$, is to be changed from the initial voltage $v(0)$ to within 1 percent of its final value within 0.2 s. Determine the required C and sketch $v(t)$ for $t \geq 0$.

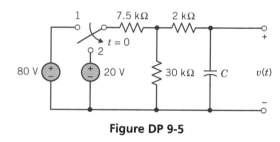

Figure DP 9-5

DP 9-6 A laser trigger circuit is shown in Figure DP 9-6. In order to trigger the laser, we require 60 mA $< |i| <$ 180 mA for $0 < t < 200$ μs. Determine a suitable value for R_1 and R_2.

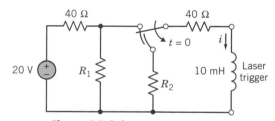

Figure DP 9-6 Laser trigger circuit.

DP 9-7 The initial voltage of the capacitor in the circuit of Figure DP 9-7 is -10 V. Select C so that $v(t) = 0$ at $t = 4.0$ s when $v_s = 20$ V. Sketch $v(t)$ for $0 < t < 5$ s.

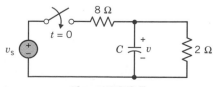

Figure DP 9-7

DP 9-8 Fuses are used to open a circuit when excessive current flows (Wright, 1990). One fuse is designed to open when the power absorbed by R exceeds 10 W for 0.5 s. The source represents the turn-on condition for the load where $v_s = A[u(t) - u(t - 0.75)]$ V. Assume that $i_L(0^-) = 0$. The goal is to achieve the maximum current while not opening the fuse. Determine an appropriate value of A and sketch the current waveform. The circuit is shown in Figure DP 9-8.

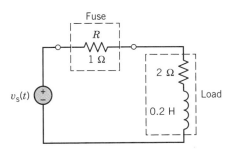

Figure DP 9-8 Fuse circuit.

DP 9-9 Repeat Design Problem 9-7 when $v_s(t) = 10e^{-t/2}u(t)$ V.

DP 9-10 An *RL* circuit as shown in Figure DP 9-10 is used to provide an actuating pulse for a power laser. The circuit is at steady state at $t = 0^-$ and $i(0^-) = 0$. The voltage source $v_s = V_0 e^{-bt}$ V is connected at $t = 0$. Select V_0 and b so that the peak magnitude of the current pulse is greater than 0.6 A. Determine $i(t)$ and plot the current pulse.

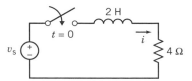

Figure DP 9-10 Laser pulse circuit.

DP 9-11 The adjustable valve in an oil refinery is actuated by a control voltage v_s that consists of a pulse signal. The model of the actuator circuit is shown in Figure DP 9-11 where v is the voltage across the valve actuator and the actuator load is $R = 2\ \Omega$. Determine the circuit L and select the required pulse when it is desired that

$$v = \begin{cases} 10(1 - e^{-4t})\ \text{V} & 0 < t < \dfrac{1}{2}\ \text{s} \\ 8.65e^{-4t}\ \text{V} & t > \dfrac{1}{2}\ \text{s} \end{cases}$$

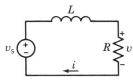

Figure DP 9-11 The circuit for actuating an oil refinery valve.

CHAPTER 10

THE COMPLETE RESPONSE OF CIRCUITS WITH TWO ENERGY STORAGE ELEMENTS

PREVIEW

In the preceding chapter we determined the natural response and the forced response of circuits with one energy storage element. In this chapter we proceed to determine the complete response $x(t)$ of a circuit with two energy storage elements. A circuit with two energy storage elements is described by a second-order differential equation in terms of $x(t)$.

We describe three methods of obtaining the second-order differential equation: (1) the substitution method, (2) the operator method, and (3) the state variable method. Then, using the differential equation, we proceed to find the natural response x_n and the forced response x_f.

Although we focus on circuits with two energy storage elements, the methods described can be used for circuits with three or more energy storage elements.

10-1 DESIGN CHALLENGE

AUTO AIRBAG IGNITER

Problem

Airbags are now widely used for driver protection in automobiles. A pendulum is used to switch a charged capacitor to the inflation ignition device, as shown in Figure 10D-1. The automobile airbag is inflated by an explosive device that is ignited by the energy

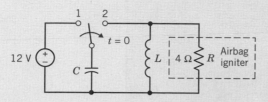

Figure 10D-1 An automobile airbag ignition device.

absorbed by the resistive device represented by R. In order to inflate, it is required that the energy dissipated in R is at least 1 J. It is required that the ignition device trigger within 0.1 s. Select the L and C that meet the specifications.

This problem involves a circuit that contains *two* energy storage elements. This type of circuit is the subject of this chapter. We will return to the problem at the end of the chapter.

10-2 COMMUNICATIONS AND POWER SYSTEMS

The objective of an electrical circuit is to transmit an electrical signal or electrical power. An *electrical system* is an interconnection of electrical elements and circuits to achieve a desired objective. Consider the electrical system shown in Figure 10-1*a*. This system has a source and transmits the source power or signal to a receiver and ultimately to the end user.

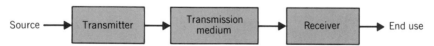

Figure 10-1a Electrical system.

In a communication system the source is an input signal such as a voltage signal. The transmitter converts the signal to a suitable form for the transmission medium. Then the output of the transmitter proceeds through the medium until it reaches the receiver, where it is converted into a form that is useful for the end user. The Marconi receiver is shown in Figure 10-1*b*.

An example of a well-known communication system is the early railway electric telegraph for communication from one railway station to another. The message is converted to Morse code, and the transmitter key is opened and closed to generate a series of short and long pulses. These electrical pulses travel along the telegraph wire until they reach a telegraph receiver at another railway station, where they are received and decoded for use by the reader.

Another example of a communication system is the radio, which converts spoken words into electrical waveforms. The transmitter then couples these waveforms to the transmission medium, in this case the earth's atmosphere. At the receiver, an antenna couples the medium to the receiver and eventually converts these electrical waveforms back to audio signals for the listener.

The goal of an efficient communication system is to achieve the best transmission and reception of the original signal so that the user may accurately discern the message. Thus,

Figure 10-1*b* Marconi with the receiving apparatus used at Signal Hill, 1901. Courtesy of the IEEE Center for the History of Electrical Engineering.

the goal is to provide the user with a faithful, undistorted output waveform that is true to the input signal. The chronology of electrical communication developments is provided in Table 10-1.

Table 10-1
Twentieth-Century Chronology of Electrical Communication

1920	Birth of commercial radio. KDKA, Pittsburgh, the first radio broadcasting station.
1923–1938	Birth of television. Philo T. Farnsworth and Vladimir Zworykin propose television systems. Experimental broadcasts begin.
1939	First commercial television broadcast in New York City.
1938–1945	World War II. Radar and microwave systems developed.
1955	John R. Pierce proposes satellite communication systems.
1962	Telstar I launched as first communication satellite.
1983	First optical fiber telephone communication system installed in Chicago.
1984	Cellular mobile telephone systems introduced.

Now let us consider the goal of an electric power system. A power plant, such as a hydroelectric power station, generates large amounts of electric power. It is common to generate as much as 30 MW to 70 MW continuously at a hydroelectric plant. Since these plants are usually distant from customers in the cities, the electrical power must be transmitted over long pairs of wires to the receiver (see Figure 10-1*a*). The transmitter, in this case, must efficiently couple the power to the transmission medium (wires). Similarly, the receiver must efficiently couple the power to local distribution circuits for ultimate use by the end user.

Thus, the goal of the communication system is faithful undistorted transmission, while the goal of the electric power system is to transmit the energy efficiently, with minimum loss of energy in the transmission medium (the wires).

Both communication and power systems contain capacitance and inductance throughout their circuits. Thus, with both energy storage elements present it is necessary to consider circuits that have both elements and determine the forced and natural responses of the circuits. It is the purpose of this chapter to consider the behavior of circuits that contain both capacitors and inductors.

10-3 DIFFERENTIAL EQUATION FOR CIRCUITS WITH TWO ENERGY STORAGE ELEMENTS

In Chapter 9 we considered circuits that contained only one energy storage element, and these could be described by a first-order differential equation. In this section we consider the description of circuits with two irreducible energy storage elements that are described by a second-order differential equation. Later, we will consider circuits with three or more irreducible energy storage elements that are described by a third-order (or higher) differential equation. We use the term "irreducible" to indicate that all parallel or series connections or other reducible combinations of like storage elements have been reduced to their irreducible form. Thus, for example, all parallel capacitors have been reduced to one capacitor, C_p.

In the following paragraphs we use two methods to obtain the second-order differential equation for circuits with two energy storage elements. Then in the next section we obtain the solution to these second-order differential equations.

First, let us consider the circuit shown in Figure 10-2, which consists of a parallel combination of a resistor, an inductor, and a capacitor. Writing the nodal equation at the top node, we have

$$\frac{v}{R} + i + C\frac{dv}{dt} = i_s \tag{10-1}$$

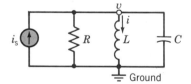

Ground   **Figure 10-2** An *RLC* circuit with a current source.

Then we write the equation for the inductor as

$$v = L\frac{di}{dt} \tag{10-2}$$

Substitute Eq. 10-2 into Eq. 10-1, obtaining

$$\frac{L}{R}\frac{di}{dt} + i + C\frac{d^2i}{dt^2} = i_s \tag{10-3}$$

which is the second-order differential equation we seek. Solve this equation for $i(t)$. If $v(t)$ is required, use Eq. 10-2 to obtain it.

This method of obtaining the second-order differential equation may be called the *direct method* and is summarized in Table 10-2.

In Table 10-2 the circuit variables are called x_1 and x_2. In any example, x_1 and x_2 will be specific element currents and/or voltages. When we analyzed the circuit of Figure 10-2

Table 10-2
The Direct Method for Obtaining the Second-order Differential Equation of a Circuit

Step 1   Identify the first and second variables x_1 and x_2. These variables are capacitor voltages and/or inductor currents.

Step 2   Write one first-order differential equation, obtaining $\dfrac{dx_1}{dt} = f(x_1, x_2)$.   (Eqn. 1)

Step 3   Obtain an additional first-order differential equation in terms of the second variable so that $\dfrac{dx_2}{dt} = Kx_1$ or $x_1 = \dfrac{1}{K}\dfrac{dx_2}{dt}$.   (Eqn. 2)

Step 4   Substitute the equation of Step 3 into the equation of Step 2, thus obtaining a second-order differential equation in terms of x_2.

we used $x_1 = v$ and $x_2 = i$. In contrast, to analyze the circuit of Figure 10-3 we will use $x_1 = i$ and $x_2 = v$, where i is the inductor current and v is the capacitor voltage.

Now let us consider the *RLC* series circuit shown in Figure 10-3 and use the direct method to obtain the second-order differential equation. We chose $x_1 = i$ and $x_2 = v$.

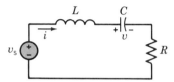

Figure 10-3 An *RLC* series circuit.

First we seek an equation for $\dfrac{dx_1}{dt} = \dfrac{di}{dt}$. Writing KVL around the loop, we have

$$L\frac{di}{dt} + v + Ri = v_s \qquad (10\text{-}4)$$

where v is the capacitor voltage. This equation may be written as

$$\frac{di}{dt} + \frac{v}{L} + \frac{R}{L}i = \frac{v_s}{L} \qquad (10\text{-}5)$$

Recall $v = x_2$, and obtain an equation in terms of $\dfrac{dx_2}{dt}$. Since

$$C\frac{dv}{dt} = i \qquad (10\text{-}6)$$

or

$$C\frac{dx_2}{dt} = x_1 \qquad (10\text{-}7)$$

substitute Eq. 10-6 into Eq. 10-5 to obtain the desired second-order differential equation:

$$C\frac{d^2v}{dt^2} + \frac{v}{L} + \frac{RC}{L}\frac{dv}{dt} = \frac{v_s}{L} \qquad (10\text{-}8)$$

Equation 10-8 may be rewritten as

$$\frac{d^2v}{dt^2} + \frac{R}{L}\frac{dv}{dt} + \frac{1}{LC}v = \frac{v_s}{LC} \tag{10-9}$$

Another method of obtaining the second-order differential equation describing a circuit is called the operator method. First, we obtain differential equations describing node voltages or circuit meshes and use operators to obtain the differential equation for the circuit.

As a more complicated example of a circuit with two energy storage elements, consider the circuit shown in Figure 10-4. This circuit has two inductors and can be described by the mesh currents as shown in Figure 10-4. The mesh equations are

$$L_1\frac{di_1}{dt} + R(i_1 - i_2) = v_s \tag{10-10}$$

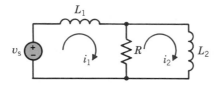

Figure 10-4 Circuit with two inductors.

and

$$R(i_2 - i_1) + L_2\frac{di_2}{dt} = 0 \tag{10-11}$$

Now, let us use $R = 1\,\Omega$, $L_1 = 1$ H, and $L_2 = 2$ H. Then we have

$$\frac{di_1}{dt} + i_1 - i_2 = v_s$$

and

$$i_2 - i_1 + 2\frac{di_2}{dt} = 0 \tag{10-12}$$

In terms of i_1 and i_2, we may rearrange these equations as

$$\frac{di_1}{dt} + i_1 - i_2 = v_s \tag{10-13}$$

and

$$-i_1 + i_2 + 2\frac{di_2}{dt} = 0 \tag{10-14}$$

It remains to obtain one second-order differential equation. A useful process is to use the operator s, where $s = d/dt$, which enables us to transform a differential equation into an algebraic equation. (See Section 8-5.) This method will be called the *operator method*. The choice of the symbol for the operator is arbitrary, and others use p or D instead of s. Then Eqs. 10-13 and 10-14 become

$$si_1 + i_1 - i_2 = v_s$$

and

$$-i_1 + i_2 + 2si_2 = 0$$

These two equations may be rewritten as

$$(s + 1)i_1 - i_2 = v_s$$

and

$$-i_1 + (2s + 1)i_2 = 0$$

We may use Cramer's rule (see Appendix B) to solve for i_2, obtaining

$$i_2 = \frac{1v_s}{(s + 1)(2s + 1) - 1}$$

Therefore,

$$(2s^2 + 3s)i_2 = v_s$$

and the differential equation is

$$2\frac{d^2i_2}{dt^2} + 3\frac{di_2}{dt} = v_s \tag{10-15}$$

The operator method for obtaining the second-order differential equation is summarized in Table 10-3.

Table 10-3
Operator Method for Obtaining the Second-Order Differential Equation of a Circuit

Step 1	Identify the variable x_1 for which the solution is desired.
Step 2	Write one differential equation in terms of the desired variable x_1 and a second variable x_2.
Step 3	Obtain an additional equation in terms of the second variable and the first variable.
Step 4	Use the operator $s = d/dt$ and $1/s = \int dt$ to obtain two algebraic equations in terms of s and the two variables x_1 and x_2.
Step 5	Using Cramer's rule, solve for the desired variable so that $x_1 = f(s, \text{sources}) = P(s)/Q(s)$, where $P(s)$ and $Q(s)$ are polynomials in s.
Step 6	Rearrange the equation of Step 5 so that $Q(s)x_1 = P(s)$.
Step 7	Convert the operators back to derivatives for the equation of Step 6 to obtain the second-order differential equation.

Example 10-1
Find the differential equation for the current i_2 for the circuit of Figure 10-5.

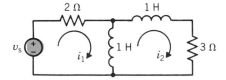

Figure 10-5 Circuit for Example 10-1.

Solution
Write the two mesh equations using KVL to obtain

$$2i_1 + \frac{di_1}{dt} - \frac{di_2}{dt} = v_s$$

$$-\frac{di_1}{dt} + 3i_2 + 2\frac{di_2}{dt} = 0$$

Using the operator $s \equiv d/dt$, we have

$$(2 + s)i_1 - si_2 = v_s$$

and

$$-si_1 + (3 + 2s)i_2 = 0$$

Using Cramer's rule to solve for i_2, we obtain

$$i_2 = \frac{sv_s}{(2 + s)(3 + 2s) - s^2} = \frac{sv_s}{s^2 + 7s + 6} \tag{10-16}$$

Rearranging Eq. 10-16, we obtain

$$(s^2 + 7s + 6)i_2 = sv_s \tag{10-17}$$

Therefore, the differential equation for i_2 is

$$\frac{d^2i_2}{dt^2} + 7\frac{di_2}{dt} + 6i_2 = \frac{dv_s}{dt} \tag{10-18}$$

Example 10-2

Find the differential equation for the voltage v for the circuit of Figure 10-6.

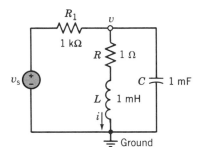

Figure 10-6 The *RLC* circuit for Example 10-2.

Solution

The KCL node equation at the upper node is

$$\frac{v - v_s}{R_1} + i + C\frac{dv}{dt} = 0 \tag{10-19}$$

Since we wish to determine the equation in terms of v, we need a second equation in terms of the current i. Write the equation for the current through the branch containing the inductor as

$$Ri + L\frac{di}{dt} = v \tag{10-20}$$

Using the operator $s = d/dt$, we have the two equations

$$\frac{v}{R_1} + Csv + i = \frac{v_s}{R_1}$$

and

$$-v + Ri + Lsi = 0$$

Substituting the parameter values and rearranging we have

$$(10^{-3} + 10^{-3}s)v + i = 10^{-3} v_s$$

and

$$-v + (10^{-3}s + 1)i = 0$$

Using Cramer's rule, solve for v to obtain

$$v = \frac{(s + 1000)v_s}{(s + 1)(s + 1000) + 10^6}$$

$$= \frac{(s + 1000)v_s}{s^2 + 1001s + 1001 \times 10^3}$$

Therefore, we have

$$(s^2 + 1001s + 1001 \times 10^3)v = (s + 1000)v_s$$

or the differential equation we seek is

$$\frac{d^2v}{dt^2} + 1001 \frac{dv}{dt} + 1001 \times 10^3 \, v = \frac{dv_s}{dt} + 1000v_s \qquad (10\text{-}21)$$

EXERCISE 10-1

Find the second-order differential equation for the circuit shown in Figure E 10-1 in terms of i using the direct method.

Answer: $\dfrac{d^2i}{dt^2} + \dfrac{1}{2}\dfrac{di}{dt} + i = \dfrac{1}{2}\dfrac{di_s}{dt}$

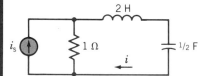

Figure E 10-1

EXERCISE 10-2

Find the second-order differential equation for the circuit shown in Figure E 10-2 in terms of v using the operator method.

Answer: $\dfrac{d^2v}{dt^2} + 2\dfrac{dv}{dt} + 2v = 2\dfrac{di_s}{dt}$

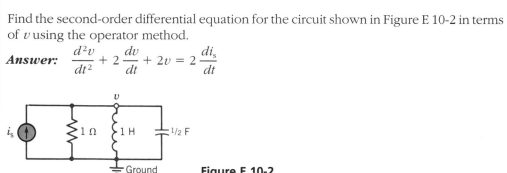

Figure E 10-2

EXERCISE 10-3

Find the second differential equation for i_2 for the circuit of Figure E 10-3 using the operator method. Recall that the operator for the integral is $1/s$.

Answer: $3\dfrac{d^2 i_2}{dt^2} + 4\dfrac{di_2}{dt} + 2i_2 = \dfrac{d^2 v_s}{dt^2}$

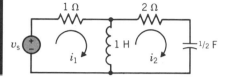

Figure E 10-3

10-4   SOLUTION OF THE SECOND-ORDER DIFFERENTIAL EQUATION—THE NATURAL RESPONSE

In the preceding section we found that a circuit with two irreducible energy storage elements can be represented by a second-order differential equation of the form

$$a_2 \frac{d^2 x}{dt^2} + a_1 \frac{dx}{dt} + a_0 x = f(t)$$

where the constants a_2, a_1, a_0 are known and the forcing function $f(t)$ is specified.
The complete response $x(t)$ is given by

$$x = x_n + x_f \tag{10-22}$$

where x_n is the natural response and x_f is the forced response. The natural response satisfies the unforced differential equation when $f(t) = 0$. The forced response x_f satisfies the differential equation with the forcing function present.
The natural response of a circuit, x_n, will satisfy the equation

$$a_2 \frac{d^2 x_n}{dt^2} + a_1 \frac{dx_n}{dt} + a_0 x_n = 0 \tag{10-23}$$

Since x_n and its derivatives must satisfy the equation, we postulate the exponential solution

$$x_n = A e^{st} \tag{10-24}$$

where A and s are to be determined. The exponential is the only function that is proportional to all of its derivatives and integrals and therefore is the natural choice for the solution of a differential equation with constant coefficients. Substituting Eq. 10-24 in Eq. 10-23 and differentiating where required, we have

$$a_2 A s^2 e^{st} + a_1 A s e^{st} + a_0 A e^{st} = 0 \tag{10-25}$$

Since $x_n = A e^{st}$, we may rewrite Eq. 10-25 as

$$a_2 s^2 x_n + a_1 s x_n + a_0 x_n = 0$$

or

$$(a_2 s^2 + a_1 s + a_0) x_n = 0$$

Since we do not accept the trivial solution, $x_n = 0$, it is required that

$$(a_2 s^2 + a_1 s + a_0) = 0 \qquad (10\text{-}26)$$

This equation, in terms of s, is called a *characteristic equation*. It is readily obtained by replacing the derivative by s and the second derivative by s^2. Clearly, we have returned to the familiar operator

$$s^n \equiv \frac{d^n}{dt^n}$$

The **characteristic equation** is derived from the governing differential equation for a circuit by setting all independent sources to zero value and assuming an exponential solution.

Oliver Heaviside (1850–1925), shown in Figure 10-7, advanced the theory of operators for the solution of differential equations.

Figure 10-7 Oliver Heaviside (1850–1925). Photograph courtesy of the Institution of Electrical Engineers.

The solution of the quadratic equation (10-26) has two roots s_1 and s_2, where

$$s_1 = \frac{-a_1 + \sqrt{a_1^2 - 4a_2 a_0}}{2a_2} \qquad (10\text{-}27)$$

and

$$s_2 = \frac{-a_1 - \sqrt{a_1^2 - 4a_2 a_0}}{2a_2} \qquad (10\text{-}28)$$

When there are two distinct roots, there are two solutions such that

$$x_n = A_1 e^{s_1 t} + A_2 e^{s_2 t} \qquad (10\text{-}29)$$

While there are indeed two solutions to the second-order differential equation, their sum is also a solution, *since the equation is linear.* Further, the *general solution* must consist of as many terms, each with an arbitrary coefficient, as the order of the equation to satisfy the fundamental theorem of differential equations. We will delay considering the special case when $s_1 = s_2$.

The **roots** of the characteristic equation contain all the information necessary for determining the character of the natural response.

Example 10-3

Find the natural response of the circuit current i_2 shown in Figure 10-8. Use operators to formulate the differential equation and obtain the response in terms of two arbitrary constants.

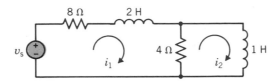

Figure 10-8 Circuit of Example 10-3.

Solution

Writing the two mesh equations, we have

$$12i_1 + 2\frac{di_1}{dt} - 4i_2 = v_s$$

and

$$-4i_1 + 4i_2 + 1\frac{di_2}{dt} = 0$$

Using the operator $s = d/dt$, we obtain

$$(12 + 2s)i_1 - 4i_2 = v_s \tag{10-30}$$

$$-4i_1 + (4 + s)i_2 = 0 \tag{10-31}$$

Using Cramer's rule and solving for i_2, we have

$$i_2 = \frac{4v_s}{(12 + 2s)(4 + s) - 16}$$

$$= \frac{4v_s}{2s^2 + 20s + 32}$$

$$= \frac{2v_s}{s^2 + 10s + 16}$$

Therefore,

$$(s^2 + 10s + 16)i_2 = 2v_s$$

Note that $(s^2 + 10s + 16) = 0$ is the characteristic equation and may be determined directly by evaluating the determinant for Eqs. 10-30 and 10-31. Thus, the roots of the characteristic equation are $s_1 = -2$ and $s_2 = -8$. Therefore, the natural response is

$$x_n = A_1e^{-2t} + A_2e^{-8t}$$

where $x = i_2$. The roots s_1 and s_2 are the *characteristic roots* and are often called the *natural frequencies*. The reciprocals of the magnitude of the real characteristic roots are the *time constants*. The time constants of this circuit are $1/2$ s and $1/8$ s.

EXERCISE 10-4

Find the characteristic equation and the natural frequencies for the circuit shown in Figure E 10-4.

Answer: $s^2 + 7s + 10 = 0$
$$s_1 = -2$$
$$s_2 = -5$$

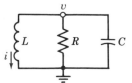

Figure E 10-4

EXERCISE 10-5

German automaker Volkswagen, in its bid to make more efficient cars, has come up with an auto whose engine saves energy by shutting itself off at stoplights. The "stop–start" system springs from a campaign to develop cars in all its world markets that use less fuel and pollute less than vehicles now on the road. The stop-start transmission control has a mechanism that senses when the car does not need fuel: coasting downhill and idling at an intersection. The engine shuts off, but a small starter flywheel keeps turning so that power can be quickly restored when the driver touches the accelerator.

A model of the stop–start circuit is shown in Figure E 10-5. Determine the characteristic equation and the natural frequencies for the circuit.

Answer: $s^2 + 20s + 400 = 0$
$$s = -10 \pm j17.3$$

Figure E 10-5 Stop-start circuit.

10-5 NATURAL RESPONSE OF THE UNFORCED PARALLEL *RLC* CIRCUIT

In this section we consider the (unforced) natural response of the parallel *RLC* circuit shown in Figure 10-9. We choose to examine the parallel *RLC* circuit to illustrate the three forms of the natural response. An analogous discussion of the series *RLC* circuit could be presented, but it is omitted since the purpose is not to obtain the solution to specific circuits but rather to illustrate the general method.

Figure 10-9 Parallel *RLC* circuit.

Write the KCL at the node to obtain

$$\frac{v}{R} + \frac{1}{L} \int_0^t v \, d\tau + i(0) + C \frac{dv}{dt} = 0 \tag{10-32}$$

Taking the derivative of Eq. 10-32, we have

$$C \frac{d^2v}{dt^2} + \frac{1}{R} \frac{dv}{dt} + \frac{1}{L} v = 0 \tag{10-33}$$

A **second-order circuit** has a homogeneous differential equation containing a second-degree term due to the presence of two independent energy storage elements.

Using the operator s, we obtain the characteristic equation

$$s^2 + \frac{1}{RC} s + \frac{1}{LC} = 0 \tag{10-34}$$

The two roots of the characteristic equation are

$$s_1 = -\frac{1}{2RC} + \left[\left(\frac{1}{2RC} \right)^2 - \frac{1}{LC} \right]^{1/2} \tag{10-35}$$

$$s_2 = -\frac{1}{2RC} - \left[\left(\frac{1}{2RC} \right)^2 - \frac{1}{LC} \right]^{1/2} \tag{10-36}$$

When s_1 is not equal to s_2, the solution to the second-order differential equation 10-33 is

$$v_n = A_1 e^{s_1 t} + A_2 e^{s_2 t} \tag{10-37}$$

The roots of the characteristic equation may be rewritten as

$$s_1 = -\alpha + \sqrt{\alpha^2 - \omega_0^2} \tag{10-38}$$

$$s_2 = -\alpha - \sqrt{\alpha^2 - \omega_0^2} \tag{10-39}$$

where $\alpha = 1/2RC$ and $\omega_0^2 = 1/LC$. Normally, ω_0 is called the *resonant frequency*. The concept of resonant frequency is expanded in subsequent chapters.

The roots of the characteristic equation assume three possible conditions:

1 Two real and distinct roots when $\alpha^2 > \omega_0^2$.
2 Two real equal roots when $\alpha^2 = \omega_0^2$.
3 Two complex roots when $\alpha^2 < \omega_0^2$.

When the two roots are real and distinct, the circuit is said to be *overdamped*. When the roots are both real and equal, the circuit is *critically damped*. When the two roots are complex conjugates, the circuit is said to be *underdamped*.

Let us determine the natural response for the overdamped *RLC* circuit of Figure 10-9 when the initial conditions are $v(0)$ and $i(0)$ for the capacitor and the inductor, respectively. Notice that because the circuit in Figure 10-9 has no input, $v_n(0)$ and $v(0)$ are both names for the same voltage. Then, at $t = 0$ for Eq. 10-37, we have

$$v_n(0) = A_1 + A_2 \tag{10-40}$$

Since A_1 and A_2 are both unknown, we need one more equation at $t = 0$. Rewriting Eq. 10-32 at $t = 0$, we have[1]

$$\frac{v(0)}{R} + i(0) + C\frac{dv(0)}{dt} = 0$$

Since $i(0)$ and $v(0)$ are known, we have

$$\frac{dv(0)}{dt} = -\frac{v(0)}{RC} - \frac{i(0)}{C} \qquad (10\text{-}41)$$

Thus, we now know the initial value of the derivative of v. Taking the derivative of Eq. 10-37 and setting $t = 0$, we obtain

$$\frac{dv_\mathrm{n}(0)}{dt} = s_1 A_1 + s_2 A_2 \qquad (10\text{-}42)$$

Equating Eqs. 10-41 and 10-42, we obtain a second equation in terms of the two constants as

$$s_1 A_1 + s_2 A_2 = -\frac{v(0)}{RC} - \frac{i(0)}{C} \qquad (10\text{-}43)$$

Using Eqs. 10-40 and 10-43, we may obtain A_1 and A_2.

Example 10-4

Find the natural response of $v(t)$ for $t > 0$ for the parallel RLC circuit shown in Figure 10-9 when $R = 2/3\ \Omega$, $L = 1$ H, $C = 1/2$ F, $v(0) = 10$ V, and $i(0) = 2$ A.

Solution

The characteristic equation is

$$s^2 + \frac{1}{RC}s + \frac{1}{LC} = 0$$

or

$$s^2 + 3s + 2 = 0$$

Therefore, the roots of the characteristic equation are

$$s_1 = -1$$

and

$$s_2 = -2$$

Then the natural response is

$$v_\mathrm{n} = A_1 e^{-t} + A_2 e^{-2t} \qquad (10\text{-}44)$$

The initial capacitor voltage is $v(0) = 10$, so we have

$$v_\mathrm{n}(0) = A_1 + A_2$$

or

$$10 = A_1 + A_2 \qquad (10\text{-}45)$$

---

[1] Note: $\dfrac{dv(0)}{dt}$ means $\dfrac{dv(t)}{dt}\Big|_{t=0}$

We use Eq. 10-43 to obtain the second equation for the unknown constants. Then

$$s_1 A_1 + s_2 A_2 = -\frac{v(0)}{RC} - \frac{i(0)}{C}$$

or

$$-A_1 - 2A_2 = -\frac{10}{1/3} - \frac{2}{1/2}$$

Therefore, we have

$$-A_1 - 2A_2 = -34 \tag{10-46}$$

Solving Eqs. 10-45 and 10-46 simultaneously, we obtain $A_2 = 24$ and $A_1 = -14$. Therefore, the natural response is

$$v_n = (-14e^{-t} + 24e^{-2t}) \text{ V}$$

The natural response of the circuit is shown in Figure 10-10.

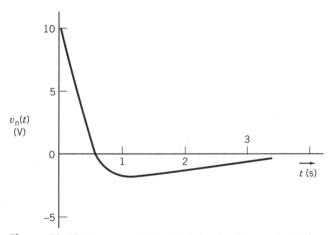

Figure 10-10 Response of the *RLC* circuit of Example 10-4.

EXERCISE 10-6

Find the natural response of the *RLC* circuit of Figure 10-9 when $R = 6\ \Omega$, $L = 7$ H, and $C = 1/42$ F. The initial conditions are $v(0) = 0$ and $i(0) = 10$ A.
Answer: $v_n(t) = -84(e^{-t} - e^{-6t})$ V

EXERCISE 10-7

The circuit shown in Figure E 10-7 is used to detect smokers in airplanes who surreptitiously light up in nonsmoking areas before they can take a single puff. The sensor activates the switch, and the change in the voltage $v(t)$ activates a light at the flight attendant's station. Determine the natural response $v(t)$.
Answer: $v(t) = -1.16e^{-2.7t} + 1.16e^{-37.3t}$ V

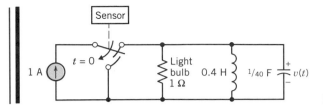

Figure E 10-7 Smoke detector.

10-6 NATURAL RESPONSE OF THE CRITICALLY DAMPED UNFORCED PARALLEL *RLC* CIRCUIT

Again we consider the parallel *RLC* circuit, and here we will determine the special case when the characteristic equation has two equal real roots. Two real equal roots occur when $\alpha^2 = \omega_0^2$ where

$$\alpha = \frac{1}{2RC} \quad \text{and} \quad \omega_0^2 = \frac{1}{LC}$$

Let us assume that $s_1 = s_2$ and proceed to find $v_n(t)$. We write the natural response as the sum of two exponentials as

$$\begin{aligned} v_n &= A_1 e^{s_1 t} + A_2 e^{s_1 t} \\ &= A_3 e^{s_1 t} \end{aligned} \tag{10-47}$$

where $A_3 = A_1 + A_2$. Since the two roots are equal, we have only one undetermined constant, but we still have two initial conditions to satisfy. Clearly, Eq. 10-47 is not the total solution for the natural response of a critically damped circuit. We need the solution that will contain two arbitrary constants, so with some foreknowledge we try the solution

$$x_n = g(t)e^{s_1 t}$$

where $g(t)$ is a polynomial in t. Let us try

$$g(t) = A_2 + A_1 t$$

Substituting $x_n = g(t)e^{s_1 t}$ into the original differential equation and applying the initial conditions, we will obtain the solution for A_1 and A_2.

Therefore, for two simultaneous equal roots of the characteristic equation, we try the solution

$$v_n = e^{s_1 t}(A_1 t + A_2) \tag{10-48}$$

Let us consider a parallel *RLC* circuit where $L = 1$ H, $R = 1$ Ω, $C = 1/4$ F, $v(0) = 5$ V, and $i(0) = -6$ A. The characteristic equation for the circuit is

$$s^2 + \frac{1}{RC}s + \frac{1}{LC} = 0$$

or

$$s^2 + 4s + 4 = 0$$

The two roots are then $s_1 = s_2 = -2$. Using Eq. 10-48 for the natural response, we have

$$v_n = e^{-2t}(A_1 t + A_2) \qquad (10\text{-}49)$$

Since $v_n(0) = 5$, we have at $t = 0$

$$5 = A_2$$

Now, to obtain A_1, we proceed to find the derivative of v_n and evaluate it at $t = 0$. The derivative of v_n is found by differentiating Eq. 10-49 to obtain

$$\frac{dv}{dt} = -2A_1 t e^{-2t} + A_1 e^{-2t} - 2A_2 e^{-2t} \qquad (10\text{-}50)$$

Evaluating Eq. 10-50 at $t = 0$, we have

$$\frac{dv(0)}{dt} = A_1 - 2A_2$$

Again, we may use Eq. 10-41 so that

$$\frac{dv(0)}{dt} = -\frac{v(0)}{RC} - \frac{i(0)}{C}$$

or

$$\frac{dv(0)}{dt} = \frac{-5}{1/4} - \frac{-6}{1/4}$$

$$= 4$$

Therefore, $A_1 = 14$ and the natural response is

$$v_n = e^{-2t}(14t + 5) \text{ V}$$

The critically damped natural response of this *RLC* circuit is shown in Figure 10-11.

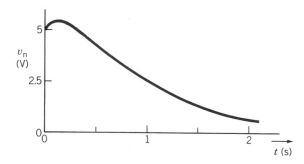

Figure 10-11 Critically damped response of the parallel *RLC* circuit.

EXERCISE 10-8

A parallel *RLC* circuit has $R = 10 \ \Omega$, $C = 1$ mF, $L = 0.4$ H, $v(0) = 8$ V, and $i(0) = 0$. Find the natural response $v_n(t)$ for $t > 0$.
Answer: $v_n(t) = e^{-50t}(8 - 400t)$ V

10-7 NATURAL RESPONSE OF AN UNDERDAMPED UNFORCED PARALLEL *RLC* CIRCUIT

The characteristic equation of the parallel *RLC* circuit will have two complex conjugate roots when $\alpha^2 < \omega_0^2$. This condition is met when

$$LC < (2RC)^2$$

or when

$$L < 4R^2 C$$

Recall that

$$v_n = A_1 e^{s_1 t} + A_2 e^{s_2 t} \tag{10-51}$$

where

$$s_{1,2} = -\alpha \pm \sqrt{\alpha^2 - \omega_0^2}$$

When

$$\omega_0^2 > \alpha^2$$

we have

$$s_{1,2} = -\alpha \pm j\sqrt{\omega_0^2 - \alpha_0^2}$$

where

$$j = \sqrt{-1}$$

See Appendix C for a review of complex numbers.

The complex roots lead to an oscillatory-type response. We define the square root $\sqrt{\omega_0^2 - \alpha^2}$ as ω_d, which we will call the *damped resonant frequency*. The factor α, called the *damping coefficient,* determines how quickly the oscillations subside. Then the roots are

$$s_{1,2} = -\alpha \pm j\omega_d$$

Therefore, the natural response is

$$v_n = A_1 e^{-\alpha t} e^{j\omega_d t} + A_2 e^{-\alpha t} e^{-j\omega_d t}$$

or

$$v_n = e^{-\alpha t}(A_1 e^{j\omega_d t} + A_2 e^{-j\omega_d t}) \tag{10-52}$$

Let us use the Euler identity[2]

$$e^{\pm j\omega t} = \cos \omega t \pm j \sin \omega t \tag{10-53}$$

Let $\omega = \omega_d$ in Eq. 10-53 and substitute into Eq. 10-52 to obtain

$$\begin{aligned}v_n &= e^{-\alpha t}(A_1 \cos \omega_d t + jA_1 \sin \omega_d t + A_2 \cos \omega_d t - jA_2 \sin \omega_d t) \\ &= e^{-\alpha t}[(A_1 + A_2) \cos \omega_d t + j(A_1 - A_2) \sin \omega_d t] \end{aligned} \tag{10-54}$$

Since the unknown constants A_1 and A_2 remain arbitrary, we replace $(A_1 + A_2)$ and $j(A_1 - A_2)$ with new arbitrary (yet unknown) constants B_1 and B_2. A_1 and A_2 must be

---

[2] See Appendix E for a discussion of Euler's identity.

complex conjugates so that B_1 and B_2 are real numbers. Therefore, Eq. 10-54 becomes

$$v_n = e^{-\alpha t}(B_1 \cos \omega_d t + B_2 \sin \omega_d t) \tag{10-55}$$

where B_1 and B_2 will be determined by the initial conditions, $v(0)$ and $i(0)$.

The natural underdamped response is oscillatory with a decaying magnitude. The rapidity of decay depends on α, and the frequency of oscillation depends on ω_d.

Let us find the general form of the solution for B_1 and B_2 in terms of the initial conditions when the circuit is unforced. Then at $t = 0$ we have

$$v_n(0) = B_1$$

In order to find B_2, we evaluate the first derivative of v_n at $t = 0$. Therefore, the derivative is

$$\frac{dv_n}{dt} = e^{-\alpha t}[(\omega_d B_2 - \alpha B_1) \cos \omega_d t - (\omega_d B_1 + \alpha B_2) \sin \omega_d t]$$

and at $t = 0$ we obtain

$$\frac{dv_n(0)}{dt} = \omega_d B_2 - \alpha B_1 \tag{10-56}$$

Recall that we found earlier that Eq. 10-41 provides $dv(0)/dt$ for the parallel *RLC* circuit as

$$\frac{dv_n(0)}{dt} = -\frac{v(0)}{RC} - \frac{i(0)}{C} \tag{10-57}$$

Therefore, we use Eqs. 10-56 and 10-57 to obtain

$$\omega_d B_2 = \alpha B_1 - \frac{v(0)}{RC} - \frac{i(0)}{C} \tag{10-58}$$

Example 10-5

Consider the parallel *RLC* circuit when $R = 25/3 \ \Omega$, $L = 0.1$ H, $C = 1$ mF, $v(0) = 10$ V, and $i(0) = -0.6$ A. Find the natural response $v_n(t)$ for $t > 0$.

Solution

First, we determine α^2 and ω_0^2 to determine the form of the response. Consequently, we obtain

$$\alpha = \frac{1}{2RC} = 60$$

and

$$\omega_0^2 = \frac{1}{LC} = 10^4$$

Therefore, $\omega_0^2 > \alpha^2$ and the natural response is underdamped. We proceed to determine the damped resonant frequency ω_d as

$$\omega_d = (\omega_0^2 - \alpha^2)^{1/2} = (10^4 - 3.6 \times 10^3)^{1/2} = 80 \text{ rad/s}$$

Hence, the characteristic roots are

$$s_1 = -\alpha + j\omega_d = -60 + j80$$

and

$$s_2 = -\alpha - j\omega_d$$

Consequently, the natural response is obtained from Eq. 10-55 as

$$v_n(t) = B_1 e^{-60t} \cos 80t + B_2 e^{-60t} \sin 80t$$

Since $v(0) = 10$, we have

$$B_1 = v(0) = 10$$

We can use Eq. 10-58 to obtain B_2 as

$$B_2 = \frac{\alpha}{\omega_d} B_1 - \frac{v(0)}{\omega_d RC} - \frac{i(0)}{\omega_d C}$$

$$= \frac{60 \times 10}{80} - \frac{10}{80 \times 25/3000} - \frac{-0.6}{80 \times 10^{-3}}$$

$$= 7.5 - 15.0 + 7.5$$

$$= 0$$

Therefore, the natural response is

$$v_n(t) = 10e^{-60t} \cos 80t \quad V$$

This response is shown in Figure 10-12. While the response is oscillatory in form because of the cosine function, it is damped by the exponential function, e^{-60t}.

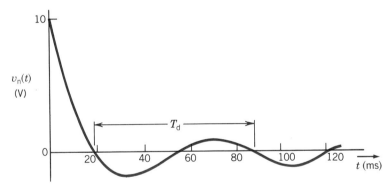

Figure 10-12 Natural response of the underdamped parallel RLC circuit.

The *period of the damped oscillation* is the time interval, denoted as T_d, expressed as

$$T_d = \frac{2\pi}{\omega_d} \text{ s}$$

The natural response of an underdamped circuit is not a pure oscillatory response, but it does exhibit the form of an oscillatory response. Thus we may approximate T_d by the period between the first and third zero-crossings, as shown in Figure 10-12. Therefore, the frequency in hertz is

$$f_d = \frac{1}{T_d} \text{ Hz}$$

The period of the oscillation of the circuit of Example 10-5 is

$$T_d = \frac{2\pi}{80} = 79 \text{ ms}$$

EXERCISE 10-9

A parallel *RLC* circuit has $R = 62.5 \, \Omega$, $L = 10$ mH, $C = 1 \, \mu$F, $v(0) = 10$ V, and $i(0) = 80$ mA. Find the natural response $v_n(t)$.

Answer: $v_n(t) = e^{-8000t}[10 \cos 6000t - 26.7 \sin 6000t]$ V

10-8 FORCED RESPONSE OF AN *RLC* CIRCUIT

The forced response of an *RLC* circuit described by a second-order differential equation must satisfy the differential equation and contain no arbitrary constants. As we noted earlier, the response to a forcing function will often be of the same form as the forcing function. Again, we consider the differential equation for the second-order circuit as

$$\frac{d^2x}{dt^2} + a_1 \frac{dx}{dt} + a_0 x = f(t) \tag{10-59}$$

The forced response x_f must satisfy Eq. 10-59. Therefore, substituting x_f, we have

$$\frac{d^2x_f}{dt^2} + a_1 \frac{dx_f}{dt} + a_0 x_f = f(t) \tag{10-60}$$

We need to determine x_f so that x_f and its first and second derivatives all satisfy Eq. 10-60.

If the forcing function is a constant, we expect the forced response also to be a constant since the derivatives of a constant are zero. If the forcing function is of the form $f(t) = Be^{-at}$, then the derivatives of $f(t)$ are all exponentials of the form Qe^{-at} and we expect

$$x_f = De^{-at}$$

If the forcing function is a sinusoidal function, we can expect the forced response to be a sinusoidal function. If $f(t) = A \sin \omega_0 t$, we will try

$$x_f = M \sin \omega_0 t + N \cos \omega_0 t$$
$$= Q \sin(\omega_0 t + \theta)$$

Table 10-4 summarizes selected forcing functions and their associated assumed solutions.

Table 10-4

Forcing Function	Assumed Solution
K	A
Kt	$At + B$
Kt^2	$At^2 + Bt + C$
$K \sin \omega t$	$A \sin \omega t + B \cos \omega t$
Ke^{-at}	Ae^{-at}

Example 10-6

Find the forced response for the inductor current i_f for the parallel *RLC* circuit shown in Figure 10-13 when $i_s = 8e^{-2t}$ A. Let $R = 6 \, \Omega$, $L = 7$ H, and $C = 1/42$ F.

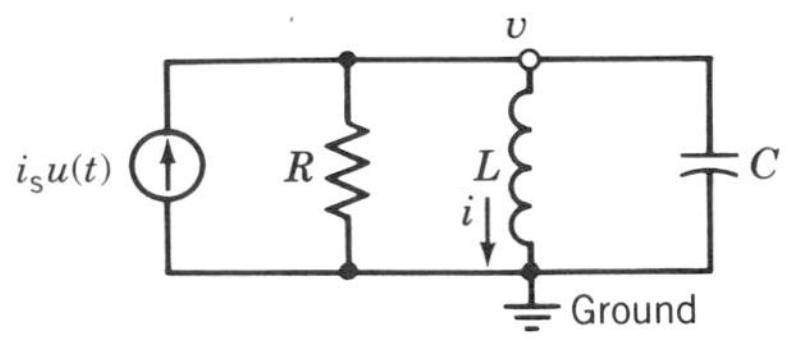

Figure 10-13 Circuit for Examples 10-6 and 10-7.

Solution

The source current is applied at $t = 0$ as indicated by the unit step function $u(t)$. The KCL equation at the upper node is

$$i + \frac{v}{R} + C\frac{dv}{dt} = i_s \tag{10-61}$$

We wish to obtain the second-order differential equation in terms of i using Eq. 10-61. We note that

$$v = L\frac{di}{dt} \tag{10-62}$$

and

$$\frac{dv}{dt} = L\frac{d^2i}{dt^2} \tag{10-63}$$

Substituting Eqs. 10-62 and 10-63 into Eq. 10-61, we have

$$i + \frac{L}{R}\frac{di}{dt} + CL\frac{di^2}{dt^2} = i_s$$

Then we divide by LC and rearrange to obtain the familiar second-order differential equation

$$\frac{d^2i}{dt^2} + \frac{1}{RC}\frac{di}{dt} + \frac{1}{LC}i = \frac{i_s}{LC} \tag{10-64}$$

Substituting the component values and the source i_s, we obtain

$$\frac{d^2i}{dt^2} + 7\frac{di}{dt} + 6i = 48e^{-2t} \tag{10-65}$$

We wish to obtain the forced response, so we assume that the response will be

$$i_f = Be^{-2t} \tag{10-66}$$

where B is to be determined. Substituting the assumed solution, Eq. 10-66, into the differential equation, we have

$$4Be^{-2t} + 7(-2Be^{-2t}) + 6Be^{-2t} = 48e^{-2t}$$

or

$$(4 - 14 + 6)Be^{-2t} = 48e^{-2t}$$

Therefore, $B = -12$ and

$$i_f = -12e^{-2t}\,\text{A}$$

Example 10-7

Find the forced response i_f of the circuit of Example 10-6 when $i_s = I_0$, where I_0 is a constant.

Solution

Since the source is a constant applied at $t = 0$, we expect the forced response to be a constant also. As a first method, we will use the differential equation to find the forced response. Second, we will demonstrate the alternative method that uses the steady-state behavior of the circuit to find i_f.

The differential equation with the constant source is obtained from Eq. 10-64 as

$$\frac{d^2i}{dt^2} + 7\frac{di}{dt} + 6i = 6I_0$$

Again, we assume that the forced response is $i_f = D$, a constant. Since the first and second derivatives of the assumed forced response are zero, we have

$$6D = 6I_0$$

or

$$D = I_0$$

Therefore,

$$i_f = I_0$$

Another approach is to determine the steady-state response i_f of the circuit of Figure 10-13 by drawing the steady-state circuit model. The inductor is represented by a short circuit and the capacitor is represented by an open circuit, as shown in Figure 10-14. Clearly, since the steady-state model of the inductor is a short circuit, all the source current flows through the inductor in steady state and

$$i_f = I_0$$

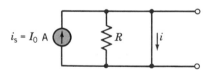

Figure 10-14 Parallel *RLC* circuit in steady state for a constant input.

The two previous examples showed that it is relatively easy to obtain the response of the circuit to a forcing function. However, we are sometimes confronted with a special case where the form of the forcing function is the same as the form of one of the components of the natural response.

Again, consider the circuit of Examples 10-6 and 10-7 (Figure 10-13) when the differential equation is

$$\frac{d^2i}{dt^2} + 7\frac{di}{dt} + 6i = 6i_s \tag{10-67}$$

The characteristic equation of the circuit is

$$s^2 + 7s + 6 = 0$$

or

$$(s + 1)(s + 6) = 0$$

Thus, the natural response is

$$i_n = A_1 e^{-t} + A_2 e^{-6t} \tag{10-68}$$

Consider the special case where $i_s = 3e^{-6t}$. Then we at first expect the forced response to be

$$i_f = Be^{-6t} \tag{10-69}$$

However, the forced response and one component of the natural response would then both have the form De^{-6t}. Will this work? Let's try substituting Eq. 10-69 into the differential equation 10-67. We then obtain

$$36Be^{-6t} - 42Be^{-6t} + 6Be^{-6t} \neq 18e^{-6t}$$

or

$$0 \neq 18e^{-6t}$$

which is an impossible solution. Therefore, we need another form of the forced response when one of the natural response terms has the same form as the forcing function.

Let us try the forced response

$$i_f = Bte^{-6t} \tag{10-70}$$

Then, substituting Eq. 10-70 into Eq. 10-67, we have

$$B(-6g - 6g + 36tg) + 7B(g - 6tg) + 6Btg = 18g \tag{10-71}$$

where $g = g(t) = e^{-6t}$. Simplifying Eq. 10-71, we have

$$B = -\frac{18}{5}$$

Therefore,

$$i_f = -\frac{18}{5} te^{-6t}$$

In general, if the forcing function is of the same form as one of the components of the natural response, x_{n1}, we will use

$$x_f = t^p x_{n1}$$

where the integer p is selected so that the x_f is not duplicated in the natural response. Use the lowest power, p, of t that is not duplicated in the natural response.

EXERCISE 10-10

A circuit is described for $t > 0$ by the equation

$$\frac{d^2v}{dt^2} + 5\frac{dv}{dt} + 6v = v_s$$

Find the forced response v_f for $t > 0$ when (a) $v_s = 8$ V, (b) $v_s = 3e^{-4t}$ V, and (c) $v_s = 2e^{-2t}$ V.

Answer: (a) $v_f = 8/6$ V
(b) $v_f = \frac{3}{2} e^{-4t}$ V
(c) $v_f = 2te^{-2t}$ V

EXERCISE 10-11

A circuit is described for $t > 0$ by the equation

$$\frac{d^2i}{dt^2} + 9\frac{di}{dt} + 20i = 6i_s$$

where $i_s = 6 + 2t$ A. Find the forced response i_f.
Answer: $i_f = 1.53 + 0.6t$ A

10-9 ∥ COMPLETE RESPONSE OF AN *RLC* CIRCUIT

We have succeeded in finding the natural response and the forced response of a circuit described by a second-order differential equation. We wish to proceed to determine the complete response for the circuit.

We know that the *complete response* is the sum of the natural response and the forced response, thus

$$x = x_n + x_f$$

We may then obtain the complete response along with its unspecified constants by evaluating $x(t)$ at $t = 0$ and dx/dt at $t = 0$ to determine these constants.

Let us consider the series *RLC* circuit of Figure 10-3 with a differential equation (10-7) as

$$LC\frac{d^2v}{dt^2} + RC\frac{dv}{dt} + v = v_s$$

When $L = 1$ H, $C = \frac{1}{6}$ F, and $R = 5$ Ω we obtain

$$\frac{d^2v}{dt^2} + 5\frac{dv}{dt} + 6v = 6v_s \qquad (10\text{-}72)$$

We let $v_s = \dfrac{2e^{-t}}{3}$ V, $v(0) = 10$ V, and $dv(0)/dt = -2$ V/s.

We will first determine the form of the natural response and then determine the forced response. Adding these responses, we have the complete response with two unspecified constants. We will then use the initial conditions to specify these constants to obtain the complete response.

To obtain the natural response, we write the characteristic equation using operators as

$$s^2 + 5s + 6 = 0$$

or

$$(s + 2)(s + 3) = 0$$

Therefore, the natural response is

$$v_n = A_1 e^{-2t} + A_2 e^{-3t}$$

The forced response is obtained by examining the forcing function and noting that its exponential response has a different time constant than the natural response, so we may write

$$v_f = Be^{-t} \tag{10-73}$$

We can determine B by substituting Eq. 10-73 into Eq. 10-72. Then we have

$$Be^{-t} + 5(-Be^{-t}) + 6(Be^{-t}) = 4e^{-t}$$

or

$$B = 2$$

The complete response is then

$$v = v_n + v_f$$
$$= A_1 e^{-2t} + A_2 e^{-3t} + 2e^{-t}$$

In order to find A_1 and A_2, we use the initial conditions. At $t = 0$ we have $v(0) = 10$, so we obtain

$$10 = A_1 + A_2 + 2 \tag{10-74}$$

From the fact that $dv/dt = -2$ at $t = 0$, we have

$$-2A_1 - 3A_2 - 2 = -2 \tag{10-75}$$

Solving Eqs. 10-74 and 10-75 by Cramer's rule, we have $A_1 = 24$ and $A_2 = -16$. Therefore,

$$v = 24e^{-2t} - 16e^{-3t} + 2e^{-t} \text{ V}$$

Example 10-8

Find the complete response $v(t)$ for $t > 0$ for the circuit of Figure 10-15. Assume the circuit is in steady state at $t = 0^-$.

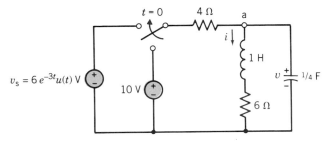

Figure 10-15 Circuit of Example 10-8.

Solution

First, we determine the initial conditions of the circuit. At $t = 0^-$ we have the circuit model shown in Figure 10-16, where we replace the capacitor with an open circuit and the inductor with a short circuit. Then the voltage is

$$v(0^-) = 6 \text{ V}$$

and the inductor current is

$$i(0^-) = 1 \text{ A}$$

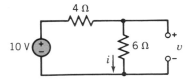

Figure 10-16 Circuit of Example 10-8 at $t = 0^-$.

After the switch is thrown, we can write the KVL for the right-hand mesh of Figure 10-15 to obtain

$$-v + \frac{di}{dt} + 6i = 0 \tag{10-76}$$

The KCL equation at node a will provide a second equation in terms of v and i as

$$\frac{v - v_s}{4} + i + \frac{1}{4}\frac{dv}{dt} = 0 \tag{10-77}$$

Equations 10-76 and 10-77 may be rearranged as

$$\left(\frac{di}{dt} + 6i\right) - v = 0 \tag{10-78}$$

$$i + \left(\frac{v}{4} + \frac{1}{4}\frac{dv}{dt}\right) = \frac{v_s}{4} \tag{10-79}$$

We will use operators so that $s = d/dt$, $s^2 = d^2/dt^2$, and $1/s = \int dt$. Then we obtain

$$(s + 6)i - v = 0 \tag{10-80}$$

$$i + \frac{1}{4}(s + 1)v = v_s/4 \tag{10-81}$$

The characteristic equation is obtained from Cramer's rule as the determinant Δ where

$$\Delta = \frac{1}{4}(s + 6)(s + 1) + 1$$

Set the determinant to zero to obtain

$$(s + 6)(s + 1) + 4 = 0$$

or

$$s^2 + 7s + 10 = 0$$

Therefore, the roots of the characteristic equation are

$$s_1 = -2 \quad \text{and} \quad s_2 = -5$$

To find the second-order differential equation describing the circuit, we use Cramer's rule for Eqs. 10-80 and 10-81 to solve for v in order to obtain

$$v = \frac{(s + 6)(v_s/4)}{\Delta}$$

$$= \frac{(s + 6)v_s}{s^2 + 7s + 10}$$

Of course, this equation can be rewritten as

$$(s^2 + 7s + 10)v = (s + 6)v_s$$

and hence the second-order differential equation is

$$\frac{d^2v}{dt^2} + 7\frac{dv}{dt} + 10v = \frac{dv_s}{dt} + 6v_s \tag{10-82}$$

The natural response v_n is

$$v_n = A_1 e^{-2t} + A_2 e^{-5t}$$

The forced response is assumed to be of the form

$$v_f = Be^{-3t} \qquad (10\text{-}83)$$

Substituting v_f into the differential equation, we have

$$9Be^{-3t} - 21Be^{-3t} + 10Be^{-3t} = -18e^{-3t} + 36e^{-3t}$$

Therefore,

$$B = -9$$

and

$$v_f = -9e^{-3t}$$

The complete response is then

$$\begin{aligned} v &= v_n + v_f \\ &= A_1 e^{-2t} + A_2 e^{-5t} - 9e^{-3t} \end{aligned} \qquad (10\text{-}84)$$

Since $v(0) = 6$, we have

$$v(0) = 6 = A_1 + A_2 - 9$$

or

$$A_1 + A_2 = 15 \qquad (10\text{-}85)$$

We also know that $i(0) = 1$ A. We can use Eq. 10-79 to determine $dv(0)/dt$ and then evaluate the derivative of Eq. 10-84 at $t = 0$. Equation 10-79 states that

$$\frac{dv}{dt} = -4i - v + v_s$$

At $t = 0$ we have

$$\begin{aligned} \frac{dv(0)}{dt} &= -4i(0) - v(0) + v_s(0) \\ &= -4 - 6 + 6 \\ &= -4 \end{aligned}$$

Let us take the derivative of Eq. 10-84 to obtain

$$\frac{dv}{dt} = -2A_1 e^{-2t} - 5A_2 e^{-5t} + 27e^{-3t}$$

At $t = 0$ we obtain

$$\frac{dv(0)}{dt} = -2A_1 - 5A_2 + 27$$

Since $dv(0)/dt = -4$, we have

$$2A_1 + 5A_2 = 31 \qquad (10\text{-}86)$$

Solving Eqs. 10-86 and 10-85 simultaneously, we obtain

$$A_1 = \frac{44}{3} \quad \text{and} \quad A_2 = \frac{1}{3}$$

Therefore,

$$v = \frac{44}{3} e^{-2t} + \frac{1}{3} e^{-5t} - 9e^{-3t} \text{ V}$$

Note that we used the capacitor voltage and the inductor current as the unknowns. This is very convenient, since you will normally have the initial conditions of these variables. These variables, v_c and i_L, are known as the *state variables*. We will consider this approach more fully in the next section.

EXERCISE 10-12

For the circuit of Figure 10-15, repeat Example 10-8 when $v_s = 4\ u(t)$ V and all other parameters remain unchanged. Find $v(t)$ for $t > 0$.
Answer: $v(t) = (4e^{-2t} - 0.4e^{-5t} + 2.4)$ V

EXERCISE 10-13

Obtain the differential equation for $v_2(t)$ describing the circuit shown in Figure E 10-13 using operators. (*Hint:* Write the two node equations.)
Answer: $4\dfrac{d^2v_2}{dt^2} + \dfrac{3}{2}\dfrac{dv_2}{dt} + \dfrac{1}{4}v_2 = \dfrac{1}{2}\dfrac{dv_s}{dt}$

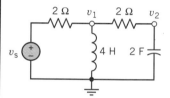

Figure E 10-13

10-10 STATE VARIABLE APPROACH TO CIRCUIT ANALYSIS

The *state variables* of a circuit are a set of variables associated with the energy of the energy storage elements of the circuit. Thus they describe the complete response of a circuit to a forcing function and the circuit's initial conditions. Here the word "state" means condition, as in "state of the union." We will choose as the state variables those variables that describe the energy storage of the circuit. Thus, we will use the independent capacitor voltages and the independent inductor currents.

Consider the circuit shown in Figure 10-17. The two energy storage elements are C_1 and C_2, and the two capacitors cannot be reduced to one. We expect the circuit to be described by a second-order differential equation. However, let us first obtain the two first-order differential equations that describe the response for $v_1(t)$ and $v_2(t)$, which are the state variables of the circuit. If we know the value of the state variables at one time and the value of the input variables thereafter, we can find the value of any state variable for any subsequent time.

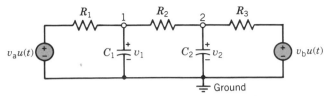

Figure 10-17 Circuit with two energy storage elements.

Writing the KCL at nodes 1 and 2, we have

$$node\ 1: \quad C_1 \frac{dv_1}{dt} = \frac{v_a - v_1}{R_1} + \frac{v_2 - v_1}{R_2} \tag{10-87}$$

$$node\ 2: \quad C_2 \frac{dv_2}{dt} = \frac{v_b - v_2}{R_3} + \frac{v_1 - v_2}{R_2} \tag{10-88}$$

Equations 10-87 and 10-88 can be rewritten as

$$\frac{dv_1}{dt} + \frac{v_1}{C_1 R_1} + \frac{v_1}{C_1 R_2} - \frac{v_2}{C_1 R_2} = \frac{v_a}{C_1 R_1} \tag{10-89}$$

$$\frac{dv_2}{dt} + \frac{v_2}{C_2 R_3} + \frac{v_2}{C_2 R_2} - \frac{v_1}{C_2 R_2} = \frac{v_b}{C_2 R_3} \tag{10-90}$$

Assume that $C_1 R_1 = 1$, $C_1 R_2 = 1$, $C_2 R_3 = 1$, and $C_2 R_2 = 1/2$. Then we have

$$\frac{dv_1}{dt} + 2v_1 - v_2 = v_a \tag{10-91}$$

and

$$-2v_1 + \frac{dv_2}{dt} + 3v_2 = v_b \tag{10-92}$$

Using operators, we have

$$(s + 2)v_1 - v_2 = v_a$$
$$-2v_1 + (s + 3)v_2 = v_b$$

If we wish to solve for v_1, we use Cramer's rule to obtain

$$v_1 = \frac{(s + 3)v_a + v_b}{(s + 2)(s + 3) - 2} \tag{10-93}$$

The characteristic equation is obtained from the denominator and has the form

$$s^2 + 5s + 4 = 0$$

The characteristic roots are $s = -4$ and $s = -1$. The second-order differential equation can be obtained by rewriting Eq. 10-93 as

$$(s^2 + 5s + 4)v_1 = (s + 3)v_a + v_b$$

Then the differential equation for v_1 is

$$\frac{d^2 v_1}{dt^2} + 5\frac{dv_1}{dt} + 4v_1 = \frac{dv_a}{dt} + 3v_a + v_b \tag{10-94}$$

We now proceed to obtain the natural response

$$v_{1n} = A_1 e^{-t} + A_2 e^{-4t}$$

and the forced response, which depends on the form of the forcing function. For example, if $v_a = 10$ V and $v_b = 6$ V, v_{1f} will be a constant (see Table 10-4). We obtain v_{1f} by substituting v_a and v_b into Eq. 10-94, obtaining

$$4v_{1f} = 3v_a + v_b$$

or

$$4v_{1f} = 30 + 6$$
$$= 36$$

Therefore,

$$v_{1f} = 9$$

Then

$$v_1 = v_{1n} + v_{1f}$$
$$= A_1 e^{-t} + A_2 e^{-4t} + 9 \qquad (10\text{-}95)$$

We will usually know the initial conditions of the energy storage elements. For example, if we know that $v_1(0) = 5$ V and $v_2(0) = 10$ V, we first use $v_1(0) = 5$ along with Eq. 10-95 to obtain

$$v_1(0) = A_1 + A_2 + 9$$

and, therefore,

$$A_1 + A_2 = -4 \qquad (10\text{-}96)$$

Now we need the value of dv_1/dt at $t = 0$. Referring back to Eq. 10-91, we have

$$\frac{dv_1}{dt} = v_a + v_2 - 2v_1$$

Therefore, at $t = 0$ we have

$$\frac{dv_1(0)}{dt} = v_a(0) + v_2(0) - 2v_1(0)$$
$$= 10 + 10 - 2(5)$$
$$= 10$$

The derivative of the complete solution, Eq. 10-95, at $t = 0$ is

$$\frac{dv_1(0)}{dt} = -A_1 - 4A_2$$

Therefore,

$$A_1 + 4A_2 = -10 \qquad (10\text{-}97)$$

Solving Eqs. 10-96 and 10-97, we have

$$A_1 = -2 \quad \text{and} \quad A_2 = -2$$

Therefore,

$$v_1(t) = -2e^{-t} - 2e^{-4t} + 9 \text{ V}$$

As you encounter circuits with two or more energy storage elements, you should consider using the state variable method of describing a set of first-order differential equations.

Table 10-5
State Variable Method of Circuit Analysis

1. Identify the state variables as the independent capacitor voltages and inductor currents.
2. Determine the initial conditions at $t = 0$ for the capacitor voltages and the inductor currents.
3. Obtain a first-order differential equation for each state variable using KCL or KVL.
4. Use the operator s to substitute for d/dt.
5. Obtain the characteristic equation of the circuit by noting that it can be obtained by setting the determinant of Cramer's rule equal to zero.
6. Determine the roots of the characteristic equation, which then determines the form of the natural response.
7. Obtain the second-order (or higher order) differential equation for the selected variable x by Cramer's rule.
8. Determine the forced response x_f by assuming an appropriate form of x_f and determining the constant by substituting the assumed solution in the second-order differential equation.
9. Obtain the complete solution $x = x_n + x_f$.
10. Use the initial conditions on the state variables along with the set of first-order differential equations (step 3) to obtain $dx(0)/dt$.
11. Using $x(0)$ and $dx(0)/dt$ for each state variable, find the arbitrary constants $A_1, A_2, \ldots A_n$ to obtain the complete solution $x(t)$.

The **state variable method** uses a first-order differential equation for each state variable to determine the complete response of a circuit.

A summary of the state variable method is given in Table 10-5. We will use this method in Example 10-9.

Example 10-9
Find $i(t)$ for $t > 0$ for the circuit shown in Figure 10-18 when $R = 3\ \Omega$, $L = 1\ \text{H}$, $C = 1/2\ \text{F}$, and $i_s = 2e^{-3t}\ \text{A}$. Assume steady state at $t = 0^-$.

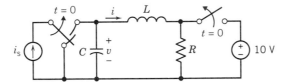

Figure 10-18 Circuit of Example 10-9.

Solution
First, we identify the state variables as i and v. The initial conditions at $t = 0$ are obtained by considering the circuit with the 10-V source connected for a long time at $t = 0^-$. Then $v(0) = 10\ \text{V}$ and $i(0) = 0\ \text{A}$. At $t = 0$, the voltage source is disconnected and the current source is connected.

The first differential equation is obtained by using KVL around the RLC mesh to obtain

$$L\frac{di}{dt} + Ri = v$$

The second differential equation is obtained at the node at the top of the capacitor as

$$C\frac{dv}{dt} + i = i_s$$

We may rewrite these two first-order differential equations as

$$\frac{di}{dt} + \frac{R}{L}i - \frac{v}{L} = 0$$

and

$$\frac{dv}{dt} + \frac{i}{C} = \frac{i_s}{C}$$

Substituting the component values, we have

$$\frac{di}{dt} + 3i - v = 0 \tag{10-98}$$

and

$$\frac{dv}{dt} + 2i = 2i_s \tag{10-99}$$

Using the operator $s = d/dt$, we have

$$(s + 3)i - v = 0 \tag{10-100}$$

$$2i + sv = 2i_s \tag{10-101}$$

Therefore, the characteristic equation obtained from the determinant is

$$(s + 3)s + 2 = 0$$

or

$$s^2 + 3s + 2 = 0$$

Thus, the roots of the characteristic equation are

$$s_1 = -2 \quad \text{and} \quad s_2 = -1$$

Since we wish to solve for $i(t)$ for $t > 0$, we use Cramer's rule to solve Eqs. 10-100 and 10-101 for i, obtaining

$$i = \frac{2i_s}{s^2 + 3s + 2}$$

Therefore, the differential equation is

$$\frac{d^2i}{dt^2} + 3\frac{di}{dt} + 2i = 2i_s \tag{10-102}$$

The natural response is

$$i_n = A_1 e^{-t} + A_2 e^{-2t}$$

We assume the forced response is of the form

$$i_f = Be^{-3t}$$

Substituting i_f into Eq. 10-102, we have

$$(9Be^{-3t}) + 3(-3Be^{-3t}) + 2Be^{-3t} = 2(2e^{-3t})$$

or

$$9B - 9B + 2B = 4$$

Therefore, $B = 2$ and

$$i_f = 2e^{-3t}$$

The complete response is

$$i = A_1e^{-t} + A_2e^{-2t} + 2e^{-3t}$$

Since $i(0) = 0$,

$$0 = A_1 + A_2 + 2 \qquad (10\text{-}103)$$

We need to obtain $di(0)/dt$ from Eq. 10-98, which we repeat here as

$$\frac{di}{dt} + 3i - v = 0$$

Therefore, at $t = 0$ we have

$$\frac{di(0)}{dt} = -3i(0) + v(0)$$

$$= 10$$

The derivative of the complete response at $t = 0$ is

$$\frac{di(0)}{dt} = -A_1 - 2A_2 - 6$$

Since $di(0)/dt = 10$, we have

$$-A_1 - 2A_2 = 16$$

and repeating Eq. 10-103, we have

$$A_1 + A_2 = -2$$

Adding these two equations, we determine that $A_1 = 12$ and $A_2 = -14$. Then we have the complete solution for i as

$$i = 12e^{-t} - 14e^{-2t} + 2e^{-3t} \text{ A}$$

We recognize that the state variable method is particularly powerful for finding the response of energy storage elements in a circuit. This is also true if we encounter higher order circuits with three or more energy storage elements. For example, consider the circuit shown in Figure 10-19. The state variables are v_1, v_2, and i. Two first-order differential equations are obtained by writing the KCL equations at node a and node b. Then a third first-order differential equation is obtained by writing the KVL around the middle mesh containing i. The solution for one or more of these variables can then be obtained by proceeding with the state variable method summarized in Table 10-5.

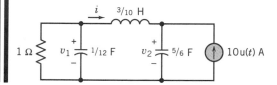

Figure 10-19 Circuit with three energy storage elements.

EXERCISE 10-14

Find $v_2(t)$ for $t > 0$ for the circuit of Figure E 10-14. Assume there is no initial stored energy.

Answer: $v_2(t) = -15e^{-2t} + 6e^{-4t} - e^{-6t} + 10$ V

Figure E 10-14

10-11 ROOTS IN THE COMPLEX PLANE

We have observed that the character of the natural response of a second-order system is determined by the roots of the characteristic equation. Let us consider the roots of a parallel RLC circuit. The characteristic equation (10-34) is

$$s^2 + \frac{s}{RC} + \frac{1}{LC} = 0$$

and the roots are (10-38)

$$s = -\alpha \pm \sqrt{\alpha^2 - \omega_0^2}$$

where $\alpha = 1/2RC$ and $\omega_0^2 = 1/LC$. When $\omega_0 > \alpha$, the roots are complex and

$$s = -\alpha \pm j\sqrt{\omega_0^2 - \alpha^2} = -\alpha \pm j\omega_d \qquad (10\text{-}104)$$

In general, roots are located in the complex plane, the location being defined by coordinates measured along the real or σ axis and the imaginary or $j\omega$-axis. This is referred to as the s-plane or, since s has the units of frequency, as the *complex frequency plane*. Where the roots are real, negative, and distinct, the response is the sum of two decaying exponentials and is said to be overdamped. Where the roots are complex conjugates, the natural response is an exponentially decaying sinusoid and is said to be underdamped or oscillatory.

Now, let us show the location of the roots of the characteristic equation for the four conditions: (a) undamped, $\alpha = 0$; (b) underdamped, $\alpha < \omega_0$; (c) critically damped, $\alpha = 0$; and (d) overdamped, $\alpha > \omega$. These four conditions lead to root locations on the s-plane as shown in Figure 10-20. When $\alpha = 0$, the two complex roots are $\pm j\omega_0$. When $\alpha < \omega_0$, the roots are $s = -\alpha \pm j\omega_d$. When $\alpha = \omega_0$, there are two roots at $s = -\alpha$. Finally, when $\alpha > \omega_0$, there are two real roots, $s = -\alpha \pm \sqrt{\alpha^2 - \omega_0^2}$.

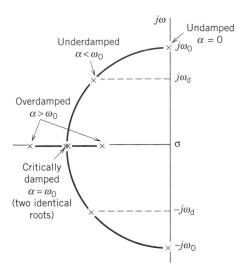

Figure 10-20 The complex s-plane showing the location of the two roots, s_1 and s_2, of the characteristic equation in the left-hand portion of the s-plane. The roots are designated by the $\times$ symbol.

Table 10-6
The Natural Response of a Parallel RLC Circuit

Root Location	Type of Response	Form of Response for $v(0) = 1$ V and $i(0) = 0$
$s = -r_1, -r_2$ two real roots	Overdamped	
$s = -r_1, -r_1$ two equal roots	Critically Damped	
$s = -\alpha \pm j\omega_d$	Underdamped	

$s = \pm j\omega_o$ Undamped

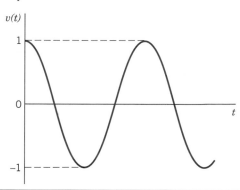

Note: $v(t)$ is the capacitor voltage

A summary of the root locations, the type of response, and the form of the response $v(t)$ for $v(0) = 1$ V and $i(0) = 0$ is shown in Table 10-6.

EXERCISE 10-15

A parallel *RLC* circuit has $L = 0.1$ H and $C = 100$ mF. Determine the roots of the characteristic equation and plot them on the *s*-plane when (a) $R = 0.4\ \Omega$ and (b) $R = 1.0\ \Omega$.
Answer: (a) $s = -5, -20$ (Figure E 10-15)

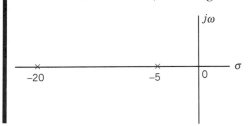

Figure E 10-15

10-12 PSpice ANALYSIS OF THE *RLC* CIRCUIT

In this section we use PSpice analysis methods to obtain the transient response of *RLC* circuits. First, let us consider the *RLC* series circuit shown in Figure 10-3 and redrawn in PSpice format in Figure 10-21 with $v_s(t) = 0$. Let us consider that $i(0) = 0$ A and $v(0) = 10$ V are established at $t = 0$. Using Eq. 10-7, we have

$$LC\frac{d^2v}{dt^2} + RC\frac{dv}{dt} + v = v_s$$

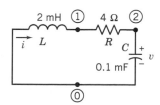

Figure 10-21 The *RLC* series circuit.

Let $s = d/dt$ and substitute the values of the parameters to obtain the unforced equation

$$(2 \times 10^{-7}s^2 + 4 \times 10^{-4}s + 1)v = 0$$

or

$$s^2 + 2000s + 5 \times 10^6 = 0$$

The roots of the characteristic equation are complex conjugates, and the circuit response is underdamped and oscillatory. We wish to determine the circuit response.

The PSpice program is shown in Figure 10-22, incorporating the initial conditions. The transient response, $v(t)$, is obtained for 20 different values of time (Figure 10-23a) by using the .PLOT statement. The .PROBE statement is used to obtain a high-resolution graph of $v(t)$, as shown in Figure 10-23b. Note that the voltage $v(t)$ attains a negative value and eventually decays to zero after 4 ms.

```
**      RLC  SERIES  CIRCUIT

R        1       2        4
L        0       1        2M   IC=O
C        2       0        0.1M IC=10
 .TRAN           0.2M     4M   UIC
 .PLOT           TRAN     V(2)
 .PROBE
 .END
```

Figure 10-22 The PSpice program for the *RLC* series circuit.

As an example of the solution of an *RLC* circuit using PSpice with a forcing function and initial conditions, reconsider Example 10-9. Reviewing Figure 10-18, we find that $v(0) = 10$ V and $i(0) = 0$. Figure 10-18 is redrawn for $t \geq 0$ in a PSpice format as shown in Figure 10-24. The input signal is $i_s = 2e^{-3t}$ A. The EXP input is used in the PSpice program shown in Figure 10-25. We plot 25 values of the capacitor voltage as shown in Figure 10-26a, and the high-resolution plot is shown in Figure 10-26b.

```
   TIME        V(2)
(*)----------   -5.0000E+00    0.0000E+00    5.0000E+00    1.0000E+01    1.5000E+01
                      - - - - - - - - - - - - - - - - - - - - - - - - - - - - - -
 0.000E+00   1.000E+01  ·              ·              ·              *           ·
 2.000E-04   9.127E+00  ·              ·              ·           *  ·           ·
 4.000E-04   7.085E+00  ·              ·              ·    *         ·           ·
 6.000E-04   4.570E+00  ·              ·              * ·            ·           ·
 8.000E-04   2.139E+00  ·              ·         *    ·              ·           ·
 1.000E-03   1.577E-01  ·              *              ·              ·           ·
 1.200E-03  -1.198E+00  ·         *    ·              ·              ·           ·
 1.400E-03  -1.918E+00  ·       *      ·              ·              ·           ·
 1.600E-03  -2.090E+00  ·       *      ·              ·              ·           ·
 1.800E-03  -1.870E+00  ·        *     ·              ·              ·           ·
 2.000E-03  -1.418E+00  ·          *   ·              ·              ·           ·
 2.200E-03  -8.859E-01  ·           *  ·              ·              ·           ·
 2.400E-03  -3.846E+01  ·            *·               ·              ·           ·
 2.600E-03   1.490E-02  ·            *                ·              ·           ·
 2.800E-03   2.807E-01  ·            ·*               ·              ·           ·
 3.000E-03   4.145E-01  ·            ·*               ·              ·           ·
 3.200E-03   4.370E-01  ·            ·*               ·              ·           ·
 3.400E-03   3.819E-01  ·            ·*               ·              ·           ·
 3.600E-03   2.829E-01  ·            ·*               ·              ·           ·
 3.800E-03   1.707E-01  ·            *                ·              ·           ·
 4.000E-03   6.721E-02  ·            *                ·              ·           ·
                      - - - - - - - - - - - - - - - - - - - - - - - - - - - - - -
```

(a)

Figure 10-23 The (a) PLOT and (b) PROBE outputs for the unforced *RLC* series circuit.

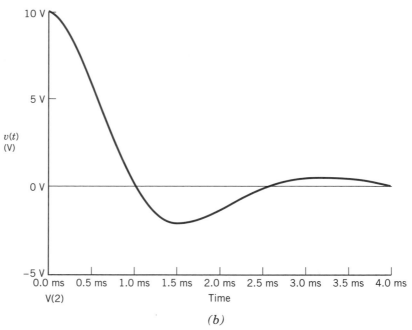

(b)

Figure 10-23 *Continued.*

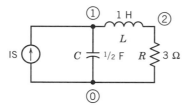

Figure 10-24 The RLC circuit of Example 10-9 redrawn for PSpice for $t \geq 0$ with a current source input.

```
**       FORCED RLC CIRCUIT
R   2    0   3
C   1    0   0.5    IC=10
L   1    2   1      IC=0
IS 0    1 EXP  (0   2 0   1E-6   0   .333)
.TRAN 0.2   5   UIC
.PLOT  TRAN V(1)
.PROBE
.END
```

Figure 10-25 The PSpice program for the *RLC* circuit with an exponential input current waveform.

```
     TIME        V(1)
  (*)----------     0.0000E+00    5.0000E+00    1.0000E+01    1.5000E+01    2.0000E+01
                  - - - - - - - - - - - - - - - - - - - - - - - - - - -
  0.000E+00   1.000E+01  ·                          *
  2.000E-01   1.046E+01  ·                         ·*
  4.000E-01   1.024E+01  ·                         ·*
  6.000E-01   9.528E+00  ·                       * ·
  8.000E-01   8.566E+00  ·                    *
  1.000E+00   7.525E+00  ·                 *
  1.200E+00   6.503E+00  ·              *
  1.400E+00   5.554E+00  ·            ·*
  1.600E+00   4.700E+00  ·          *·
  1.800E+00   3.951E+00  ·        *
  2.000E+00   3.303E+00  ·      *
  2.200E+00   2.750E+00  ·     *
  2.400E+00   2.282E+00  ·    *
  2.600E+00   1.888E+00  ·   *
  2.800E+00   1.560E+00  ·  *
  3.000E+00   1.286E+00  ·  *
  3.200E+00   1.059E+00  · *
  3.400E+00   8.708E-01  · *
  3.600E+00   7.156E-01  · *
  3.800E+00   5.876E-01  · *
  4.000E+00   4.822E-01  ·*
  4.200E+00   3.955E-01  ·*
  4.400E+00   3.243E-01  ·*
  4.600E+00   2.659E-01  ·*
  4.800E+00   2.179E-01  ·*
  5.000E+00   1.783E-01  *
                  - - - - - - - - - - - - - - - - - - - - - - - - - - -
```

(a)

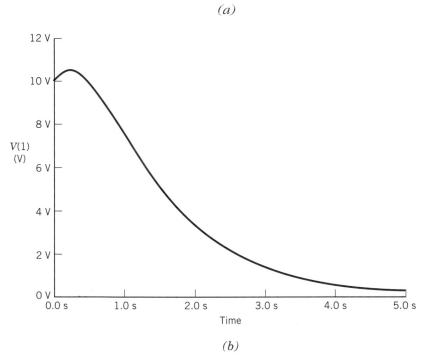

(b)

Figure 10-26 The response of the *RLC* circuit to a current source input *(a)* PLOT and *(b)* PROBE outputs.

10-13 ∥ VERIFICATION EXAMPLE

An RLC circuit is shown in Figure 10 V.1. An engineer's report states that

$$v(t) = 16e^{-500t} - 1e^{-8000t} \text{ V}, \qquad t \geq 0$$

Verify this result.

Solution

First, we check the final value of $v(t)$ as $t \to \infty$, which is $v(t) = 0$ at $t = \infty$. This checks since the circuit is unforced and we expect the steady state to be $v(\infty) = 0$. The initial value $v(0) = 15$ V due to the source connected to C prior to the switch opening. Checking at $t = 0$ we have

$$v(0) = 16 - 1 = 15$$

which checks.

The solution for $v(t)$ must satisfy the differential equations. Using the state variable formulation of Section 10-10, we obtain the two differential equations

$$L\frac{di}{dt} + Ri = v$$

and

$$C\frac{dv}{dt} + i = 0$$

Using the values of R, L, and C and substituting v we find that the equations are identically satisfied. Thus, the answer provided by the engineer is correct.

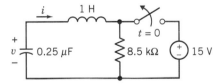

Figure 10 V.1

10-14 DESIGN CHALLENGE SOLUTION

AUTO AIRBAG IGNITER

Problem

Airbags are now widely used for driver protection in automobiles. A pendulum is used to switch a charged capacitor to the inflation ignition device, as shown in Figure 10D-1. The automobile airbag is inflated by an explosive device that is ignited by the energy absorbed by the resistive device represented by R. In order to inflate, it is required that the energy dissipated in R is at least 1 J. It is required that the ignition device trigger within 0.1 s. Select the L and C that meet the specifications.

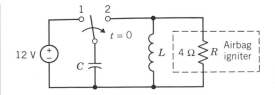

Figure D 10-1 An automobile airbag ignition device.

Define the Situation

1 The switch is changed from position 1 to position 2 at $t = 0$.
2 The switch was connected to position 1 for a long time.
3 A parallel RLC circuit occurs for $t \geq 0$.

The Goal

Select L and C so that the energy stored in the capacitor is quickly delivered to the resistive device R.

Generate a Plan

1 Select L and C so that an underdamped response is obtained with a period of less than or equal to 0.4 s ($T \leq 0.4$ s).
2 Solve for $v(t)$ and $i(t)$ for the resistor R.

Take Action Using the Plan

We assume that the initial capacitor voltage is $v(0) = 12$ V and $i_L(0) = 0$ since the switch is in position 1 for a long time prior to $t = 0$. The response of the parallel RLC circuit for an underdamped response is of the form (Eq. 10-55)

$$v(t) = e^{-\alpha t}(B_1 \cos \omega_d t + B_2 \sin \omega_d t) \qquad (10D\text{-}1)$$

as discussed in Section 10-6. This natural response is obtained when $\alpha^2 < \omega_0^2$ or $L < 4R^2 C$. We choose an underdamped response for our design but recognize that an overdamped or critically damped response may satisfy the circuit's design objectives. Furthermore, we recognize that the parameter values selected below represent only one acceptable solution.

Since we want a rapid response, we will select $\alpha = 2$ (a time constant of $1/2$ s) where $\alpha = 1/2RC$. Therefore, we have

$$C = \frac{1}{2R\alpha} = \frac{1}{16} \text{ F}$$

Recall that $\omega_0^2 = 1/LC$ and it is required that $\alpha^2 < \omega_0^2$. Since we want a rapid response, we select the natural frequency ω_0 so that (recall $T \cong 0.4$ s)

$$\omega_0 = \frac{2\pi}{T} = \frac{2\pi}{0.4} = 5\pi$$

Therefore, we obtain

$$L = \frac{1}{\omega_0^2 C} = \frac{1}{25\pi^2(1/16)} = 0.065 \text{ H}$$

Thus, we will use $C = 1/16$ F and $L = 65$ mH. We then find that $\omega_d = 15.58$ rad/s and, using Eq. 10-55, we have

$$v(t) = e^{-2t}(B_1 \cos \omega_d t + B_2 \sin \omega_d t) \qquad (10D\text{-}2)$$

Then $B_1 = v(0) = 12$ and

$$\omega_d B_2 = \alpha B_1 - \frac{B_1}{RC}$$
$$= (2 - 4)12$$
$$= -24$$

Therefore, $B_2 = -24/15.58 = -1.54$. Since $B_2 \ll B_1$, we can approximate Eq. 10D-2 as

$$v(t) \cong 12e^{-2t} \cos \omega_d t \quad V$$

The power is then

$$p = \frac{v^2}{R} = 36e^{-4t} \cos^2 \omega_d t \quad W$$

The actual voltage and current for the resistor R are shown in Figure 10D-2 for the first 100 ms. If we sketch the product of v and i for the first 100 ms, we obtain a linear approximation declining from 36 W at $t = 0$ to 0 W at $t = 95$ ms. The energy absorbed by the resistor over the first 100 ms is then

$$w \cong \tfrac{1}{2}(36)(0.1 \text{ s})$$
$$= 1.8 \text{ J}$$

Therefore, the airbag will trigger in less than 0.1 s and our objective is achieved.

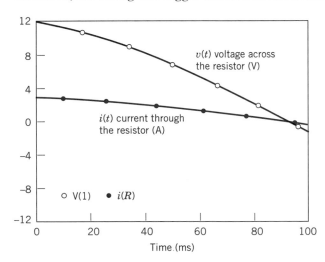

Figure 10D-2 The response of the RLC circuit.

SUMMARY

Many circuits contain two or more energy storage elements, and we need to obtain the complete solution for the variables of these circuits. Thus, we require both the natural and the forced responses of a circuit that contains inductors and capacitors. The first step in this process is to obtain the differential equation that describes the circuit in terms of the desired variable x.

We can obtain the desired differential equation by the substitution method or the operator method. Both of these methods use the node voltage or mesh current equations and enable us to proceed to obtain the differential equation in terms of the one variable.

The next step in the process is to determine the characteristic equation and the roots of this equation. The roots of the characteristic equation dictate the natural response x_n. We then proceed to find the forced response x_f that will satisfy the differential equation. Finally, the complete response is obtained as $x = x_n + x_f$, and the arbitrary constants are evaluated by utilizing the initial conditions.

Another procedure for obtaining the desired variable $x(t)$ of a circuit is the state variable method, which identifies the variables as the capacitor voltages and the inductor currents. We then obtain a set of first-order differential equations and use operators and Cramer's rule to solve for the desired variable x. We then have the desired differential equation in terms of x and proceed to determine $x(t)$. The state variable method is attractive because we usually know the initial conditions for the energy storage variables.

TERMS AND CONCEPTS

Characteristic Equation   Equation obtained by setting $s^n = d^n/dt^n$ in the unforced differential equation; the second-order characteristic equation is $s^2 + a_1 s + a_0 = 0$. Equation for determining the character of the natural response of a circuit.

Characteristic Roots   Roots of the characteristic equation. Also called natural frequencies.

Critically Damped   Condition that exists when the characteristic roots are equal.

Damped Resonant Frequency $\omega_d = \sqrt{\omega_0^2 - \alpha^2}$ of the second-order circuit with the characteristic equation $s^2 + 2\alpha s + \omega_0^2 = 0$. The oscillation frequency of a damped sinusoid.

Electric Power System   System that generates, transmits, and distributes electric power to the end user.

Natural Frequencies   Roots of the characteristic equation, s_1 and s_2.

Natural Response   Unforced response x where $x = A_1 e^{s_1 t} + A_2 e^{s_2 t}$ when s_1 and s_2 are distinct. Response of an RL, RC, or RLC circuit that depends only on the nature of the circuit and not on external sources.

Operator Method   Method of obtaining the second-order differential equation that uses the operator $s^n = d^n/dt^n$ and Cramer's rule to obtain the complete solution.

Overdamped   Condition that exists when the roots of the characteristic equation are real and distinct.

Parallel *RLC* Circuit   Circuit with three parallel branches where the branches contain a resistor, a capacitor, and an inductor, respectively.

Period of Oscillation   Time for an underdamped response to proceed through one cycle of oscillation, denoted as T_d. T_d is the period of oscillation of the periodic functions.

Resonant Frequency $\omega_0 = 1/\sqrt{LC}$ for the parallel *RLC* circuit.

Second-Order Circuit   Circuit with a homogeneous differential equation containing a second-degree term due to the presence of two independent energy storage elements.

State Variable Method   Method of identifying the state variables and obtaining a set of first-order differential equations, then proceeding to obtain the second-order differential equation in terms of one of the state variables, and then solving for the complete solution for one or more of the state variables.

State Variables   Set of variables describing the energy of the storage elements of a circuit; the capacitor voltages and the inductor currents.

Substitution Method   Method of obtaining the second-order, or higher order, differential equation in terms of a selected variable x that substitutes equations for the desired variable in order to eliminate other variables.

System Interconnection of electrical elements and circuits to achieve a desired objective.

Underdamped Condition that exists when two roots of the characteristic equation are complex conjugates.

PROBLEMS

Section 10-3 Differential Equation for Circuits with Two Energy Storage Elements

P 10.3-1 Find the differential equation for the circuit shown in Figure P 10.3-1 using the substitution method.

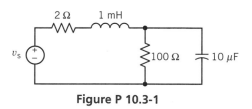

Figure P 10.3-1

P 10.3-2 Find the differential equation for the circuit shown in Figure P 10.3-2 using the operator method.

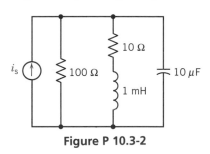

Figure P 10.3-2

P 10.3-3 Find the differential equation for $i_L(t)$ for $t > 0$ for the circuit of Figure P 10.3-3.

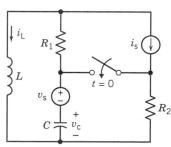

Figure P 10.3-3

Section 10-4 Solution of the Second-Order Differential Equation—The Natural Response

P 10.4-1 Find the characteristic equation and its roots for the circuit of Figure P 10.3-2.

P 10.4-2 Find the characteristic equation and its roots for the circuit of Figure P 10.4-2.
Answer: $s^2 + 400s + 3 \times 10^4 = 0$
roots: $s = -300, -100$

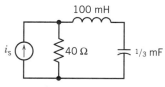

Figure P 10.4-2

P 10.4-3 Find the characteristic equation and its roots for circuit shown in Figure P 10.4-3.

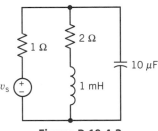

Figure P 10.4-3

Section 10-5 Natural Response of the Unforced Parallel *RLC* Circuit

P 10.5-1 For the circuit of Figure P 10.5-1 determine the characteristic equation and its roots and the initial conditions $v_c(0)$ and $i_L(0)$ when $v_S = 16$ V, $i_s = 2$ mA, $R_1 = 6$ kΩ, $R_2 = 6$ kΩ, $L = 600$ mH, and $C = 1/6$ μF.
Answer: $s^2 + 11 \times 10^3 s + 2 \times 10^7 = 0$
$v_c(0) = -16$ V, $i_L(0) = -2$ mA

P 10.5-2 Determine $v(t)$ for the circuit of Figure P 10.5-2 when $L = 1$ H and $v_s = 0$ for $t \geq 0$. The initial conditions are $v(0) = 6$ V and $dv/dt(0) = -3000$ V/s.

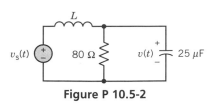

Figure P 10.5-2

P 10.5-3 An *RLC* circuit is shown in Figure P 10.5-3 where $v(0) = 2$ V. The switch has been open for a long time before closing at $t = 0$. Determine and plot $v(t)$.

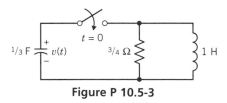

Figure P 10.5-3

P 10.5-4 Determine $i_1(t)$ and $i_2(t)$ for the circuit of Figure P 10.5-4 when $i_1(0) = i_2(0) = 11$ A.

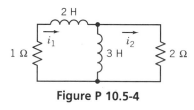

Figure P 10.5-4

P 10.5-5 Find $v_1(t)$ for $t > 0$ using the method of substitution for the circuit of Figure P 10.5-5. Assume the circuit is in steady state at $t = 0^-$.
Answer: $v_1(t) = 18e^{-4t} - 12e^{-6t}$ V

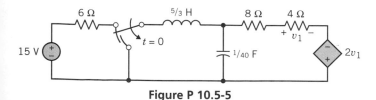

Figure P 10.5-5

Section 10-6 Natural Response of the Critically Damped Unforced Parallel *RLC* Circuit

P 10.6-1 Find $v_c(t)$ for $t > 0$ for the circuit shown in Figure P 10.6-1.
Answer: $v_c(t) = (3 + 6000t)e^{-2000t}$ V

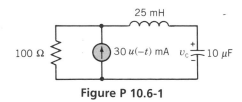

Figure P 10.6-1

P 10.6-2 Find $v_c(t)$ for $t > 0$ for the circuit of Figure P 10.6-2. Assume steady-state conditions exist at $t = 0^-$.
Answer: $v_c(t) = -8te^{-2t}$ V

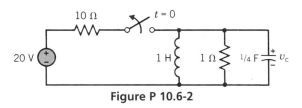

Figure P 10.6-2

P 10.6-3 Police often use stun guns to incapacitate potentially dangerous felons. The hand-held device provides a series of high-voltage, low-current pulses. The power of the pulses is far below lethal levels, but it is enough to cause muscles to contract and put the person out of action. The device provides a pulse of up to 50,000 V, and a current of 1 mA flows through an arc. A model of the circuit for one period is shown in Figure P 10.6-3. Find $v(t)$ for $0 < t < 1$ ms. The resistor R represents the spark gap. Select C so that the response is critically damped.

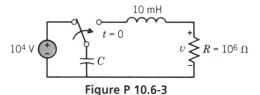

Figure P 10.6-3

P 10.6-4 Reconsider Problem 10.5-2 when $L = 640$ mH and the other parameters and conditions remain the same.

P 10.6-5 An automobile ignition uses an electromagnetic trigger. The *RLC* trigger circuit shown in Figure P 10.6-5 has a step input of 6 V, and $v(0) = 2$ V and $i(0) = 0$. The resistance R must be selected from $2\ \Omega < R < 7\ \Omega$ so that the current $i(t)$ exceeds 0.6 A for greater than 0.5 s in order to activate the trigger. A critically damped response $i(t)$ is required to avoid oscillations in the trigger current. Select R and determine and plot $i(t)$.

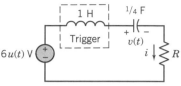

Figure P 10.6-5

Section 10-7   Natural Response of an Underdamped Unforced Parallel *RLC* Circuit

P 10.7-1   A communication system from a space station uses short pulses to control a robot operating in space. The transmitter circuit is modeled in Figure P 10.7-1. Find the output voltage $v_c(t)$ for $t > 0$. Assume steady-state conditions at $t = 0^-$.

Answer: $v_c(t) = e^{-400t}[3\cos 300t + 4\sin 300t]$ V

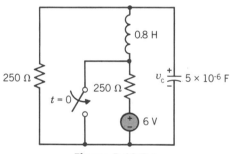

Figure P 10.7-1

P 10.7-2   The switch in Figure P 10.7-2 has been connected to position 1 for a long time before moving instantaneously to position 2 at $t = 0$. Determine $v(t)$.

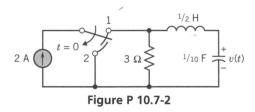

Figure P 10.7-2

P 10.7-3   The switch of the circuit shown in Figure P 10.7-3 is opened at $t = 0$. Determine and plot $v(t)$ when $C = 1/4$ F. Assume steady state at $t = 0^-$.

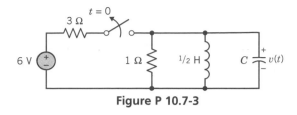

Figure P 10.7-3

P 10.7-4   The initial voltage of the capacitor, $v(0)$, shown in Figure P 10.7-4 is equal to 10 V. Determine the peak positive voltage $v(t_1)$ and the time of occurrence, t_1, when $i(0) = 90$ mA.

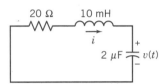

Figure P 10.7-4

P 10.7-5   A 240-watt power supply circuit is shown in Figure P 10.7-5*a*. This circuit employs a large inductor and a large capacitor. The model of the circuit is shown in Figure P 10.7-5*b*. Find $i_L(t)$ for $t > 0$ for the circuit of Figure P 10.7-5*b*. Assume steady-state conditions exist at $t = 0^-$.

Answer: $i_L(t) = e^{-2t}(-4\cos t + 2\sin t)$ A

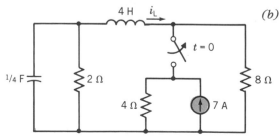

Figure P 10.7-5   (*a*) A 240-W power supply. Courtesy of Kepco, Inc. (*b*) Model of the power supply circuit.

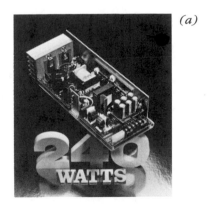

P 10.7-6 The natural response of a parallel *RLC* circuit is measured and plotted as shown in Figure P 10.7-6.

Using this chart, determine an expression for $v(t)$ for the circuit of Figure 10-9.

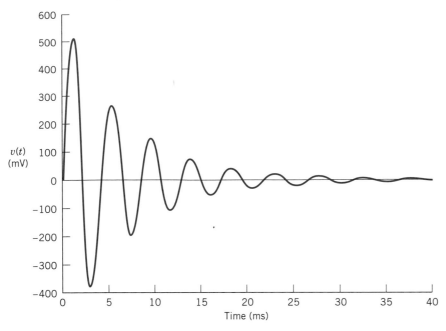

Figure P 10.7-6 The natural response of a parallel *RLC* circuit.

P 10.7-7 The photovoltaic cells of the proposed space station shown in Figure P 10.7-7*a* provide the voltage $v(t)$ of the circuit shown in Figure P 10.7-7*b*. The space station passes behind the shadow of Earth (at $t = 0$) with $v(0) = 2$ V and $i(0) = 1/10$ A. Determine and sketch $v(t)$ for $t > 0$.

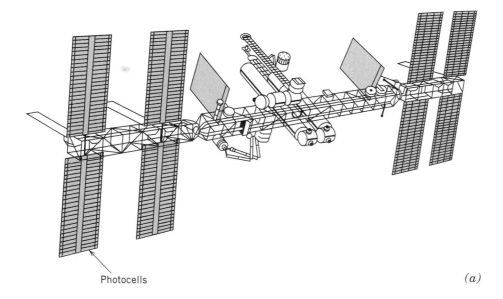

Photocells

(a)

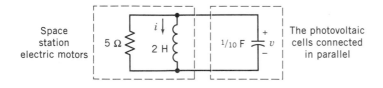

Figure P 10.7-7 *(a)* Photocells on space station. *(b)* Circuit with photocells.

(b)

Section 10-8   Forced Response of an *RLC* Circuit

P 10.8-1   Determine the forced response for the inductor current i_f when (a) $i_s = 1$ A, (b) $i_s = 0.5\ t$ A, and (c) $i_s = 2e^{-250t}$ A for the circuit of Figure P 10.8-1.

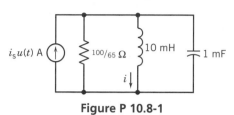

Figure P 10.8-1

P 10.8-2   Determine the forced response for the capacitor voltage, v_f, for the circuit of Figure P 10.8-2 when (a) $v_s = 2$ V, (b) $v_s = 0.2\ t$ V, and (c) $v_s = 1e^{-30t}$ V.

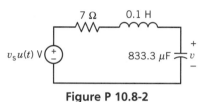

Figure P 10.8-2

Section 10-9   Complete Response of an *RLC* Circuit

P 10.9-1   Find $i(t)$ for $t > 0$ using the method of substitution for the circuit of Figure P 10.9-1.

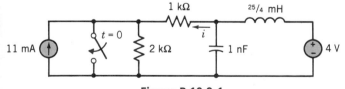

Figure P 10.9-1

P 10.9-2   With $v_c(0^-) = -10$ V and $i_L(0^-) = 2/3$ A, find $v_1(t)$ for $t > 0$ for the circuit of Figure P 10.9-2. Use the method of substitution.
Answer: $v_1(t) = -17e^{-t} + 7e^{-3t}$ V

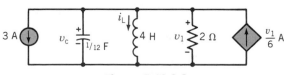

Figure P 10.9-2

P 10.9-3   Find $v(t)$ for $t > 0$ for the circuit of Figure P 10.9-3. Assume that the circuit is in steady state at $t = 0^-$. Use the method of substitution.

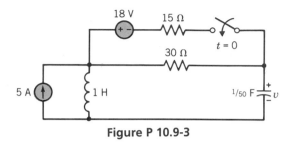

Figure P 10.9-3

P 10.9-4   Find $v(t)$ for $t > 0$ for the circuit shown in Figure P 10.9-4. Use the method of operators. Assume steady-state conditions exist at $t = 0^-$.

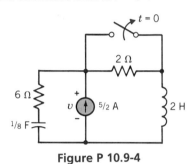

Figure P 10.9-4

P 10.9-5   Find $v(t)$ for $t > 0$ for the circuit shown in Figure P 10.9-5. Assume steady-state conditions exist at $t = 0^-$.

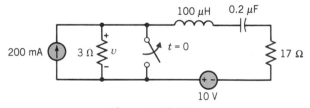

Figure P 10.9-5

P 10.9-6   Find $v_c(t)$ for $t > 0$ using the method of operators for the circuit of Figure P 10.9-6. Assume steady-state conditions at $t = 0^-$.

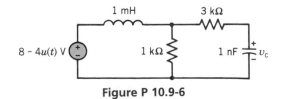

Figure P 10.9-6

P 10.9-7 Find $i_1(t)$ and $i_2(t)$ for $t > 0$ for the circuit shown in Figure P 10.9-7. Assume steady-state conditions at $t = 0^-$.

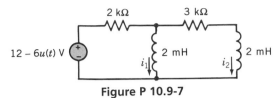

Figure P 10.9-7

P 10.9-8 Find $v_1(t)$ for $t > 0$ for the circuit shown in Figure P 10.9-8. Assume steady-state conditions at $t = 0^-$.

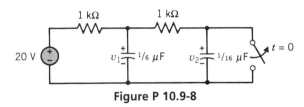

Figure P 10.9-8

P 10.9-9 Find $v(t)$ for $t > 0$ for the circuit shown in Figure P 10.9-9 when $v(0) = 1$ V and $i_L(0) = 0$.
Answer: $v = 25e^{-3t} - \frac{1}{17}[429e^{-4t} - 21\cos t + 33\sin t]$ V

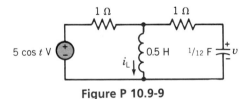

Figure P 10.9-9

P 10.9-10 Find $i(t)$ for $t > 0$ by the method of operators for the circuit of Figure P 10.9-10. Assume steady-state conditions at $t = 0^-$.

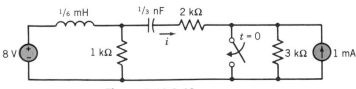

Figure P 10.9-10

P 10.9-11 Find $v(t)$ for $t > 0$ for the circuit of Figure P 10.9-11 using the method of operators.
Answer: $v(t) = [-16e^{-t} + 16e^{-3t} + 8]u(t)$
$\qquad + [16e^{-(t-2)} - 16e^{-3(t-2)} - 8]u(t - 2)$ V

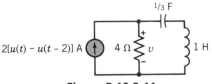

Figure P 10.9-11

P 10.9-12 An experimental space station power supply system is modeled by the circuit shown in Figure P 10.9-12. Find $v_c(t)$ for $t > 0$. Assume steady-state conditions at $t = 0^-$.

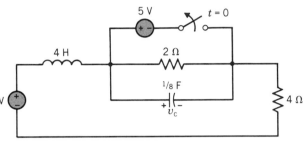

Figure P 10.9-12

P 10.9-13 The switch in the circuit shown in Figure P 10.9-13 is open a long time before $t = 0$. Find $i(t)$ for $t > 0$ by the method of operators.

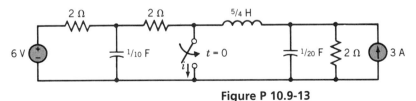

Figure P 10.9-13

P 10.9-14 Find $v_c(t)$ for $t > 0$ in the circuit of Figure P 10.9-14 when (a) $C = 1/18$ F, (b) $C = 1/10$ F, and (c) $C = 1/20$ F.

Answer: (a) $v_c(t) = 8e^{-3t} + 24te^{-3t} - 8$ V
(b) $v_c(t) = 10e^{-t} - 2e^{-5t} - 8$ V
(c) $v_c(t) = e^{-3t}(8 \cos t + 24 \sin t) - 8$ V

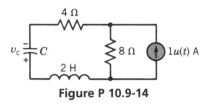

Figure P 10.9-14

P 10.9-15 For the circuit of Figure P 10.9-15 find $i(t)$ for $t > 0$. Assume steady state at $t = 0^-$.
Answer: $i(t) = 3e^{-5 \times 10^9 t} - e^{-2000t} - 6$ mA

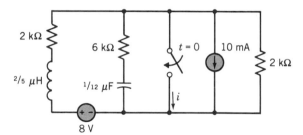

Figure P 10.9-15

P 10.9-16 An *RLC* circuit is shown in Figure P 10.9-16.
(a) Find v and i for $t > 0$.
(b) Plot v and i.

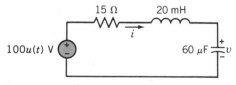

Figure P 10.9-16

P 10.9-17 In Figure P 10.9-17, determine the in tor current $i(t)$ when $i_s = 5u(t)$ A. Assume that $i($ $v_c(0) = 0$.

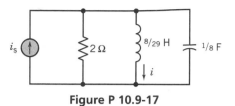

Figure P 10.9-17

P 10.9-18 Find the complete response $v(t)$ for $t > 0$ for the circuit of Figure P 10.9-18 when $v_s = 8e^{-4t}u(t)$. Assume that the circuit is in steady state at $t = 0^-$.

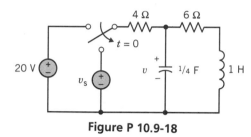

Figure P 10.9-18

P 10.9-19 Determine $v(t)$ for the circuit shown in Figure P 10.9-19 when $v_1(0) = v(0) = 0$ and $v_s = 5 \cos 2000t$ V. Assume an ideal op amp, $C_1 = 1$ μF, and $C_2 = 1/8$ μF.

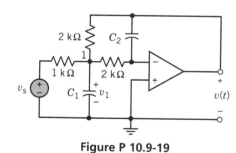

Figure P 10.9-19

P 10.9-20 Determine $v(t)$ for the circuit shown in Figure P 10.9-20 when

$$v_s = \begin{cases} 0 & t < 0 \\ t & t \geq 0 \end{cases}$$

Determine the difference between $v_s(t)$ and $v(t)$ at $t = 20$ s.

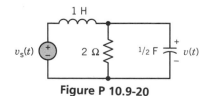

Figure P 10.9-20

P 10.9-21 A model of a circuit with a MOSFET transistor amplifier is shown in Figure P 10.9-21. The source voltage is $v_s = 10 \sin \omega t$ V. Find $v(t)$ for $t > 0$ when $\omega = 10^3$ rad/s and $C = 10 \mu$F. The initial capacitor voltages are zero at $t = 0^-$.

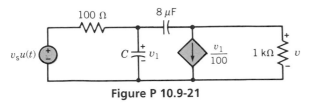

Figure P 10.9-21

P 10.9-22 A circuit is shown in Figure P 10.9-22 where switch 1 is closed at $t = 0$ and switch 2 is opened at $t = 40$ ms.

(a) Find $i(t)$ and $v(t)$ for $0 < t < 100$ ms.
(b) Plot v and i.

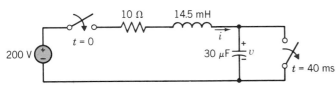

Figure P 10.9-22

P 10.9-23 Find $i_L(t) > 0$ for the circuit of Figure P 10.9-23. Hint: Use the same method of operators as you used with a second-order circuit. Assume steady-state conditions at $t = 0^-$.
Answer: $i_L(t) = 8e^{-t} - 10e^{-2t} - 12te^{-2t} + 12$ A

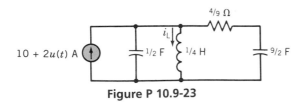

Figure P 10.9-23

P 10.9-24 Consider the circuit shown in Figure P 10.9-24.

(a) Derive the differential equation that represents $i_c(t)$ for $t > 0$.
(b) Assume that $C = 1$ F, and find the value of the inductor L such that the natural response of $i_c(t)$ has a double root at $-1/2$.
(c) Using the value of L found in (b), find the total response of $i_c(t)$ for $t > 0$.
(d) Derive the expressions for $v_c(t)$, $i_L(t)$ for $t > 0$ using previous results.
(e) A circuit is defined to be stable if the total response is finite as time approaches infinity. Is this circuit stable? If so, find the steady-state value $v_c(\infty)$.
(f) Using your results from (d) and (e), roughly sketch $v_c(t)$ for $t > 0$.

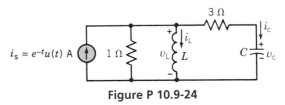

Figure P 10.9-24

P 10.9-25 For the circuit shown in Figure P 10.9-25, find $v_0(t)$ for $t > 0$ if $v_s = 10u(t)$ volts, $i_L(0) = 0$, and $v_c(0) = 0$ V.
Answer: $v_0 = 1200te^{-4000t}$ V

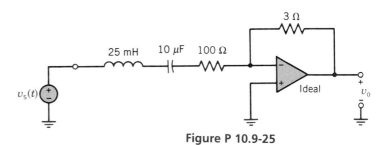

Figure P 10.9-25

P 10.9-26 Railroads widely use automatic identification of railcars. When a train passes a tracking station, a wheel detector activates a radio-frequency module. The module's antenna, as shown in Figure P 10.9-26a, transmits and receives a signal that bounces off a transponder on the locomotive. A trackside processor turns the received signal into useful information consisting of the train's location, speed, and direction of travel. The railroad uses this information to schedule locomotives, trains, crews, and equipment more efficiently.

One proposed transponder circuit is shown in Figure P 10.9-26b with a large transponder coil of $L = 5$ H. Determine $i(t)$ and $v(t)$. The received signal is $i_s = 9 + 3e^{-2t}u(t)$ A.

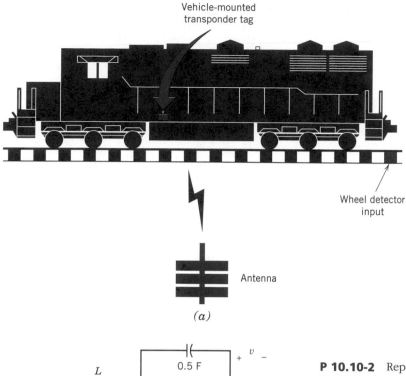

Figure P 10.9-26 *(a)* Railroad identification system. *(b)* Transponder circuit.

Section 10-10 State Variable Approach to Circuit Analysis

P 10.10-1   Find $v(t)$ for $t > 0$ using the state variable method of Section 10-10 when $C = 1/5$ F in the circuit of Figure P 10.10-1. Sketch the response for $v(t)$ for $0 < t < 10$ s.

P 10.10-2   Repeat Problem 10.10-1 when $C = 1/10$ F. Sketch the response for $v(t)$ for $0 < t < 3$ s.
Answer: $v(t) = e^{-3t}(-24 \cos t - 32 \sin t) + 24$ V

P 10.10-3   A circuit is shown in Figure P 10.10-3*a* where $R_2 = -2\ \Omega$ and R_1 is to be determined. Find the value of R_1 that will cause $i(t)$ to be a constant amplitude oscillation, as shown in Figure P 10.10-3*b*. Let $C = 1$ F and $L = 1$ H.
Answer: $R_1 = 1/2\ \Omega$

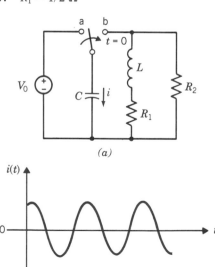

Figure P 10.10-3

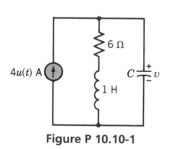

Figure P 10.10-1

P 10.10-4 Determine the current $i(t)$ and the voltage $v(t)$ for the circuit of Figure P 10.10-4.

Answer: $i(t) = (3.08e^{-2.57t} - 0.08e^{-97.4t} - 6)$ A

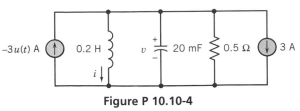

Figure P 10.10-4

P 10.10-5 Determine the two state variable equations for the circuit of Figure P 10.10-5.

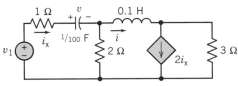

Figure P 10.10-5

P 10.10-6 For the circuit of Figure P 10.10-6, find:

(a) $i(t)$ for $0 < t \le 0.05$ s

(b) $i(t)$ for $t \ge 0.05$ s.

Assume steady-state conditions at $t = 0^-$.

Answer:

(a) $i(t) = e^{-t}(3 \cos 3t - \sin 3t)$ A

(b) $i(t) = -2.26e^{-2x} + 4.94e^{-5x}$ A, $x = t - 0.05$

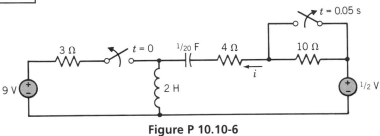

Figure P 10.10-6

P 10.10-7 Clean-air laws are pushing the auto industry toward the development of electric cars. One proposed vehicle using an ac motor is shown in Figure P 10.10-7a. The motor-controller circuit is shown in Figure P 10.10-7b with $L = 100$ mH and $C = 10$ mF. Using the state equation approach, determine $i(t)$ and $v(t)$ where $i(t)$ is the motor-control current. The initial conditions are $v(0) = 10$ V and $i(0) = 0$.

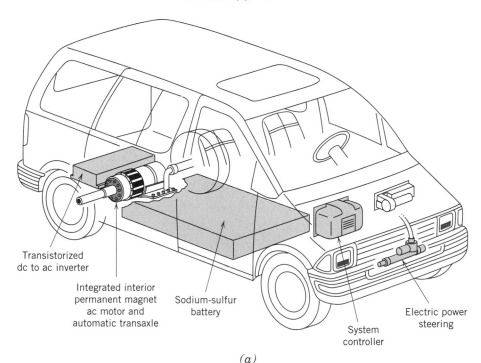

(a)

Figure P 10.10-7a Electric vehicle.

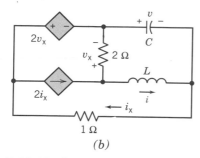

P 10.10-7b Motor controller circuit.

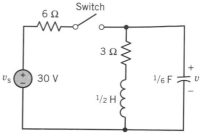

Figure P 10.10-9 Newton circuit model.

Section 10-11 Roots in the Complex Plane

P 10.11-1 For the circuit of Figure P 10.9-7 determine the roots of the characteristic equation and plot the roots on the *s*-plane.

P 10.11-2 For the circuit of Figure P 10.7-1 determine the roots of the characteristic equation and plot the roots on the *s*-plane.

P 10.11-3 For the circuit of Figure P 10.11-3 determine the roots of the characteristic equation and plot the roots on the *s*-plane.

P 10.10-8 The switch in the circuit of Figure P 10.10-8 has been closed for a long time during a laboratory experiment. One of the students, seeing that the source voltage is only 10 V, carelessly lets his hands roam across the inductor terminals just as (at $t = 0$) his lab partner opens the switch. What is the maximum voltage to which he will be subjected? At what time will this maximum voltage occur? Sketch the voltage, $v(t)$, versus time.

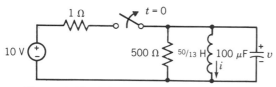

Figure P 10.10-8 A circuit in a laboratory.

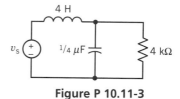

Figure P 10.11-3

P 10.11-4 An *RLC* circuit is shown in Figure P 10.11-4.
(a) Obtain the two node voltage equations using operators.
(b) Obtain the characteristic equation for the circuit.
(c) Show the location of the roots of the characteristic equation in the *s*-plane.
(d) Determine $v(t)$ for $t > 0$.

P 10.10-9 Studies of an artificial insect are being used to understand the nervous system of animals (Beer, 1991). A model neuron in the nervous system of the artificial insect is shown in Figure P 10.10-9. The input signal v_S is used to generate a series of pulses, called synapses. The switch generates a pulse by opening at $t = 0$ and closing at $t = 0.5$ s. Assume that the circuit is in steady state and that $v(0^-) = 10$ V. Determine the voltage $v(t)$ for $0 < t < 2$ s.

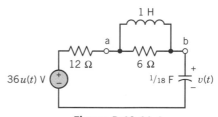

Figure P 10.11-4

PSpice PROBLEMS

SP 10-1 Determine and plot $v(t)$ for Example 10-4.

SP 10-2 A simple amplifier is represented by the circuit model shown in Figure SP 10-2. Determine $v(t)$ when $v_s = 5u(t)$.

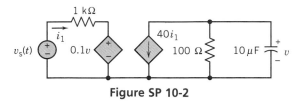

Figure SP 10-2

SP 10-3 Determine and plot $v(t)$ for Problem P 10.7-3.

SP 10-4 Determine and plot the natural response $v(t)$ of the circuit of Figure SP 10-4 when $i_s = 3e^{-6t}$ A (see Example 10-7).

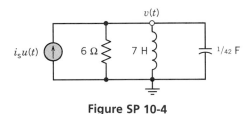

Figure SP 10-4

SP 10-5 Determine and plot $v(t)$ for Example 10-8. Determine $v(t)$ at $t = 0.5$ s, and compare the result with the analytical solution of Example 10-8.

SP 10-6 Determine and plot $i(t)$ for Problem P 10.9-17.

SP 10-7 Determine the voltage $v(t)$ for Problem P 10.7-4. Also determine the peak positive voltage $v(t_1)$ and t_1 using a PSpice program.

SP 10-8 Determine and plot the voltage $v(t)$ for Problem P 10.9-20.

SP 10-9 Obtain the solution for Problem P 10.6-1 using PSpice.

SP 10-10 Obtain the solution for Problem 10.10-9 using PSpice.

SP 10-11 Obtain $v(t)$ for the circuit shown in Figure P 10.11-4 for $0 \le t \le 0.1$ s.

SP 10-12 Determine and plot the capacitor voltage $v(t)$ for $0 < t < 300$ μs for the circuit shown in Figure

SP 10-12*a*. The sources are pulses as shown in Figure SP 10-12b and SP 10-12c.

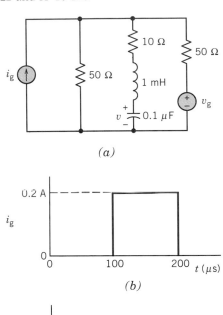

(*a*)

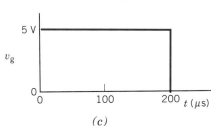

(*b*)

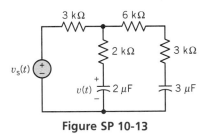

(*c*)

Figure SP 10-12 (*a*) Circuit, (*b*) current pulse, and (*c*) voltage pulse.

SP 10-13 Determine and plot $v(t)$ for the circuit of Figure SP 10-13 when $v_s(t) = 5u(t)$ V. Plot $v(t)$ for $0 < t < 0.25$ s.

Figure SP 10-13

VERIFICATION PROBLEMS

VP 10-1 A series *RLC* circuit is shown in Figure VP 10-1. A student report states that the natural response ($v_s = 0$) of this circuit will be critically damped when $R = 30 \ \Omega$. Determine if this is the correct value for critical damping.

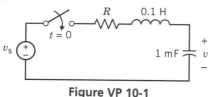

Figure VP 10-1

DESIGN PROBLEMS

DP 10-1

(a) Select the inductor *L* of the circuit shown in Figure DP 10-1 so that the response $v(t)$ is critically damped.

(b) Determine and plot the response $v(t)$ when $v(0) = 1$ V and $i(0) = 0.5$ A.

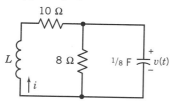

Figure DP 10-1

DP 10-2 An unforced *RLC* circuit is shown in Figure DP 10-2. The initial conditions are $i(0) = 16$ A and $v(0) = 8$ V. Select *R* so that a critically damped response, $v(t)$, is achieved. Determine $v(t)$ and sketch the response.

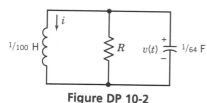

Figure DP 10-2

DP 10-3 The *RLC* circuit shown in Figure DP 10-3 may have a resistance *R* that is greater than 2 Ω and less than 10 Ω. Select *R* so that $v(t)$ attains a value of 1.0 V in less than 0.5 second but $|v(t)|$ never exceeds 1.70 V. The source is $v_s = 1 \ u(t)$ V.

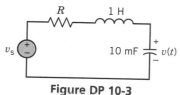

Figure DP 10-3

VP 10-2 A series *RLC* circuit is shown in Figure VP 10-1. Consider the case when $v_s = 10$ V, $v(0) = 0$, and $R = 25 \ \Omega$. One laboratory reports an overdamped response. Verify the form of the response and predict $v(t)$ when $t = 0.1$ s.

DP 10-4 A circuit designer has selected a parallel *RLC* circuit, as shown in Figure DP 10-4, and now needs to specify the values of the components. The switch has been open for a long time and is closed at $t = 0$. The initial voltage on the capacitor is $v_c = 10$ V at $t = 0$. The designer desires a critically damped voltage response, $v(t)$ for $t > 0$ and $v(t) \geq 1$ V for $t \leq 0.5$ s. Determine *R*, *L*, and *C* and plot $v(t)$ for $t > 0$.

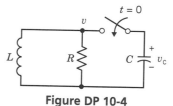

Figure DP 10-4

DP 10-5 Repeat Design Problem 10-4 when it is desired that $v(t)$ exhibit an underdamped response with a resonant frequency, ω_0, of 100 rad/s. The first negative peak should have $|v| < 6$ V. Select *R*, *L*, and *C* and sketch $v(t)$ for $t > 0$.

DP 10-6 A fluorescent light uses cathodes (coiled tungsten filaments coated with an electron-emitting substance) at each end that send current through mercury vapors sealed in the tube. Ultraviolet radiation is produced as electrons from the cathodes knock mercury electrons out of their natural orbits. Some of the displaced electrons settle back into orbit, throwing off the excess energy absorbed in the collision. Almost all of this energy is in the form of ultraviolet radiation. The ultraviolet rays, which are invisible, strike a phosphor coating on the inside of the tube. The rays energize the electrons in the phosphor atoms, and the atoms emit white light. The conversion of one kind of light into another is known as fluorescence.

One form of a fluorescent lamp is represented by the *RLC* circuit shown in Figure DP 10 6. Select *L* so that the current $i(t)$ reaches a maximum at approximately $t = 0.5$ s. Determine the maximum value of $i(t)$. Assume that the switch was in position 1 for a long time before switching to position 2 at $t = 0$.

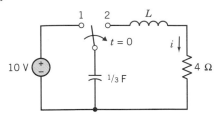

Figure DP 10-6 Fluorescent lamp circuit.

DP 10-7 A second-order circuit is given in Figure DP 10-7 with L and R unspecified. Select R and L so that the damped resonant frequency is $\omega_d = 1000$ rad/s and (a) $\alpha = 10$ rad/s, (b) $\alpha = 0$ rad/s, and (c) $\alpha = 0.1$ rad/s. Verify the design using PSpice.

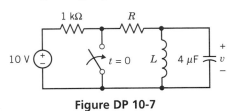

Figure DP 10-7

CHAPTER 11

SINUSOIDAL STEADY-STATE ANALYSIS

PREVIEW

The wide use of alternating current voltage sources requires us to study the analysis of circuits in the sinusoidal steady state. We are interested primarily, in this chapter, in the steady-state response of circuits to these sinusoidal sources. The need for the analysis of sinusoidal steady-state behavior led engineers to develop the concept of the phasor and impedance, which linearly relates the phasor current and phasor voltage of a circuit element. With this concept we can proceed to use all the circuit analysis methods we developed carefully in Chapter 5 for resistive circuits.

11-1 DESIGN CHALLENGE

OP AMP CIRCUIT

Figure 11D-1a shows two sinusoidal voltages, one labeled as input and the other labeled as output. We want to design a circuit that will transform the input sinusoid into the output sinusoid. Figure 11D-1b shows a candidate circuit. We must first determine if this circuit can do the job. Then, if it can, we will design the circuit, that is, specify the required values of R_1, R_2, and C.

In this chapter we will learn how to find the steady-state response of a linear circuit when the input is a sinusoid. We will see that phasors can be used to describe the way a circuit transforms a sinusoidal input into a sinusoidal output. With these tools we will be able to solve this problem, to which we will return at the end of this chapter.

490

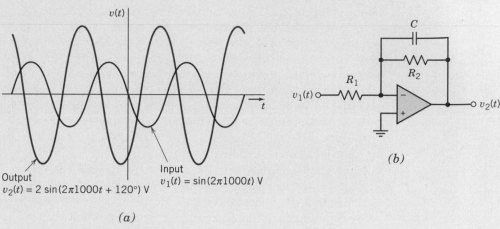

$$v_2(t) = 2\sin(2\pi 1000t + 120°)\ \text{V}$$

Output

Input
$$v_1(t) = \sin(2\pi 1000t)\ \text{V}$$

(a)

(b)

Figure 11D-1 *(a)* Input and output voltages. *(b)* Proposed circuit.

11-2 | ALTERNATING CURRENT BECOMES STANDARD

In the late 1800s the primary use for electricity was illumination, with Edison's direct current system operating in New York and New Jersey. As electricity became an important energy source, inventors such as Michael Faraday and Joseph Henry demonstrated small electric motors. However, it was clear that reliance on batteries had hindered the development of widespread use of electric power. Thus began the battle of alternating current (ac) versus direct current (dc).

Alternating current was generated in a power plant by a rotating machine called an alternator. Direct current was generated by a machine called a dynamo. Thus, it became a significant task for electrical engineers in the late 1800s to decide which system would be superior for future use.

Direct current seemed to offer some advantages. Batteries could be used as emergency backup when dynamos failed or could supply power during periods of low demand. Also, dynamos could be run in parallel to add supply as power demand increased. In 1890 the parallel use of alternators was thought to be difficult because of synchronization problems.

The Frankfurt Exhibition of 1891 was largely responsible for thrusting the case for ac transmission into the consciousness of European and American electrical engineers. Frankfurt had been ready to establish a municipal dc system, whereas Cologne had just become the first German city to adopt an ac system.

The battle of the systems was in full swing, and ac was demonstrated to be worthy of consideration at the Frankfurt Exhibition. When the exhibition opened, lights shone and motors ran on power generated by alternators at a dam on the Neckar River, 176 km away. The power was transmitted at 15 kV, and then a transformer was used to reduce the voltage to the 110 V used at the exhibition. A transformer is a device with two or more coils (windings) used to transfer ac power between circuits, usually with changed values of voltage and current. Transformers are considered in Chapter 12.

In March 1892 the Cataract Construction Company issued a request for proposals to build a large power plant at Niagara Falls, New York, and transmit the power to Buffalo, New York, some 20 miles distant. However, by 1893 the Cataract directors had decided to proceed with their own plan. But they had to choose: Was it to be an ac or a dc system?

The main advantage of ac over dc is efficiency of transmission. Alternating voltages could be transformed to high voltage, thus reducing the loss on the transmission line. If a line has a wire resistance R and the power transmitted is VI, then power lost in the lines is I^2R. Thus, if the transmitted voltage could be set at a high level and the current I kept low, the line losses would be minimized.

Many engineers defended the dc system. Thomas Edison, for example, was firmly against the adoption of ac systems. He emphasized the greater reliability of dc and the potential hazard of high ac voltages. However, after 1891 the safety issue died with the safe operation of a 19-km ac transmission line from Willamette Falls to Portland, Oregon, which operated at 4 kV. The Frankfurt line mentioned earlier operated at 30 kV. Thus the Cataract Company began the construction of an ac power system at Niagara, and the ac versus dc issue was settled.

In 1893, when General Electric Company acquired a small firm, it acquired the services of a young electrical engineer, Charles P. Steinmetz, shown in Figure 11-1. Steinmetz had completed his doctoral dissertation in mathematics at the University of Breslau but had left for Switzerland and eventually the United States.

Figure 11-1 Charles P. Steinmetz (1865–1923), the developer of the mathematical analytical tools for studying ac circuits. Courtesy of General Electric Co.

Steinmetz's claim to fame was his symbolic method of steady-state ac calculation—the $j\omega$ notation—and the theory of electrical transients. McGraw-Hill published the first of his many textbooks, *Theory and Calculation of AC Phenomena,* in 1897. Four years later, he was elected president of the American Institute of Electrical Engineers in recognition of his leadership.

One of Steinmetz's greatest insights occurred in 1893, when he discussed a paper by A. E. Kennelly on impedance. Steinmetz, noting that $a + jb = r(\cos \phi + j \sin \phi)$, showed the equivalence of the complex notation and the polar form in ac circuit calculations. Thus, the use of complex numbers to solve ac circuit problems was born.

The subject of this chapter is the analysis of the steady-state response using complex numbers as originally introduced by Steinmetz in 1893.

11-3 ‖ SINUSOIDAL SOURCES

In electrical engineering, sinusoidal forcing functions are particularly important since power sources and communication signals are usually transmitted as sinusoids or modified sinusoids. The forcing function causes the forced response, and the natural response is caused by the internal dynamics of the circuit. The natural response will normally decay after some period of time, but the forced response continues indefinitely. Therefore, in this chapter we are primarily interested in the steady-state response of a circuit to the sinusoidal forcing function.

We consider the forcing function

$$v_s = V_m \sin \omega t \tag{11-1}$$

or, in the case of a current source,

$$i_s = I_m \sin \omega t \tag{11-2}$$

The amplitude of the sinusoid is V_m and the radian frequency is ω (rad/s). The sinusoid is a *periodic function* defined by the property

$$x(t + T) = x(t)$$

for all t and where T is the period of oscillation.

The reciprocal of T defines the *frequency* or number of cycles per second, denoted by f, where

$$f = \frac{1}{T}$$

The frequency f is in cycles per second, more commonly referred to as hertz (Hz) in honor of the scientist Heinrich Hertz, shown in Figure 11-2. Therefore, the angular

Figure 11-2 Heinrich R. Hertz (1857–1894). Courtesy of the Institution of Electrical Engineers.

(radian) frequency of the sinusoidal function is

$$\omega = 2\pi f = \frac{2\pi}{T} \text{ (rad/s)}$$

For the voltage source of Eq. 11-1, the maximum value is V_m. If the sinusoidal voltage has an associated *phase shift* ϕ, in radians, the voltage source is

$$v_s = V_m \sin(\omega t + \phi) \tag{11-3}$$

The sinusoidal voltage of Eq. 11-3 is represented by Figure 11-3. Normally, we consider ωt and ϕ in radians, but many use degrees as the convention for the total angle. Either approach is acceptable as long as it is clear which notation you are using.

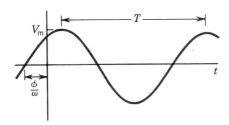

Figure 11-3 Sinusoidal voltage source $v_s = V_m \sin(\omega t + \phi)$.

Since, conventionally, the angle ϕ may be expressed in degrees, you will encounter the notation

$$v_s = V_m \sin(4t + 30°)$$

or alternatively

$$v_s = V_m \sin\left(4t + \frac{\pi}{6}\right)$$

This angular inconsistency will not deter us as long as we recognize that in the actual calculation of $\sin \theta$, θ must be in degrees or radians as our calculator or math tables require.

In addition it is worth noting that

$$V_m \sin(\omega t + 30°) = V_m \cos(\omega t - 60°)$$

This relationship can be deduced using the trigonometric formulas summarized in Toolbox 11-1.

TOOLBOX 11-1

TRIGONOMETRIC FORMULAS—SINE AND COSINE RELATIONSHIPS

a. $\sin \omega t = \cos(\omega t - 90°)$
b. $\cos \omega t = \sin(\omega t + 90°)$
c. $\sin(\omega t \pm \theta) = \cos \theta \sin \omega t \pm \sin \theta \cos \omega t$
d. $\cos(\omega t \pm \theta) = \cos \theta \cos \omega t \mp \sin \theta \sin \omega t$
e. $A \cos \omega t + B \sin \omega t = \sqrt{A^2 + B^2} \cos(\omega t - \theta)$ where $\theta = \tan^{-1}(B/A)$ when $A > 0$ and $\theta = 180° - \tan^{-1}(B/-A)$ when $A < 0$.

Note that ω is the same for all sinusoids.

If a circuit has a voltage across an element as

$$v = V_m \sin \omega t$$

and a current flows through the element

$$i = I_m \sin(\omega t + \phi)$$

we have the v and the i shown in Figure 11-4. We say that the current *leads* the voltage by ϕ radians. Examining Figure 11-4, we note that the current reaches its peak value before

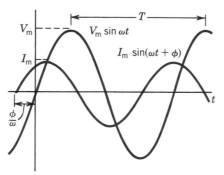

Figure 11-4 Voltage and current of a circuit element.

the voltage and thus is said to lead the voltage. Thus, a point on $i(t)$ is reached in time ahead of the corresponding point of $v(t)$. An alternative form is to state that the voltage lags the current by ϕ radians.

Consider a sine waveform with

$$v = 2 \sin(3t + 20°)$$

and the associated current waveform

$$i = 4 \sin(3t - 10°)$$

Clearly, the voltage v leads the current i by 30°, or $\pi/6$ radians.

Example 11-1

The voltage across an element is $v = 3 \cos 3t$ V, and the associated current through the element is $i = -2 \sin(3t + 10°)$ A. Determine the angle relationship between the voltage and current.

Solution

First, we need to convert the current to a cosine form with a positive magnitude so that it can be contrasted with the voltage. In order to determine a phase relationship it is necessary to express both waveforms in a consistent form.

Since $-\sin \omega t = \sin(\omega t + \pi)$, we have

$$i = 2 \sin(3t + 180° + 10°)$$

Also, we note that

$$\sin \theta = \cos(\theta - 90°)$$

Therefore,

$$i = 2 \cos(3t + 180° + 10° - 90°)$$
$$= 2 \cos(3t + 100°)$$

Recall that $v = 3 \cos 3t$. Therefore, the current leads the voltage by 100°.

In the preceding chapter we learned that the response of a circuit to a forcing function

$$v_s = V_0 \cos \omega t$$

will result in a forced response of the form

$$v_f = A \cos \omega t + B \sin \omega t \tag{11-4}$$

Equation 11-4 may also be written as

$$v_f = \sqrt{A^2 + B^2} \left(\frac{A}{\sqrt{A^2 + B^2}} \cos \omega t + \frac{B}{\sqrt{A^2 + B^2}} \sin \omega t \right)$$

Consider the triangle shown in Figure 11-5 and note that

$$\sin \theta = \frac{B}{\sqrt{A^2 + B^2}}$$

and

$$\cos \theta = \frac{A}{\sqrt{A^2 + B^2}}$$

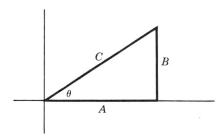

Figure 11-5 Triangle for A and B of Eq. 11-4, where $C = \sqrt{A^2 + B^2}$.

Then we have for v_f

$$v_f = \sqrt{A^2 + B^2} (\cos \theta \cos \omega t + \sin \theta \sin \omega t) \tag{11-5}$$

However, we can also write Eq. 11-5 as

$$v_f = \sqrt{A^2 + B^2} \cos(\omega t - \theta) \tag{11-6}$$

where $\theta = \tan^{-1} B/A$ since A > 0. (See Toolbox 11-1)

Therefore, if a circuit has a forcing function $V_0 \cos \omega t$, the resulting forced response is

$$v_f = A \cos \omega t + B \sin \omega t$$
$$= \sqrt{A^2 + B^2} \cos(\omega t - \theta)$$

where $\theta = \tan^{-1} B/A$ and B and A are as they appear in Figure 11-5. We will use this result in the next section.

Example 11-2

A current has the form $i = -6 \cos 2t + 8 \sin 2t$. Find the current restated in the form of Eq. 11-6.

Solution

The triangle for A and B is shown in Figure 11-6. Since the coefficient A is equal to -6 and

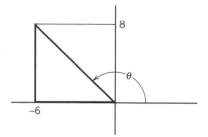

Figure 11-6 The *A-B* triangle for Example 11-2.

B is $+8$, we have the angle θ shown. Therefore, using Toolbox 11-1

$$\theta = \tan^{-1}\left(\frac{8}{-6}\right)$$

$$= 180° - 53.1°$$
$$= 126.9°$$

Hence

$$i = 10 \cos(2t - 126.9°)$$

EXERCISE 11-1

A voltage is $v = 6 \cos(4t + 30°)$. (a) Find the period of oscillation. (b) State the phase relation to the associated current $i = 8 \cos(4t - 70°)$.
Answer: (a) $T = 2\pi/4$.
             (b)   The voltage leads the current by 100°.

EXERCISE 11-2

A voltage is $v = 3 \cos 4t + 4 \sin 4t$. Find the voltage in the form of Eq. 11-6.
Answer: $v = 5 \cos(4t - 53°)$

EXERCISE 11-3

A current is $i = 12 \sin 5t - 5 \cos 5t$. Find the current in the form of Eq. 11-6.
Answer: $i = 13 \cos(5t - 112.6°)$

11-4   STEADY-STATE RESPONSE OF AN *RL* CIRCUIT FOR A SINUSOIDAL FORCING FUNCTION

As an example of the task of determining the steady-state response of a circuit, let us consider the *RL* circuit shown in Figure 11-7. We will use the forcing function

$$v_s = V_m \cos \omega t$$

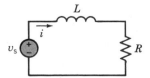

Figure 11-7 An *RL* circuit.

We wish to determine only the forced response i_f since we will take the long-run view and look for the steady-state response. Irrespective of the initial value of the current and provided that the root of the characteristic equation is in the left-hand s-plane, the response will become purely sinusoidal as $t \to \infty$. This response is called the sinusoidal steady-state response.

The governing differential equation of the RL circuit is given by

$$L\frac{di}{dt} + Ri = V_m \cos \omega t \tag{11-7}$$

Following the method of the previous chapter, we assume that

$$i_f = A \cos \omega t + B \sin \omega t \tag{11-8}$$

At this point, since we are only solving for the forced response, we drop the subscript f notation. Substituting the assumed solution of Eq. 11-8 into the differential equation and completing the derivative we have

$$L(-\omega A \sin \omega t + \omega B \cos \omega t) + R(A \cos \omega t + B \sin \omega t) = V_m \cos \omega t$$

Equating the coefficients of $\cos \omega t$, we obtain

$$\omega LB + RA = V_m$$

Next, equating the coefficients of $\sin \omega t$, we obtain

$$-\omega LA + RB = 0$$

Solving for A and B, we have

$$A = \frac{RV_m}{R^2 + \omega^2 L^2}$$

and

$$B = \frac{\omega L V_m}{R^2 + \omega^2 L^2}$$

The response to the sinusoidal input is then

$$i = A \cos \omega t + B \sin \omega t$$

or

$$i = \frac{V_m}{Z} \cos(\omega t - \beta)$$

where

$$Z = \sqrt{R^2 + \omega^2 L^2}$$

and

$$\beta = \tan^{-1} \frac{\omega L}{R}$$

Thus, the forced (steady-state) response is of the form

$$i = I_m \cos(\omega t + \phi)$$

where

$$I_m = \frac{V_m}{Z}$$

and

$$\phi = -\beta$$

In this case, we have found only the steady-state response of a circuit with one energy storage element. Clearly, this approach can be quite complicated if the circuit has several storage elements.

Steinmetz saw a way beyond this differential equation approach. The method is discussed in the next section. It was an insight that, in many ways, enabled electrical engineering to leap forward to new achievements.

EXERCISE 11-4

Find the forced response v for the RC circuit shown in Figure E 11-4 when $i_s = I_m \cos \omega t$.

Answer: $v = (RI_m/P) \cos(\omega t - \theta)$, $P = \sqrt{1 + \omega^2 R^2 C^2}$, $\theta = \tan^{-1}(\omega RC)$

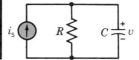

Figure E 11-4

EXERCISE 11-5

Find the forced response $i(t)$ for the RL circuit of Figure 11-7 when $R = 2\,\Omega$, $L = 1\,$H, and $v_s = 10 \cos 3t$ V.

Answer: $i = 2.77 \cos(3t - 56.3°)$ A

11-5 COMPLEX EXPONENTIAL FORCING FUNCTION

Upon reviewing the preceding section, we see that the forcing function is of the form

$$v_s = V_m \cos \omega t$$

and the response was

$$i = \frac{V_m}{Z} \cos(\omega t - \beta)$$

Thus, the response has a different magnitude and phase than the original voltage source.

TOOLBOX 11-2

COMPLEX NUMBERS AND EULER'S FORMULA

A complex number is defined as the sum of a real number and an imaginary number so that

$$c = a + jb$$

where a and b are real numbers and j is the imaginary unit ($j = \sqrt{-1}$). We designate a as the real part of the complex number and b as the imaginary part and use the notation

$$\text{Re}\{c\} = a \quad \text{and} \quad \text{Im}\{c\} = b$$

The exponential form of a complex number is written as

$$c = re^{j\theta}$$

Euler's formula is

$$e^{j\omega t} = \cos \omega t + j \sin \omega t$$

or the alternate form is

$$e^{-j\omega t} = \cos \omega t - j \sin \omega t$$

See Appendices C and E for further discussion.

Let us now consider the exponential signal

$$v_s = V_m e^{j\omega t} \tag{11-9}$$

Of course, using Euler's equation we recognize that v_s is

$$v_s = V_m \cos \omega t = \text{Re}\{V_m e^{j\omega t}\}$$

The notation $\text{Re}\{a + jb\}$ is read as the real part of the complex number $(a + jb)$. Our aim is to find i, where

$$i = \text{Re}\{I_m e^{j\omega t}\}$$

where $\text{Re}\{\ \}$ means the real part of the complex number. For example,

$$\text{Re}\{a + jb\} = a$$

Let us try the exponential source v_s of Eq. 11-9 with the differential equation of the RL circuit as described by

$$L\frac{di}{dt} + Ri = v_s \tag{11-10}$$

Since the source is an exponential, we try the solution

$$i = Ae^{j\omega t}$$

and substitute into Eq. 11-10 to obtain

$$(j\omega L + R)Ae^{j\omega t} = V_m e^{j\omega t}$$

Hence,[1]

$$A = \frac{V_m}{R + j\omega L}$$

$$= \frac{V_m}{Z} e^{-j\beta}$$

where

$$\beta = \tan^{-1} \frac{\omega L}{R}$$

and

$$Z = \sqrt{R^2 + \omega^2 L^2}$$

Therefore, substituting for A, we have

$$i = \frac{V_m}{Z} e^{-j\beta} e^{j\omega t} \qquad (11\text{-}11)$$

Again noting that the original forcing function was

$$v_s = \text{Re}\{V_m e^{j\omega t}\} = V_m \cos \omega t$$

we expect that

$$i = \text{Re}\{i \text{ of Eq. 11-11}\}$$

Accordingly,

$$i = \frac{V_m}{Z} \text{Re}\{e^{-j\beta} e^{j\omega t}\}$$

$$= \frac{V_m}{Z} \text{Re}\{e^{j(\omega t - \beta)}\}$$

$$= \frac{V_m}{Z} \cos(\omega t - \beta)$$

We are actually interested in using the sinusoidal excitation

$$v_s = \text{Re}\{V_m e^{j\omega t}\} = V_m \cos \omega t$$

to determine the forced response

$$i = I_m \cos(\omega t - \beta)$$

However, we have learned that this same response

$$i = \text{Re}\left\{\frac{V_m}{Z} e^{j(\omega t - \beta)}\right\}$$

is readily obtained by using the complex exponential excitation, $\text{Re}\{V_m e^{j\omega t}\}$.
Thus, if

$$v_s = \text{Re}\{V_m e^{j\omega t}\} \qquad (11\text{-}12)$$

---

[1] Note: See Appendix C for a review of complex numbers.

then

$$i = \text{Re} \left\{ \frac{V_\text{m}}{Z} e^{j(\omega t - \beta)} \right\} \tag{11-13}$$

As an example, let us try these two defining equations for the RLC circuit represented by the differential equation

$$\frac{d^2 i}{dt^2} + \frac{di}{dt} + 12i = 12 \cos 3t \tag{11-14}$$

First, replace the real excitation by the complex exponential excitation

$$v = 12e^{j3t}$$

Then we have Eq. 11-14 restated as

$$\frac{d^2 i}{dt^2} + \frac{di}{dt} + 12i = 12e^{j3t} \tag{11-15}$$

and it is clear that we must assume the current is of the form

$$i = Ae^{j3t} \tag{11-16}$$

The first and second derivatives of the i of Eq. 11-16 are

$$\frac{di}{dt} = j3Ae^{j3t}$$

and

$$\frac{d^2 i}{dt^2} = -9Ae^{j3t}$$

Substituting into Eq. 11-15, we have

$$(-9 + j3 + 12)Ae^{j3t} = 12e^{j3t}$$

Solving for A, we obtain

$$A = \frac{12}{3 + j3}$$
$$= \frac{12(3 - j3)}{(3 + j3)(3 - j3)}$$
$$= \frac{12(3 - j3)}{18}$$
$$= 2\sqrt{2}\underline{/-45°}$$

Therefore,

$$i = Ae^{j3t}$$
$$= 2\sqrt{2}e^{-j(\pi/4)}e^{j3t}$$
$$= 2\sqrt{2}e^{j(3t - \pi/4)}$$

Recall that Euler's identity is $e^{j\phi} = \cos \phi + j \sin \phi$. Thus, the desired answer for the steady-state current is[2]

$$i(t) = \text{Re}\{i\}$$
$$= 2\sqrt{2}\,\text{Re}\{\cos(3t - 45°) + j\sin(3t - 45°)\}$$
$$= 2\sqrt{2}\,\cos(3t - 45°)$$

Note we have changed from $\pi/4$ radians to 45°, which are interchangeable and equivalent. Both degree and radian notation are acceptable and interchangeable.

We have developed a straightforward method for determining the steady-state response of a circuit to a sinusoidal excitation. The process is as follows: (1) instead of applying the actual forcing function, we apply a complex exponential forcing function, and (2) we then obtain the complex response whose real part is the desired response. The process is summarized in Table 11-1. Let us use this process in another example.

Table 11-1
Use of the Complex Exponential Excitation to Determine a Circuit's Steady-State Response to a Sinusoidal Source

1. Write the excitation (forcing function) as a cosine waveform with a phase angle so that $y_s = Y_m \cos(\omega t + \phi)$, where y_s is a current source i_s or a voltage source v_s in by the circuit.
2. Recall Euler's identity, which is

$$e^{j\alpha} = \cos\alpha + j\sin\alpha$$

where $\alpha = \omega t + \phi$ in this case.
3. Introduce the complex excitation so that for a voltage source, for example, we have

$$v_s = \text{Re}\{V_m e^{j(\omega t + \phi)}\}$$

where $V_m e^{j(\omega t + \phi)}$ is a complex exponential excitation.
4. Use the complex excitation and the differential equation along with the assumed response $x = A e^{j(\omega t + \phi)}$, where A is to be determined. Note that A will normally be a complex quantity.
5. Determine the constant $A = B e^{-j\beta}$, so that

$$x = A e^{j(\omega t + \phi)}$$
$$= B e^{j(\omega t + \phi - \beta)}$$

6. Recognize that the desired response is

$$x(t) = \text{Re}\{x\}$$
$$= B\cos(\omega t + \phi - \beta)$$

Example 11-3
Find the response i of the RL circuit of Figure 11-7 when $R = 2\,\Omega$, $L = 1$ H, and $v_s = 10\sin 3t$ V.

Solution
First, we will rewrite the voltage source so that it is expressed as a cosine waveform as follows:

$$v_s = 10\sin 3t$$
$$= 10\cos(3t - 90°)$$

---
[2] See Appendix E for a discussion of Euler's equation.

Using the complex excitation, we have

$$v_s = 10e^{j(3t-90°)}$$

Introduce the complex excitation into the circuit's differential equation, which is

$$L\frac{di}{dt} + Ri = v_s$$

obtaining

$$\frac{di}{dt} + 2i = 10e^{j(3t-90°)}$$

Assume that the response is

$$i = Ae^{j(3t-90°)} \tag{11-17}$$

where A is a complex quantity to be determined. Substituting the assumed solution, Eq. 11-17, into the differential equation and taking the derivative, we have

$$j3Ae^{j(3t-90°)} + 2Ae^{j(3t-90°)} = 10e^{j(3t-90°)}$$

Therefore,

$$j3A + 2A = 10$$

or

$$A = \frac{10}{j3+2}$$

$$= \frac{10}{\sqrt{9+4}}e^{-j\beta}$$

where

$$\beta = \tan^{-1}\frac{3}{2}$$

$$= 56.3°$$

Then the solution is

$$i = Ae^{j(3t-90°)}$$

$$= \frac{10}{\sqrt{13}}e^{j(3t-90°-56.3°)}$$

$$= \frac{10}{\sqrt{13}}e^{j(3t-90°-56.3°)}$$

$$= \frac{10}{\sqrt{13}}e^{j(3t-146.3°)}$$

Consequently, the actual response is

$$i(t) = \text{Re}\{i\}$$

$$= \frac{10}{\sqrt{13}}\cos(3t - 146.3°) \text{ A}$$

Steinmetz observed the process we just utilized and decided to formulate a method for solving the sinusoidal steady-state response of circuits using complex number notation. The development of this approach is the subject of the next section.

EXERCISE 11-6

Find a and b when

$$\frac{10}{a + jb} = 2.36e^{j45}$$

Answer: $a = 3, b = -3$

EXERCISE 11-7

Find A and θ when

$$[A\underline{/\theta}](-3 + j8) = j32$$

Answer: $A = 3.75, \theta = -20.56°$

11-6 THE PHASOR CONCEPT

A sinusoidal current or voltage at a given frequency is characterized by its amplitude and phase angle. For example, the current response in the *RL* circuit considered in Example 11-3 was

$$i(t) = \text{Re}\{I_m e^{j(\omega t + \phi - \beta)}\}$$
$$= I_m \cos(\omega t + \phi - \beta)$$

The magnitude I_m and the phase angle $(\phi - \beta)$, along with knowledge of ω, completely specify the response. Thus, we may write $i(t)$ as

$$i(t) = \text{Re}\{I_m e^{j(\phi - \beta)} e^{j\omega t}\}$$

However, we note that the complex factor $e^{j\omega t}$ remained unchanged throughout all our previous calculations. Thus, the information we seek is represented by

$$\mathbf{I} = I_m e^{j(\phi - \beta)}$$
$$= I_m \underline{/\phi - \beta} \tag{11-18}$$

where **I** is called a *phasor*. A phasor is a complex number that represents the magnitude and phase of a sinusoid. The term phasor is used instead of vector because the angle is time based rather than space based. A phasor may be written in exponential form, polar form, or rectangular form.

The **phasor concept** may be used when the circuit is linear, the steady-state response is sought, and all independent sources are sinusoidal and have the same frequency.

A real sinusoidal current, where $\theta = \phi - \beta$, is written as

$$i(t) = I_m \cos(\omega t + \theta)$$

It can be represented by

$$i(t) = \text{Re}\{I_m e^{j(\omega t + \theta)}\}$$

We then decide to drop the notation Re and the redundant $e^{j\omega t}$ to obtain the phasor representation

$$\mathbf{I} = I_m e^{j\theta}$$
$$= I_m \underline{/\theta}$$

This abbreviated representation is the *phasor notation*. Phasor quantities are complex and thus are printed in boldface in this book. You may choose to use the underline notation as follows:

$$\underline{I} = I_m \underline{/\theta}$$

Although we have dropped or suppressed the complex frequency $e^{j\omega t}$, we continue to note that we are in the complex frequency form and are performing calculations in the *frequency domain*. We have transformed the problem from the time domain to the frequency domain by the use of phasor notation. A transform is a means of encoding to simplify a calculation process. One example of a mathematical transform is the logarithmic transform.

A **transform** is a change in the mathematical description of a physical variable to facilitate computation.

EXERCISE 11-8

Using the method of complex excitation as described in the preceding section, find the response v for the circuit shown in Figure E 11-8 when $i_s = 10 \cos 4t$ A.

Answer: $v = \dfrac{40}{\sqrt{17}} \cos(4t - 76°)$ V

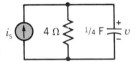

Figure E 11-8

EXERCISE 11-9

Find the current $i(t)$ for the *RLC* circuit of Figure E 11-9 when $v_s = 4 \cos 100t$ V.
Answer: $i(t) = 2\sqrt{2} \cos(100t + 45°)$ A

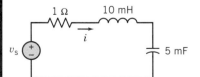

Figure E 11-9

The actual steps involved in transforming a function in the time domain to the frequency domain are summarized in Table 11-2. Since it is easy to move through these steps, we usually jump directly from step 1 to step 4.

Table 11-2

Transformation from the Time Domain to the Frequency Domain

1. Write the function in the time domain, $y(t)$, as a cosine waveform with a phase angle ϕ as

$$y(t) = Y_m \cos(\omega t + \phi)$$

2. Express the cosine waveform as the real part of a complex quantity by using Euler's identity so that

$$y(t) = \text{Re}\{Y_m e^{j(\omega t + \phi)}\}$$

3. Drop the real part notation.
4. Suppress the $e^{j\omega t}$, while noting the value of ω for later use, obtaining the phasor

$$\mathbf{Y} = Y_m e^{j\phi}$$
$$= Y_m \underline{/\phi}$$

For example, let us determine the phasor notation for

$$i = 5 \sin(100t + 120°)$$

We have chosen to use cosine functions as the standard for phasor notation. Thus, we express the current as a cosine waveform:

$$i = 5 \cos(100t + 30°)$$

At this point, it is easy to see that the information we require is the amplitude and the phase. Thus, the phasor is

$$\mathbf{I} = 5\underline{/30°}$$

Of course, the reverse process from phasor notation to time notation is exactly the reverse of the steps required to go from the time to the phasor notation. Thus, if we have a voltage in phasor notation:

$$\mathbf{V} = 25\underline{/125°}$$

the time domain notation is

$$v(t) = 25 \cos(\omega t + 125°)$$

where the frequency ω was noted in the original statement of the circuit formulation. This transformation from the frequency domain to the time domain is summarized in Table 11-3.

A **phasor** is a transformed version of a sinusoidal voltage or current waveform and consists of the magnitude and phase angle information of the sinusoid.

The phasor method uses the transformation from the time domain to the frequency domain to more easily obtain the sinusoidal steady-state solution of the differential equation. Consider the RL circuit of Figure 11-8. We wish to find the solution for the steady-state current i when the voltage source is $v_s = V_m \cos \omega t$ V and $\omega = 100$ rad/s.

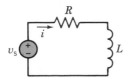

Figure 11-8 RL circuit.

Table 11-3
Transformation from the Frequency Domain to the Time Domain

1. Write the phasor in exponential form as

$$\mathbf{Y} = Y_m e^{j\beta}$$

2. Reinsert the factor $e^{j\omega t}$ so that you have

$$Y_m e^{j\beta} e^{j\omega t}$$

3. Reinsert the real part operator Re as

$$\text{Re}\{Y_m e^{j\beta} e^{j\omega t}\}$$

4. Use Euler's identity to obtain the time function

$$y(t) = \text{Re}\{Y_m e^{j(\omega t + \beta)}\}$$
$$= Y_m \cos(\omega t + \beta)$$

Also, for this circuit let $R = 200 \; \Omega$ and $L = 2$ H. Then we may write the differential equation as

$$L\frac{di}{dt} + Ri = v_s \tag{11-19}$$

Since

$$v_s = V_m \cos(\omega t + \phi)$$
$$= \text{Re}\{V_m e^{j(\omega t + \phi)}\} \tag{11-20}$$

we will use the assumed solution

$$i = I_m \cos(\omega t + \beta)$$
$$= \text{Re}\{I_m e^{j(\omega + \phi)}\} \tag{11-21}$$

Therefore, we may substitute Eqs. 11-20 and 11-21 into Eq. 11-19 and suppress the Re notation to obtain

$$(j\omega L I_m + RI_m)e^{j(\omega t + \beta)} = V_m e^{j(\omega t + \phi)}$$

Suppress the $e^{j\omega t}$ to obtain

$$(j\omega L + R)I_m e^{j\beta} = V_m e^{j\phi}$$

Now we recognize the phasors

$$\mathbf{I} = I_m e^{j\beta}$$

and

$$\mathbf{V} = V_m e^{j\phi}$$

Therefore, in phasor notation we have

$$(j\omega L + R)\mathbf{I} = \mathbf{V}$$

Solving for $\mathbf{I}$, we have

$$\mathbf{I} = \frac{\mathbf{V}}{j\omega L + R}$$

$$= \frac{\mathbf{V}}{j200 + 200}$$

for $\omega = 100$, $L = 2$, and $R = 200$. Therefore, since

$$\mathbf{V} = V_m\underline{/0°}$$

we have

$$\mathbf{I} = \frac{V_m}{283\underline{/45°}}$$

$$= \frac{V_m}{283}\underline{/-45°}$$

Using the method of Table 11-3, we may transform this result back to the time domain to obtain the steady-state time solution as

$$i(t) = \frac{V_m}{283}\cos(100t - 45°) \text{ A}$$

It is clear that we can use phasors directly to obtain a linear algebraic equation expressed in terms of the phasors and complex numbers and then solve for the phasor variable of interest. After obtaining the phasor we desire, we simply transform it back to the time domain to obtain the steady-state solution.

Example 11-4
Find the steady-state voltage v for the RC circuit shown in Figure 11-9 when $i = 10\cos\omega t$ A, $R = 1\ \Omega$, $C = 10$ mF, and $\omega = 100$ rad/s.

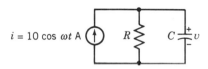

$i = 10\cos\omega t$ A

Figure 11-9 An RC circuit with a sinusoidal current source.

Solution
First, we find the phasor representation of the source current as

$$\mathbf{I} = I_m\underline{/0°}$$
$$= 10\underline{/0°} \tag{11-22}$$

We seek to find the voltage v by first obtaining the phasor $\mathbf{V}$.

Write the node voltage differential equation for the circuit to obtain

$$\frac{v}{R} + C\frac{dv}{dt} = i \tag{11-23}$$

Since

$$i = 10\ \text{Re}\{e^{j\omega t}\}$$

and

$$v = V_m\ \text{Re}\{e^{j(\omega t + \phi)}\}$$

we substitute into Eq. 11-23 and suppress the Re notation to obtain

$$\frac{V_m}{R}e^{j(\omega t + \phi)} + j\omega C V_m e^{j(\omega t + \phi)} = 10e^{j\omega t}$$

We now suppress the $e^{j\omega t}$ and obtain

$$\left(\frac{1}{R} + j\omega C\right) V_m e^{j\phi} = 10 e^{j0°}$$

Recalling the phasor representation of Eq. 11-22, we have

$$\left(\frac{1}{R} + j\omega C\right) \mathbf{V} = \mathbf{I}$$

Since $R = 1$, $C = 10^{-2}$, and $\omega = 100$, we have

$$(1 + j1)\mathbf{V} = \mathbf{I}$$

or

$$\mathbf{V} = \frac{\mathbf{I}}{1 + j1}$$

Therefore,

$$\mathbf{V} = \frac{10}{\sqrt{2}\underline{/45°}}$$

$$= \frac{10}{\sqrt{2}} \underline{/-45°}$$

Transforming from the phasor notation back to the steady-state time solution, we have

$$v = \frac{10}{\sqrt{2}} \cos(100t - 45°) \text{ V}$$

EXERCISE 11-10

Express the current i as a phasor.
(a) $i = 4 \cos(\omega t - 80°)$ (b) $i = 10 \cos(\omega t + 20°)$
(c) $i = 8 \sin(\omega t - 20°)$
Answers: (a) $4\underline{/-80°}$
                     (b) $10\underline{/+20°}$
                     (c) $8\underline{/-110°}$

EXERCISE 11-11

Find the steady-state voltage v represented by the phasor
(a) $\mathbf{V} = 10\underline{/-140°}$ (b) $\mathbf{V} = 80 + j75$
Answer: (a) $v = 10 \cos(\omega t - 140°)$
                     (b) $109.7 \cos(\omega t + 43.2°)$

11-7 | PHASOR RELATIONSHIPS FOR R, L, AND C ELEMENTS

In the preceding section we found that the phasor representation is actually a transformation from the time domain into the frequency domain. With this transform, we have converted the solution of a differential equation into the solution of an algebraic equation.

In this section we determine the relationship between the phasor voltage and the phasor current of the elements: R, L, and C. We use the transformation from the time to the frequency domain and then solve for the phasor relationship for a specified element. We are using the method of Table 11-2 as recorded in the last section.

Let us begin with the resistor, as shown in Figure 11-10a. The voltage–current relationship in the time domain is

$$v = Ri \qquad (11\text{-}24)$$

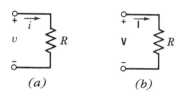

Figure 11-10 (a) The v-i time domain relationship for R. (b) The phasor relationship for R.

Now consider the steady-state voltage

$$v = V_{\mathrm{m}} \cos(\omega t + \phi)$$

Then

$$v = \mathrm{Re}\{V_{\mathrm{m}} e^{j(\omega t + \phi)}\} \qquad (11\text{-}25)$$

Assume that the current source is of the form

$$i = \mathrm{Re}\{I_{\mathrm{m}} e^{j(\omega t + \beta)}\} \qquad (11\text{-}26)$$

Then substitute Eqs. 11-25 and 11-26 into Eq. 11-24 and suppress the Re notation to obtain

$$V_{\mathrm{m}} e^{j(\omega t + \phi)} = R I_{\mathrm{m}} e^{j(\omega t + \beta)}$$

Suppress $e^{j\omega t}$ to obtain

$$V_{\mathrm{m}} e^{j\phi} = R I_{\mathrm{m}} e^{j\beta}$$

Therefore, we note that $\beta = \phi$ and

$$\mathbf{V} = R\mathbf{I} \qquad (11\text{-}27)$$

Since $\beta = \phi$, the current and voltage waveforms are in phase. This phasor relationship is shown in Figure 11-10b.

For example, if the voltage across a resistor is $v = 10 \cos 10t$, we know that the current will be

$$i = \frac{10}{R} \cos 10t$$

in the time domain.

In the frequency domain, we first note that the voltage is

$$\mathbf{V} = 10\underline{/0°}$$

Then, using the phasor relationship of the resistor, Eq. 11-27, we have

$$\mathbf{I} = \frac{\mathbf{V}}{R}$$
$$= \frac{10\underline{/0°}}{R}$$

Then, obtaining the time domain relationship for i, we have

$$i = \frac{10}{R} \cos 10t$$

Now consider the inductor as shown in Figure 11-11a. The time domain voltage-current

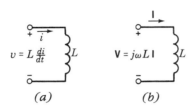

Figure 11-11 (*a*) The time domain v-i relationship for an inductor. (*b*) The phasor relationship for an inductor.

relationship is

$$v = L\frac{di}{dt} \tag{11-28}$$

Again, we use the complex voltage as

$$v = \text{Re}\{V_m e^{j(\omega t + \phi)}\} \tag{11-29}$$

and assume that the current is

$$i = \text{Re}\{I_m e^{j(\omega t + \beta)}\} \tag{11-30}$$

Substituting Eqs. 11-29 and 11-30 into Eq. 11-28 and suppressing the Re notation, we have

$$V_m e^{j\phi}e^{j\omega t} = L\frac{d}{dt}\{I_m e^{j\omega t}e^{j\beta}\}$$

Taking the derivative, we have

$$V_m e^{j\phi}e^{j\omega t} = j\omega L I_m e^{j\omega t}e^{j\beta}$$

Now suppressing the $e^{j\omega t}$, we have

$$V_m e^{j\phi} = j\omega L I_m e^{j\beta} \tag{11-31}$$

or

$$\mathbf{V} = j\omega L\mathbf{I} \tag{11-32}$$

This phasor relationship is shown in Figure 11-11b. Since $j = e^{j90°}$, Eq. 11-31 can also be written as

$$V_m e^{j\phi} = \omega L I_m e^{j90°}e^{j\beta}$$

Therefore,

$$\phi = \beta + 90°$$

Thus, the voltage leads the current by exactly 90°.

As an illustration, consider an inductor of 2 H with $\omega = 100$ rad/s and with voltage $v = 10 \cos(\omega t + 50°)$ V. Then the phasor voltage is

$$\mathbf{V} = 10\underline{/50°}$$

and the phasor current is

$$\mathbf{I} = \frac{\mathbf{V}}{j\omega L}$$

Since $\omega L = 200$, we have

$$\mathbf{I} = \frac{V}{j200}$$
$$= \frac{10\underline{/50°}}{200\underline{/90°}}$$
$$= 0.05\underline{/-40°} \text{ A}$$

Then the current expressed in the time domain is

$$i = 0.05 \cos(100t - 40°) \text{ A}$$

Therefore, the current lags the voltage by 90°.

Finally, let us consider the case of the capacitor, as shown in Figure 11-12a. The

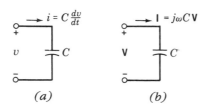

(a) *(b)*

Figure 11-12 *(a)* The time domain v-i relationship for a capacitor. *(b)* The phasor relationship for a capacitor.

current–voltage relationship is

$$i = C\frac{dv}{dt} \tag{11-33}$$

We assume that the voltage is

$$v = V_m \cos(\omega t + \phi)$$
$$= \text{Re}\{V_m e^{j(\omega t + \phi)}\} \tag{11-34}$$

and the current is of the form

$$i = \text{Re}\{I_m e^{j(\omega t + \beta)}\} \tag{11-35}$$

Suppress the Re notation in Eqs. 11-34 and 11-35 and substitute them into Eq. 11-33 to obtain

$$I_m e^{j(\omega t + \beta)} = C\frac{d}{dt}(V_m e^{j(\omega t + \phi)})$$

Taking the derivative, we have

$$I_m e^{j\omega t} e^{j\beta} = j\omega C V_m e^{j\omega t} e^{j\phi}$$

Suppressing the $e^{j\omega t}$, we obtain

$$I_m e^{j\beta} = j\omega C V_m e^{j\phi}$$

or

$$\mathbf{I} = j\omega C \mathbf{V} \qquad (11\text{-}36)$$

This phasor relationship is shown in Figure 11-12b. Since $j = e^{j90°}$, the current leads the voltage by 90°. As an example, consider a voltage $v = 100 \cos \omega t$ V and let us find the current when $\omega = 1000$ rad/s and $C = 1$ mF. Since

$$\mathbf{V} = 100\underline{/0°}$$

we have

$$\begin{aligned}
\mathbf{I} &= j\omega C \mathbf{V} \\
&= (\omega C e^{j90°})100 e^{j0} \\
&= (1 e^{j90°})100 \\
&= 100\underline{/90°}
\end{aligned}$$

Therefore, transforming this phasor into the time domain, we have

$$i = 100 \cos(\omega t + 90°) \text{ A}$$

We can rewrite Eq. 11-36 as

$$\mathbf{V} = \frac{1}{j\omega C} \mathbf{I} \qquad (11\text{-}37)$$

Using this form, we summarize the phasor equations for sources and the resistor, inductor, and capacitor in Table 11-4, where the phasor voltage is expressed in its relationship to the phasor current.

EXERCISE 11-12

A current in an element is $i = 5 \cos 100t$ A. Find the steady-state voltage $v(t)$ across the element for (a) a resistor of 10 Ω, (b) an inductor $L = 10$ mH, and (c) a capacitor $C = 1$ mF.

Answer: (a)   50 cos 100t V
                (b)   5 cos(100t + 90°) V
                (c)   50 cos(100t − 90°) V

EXERCISE 11-13

A capacitor $C = 10$ μF has a steady-state voltage across it of $v = 100 \cos(500t + 30°)$ V. Find the steady-state current in the capacitor.

Answer: $i = 0.5 \cos(500t + 120°)$ A

Table 11-4

Time Domain and Phasor Relationships

Element	Time Domain	Frequency Domain
Current source	$i(t) = A \cos(\omega t + \theta)$	$\mathbf{I}(\omega) = Ae^{j\theta}$
Voltage source	$v(t) = B \cos(\omega t + \phi)$	$\mathbf{V}(\omega) = Be^{j\phi}$
Resistor	R $v(t)$ $v(t) = Ri(t)$	R $\mathbf{V}(\omega)$ $\mathbf{V}(\omega) = R\mathbf{I}(\omega)$
Capacitor	C $v(t)$ $v(t) = \dfrac{1}{C}\displaystyle\int_{-\infty}^{t} i(\tau)\,d\tau$	$\dfrac{1}{j\omega C}$ $\mathbf{V}(\omega)$ $\mathbf{V}(\omega) = \dfrac{1}{j\omega C}\,\mathbf{I}(\omega)$
Inductor	L $v(t)$ $v(t) = L\dfrac{d}{dt}i(t)$	$j\omega L$ $\mathbf{V}(\omega)$ $\mathbf{V}(\omega) = j\omega L\mathbf{I}(\omega)$
CCVS	$i_c(t)$ $i(t)$ $v(t) = Ki_c(t)$	$\mathbf{I}_c(\omega)$ $\mathbf{I}(\omega)$ $\mathbf{V}(\omega) = K\mathbf{I}_c(\omega)$
Ideal Op-amp	$i_1 = 0$ $i(t)$ $i_2 = 0$ $v(t)$	$I_1 = 0$ $\mathbf{I}(\omega)$ $I_2 = 0$ $\mathbf{V}(\omega)$

EXERCISE 11-14

The voltage $v(t)$ and current $i(t)$ for an element are shown in Figure E 11-14. Determine if the element is an inductor or a capacitor.

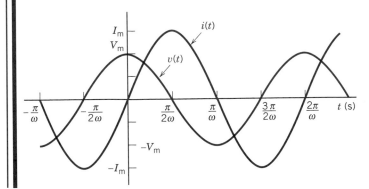

Figure E 11-14

11-8 IMPEDANCE AND ADMITTANCE

The relationships in the frequency domain for the phasor current and phasor voltage of a capacitor, inductor, and resistor are summarized in Table 11-4. These relationships appear similar to Ohm's law for resistors.

We will define the *impedance* of an element as the ratio of the phasor voltage to the phasor current, which we denote by **Z**. Therefore,

$$\mathbf{Z} = \frac{\mathbf{V}}{\mathbf{I}} \tag{11-38}$$

This is called Ohm's law in phasor notation.

TOOLBOX 11-3

MULTIPLICATION AND DIVISION OF COMPLEX NUMBERS

The complex number $c = a + jb$ may be written in polar form as

$$c = |c|\underline{/\theta}$$
$$= r\underline{/\theta}$$

where $r = (a^2 + b^2)^{1/2}$ and $\theta = \tan^{-1}(b/a)$. For two complex numbers $c_1 = r_1\underline{/\theta_1}$ and $c_2 = r_2\underline{/\theta_2}$ we may obtain multiplication as

$$c_1 c_2 = r_1 r_2\underline{/\theta_1 + \theta_2}$$

Division of c_1 by c_2 is obtained as

$$\frac{c_1}{c_2} = \frac{r_1}{r_2}\underline{/\theta_1 - \theta_2}$$

See Appendix C for examples.

Since $\mathbf{V} = V_m\underline{/\phi}$ and $\mathbf{I} = I_m\underline{/\beta}$, we have

$$\mathbf{Z} = \frac{V_m\underline{/\phi}}{I_m\underline{/\beta}}$$

$$= \frac{V_m}{I_m}\underline{/\phi - \beta} \tag{11-39}$$

Thus, the impedance is said to have a magnitude $|\mathbf{Z}|$ and a phase angle $\theta = \phi - \beta$. Therefore,

$$|\mathbf{Z}| = \frac{V_m}{I_m} \tag{11-40}$$

and

$$\theta = \phi - \beta \tag{11-41}$$

Impedance has a role similar to the role of resistance in resistive circuits. Also, since it is a ratio of volts to amperes, impedance has units of ohms. Impedance is the ratio of two phasors; however, it is not a phasor itself. Impedance is a complex number that relates one phasor $\mathbf{V}$ to the other phasor $\mathbf{I}$ as

$$\mathbf{V} = \mathbf{ZI} \tag{11-42}$$

The phasors $\mathbf{V}$ and $\mathbf{I}$ may be transformed to the time domain to yield the steady-state voltage or current, respectively. Impedance has no meaning in the time domain, however.

With the concept of impedance we can solve for the behavior of sinusoidally excited circuits, using complex algebra, in the same way we solved resistive circuits.

Since the impedance is a complex number, it may be written in several forms, as follows:

$$\begin{aligned}\mathbf{Z} &= |\mathbf{Z}|\underline{/\theta} \longrightarrow \text{polar form}\\ &= Ze^{j\theta} \quad\;\; \longrightarrow \text{exponential form}\\ &= R + jX \longrightarrow \text{rectangular form}\end{aligned} \tag{11-43}$$

where R is the real part and X is the imaginary part of the complex number $\mathbf{Z}$. We introduce the notation, in Eq. 11-43, $|\mathbf{Z}| = Z$. Thus, the magnitude of the impedance can be written as Z (not boldface). The $R = \text{Re } \mathbf{Z}$ is called the resistive part of the impedance, and $X = \text{Im } \mathbf{Z}$ is called the reactive part of the impedance. Both R and X are measured in ohms.

We also note that the magnitude of the impedance is

$$Z = \sqrt{R^2 + X^2} \tag{11-44}$$

and the phase angle is

$$\theta = \tan^{-1}\frac{X}{R} \tag{11-45}$$

These relationships are summarized graphically in the complex plane, in Figure 11-13. As

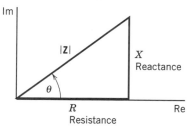

Figure 11-13 Graphical representation of impedance.

an example, let us consider

$$\mathbf{Z} = 2 + j2$$

Then,

$$Z = \sqrt{8}$$

and

$$\theta = 45°$$

The three elements R, L, and C are uniquely represented by an impedance that follows from their **V–I** relationship. For a resistor we have

$$\mathbf{V} = R\mathbf{I}$$

and, therefore,

$$\mathbf{Z} = R \tag{11-46}$$

For the inductor we have

$$\mathbf{V} = j\omega L\mathbf{I}$$

and, therefore,

$$\mathbf{Z} = j\omega L \tag{11-47}$$

Finally, for the capacitor we have

$$\mathbf{V} = \frac{\mathbf{I}}{j\omega C}$$

so that

$$\mathbf{Z} = \frac{1}{j\omega C} \tag{11-48}$$

The impedances for R, L, and C are summarized in Table 11-5. The unit for an impedance is ohms.

Table 11-5
Impedances of R, L, and C

Element	Impedance
Resistor	$\mathbf{Z} = R$
Inductor	$\mathbf{Z} = j\omega L$
Capacitor	$\mathbf{Z} = \dfrac{1}{j\omega C}$

The reciprocal of impedance is called the *admittance* and is denoted by **Y**:

$$\mathbf{Y} = \frac{1}{\mathbf{Z}} \tag{11-49}$$

Admittance is analogous to conductance for resistive circuits. The units of admittance are siemens, abbreviated as S. Recalling from Eq. 11-43 that $\mathbf{Z} = Z\,\underline{/\theta}$, we have

$$\mathbf{Y} = \frac{1}{|\mathbf{Z}|\,\underline{/\theta}}$$
$$= |\mathbf{Y}|\,\underline{/-\theta} \tag{11-50}$$

Therefore, $|\mathbf{Y}| = 1/|\mathbf{Z}|$ and the angle of $\mathbf{Y}$ is $-\theta$. We may also write the magnitude relation as $Y = 1/Z$.

Using the form

$$\mathbf{Z} = R + jX$$

we obtain

$$\mathbf{Y} = \frac{1}{R + jX}$$
$$= \frac{R - jX}{R^2 + X^2}$$
$$= G + jB \tag{11-51}$$

Note that G is not simply the reciprocal of R, nor is B the reciprocal of X. The real part of admittance, G, is called the *conductance,* and the imaginary part, B, is called the *susceptance.* The units of G and B are siemens.

The impedance of an element is $\mathbf{Z} = R + jX$. The element is inductive if the reactive part X is positive, capacitive if X is negative. Since $\mathbf{Y}$ is the reciprocal of $\mathbf{Z}$ and $\mathbf{Y} = G + jB$, one can also say that if B is positive the element is capacitive and that a negative B indicates an inductive element.

Let us consider the capacitor $C = 1$ mF and find its impedance and admittance. The impedance of a capacitor is

$$\mathbf{Z} = \frac{1}{j\omega C}$$

Therefore, in addition to the value of $C = 1$ mF, we need the frequency ω. If we consider the case $\omega = 100$ rad/s, we obtain

$$\mathbf{Z} = \frac{1}{j0.1}$$
$$= \frac{10}{j}$$
$$= -10j$$
$$= 10\,\underline{/-90°}\ \Omega$$

To find the admittance, we note that

$$\mathbf{Y} = \frac{1}{\mathbf{Z}}$$
$$= j\omega C$$
$$= j0.1$$
$$= 0.1\,\underline{/90°}\ S$$

EXERCISE 11-15

Find the impedance and admittance for an inductor when $L = 10$ mH and $\omega = 1000$.
Answer: $\mathbf{Z} = j10\ \Omega$
$\mathbf{Y} = \frac{1}{10}\ \underline{/-90°}\ \text{S}$

EXERCISE 11-16

Find the impedance and admittance at terminals a–b for the circuit of Figure E 11-16 when $\omega = 1000$ rad/s. Hint: first show that $C_{eq} = 12/7$ mF.
Answer: $\mathbf{Z} = -0.583j\ \Omega$
$\mathbf{Y} = 1.71j\ \text{S}$

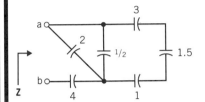

Figure E 11-16 All capacitors are in millifarads.

11-9 KIRCHHOFF'S LAWS USING PHASORS

Kirchhoff's current law and voltage law were considered earlier in the time domain. Consider the KVL around a closed path, which requires that

$$v_1 + v_2 + v_3 + \cdots + v_n = 0 \qquad (11\text{-}52)$$

For sinusoidal steady-state voltages, we may write the equation in terms of cosine waveforms as

$$V_{m_1}\cos(\omega t + \theta_1) + V_{m_2}\cos(\omega t + \theta_2) + \cdots + V_{m_n}\cos(\omega t + \theta_n) = 0 \quad (11\text{-}53)$$

All the information concerning each voltage v_n is incorporated in the magnitude and phase, V_{m_n} and θ_n (assuming we keep note of ω, which is the same for each term). Equation 11-53 can be rewritten, using Euler's identity, as

$$\text{Re}\{V_{m_1}e^{j\theta_1}e^{j\omega t}\} + \cdots + \text{Re}\{V_{m_n}e^{j\theta_n}e^{j\omega t}\} = 0$$

or

$$\text{Re}\{V_{m_1}e^{j\theta_1}e^{j\omega t} + \cdots + V_{m_n}e^{j\theta_n}e^{j\omega t}\} = 0$$

We can factor out the $e^{j\omega t}$ to obtain

$$\text{Re}\{(V_{m_1}e^{j\theta_1} + \cdots + V_{m_n}e^{j\theta_n})e^{j\omega t}\} = 0$$

Writing $V_{m_p}e^{j\theta_p}$ as $\mathbf{V}_p$, we have

$$\text{Re}(\mathbf{V}_1 + \mathbf{V}_2 + \cdots + \mathbf{V}_n)e^{j\omega t} = 0$$

Since $e^{j\omega t}$ cannot equal zero, we require that

$$\mathbf{V}_1 + \mathbf{V}_2 + \cdots + \mathbf{V}_n = 0 \qquad (11\text{-}54)$$

Therefore, we have the important result that the sum of the *phasor voltages* in a closed path is zero. Thus, Kirchhoff's voltage law holds in the frequency domain with phasor voltages. Using a similar process, one can show that Kirchhoff's current law holds in the frequency domain for phasors, so that at a node we have

$$\mathbf{I}_1 + \mathbf{I}_2 + \cdots + \mathbf{I}_n = 0 \tag{11-55}$$

Since both the KVL and the KCL hold in the frequency domain, it is easy to conclude that all the techniques of analysis we developed for resistive circuits hold for phasor currents and voltages. For example, we can use the principle of superposition, source transformations, Thévenin and Norton equivalent circuits, and node voltage and mesh current analysis. All these methods apply as long as the circuit is linear.

TOOLBOX 11-4

ADDITION OF COMPLEX NUMBERS

To add two or more complex numbers we add their real parts and their imaginary parts. Therefore, if $c_1 = a_1 + jb_1$ and $c_2 = a_2 + jb_2$, we have

$$c_1 + c_2 = (a_1 + a_2) + j(b_1 + b_2)$$

In general, for n complex numbers,

$$\sum_{i=1}^{n} c_i = \sum_{i=1}^{n} a_i + j \sum_{i=1}^{n} b_i$$

See Appendix C-3 for examples.

First, let us consider impedances connected in series, as shown in Figure 11-14. The

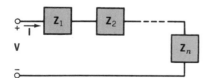

Figure 11-14 Impedances in series.

phasor current **I** flows through each impedance. Applying KVL, we can write

$$\mathbf{V}_1 + \mathbf{V}_2 + \cdots + \mathbf{V}_n = \mathbf{V}$$

Since $\mathbf{V}_j = \mathbf{Z}_j \mathbf{I}_j$, we have

$$(\mathbf{Z}_1 + \mathbf{Z}_2 + \cdots + \mathbf{Z}_n)\mathbf{I} = \mathbf{V}$$

Therefore, the equivalent impedance seen at the input terminals is

$$\mathbf{Z}_{eq} = \mathbf{Z}_1 + \mathbf{Z}_2 + \cdots + \mathbf{Z}_n \tag{11-56}$$

Thus, the equivalent impedance for a series of impedances is the sum of the individual impedances.

Consider the set of parallel admittances shown in Figure 11-15. It can easily be shown that the equivalent admittance $\mathbf{Y}_{eq}$ is

$$\mathbf{Y}_{eq} = \mathbf{Y}_1 + \mathbf{Y}_2 + \cdots + \mathbf{Y}_n \tag{11-57}$$

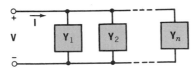

Figure 11-15 Admittances in parallel.

In the case of two parallel admittances, we have

$$\mathbf{Y}_{eq} = \mathbf{Y}_1 + \mathbf{Y}_2$$

and the corresponding equivalent impedance is

$$\mathbf{Z}_{eq} = \frac{1}{\mathbf{Y}_{eq}} = \frac{1}{\mathbf{Y}_1 + \mathbf{Y}_2} = \frac{\mathbf{Z}_1\mathbf{Z}_2}{\mathbf{Z}_1 + \mathbf{Z}_2} \tag{11-58}$$

Similarly, the current divider and voltage divider rules hold for phasor currents and voltages.

Example 11-5

Consider the *RLC* circuit shown in Figure 11-16 where $R = 9\ \Omega$, $L = 10$ mH, and $C = 1$ mF. Find the steady-state current i using phasors.

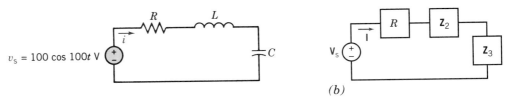

(b)

Figure 11-16 *(a)* An *RLC* series circuit. *(b)* Circuit in phasor form.

Solution

First, we redraw the circuit in phasor form as shown in Figure 11-16*b*. Then, we write KVL around the mesh to obtain

$$R\mathbf{I} + \mathbf{Z}_2\mathbf{I} + \mathbf{Z}_3\mathbf{I} = \mathbf{V}_s \tag{11-59}$$

where

$$\mathbf{Z}_2 = j\omega L$$

and

$$\mathbf{Z}_3 = \frac{1}{j\omega C}$$

Since $\omega = 100$, $L = 10$ mH, and $C = 1$ mF, we have

$$\mathbf{Z}_2 = j1$$

and

$$\mathbf{Z}_2 = -j10$$

Therefore, Eq. 11-59 becomes

$$(9 + j1 - j10)\mathbf{I} = \mathbf{V}_s \tag{11-60}$$

or

$$I = \frac{V_s}{9 - j9}$$

Since $V_s = 100\underline{/0°}$, we obtain the phasor current as

$$I = \frac{100\underline{/0°}}{9\sqrt{2}\underline{/-45°}}$$

$$= 7.86\underline{/45°}$$

Therefore, the steady-state current in the time domain is

$$i = 7.86 \cos(100t + 45°) \text{ A}$$

Example 11-6

Find the steady-state voltage v for the circuit of Figure 11-17 when $i_s = 10 \cos 1000t$ A, $R = 10 \ \Omega$, $L = 10$ mH, and $C = 100 \ \mu$F.

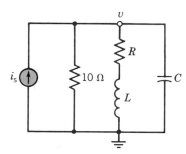

Figure 11-17 Circuit for Example 11-6.

Solution

The phasor voltage **V** can be obtained from Kirchhoff's current law at the top node as

$$(\mathbf{Y}_1 + \mathbf{Y}_2 + \mathbf{Y}_3)\mathbf{V} = \mathbf{I}_s \tag{11-61}$$

The source is represented by the phasor current

$$\mathbf{I}_s = 10\underline{/0°}$$

The first admittance, moving from left to right in the circuit, is

$$\mathbf{Y}_1 = \frac{1}{\mathbf{Z}_1} = \frac{1}{10}$$

The second admittance is evaluated for the R and L in series. Therefore, we have $\mathbf{Z}_2 = R + j\omega L$ and thus

$$\mathbf{Y}_2 = \frac{1}{\mathbf{Z}_2} = \frac{1}{10 + j10}$$

Finally, for the capacitor branch, $\mathbf{Z}_3 = 1/j\omega C$ and, hence,

$$\mathbf{Y}_3 = \frac{1}{\mathbf{Z}_3} = j\omega C = \frac{j}{10}$$

Substituting the admittances and the source current into Eq. 11-61, we have

$$\left(\frac{1}{10} + \frac{1}{10 + j10} + \frac{j}{10} \right) \mathbf{V} = 10$$

or

$$(0.15 + j0.05)\mathbf{V} = 10$$

Solving for $\mathbf{V}$, we have

$$\mathbf{V} = \frac{10\underline{/0°}}{0.158\underline{/18.4°}}$$
$$= 63.3\underline{/-18.4°}$$

Therefore, we have the steady-state voltage as

$$v = 63.3 \cos(1000t - 18.4°) \text{ V}$$

EXERCISE 11-17

Find the steady-state voltage v when $i_s = 10 \cos \omega t$ A and $\omega = 10^5$ rad/s for the circuit of Figure E 11-17.
Answer: $v = 50 \cos(\omega t - 36.9°)$ V where $\omega = 10^5$

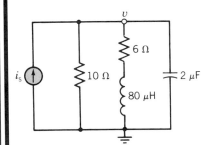

Figure E 11-17

EXERCISE 11-18

Find the steady-state current i when $v_s = 10 \cos 3t$ V for the circuit of Figure E 11-18.
Answer: $i = 2 \cos(3t + 36.9°)$ A

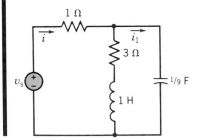

Figure E 11-18

EXERCISE 11-19

For the circuit of Figure E 11-18 find the current i_1 by the current divider principle using phasors.
Answer: $i_1 = 2\sqrt{2} \cos(3t + 81.9°)$ A

EXERCISE 11-20

Determine the equivalent impedance and admittance at terminals a–b of the circuit shown in Figure E 11-20 when $\omega = 100$ rad/s.

Answer: $\mathbf{Z} = 4 + j6 \; \Omega$
$\mathbf{Y} = 0.14 \; \underline{/-56.3°} \; S$

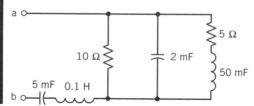

Figure E 11-20

11-10 NODE VOLTAGE AND MESH CURRENT ANALYSIS USING PHASORS

Circuit analysis in the frequency domain follows the same procedure as we utilized for resistive circuits; however, we use impedances and phasors instead of resistances and time functions. Since Ohm's law can be used in the frequency domain, we use the relationship $\mathbf{V} = \mathbf{ZI}$ for the passive elements and proceed to use the node voltage and mesh current techniques.

As an example of the node voltage method using phasors, consider the circuits of Figure 11-18 when $i_s = I_m \cos \omega t$. For a specified ω and for specified L and C, we can obtain the

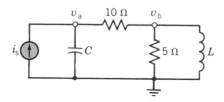

Figure 11-18 Circuit for which we wish to determine v_a and v_b.

impedance for the L and C elements. When $\omega = 1000$ rad/s and $C = 100 \; \mu F$, we obtain

$$\mathbf{Z}_1 = \frac{1}{j\omega C} = -j10$$

When $L = 5$ mH for the inductor, we have the impedance

$$\mathbf{Z}_L = j\omega L = j5$$

Then, we may redraw the circuit shown in Figure 11-18 using the phasor format shown in Figure 11-19. Clearly, $\mathbf{Z}_3 = 10 \; \Omega$ and $\mathbf{Z}_2$ is obtained from the parallel combination of the

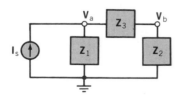

Figure 11-19 Circuit equivalent to that of Figure 11-18 in phasor form.

5-Ω resistor and the inductor's impedance $\mathbf{Z}_L$. Rather than obtaining $\mathbf{Z}_2$, let us determine $\mathbf{Y}_2$, which is readily found by adding the two parallel admittances as follows:

$$\mathbf{Y}_2 = \frac{1}{5} + \frac{1}{\mathbf{Z}_L}$$
$$= \frac{1}{5} + \frac{1}{j5}$$
$$= \frac{1}{5}(1 - j)$$

Using KCL at node a, we have

$$\frac{\mathbf{V}_a}{\mathbf{Z}_1} + \frac{\mathbf{V}_a - \mathbf{V}_b}{\mathbf{Z}_3} = \mathbf{I}_s \tag{11-62}$$

At node b, we have

$$\frac{\mathbf{V}_b}{\mathbf{Z}_2} + \frac{\mathbf{V}_b - \mathbf{V}_a}{\mathbf{Z}_3} = 0 \tag{11-63}$$

Rearranging Eqs. 11-62 and 11-63, we obtain

$$(\mathbf{Y}_1 + \mathbf{Y}_3)\mathbf{V}_a + (-\mathbf{Y}_3)\mathbf{V}_b = \mathbf{I}_s \tag{11-64}$$

$$(-\mathbf{Y}_3)\mathbf{V}_a + (\mathbf{Y}_2 + \mathbf{Y}_3)\mathbf{V}_b = 0 \tag{11-65}$$

where we use the admittance $\mathbf{Y}_n = 1/\mathbf{Z}_n$ and $\mathbf{I}_s = I_m \underline{/0°}$.

We find that Eqs. 11-64 and 11-65 are similar to the node voltage equations we found in Chapter 4 for resistive circuits. In this case, however, we obtain the node voltage equations in terms of phasor currents, phasor voltages, and complex impedances and admittances.

In general, we may state that for circuits containing only admittances and independent sources, KCL at node k requires that the coefficient of $\mathbf{V}_k$ be the sum of the admittances at node k and the coefficients of the other terms are the negative of the admittance between those nodes and the kth node.

Let us proceed to solve for $\mathbf{V}_a$ for the circuit shown in Figures 11-18 and 11-19 when $I_m = 10$ A. Substituting the admittances into Eqs. 11-64 and 11-65, we have

$$\left(\frac{1}{-j10} + \frac{1}{10}\right)\mathbf{V}_a + \frac{-1}{10}\mathbf{V}_b = 10 \tag{11-66}$$

$$\frac{-1}{10}\mathbf{V}_a + \left[\frac{1}{5}(1 - j) + \frac{1}{10}\right]\mathbf{V}_b = 0 \tag{11-67}$$

We then use Cramer's rule to solve for $\mathbf{V}_a$, obtaining

$$\mathbf{V}_a = \frac{100(3 - 2j)}{4 + j}$$
$$= \frac{100(3 - 2j)(4 - j)}{17}$$
$$= \frac{100}{17}(10 - 11j)$$
$$= 87.5\underline{/-47.7°}$$

Therefore, we have the steady-state voltage v_a:

$$v_a = 87.5 \cos(1000t - 47.7°) \text{ V}$$

The general nodal analysis methods of Chapter 4 may be utilized here, where we are careful to note that we use complex impedances and admittances and phasor voltages and currents. Once we have determined the desired phasor currents or voltages, we transform them back to the time domain to obtain the steady-state sinusoidal current or voltage desired. We use the concept of a supernode, if necessary, and include the effect of a dependent source, if required.

Example 11-7

A circuit is shown in Figure 11-20 with $\omega = 10$ rad/s, $L = 0.5$ H, and $C = 10$ mF. Find the node voltage v in its sinusoidal steady form when $v_s = 10 \cos \omega t$ V.

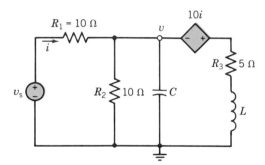

Figure 11-20 Circuit for Example 11-7.

Solution

The circuit has a dependent source between two nodes, so we identify a supernode as shown in Figure 11-21, where we also show the impedance for each element in phasor

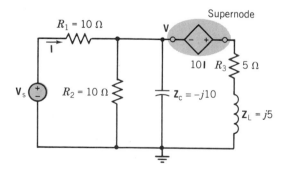

Figure 11-21 Circuit for Example 11-7 in phasor notation.

form. For example, the impedance of the inductor is $\mathbf{Z}_L = j\omega L = j5$. Similarly, the impedance for the capacitor is

$$\mathbf{Z}_c = \frac{1}{j\omega C} = \frac{10}{j} = -j10$$

First, we note that $\mathbf{Y}_1 = 1/R_1 = 1/10$. We now wish to bring together the two parallel admittances for R_2 and C to yield one admittance $\mathbf{Y}_2$ as shown in Figure 11-22. We then

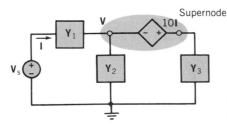

Figure 11-22 Circuit for Example 11-7 with three admittances and the supernode identified.

obtain

$$\mathbf{Y}_2 = \frac{1}{R_2} + \frac{1}{\mathbf{Z}_c}$$

$$= \frac{1}{10} + \frac{j}{10}$$

$$= \frac{1}{10}(1 + j)$$

We may obtain $\mathbf{Y}_3$ for the series resistance and inductance as

$$\mathbf{Y}_3 = \frac{1}{\mathbf{Z}_3}$$

where $\mathbf{Z}_3 = R_3 + \mathbf{Z}_L = 5 + j5$. Therefore, we have

$$\mathbf{Y}_3 = \frac{1}{5 + j5} = \frac{1}{50}(5 - j5)$$

Writing the KCL at the supernode of Figure 11-22, we have

$$\mathbf{Y}_1(\mathbf{V} - \mathbf{V}_s) + \mathbf{Y}_2\mathbf{V} + \mathbf{Y}_3(\mathbf{V} + 10\mathbf{I}) = 0 \tag{11-68}$$

Furthermore, we note that

$$\mathbf{I} = \mathbf{Y}_1(\mathbf{V}_s - \mathbf{V}) \tag{11-69}$$

Substituting Eq. 11-69 into Eq. 11-68, we obtain

$$\mathbf{Y}_1(\mathbf{V} - \mathbf{V}_s) + \mathbf{Y}_2\mathbf{V} + \mathbf{Y}_3[\mathbf{V} + 10\mathbf{Y}_1(\mathbf{V}_s - \mathbf{V})] = 0$$

Rearranging, we have

$$(\mathbf{Y}_1 + \mathbf{Y}_2 + \mathbf{Y}_3 - 10\mathbf{Y}_1\mathbf{Y}_3)\mathbf{V} = (\mathbf{Y}_1 - 10\mathbf{Y}_1\mathbf{Y}_3)\mathbf{V}_s$$

Therefore

$$\mathbf{V} = \frac{(\mathbf{Y}_1 - 10\mathbf{Y}_1\mathbf{Y}_3)\mathbf{V}_s}{\mathbf{Y}_1 + \mathbf{Y}_2 + \mathbf{Y}_3 - 10\mathbf{Y}_1\mathbf{Y}_3}$$

Since $\mathbf{V}_s = 10\underline{/0°}$, we have

$$\mathbf{V} = \frac{(\frac{1}{10} - \frac{1}{50}(5 - j5))10}{\frac{1}{10} + \frac{1}{10}(1 + j)}$$

$$= \frac{1 - (1 - j)}{\frac{1}{10}(2 + j)}$$

$$= \frac{10j}{2 + j}$$

Therefore, we obtain

$$v = \frac{10}{\sqrt{5}} \cos(10t + 63.4°) \text{ V}$$

The processes of node voltage and mesh current analysis using phasors for determining the steady-state sinusoidal response of a circuit are recorded in Tables 11-6 and 11-7, respectively.

Mesh current analysis, using the method of Table 11-7, is relatively straightforward. Once you have the impedance of each element, you may readily write the KVL equations for each mesh.

Example 11-8

Find the steady-state sinusoidal current i_1 for the circuit of Figure 11-23 when $v_s = 10\sqrt{2} \cos(\omega t + 45°)$ V and $\omega = 100$ rad/s. Also, $L = 30$ mH and $C = 5$ mF.

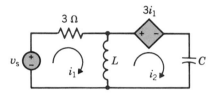

Figure 11-23 Circuit of Example 11-8.

Solution

First, we transform the source voltage to phasor form to obtain

$$\mathbf{V}_s = 10\sqrt{2}\underline{/45°}$$
$$= 10 + 10j$$

We then select the two mesh currents as $\mathbf{I}_1$ and $\mathbf{I}_2$, as shown in Figure 11-24. Since the

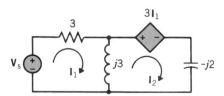

Figure 11-24 Circuit of Example 11-8 with phasors and impedances.

frequency of the source is $\omega = 100$, we find that the inductance has an impedance of

$$\mathbf{Z}_L = j\omega L$$
$$= j3$$

The capacitor has an impedance of

$$\mathbf{Z}_c = \frac{1}{j\omega C}$$
$$= \frac{1}{j(\frac{1}{2})}$$
$$= -j2$$

We can then summarize the circuit's phasor currents and the impedance of each element by redrawing the circuit in terms of phasors, as shown in Figure 11-24.

Table 11-6
Node Voltage Analysis Using the Phasor Concept to Find the Sinusoidal Steady-State Node Voltages

1. Convert the independent sources to phasor form.
2. Select the nodes and the reference node and label the node voltages in the time domain, v_n, and the corresponding phasor voltage, $\mathbf{V}_n$.
3. If the circuit contains only independent current sources, proceed to step 5; otherwise proceed to step 4.
4. If the circuit contains a voltage source, select one of the following three cases and the associated method:

Case	Method
a. The voltage source connects node q and the reference node.	Set $\mathbf{V}_q = \mathbf{V}_s$ and proceed.
b. The voltage source lies between two nodes.	Create a supernode including both nodes.
c. The voltage source in series with an impedance lies between node d and the ground with its positive terminal at node d.	Replace the voltage source and series impedance with a parallel combination of an admittance $\mathbf{Y}_1 = 1/\mathbf{Z}_1$ and a current source $\mathbf{I}_1 = \mathbf{V}_s\mathbf{Y}_1$ entering node d.

5. Using the known frequency of the sources, ω, find the impedance of each element in the circuit.
6. For each branch at a given node, find the equivalent admittance of that branch, $\mathbf{Y}_n$.
7. Write KCL at each node.
8. Solve for the desired node voltage $\mathbf{V}_a$ using Cramer's rule.
9. Convert the phasor voltage $\mathbf{V}_a$ back to the time domain form.

Table 11-7
Mesh Current Analysis Using the Phasor Concept to Find the Sinusoidal Steady-State Mesh Currents

1. Convert the independent sources to phasor form.
2. Select the mesh currents and label the currents in the time domain, i_n, and the corresponding phasor currents, $\mathbf{I}_n$.
3. If the circuit contains only independent voltage sources, proceed to step 5; otherwise proceed to step 4.
4. If the circuit contains a current source, select one of the following two cases and the associated method:

Case	Method
a. The current source appears as an element of only one mesh, n.	Equate the mesh current $\mathbf{I}_n$ to the current of the source, accounting for the direction of the current source.
b. The current source is common to two meshes.	Create a supermesh as the periphery of the two meshes. In step 5 write one KVL equation around the periphery of the supermesh. Also record the constraining equation incurred by the current source.

5. Using the known frequency of the sources, ω, find the impedance of each element in the circuit.
6. Write KVL for each mesh.
7. Solve for the desired mesh current $\mathbf{I}_n$ using Cramer's rule.
8. Convert the phasor current $\mathbf{I}_n$ back to the time domain form.

Now we can write the KVL equations for each mesh, obtaining

$$\text{mesh 1:} \qquad (3 + j3)\mathbf{I}_1 - j3\mathbf{I}_2 = \mathbf{V}_s$$

$$\text{mesh 2:} \qquad (3 - j3)\mathbf{I}_1 + (j3 - j2)\mathbf{I}_2 = 0$$

Solving for $\mathbf{I}_1$ using Cramer's rule, we have

$$\mathbf{I}_1 = \frac{(10 + j10)j}{\Delta}$$

where the determinant is

$$\Delta = (3 + j3)(j) + j3(3 - j3)$$
$$= 6 + 12j$$

Therefore, we have

$$\mathbf{I}_1 = \frac{10j - 10}{6 + 12j}$$

Continuing, we obtain

$$\mathbf{I}_1 = \frac{10(j - 1)}{6(1 + 2j)}$$
$$= \frac{10(\sqrt{2}\,\underline{/135°})}{6(\sqrt{5}\,\underline{/63.4°})}$$
$$= 1.05\,\underline{/71.6°}$$

Thus, the steady-state time response is

$$i_1 = 1.05 \cos(100t + 71.6°) \text{ A}$$

Example 11-9

Find the steady-state current i_1 when the voltage source is $v_s = 10\sqrt{2} \cos(\omega t + 45°)$ V and the current source is $i_s = 3 \cos \omega t$ A for the circuit of Figure 11-25. The circuit of the figure provides the impedance in ohms for each element at the specified ω.

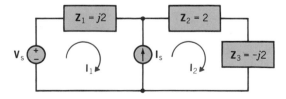

Figure 11-25 Circuit of Example 11-9.

Solution

First, we transform the independent sources into phasor form. The voltage source is

$$\mathbf{V}_s = 10\sqrt{2}\,\underline{/45°}$$
$$= 10(1 + j)$$

and the current source is

$$\mathbf{I}_s = 3\,\underline{/0°}$$

We note that the current source connects the two meshes and provides a constraining equation

$$\mathbf{I}_2 - \mathbf{I}_1 = \mathbf{I}_s \tag{11-70}$$

Creating a supermesh around the periphery of the two meshes, we write one KVL equation, obtaining

$$\mathbf{I}_1\mathbf{Z}_1 + \mathbf{I}_2(\mathbf{Z}_2 + \mathbf{Z}_3) = \mathbf{V}_s \tag{11-71}$$

Since we wish to solve for $\mathbf{I}_1$, we will use $\mathbf{I}_2$ from Eq. 11-70 and substitute it into Eq. 11-71, obtaining

$$\mathbf{I}_1\mathbf{Z}_1 + (\mathbf{I}_s + \mathbf{I}_1)(\mathbf{Z}_2 + \mathbf{Z}_3) = \mathbf{V}_s$$

Rearranging, we have

$$(\mathbf{Z}_1 + \mathbf{Z}_2 + \mathbf{Z}_3)\mathbf{I}_1 = \mathbf{V}_s - (\mathbf{Z}_2 + \mathbf{Z}_3)\mathbf{I}_s$$

Therefore, we have

$$\mathbf{I}_1 = \frac{\mathbf{V}_s - (\mathbf{Z}_2 + \mathbf{Z}_3)\mathbf{I}_s}{\mathbf{Z}_1 + \mathbf{Z}_2 + \mathbf{Z}_3}$$

Substituting the impedances and the sources, we have

$$\mathbf{I} = \frac{(10 + j10) - (2 - j2)3}{2}$$
$$= 2 + j8$$
$$= 8.25\underline{/76°}$$

Thus, we obtain

$$i_1 = 8.25 \cos(\omega t + 76°) \text{ A}$$

EXERCISE 11-21

A circuit has the form shown in Figure E 11-21 when $i_{s1} = 1 \cos 100t$ A and $i_{s2} = 0.5 \cos(100t - 90°)$ A. Find the voltage v_a in the time domain.

Answer: $v_a = \sqrt{5} \cos(100t - 63.5°)$ V

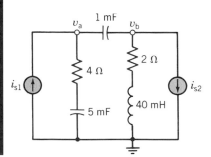

Figure E 11-21

EXERCISE 11-22

Use mesh current analysis for the circuit of Figure E 11-22 to find the steady-state voltage across the inductor, v_L, when $v_{s1} = 20 \cos \omega t$ V, $v_{s2} = 30 \cos(\omega t - 90°)$ V, and $\omega = 1000$ rad/s.

Answer: $v_L = 24\sqrt{2} \cos(\omega t + 82°)$ V

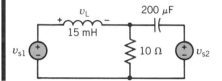

Figure E 11-22

EXERCISE 11-23

Determine the node phasor voltages at terminals a and b for the circuit of Figure E 11-23 when $\mathbf{V}_s = j50$ V and $\mathbf{V}_1 = j30$ V.

Answer: $\mathbf{V}_a = 14.33\underline{/-71.75°}$ V, $\mathbf{V}_b = 36.67\underline{/83°}$ V

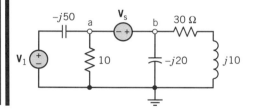

Figure E 11-23 All impedances in ohms.

11-11	## SUPERPOSITION, THÉVENIN AND NORTON EQUIVALENTS, AND SOURCE TRANSFORMATIONS

Circuits in the frequency domain with phasor currents and voltages and impedances are analogous to the resistive circuits we considered earlier. Since they are linear, we expect that the principle of superposition and the source transformation method will hold. Furthermore, we can define Thévenin and Norton equivalent circuits in terms of impedance or admittance.

First, let us consider the *superposition principle,* which may be restated as follows: For a linear circuit containing two or more independent sources, any circuit voltage or current may be calculated as the algebraic sum of all the individual currents or voltages caused by each independent source acting alone.

If a linear circuit is excited by several sinusoidal sources all having the same frequency, ω, then superposition *may* be used. If a linear circuit is excited by several sources all having different frequencies, then superposition *must* be used.

The superposition principle is particularly useful if a circuit has two or more sources acting at different frequencies. Clearly, the circuit will have one set of impedance values at one frequency and a different set of impedance values at another frequency. We can determine the phasor response at each frequency. Then we find the time response corresponding to each phasor response and add them. Note that superposition, in the case of sources operating at two or more frequencies applies to time responses only. We cannot superpose the phasor responses.

Example 11-10
Using the superposition principle, find the steady-state current i for the circuit shown in

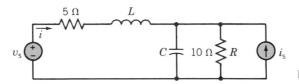

Figure 11-26 Circuit of Example 11-10.

Figure 11-26 when $v_s = 10 \cos 10t$ V, $i_s = 3$ A, $L = 1.5$ H, and $C = 10$ mF.

Solution
The first step is to convert the independent sources into phasor form noting that the sources operate at different frequencies. For the voltage source operating at $\omega = 10$, we have

$$\mathbf{V}_s = 10\underline{/0°}$$

We note that the current source is a direct current, so we can state that $\omega = 0$ for the current source. The phasor form of the current source is then

$$\mathbf{I}_s = 3\underline{/0°}$$

The second step is to convert the circuit to phasor form with the impedance of each element shown as in Figure 11-27.

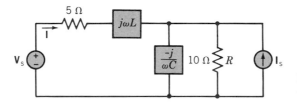

Figure 11-27 Circuit of Example 11-10 with phasors and impedances.

Now let us determine the phasor current $\mathbf{I}_1$, which is the component of current $\mathbf{I}$ due to the voltage source. We remove the current source, replacing it with an open circuit across the 10-Ω resistor. Then we may find the current $\mathbf{I}_1$ due to the first source as

$$\mathbf{I}_1 = \frac{\mathbf{V}_s}{5 + j\omega L + \mathbf{Z}_p} \tag{11-72}$$

where $\mathbf{Z}_p$ is the impedance of the capacitor and the 10-Ω resistance in parallel. Recall that $\omega = 10$ and $C = 10$ mF. Therefore, since $\mathbf{Z}_c = -j10$, we have

$$\begin{aligned}
\mathbf{Z}_p &= \frac{\mathbf{Z}_c R}{R + \mathbf{Z}_c} \\
&= \frac{(-j10)10}{10 - j10} \\
&= 5(1 - j)
\end{aligned}$$

Substituting $\mathbf{Z}_p$ and $\omega L = 15$ into Eq. 11-72, we have

$$\mathbf{I}_1 = \frac{10\underline{/0°}}{5 + j15 + (5 - j5)}$$

$$= \frac{10}{10 + j10}$$

$$= \frac{10}{\sqrt{200}}\underline{/-45°}$$

Therefore, the time domain current resulting from the voltage source is

$$i_1 = 0.71 \cos(10t - 45°) \text{ A}$$

Now, let us consider the situation for the current source with the voltage source deactivated. Setting the voltage source to zero results in a short circuit. Since $\omega = 0$ for the dc source, the capacitor impedance becomes an open circuit because $\mathbf{Z} = 1/j\omega C = \infty$. The inductor's impedance becomes a short circuit because $\mathbf{Z} = j\omega L = 0$. Hence, we obtain the circuit shown in Figure 11-28. We see that we have returned to a

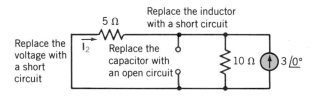

Figure 11-28 Circuit of Example 11-10 for the dc current source acting alone.

familiar resistive circuit for a dc source. Then the response due to the current source is

$$\mathbf{I}_2 = -\frac{10}{15}3$$

$$= -2 \text{ A}$$

Therefore, using the principle of superposition, the total steady-state current is $i = i_1 + i_2$ or

$$i = 0.71 \cos(10t - 45°) - 2 \text{ A}$$

Now let us consider the *source transformations* for frequency domain (phasor) circuits. The techniques considered for resistive circuits discussed in Chapter 5 can readily be extended. The source transformation is concerned with transforming a voltage source and its associated series impedance to a current source and its associated parallel impedance,

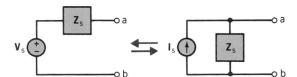

Figure 11-29 Two equivalent sources.

or vice versa, as shown in Figure 11-29. The method of transforming from one source to another source is summarized in Figure 11-30.

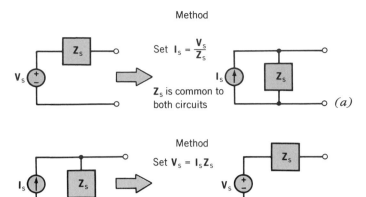

Figure 11-30 Method of source transformations. *(a)* Converting a voltage source to a current source. *(b)* Converting a current source to a voltage source.

Example 11-11

A circuit has a voltage source v_s in series with two elements, as shown in Figure 11-31. Determine the phasor equivalent current source form when $v_s = 10 \cos(\omega t + 45°)$ V and $\omega = 100$ rad/s.

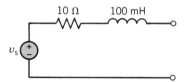

Figure 11-31 Circuit of Example 11-11.

Solution

First, we determine the equivalent current source as

$$\mathbf{I}_s = \frac{\mathbf{V}_s}{\mathbf{Z}_s}$$

Since $\mathbf{Z}_s = 10 + j10$ and $\mathbf{V}_s = 10\underline{/45°}$, we obtain

$$\mathbf{I}_s = \frac{10\underline{/45°}}{\sqrt{200}\underline{/45°}}$$

$$= \frac{10}{\sqrt{200}}\underline{/0°} \text{ A}$$

The equivalent current source circuit is shown in Figure 11-32.

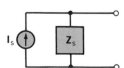

Figure 11-32 Circuit of Example 11-11 transformed to a current source where $\mathbf{Z}_s = 10 + j10$ Ω and $\mathbf{I}_s = 1/\sqrt{2}$ A.

Thévenin's and Norton's theorems apply to phasor current or voltages and impedances in the same way that they do for resistive circuits. The Thévenin theorem is used to obtain an equivalent circuit as discussed in Chapter 5. The Thévenin equivalent circuit is shown in Figure 11-33.

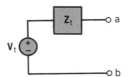

b   **Figure 11-33** The Thévenin equivalent circuit.

A procedure for determining the Thévenin equivalent circuit is as follows:

1   Identify a separate circuit portion of a total circuit.
2   Determine the Thévenin voltage $\mathbf{V}_t = \mathbf{V}_{oc}$, the open-circuit voltage at the terminals.
3   (a) Find $\mathbf{Z}_t$ by deactivating all the independent sources and reducing the circuit to an equivalent impedance; (b) if the circuit has one or more dependent sources, then either short circuit the terminals and determine $\mathbf{I}_{sc}$ from which $\mathbf{Z}_t = \mathbf{V}_{oc}/\mathbf{I}_{sc}$; or (c) deactivate the independent sources, attach a voltage or current source at the terminals, and determine both $\mathbf{V}$ and $\mathbf{I}$ at the terminals from which $\mathbf{Z}_t = \mathbf{V}/\mathbf{I}$.

Example 11-12

Find the Thévenin equivalent circuit for the circuit shown in Figure 11-34 when $\mathbf{Z}_1 = 1 + j$ and $\mathbf{Z}_2 = -j1$.

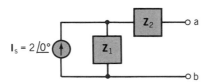

Figure 11-34 Circuit of Example 11-12.

Solution

The open-circuit voltage is

$$\mathbf{V}_{oc} = \mathbf{I}_s \mathbf{Z}_1$$
$$= (2\underline{/0°})(1 + j)$$
$$= 2\sqrt{2}\underline{/45°}$$

The impedance $\mathbf{Z}_t$ is found by deactivating the current source by replacing it with an open circuit. Then we have $\mathbf{Z}_1$ in series with $\mathbf{Z}_2$, so that

$$\mathbf{Z}_t = \mathbf{Z}_1 + \mathbf{Z}_2$$
$$= (1 + j) - j$$
$$= 1\ \Omega$$

Example 11-13

Find the Thévenin equivalent circuit of the circuit shown in Figure 11-35 in phasor form.

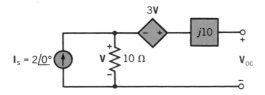

Figure 11-35 Circuit of Example 11-13.

Solution

The Thévenin voltage $\mathbf{V}_t = \mathbf{V}_{oc}$, so we first determine $\mathbf{V}_{oc}$. Note that with the open circuit,

$$\mathbf{V} = 10\mathbf{I}_s$$
$$= 20$$

Then, for the mesh on the right, using KVL, we have

$$\mathbf{V}_{oc} = 3\mathbf{V} + \mathbf{V}$$
$$= 4\mathbf{V}$$
$$= 80\underline{/0^\circ}$$

Examining the circuit of Figure 11-35, we transform the current source and 10-Ω resistance to the voltage source and 10-Ω series resistance as shown in Figure 11-36a. When the

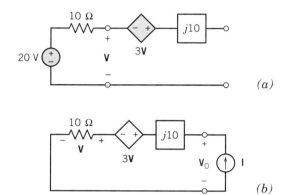

(a)

(b)

Figure 11-36 *(a)* Circuit of Example 11-13 with an open circuit at the output and the current source transformed to a voltage source. *(b)* Circuit with a test current source connected at the output terminal.

voltage source is deactivated and a current source is connected at the terminals as shown in Figure 11-36b, KVL gives

$$\mathbf{V}_o = j10\mathbf{I} + 4\mathbf{V}$$
$$= (j10 + 40)\mathbf{I}$$

Therefore,

$$\mathbf{Z}_t = 40 + j10 \ \Omega$$

Now let us consider the procedure for finding the Norton equivalent circuit. The steps are similar to those used for the Thévenin equivalent since $\mathbf{Z}_t$ in series with the Thévenin voltage is equal to the Norton impedance in parallel with the Norton current source. The Norton equivalent circuit is shown in Figure 11-37.

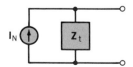

Figure 11-37 The Norton equivalent circuit expressed in terms of a phasor current and an impedance.

In order to determine the Norton circuit, we follow the procedure as follows:

1 Identify a separate circuit portion of a total circuit.
2 The Norton current $\mathbf{I}_N$ is the current through a short circuit at the terminals, so $\mathbf{I}_N = \mathbf{I}_{sc}$.

3 Find $\mathbf{Z}_t$ by (a) deactivating all the independent sources and reducing the circuit to an equivalent impedance, *or* (b) if the circuit has one or more dependent sources, find the open-circuit voltage at the terminals, $\mathbf{V}_{oc}$, so that

$$\mathbf{Z}_t = \frac{\mathbf{V}_{oc}}{\mathbf{I}_{sc}}$$

Example 11-14

Find the Norton equivalent of the circuit shown in Figure 11-38 in phasor and impedance forms. Assume that $\mathbf{V}_s = 100\underline{/0°}$ V.

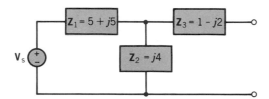

Figure 11-38 Circuit of Example 11-14.

Solution

First, let us find the equivalent impedance by deactivating the voltage source by replacing it with a short circuit. Since $\mathbf{Z}_1$ appears in parallel with $\mathbf{Z}_2$, we have

$$\mathbf{Z}_t = \mathbf{Z}_3 + \frac{\mathbf{Z}_1\mathbf{Z}_2}{\mathbf{Z}_1 + \mathbf{Z}_2}$$

$$= (1 - j2) + \frac{(5 + j5)(j4)}{(5 + j5) + (j4)}$$

$$= (1 - j2) + \frac{20}{53}(2 + j7)$$

$$= \frac{93}{53} + j\frac{34}{53}$$

$$= \frac{1}{53}(93 + j34)$$

We now proceed to determine the Norton equivalent current source by determining the current flowing through a short circuit connected at terminals a–b, as shown in Figure 11-39.

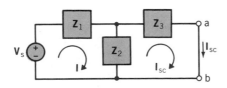

Figure 11-39 Circuit of Example 11-14 with a short circuit at terminals a–b.

We will use mesh currents to find $\mathbf{I}_{sc}$ as shown in Figure 11-39. The two mesh KVL equations are

$$\text{mesh 1:} \qquad (\mathbf{Z}_1 + \mathbf{Z}_2)\mathbf{I} + (-\mathbf{Z}_2)\mathbf{I}_{sc} = \mathbf{V}_s$$

$$\text{mesh 2:} \qquad (-\mathbf{Z}_2)\mathbf{I} + (\mathbf{Z}_2 + \mathbf{Z}_3)\mathbf{I}_{sc} = 0$$

Using Cramer's rule, we find $\mathbf{I}_N = \mathbf{I}_{sc}$ as follows:

$$
\begin{aligned}
\mathbf{I}_{sc} &= \frac{\mathbf{Z}_2\mathbf{V}_s}{(\mathbf{Z}_1 + \mathbf{Z}_2)(\mathbf{Z}_2 + \mathbf{Z}_3) - \mathbf{Z}_2^2} \\
&= \frac{(j4)100}{(5 + j9)(1 + j2) - (-16)} \\
&= \frac{j400}{3 + j19} \\
&= \frac{400}{370}(19 + 3j) \ \text{A}
\end{aligned}
$$

EXERCISE 11-24

Find the Thévenin equivalent circuit for the circuit of Figure E 11-24 when $b = 0.1$.
Answer: $\mathbf{V}_t = 8.9\ \underline{/63.4°}$ V
$\mathbf{Z}_t = 2 + j4\ \Omega$

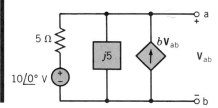

Figure E 11-24 Impedances in ohms.

EXERCISE 11-25

Find the Norton equivalent circuit for the circuit of Figure E 11-24 when $b = 0.1$.
Answer: $\mathbf{I}_N = 2\underline{/0°}$ A
$\mathbf{Z}_t = 2 + j4\ \Omega$

EXERCISE 11-26

Using the principle of superposition, determine $i(t)$ of the circuit shown in Figure E 11-26 when $v_1 = 10 \cos 10t$ V.
Answer: $i = -2 + 0.71 \cos(10t - 45°)$ A

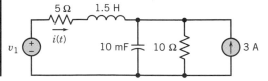

Figure E 11-26

11-12 | PHASOR DIAGRAMS

Phasors representing the voltage or current of a circuit are time quantities transformed or converted into the frequency domain. Phasors are complex numbers and can be portrayed in a complex plane. The relationship of phasors on a complex plane is called a *phasor diagram*.

Let us consider an *RLC* series circuit as shown in Figure 11-40. The impedance of each

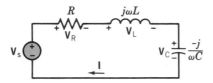

Figure 11-40 An *RLC* circuit.

element is also identified in the diagram. Since the current flows through all elements and is common to all, we take **I** as the reference phasor.

$$\mathbf{I} = I\underline{/0°}$$

Then the voltage phasors are

$$\mathbf{V}_R = R\mathbf{I} = RI\underline{/0°} \tag{11-73}$$

$$\mathbf{V}_L = j\omega L\mathbf{I} = \omega LI\underline{/90°} \tag{11-74}$$

$$\mathbf{V}_c = \frac{-j\mathbf{I}}{\omega C} = \frac{I}{\omega C}\underline{/-90°} \tag{11-75}$$

These phasors are shown in the phasor diagram of Figure 11-41. Note that KVL for this

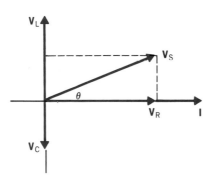

Figure 11-41 Phasor diagram for the *RLC* circuit of Figure 11-40.

circuit requires that

$$\mathbf{V}_s = \mathbf{V}_R + \mathbf{V}_L + \mathbf{V}_c$$

A **phasor diagram** is a graphical representation of phasors and their relationship on the complex plane.

The current **I** and the voltage across the resistor are in phase. The inductor voltage leads the current by 90°, and the capacitor voltage lags the current by 90°. For a given *L* and *C* there will be a frequency ω that results in

$$| \mathbf{V}_L | + | \mathbf{V}_c |$$

Referring to Eqs. 11-74 and 11-75, this equality of voltage magnitudes occurs when

$$\omega L I = \frac{I}{\omega C}$$

or

$$\omega^2 = \frac{1}{LC}$$

When $\omega^2 = 1/LC$, the magnitudes of the inductor voltage and capacitor voltage are equal. Since they are out of phase by 180°, they cancel, and the resulting condition is

$$\mathbf{V}_s = \mathbf{V}_R$$

and then both $\mathbf{V}_s$ and $\mathbf{V}_R$ are in phase with $\mathbf{I}$. This condition is called *resonance*.

EXERCISE 11-27

Consider the *RLC* series circuit of Figure 11-40 when $L = 1$ mH and $C = 1$ mF. Find the frequency ω when the current, source voltage, and $\mathbf{V}_R$ are all in phase.
Answer: $\omega = 1000$ rad/s.

EXERCISE 11-28

Draw the phasor diagram for the circuit of Figure E 11-28 when $\mathbf{V} = V\underline{/0°}$. Show each current on the diagram.

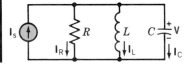

Figure E 11-28

EXERCISE 11-29

The circuit shown in Figure E 11-29 contains a sinusoidal current source of $25\underline{/0°}$ A. An ammeter reads the magnitude of the current. Ammeter A_1 reads 15 A, and ammeter A_2 reads 6 A. Find the reading of ammeter A_3.

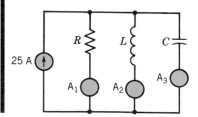

Figure E 11-29

11-13 SINUSOIDAL STEADY-STATE ANALYSIS USING PSpice

We can readily determine the sinusoidal steady-state response of a circuit using the **.AC** analysis command of PSpice (see Appendix G-7). This is often called ac analysis and the calculation is obtained for a specific frequency. The form of the **.AC** statement is

.AC <sweep type> <n> <start freq> <end freq>

For phasor analysis, start freq = end freq and we choose LIN (linear) for the sweep type. Then n is the number of frequency points that will be output, which is one for the single frequency chosen. Thus, if we wish to determine the phasor calculation at a frequency of 100 Hz, we use

.AC LIN 1 100 100

This statement requires the frequency stated in hertz. The output of the calculation is obtained by using the statement

.PRINT AC <ac output variables>

Let us use PSpice to determine the solution of Example 11-6. The circuit of Figure 11-17 is redrawn in Figure 11-42 where $i_s = 10 \cos 1000t$. The PSpice program is shown in

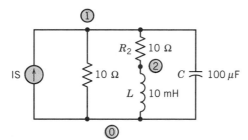

Figure 11-42 The circuit of Example 11-6 formatted for PSpice.

Figure 11-43. The current source is designated as IS and has a magnitude of 10 and a phase of 0°. PSpice uses $M \sin(\omega t + \phi)$ as the sinusoidal input signal for transient analysis. However, for **.AC** analysis, we can designate the input source as a cosine or sine function and PSpice determines the angle of the output variable, given the assumed form of the input source. Note that the **.AC** statement calculates the response at $f = 1000/2\pi = 159.15$ Hz as required by PSpice.

We then call for a printout of the magnitude and phase of the voltage at node 1, V(1). The simulation results are provided and we have VM(1) = 63.28 V and VP(1) = − 18.4°. Therefore, the capacitor voltage is

$$v = 63.3 \cos(1000t - 18.4°)$$

as we determined in Example 11-6.

```
                EXAMPLE 11-6 BY SPICE


    IS        0     1      AC        10    0

    R1        1     0      10

    R2        1     2      10

    L1        2     0      10M

    C1        1     0      0.1M

    .AC      LIN    1     159.15        159.15

    .PRINT    AC           VM(1)       VP(1)

    .END
```

Figure 11-43 The PSpice program for Example 11-6.

11-14 ‖ PHASOR CIRCUITS AND THE OPERATIONAL AMPLIFIER

The discussion in the prior sections considered the behavior of operational amplifiers and their associated circuits in the time domain. In this section we consider the behavior of operational amplifiers and associated *RLC* circuits in the frequency domain using phasors.

Figure 11-44 shows two frequently used operational amplifier circuits, the inverting amplifier, and the noninverting amplifier. These circuits are represented using imped-ances and phasors. This representation is appropriate when the input is sinusoidal and the circuit is at steady state. $\mathbf{V}_s$ is the phasor corresponding to a sinusoidal input voltage and $\mathbf{V}_o$ is the phasor representing the resulting sinusoidal output voltage. Both circuits involve two impedances, $\mathbf{Z}_1$ and $\mathbf{Z}_2$.

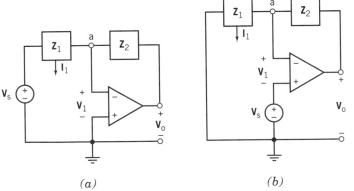

(a)   (b)

Figure 11-44 (a) An inverting amplifier and (b) a noninverting amplifier.

Now let us proceed to determine the ratio of output-to-input voltage, $\mathbf{V}_o/\mathbf{V}_s$, for the inverting amplifier shown in Figure 11-44a. This circuit can be analyzed by writing the node equation at node a as

$$\frac{\mathbf{V}_s - \mathbf{V}_1}{\mathbf{Z}_1} + \frac{\mathbf{V}_o - \mathbf{V}_1}{\mathbf{Z}_2} - \mathbf{I}_1 = 0 \qquad (11\text{-}76)$$

When the operational amplifier is ideal, $\mathbf{V}_1$ and $\mathbf{I}_1$ are both 0. Then

$$\frac{\mathbf{V}_s}{\mathbf{Z}_1} + \frac{\mathbf{V}_o}{\mathbf{Z}_2} = 0 \qquad (11\text{-}77)$$

Finally,

$$\frac{\mathbf{V}_o}{\mathbf{V}_s} = -\frac{\mathbf{Z}_2}{\mathbf{Z}_1} \qquad (11\text{-}78)$$

Next, we will determine the ratio of output-to-input voltage, $\mathbf{V}_o/\mathbf{V}_s$, for the noninverting amplifier shown in Figure 11-44b. This circuit can be analyzed by writing the node equation at node a as

$$\frac{(\mathbf{V}_s + \mathbf{V}_1)}{\mathbf{Z}_1} - \frac{\mathbf{V}_o - (\mathbf{V}_s - \mathbf{V}_1)}{\mathbf{Z}_2} + \mathbf{I}_1 = 0 \qquad (11\text{-}79)$$

When the operational amplifier is ideal, $\mathbf{V}_1$ and $\mathbf{I}_1$ are both 0. Then

$$\frac{\mathbf{V}_s}{\mathbf{Z}_1} - \frac{\mathbf{V}_o - \mathbf{V}_s}{\mathbf{Z}_2} = 0$$

Finally,

$$\frac{\mathbf{V}_o}{\mathbf{V}_s} = \frac{\mathbf{Z}_1 + \mathbf{Z}_2}{\mathbf{Z}_1} \qquad (11\text{-}80)$$

Typically, impedances $\mathbf{Z}_1$ and $\mathbf{Z}_2$ are obtained using only resistors and capacitors. Of course, in theory we could use inductors, but their cost and size relative to capacitors result in little use of inductors with operational amplifiers.

An example of the inverting amplifier is shown in Figure 11-45. It is possible later to set a selected R and C equal to zero, so we will consider the more general case as shown.

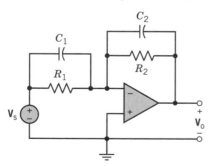

Figure 11-45 Operational amplifier with two RC circuits connected.

The impedance $\mathbf{Z}_n$, where n is equal to 1 or 2, is a parallel $R_n C_n$ impedance so that

$$\mathbf{Z}_n = \frac{R_n(-jX_n)}{R_n - jX_n}$$

$$= \frac{R_n X_n(X_n - jR_n)}{R_n^2 + X_n^2} \qquad (11\text{-}81)$$

where $X_n = 1/\omega C_n$. Using Eqs. 11-80 and 11-81, one may obtain the ratio $\mathbf{V}_o/\mathbf{V}_s$.

Example 11-15

Find the ratio $\mathbf{V_o}/\mathbf{V_s}$ for the circuit of Figure 11-45 when $R_1 = 1\text{ k}\Omega$, $R_2 = 10\text{ k}\Omega$, $C_1 = 0$, and $C_2 = 0.1\ \mu\text{F}$ for $\omega = 1000$ rad/s.

Solution

Since $C_1 = 0$, $\mathbf{Z}_1 = R_1 = 1\text{ k}\Omega$. The impedance $\mathbf{Z}_2$ is

$$\mathbf{Z}_2 = \frac{R_2 X_2 (X_2 - jR_2)}{R_2^2 + X_2^2}$$

The reactance X_2 is

$$X_2 = \frac{1}{\omega C_2}$$
$$= 10^4$$

Therefore, since $R_2 = 10^4$, we have

$$\mathbf{Z}_2 = \frac{(10^4)(10^4)(10^4 - j10^4)}{(10^4)^2 + (10^4)^2}$$
$$= \frac{10^4(1 - j)}{2}$$

Therefore,

$$\frac{\mathbf{V_o}}{\mathbf{V_s}} = -\frac{\mathbf{Z}_2}{\mathbf{Z}_1}$$
$$= \frac{-10^4(1 - j)}{2(10^3)}$$
$$= -5(1 - j)$$

EXERCISE 11-30

Find the ratio $\mathbf{V_o}/\mathbf{V_s}$ for the circuit shown in Figure 11-45 when $R_1 = R_2 = 1\text{ k}\Omega$, $C_2 = 0$, $C_1 = 1\ \mu\text{F}$, and $\omega = 1000$ rad/s.
Answer: $\mathbf{V_o}/\mathbf{V_s} = -1 - j$

11-15 VERIFICATION EXAMPLES

Example 11V-1

It is known that

$$\frac{10}{R - j4} = A\underline{/53°}$$

A computer program states that $A = 2$. Verify this result. (Notice that values are given to only two significant figures.)

Solution

The equation for the angle is

$$-\tan^{-1}\left(\frac{-4}{R}\right) = 53°$$

Then, we have

$$R = \frac{-4}{\tan(-53°)} = 3.014$$

Solving for A in terms of R we obtain

$$A = \frac{10}{(R^2 + 16)^{1/2}} = 1.997$$

Therefore $A = 2$ is correct to two significant figures.

Example 11V-2

Consider the circuit shown in Figure 11V-1. Suppose we know that the capacitor

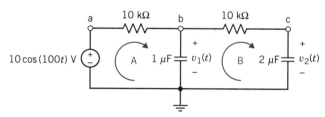

Figure 11V-1 An example circuit.

voltages are

$$1.96 \cos(100t - 101.3°) \text{ V} \quad \text{and} \quad 4.39 \cos(100t - 37.88°) \text{ V}$$

but we do not know which voltage is $v_1(t)$ and which is $v_2(t)$. How can we match the voltages with the capacitors?

Solution

Let us guess that

$$v_1(t) = 1.96 \cos(100t - 101.3°)$$

and

$$v_2(t) = 4.39 \cos(100t - 37.88°)$$

and then check to see if this choice satisfies the node equations representing the circuit. These node equations are

$$\frac{10 - \mathbf{V}_1}{R_1} = j\omega C \mathbf{V}_1 + \frac{\mathbf{V}_1 - \mathbf{V}_2}{R_2}$$

and

$$j\omega C \mathbf{V}_2 = \frac{\mathbf{V}_1 - \mathbf{V}_2}{R_2}$$

where $\mathbf{V}_1$ and $\mathbf{V}_2$ are the phasors corresponding to $v_1(t)$ and $v_2(t)$. That is,

$$\mathbf{V}_1 = 1.96e^{-j101.3°} \quad \text{and} \quad \mathbf{V}_2 = 4.39e^{-j37.88°}$$

Substituting the phasors $\mathbf{V}_1$ and $\mathbf{V}_2$ into the left-hand side of first node equation gives

$$\frac{10 - 1.96e^{-j101.3}}{10 \times 10^3} = 0.0001 + j1.92 \times 10^{-4}$$

Substituting the phasors $\mathbf{V}_1$ and $\mathbf{V}_2$ into the right-hand side of first node equation gives

$$j \cdot 100 \times 10^{-6} \cdot 1.96e^{-j101.3} + \frac{1.96e^{-j101.3} - 4.39e^{-j37.88}}{10 \times 10^3}$$

$$= -19.3 \times 10^{-4} + j3.89 \times 10^{-5}$$

Since the right-hand side is not equal to the left-hand side, $\mathbf{V}_1$ and $\mathbf{V}_2$ do not satisfy the node equation. That means that the selected order of $v_1(t)$ and $v_2(t)$ are not correct. Instead, use the reverse order so that

$$v_1(t) = 4.39 \cos(100t - 37.88°)$$

and

$$v_2(t) = 1.96 \cos(100t - 101.3°)$$

Now the phasors $\mathbf{V}_1$ and $\mathbf{V}_2$ will be

$$\mathbf{V}_1 = 4.39e^{-j37.88°} \quad \text{and} \quad \mathbf{V}_2 = 1.96e^{-j101.3°}$$

Substituting the new values of the phasors $\mathbf{V}_1$ and $\mathbf{V}_2$ into the left-hand side of the first node equation gives

$$\frac{10 - 4.39e^{-j101.3}}{10 \times 10^3} = 6.335 \times 10^{-4} + j2.696 \times 10^{-4}$$

Substituting the new values for the phasors $\mathbf{V}_1$ and $\mathbf{V}_2$ into the right-hand side of the first node equation gives

$$j \cdot 100 \cdot 10^{-6} \cdot 4.39e^{-j37.88} + \frac{4.39e^{-j37.88} - 1.96e^{-j101.3}}{10 \times 10^3}$$

$$= +6.545 \times 10^{-4} + j2.69 \times 10^{-4}$$

Since the right-hand side is very close to equal to the left-hand side, $\mathbf{V}_1$ and $\mathbf{V}_2$ satisfy the first node equation. That means that $v_1(t)$ and $v_2(t)$ are probably correct. To be certain, we will also check the second node equation. Substituting the phasors $\mathbf{V}_1$ and $\mathbf{V}_2$ into the left-hand side of the second node equation gives

$$j \cdot 100 \cdot 2 \times 10^{-6} \cdot 1.96e^{-j101.3} = +3.84 \times 10^{-4} - j7.681 \times 10^{-5}$$

Substituting the phasors $\mathbf{V}_1$ and $\mathbf{V}_2$ into the right-hand side of the second node equation gives

$$\frac{4.39e^{-j37.88} - 1.96e^{-j101.3}}{10 \times 10^3} = 3.85 \times 10^{-4} - j7.735 \times 10^{-5}$$

Since the right-hand side is equal to the left-hand side, $\mathbf{V}_1$ and $\mathbf{V}_2$ satisfy the second node equation. Now we are certain that

$$v_1(t) = 4.39 \cos(100t - 37.88°) \text{ V}$$

and

$$v_2(t) = 1.96 \cos(100t - 101.3°) \text{ V}$$

11-16 DESIGN CHALLENGE SOLUTION

OP AMP CIRCUIT

Figure 11D-1*a* shows two sinusoidal voltages, one labeled as input and the other labeled as output. We want to design a circuit that will transform the input sinusoid into the output sinusoid. Figure 11D-1*b* shows a candidate circuit. We must first determine if this circuit can do the job. Then, if it can, we will design the circuit, that is, specify the required values of R_1, R_2, and C.

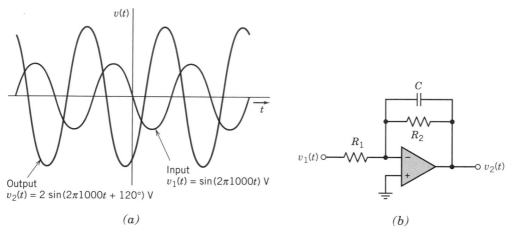

(a)   (b)

Figure 11D-1 (*a*) Input and output voltages. (*b*) Proposed circuit.

Define the Situation

The input and output sinusoids have different amplitudes and phase angles but the same frequency

$$f = 1000 \text{ Hz}$$

or, equivalently,

$$\omega = 2\pi 1000 \text{ rad/s}$$

We now know that this must be the case. When the input to a linear circuit is a sinusoid, the steady-state output will also be a sinusoid having the same frequency.

In this case, the input sinusoid is

$$v_1(t) = \sin(2\pi 1000 t) = \cos(2\pi 1000 t - 90°)$$

and the corresponding phasor is

$$\mathbf{V}_1 = 1\, e^{-j90°}$$
$$= 1\, \underline{/-90°}$$

The output sinusoid is

$$v_2(t) = 2\sin(2\pi 1000 t + 120°) = 2\cos(2\pi 1000 t + 30°)$$

and the corresponding phasor is

$$\mathbf{V}_2 = 2\, e^{j30°}$$

The ratio of these phasors is

$$\frac{\mathbf{V}_2}{\mathbf{V}_1} = \frac{2\ e^{j30°}}{1\ e^{-j90°}}$$

$$= 2\ e^{j120°}$$

The magnitude of this ratio is called the gain, G, of the circuit used to transform the input sinusoid into the output sinusoid:

$$G = \left| \frac{\mathbf{V}_2}{\mathbf{V}_1} \right|$$

$$= 2$$

The angle of this ratio is called the phase shift, θ, of the required circuit.

$$\theta = \angle \frac{\mathbf{V}_2}{\mathbf{V}_1}$$

$$= 120°$$

Therefore, we need a circuit that has a gain of 2 and a phase shift of 120°.

State the Goal

Determine if it is possible to design the circuit shown in Figure 11D-1b to have a gain of 2 and a phase shift of 120°. If it is possible, specify the appropriate values of R_1, R_2, and C.

Generate a Plan

Analyze the circuit shown in Figure 11D-1b to determine the ratio of the output phasor to the input phasor, $\mathbf{V}_2/\mathbf{V}_1$. Determine if this circuit can have a gain of 2 and a phase shift of 120°. If so, determine the required values of R_1, R_2, and C.

Take Action on the Plan

The circuit in Figure 11D-1b is a special case of the circuit shown in Figure 11-44. The impedance $\mathbf{Z}_1$ in Figure 11-44 corresponds to the resistor R_1 in Figure 11D-1b and impedance $\mathbf{Z}_2$ corresponds to the parallel combination of resistor R_2 and capacitor C. That is,

$$\mathbf{Z}_1 = R_1$$

and

$$\mathbf{Z}_2 = \frac{R_2(1/j\omega C)}{R_2 + 1/j\omega C}$$

$$= \frac{R_2}{1 + j\omega C R_2}$$

Then, using Eq. 11-79,

$$\frac{\mathbf{V}_2}{\mathbf{V}_1} = -\frac{\mathbf{Z}_2}{\mathbf{Z}_1}$$

$$= -\frac{R_2/(1 + j\omega C R_2)}{R_1}$$

$$= -\frac{R_2/R_1}{1 + j\omega C R_2}$$

The phase shift of the circuit in Figure 11D-1b is given by

$$\theta = \left/ \frac{\mathbf{V}_2}{\mathbf{V}_1} \right.$$

$$= \left/ \left(-\frac{R_2/R_1}{1 + j\omega CR_2} \right) \right. \tag{11D-1}$$

$$= 180° - \tan^{-1}\omega CR_2$$

What values of phase shift are possible? Notice that ω, C, and R_2 are all positive, which means that

$$0° \leq \tan^{-1}\omega CR_2 \leq 90°$$

Therefore, the circuit shown in Figure 11D-1b can be used to obtain phase shifts between 90° and 180°. Hence, we can use this circuit to produce a phase shift of 120°.

The gain of the circuit in Figure 11D-1b is given by

$$G = \left| \frac{\mathbf{V}_2}{\mathbf{V}_1} \right|$$

$$= \left| -\frac{R_2/R_1}{1 + j\omega CR_2} \right|$$

$$= \frac{R_2/R_1}{\sqrt{1 + \omega^2 C^2 R_2^2}}$$

$$= \frac{R_2/R_1}{\sqrt{1 + \tan^2(180° - \theta)}} \tag{11D-2}$$

Next, first solve Eq. 11D-1 for R_2 and then Eq. 11D-2 for R_1 to get

$$R_2 = \frac{\tan(180° - \theta)}{\omega C}$$

and

$$R_1 = \frac{R_2/G}{\sqrt{1 + \tan^2(180° - \theta)}}$$

These equations can be used to design the circuit. First, pick a convenient, readily available and inexpensive, value of the capacitor, say

$$C = 0.02 \ \mu F$$

Next, calculate values of R_1 and R_2 from the values of ω, C, G, and θ. For $\omega = 1000$, $C = 0.02 \ \mu F$, $G = 2$, and $\theta = 120°$ we calculate

$$R_1 = 3446 \ \Omega \quad \text{and} \quad R_2 = 13.78 \ k\Omega$$

and the design is complete.

SUMMARY

With the pervasive use of ac electric power in the home and industry, it is important for engineers to analyze circuits with sinusoidal independent sources. In this chapter we focused on developing a method of analyzing the steady-state response of a circuit to one or more sinusoidal forcing functions.

We introduced the complex exponential forcing function $e^{j\omega t}$ and noted that a sinusoidal source could be represented by the real part of the exponential function. This approach led to the insight that the unknown response is characterized by an amplitude and a phase angle. Thus, we proposed the phasor concept, where the phasor quantity is a complex number.

With a phasor voltage $\mathbf{V} = V\underline{/\phi}$ and a phasor current $\mathbf{I} = I\underline{/\beta}$, we define the ratio of these two phasor quantities, $\mathbf{V}/\mathbf{I}$, as an impedance, $\mathbf{Z}$. Thus, with the use of phasor voltages and currents and the concept of impedance and admittance, we have converted the circuit analysis problem from the solution of differential equations in the time domain to the solution of algebraic equations in the frequency domain. Then the final time domain solution can be obtained by converting the desired phasor voltage (or current) to the time domain by obtaining the inverse transform of the phasor.

We then find that we can use all the circuit analysis techniques we developed in Chapter 5 for resistive circuits. We can use the principle of superposition, Norton's and Thévenin's theorems, and mesh and node analysis methods. Therefore, we are in a position to analyze the steady-state performance of complex linear circuits excited with one or more independent sinusoidal sources.

TERMS AND CONCEPTS

Admittance   Ratio of the phasor current $\mathbf{I}$ to the phasor voltage $\mathbf{V}$ of an element or set of elements so that $\mathbf{Y} = \mathbf{I}/\mathbf{V}$. The units are siemens.

Complex Exponential Forcing Function $V_m e^{j\omega t}$ for a voltage source, or $I_m e^{j\omega t}$ for a current source.

Conductance   Real part of admittance, denoted as G. The units are siemens.

Frequency   Radian frequency ω in the sinusoidal waveform $x = \cos \omega t$; also ordinary frequency f, where $\omega = 2\pi f$ (f has units of hertz).

Frequency Domain   Mathematical domain where the set of possible values of a variable is expressed in terms of frequency.

Impedance   Ratio of the phasor voltage $\mathbf{V}$ to the phasor current $\mathbf{I}$ for a circuit element or set of elements so that $\mathbf{Z} = \mathbf{V}/\mathbf{I}$. The units are ohms.

Periodic Function   Function defined by the property $x(t + T) = x(t)$ so that it repeats every T seconds.

Period of Oscillation T is the period of oscillation of the periodic function, the time between two identical maximum points in the waveform.

Phase Shift   Phase angle ϕ associated with a variable x so that $x = V_m \sin(\omega t + \phi)$, or phase angle between two sinusoidal waveforms.

Phasor   Complex number associated with a circuit variable—for example, the phasor voltage $\mathbf{V}$. The transform of a sinusoidal voltage or current that contains the magnitude and the phase angle information.

Phasor Diagram   Relationship of phasors on the complex plane.

Reactance   Imaginary part of impedance, denoted as X. The units are ohms.

Resistance   Real part of impedance, denoted as R. The units are ohms.

Susceptance   Imaginary part of admittance, denoted as B. The units are siemens.

Transform   Change in the mathematical description of a physical variable to facilitate computation.

REFERENCES Chapter 11

Dana Gardner, "The Walking Piano," *Design News,* December 11, 1988, pp. 60–65.
James E. Lenz, "A Review of Magnetic Sensors," *Proc. IEEE,* June 1990, pp. 973–989.
Lyle H. McCarty, "Catheter Clears Coronary Arteries," *Design News,* September 23, 1991, pp. 88–92.
Henry Petroski, "Images of an Engineer," *American Scientist,* August 1991, pp. 300–303.
E. D. Smith, "Electric Shark Barrier," *Power Engineering Journal,* July 1991, pp. 167–177.

PROBLEMS

Section 11-3   Sinusoidal Sources

P 11.3-1 Express the following summations of sinusoids in the general form $A \sin(\omega t + \theta)$ by using trigonometric identities.
(a) $i(t) = 2 \cos(6t + 120°) + 4 \sin(6t - 60°)$
(b) $v(t) = 5\sqrt{2} \cos 8t + 10 \sin(8t + 45°)$

P 11.3-2 A sinusoidal voltage has a maximum amplitude of 100 V, and the magnitude is 10 V at $t = 0$. Determine $v(t)$ when the period is $T = 1$ ms.

P 11.3-3 A sinusoidal current is given as $i = 300 \cos(1200\pi t + 55°)$ mA. Determine the frequency f and the value of the current at $t = 2$ ms.

P 11.3-4 Plot a graph of the voltage signal

$$v(t) = 15 \cos(628t + 45°) \text{ mV}$$

P 11.3-5 Represent each of the signals shown in Figure P 11.3-5 by a function of the form $A \cos(\omega t + \theta)$.

Section 11-4   Steady-State Response of an *RL* Circuit for a Sinusoidal Forcing Function

P 11.4-1 Find the forced response i for the circuit of Figure P 11.4-1 when $v_s(t) = 10 \cos(300t)$ V.

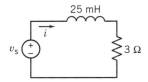

Figure P 11.4-1

P 11.4-2 Find the forced response v for the circuit of Figure P 11.4-2 when $i_s(t) = 0.5 \cos \omega t$ A and $\omega = 1000$ rad/s.

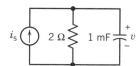

Figure P 11.4-2

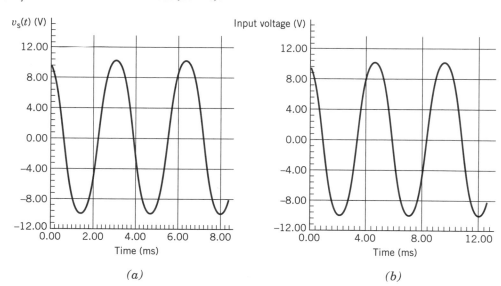

(a)   (b)

Figure P 11.3-5

P 11.4-3 Find the forced response $i(t)$ for the circuit of Figure P 11.4-3.

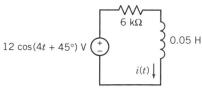

Figure P 11.4-3

Section 11-5 Complex Exponential Forcing Function
(a) Complex Numbers

P 11.5-1 Determine the polar form of the quantity

$$\frac{(5\underline{/36.9°})(10\underline{/-53.1°})}{(4 + j3) + (6 - j8)}$$

Answer: $2\sqrt{5}\ \underline{/10.36°}$

P 11.5-2 Determine the polar and rectangular form of the expression

$$5\underline{/+81.87°}\left(4 - j3 + \frac{3\sqrt{2}\underline{/-45°}}{7 - j1}\right)$$

Answer: $28\underline{/+45°} = 14\sqrt{2} + j14\sqrt{2}$

P 11.5-3 Given $\mathbf{A} = 3 + j7$, $\mathbf{B} = 6\underline{/15°}$, and $\mathbf{C} = 5e^{j2.3}$, find $(\mathbf{A^*C^*})/\mathbf{B}$.
Answer: $-6.016 + j\,2.022$

P 11.5-4 Determine a and b when (angles in degrees)

$$(6\underline{/120°})(-4 + j3 + 2\,e^{j15}) = a + jb$$

P 11.5-5 Find a, b, A, and θ as required. (angles given in degrees)
(a) $A\,e^{j120} + jb = -4 + j3$
(b) $6\,e^{j120}(-4 + jb + 8\,e^{j\theta}) = 18$
(c) $(a + j4)j2 = 2 + A\,e^{j60}$

Section 11-5 Complex Exponential Forcing Function
(b) Response of a Circuit

P 11.5-6 For the circuit of Figure P 11.5-6 find $i(t)$ when $v_s(t) = 0.1 \cos(\omega t + 90°)$ V and $\omega = 10^7$ rad/s.
Answer: $i(t) = 1 \cos(\omega t + 90°)$ mA

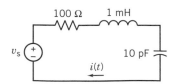

Figure P 11.5-6

P 11.5-7 For the circuit of Figure P11.5-7 find $i(t)$ when $v_s(t) = 20 \cos(\omega t + 45°)$ V when $\omega = 25$ Mrad/s.

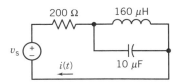

Figure P 11.5-7

P 11.5-8 For the circuit of Figure P 11.5-8 find $i(t)$ when $v_1 = 12 \cos(4000t + 45°)$ V and $v_2 = 5 \cos 3000t$ V. Hint: Use superposition.

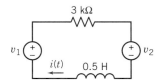

Figure P 11.5-8

Section 11-6 The Phasor Concept

P 11.6-1 For the circuit of Figure P 11.6-1 find $v(t)$ when $i_s(t) = 5 \cos(\omega t - 120°)$ mA when $\omega = 10^5$ rad/s.
Answer: $v(t) = 1.5 \cos(\omega t + 60°)$ V

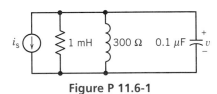

Figure P 11.6-1

P 11.6-2 For the circuit of Figure P 11.6-2 find $i(t)$ when $i_s = 25 \cos(\omega t - 120°)$ mA and $\omega = 10^3$ rad/s.

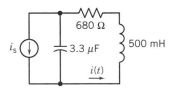

Figure P 11.6-2

P 11.6-3 For the circuit of Figure P 11.6-3, find $v(t)$ when $v_s = 2 \sin 500t$ V.

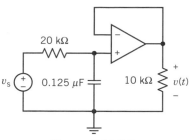

Figure P 11.6-3

Section 11-7 Phasor Relationships for R, L, and C Elements

P 11.7-1 Given the phasor $\mathbf{I} = 6 + j8$, rotate it in the phasor plane by the following amounts and express the final result in rectangular coordinates: (a) $-45°$; (b) $90°$.
Answer: (a) $\mathbf{I} = 7\sqrt{2} + j\sqrt{2}$
(b) $\mathbf{I} = -8 + j6$

P 11.7-2 Add the following voltage waveforms using phasors, obtaining the resultant voltage in the form $v_1 + v_2 = A \cos(\omega t + \phi)$.
(a) $v_1 = 3 \cos(2t + 60°)$; $v_2 = 8 \cos(2t - 22.5°)$
(b) $v_1 = 2\sqrt{2} \sin 4t$; $v_2 = 10 \cos(4t + 30°)$

P 11.7-3 Find the two phasors $\mathbf{A}$ and $\mathbf{B}$ so that $|A| = 5\sqrt{2}$, $|B| = 4$ and so that $2\mathbf{A} + 5\mathbf{B} = j10(1 + \sqrt{3})$ and $\mathbf{B}$ leads $\mathbf{A}$ by $75°$.

P 11.7-4 For the following voltage and current expressions, indicate whether the element is capacitive, inductive, or resistive and find the element value.
(a) $v = 15 \cos(400t + 30°)$; $i = 3 \sin(400t + 30°)$
(b) $v = 8 \sin(900t + 50°)$; $i = 2 \sin(900t + 140°)$
(c) $v = 20 \cos(250t + 60°)$; $i = 5 \sin(250t + 150°)$
Answer: (a) $L = 12.5$ mH
(b) $C = 277.77$ μF
(c) $R = 4$ Ω

P 11.7-5 Two voltages appear in series so $v = v_1 + v_2$. Find v when $v_1 = 150 \cos(377t - \pi/6)$ V and $\mathbf{V}_2 = 200\underline{/+ 60°}$ V.

Section 11-8 Impedance and Admittance

P 11.8-1 Find $\mathbf{Z}$ and $\mathbf{Y}$ for the circuit of Figure P 11.8-1 operating at 10 kHz.

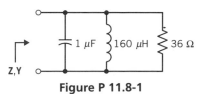

Figure P 11.8-1

P 11.8-2 Find R and L of the circuit of Figure P 11.8-2 when $v(t) = 10 \cos(\omega t + 40°)$ V, $i(t) = 2 \cos(\omega t + 195°)$ mA, and $\omega = 2 \times 10^6$ rad/s.
Answer: 4.532 kΩ, $L = 1.057$ mH

Figure P 11.8-2

P 11.8-3 Consider the circuit of Figure P 11.8-3 when $R = 6$ Ω, $L = 27$ μH, and $C = 22$ μF. Determine the frequency f when the impedance $\mathbf{Z}$ is purely resistive, and find the input resistance at that frequency.

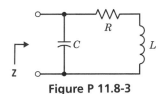

Figure P 11.8-3

P 11.8-4 Consider the circuit of Figure P 11.8-4 when $R = 10$ kΩ and $f = 1$ kHz. Find L and C so that $\mathbf{Z} = 100 + j0$ Ω.

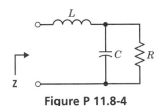

Figure P 11.8-4

P 11.8-5 For the circuit of Figure P 11.8-5 find the value of C required so that $\mathbf{Z} = 590.7$ Ω when $f = 1$ MHz.

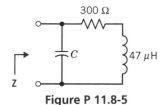

Figure P 11.8-5

P 11.8-6 The source voltage, v_s, shown in the circuit of Figure P 11.8-6a is a sinusoid having a frequency of 500 Hz and an amplitude of 8 V. The circuit is in steady state. The oscilloscope traces show the input and output waveforms as shown in Figure P 11.8-6b.
(a) Determine the gain and phase shift of the circuit at 500 Hz. Gain and phase shift are defined in the Design Challenge Solution. (page 550).

(b) Determine the value of the capacitor.
(c) If the frequency of the input is changed, then the gain and phase shift of the circuit will change. What are the values of the gain and phase shift at the frequency 200 Hz? At 2000 Hz? At what frequency will the phase shift be − 45°? At what frequency will the phase shift be − 135°?
(d) What value of capacitance would be required to make the phase shift at 500 Hz be − 60°? What value of capacitance would be required to make the phase shift at 500 Hz be − 300°?
(e) Suppose the phase shift had been − 120° at 500 Hz. What would be the value of the capacitor?

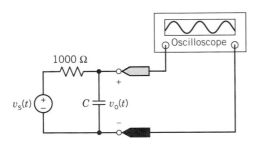

(a)

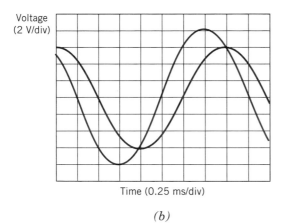

Time (0.25 ms/div)

(b)

Figure P 11.8-6

Section 11-9 Kirchhoff's Laws Using Phasors

P11.9-1 For the circuit shown in Figure P 11.9-1, find (a) the impedances $\mathbf{Z}_1$ and $\mathbf{Z}_2$ in polar form, (b) the total combined impedance in polar form, and (c) the steady-state current $i(t)$.

Answer: (a) $\mathbf{Z}_1 = 5\underline{/53.1°}$; $\mathbf{Z}_2 = 8\sqrt{2}\underline{/-45°}$
(b) $\mathbf{Z}_1 + \mathbf{Z}_2 = 11.7\underline{/-20°}$
(c) $i(t) = (8.55)\cos(1250t + 20°)$ A

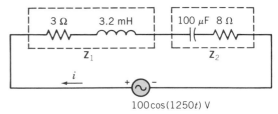

Figure P 11.9-1

P 11.9-2 Given the phasor circuit shown in Figure P 11.9-2, find (a) $\mathbf{Z}_{eq}$, (b) $\mathbf{I}$, (c) $\mathbf{V}_{ef}$, (d) $\mathbf{Z}_{fg}$, and (e) $\mathbf{I}_{cd}$. The $\mathbf{Z}_{eq}$ is the equivalent impedance found by disconnecting the source and determining the impedance at terminals e−g.

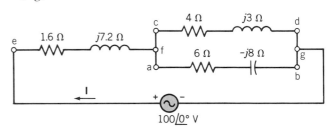

Figure P 11.9-2

P 11.9-3 Spinal cord injuries result in paralysis of the lower body and can cause loss of bladder control. Numerous electrical devices have been proposed to replace the normal nerve pathway stimulus for bladder control. Figure P 11.9-3 shows the model of a bladder control system where $v_s = 20 \cos \omega t$ V and $\omega = 100$ rad/s. Find the steady-state voltage across the 10-Ω load resistor.

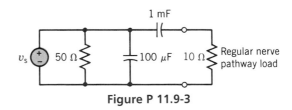

Figure P 11.9-3

P 11.9-4 There are 500 to 1000 deaths each year in the United States from electric shock. If a person makes a good contact with his hands, the circuit can be represented by Figure P 11.9-4, where $v_s = 160 \cos \omega t$ V and $\omega = 2\pi f$. Find the steady-state current i flowing through the body when (a) $f = 60$ Hz and (b) $f = 400$ Hz.

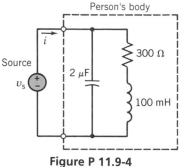

Figure P 11.9-4

P 11.9-5 For $t > 0$, find the total capacitor voltage, $v(t)$, for the circuit shown in Figure P 11.9-5 when

$$v_s = \begin{cases} 50 \cos 3t \text{ V} & t < 0 \\ 20 \text{ V} & t > 0 \end{cases}$$

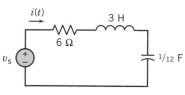

Figure P 11.9-5

P 11.9-6 Determine $i(t)$ of the *RLC* circuit shown in Figure P 11.9-6 when $v_s = 2 \cos(4t + 30°)$ V.

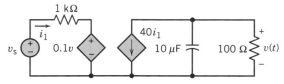

Figure P 11.9-6

P 11.9-7 An amplifier circuit is shown in Figure P 11.9-7 with an input $v_s = 5 \cos 200t$ V. Determine the output voltage $v(t)$.

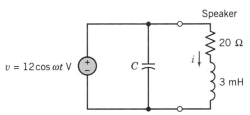

Figure P 11.9-7 Amplifier circuit.

P 11.9-8 The circuit of Figure P 11.9-8 contains a voltage source operating at $\omega = 100$. Find the current $i(t)$.
Answer: $i(t) = (1/\sqrt{2}) \cos(100t - 45°)$

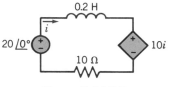

Figure P 11.9-8

P 11.9-9 The big toy from the hit movie *Big* is a child's musical fantasy come true—a sidewalk-sized piano. Like a hopscotch grid, this Christmas's hot toy invites anyone who passes to jump on, move about, and make music. The developer of the "toy" piano used a tone synthesizer and stereo speakers as shown in Figure P 11.9-9 (Gardner, 1988). Determine the current $i(t)$ for a tone at 796 Hz when $C = 10 \ \mu$F.

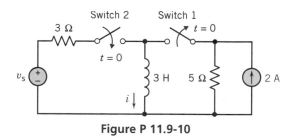

Figure P 11.9-9 Tone synthesizer.

P 11.9-10 Determine the total inductor current $i(t)$ for the circuit shown in Figure P 11.9-10 when $v_s = 3 \sin 2t$ V. Assume that switch 1 was closed for a long time prior to $t = 0$.

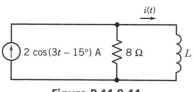

Figure P 11.9-10

P 11.9-11 Determine B and L for the circuit of Figure P 11.9-11 when $i(t) = B \cos(3t - 51.87°)$ A.

Figure P 11.9-11

Section 11-10 Node Voltage and Mesh Current Analysis Using Phasors
(a) Node Voltage Analysis

P 11.10-1 Find the phasor voltage $\mathbf{V}_c$ for the circuit shown in Figure P 11.10-1.

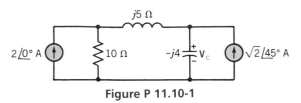

Figure P 11.10-1

P 11.10-2 For the circuit shown in Figure P 11.10-2, determine the phasor currents $\mathbf{I}_s$, $\mathbf{I}_c$, $\mathbf{I}_L$, and $\mathbf{I}_R$ if $\omega = 1000$ rad/s.

Answer: $\mathbf{I}_s = 0.347\underline{/-25.5°}$A
$\mathbf{I}_c = 0.461\underline{/112.9°}$ A
$\mathbf{I}_L = 0.720\underline{/-67.1°}$ A
$\mathbf{I}_R = 0.230\underline{/22.9°}$ A

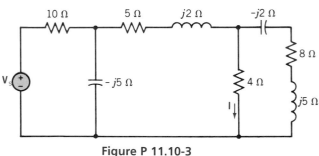

Figure P 11.10-2

P 11.10-3 Consider the circuit of Figure P 11.10-3. Using nodal analysis, find $\mathbf{V}_s$ when $\mathbf{I}$ in the 4-Ω resistor is $3\underline{/45°}$ A.

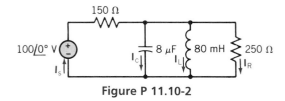

Figure P 11.10-3

P 11.10-4 Find the steady-state response v_x for the circuit in Figure P 11.10-4 using nodal analysis.
Answer: $v_x(t) = 1168 \cos(500t - 66.3°)$ V

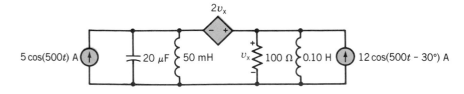

Figure P 11.10-4

P 11.10-5 Find the node voltages $v_1(t)$ and $v_2(t)$ for the circuit shown in Figure P 11.10-5 when (a) $i_1 = 1 \cos 100t$ A and (b) $i_1 = 1$ A (a constant). The second current source is $i_2 = 0.5 \sin 100t$.

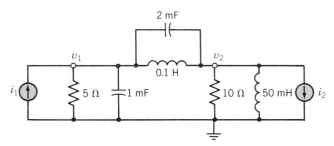

Figure P 11.10-5

P 11.10-6 Find the two node voltages $v_a(t)$ and $v_b(t)$ for the circuit of Figure P 11.10-6 when $v_s(t) = 1.2 \cos 4000t$.

Answer: $v_a(t) = 1.97 \cos(4000t - 171°)$ V
$v_b(t) = 2.21 \cos(4000t - 144°)$ V

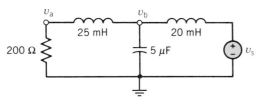

Figure P 11.10-6

P 11.10-7 The circuit shown in Figure P 11.10-7 has two sources: $v_s = 20 \cos(\omega_0 t + 90°)$ V and $i_s = 6 \cos \omega_0 t$ A where $\omega_0 = 10^5$ rad/s. Determine $v_0(t)$.

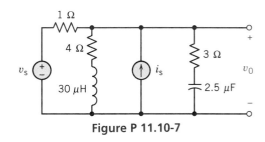

Figure P 11.10-7

P 11.10-8 Determine the voltage v_a for the circuit in Figure P 11.10-8 when $i_s = 20 \cos(\omega t + 53.13°)$ A and $\omega = 10^4$ rad/s.

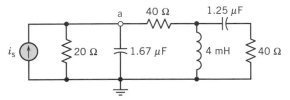

Figure P 11.10-8

P 11.10-9 Determine $v(t)$ for the circuit of Figure P 11.10-9 when $v_s = \sqrt{2} \sin(\omega t + 135°)$ V, $i_s = 4 \cos(\omega t + 30°)$ A, and $\omega = 100$ rad/s.

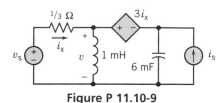

Figure P 11.10-9

P 11.10-10 A commercial airliner has sensing devices to indicate to the cockpit crew that each door and baggage hatch is closed. A device called a search coil magnetometer, also known as a proximity sensor, provides a signal indicative of the proximity of metal or other conducting material to an inductive sense coil. The inductance of the sense coil changes as the metal gets closer to the sense coil. The sense coil inductance is compared to a reference coil inductance with a circuit called a balanced inductance bridge (see Figure P 11.10-10). In the inductance bridge, a signal indicative of proximity is observed between terminals A and B by subtracting the voltage at B, v_B, from the voltage at A, v_A (Lenz, 1990).

The bridge circuit is excited by a sinusoidal voltage source $v_s = \sin(800 \pi t)$ V. The two resistors, $R = 100$ Ω, are of equal resistance. When the door is open (no metal is present), the sense coil inductance, L_s, is equal to the reference coil inductance, $L_r = 40$ mH. In this case, what is the magnitude of the signal $V_A - V_B$?

When the airliner door is completely closed, $L_s = 60$ mH. With the door closed, what is the phasor representation of the signal $V_A - V_B$?

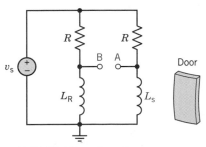

Figure P 11.10-10 Airline door sensing circuit.

P 11.10-11 Using a tiny diamond-studded burr operating at 190,000 rpm, cardiologists can remove life-threatening plaque deposits in coronary arteries. The procedure is fast, uncomplicated, and relatively painless (McCarty, 1991). The Rotablator, an angioplasty system, consists of an advancer/catheter, a guide wire, a console, and a power source. The advancer/catheter contains a tiny turbine that drives the flexible shaft that rotates the catheter burr. The model of the operational and control circuit is shown in Figure P 11.10-11. Determine $v(t)$, the voltage that drives the tip, when $v_s = \sqrt{2} \cos(40t + 135°)$ V.

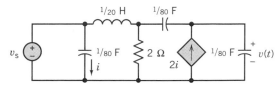

Figure P 11.10-11 Control circuit for Rotablator.

P 11.10-12 For the circuit of Figure P 11.10-12, it is known that

$$v_2(t) = 0.7571 \cos(2t + 66.7°) \text{ V}$$
$$v_3(t) = 0.6064 \cos(2t - 69.8°) \text{ V}$$

Determine $i_1(t)$.

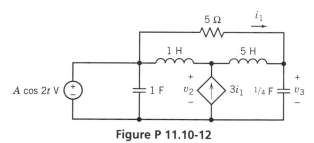

Figure P 11.10-12

Section 11-10 Node Voltage and Mesh Current Analysis Using Phasors
(b) Mesh Current Analysis

P 11.10-13 Determine $\mathbf{I}_1$, $\mathbf{I}_2$, $\mathbf{V}_L$, and $\mathbf{V}_c$ for the circuit of Figure P 11.10-13 using KVL and mesh analysis.
Answer: $\mathbf{I}_1 = 2.5\underline{/29.0°}$ A
$\mathbf{I}_2 = 1.8\underline{/105°}$ A
$\mathbf{V}_L = 16.3\underline{/78.7°}$ V
$\mathbf{V}_c = 7.2\underline{/15°}$ V

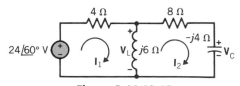

Figure P 11.10-13

P 11.10-14 The model of a high-frequency transistor amplifier is shown in Figure P 11.10-14 with a source voltage and a load resistor. The source voltage is $v_s = 10 \cos \omega t$ where $\omega = 10^8$ rad/s. Find the voltage across the load resistor.

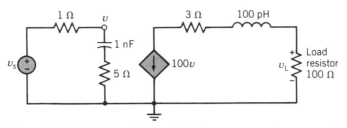

Figure P 11.10-14 Model of a high-frequency transis- tor amplifier.

P 11.10-15 Using a supermesh and mesh analysis, find the steady-state response i_x for the circuit shown in Figure P 11.10-15.

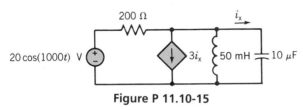

Figure P 11.10-15

P 11.10-16 Determine the current $i(t)$ for the circuit of Figure P 11.10-16 using mesh currents when $\omega = 1000$ rad/s.

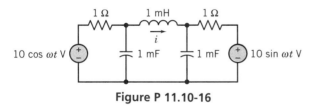

Figure P 11.10-16

P 11.10-17 The idea of using an induction coil in a lamp isn't new, but applying it in a commercially available product is. An induction coil in a bulb induces a high-frequency energy flow in mercury vapor to produce light.

The lamp uses about the same amount of energy as a fluorescent bulb but lasts six times longer, with 60 times the life of a conventional incandescent bulb. The circuit model of the bulb and its associated circuit is shown in Figure P 11.10-17. Determine the voltage $v(t)$ across the 2 Ω resistor when $C = 40$ μF, $L = 40$ μH, $v_s = 10 \cos(\omega_0 t + 30°)$, and $\omega_0 = 10^5$ rad/s.

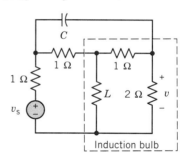

Figure P 11.10-17 Induction bulb circuit.

P 11.10-18 The development of coastal hotels in various parts of the world is a rapidly growing enterprise. The need for environmentally acceptable shark protection is manifest where these developments take place alongside shark-infested waters (Smith, 1991). One concept is to use an electrified line submerged in the water in order to deter the sharks, as shown in Figure P 11.10-18a. The circuit model of the electric fence is shown in Figure P 11.10-18b where the shark is represented by an equivalent resistance of 100 Ω. Determine the current flowing through the shark's body, $i(t)$, when $v_s = 375 \cos 400t$ V.

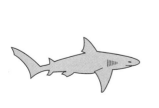

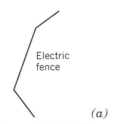

(a)

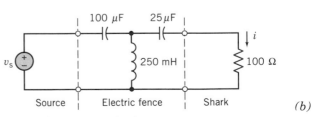

(b)

Figure P 11.10-18 Electric fence for repelling sharks.

Section 11-11 Superposition, Thévenin and Norton Equivalents, and Source Transformations
(a) Superposition Principle

P 11.11-1 Determine $i_3(t)$ for the circuit of Figure P 11.11-1 when $i_1 = 3.3 \cos(800t - 145°)$ mA and $i_2 = 1.6 \sin(800t + 16°)$ mA.
Answer: $i_3 = 2.94 \cos(800t + 42.3°)$ mA

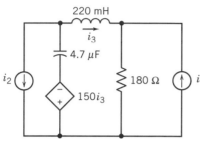

Figure P 11.11-1

P 11.11-2 Determine $i(t)$ for the circuit of Figure P 11.11-2.

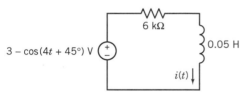

Figure P 11.11-2

P 11.11-3 Determine $i(t)$ for the circuit of Figure P 11.11-3.

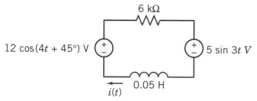

Figure P 11.11-3

P 11.11-4 Find the current in the inductor for the circuit of Figure P 11.11-4 when $i_1 = 10 \cos 100t$ A and $v_1 = 100 \cos 1000t$ V.

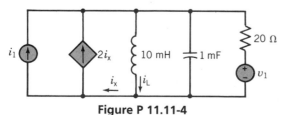

Figure P 11.11-4

Section 11-11 Superposition, Thévenin and Norton Equivalents, and Source Transformations
(b) Thévenin and Norton Equivalent Circuits

P 11.11-5 Determine the Thévenin equivalent circuit for the circuit shown in Figure P11.11-5 when $v_s = 5 \cos(4000t - 30°)$.
Answer: $\mathbf{V}_t = 5.7\underline{/-21.9°}$ V
$\mathbf{Z}_t = 23\underline{/-81.9°}$ Ω

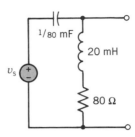

Figure P 11.11-5

P 11.11-6 Find $v_x(t)$ by first replacing the circuit to the left of terminals a–b in Figure P 11.11-6 with its Thévenin equivalent circuit when $C = 10$ mF.
Answer: $v_x(t) = 33.13 \cos(20t - 83.66°)$ V

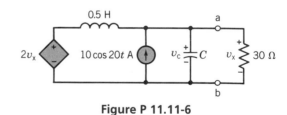

Figure P 11.11-6

P 11.11-7 Determine the Thévenin equivalent at terminals a–b for the circuit shown in Figure P 11.11-7 when $\omega = 100$ rad/s.

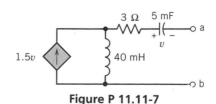

Figure P 11.11-7

P 11.11-8 Find the Thévenin equivalent circuit for the circuit shown in Figure P 11.11-8 using the mesh current method.

Answer: $\mathbf{V}_t = 3.71/\!\!-16°$ V
$\mathbf{Z}_t = 247/\!\!-16°\ \Omega$

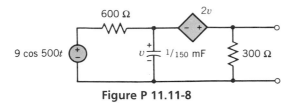

Figure P 11.11-8

P 11.11-9 For the circuit shown in Figure P 11.11-9, find the capacitance, C, if $v(t) = 100\sqrt{2}\cos(100t - 45°)$ V.

Answer: $C = 1/800$ F

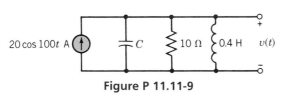

Figure P 11.11-9

P 11.11-10 Consider the circuit shown in Figure P 11.11-10. Determine the required value of X_c and X_L in order to make the impedance $\mathbf{Z}$ seen at the terminals equal to aR, where $0 \le a \le 1$. Find X_c and X_L in terms of a and R.

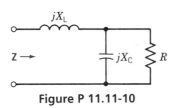

Figure P 11.11-10

P 11.11-11 Find the equivalent input impedance $\mathbf{Z}_{in}$ for the circuit shown in Figure P 11.11-11.

Answer: $6\sqrt{5}/\!\!26.6°\ \Omega$

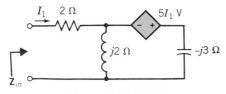

Figure P 11.11-11

P 11.11-12 A pocket-sized mini-disk CD player system has an amplifier circuit shown in Figure P 11.11-12 with a signal $v_s = 10\cos(\omega t + 53.1°)$ at $\omega = 10,000$ rad/s. Determine the Thévenin equivalent at the output terminals a–b.

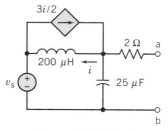

Figure P 11.11-12

P 11.11-13 In the analysis of guided waves on transmission lines, circuit theory is often used. One of the simplest unit-length circuit models of a transmission line is shown in Figure P 11.11-13. The transmission line is represented by L and C and the load by R. A line has $L = 97.5$ nH and $C = 39$ pF and operates at 100 MHz. Determine the input impedance when $R = 25\ \Omega$ and $50\ \Omega$. Note that this circuit translates one of the resistances almost exactly to the input impedance. This resistance is the characteristic impedance or resistance of the transmission line.

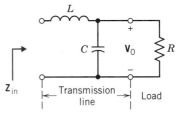

Figure P 11.11-13 Circuit model of transmission line.

P 11.11-14 An AM radio receiver uses the parallel RLC circuit shown in Figure P 11.11-14. Determine the frequency, f_0, at which the admittance $\mathbf{Y}$ is a pure conductance. What is the "number" of this station on the AM radio dial?

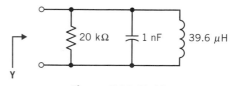

Figure P 11.11-14

P 11.11-15 A linear circuit is placed within a black box with only the terminals a–b available as shown in Figure P 11.11-15. There are three elements available in the laboratory: (1) a 50-Ω resistor, (2) a 2.5-μF capacitor, and (3) a 50-mH inductor. These three elements are placed across terminals a–b as the load $\mathbf{Z}_L$, and the magnitude of $\mathbf{V}$ is measured as (1) 25 V, (2) 100 V, and (3) 50 V, respectively. It is known that the sources within the box are sinusoidal with $\omega = 2 \times 10^3$ rad/s. Determine the Thévenin equivalent for the circuit in the box as shown in Figure P 11.11-15.

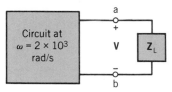

Figure P 11.11-15 A circuit within a black box is connected to a selected impedance $\mathbf{Z}_n$.

Section 11-11 Superposition, Thévenin and Norton Equivalents, and Source Transformations
(c) Source Transformations

P 11.11-16 Consider the circuit of Figure P 11.11-16, where we wish to determine the current $\mathbf{I}$. Use a series of source transformations to find a current source in parallel with an equivalent impedance, and then find the current in the 2-Ω resistor by current division.

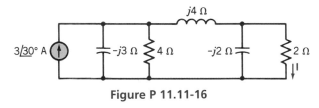

Figure P 11.11-16

P 11.11-17 For the circuit of Figure P 11.11-17 determine the current $\mathbf{I}$ using a series of source transformations. The source has $\omega = 25 \times 10^3$ rad/s.

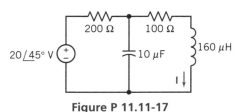

Figure P 11.11-17

Section 11-12 Phasor Diagrams

P 11.12-1 Using a phasor diagram determine $\mathbf{V}$ when $\mathbf{V} = \mathbf{V}_1 - \mathbf{V}_2 + \mathbf{V}_3^*$ and $\mathbf{V}_1 = 3 + j3$, $\mathbf{V}_2 = 4 + j2$, and $\mathbf{V}_3 = -3 - j2$. (Units are volts.)
Answer: $\mathbf{V} = 5\underline{/143.1°}$ V

P 11.12-2 Consider the series RLC circuit of Figure P 11.12-2 when $R = 10$ Ω, $L = 1$ mH, $C = 100$ μF, and $\omega = 10^3$ rad/s. Find $\mathbf{I}$ and plot the phasor diagram.

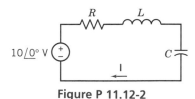

Figure P 11.12-2

P 11.12-3 Consider the signal

$$i(t) = 72\sqrt{3}\cos 8t + 36\sqrt{3}\sin(8t + 140°) \\ + 144\cos(8t + 210°) + 25\cos(8t + \phi)$$

Using the phasor plane, for what value of ϕ does the $|I|$ attain its maximum?

Section 11-14 Phasor Circuits and the Operational Amplifier

P 11.14-1 Find the steady-state response $v_0(t)$ if $v_s(t) = \sqrt{2}\cos 1000t$ for the circuit of Figure P 11.14-1. Assume an ideal op amp.
Answer: $v_0(t) = 10\cos(1000t - 225°)$

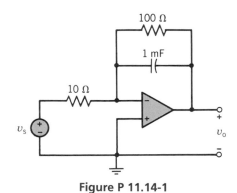

Figure P 11.14-1

P 11.14-2 Determine $\mathbf{V}_0/\mathbf{V}_s$ for the op amp circuit shown in Figure P 11.14-2. Assume an ideal op amp.

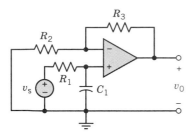

Figure P 11.14-2 Amplifier circuit for disk player.

P 11.14-3 Determine $\mathbf{V}_0/\mathbf{V}_s$ for the op amp circuit shown in Figure P 11.14-3. Assume an ideal op amp.

Answer: $\dfrac{\mathbf{V}_o}{\mathbf{V}_s} = \dfrac{j\omega R_1 C_1(1 + R_3/R_2)}{1 + j\omega R_1 C_1}$

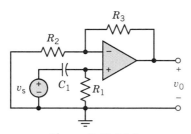

Figure P 11.14-3

P 11.14-4 Determine $v_0(t)$ for the op amp circuit of Figure P 11.14-4. Assume an ideal op amp and that $v_s = 1 \sin(3t + 60°)$ V.

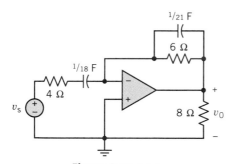

Figure P 11.14-4

P 11.14-5 For the circuit of Figure P 11.14-5 determine $v_0(t)$ when $v_s = 5 \cos \omega t$ mV and $f = 10$ kHz.
Answer: $v_0 = 0.5 \cos(\omega t - 89.5°)$ mV

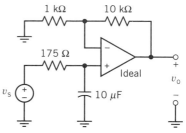

Figure P 11.14-5

P 11.14-6 Determine $v(t)$ for the circuit of Figure P 11.14-6.

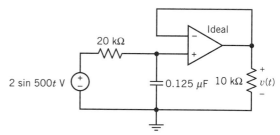

Figure P 11.14-6

PSpice PROBLEMS

SP 11-1 Determine the current $i(t)$ for the circuit of Figure P 11.10-14 when $\omega = 1000$ rad/s.

SP 11-2 Determine $v(t)$ for the circuit of Figure P 11.10-9 when $v_s = \sqrt{2} \sin(\omega t + 135°)$ V, $i_s = 4 \cos(\omega t + 30°)$ A, and $\omega = 100$ rad/s.

SP 11-3 Determine $i(t)$ of the circuit of Problem P 11.9-6.

SP 11-4 Determine $v(t)$ for Problem P 11.9-7.

SP 11-5 Determine $\mathbf{V}_0$ for Problem P 11.10-7.

SP 11-6 Determine the current $i(t)$ for the circuit of Figure SP 11-6 when $v_s = 10 \cos(6t + 45°)$ V and $i_s = 2 \cos(6t + 60°)$ A.

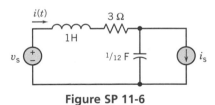

Figure SP 11-6

SP 11-7 Determine $v_0(t)$ for the circuit of Figure SP 11-7 when $v_s = 5 \cos(3t - 30°)$ V.

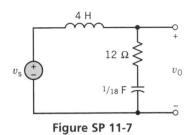

Figure SP 11-7

SP 11-8 Determine the current $i(t)$ for the circuit shown in Figure SP 11-8 when $v_s = 200 \cos \omega t$ V, $i_s = 8 \cos(\omega t + 90°)$ A, and $\omega = 1000$ rad/s.

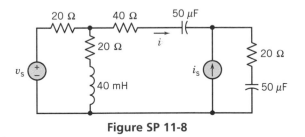

Figure SP 11-8

SP 11-9 Determine the output voltage, v, for the circuit shown in Figure SP 11-9 when $v_s = 4 \cos \omega t$ V and $\omega = 2\pi \times 1000$ rad/s.

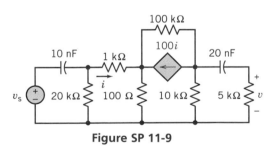

Figure SP 11-9

SP 11-10 Determine $i(t)$ for the circuit of Figure SP 11-10 when $i_s = 2 \cos(3t + 10°)$ A and $v_s = 3 \cos(2t + 30°)$ V.

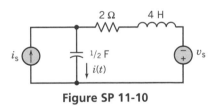

Figure SP 11-10

SP 11-11 Determine the impedance at terminals a–b for Design Problem 11-7 for a suitable design.

SP 11-12 Determine the voltage v_a for Problem P 11.10-8.

SP11-13 Determine $v_a(t)$ and $i(t)$ for the circuit shown in Figure SP 11-13 when $v_s = 5 \cos 2t$ V and $i_s = 5 \cos 2t$ A.

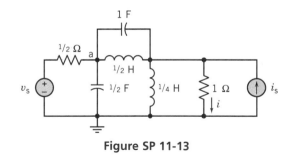

Figure SP 11-13

SP 11-14 Determine $i(t)$ for the circuit shown in Figure SP 11-14 when $v_s = 4 \cos 5000t$ V.

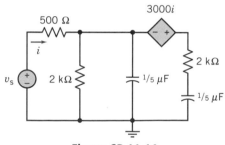

Figure SP 11-14

SP 11-15 Determine the impedance **Z** for the circuit as shown in Figure SP 11-15 at a frequency of 60 Hz.

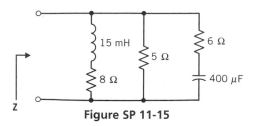

Figure SP 11-15

SP 11-16 Determine the current $i(t)$ for the circuit shown in Figure SP 11-16 when $v_s(t) = 120 \sin(\omega t + 30°)$ V and $f = 10$ kHz.

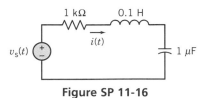

Figure SP 11-16

SP 11-17 Consider the RLC circuit of Figure 11-16a with $v_s = 1 \sin \omega t$ at 1023 Hz. Determine and plot $v_s(t)$ and $v_c(t)$ when $R = 33$ kΩ, (a) $C = 4.7$ μF and $L = 0$ and (b) $L = 1$ mH and $C = 20$ μF.

VERIFICATION PROBLEMS

VP 11-1 It is known that

$$\frac{10}{6 + jX} = A\underline{/-53.1°}$$

It is suggested that $A = 2$. Verify if this is correct (for 3 significant figures).

VP 11-2 A student report states that for the circuit of Figure VP 11-2, the impedance $\mathbf{Z}_{ab}$ is $\mathbf{Z}_{ab} = (90 + j120)\,\Omega$ when $\omega = 5000$ rad/s. Check this result.

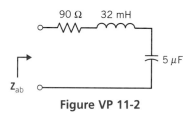

Figure VP 11-2

VP 11-3 A computer solution of $\mathbf{Z}_{ab}$ of Figure VP 11-2 reports that $\mathbf{Z}_{ab} = 95\underline{/18°}\,\Omega$ when $\omega = 3000$ rad/s. Check this result and its accuracy.

VP 11-4 A computer program reports that the currents of the circuit of Figure VP 11-4 are $\mathbf{I} = 0.2\underline{/53.1°}$ A, $\mathbf{I}_1 = 632\underline{/-18.4°}$ mA, and $\mathbf{I}_2 = 190\underline{/71.6°}$ mA. Verify this result.

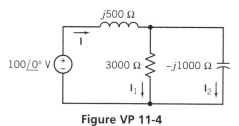

Figure VP 11-4

VP 11-5 An electrical engineering textbook states that $\mathbf{Z}_{in}$ of the circuit of Figure VP 11-5 is 10 kΩ for all frequencies. Check this statement.

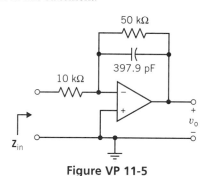

Figure VP 11-5

DESIGN PROBLEMS

DP 11-1 A circuit with an unspecified R, L, and C is shown in Figure DP 11-1. The input source is $i_s = 10 \cos 1000t$ A, and the goal is to select the R, L, and C so that the node voltage is $v = 80 \cos(1000t - \theta)$ V where $-40° < \theta < 40°$.

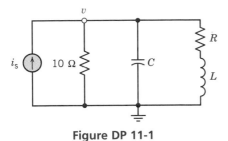

Figure DP 11-1

DP 11-2 The source $v_s = (15/\sqrt{2}) \cos 3t$ V in the circuit of Figure DP 11-2. Select the capacitance C so that the output voltage is $v_0 = V_0 \cos(3t + \phi)$ V and $V_0 = 9$ V. Determine the resulting ϕ.

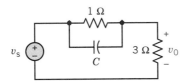

Figure DP 11-2

DP 11-3 An RLC circuit shown in Figure DP 11-3 has a source voltage $v_s = 10 \cos \omega_0 t$ V where $\omega_0 = 10^4$ rad/s. It is desired to adjust the capacitor from 0.5 μF to 3.5 μF in steps of 0.5 μF to achieve the maximum steady-state magnitude $| V_0 |$. Determine the required value of C and the maximum value of $| V_0 |$.

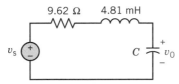

Figure DP 11-3

DP 11-4 The input impedance of the circuit of Figure DP 11-4 can be adjusted to yield $Z = aR$ where $0 \le a \le 1$.
(a) Determine the required values for $X_L = \omega L$ and $X_C = 1/\omega C$ in terms of R and a.
(b) Select L and C when $\omega = 1000$, $a = 0.2$, and $R = 100$ Ω.

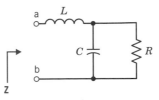

Figure DP 11-4

DP 11-5 Select L so that the impedance $\mathbf{Z}$ is $10\underline{/0°}$ at $\omega = 400$ for the circuit shown in Figure DP 11-5.

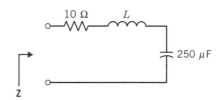

Figure DP 11-5

DP 11-6 A power circuit shown in Figure DP 11-6 takes the input signal $v_s = 12 \cos 1000t$ V and provides an output $v_0 = 6 \cos 1000t$ V. Determine the required value of b in order to achieve this output voltage.

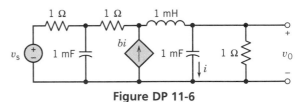

Figure DP 11-6

DP 11-7 Determine the impedance at terminals a–b for the circuit shown in Figure DP 11-7. One resistor, one inductor, and one capacitor are available and $b = 2$.
(a) Select suitable values for the components and place in the appropriate positions in the circuit so that $\mathbf{Z}_{ab} = 14.1\underline{/- 45°}$ Ω. Assume $\omega = 377$ rad/s.
(b) Determine the effect on $\mathbf{Z}_{ab}$ of b changing by $\pm 10\%$.

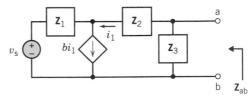

Figure DP 11-7

CHAPTER 12

AC STEADY-STATE POWER

PREVIEW

We have seen how sinusoidal steady-state currents and voltages, described as ac, have become the predominant form of power delivery to the home and industry. In this chapter we develop appropriate relationships for describing the power delivered by ac current and voltage to a load impedance. We also reexamine the concepts of superposition and the maximum power theorem in this chapter. Finally, we consider the concept of mutual inductance associated with an electric transformer. We describe how a transformer enables us to increase or decrease a voltage or current flowing through a load.

12-1 DESIGN CHALLENGE

MAXIMUM POWER TRANSFER

The matching network in Figure 12D-1 is used to interface the source with the load, which means that the matching network is used to connect the source to the load in a desirable way. In this case, the purpose of the matching network is to transfer as much power as possible to the load. This problem occurs frequently enough that it has been given a name, the maximum power transfer problem.

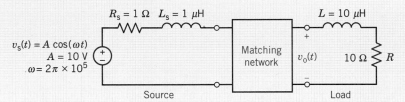

Figure 12D-1 Design the matching network to transfer maximum power to the load where the load is the model of an antenna of a wireless communication system.

An important example of the application of maximum power transfer is the connection of a cellular phone or wireless radio transmitter to the cell's antenna. For example, the input impedance of a practical cellular telephone antenna is $\mathbf{Z} = (10 + j6.28)\ \Omega$ (Dorf, 1993).

In this chapter we continue our study of circuits that are excited by sinusoidal sources. In particular, we will learn that phasors can be used to calculate the average power delivered to an element. We will return to the problem of designing the matching network at the end of the chapter. Then we will be ready to design this network to transfer maximum power to the load.

12-2 | ELECTRIC POWER

Human civilization's progress has been enhanced by society's ability to control and distribute energy. Electricity serves as a carrier of energy to the user. Energy present in a fossil fuel or a nuclear fuel is converted to electric power in order to transport and readily distribute it to customers. By means of transmission lines, electric power is transmitted and distributed to essentially all the residences, industries, and commercial buildings in the United States and Canada.

Electric power may be readily transported with low attendant losses, and improved methods for safe handling of electric power have been developed over the past 80 years. Furthermore, methods of converting fossil fuels to electric power are well developed, economical, and safe. Means of converting solar and nuclear energy to electric power are currently in various stages of development or of proven safety. Geothermal energy, tidal energy, and wind energy may also be converted to electric power. The kinetic energy of falling water may readily be used to generate hydroelectric power.

The United States consumed about 3.7 billion kWh of electric energy in 1900, for a per capita consumption of 49 kWh per person per year. By 1985 total consumption had grown to 2306 billion kWh, with a per capita consumption of 8906 kWh per year. The growth in the consumption of electric energy is shown in Table 12-1. The annual growth rate averaged 7.5 percent per year over the period 1900 to 1970. This growth rate decreased over the next 25 years as population leveled off, conservation practices were developed, and more efficient use of electricity was achieved.

As we noted earlier, the necessity of transmitting electrical power over long distances fostered the development of ac high-voltage power lines from power plant to end user.

George Westinghouse and his company developed the transformer, which permitted the efficient transformation of ac voltages to high magnitudes. Westinghouse, who is shown in Figure 12-1, designed a transformer that could be produced in quantities by ordinary manufacturing processes. The first large-scale installation was made in Buffalo in 1886.

Frequencies were generally rather high; 133 Hz was common in the early days of ac power transmission. Nikola Tesla, the inventor of the ac induction motor, who is shown in

Table 12-1
Electric Energy Use in the United States

Year	Population (millions)	Generating Capacity (million kW)	Consumption (billion kWh)	Consumption per Person per Year (kWh)
1900	76	1	4	49
1920	106	20	58	540
1940	132	50	142	1074
1960	181	168	753	4169
1980	230	570	2126	9025
1990	242	699	2550	9481

Figure 12-1 George Westinghouse, 1846–1914, was one of the persons who championed the development of ac power and fostered the development and use of the practical ac transformer. Courtesy of Westinghouse Corporation.

Figure 12-2, found that 133 Hz was too high for his motor. By 1910, Tesla had convinced electrical engineers that 60 Hz should be adopted in the United States.

Transmission over long distances required higher ac voltages. A 155-mile line at 110,000 V was put into operation in California in 1908. As needs grew, transmission voltages increased to 230,000 V, and the Hoover Dam project of the 1930s used 287 kV on a 300-mile transmission line. By the 1970s, voltages over 600 kV were used. A modern transmission line is shown in Figure 12-3.

Figure 12-2 Nikola Tesla, 1856–1943, joined George Westinghouse in the promotion of ac power in the late 1800s. Tesla was responsible for many inventions, including the ac induction motor, and was a contributor to the selection of 60 Hz as the standard ac frequency in the United States. Courtesy of Burndy Library.

Figure 12-3 AC power high-voltage transmission lines. Courtesy of Pacific Gas and Electric Company.

Figure 12-4 A large hydroelectric power plant. Courtesy of Hydro Quebec.

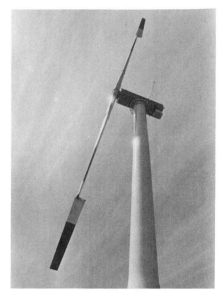

Figure 12-5 A large wind-power turbine and generator. Courtesy of *EPRI Journal.*

Electric energy generation uses original sources such as hydropower, coal, and nuclear. An example of a large hydroelectric power project is shown in Figure 12-4. A typical hydroelectric power plant can generate 1000 MW. On the other hand, many regions are turning to small generators such as the wind power device shown in Figure 12-5. A typical wind power machine may be capable of generating 75 kW.

During the period 1920 to 1960, there was extensive development of transmission tower equipment, insulators, and power lines. An example of an early high-voltage power laboratory is shown in Figure 12-6.

A unique element of the American power system is its interconnectedness. Although the power system of the United States consists of many independent companies, it is interconnected by large transmission facilities. An electric utility is often able to save

Figure 12-6 High-voltage laboratory of the Ohio Insulator Co. (Ohio Brass Co.): outdoor impulse gaps (left and right); test section of 220-V transmission towers (left). From A. O. Austin, "A Laboratory for Making Lightning" (paper presented at the International High Tension Congress, Paris, June 6–15, 1929). Courtesy of the Ohio Brass Co.

money by buying electricity from another utility and by transmitting the energy over the transmission lines of a third utility.

Coal fueled the largest percentage of the 2485×10^{12} watt-hours produced in the United States in 1986, as shown in Figure 12-7. Nuclear sources are projected to grow to 22 percent of the total by 1997.

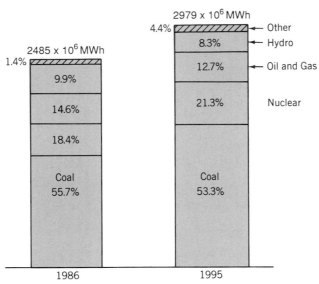

Figure 12-7 Sources of electric energy in 1986 and projected for 1995. Source: *IEEE Spectrum*, January 1987, page 53.

12-3 ‖ INSTANTANEOUS POWER

We are interested in determining the power generated and absorbed in a circuit under study. We shall start with an examination of the instantaneous power, which is the product of the time domain voltage and current associated with one or more circuit elements. The instantaneous power will permit us to determine the maximum value of the power over time.

The *instantaneous power* delivered to a circuit element is the product of the voltage $v(t)$ and the current $i(t)$, so that

$$p = vi \qquad (12-1)$$

The unit of power is watts (W). Let us consider the sinusoidal source in the circuit shown in Figure 12-8, where $v = V_m \cos \omega t$. The time domain steady-state response for the current is

$$i = I_m(\cos \omega t + \theta)$$

where

$$I_m = \frac{V_m}{\sqrt{R^2 + \omega^2 L^2}}$$

and

$$\theta = -\tan^{-1}\left(\frac{\omega L}{R}\right)$$

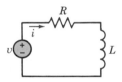

Figure 12-8 An *RL* circuit with a sinusoidal voltage source.

The instantaneous power delivered to the entire circuit is

$$p = vi$$
$$= V_m I_m \cos(\omega t + \theta) \cos \omega t$$

TOOLBOX 12-1

 TRIGONOMETRIC FORMULA FOR THE PRODUCT OF TWO SINUSOIDAL FUNCTIONS

(1) $\cos \alpha \cos(\alpha + \beta) = \cos \beta + \cos(2\alpha + \beta)$
(2) $\sin \alpha \cos \alpha = \frac{1}{2}\sin 2\alpha$

Using the trigonometric identity for the product of two cosine functions,

$$p = \frac{V_m I_m}{2}[\cos \theta + \cos(2\omega t + \theta)] \qquad (12\text{-}2)$$

We see that the instantaneous power has two terms. The first term is independent of time, and the second term varies sinusoidally over time at twice the radian frequency. The second term averages zero over one cycle, and the average power delivered to the circuit is given by the first term.

For example, let $v(t) = 10 \cos \omega t$ V, $\omega = 10^4$ rad/s, $L = 12$ mH, and $R = 50$ Ω. Then $i = I_m \cos(\omega t + \theta)$ where $I_m = 10/130 = 0.077$ A and $\theta = -\tan^{-1}(120/50) = -67.4°$. Therefore, the instantaneous power delivered by the source is

$$p = 0.15 + 0.38 \cos(2\omega t - 67.4°) \text{ W}$$

In the next section we proceed to determine the average power of an element.

EXERCISE 12-1

An *RLC* circuit is shown in Figure E 12-1 with a voltage source $v_s = 7 \cos 10t$ V.
(a) Determine the instantaneous power delivered to the circuit by the voltage source.
(b) Find the instantaneous power delivered to the inductor.
Answer: (a) $p = 7.54 + 15.2 \cos(20t - 60.3°)$ W
(b) $p = 28.3 \cos(20t - 30.6°)$ W

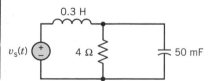

Figure E 12-1

EXERCISE 12-2

Determine the instantaneous power delivered to an element and sketch $p(t)$ when the element is (a) a resistance R and (b) an inductor L. The voltage across the element is $v(t) = V_m \cos(\omega t + \theta)$ V.

Answer: (a) $P_R = \dfrac{V_m^2}{2R}[1 + \cos(2\omega t + 2\theta)]$ W

(b) $P_L = \dfrac{V_m^2}{2\omega L} \cos(2\omega t + 2\theta - 90°)$ W

12-4 AVERAGE POWER

The usefulness of an electrical device is measured by its ability to deliver or absorb energy. In this section we consider the average power of an element having a sinusoidal voltage and sinusoidal current under steady-state condition.

Let us calculate the average power of a sinusoidal waveform of period T. When a source voltage is periodic, we have

$$v(t + T) = v(t)$$

since the voltage repeats every T seconds. Then, for a linear circuit, the current will also be a periodic function, so

$$i(t) = i(t + T)$$

Therefore, the power is

$$p = vi$$
$$= v(t + T)i(t + T)$$

In the preceding section, we found in Eq. 12-2 that

$$p = \frac{V_m I_m}{2}[\cos\theta + \cos(2\omega t + \theta)] \tag{12-3}$$

The first term within the brackets is a constant, $\cos\theta$, and the second term is periodic. The period T_2 of the second term is found when

$$2\omega T_2 = 2\pi$$

or

$$T_2 = \frac{\pi}{\omega}$$

Therefore, while the original source voltage has a period of $T = 2\pi/\omega$, the second term in the power expression has a period of $T_2 = \pi/\omega$.

The *average value* of a periodic function is the integral of the time function over a complete period, divided by the period. We use a capital P to denote average power and lowercase p to denote instantaneous power. Therefore, the average power P is given by

$$P = \frac{1}{T}\int_{t_0}^{t_0 + T} p(t)\, dt$$

where t_0 is an arbitrary starting point in time. Since $T = 2T_2$, we may also use

$$P = \frac{1}{2T_2} \int_{t_0}^{t_0 + 2T_2} p(t)\ dt \qquad (12\text{-}4)$$

Integrating over two periods, $2T_2$, will give the same value as integrating over one period of T_2.

Let us find the average power delivered to the RL circuit of Figure 12-8. Using Eqs. 12-3 and 12-4, we have

$$P = \frac{1}{T} \int_0^T \frac{V_m I_m}{2} [\cos \theta + \cos(2\omega t + \theta)]\ dt$$

where we have chosen $t_0 = 0$. Then we have

$$P = \frac{1}{T} \int_0^T k_1 \cos \theta\ dt + \frac{1}{T} \int_0^T k_1 \cos(2\omega t + \theta)\ dt$$

$$= \frac{k_1 \cos \theta}{T} \int_0^T dt + \frac{k_1}{T} \int_0^T \cos(2\omega t + \theta)\ dt$$

where $k_1 = V_m I_m / 2$. The integral of the second term is zero, since the average value of the periodic cosine function is zero. Then we have

$$P = k_1 \cos \theta$$

$$= \frac{V_m I_m}{2} \cos \theta \qquad (12\text{-}5)$$

Example 12-1

Find the average power delivered to a resistor R when the current through the resistor is $i(t)$, as shown in Figure 12-9.

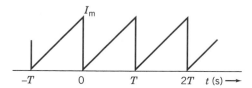

Figure 12-9 Current through a resistor in Example 12-1.

Solution

The current waveform repeats every T seconds and attains a maximum value of I_m. Using the period from $t = 0$ to $t = T$, we have

$$i = \frac{I_m}{T} t \qquad 0 < t \le T$$

Then the instantaneous power is

$$p = i^2 R$$

$$= \frac{I_m^2 t^2}{T^2} R \qquad 0 < t \le T$$

It is sufficient to find the average power over $0 < t < T$ since the power is periodic with period T. Then the average power is

$$P = \frac{1}{T} \int_0^T k_2 t^2\ dt$$

where

$$k_2 = \frac{I_m^2 R}{T^2}$$

Integrating, we have

$$P = \frac{k_2}{T} \int_0^T t^2 \, dt$$

$$= \frac{k_2}{T} \frac{T^3}{3}$$

$$= \frac{I_m^2 R}{3} \text{ W}$$

Now let us determine the average power absorbed by an impedance **Z** for a circuit excited by a sinusoidal source. The phasor voltage for the impedance is related to the phasor current by

$$\mathbf{V} = \mathbf{ZI}$$

and

$$\mathbf{Z} = Z\underline{/\theta}$$

Then, if $v = V_m \cos \omega t$, we have

$$i = I_m \cos(\omega t - \theta)$$

The average power delivered to the impedance is

$$P = \frac{V_m I_m}{T} \int_0^T \cos \omega t \cos(\omega t - \theta) \, dt$$

Referring to Table 12-2, which summarizes the integrals of sinusoidal functions, we find that

$$P = \frac{V_m I_m}{T} \left(\frac{T}{2} \cos \theta \right)$$

$$= \frac{V_m I_m}{2} \cos \theta \tag{12-6}$$

Table 12-2
Integrals of Sinusoidal Functions of Period $T = 2\pi/\omega$

$y(t)$	$\int_0^T y(t) \, dt$
$\sin(n\omega t + \phi)$; $\cos(n\omega t + \phi)$	0
$\sin^2(\omega t + \phi)$; $\cos^2(\omega t + \phi)$	$T/2$
$\sin(n\omega t + \phi) \cos(m\omega t + \theta)$	$\begin{cases} 0, \ m = n \\ \dfrac{T \sin(\phi - \theta)}{2}, \ m \neq n \end{cases}$
$\cos(n\omega t + \phi) \cos(m\omega t + \theta)$	$\begin{cases} 0, \ m \neq n \\ \dfrac{T \cos(\phi - \theta)}{2}, \ m = n \end{cases}$

Note: n and m are integers.

Therefore, the average power absorbed by an impedance is determined by the amplitude of the voltage and the amplitude of the current multiplied by the cosine of the angle of the impedance. The power of Eq. 12-6 also represents the power delivered (or supplied) by a source.

Consider the case where the impedance is that of a pure resistor R. Then $\theta = 0°$ and the average power is

$$P_R = \frac{V_m I_m}{2}$$

If the impedance is that of an inductor, we have

$$\mathbf{Z}_L = \omega L \underline{/90°}$$

Therefore, the average power is

$$P_L = \frac{V_m I_m}{2} \cos 90°$$
$$= 0$$

Similarly, for a capacitor

$$\mathbf{Z}_c = \frac{1}{\omega C} \underline{/-90°}$$

or

$$P_c = 0$$

The average power supplied by a sinusoidal source to a capacitor or inductor is zero.

Now let us consider an impedance $\mathbf{Z}$ that represents several circuit elements in series or parallel, so that

$$\mathbf{Z} = R + jX$$
$$= Z\underline{/\theta}$$

Referring back to Chapter 2, we know that a passive impedance requires that the net energy delivered be nonnegative. Therefore, we require that the average power is

$$P \geq 0$$

Since the average power is

$$P = \frac{V_m I_m}{2} \cos \theta$$

we require for a passive impedance that

$$-\frac{\pi}{2} \leq \theta \leq \frac{\pi}{2}$$

Example 12-2
Find the average power (a) absorbed by the resistor and (b) supplied by the source for the circuit of Figure 12-10 when $v_s = 100 \cos 1000t$ V.

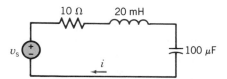

Figure 12-10 Circuit for Example 12-2.

Solution

First, let us determine the impedance of the series circuit. We have

$$\mathbf{Z} = R + j\left(\omega L - \frac{1}{\omega C}\right)$$
$$= 10 + j10$$
$$= \sqrt{200}\underline{/45°}$$

Therefore, the power absorbed by the series impedance (or supplied by the source) is

$$P = \frac{V_m I_m}{2}\cos\theta$$

where $\theta = 45°$. Since the source is $\mathbf{V}_s = 100\underline{/0°}$, we have $V_m = 100$ V. To find the current, we note that

$$\mathbf{I} = \frac{\mathbf{V}_s}{\mathbf{Z}}$$
$$= \frac{100\underline{/0°}}{10\sqrt{2}\underline{/45°}}$$
$$= \frac{10}{\sqrt{2}}\underline{/-45°}$$

Therefore, $I_m = 10/\sqrt{2}$ A and the average power is

$$P = \frac{100}{2}\left(\frac{10}{\sqrt{2}}\right)\cos 45°$$
$$= \frac{1000}{2\sqrt{2}}\left(\frac{1}{\sqrt{2}}\right)$$
$$= 250 \text{ W}$$

Another method for finding the power delivered to the impedance is to note that for the impedance

$$\cos\theta = \frac{R}{Z}$$

where $\mathbf{Z} = R + jX$ and $Z = |\mathbf{Z}|$. Also, since

$$V_m = ZI_m$$

we substitute into Eq. 12-6 as follows:

$$P = \frac{V_m I_m}{2}\cos\theta$$
$$= \frac{(ZI_m)I_m}{2}\left(\frac{R}{Z}\right)$$
$$= \frac{I_m^2}{2}R$$

Remember that R is the real part of the impedance $\mathbf{Z}$ and that the power delivered to R is the power delivered to the impedance since $P_L = P_c = 0$. Therefore, the average power delivered to the circuit of Figure 12-10 may be obtained by noting that $I_m = 10/\sqrt{2}$ A and $R = 10\ \Omega$, so that

$$P = \frac{1}{2}\left(\frac{10}{\sqrt{2}}\right)^2 10$$

$$= 250\ \text{W}$$

EXERCISE 12-3

(a) Find the average power delivered by the source to the circuit shown in Figure E 12-3.
(b) Find the power absorbed by resistor R_1.
Answer: (a) 30 W
(b) 20 W

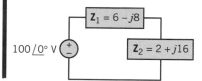

$R_1 = 1\ \Omega$

10 cos t V

$1\ \Omega$ 1 H

Figure E 12-3

EXERCISE 12-4

Find the average power delivered to each of the impedances in the network shown in Figure E 12-4.
Answer: $P_1 = 234.4$ W
$P_2 = 78.1$ W

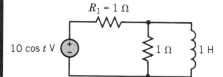

$\mathbf{Z}_1 = 6 - j8$

$100\underline{/0°}$ V

$\mathbf{Z}_2 = 2 + j16$

Figure E 12-4 Impedances in ohms.

EXERCISE 12-5

(a) Determine the instantaneous power of an element, and sketch it, when the voltage across it and the current through it are

$$v(t) = \begin{cases} 2\ \text{V} & 0 < t \le 1.5 \\ 1\ \text{V} & 1.5 < t < 2 \end{cases}$$

$$i(t) = \begin{cases} 2t\ \text{A} & 0 < t \le 1 \\ 2\ \text{A} & 1 < t < 2 \end{cases}$$

The period of the voltage and current is 2 s.
(b) Calculate the average power of the element.
Answer: (b) $P = 2.5$ W

EXERCISE 12-6

For the circuit of Figure E 12-6, determine the average power of each element and verify that the average power delivered by the source is equal to the average power absorbed by the circuit elements. The source is $v_s = 5 \cos 2t$ V.

Answer: $P_s = P_{R_1} + P_{R_2} = 4.6$ W

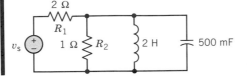

Figure E 12-6

12-5 THE POWER SUPERPOSITION PRINCIPLE AND THE MAXIMUM POWER THEOREM

In this section, let us consider the case where the circuit contains two or more sources. For example, consider the circuit shown in Figure 12-11 with two sinusoidal voltage sources. If we take each source individually, deactivating the other source in turn, we can use the principle of superposition. Then we have the response i_1 to source 1 and the response i_2 to source 2. The total response is

$$i = i_1 + i_2$$

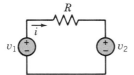

Figure 12-11 Circuit with two sources.

The instantaneous power is

$$
\begin{aligned}
p &= i^2 R \\
&= R(i_1 + i_2)^2 \\
&= R(i_1^2 + i_2^2 + 2i_1 i_2)
\end{aligned}
$$

where R is the resistance of the circuit. Then the average power is

$$
\begin{aligned}
P &= \frac{1}{T} \int_0^T p \, dt \\
&= \frac{R}{T} \int_0^T (i_1^2 + i_2^2 + 2i_1 i_2) \, dt \\
&= \frac{R}{T} \int_0^T i_1^2 \, dt + \frac{R}{T} \int_0^T i_2^2 \, dt + \frac{2R}{T} \int_0^T i_1 i_2 \, dt \\
&= P_1 + P_2 + \frac{2R}{T} \int_0^T i_1 i_2 \, dt
\end{aligned}
$$

where P_1 is the average power due to v_1 and P_2 is the average power due to v_2.

Now let us consider the last term and determine under what conditions it is equal to zero. Let the radian frequency for the first source be $m\omega$, and let the radian frequency for the second source be $n\omega$. The currents can be represented by the general form

$$i_1 = I_1 \cos(m\omega t + \phi)$$

and

$$i_2 = I_2 \cos(n\omega t + \theta)$$

Then the average power P_{12} for the product of the two currents (the third term of the expression for P) is

$$P_{12} = \frac{2R}{T} \int_0^T I_1 I_2 \cos(m\omega t + \phi) \cos(n\omega t + \theta) \, dt \qquad (12\text{-}7)$$

Referring back to Table 12-2, we note that the integral is equal to zero when $m \neq n$ and is equal to a nonzero quantity when $m = n$, where m and n are integers.

Therefore, we may state that the total average power delivered to a load is equal to the sum of the average power delivered by each source when the radian frequency of each source is an integer multiple of the other sources. However, when two or more sources have the same radian frequency, the total average power is not the sum of the average power due to each source.

Now let us further consider Eq. 12-7 when m and n are not integers. For convenience, let $m = 1$, $n = 1.5$, and $\phi = \theta = 0°$. Furthermore, since we wish to determine the average power when one of the cosine functions has a period that is not an integer multiple of period T, we must revert to the definition of average power over all time as

$$P_{12} = \lim_{\tau \to \infty} \frac{1}{\tau} \int_{-\tau/2}^{\tau/2} p \, dt$$

$$= \lim_{\tau \to \infty} \frac{1}{\tau} \int_{-\tau/2}^{\tau/2} 2RI_1 I_2 \cos \omega t \cos(1.5\omega t) \, dt$$

$$= \lim_{\tau \to \infty} \int_{-\tau/2}^{\tau/2} RI_1 I_2 (\cos 0.5\omega t + \cos 2.5\omega t) \, dt$$

$$= 0$$

since the integral of each cosine wave within the final form of the integral is zero.

Therefore, in summary, we may state that the *superposition of average power* due to multiple sinusoidal sources holds as long as no two of the sources have the same radian frequency.

Thus, superposition of average power is not applicable when two sources are coherent (have the same frequency ω), including the case when we have two dc sources ($\omega = 0$).

If two or more sources are operating at the same frequency, use the principle of superposition to find each phasor current and then add the currents to obtain the total phasor current

$$\mathbf{I} = \mathbf{I}_1 + \mathbf{I}_2 + \cdots + \mathbf{I}_N$$

for N sources. Then we have the average power

$$P = \frac{I_m^2 R}{2} \qquad (12\text{-}8)$$

where $\mathbf{I} = I_m \cos \omega t$.

Remember that phasors resulting from sources with different frequencies cannot be superposed.

Example 12-3

Find the average power in the circuit of Figure 12-11 when $v_1 = 10 \cos 10t$ V, $R = 1\ \Omega$, and (a) $v_2 = 2 \cos 20t$ V; (b) $v_2 = 2 \cos 10t$ V.

Solution

(a) First, for v_1 and v_2, which have two different frequencies, the principle of power superposition holds and

$$P = P_1 + P_2$$

Then we have

$$P_1 = \frac{I_1^2 R}{2}$$
$$= \frac{(10)^2}{2}$$
$$= 50\text{ W}$$

and

$$P_2 = \frac{I_2^2 R}{2}$$
$$= \frac{(2)^2}{2}$$
$$= 2\text{ W}$$

Therefore, $P = P_1 + P_2 = 52$ W.

(b) When both sources have the same frequency, the principle of *power* superposition does not hold. The circuit is linear so we can use superposition to calculate the *current* in the resistor. We find $\mathbf{I}_1 = 10$ and $\mathbf{I}_2 = -2$. Since the phasors $\mathbf{I}_1$ and $\mathbf{I}_2$ correspond to sinusoids having the same frequency, $\omega = 10$, we can add them,

$$\mathbf{I} = \mathbf{I}_1 + \mathbf{I}_2$$
$$= 8\text{ A}$$

The total average power can be obtained as

$$P = \frac{I^2 R}{2}$$
$$= \frac{(8)^2}{2}$$
$$= 32\text{ W}$$

In Chapter 5, we proved that for a resistive network, maximum power is transferred from a source to a load when the load resistance is set equal to the Thévenin resistance of the Thévenin equivalent source. Now let us consider a circuit represented by a Thévenin equivalent circuit for a sinusoidal steady source, as shown in Figure 12-12, when the load is $\mathbf{Z}_L$.

We then have

$$\mathbf{Z}_t = R_t + jX_t$$

and

$$\mathbf{Z}_L = R_L + jX_L$$

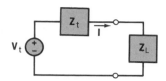

Figure 12-12 The Thévenin equivalent circuit with a load impedance.

The average power delivered to the load is

$$P = \frac{I_m^2}{2} R_L$$

The phasor current $\mathbf{I}$ is given by

$$\mathbf{I} = \frac{\mathbf{V}_t}{\mathbf{Z}_t + \mathbf{Z}_L}$$

$$= \frac{\mathbf{V}_t}{(R_t + jX_t) + (R_L + jX_L)}$$

where we may select the values of R_L and X_L. The average power delivered to the load is

$$P = \frac{I^2 R_L}{2} = \frac{|\mathbf{V}_t|^2 R_L/2}{(R_t + R_L)^2 + (X_t + X_L)^2}$$

and we wish to maximize P. The term $(X_t + X_L)^2$ can be eliminated by setting $X_L = -X_t$. We have

$$P = \frac{|\mathbf{V}_t|^2 R_L}{2(R_t + R_L)^2}$$

The maximum is determined by taking the derivative dP/dR_L and setting it equal to zero. Then we find that $dP/dR_L = 0$ when $R_L = R_t$.

Consequently we have

$$\mathbf{Z}_L = R_t - jX_t$$

Thus, the *maximum power* of a circuit with a Thévenin equivalent circuit with an impedance $\mathbf{Z}_t$ is obtained when $\mathbf{Z}_L$ is set equal to the complex conjugate of $\mathbf{Z}_t$.[1]

Example 12-4

Find the load impedance that transfers maximum power to the load and determine the maximum power quantity obtained for the circuit shown in Figure 12-13.

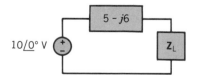

Figure 12-13 Circuit for Example 12-4. Impedances in ohms.

Solution

We select the load to be the complex conjugate of $\mathbf{Z}_t$ so that

$$\mathbf{Z}_L = \mathbf{Z}_t^*$$
$$= 5 + j6$$

---

[1] See Appendix C for a review of complex numbers.

Then the maximum power transferred can be obtained by noting that

$$\mathbf{I} = \frac{10\underline{/0°}}{5+5}$$
$$= 1\underline{/0°}$$

Therefore, the average power transferred to the load is

$$P = \frac{I_m^2}{2} R_L$$
$$= \frac{(1)^2}{2} 5$$
$$= 2.5 \text{ W}$$

EXERCISE 12-7

Find the average power delivered to the resistor of the circuit of Figure 12-11 when $v_1 = 100 \cos 100t$ V, $R = 10$ Ω, and (a) $v_2 = 50 \cos 25t$ V, (b) $v_2 = 50 \cos 100t$ V.

EXERCISE 12-8

For the circuit of Figure 12-12, find $\mathbf{Z}_L$ to obtain the maximum power transferred when the Thévenin equivalent circuit has $\mathbf{V}_t = 100\underline{/0°}$ V and $\mathbf{Z}_t = 10 + j14$ Ω. Also determine the maximum power transferred to the load.
Answer: $\mathbf{Z}_L = 10 - j14$ Ω
$\qquad P = 125$ W

EXERCISE 12-9

A television receiver uses a cable to connect the antenna to the TV, as shown in Figure E 12-9, with $v_s = 4 \cos \omega t$ mV. The TV station is received at 52 MHz. Determine the average power delivered to each TV set if (a) the load impedance is $\mathbf{Z} = 300$ Ω; (b) two identical TV sets are connected in parallel with $\mathbf{Z} = 300$ Ω for each set; (c) two identical sets are connected in parallel and $\mathbf{Z}$ is to be selected so that maximum power is delivered at each set.
Answer: (a) 9.6 nW   (b) 4.9 nW   (c) 5 nW

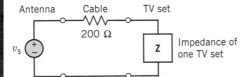

Figure E 12-9

12-6 | EFFECTIVE VALUE OF A SINUSOIDAL WAVEFORM

The voltage available from a wall plug in a residence is said to be 110 V. Of course, this is not the average value of the sinusoidal voltage since we know that the average would be zero. It is also not the instantaneous value or the maximum value, V_m, of the voltage $v = V_m \cos \omega t$.

The effective value of a voltage is a measure of its effectiveness in delivering power to a load resistor. The concept of an *effective value* comes from a desire to have a sinusoidal voltage (or current) deliver to a load resistor the same average power as an equivalent dc voltage (or current). The goal is to find a dc V_{eff} (or I_{eff}) that will deliver the same average power to the resistor as would be delivered by a periodically varying source, as shown in Figure 12-14. The energy delivered in a period T is $w = PT$, where P is the average power.

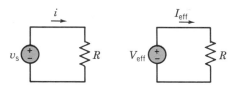

Figure 12-14 The goal is to find V_{eff} for a specified $v_s(t)$ that will deliver the same energy to R as would be delivered by a dc source.

The average power delivered to the resistor R by a current is

$$P = \frac{1}{T} \int_0^T i^2 R \, dt \tag{12-9}$$

We select the period of the time-varying current, T, as the integration interval.

The power delivered by a direct current is

$$P = I_{eff}^2 R \tag{12-10}$$

where I_{eff} is the dc current that will deliver the same power as the periodically varying current. That is, I_{eff} is defined as the steady (constant) current that is as *effective* in delivering power as the average power of the varying current.

We equate Eqs. 12-9 and 12-10, obtaining

$$I_{eff}^2 R = \frac{R}{T} \int_0^T i^2 \, dt$$

Solving for I_{eff}, we have

$$I_{eff} = \left(\frac{1}{T} \int_0^T i^2 \, dt \right)^{1/2} \tag{12-11}$$

We see that I_{eff} is the square root of the mean of the squared value. Thus, the effective current I_{eff} is commonly called the root-mean-square current I_{rms}.

Now let us find the I_{rms} of a sinusoidally varying current $i = I_m \cos \omega t$. Using Eq. 12-11 and Toolbox 12-1 we have

$$I_{rms} = \left(\frac{1}{T} \int_0^T I_m^2 \cos^2 \omega t \, dt \right)^{1/2}$$

$$= \left(\frac{I_m^2}{T} \int_0^T \frac{1}{2} (1 + \cos 2\omega t) \, dt \right)^{1/2}$$

$$= \frac{I_m}{\sqrt{2}} \tag{12-12}$$

since the integral of cos $2\omega t$ is zero over the period T. Be sure to remember that Eq. 12-12 is true only for sinusoidal currents.

The **effective value** of a current is the steady current (dc) that transfers the same average power as the given varying current.

Of course, the effective value of the voltage in a circuit is similarly found from the equation

$$V^2_{\text{eff}} = V^2_{\text{rms}} = \frac{1}{T}\int_0^T v^2 \, dt$$

Thus

$$V_{\text{rms}} = \left(\frac{1}{T}\int_0^T v^2 \, dt\right)^{1/2}$$

Example 12-5
Find the effective value of the current for the sawtooth waveform shown in Figure 12-15.

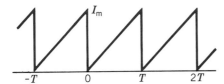

Figure 12-15 A sawtooth current waveform.

Solution
First, we will express the current waveform over the period $0 < t \le T$. The current is then

$$i = \frac{I_m}{T} t \qquad 0 < t \le T$$

The effective value is found from

$$I^2_{\text{eff}} = \frac{1}{T}\int_0^T i^2 \, dt$$

$$= \frac{1}{T}\int_0^T \frac{I^2_m}{T^2} t^2 \, dt$$

$$= \frac{I^2_m}{T^3}\left[\frac{t^3}{3}\right]_0^T$$

$$= \frac{I^2_m}{3}$$

Therefore, solving for I_{eff}, we have

$$I_{\text{eff}} = \frac{I_m}{\sqrt{3}}$$

In practice, we must be careful to determine whether a sinusoidal voltage is expressed in terms of its effective value or its maximum value I_m. In the case of power transmission and use in the home, the voltage is said to be 110 V or 220 V, and it is understood that the voltage is the rms or effective value.

In electronics or communications circuits the voltage could be described as 10 V, and the person is typically indicating the maximum or peak amplitude, V_m. Henceforth, we will use V_m as the peak value and V as the rms value (which could be written as V_{rms}). Thus, we choose normally to drop the subscript rms or eff. We will periodically remind you of this notation. However, you should be able to distinguish V from V_m, which always implies a peak amplitude value.

Remember that since the average power was used to derive the rms value, I, we can state that the average power is

$$P = (I_1^2 + I_2^2 + \cdots + I_N^2)R \tag{12-13}$$

if, and only if, each effective current has a different frequency ω_n. If two or more of the currents have the same frequency ω, the simple additive condition does not hold. For example, if we have a current of 5 A (effective value) at 50 Hz and another of 2 A at 60 Hz, then the absorbed power in a 3-Ω resistor is

$$
\begin{aligned}
P &= (5^2 + 2^2)R \\
&= (29)3 \\
&= 87 \text{ W}
\end{aligned}
$$

However, if we have a current of 5 A at 60 Hz and another current of 2 A at 60 Hz, the resulting power will depend on the phase between the two sinusoidal currents. The effective value of these two currents may then have a value between 3 A and 7 A, depending on the phase difference.

If you wish to find the effective value of a current I that consists of a number of sinusoidal currents of different frequencies, you can readily use Eq. 12-13 to prove that the square of the effective value I is

$$I^2 = I_1^2 + I_2^2 + \cdots + I_N^2 \tag{12-14}$$

EXERCISE 12-10

Find the effective value of the current waveform shown in Figure E 12-10.
Answer: $I_{eff} = 8.66$

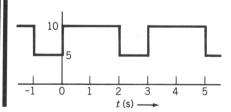

Figure E 12-10

EXERCISE 12-11

Find the effective value of the following currents: (a) $\cos 3t + \cos 3t$; (b) $\sin 3t + \cos(3t + 60°)$; (c) $2 \cos 3t + 3 \cos 5t$
Answer: (a) $\sqrt{2}$ (b) 0.366   (c) 2.55

EXERCISE 12-12

Calculate the effective value of the voltage across the resistance R of the circuit shown in Figure E 12-12 when $\omega = 100$ rad/s.
Answer: $V_{eff} = 6.18$ V

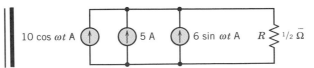

Figure E 12-12

12-7 | POWER FACTOR

The average power delivered by a sinusoidal current is related to the effective value of the current. Recall that we showed in Section 12-3 (Eq. 12-6) that the average power absorbed by an impedance is

$$P = \frac{V_m I_m}{2} \cos \theta$$

Since $V_{\text{eff}} = V_m/\sqrt{2}$ and $I_{\text{eff}} = I_m/\sqrt{2}$, we may rewrite Eq. 12-6 as

$$P = V_{\text{eff}} I_{\text{eff}} \cos \theta$$
$$= VI \cos \theta$$

where we use $V = V_{\text{eff}}$ and $I = I_{\text{eff}}$, dropping the subscript for effective.

The product VI is called the *apparent power* and has units of voltamperes (VA).

Power utility companies are concerned with the ratio of the average power to the apparent power and define the power factor (pf) as

$$\text{pf} = \frac{P}{VI} = \cos \theta \qquad (12\text{-}15)$$

The angle θ is often referred to as the *power factor angle.*

The power utility company desires an efficient transmission and distribution system for delivery of its power to a customer. Consider the model of the power source provided by a utility and a customer's load as shown in Figure 12-16. If the source current is $i_s = I_m \cos \omega t$ and the load impedance is $Z\underline{/\theta}$, then the phasor voltage is

$$\mathbf{V} = I_m Z\underline{/\theta}$$

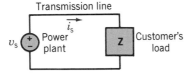

Figure 12-16 Power plant supplying a customer's electrical load. A transmission line connects the power plant to the customer's terminals.

Therefore, the average power is

$$P = \frac{V_m I_m}{2} \cos \theta$$

$$= VI \cos \theta$$

since $V = V_m/\sqrt{2}$ and $I = I_m/\sqrt{2}$. For a purely resistive load, the angle θ is zero and the power factor is unity. A power factor of 0.5 indicates an angle of either 60° or − 60°.

However, recall that

$$\mathbf{Z}_L = R + jX$$

and

$$\tan \theta = \frac{X}{R}$$

When the load is inductive, we have a positive angle θ. Conversely, when the load is capacitive, the angle θ is negative. Power engineers refer to the inductive case where the angle is positive as lagging, since the current lags the voltage. Similarly, they define the capacitive case with a negative angle as leading. The inductive case is said to have a lagging power factor and capacitive case a leading power factor. Of course, the power factor does not actually lag; it is the current that lags or leads the voltage.

The **power factor** is the ratio of average power P to apparent power VI.

The power company must send the current down the power lines to the load. Since the customer will wish to pay only for actual power received, the customer desires to pay for $P = VI \cos \theta$. For example, if the load has a $\theta = 60°$, the customer will wish to pay for $VI \cos 60° = 0.5\,VI$. Nevertheless, the power company must transmit I and suffer the power line losses I^2R, where R is the resistance of the line. Therefore, the power company is very interested in having the customer keep the power factor of the load as close to unity as possible. Look again at Eq. 12-15, where

$$\text{pf} = \frac{P}{VI} = \cos \theta$$

It is desirable to have pf $= 1$ so that $VI = P$.

For example, let us assume that the power plant source of Figure 12-16 provides an effective voltage $V = 200$ V and that the load is $Z\underline{/\theta}$, where $\theta = 60°$. If the load requires an average power of 1 kW, the current required can be calculated. Since

$$P = VI \cos \theta$$

we solve for I to obtain

$$I = \frac{P}{V \cos \theta}$$

When $P = 1$ kW, $\theta = 60°$, and $V = 200$, we have

$$I = \frac{1000}{200(0.5)} = 10 \text{ A}$$

On the other hand, if the load had a power factor of 1, the required current I would be only 5 A. Now if the power line to reach the location of the load is such that the total resistance along the line is 5 Ω, the power lost on the line is 125 W for the case when the pf is 1 and is 500 W when the pf is 0.5. Clearly, since the power company wants to minimize the power lost on the lines, they want the customer to have a power factor as close to 1 as possible and for this reason they will provide price incentives or penalties to help encourage the user to meet their objectives.

It is clearly to the advantage of the power company and the user to keep the power factor of a load as close to unity as feasible. If the load has an impedance

$$\mathbf{Z} = R + jX$$

then we may add a parallel impedance $\mathbf{Z}_1$ as shown in Figure 12-17. Since $\mathbf{Z}$ is connected to draw a current $\mathbf{I}$, the power delivered to $\mathbf{Z}$ will remain P. The benefit of the parallel impedance is that it is the parallel combination that appears as the load to the source and $\mathbf{I}_1$ is the current that flows through the transmission line. Therefore, we use $\mathbf{Z}_1$ to *correct* the power factor so that pf $= 1$ at the customer's terminals.

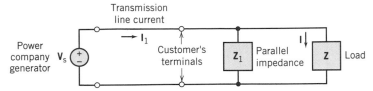

Figure 12-17 Use of an added parallel impedance $\mathbf{Z}_1$ to correct the customer's power factor.

The impedance of the parallel combination, $\mathbf{Z}_P$, is

$$\mathbf{Z}_P = \frac{\mathbf{Z}\mathbf{Z}_1}{\mathbf{Z} + \mathbf{Z}_1}$$

We wish that $\mathbf{Z}_1$ absorb no average power. Therefore, we choose a reactive element so that $\mathbf{Z}_1 = jX_1$.

The parallel impedance combination may be written as

$$\mathbf{Z}_P = R_P + jX_P = Z_P \underline{/\theta_P}$$

and the power factor of the new combination is

$$\text{pfc} = \cos \theta_c$$
$$= \cos \left(\tan^{-1} \frac{X_P}{R_P} \right) \tag{12-16}$$

where pfc is the corrected power factor and the corrected phase $\theta_c = \theta_P$. The relationship for $\mathbf{Z}_P$ is then obtained from the specification that $\mathbf{Z}_1 = jX_1$ so that

$$\mathbf{Z}_P = \frac{(R + jX)jX_1}{R + jX + jX_1}$$
$$= \frac{RX_1^2 + j[R^2X_1 + (X_1 + X)XX_1]}{R^2 + (X + X_1)^2}$$

Therefore, the ratio of R_P to X_P is

$$\frac{R_P}{X_P} = \frac{RX_1}{R^2 + (X_1 + X)X} \tag{12-17}$$

Since R_P/X_P is also specified by Eq. 12-16, we may proceed to find the required X_1 by noting that Eq. 12-16 may be written as

$$\frac{X_P}{R_P} = \tan(\cos^{-1} \text{pfc}) \tag{12-18}$$

Combining Eqs. 12-17 and 12-18 and solving for X_1, we have

$$X_1 = \frac{R^2 + X^2}{R \tan(\cos^{-1} \text{pfc}) - X} \tag{12-19}$$

We note that X_1 may be positive or negative depending on the required pfc and the original R and X of the load. The factor $\tan(\cos^{-1} \text{pfc})$ will be positive if pfc is specified as lagging and negative if it is specified as leading. Typically, we will find that the customer's load is inductive and we will need a capacitive impedance $\mathbf{Z}_1$. Recall that for a capacitor, we have

$$\mathbf{Z}_1 = \frac{-j}{\omega C} = jX_1$$

Note that we determine that X_1 is typically negative.

The formula, Eq. 12-19, for determining the parallel impedance $\mathbf{Z}_1$ is useful when the load may be inductive or capacitive. However, the load is inductive in most cases and, hence, we use a parallel capacitor to correct the power factor. Let us assume a lagging power factor (for an inductive load) and use a capacitor in parallel to correct it. Then $\mathbf{Z} = R + j\omega L$ and $\mathbf{Z}_1 = 1/j\omega C$. The admittance of the load is

$$\mathbf{Y} = \frac{1}{R + j\omega L} = G - jB$$

where $G = R/(R^2 + X^2)$ where $X = \omega L$. Furthermore, we have $\mathbf{Y}_1 = +j\omega C$. Then, we construct a phasor diagram using the admittances as shown in Figure 12-18. Then

$$\omega C = G \tan \theta - G \tan \theta_c$$
$$= G[\tan \theta - \tan \theta_c] \qquad (12\text{-}20)$$

where $\cos \theta$ is the uncorrected power factor and $\cos \theta_c$ is the corrected power factor.

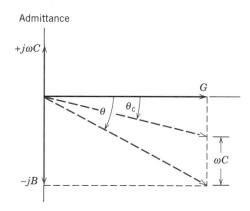

Figure 12-18 Phasor diagram for power factor correction using a shunt capacitor.

Example 12-6

A load as shown in Figure 12-17 has an impedance of $\mathbf{Z} = 100 + j100 \ \Omega$. Find the parallel capacitance required to correct the power factor to (a) 0.95 lagging and (b) 1.0. Assume that the source is operating at $\omega = 377$ rad/s.

Solution

The original load has a lagging power factor with

$$\cos \theta = \cos 45° = 0.707$$

First, we wish to correct the pf so that pfc = 0.95 lagging. Then, we use Eq. 12-19 as follows:

$$X_1 = \frac{100^2 + 100^2}{100 \tan(\cos^{-1} 0.95) - 100}$$
$$= -297.9$$

The capacitor required is determined from

$$-\frac{1}{\omega C} = X_1$$

Therefore, since $\omega = 377$ rad/s,

$$C = \frac{1}{\omega X_1} = 8.9 \ \mu F$$

If we wish to correct the load to pfc = 1, we have

$$X_1 = \frac{2 \times 10^4}{100 \tan(\cos^{-1} 1) - 100}$$
$$= -200$$

The capacitor required to correct the power factor to 1.0 is determined from

$$\frac{-1}{\omega C} = X_1$$

Therefore, since $\omega = 377$,

$$C = \frac{-1}{\omega X_1}$$
$$= 13.3 \ \mu F$$

Since the uncorrected power factor is lagging, we can alternatively use Eq. 12-20 to determine C. For example, it follows that pfc = 1. Then $\theta_c = 0°$ and $G = 100/(2 \times 10^4)$. Therefore,

$$\omega C = G(\tan \theta - \tan \theta_c) = 5 \times 10^{-3} \tan 45°$$

and

$$C = \frac{5 \times 10^{-3}}{377} = 13.3 \ \mu F$$

Generally, the uncorrected load is inductive and we have a lagging power factor. We then add a capacitor in parallel to achieve a corrected power factor between 0.90 and 1.0.

The average power to a load is measured in watts and is expressed as

$$P = VI \cos \theta \qquad (12\text{-}21)$$

where VI is called the apparent power with the units of voltamperes (VA) and V and I are the effective voltage and current values respectively. The angle θ is defined as $\theta_v - \theta_i$, the difference between the angle of the voltage and the current. Power engineers define a term *complex power* as the sum of the average power and the reactive power expressed as a complex number, so that

$$\mathbf{S} = P + jQ \qquad (12\text{-}22)$$

with the units of $\mathbf{S}$ called voltamperes (VA). Since $P = VI \cos \theta$, using the rectangular form of the complex number $\mathbf{S}$, it may be shown that

$$Q = VI \sin \theta \qquad (12\text{-}23)$$

Q is called the *reactive power* and has units called voltamperes reactive (VAR). This form of complex power is often used by power engineers to determine the complete effect of several parallel loads, as will be illustrated by Example 12-7.

TOOLBOX 12-2

THE CONJUGATE OF THE COMPLEX NUMBER

For a complex number $c = a + jb$, the conjugate is defined as c^* where

$$c^* = a - jb$$

In polar form when $c = r\underline{/\theta}$,

$$c^* = r\underline{/-\theta}$$

Since $\mathbf{S} = P + jQ$ we use Eqs. 12-21 and 12-23 to obtain

$$\begin{aligned}
\mathbf{S} &= VI(\cos\theta + j\sin\theta) \\
&= VI\underline{/\theta} \\
&= V\underline{/\theta_v}\, I\underline{/-\theta_I} \\
&= \mathbf{VI}^*
\end{aligned}$$

We can demonstrate the concept of complex power by considering an inductive circuit with an impedance

$$\mathbf{Z} = R + jX$$

with a lagging power factor $\cos\theta$. Let the current through the load impedance be the phasor $\mathbf{I} = I\underline{/0°}$, where I is the effective current. Multiplying $\mathbf{Z}$ by $\mathbf{I}$, we have the phasor voltage

$$\begin{aligned}
\mathbf{V} &= (R + jX)\mathbf{I} \\
&= RI + jXI
\end{aligned}$$

The power $\mathbf{S}$ is $\mathbf{VI}^*$, or we have

$$\begin{aligned}
\mathbf{S} &= RI^2 + jXI^2 \\
&= P + jQ
\end{aligned}$$

where P is the average power supplied to the impedance and Q is the reactive power.

We can establish a complex power triangle as shown in Figure 12-19. Again, we note

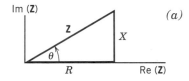

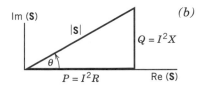

Figure 12-19 (*a*) The impedance triangle where $\mathbf{Z} = R + jX = Z\underline{/\theta}$. (*b*) The complex power triangle where $\mathbf{S} = P + jQ$.

Table 12-3
AC Power Relationships for an Impedance $\mathbf{Z} = Z\underline{/\theta}$

Quantity	Relationship	Units
Instantaneous current	$i = I_m \cos \omega t$	A
Instantaneous voltage	$v = V_m \cos(\omega t + \theta)$	V
RMS (effective) current	$I = \dfrac{I_m}{\sqrt{2}}$	A
Apparent power	VI	VA
Average power	$P = VI \cos \theta$	W
Power factor	$\cos \theta$	Dimensionless
Reactive power	$Q = VI \sin \theta$	VAR
Complex power	$\mathbf{S} = P + jQ = \mathbf{VI}^*$	VA

from the figure that $P = VI \cos \theta$ and $Q = VI \sin \theta$, where $VI = |\mathbf{S}|$. The quantity $\mathbf{S}$ is called the complex power.

A summary of the relationships for ac power is given in Table 12-3.

The complex power, $\mathbf{S}$, supplied by the source shown in Figure 12-17 is the sum of the complex powers delivered to the parallel loads. Thus, in general,

$$\mathbf{S} = \mathbf{S}_1 + \mathbf{S}_2 + \cdots + \mathbf{S}_N$$

for N loads.

Example 12-7

A customer's plant has two parallel loads connected to the power utility's distribution lines. The first load consists of 50 kW of heating and is resistive. The second load is a set of motors that operate at 0.86 lagging power factor. The motors' load is 100 kVA. Power is supplied to the plant at 10,000 volts rms. Determine the total current flowing from the utility's lines into the plant and the plant's overall power factor.

Solution

The total complex power $\mathbf{S}$ delivered to the total load is the sum of the complex power delivered to each load, hence $\mathbf{S} = \mathbf{S}_1 + \mathbf{S}_2$. We proceed to construct a table as shown below that summarizes the power factor and associated power for each load and the total plant. (Table 12-4)

Table 12-4
Summary of Power Calculations for Two Loads

Load	Power Factor	θ	VI Apparent Power (kVA)	P Average Power (kW)	Q Reactive Power (kVAR)
Heating	1.0	0°	50	50	0
Motors	0.86	30.7°	100	86	51
Plant	0.94	20.6°	145.2	136	51

The reactive power is found using Eq. 12-23, which is

$$Q = VI \sin \theta$$

For the motors, we have $\cos \theta = 0.86$ and therefore $\theta = 30.7°$. Since the apparent power of the motors is $VI = 100$ kVA, we have

$$\begin{aligned} P_1 &= VI \cos \theta \\ &= 10^5(0.86) \\ &= 86 \text{ kW} \end{aligned}$$

and

$$\begin{aligned} Q_1 &= VI \sin \theta \\ &= 10^5 \sin 30.7° \\ &= 51 \text{ kVAR} \end{aligned}$$

These results are recorded in Table 12-4.

The total complex power is

$$\mathbf{S} = \mathbf{S}_1 + \mathbf{S}_2 = P_1 + jQ_1 + P_2 + jQ_2$$

First, add the average power for each load to obtain the total average power for the plant. Then the total average power is

$$\begin{aligned} P &= P_1 + P_2 \\ &= 50 + 86 \\ &= 136 \text{ kW} \end{aligned}$$

Of course, the reactive power for the resistive load is zero. The total reactive power of the plant is

$$\begin{aligned} Q &= 51 + 0 \\ &= 51 \text{ kVAR} \end{aligned}$$

Therefore, the total plant has a complex power

$$\begin{aligned} \mathbf{S} &= P + jQ \\ &= 136 + j51 \\ &= 145.2\underline{/20.6°} \text{ kVA} \end{aligned}$$

Therefore, the power factor angle for the plant is 20.6° and the power factor is $\cos \theta = 0.94$ as summarized in Table 12-4.

Since $VI = 145.2$ kVA for the total plant and $V = 10$ kV, we have $I = 14.5$ A.

EXERCISE 12-13

A circuit has a large motor connected to the ac power lines. The impedance of the motor is determined at $\omega = 377$ rad/s. The model of the motor is a resistor of 100 Ω in series with an inductor of 5 H. Find the power factor of the motor.
Answer: pf = 0.053

EXERCISE 12-14

A circuit has a load impedance $\mathbf{Z} = 50 + j80$ Ω as shown in Figure 12-17. Determine the power factor of the uncorrected circuit. Determine the impedance $\mathbf{Z}_1$ required to obtain a corrected power factor of 1.0.
Answer: $\mathbf{Z}_1 = -j111.25$ Ω

EXERCISE 12-15

Determine the power factor for the total plant of Example 12-7 when the resistive heating load is decreased to 30 kW. The motor load and the supply voltage remain as described in Example 12-7.
Answer: pf = 0.915

EXERCISE 12-16

A 4-kW, 110-V load, as shown in Figure 12-17, has a power factor of 0.82 lagging. Find the value of the parallel capacitor that will correct the power factor to 0.95 lagging when ω = 377 rad/s.
Answer: C = 0.324 mF

12-8 THE TRANSFORMER

The concept of self-inductance was introduced in Chapter 7. We commonly use the term inductance for self-inductance, and we are familiar with circuits with inductors. In this section we consider a circuit element called a transformer, which is useful in circuits with sinusoidal steady-state (ac) voltages and currents and is also widely used in electronic circuits.

An *electric transformer* is a magnetic device that consists of two or more multiturn coils wound on a common core. One major use of transformers is in ac power distribution. Transformers possess the ability to "step up" or "step down" ac voltages or currents. Transformers are used by power utilities to raise (step up) the voltage from 10 kV at a generating plant to 200 kV or higher for transmission over long distances. Then, at a receiving plant, transformers are used to reduce (step down) the voltage to 220 or 110 V for use by the customer (Coltman, 1988).

In addition to power systems, transformers are commonly used in electronic and communication circuits. They provide the ability to raise or reduce voltages and to isolate one circuit from another.

The self-inductance of a multiturn coil behaves as

$$v = L \frac{di}{dt}$$

If the coil has N turns and the magnetic flux linking the coil is ϕ, then we have

$$v = N \frac{d\phi}{dt}$$

Now let us consider two magnetically coupled coils as shown in Figure 12-20. We assume that all the flux ϕ is contained within the magnetic core having a cross-sectional area A. If i is increasing with time, then ϕ is also increasing, since $N\phi = Li$. The voltage at the terminals of the second coil is induced by ϕ that flows through the second coil. Thus, there is a voltage v_2 with a positive sign at terminal a, since the flux ϕ flows through the second coil, where its direction of windings relative to ϕ is the same as for the first coil.

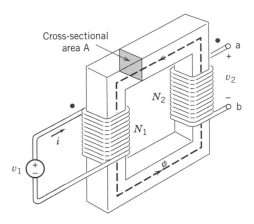

Figure 12-20 Two magnetically coupled coils mounted on a magnetic material. The flux ϕ is contained within the magnetic core.

A dot convention is used to establish the polarity of the voltage induced in the second coil relative to the first coil as shown in Figure 12-20. The dots at the appropriate end of the coils indicate that the dotted ends have a positive voltage at the same time. The number of turns of the first and second coils are N_1 and N_2 respectively.

We use the dot convention to determine the sign of the induced voltages in terms of their inducing currents. The induced voltage v_2 will be plus at its dot on the second coil L_2 if the inducing current i enters the dotted terminal for L_1.

When there is no load connected to terminals a–b in Figure 12-20, the induced voltage is

$$v_2 = N_2 \frac{d\phi}{dt}$$

However, since it can be shown that $\phi = kN_1 i$, we have

$$v_2 = kN_1 N_2 \frac{di}{dt}$$

$$= M \frac{di}{dt} \tag{12-24}$$

where M is a positive number called the *mutual inductance*. The inductances L_1 and L_2 of the first and second coils are called the self-inductances. In all cases, the unit of inductance is the henry, H.

If a current i enters the dotted end of one coil, then, for the other coil, the polarity of the induced voltage is plus at the dotted end. Conversely, if a current comes out of the dotted end of the first coil, the induced voltage at the second coil is minus at the dotted terminal.

One of the coils, typically drawn on the left of the diagram of a transformer, is designated as the *primary coil,* and the other is called the *secondary coil* or winding. The primary coil is connected to the energy source and the secondary coil is connected to the load. The circuit symbol for a transformer is shown in Figure 12-21*a* with the dots shown and the mutual inductance is identified as *M*.

The voltage at the first coil, v_1, shown in Figure 12-21*a*, is

$$v_1 = L_1 \frac{di_1}{dt} + M \frac{di_2}{dt} \tag{12-25}$$

Similarly, the voltage at the second pair of terminals is

$$v_2 = L_2 \frac{di_2}{dt} + M \frac{di_1}{dt} \tag{12-26}$$

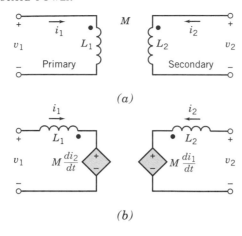

(a)

(b)

Figure 12-21 (a) Circuit symbol for the transformer. (b) Model of the transformer.

Thus, the mutual inductance can be seen to induce a voltage in a coil due to the current in the other coil. Note the assumed direction of flow of the currents i_1 and i_2 into the dotted terminals as shown in Figure 12-21a. The circuit model of the transformer is shown in Figure 12-21b using two current-controlled voltage sources.

The use of transformers is usually limited to non-dc applications since the primary and secondary windings behave as short circuits for a steady current.

When sinusoidal steady-state currents and voltages are present, the transformer shown in Figure 12-21 is represented by the phasor equations

$$\mathbf{V}_1 = j\omega L_1 \mathbf{I}_1 + j\omega M \mathbf{I}_2 \tag{12-27}$$

and

$$\mathbf{V}_2 = j\omega L_2 \mathbf{I}_2 + j\omega M \mathbf{I}_1 \tag{12-28}$$

These equations are unique and assume that the currents i_1 and i_2 enter the dotted terminals of the transformer.

Let us consider the transformer circuit as shown in Figure 12-22 when $v_1 = 100 \cos 1000t$ V. Note that i_1 enters the dotted terminal, but i_3 leaves its dotted terminal; so we have $i_3 = -i_2$, where i_2 enters the dotted terminal. The equation for the left-hand mesh is

$$(R + j10)\mathbf{I}_1 + j100\mathbf{I}_2 = 100$$

or

$$(R + j10)\mathbf{I}_1 - j100\mathbf{I}_3 = 100$$

Thus, there is a minus sign associated with the mutual inductance term since $\mathbf{I}_3$ does not enter the dotted terminal of the transformer.

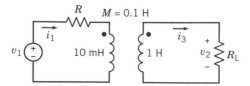

Figure 12-22 Transformer circuit with a load R_L.

The equation for the right-hand mesh is

$$-j100\mathbf{I}_1 + (R_L + j1000)\mathbf{I}_3 = 0$$

As an approximation, let $R = 0$ and solve for $\mathbf{I}_3$. Then using Cramer's rule and letting $R_L = 100\ \Omega$, we have

$$\mathbf{I}_3 = \frac{j100(100)}{j10(100 + j1000) + (100)^2}$$

$$= 10\underline{/0°}$$

Therefore,

$$\mathbf{V}_2 = R_L\mathbf{I}_3$$

$$= 1000\ \text{V}$$

The ratio $\mathbf{V}_2/\mathbf{V}_1 = 10$, so this transformer circuit results in a step-up of 10 times the primary voltage.

We can find the instantaneous power in a transformer as $v_1 i_1 + v_2 i_2$ and then proceed to determine the energy stored in the transformer using the method of Section 7-7. The energy stored in a transformer is

$$w = \tfrac{1}{2}L_1 i_1^2 + \tfrac{1}{2}L_2 i_2^2 \pm M i_1 i_2 \qquad (12\text{-}29)$$

where the sign of the last term is positive if both currents enter dotted terminals; otherwise it is negative. We can use this equation to find how large a value M can attain in terms of L_1 and L_2. Since the transformer is a passive element, the energy stored must be greater than or equal to zero. The limiting quantity for M is obtained when $w = 0$ in Eq. 12-29. Then we have

$$\tfrac{1}{2}L_1 i_1^2 + \tfrac{1}{2}L_2 i_2^2 - M i_1 i_2 = 0 \qquad (12\text{-}30)$$

as the limiting condition for the case where one current enters the dotted terminal and the other current leaves the dotted terminal. Now add and subtract the term $i_1 i_2 \sqrt{L_1 L_2}$ in the equation to generate a term that is a perfect square as follows:

$$\left(\sqrt{\frac{L_1}{2}}\, i_1 - \sqrt{\frac{L_2}{2}}\, i_2\right)^2 + i_1 i_2(\sqrt{L_1 L_2} - M) = 0$$

The perfect square term can be positive or zero. Therefore, in order to have $w \geq 0$, we require that

$$\sqrt{L_1 L_2} \geq M \qquad (12\text{-}31)$$

Thus, the maximum value of M is $\sqrt{L_1 L_2}$. Many engineers define a coefficient of coupling as

$$k = \frac{M}{\sqrt{L_1 L_2}} \qquad (12\text{-}32)$$

so it is required that $0 \leq k \leq 1$. Clearly, when $k = 0$, no coupling exists. Most power system transformers have a k that approaches one, while k is low for radio circuits.

Finally, let us consider a circuit as shown in Figure 12-23a and determine the equivalent impedance seen by the source voltage $\mathbf{V}_1$. The phasor voltage and current model of the circuit is shown in Figure 12-23b. The two mesh equations are

$$j\omega L_1 \mathbf{I}_1 - j\omega M\mathbf{I}_2 = \mathbf{V}_1$$

$$-j\omega M\mathbf{I}_1 + (j\omega L_2 + \mathbf{Z}_2)\mathbf{I}_2 = 0$$

Solving for $\mathbf{V}_1$ in terms of $\mathbf{I}_1$, we have

$$\mathbf{V}_1 = \left[j\omega L_1 - \frac{(j\omega M)^2}{(j\omega L_2 + \mathbf{Z}_2)}\right]\mathbf{I}_1 \qquad (12\text{-}33)$$

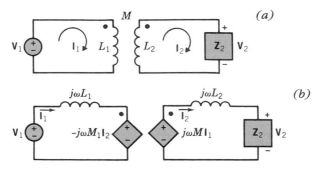

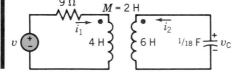

Figure 12-23 *(a)* Transformer circuit with a secondary load $\mathbf{Z}_2$. *(b)* Model of the circuit.

Therefore, the impedance seen at the terminals of the primary of the transformer is

$$\mathbf{Z}_1 = \frac{\mathbf{V}_1}{\mathbf{I}_1} = j\omega L_1 + \frac{\omega^2 M^2}{j\omega L_2 + \mathbf{Z}_2} \qquad (12\text{-}34)$$

Since $j\omega L_1$ depends on the primary coil, the second part of Eq. 12-34 is due to the secondary portion of the circuit and is called the *reflected impedance,* which is

$$\mathbf{Z}_R = \frac{\omega^2 M^2}{j\omega L_2 + \mathbf{Z}_2} \qquad (12\text{-}35)$$

Let us consider the case where $\mathbf{Z}_2 = R_2 + jX_2$. Then we have

$$\mathbf{Z}_R = \frac{\omega^2 M^2}{R_2^2 + (X_2 + \omega L_2)^2} [R_2 - j(X_2 + \omega L_2)]$$

For the special case $X_2 = -\omega L_2$, which is obtained from a capacitive reactance, we find that the reflected impedance is

$$\mathbf{Z}_R = \frac{\omega^2 M^2}{R_2} \qquad (12\text{-}36)$$

which is purely real.

EXERCISE 12-17

Consider the circuit shown in Figure E 12-17. Find the phasor $\mathbf{V}_c$ when $v = 36 \cos(3t - 60°)$ V.
Answer: $\mathbf{V}_c = 6\sqrt{2}\underline{/165°}$ V

EXERCISE 12-18

Find the ratio of phasor voltages $\mathbf{V}_2/\mathbf{V}_1$ for the circuit shown in Figure E 12-18 when $\omega = 10$.
Answer: $0.26\underline{/127°}$

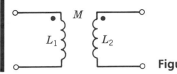

Figure E 12-18

EXERCISE 12-19

Find the maximum value of M that could be attained for the transformer of Figure E 12-19 when $L_1 = 100$ mH and $L_2 = 50$ mH.

Answer: 70.7 mH

Figure E 12-19

12-9 THE IDEAL TRANSFORMER

An *ideal transformer* is a model of a transformer with a coupling coefficient equal to unity and where the primary and secondary reactances are very large compared to the impedances connected to the transformer terminals. These characteristics are usually present for specially designed iron-core transformers. They approximate the ideal transformer for a range of frequencies.

For an ideal transformer, we have

$$\frac{L_2}{L_1} = \frac{N_2^2}{N_1^2} = n^2$$

where

$$n = \frac{N_2}{N_1}$$

The ratio of $n = N_2/N_1$ is called the *turns ratio*. The two defining equations for an ideal transformer are

$$\mathbf{V}_2 = n\mathbf{V}_1 \tag{12-37}$$

and

$$\mathbf{I}_1 = -n\mathbf{I}_2 \tag{12-38}$$

The symbol for the ideal step-up transformer is shown in Figure 12-24*a*, and the circuit model for the ideal transformer is shown in Figure 12-24*b*. The vertical bars in Figure 12-24*a* indicate the iron core, and we write ideal with the transformer to ensure recognition of the ideal case. Of course, the ideal transformer is a model, but some tightly coupled transformers ($k = 1$) approximate the behavior of the ideal. The ideal transformer is lossless since

$$\mathbf{V}_1\mathbf{I}_1 + \mathbf{V}_2\mathbf{I}_2 = 0$$

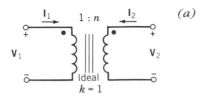

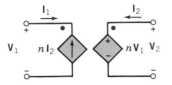

Figure 12-24 *(a)* Symbol for the ideal transformer. *(b)* Model for the ideal transformer.

Let us consider the circuit of Figure 12-25*a*, which has a load impedance $\mathbf{Z}_2$ connected to the secondary of the ideal transformer. The circuit model is shown in Figure 12-25*b*. The impedance seen at the primary of the transformer is

$$\mathbf{Z}_1 = \frac{\mathbf{V}_1}{\mathbf{I}_1}$$

However, we note that

$$\mathbf{V}_1 = \mathbf{V}_2 / n$$

and

$$\mathbf{I}_1 = n\,\mathbf{I}_2$$

where we have selected $\mathbf{I}_2$ to flow out of the dotted terminal in the model shown in Figure 12-25 so that

$$\mathbf{V}_2 = \mathbf{I}_2 \mathbf{Z}_2$$

Therefore, for $\mathbf{Z}_1$ we have

$$\mathbf{Z}_1 = \frac{\mathbf{V}_2 / n}{n\,\mathbf{I}_2}$$

$$= \frac{1}{n^2} \frac{\mathbf{V}_2}{\mathbf{I}_2}$$

$$= \frac{1}{n^2} \mathbf{Z}_2$$

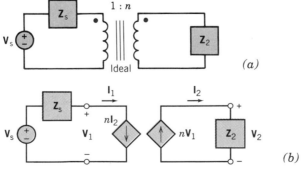

Figure 12-25 *(a)* Circuit with an ideal transformer. *(b)* Model of the circuit of Figure 12-25*a*. The direction of the dependent current source is downward due to $\mathbf{I}_2$ flowing out of the dotted terminal.

Thus the input impedance viewed from the terminals of the source, $\mathbf{V}_s$, is

$$\mathbf{Z}_{in} = \mathbf{Z}_s + \mathbf{Z}_1$$
$$= \mathbf{Z}_s + \frac{\mathbf{Z}_2}{n^2} \tag{12-39}$$

We can change the input impedance by adjusting n^2. For example, if $\mathbf{Z}_2 = R_2 = 10\ \Omega$ and $n^2 = 10$, the effective resistance seen in the primary circuit would be only $1\ \Omega$.

Example 12-8

Often we can use an ideal transformer to represent a transformer that connects the output of a stereo amplifier, $\mathbf{V}_1$, to a stereo speaker, as shown in Figure 12-26. Find the required turns ratio n when $R_L = 8\ \Omega$ and $R_s = 48\ \Omega$ if we wish to achieve maximum power transfer to the load.

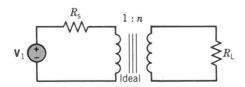

Figure 12-26 Output of an amplifier connected to a stereo speaker with resistance R_L.

Solution

The impedance seen at the primary due to R_L is

$$\mathbf{Z}_1 = \frac{R_L}{n^2}$$
$$= \frac{8}{n^2}$$

To achieve maximum power transfer, we require that

$$\mathbf{Z}_1 = R_s$$

Since $R_s = 48\ \Omega$, we require that $\mathbf{Z}_1 = 48\ \Omega$, so

$$n^2 = \frac{8}{48} = \frac{1}{6}$$

and, therefore,

$$\left(\frac{N_2}{N_1}\right)^2 = \frac{1}{6}$$

or

$$N_1 = \sqrt{6}\,N_2$$

EXERCISE 12-20

Find $\mathbf{V}_1$ and $\mathbf{I}_1$ for the circuit of Figure E 12-20 when $n = 5$. The load and source impedances are given in the figure.

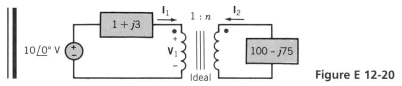

Figure E 12-20

EXERCISE 12-21

Determine v_2 and i_2 for the circuit shown in Figure E 12-21 when $n = 2$. Note that i_2 does not enter the dotted terminal.

Answer: $v_2 = 0.68 \cos(10t + 47.7°)$ V
$\qquad\quad i_2 = 0.34 \cos(10t + 42°)$ A

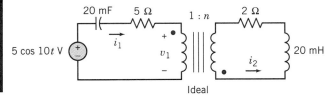

Figure E 12-21

EXERCISE 12-22

Determine the impedance $\mathbf{Z}_{ab}$ for the circuit of Figure E 12-22. All the transformers are ideal.

Answer: $\mathbf{Z}_{ab} = 4.063\,\mathbf{Z}$

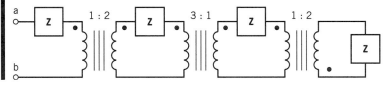

Figure E 12-22

12-10 ∥ TRANSFORMER CIRCUIT ANALYSIS USING PSpice

Transformer circuits may be modeled using PSpice by using the mutual inductor statement as

$$\text{K}<name>\text{ L}<name>\text{ L}<name><coupling\ k>$$

Therefore, the mutual inductor statement for the transformer shown in Figure 12-27 is

```
K1    L1    L2    0.5
```

where the coefficient of coupling is 0.5. This statement is used to couple the two individual inductors that model the transformer coils. Recall that the current must flow into the dotted terminals of the transformer as shown in Figure 12-27. Therefore, a dot is assumed at the first node listed on the inductor statements. Also recall that a resistance with a very high value is required to obtain a dc connection between the two parts of the circuit as required by PSpice.

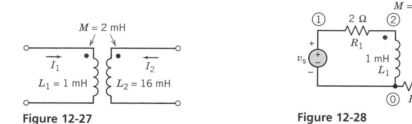

Figure 12-27   **Figure 12-28**

Consider the circuit shown in PSpice format in Figure 12-28. The source is $v_s = 110 \cos 10t$ V and the coupling coefficient is 0.5. The dummy resistance R_d will be set equal to 100 MΩ. Then we obtain the PSpice program as shown in Figure 12-29, which will provide as output the magnitude and phase of the output voltage appearing at node 3.

```
TRANSFORMER CIRCUIT
VS   1   0   AC   110   0
R1   1   2   2
L1   2   0   1M
L2   3   4   16M
R2   3   4   5
K1   L1   L2   0.5
RD   4   0   100 MEG
.AC   LIN   1   1.591   1.591
.PRINT   AC   VM (3,4)   VP (3,4)
.END
```

Figure 12-29 The PSpice program for the circuit of Figure 12-28.

Let us consider the transformer circuit shown in Figure 12-30, which is set up in PSpice format. In this case, let $\omega = 1000$ rad/s and $R_L = 100$ Ω. For these conditions

$$k = \frac{M}{\sqrt{L_1 L_2}} = 1$$

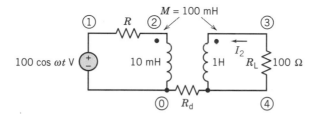

Figure 12-30 Transformer circuit with $k = 1$.

Therefore, the coupling coefficient is equal to 1. The PSpice program to calculate $\mathbf{V}_2$ is shown in Figure 12-31. Note that we use a coupling coefficient of 0.999999 since PSpice requires that $0 \le k < 1$. In this case, we set $R = 0$ Ω, since we used that value on page 601. As expected, we find an output of 1000 V with zero phase shift.

If we set $R = 1$ Ω in Figure 12-30, we can change R1 (line 3) in the PSpice program to 1. When we run this altered PSpice program, we find that the output across R_L is

$$\mathbf{V} = 499.4\underline{/2.86°} \text{ V}$$

Thus, adding the 1-Ω resistance has reduced the output voltage to one-half of the output voltage obtained when $R = 0$.

The ideal transformer is obtained when $k = 1$ and the reactances of the inductances are much greater than the other circuit impedances. The circuit of Figure 12-30 has $k = 1$ and

```
TRANSFORMER
VI   1   0   AC   100   0
R1   1   2   1E-9;   Resistance R = 0
L1   2   0   10M
L2   3   4   1
RL   4   3   100
RD   4   0   1G
K1   L1   L2   .999999
.AC   LIN   1   159.2   159.2
.PRINT   AC   VM (3,4)   VP (3,4)
.END
```

Figure 12-31 PSpice program for the circuit of Figure 12-30.

will closely satisfy the impedance requirements when ω is increased to 10,000 rad/s and $R = 1\ \Omega$. Then, $\omega L_1 = 100$ and $\omega L_2 = 10,000$. Thus, we have $\omega L_1 \gg R$ and $\omega L_2 \gg R_L$ and the transformer appears to be ideal. If we rewrite the PSpice program for these conditions, we have the program shown in Figure 12-32. The output of this program verifies that

$$\mathbf{V} = 992\underline{/0.5°}\ \text{V}$$

For a perfectly ideal transformer the output voltage across R_L is given by Eq. 12-37 as

$$\mathbf{V}_2 = n\,\mathbf{V}_1$$
$$= 10\mathbf{V}_1$$

```
TRANSFORMER
VI   1   0   AC   100   0
R1   1   2   1
L1   2   0   10M
L2   3   4   1
RL   4   3   100
RD   4   0   1G
K1   L1   L2   .999999
.AC   LIN   1   1592   1592
.PRINT   AC   VM (3,4)   VP (3,4)
.END
```

Figure 12-32 PSpice program with $R = 1\Omega$ and $\omega = 10,000$ rad/s.

since $n^2 = L_2/L_1 = 100$ and $n = 10$. Thus, we see that the actual transformer acts as a nearly ideal transformer when $\omega = 10,000$ and the impedance of each inductance is large relative to the other circuit impedances.

The ideal transformer is a model of a transformer with a coupling coefficient equal to one and with primary and secondary reactances very large compared to the impedances connected to the terminals of the transformer. This ideal transformer model fits many practical applications in ac circuits. Various components including transformers are shown in Figure 12-33. The power levels for selected electrical devices or phenomena are shown in Figure 12-34.

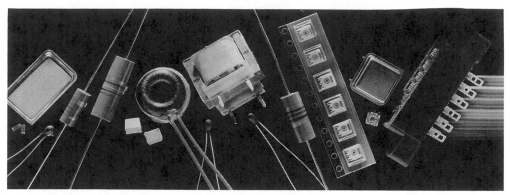

Figure 12-33 Components shown include oscillators, inductors, transformers, thermistors, chip potentiometers, and connectors. (Courtesy of Dale-Vishay Inc.)

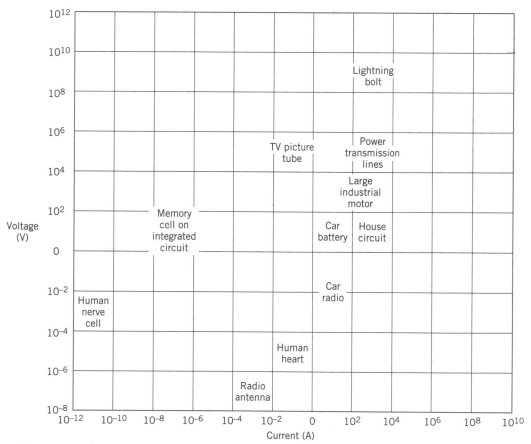

Figure 12-34 Power levels for selected electrical devices or phenomena.

12-11 | VERIFICATION EXAMPLE

The circuit shown in Figure 12 V-1*a* has been analyzed using a computer. The results are tabulated in Figure 12 V-1*b*. The labels Xp and Xs refer to the primary and secondary of the transformer. The passive convention is used for all elements, including the voltage sources, which means that

$$\frac{(30)(1.76)}{2} \cos(133° - 0) = -18.00$$

is the average power *absorbed* by the voltage source. The average power *supplied* by the voltage source is $+18.00$ W.

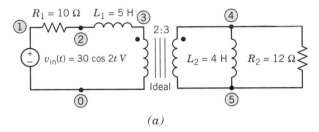

(*a*)

Element				Voltage	Current
Vin	1	0	30 ∠ 0	30 ∠ 0°	1.76 ∠ 133°
R1	1	2	10	17.6 ∠ –47°	1.76 ∠ –47°
L1	2	3	5	17.6 ∠ 43°	1.76 ∠ –47°
Xp	3	0	2	5.2 ∠ 9°	1.76 ∠ –47°
Xs	4	5	3	7.8 ∠ 9°	1.17 ∠ 133°
R2	4	5	12	7.8 ∠ 9°	0.65 ∠ 9°
L2	4	5	4	7.8 ∠ 9°	0.98 ∠ –81°

Steady-state response: $\omega = 2$ rad/s

(*b*)

Figure 12 V-1 (*a*) A circuit and (*b*) the results from computer analysis of the circuit.

How can we verify that the computer analysis of this circuit is indeed correct?

Solution

There are several things that can be easily checked.

1. The element current and voltage of each inductor should be 90° out of phase with each other so that the average power delivered to each inductor is zero. The element current and voltage of both L_1 and L_2 satisfy this condition.

2. An ideal transformer absorbs zero average power. The sum of the average power absorbed by the transformer primary and the secondary is

$$\frac{(5.2)(1.76)}{2} \cos(9° - (-47°))$$

$$+ \frac{(7.8)(1.17)}{2} \cos(133° - 9°) = 2.56 + (-2.55)$$

$$\approx 0 \text{ W}$$

so this condition is satisfied.

3. All of the power delivered to the primary of the transformer is in turn delivered to the load. In this example the load consists of the inductor L_2 and the resistor R_2. Since the average power delivered to the inductor is zero, all the power delivered to the transformer primary should be delivered by the secondary to the resistor R_2. The power delivered to the transformer primary is

$$\frac{(5.2)(1.76)}{2} \cos(9° - (-47°)) = 2.56 \text{ W}$$

The power delivered to R_2 is

$$\frac{(7.8)(0.65)}{2} \cos(0) = 2.53 \text{ W}$$

There seems to be some round-off error in the voltages and currents provided by the computer. Nonetheless, it seems reasonable to conclude that all the power delivered to the transformer primary is delivered by the secondary to the resistor R_2.

4. The average power supplied by the voltage source should be equal to the average power absorbed by the resistors. We have already calculated that the average power delivered by the voltage source is 18 W. The average power absorber by the resistors is

$$\frac{(17.6)(1.76)}{2} \cos(0) + \frac{(7.8)(0.65)}{2} \cos(0) = 15.49 + (-2.53)$$
$$= 18.02 \text{ W}$$

so this condition is satisfied.

Since these four conditions are satisfied, we are confident that the computer analysis of the circuit is correct.

12-12 DESIGN CHALLENGE SOLUTION

MAXIMUM POWER TRANSFER

The matching network in Figure 12D-1 is used to interface the source with the load, which means that the matching network is used to connect the source to the load in a desirable way. In this case, the purpose of the matching network is to transfer as much power as possible to the load. This problem occurs frequently enough that it has been given a name, the maximum power transfer problem.

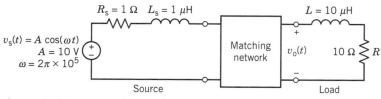

Figure 12D-1 Design the matching network to transfer maximum power to the load where the load is the model of an antenna of a wireless communication system.

An important example of the application of maximum power transfer is the connection of a cellular phone or wireless radio transmitter to the cell's antenna. For example, the input impedance of a practical cellular telephone antenna is $\mathbf{Z} = (10 + j6.28)\ \Omega$ (Dorf, 1993).

Define the Situation

The input voltage is a sinusoidal function of time. The circuit is at steady state. The matching network is to be designed to deliver as much power as possible to the load.

State the Goal

To achieve maximum power transfer, the matching network should "match" the load and source impedances. The source impedance is

$$
\begin{aligned}
\mathbf{Z}_s &= R_s + j\omega L_s \\
&= 1 + j(2 \cdot \pi \cdot 10^5)(10^{-6}) \\
&= 1 + j0.628\ \Omega
\end{aligned}
$$

For maximum power transfer, the impedance $\mathbf{Z}_{in}$, shown in Figure 12D-2, must be the complex conjugate of $\mathbf{Z}_s$. That is,

$$
\begin{aligned}
\mathbf{Z}_{in} &= \mathbf{Z}_s^{*} \\
&= 1 - j0.628
\end{aligned}
$$

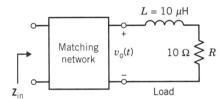

Figure 12D-2 Z_{in} is the impedance seen looking into the matching network.

Generate a Plan

Let us use a transformer for the matching network as shown in Figure 12D-3. The impedance $\mathbf{Z}_{in}$ will be a function of n, the turns ratio of the transformer. We will set $\mathbf{Z}_{in}$ equal to the complex conjugate of $\mathbf{Z}_s$ and solve the resulting equation to determine the turns ratio, n.

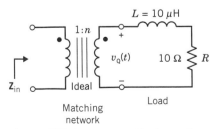

Figure 12D-3 Using an ideal transformer as the matching network.

Take Action on the Plan

$$\mathbf{Z}_{\text{in}} = \frac{1}{n^2}(R + j\omega L)$$

$$= \frac{1}{n^2}(10 + j6.28)$$

We require that

$$\frac{1}{n^2}(10 + j6.28) = 1 - j0.628$$

This requires both

$$\frac{1}{n^2}10 = 1 \qquad\qquad (12\text{D-}1)$$

and

$$\frac{1}{n^2}6.28 = -0.628 \qquad\qquad (12\text{D-}2)$$

Selecting

$$n = 3.16$$

(e.g., $N_2 = 158$ and $N_1 = 50$) satisfies Eq. 12D-1 but not Eq. 12D-2. Indeed, no positive value of n will satisfy Eq. 12D-2.

We need to modify the matching network to make the imaginary part of $\mathbf{Z}_{\text{in}}$ negative. This can be accomplished by adding a capacitor as shown in Figure 12D-4.

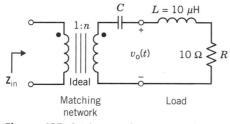

Figure 12D-4 The matching network is modified by adding a capacitor.

Then

$$\mathbf{Z}_{\text{in}} = \frac{1}{n^2}\left(R + j\omega L - j\frac{1}{\omega C}\right)$$

$$= \frac{1}{n^2}\left(10 + j6.28 - j\frac{1}{2\pi \cdot 10^5 \cdot C}\right)$$

We require that

$$\frac{1}{n^2}\left(10 + j6.28 - j\frac{1}{2\pi \cdot 10^5 \cdot C}\right) = 1 - j0.628$$

This requires both

$$\frac{1}{n^2}10 = 1 \qquad\qquad (12\text{D-}3)$$

and

$$\frac{1}{n^2}\left(6.28 - \frac{1}{2\pi \cdot 10^5 \cdot C}\right) = -0.628 \qquad (12\text{D-4})$$

First, solving Eq. 12D-3 gives

$$n = 3.16$$

Next, solving Eq. 12D-4 gives

$$C = 0.1267 \ \mu\text{F}$$

and the design is complete.

SUMMARY

With the adoption of ac power as the generally used conventional power for industry and the home, engineers became involved in analyzing ac power relationships. In this chapter we examined the representation and analysis of sinusoidal steady-state (ac) power.

The instantaneous power was examined, but we then moved to the concept of average power. The average power of a periodic source was determined, and we used this relationship to find the average ac power delivered to an impedance. Although the average ac power can be expressed in terms of V_m and I_m, we prefer to use the effective values of the voltage and current.

We found that the power superposition principle for several sources will hold as long as no two sources have the same radian frequency. Also, the maximum power is transferred to a load impedance $\mathbf{Z}_L$ when it is set equal to the conjugate of $\mathbf{Z}_t$, the Thévenin equivalent impedance of the remaining portion of the circuit.

The effective or rms value of a sinusoidal waveform is simply $I = I_m/\sqrt{2}$, where I_m is the maximum amplitude of the waveform. We choose to describe the average power as $P = VI \cos\theta$, where VI is called the apparent power and $\cos\theta$ is the power factor.

Since it is important to keep the current I as low as possible in the transmission lines, engineers strive to achieve a power factor close to one. Thus, a purely reactive impedance in parallel with the load is used to correct the power factor.

We introduced the concept of complex power $\mathbf{S} = P + jQ$, where Q is the reactive power. This concept is used to determine the effective sum of several loads in parallel.

Finally, we considered the transformer, which is used for ac power distribution as well as for electronic circuits. Transformers are based on the concept of mutual inductance, which relates the voltage in one coil to the change in current in another coil. We use the dot convention to describe the terminal that has a positive voltage when a current flows into the dotted terminal of the other coil. With the concept of mutual inductance we are able to show how one can achieve voltage multiplication or reduction between the primary and the secondary coils of a transformer. Also, a transformer may be used to reflect and alter the impedance of the secondary circuit as seen in the primary circuit.

TERMS AND CONCEPTS

Apparent Power   Product of the rms voltage and the rms current, expressed as VI with units of voltamperes (VA).

Average Power Delivered by a Sinusoidal Steady-State Source $P = (V_m I_m/2) \cos\theta$, where θ is the phase angle difference between the voltage and current and the source measured in watts.

Average Power of a Periodic Function Integral of the time function over a complete period T, divided by the period:

$$P = \frac{1}{T} \int_{t_0}^{t_0 + T} p(t) \, dt$$

Coefficient of Coupling Ratio of the mutual inductance M to the square root of the product of the two inductances L_1 and L_2:

$$k = \frac{M}{\sqrt{L_1 L_2}}$$

Complex Power Sum of the average power and the reactive power expressed as a complex number, $\mathbf{S} = P + jQ$, with units of voltamperes (VA).

Dot Convention If a current goes into the dotted end of one coil, then for the other coil the polarity of the induced voltage is plus at the dotted end.

Effective Value of a Current The steady current that is as effective in transferring power as the average power of the varying current.

$$I_{\text{eff}} = \left(\frac{1}{T} \int_0^T i^2 \, dt \right)^{1/2}$$

We normally use $I = I_{\text{eff}} = I_{\text{rms}}$, dropping the subscript.

High-Voltage Power Transmission Use of voltages greater than 100 kV to keep low resistive $I^2 R$ losses on the transmission lines.

Ideal Transformer Model of a transformer with a coupling coefficient equal to unity and with the primary and secondary reactances very large compared to the impedances connected to the transformer terminals.

Instantaneous Power Product of the voltage $v(t)$ and the current $i(t)$, $p = vi$.

Maximum Power Transfer If a circuit has a Thévenin equivalent circuit with an impedance $\mathbf{Z}_t$, the maximum power is delivered to the load when the load $\mathbf{Z}_L$ is set equal to the complex conjugate of $\mathbf{Z}_t$.

Mutual Inductance Property of two electric coils where a voltage in one is induced by a changing current in the other.

Power Factor Equal to $\cos \theta$, where θ is the phase angle difference between the sinusoidal steady-state voltage and current; ratio of average power to apparent power.

Primary Coil Coil of a transformer connected to the energy source.

Reactive Power Imaginary part Q, of the complex power with units of voltamperes reactive (VAR).

Reflected Impedance The complex impedance reflected to the primary circuit of a transformer from the secondary circuit.

RMS of a Periodic Waveform Square root of the mean of the squared value. Also called the effective value. We normally use V instead of V_{rms} to denote the rms voltage, thus dropping the subscript.

Secondary Coil Coil of a transformer connected to the load.

Superposition of the Average Power of Multiple Sources The total average power delivered to a load is equal to the sum of the average power delivered by each source when no two of the sources have the same radian frequency.

Transformer Magnetic circuit with two or more multiturn coils wound on a common core.

Turns Ratio Ratio n equal to N_2/N_1, where N_1 = number of turns in the primary coil and N_2 = number of turns in the secondary coil of an ideal transformer.

REFERENCES   Chapter 12

Adler, Jerry, "Another Bright Idea," *Newsweek,* June 15, 1992, p. 67.

Bernstein, Theodore, "Electrical Shock Hazards," *IEEE Transactions on Education,* August 1991, pp. 216–222.

Coltman, John W., "The Transformer," *Scientific American,* January 1988, pp. 86–95.

Dorf, Richard C., *Electrical Engineering Handbook,* Antennas, Chap. 36, CRC Press, Boca Raton, FL, 1993.

Mackay, Lionel, "Rural Electrification in Nepal," *Power Engineering Journal,* September 1990, pp. 223–231.

Zorpette, Glenn, "Utilities Get Serious About Efficiency," *IEEE Spectrum,* May 1991, pp. 42–43.

PROBLEMS

Section 12-3   Instantaneous Power

P 12.3-1   A large hydroelectric generating station was built on the border of Paraguay and Brazil. However, Paraguay and Brazil use 50 Hz and 60 Hz, respectively. What are the possibilities of transmitting energy from one country to the other? What is the potential for both countries adopting 60 Hz?

P 12.3-2   An *RLC* circuit is shown in Figure P 12.3-2. Find the instantaneous power delivered to the inductor when $i_s = 1 \cos \omega t$ A and $\omega = 1$ kHz.

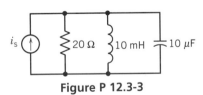

Figure P 12.3-3

Section 12-4   Average Power

P 12.4-1   Find the average power absorbed by the 0.6-kΩ resistor and the average power supplied by the current source for the circuit of Figure P 12.4-1.

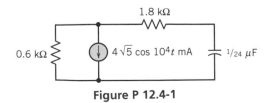

Figure P 12.4-1

P 12.4-2   For the circuit of Figure P 12.4-2, find (a) average power supplied by the independent source and (b) average power absorbed by the dependent source.

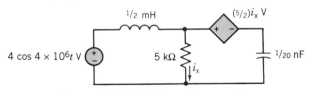

Figure P 12.4-2 Units of i_X are mA.

P 12.4-3   Use nodal analysis to find the average power absorbed by the 20-Ω resistor in the circuit of Figure P 12.4-3.

Answer: $P = 200$ W

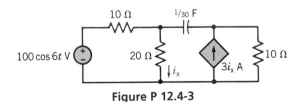

Figure P 12.4-3

P 12.4-4   Nuclear power stations have become very complex to operate, as illustrated by the training simulator for the operating room of the Pilgrim power plant shown in Figure P 12.4-4a. One control circuit has the model shown in Figure P 12.4-4b. Find the average power absorbed by each element.

Answer: $P_{\text{source current}} = -12.8$ W
$P_{8\Omega} = 6.4$ W
$P_L = 0$ W
$P_{\text{voltage source}} = 6.4$ W

(a)

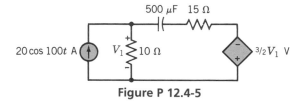

(b)

Figure P 12.4-4 (a) The simulation training room for the Pilgrim Power Station. The power station is located at Plymouth, Massachusetts, and generates 700 MW. It commenced operation in 1972. Courtesy of Boston Edison. (b) One control circuit of the reactor.

P 12.4-5 Find the average power absorbed by each element for the circuit of Figure P 12.4-5.

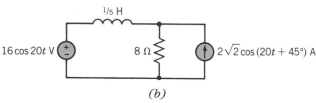

Figure P 12.4-5

P 12.4-6 Find the average power absorbed by the 6-kΩ resistor and supplied by the leftmost source in the circuit of Figure P 12.4-6.

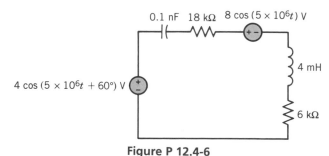

Figure P 12.4-6

P 12.4-7 A student experimenter in the laboratory encounters all types of electrical equipment. Some pieces of test equipment are battery-operated or operate at low voltage so that any hazard is minimal. Other types of equipment are isolated from electrical ground so that there is no problem if a grounded object makes contact with the circuit. There are, however, types of test equipment that are supplied by voltages that can be hazardous or can have dangerous voltage outputs. The standard power supply used in the United States for power and lighting in laboratories is the 120, grounded, 60-Hz sinusoidal supply. This supply provides power for much of the laboratory equipment, so an understanding of its operation is essential in its safe use (Bernstein, 1991).

Consider the case where the experimenter has one hand on a piece of electrical equipment and the other hand on a ground connection, as shown in the circuit diagram of Figure P 12.4-7a. The hand-to-hand resistance is 200 Ω. Shocks with an energy of 30 J are hazardous to a human. Consider the model shown in Figure P 12.4-7b, which represents the human with R. Determine the energy delivered to the human in 1 s.

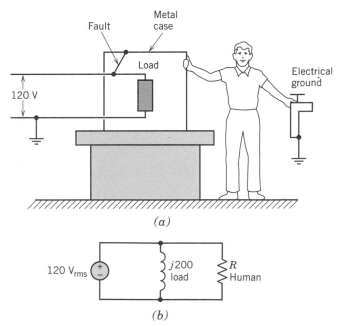

Figure P 12.4-7 Student experimenter touching an electrical device.

Section 12-5   The Superposition Principle and the Maximum Power Theorem
(a) Superposition Principle

P 12.5-1 Find the average power absorbed by the 2-Ω resistor in the circuit of Figure P 12.5-1.
Answer: $P = 413$ W

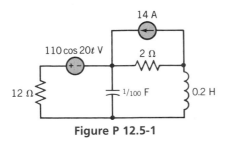

Figure P 12.5-1

P 12.5-2 Find the average power absorbed by the 8-Ω resistor in the circuit of Figure P 12.5-2.
Answer: $P = 22$ W

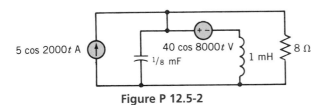

Figure P 12.5-2

P 12.5-3 For the circuit shown in Figure P 12.5-3, determine the average power absorbed by each resistor, R_1 and R_2. The voltage source is $v_s = 10 + 10 \cos(5t + 40°)$ V and the current source is $i_s = 4 \cos(5t - 30°)$ A.

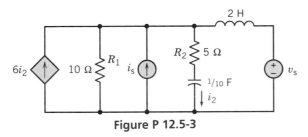

Figure P 12.5-3

P 12.5-4 For the circuit shown in Figure P 12.5-4, determine the effective value of the resistor voltage v_R and the capacitor voltage v_C.

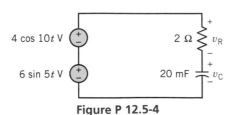

Figure P 12.5-4

Section 12-5   The Superposition Principle and the Maximum Power Theorem
(b) Maximum Power Theorem

P 12.5-5 For the circuit of Figure P 12.5-5, find the values of R and L such that R absorbs maximum power. What is this value of maximum power?
Answer: $R = 2$ kΩ, $L = 1$ mH
              $P_{\max} = 2.5$ mW

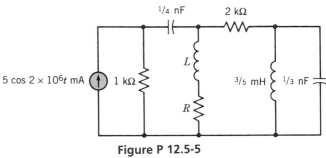

Figure P 12.5-5

P 12.5-6 Consider the circuit of Figure P 12.5-6 and find the value of L and R such that the maximum power transfer condition $\mathbf{Z}_L = \mathbf{Z}_t^*$ is met.
What power is drawn by the load in this case?

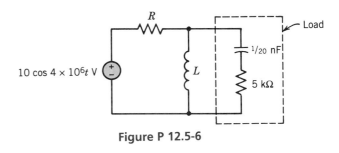

Figure P 12.5-6

P 12.5-7 The circuit shown in Figure P 12.5-7 operates at $\omega = 1000$ rad/s. Determine the appropriate values of R and C for maximum power transfer to the load. Determine the maximum power delivered to the load when $v_s = 100 \cos \omega t$ V.

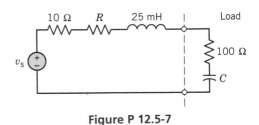

Figure P 12.5-7

P 12.5-8 It is desired to transfer maximum power to the load **Z** for the circuit shown in Figure P 12.5-8 when $i_s = 4 \cos 50t$ A. Calculate the maximum power to **Z**.

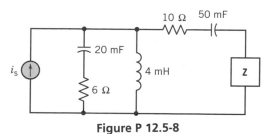

Figure P 12.5-8

P 12.5-9 For the circuit of Figure P 12.5-9, determine the load impedance that maximizes the average power to the load. Determine the maximum power delivered. The circuit operates at $\omega = 10$ Mrad/s.

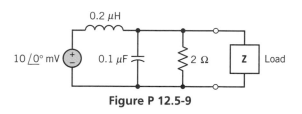

Figure P 12.5-9

P 12.5-10 Determine the Norton equivalent at terminals a–b for the circuit shown in Figure P 12.5-10. Use this information to determine the impedance at the terminals a–b in order to deliver maximum power to that impedance.

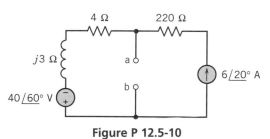

Figure P 12.5-10

P 12.5-11 For the circuit of Figure P 12.5-11, find R and α such that maximum possible power is delivered to R. Also find the maximum power delivered to R.
Answer: $\alpha = -1/3$, $R = 7 \ \Omega$, $P_{max} = 160.7$ W

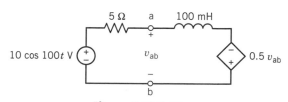

Figure P 12.5-11

P 12.5-12 Determine the Thévenin equivalent circuit at terminals a–b for the circuit shown in Figure P 12.5-12. The frequency of operation is 100 rad/s.

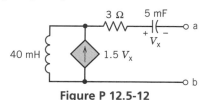

Figure P 12.5-12

P 12.5-13 Obtain the Thévenin equivalent circuit at terminals a–b for the transistor circuit shown in Figure P 12.5-13. Calculate the power delivered to the load connected at terminals a–b when the load is selected for the maximum power condition.

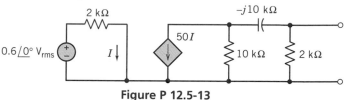

Figure P 12.5-13

P 12.5-14 Determine the required impedance **Z** in order to deliver maximum power to that impedance in Figure P 12.5-14. Calculate the power delivered to **Z** when the maximum power condition is attained. The voltage sources are $v_1 = 6 \cos(10t + 30°)$ V and $v_2 = 7 \cos(10t - 50°)$ V.

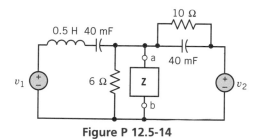

Figure P 12.5-14

P 12.5-15 (a) Determine the load impedance $\mathbf{Z}_{ab}$ that will absorb maximum power if it is connected to terminals a–b of the circuit shown in Figure P 12.5-15.
(b) Determine the maximum power absorbed by this load.
(c) Determine a model of the load and indicate the element values.

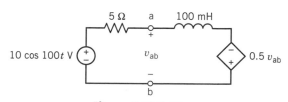

Figure P 12.5-15

Section 12-6 Effective Value of a Sinusoidal Waveform

P 12.6-1 Find the rms value of the current i for (a) $i = 2 - 4\cos 2t$ A, (b) $i = 3\sin \pi t + \sqrt{2}\cos \pi t$ A, and (c) $i = 2\cos 2t + 4\sqrt{2}\cos(2t + 45°) + 12\sin 2t$ A.
Answer: (a) $2\sqrt{3}$ A   (b) 2.35 A   (c) $5\sqrt{2}$ A

P 12.6-2 Find the rms value for each of the waveforms shown in Figure P 12.6-2.
Answer: (a) $\sqrt{7}/5$ (b) and (c) 1/2

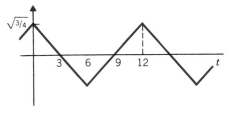

(a)

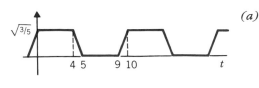

(b)

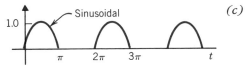

(c)

Figure P 12.6-2

P 12.6-3 Find the average and the rms value of the voltage waveform shown in Figure P 12.6-3.
Answer: $V_{\text{aver}} = 1.75$ V
$V_{\text{rms}} = 2.18$ V

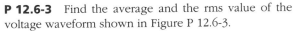

Figure P 12.6-3

P 12.6-4 Find the rms value for each of the waveforms of Figure P 12.6-4.
Answer: $V_{\text{rms}} = 1.225$ V
$I_{\text{rms}} = 5$ mA

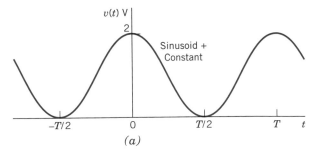

(a)

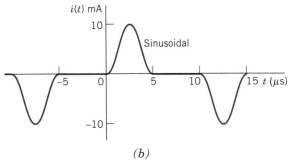

(b)

Figure P 12.6-4

P 12.6-5 Find the rms value of the voltage $v(t)$ shown in Figure P 12.6-5.
Answer: $V_{\text{rms}} = 5.196$ V

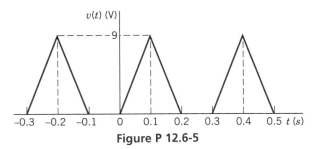

Figure P 12.6-5

Section 12-7 Power Factor

P 12.7-1 Manned space stations will require several continuously available ac power sources. Also, it is desired to keep the power factor close to 1. Consider the model of one communication circuit shown in Figure P 12.7-1. If an average power of 500 W is dissipated in the 20-Ω resistor, find (a) V_{rms}, (b) $I_{\text{s rms}}$, (c) the power factor seen by the source, and (d) $|V_s|$.

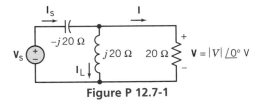

Figure P 12.7-1

P 12.7-2 Find the complex power delivered by the voltage source and the power factor seen by the voltage source for the circuit of Figure P 12.7-2.

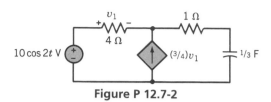

Figure P 12.7-2

P 12.7-3 For the circuit of Figure P 12.7-3, find the value of the inductor L if the complex power supplied by the voltage source is $50/3\underline{/53.13°}$ VA.
Answer: $L = 2$ H

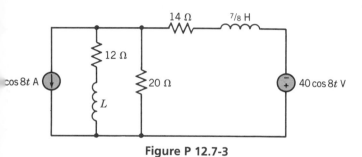

Figure P 12.7-3

P 12.7-4 Many engineers are working to develop photovoltaic power plants that provide ac power. An example of an experimental photovoltaic system is shown in Figure P 12.7-4a. A model of one portion of the energy conversion circuit is shown in Figure P 12.7-4b. (a) Find the average, reactive, and complex power delivered by the dependent source. (b) What single passive element could replace the dependent source and leave the circuit's conditions unchanged?
Answer: $\mathbf{S} = +j8/9$ VA

(a)

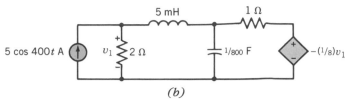

(b)

Figure P 12.7-4 (a) An experimental photovoltaic power plant. (b) Model of part of the energy conversion circuit. Courtesy of *EPRI Journal*.

P 12.7-5 Two electrical loads are connected in parallel to a 440-V rms, 60-Hz supply. The first load is 12 kVA at 0.7 lagging power factor. The second load is 10 kVA at 0.8 lagging power factor. Find the average power, the apparent power, and the power factor of the two combined loads.
Answer: Total power factor = 0.75 lagging

P 12.7-6 The source of Figure P 12.7-6 delivers 50 VA with a power factor of 0.8 lagging. Find the unknown impedance **Z**.

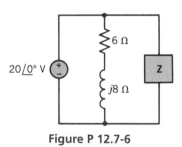

Figure P 12.7-6

P 12.7-7 Electric streetcars have been used in industrialized nations since the 1890s. A typical streetcar system will use the power distributed by overhead wires at 50 kV rms. It is desirable to keep the power factor equal to one. When 50 streetcars are running on a line, the average load is $\mathbf{Z}_L = 8 + j6\ \Omega$ and $\omega = 377$. Find (a) the complex power delivered to the load, (b) the power factor of the load, and (c) the value of a capacitor connected in parallel that will correct the total power factor to 0.95 leading.

P 12.7-8 A coil is represented by a model consisting of an inductance L in series with a resistance R. A voltmeter reading rms voltage reads 26 V when a 2-A (rms) current is supplied to the coil. A wattmeter indicates that 20 W is delivered to the coil. Determine L and R when $\omega = 377$.
Answer: $R = 5\ \Omega$, $L = 31.8$ mH

P 12.7-9 Assume the coil of Problem P12.7-8 is connected to a 26-V (rms) source. Find the complex power delivered to each element of this circuit. Verify the principle of conservation of complex power.

P 12.7-10 An industrial firm has two electrical loads connected in parallel across the power source. Power is supplied to the firm at 4000 V (rms). One load is 30 kW of heating use, and the other load is a set of motors that together operate as a load at 0.6 lagging power factor and at 150 kVA. Determine the total current and the plant power factor.

Answer: $I = 42.5$ A, pf $= 1/\sqrt{2}$

P 12.7-11 The information below is related to tests performed on the MT46 Machine Tool Relay manufactured by Furnas Controls Co.. The data are measured when the relay is energized from a 120-V, 60-Hz supply.

Inrush current	1.135 A
Seal current	0.2185 A
Inrush watts	46.0 W
Seal watts	5.0 W
Inrush VA	136.2 VA
Seal VA	26.22 VA

The inrush values describe the conditions at immediate switch-on of the relay, that is, as the relay armature is about to start moving. The seal values describe the conditions when the armature has moved to its final on position. Determine the series equivalent RL circuit representation of the relay (a) at initial switch-on and (b) with the armature sealed.

P 12.7-12 For the circuit shown in Figure P 12.7-12, determine **I** and the complex power **S** delivered by the source when $\mathbf{V} = 50\underline{/120°}\ V_{rms}$.

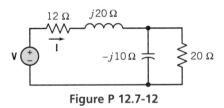

Figure P 12.7-12

P 12.7-13 For the circuit of Figure P 12.7-13, determine the complex power of the R, L, and C elements and show that the complex power delivered by the sources is equal to the complex power absorbed by the R, L, and C elements.

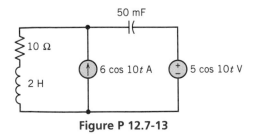

Figure P 12.7-13

P 12.7-14 A generator supplies two parallel loads at 220 V with a total power factor of 0.75 lagging. One load is 4800 VA at a power factor of 0.85 lagging. The second load absorbs 4 kW. What are the apparent power and power factor of the second load?

P 12.7-15 A 10-hp motor will draw 28 A from a 480-V line. What is the efficiency of the motor if the power factor is 0.694 lagging?

P 12.7-16 A residential electric supply three-wire circuit from a transformer is shown in Figure P 12.7-16a. The circuit model is shown in Figure P 12.7-16b. From its nameplate, the refrigerator motor is known to have a rated current of 8.5 A. It is reasonable to assume an inductive impedance angle of 45° for a small motor at rated load. Lamp and range loads are 100 W and 12 kW.

(a) Calculate the currents in line 1, line 2, and the neutral wire.
(b) Calculate: (i) P_{refrig}, Q_{refrig}, (ii) P_{lamp}, Q_{lamp}, and (iii) P_{total}, Q_{total}, S_{total}, and overall power factor.
(c) The neutral connection resistance increases, because of corrosion and looseness, to 20 Ω (this must be included as part of the neutral wire). Use mesh analysis and calculate the voltage across the lamp.

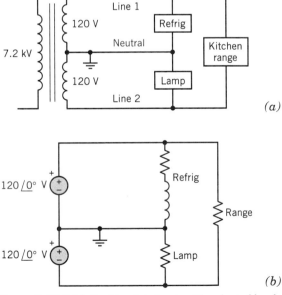

Figure P 12.7-16 Residential circuit with selected loads.

P 12.7-17 A motor draws 1-A peak current at a 0.6 lagging power factor from a 120-V-rms 60-Hz source. The motor is modeled by a series resistance R and an inductance L. Determine (a) the complex power of the motor and (b) the values of R and L.

P 12.7-18 Two impedances are supplied by $\mathbf{V} = 100\underline{/160°}$ V, as shown in Figure P 12.7-18, where $\mathbf{I} = 2\underline{/190°}$ A. The first load draws $P_1 = 23.2$ W and

$Q_1 = 50$ VAR. Calculate $\mathbf{I}_1$, $\mathbf{I}_2$, the power factor of each impedance, and the total power factor of the circuit.

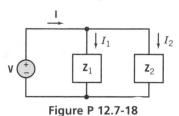

Figure P 12.7-18

P 12.7-19 To encourage conservation of electricity, many electric utility companies are offering financial incentives to their customers to purchase more efficient appliances, lamps, and insulation. One utility, Pacific Gas and Electric Company (PG&E), headquartered in San Francisco, offers a rebate on customer purchases of low-power compact fluorescent lamps (Zorpette, 1991).

One model of compact fluorescent lamp provides the same illumination as a 75-W incandescent lamp but consumes only 18 W. The equivalent resistance of the incandescent lamp is 192 Ω, and that of the compact fluorescent lamp is 800 Ω. Assume that the source resistance of the line voltage (generation source resistance plus transmission resistance) is 3 Ω. Derive expressions for power consumption in the lamp, for total power consumption (power consumed by the lamp and by the source resistance), and for power transfer efficiency (lamp power consumption divided by total power consumption). Evaluate these three expressions for the two lamps. Assume line voltage is 120 V rms. Which lamp causes the greatest power transfer to the load? Which lamp causes the greatest percentage of power delivered to the load? Which lamp causes the greatest total consumption of power? Which lamp is the most efficient? Explain what you mean by efficient.

Section 12-7   Power Factor (b) Correction

P 12.7-20 A common method of improving or correcting the power factor for a circuit was discussed in Section 12-6 for the circuit of Figure 12-17, which uses a parallel reactance. The uncorrected circuit has a power factor pf. Show that the required reactance X_1 is

$$X_1 = \frac{R_L^2 + X_L^2}{R_L \tan(\cos^{-1} \text{pf}) - X_L}$$

Discuss how the required reactance X_1 changes depending upon whether the original pf is leading or lagging.

P 12.7-21 A motor connected to a 220-V supply line from the power company has a current of 7.6 A. Both the current and the voltage are rms values. The average power delivered to the motor is 1317 W. (a) Find the apparent power, the reactive power, and the power factor

when $\omega = 377$. (b) Find the capacitance of a parallel capacitor that will result in a unity power factor of the combination. (c) Find the current in the utility lines after the capacitor is installed.

Answer: (a) pf $= 0.788$
(b) $C = 56.5 \ \mu\text{F}$
(c) $I = 6.0$ A

P 12.7-22 Two loads are connected in parallel across a 1000-V (rms), 60-Hz source. One load absorbs 500 kW at 0.6 power factor lagging, and the second load absorbs 400 kW and 600 kVAR. Determine the value of the capacitor that should be added in parallel with the two loads to improve the overall power factor to 0.9 lagging.

P 12.7-23 A voltage source with a complex internal impedance is connected to a load, as shown in Figure P 12.7-23. The load absorbs 1 kW of average power at 100 V rms with a power factor of 0.80 lagging. The source frequency is 200 rad/s. (a) Determine the source voltage $\mathbf{V}_1$. (b) Find the type of value of the element to be placed in parallel with the load so that maximum power is transferred to the load.

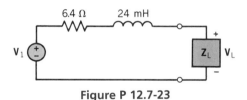

Figure P 12.7-23

P 12.7-24 A 100-kW induction motor, shown in Figure P 12.7-24, is receiving 100 kW at 0.8 power factor lagging. Determine the additional capacity in kVA that is made available by improving the power factor to (a) 0.95 lagging and (b) 1.0. (c) Find the required reactive capacity in kVAR provided by a set of parallel capacitors for parts a and b. (d) Determine the ratio of kVA released to the kVAR of capacitors required for parts a and b alone. Set up a table recording the results of this problem for the two values of power factor attained.

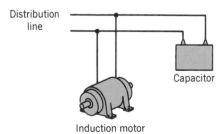

Figure P 12.7-24 Induction motor with parallel capacitor.

P 12.7-25 A resistor of 150 Ω and a parallel capacitor must draw 0.2 A from a 24-V source operating at 400 Hz. The power factor is measured as 0.8 leading. Determine the required parallel capacitor C.

P 12.7-26 A 10-hp motor operates from a 480-V, 60-Hz source with 0.8 lagging power factor and with 85 percent efficiency. Determine the parallel capacitor C required so that the power factor is changed to 0.922 lagging.

P 12.7-27 Two loads are connected in parallel and supplied from a 7.2-kV source. The first load is 50 kVA at 0.9 lagging power factor, and the second load is 45 kW at 0.91 lagging power factor. Determine the kVAR rating and capacitance required to correct the overall power factor to 0.97 lagging. Determine the resulting efficiency for the power system.

P 12.7-28 An industrial plant operates from a 500-V, 60 Hz source and has two loads connected in parallel to the source. The first load draws 48 kW at 0.6 lagging power factor. The second load draws 24 kW at 0.96 leading power factor. Select a parallel impedance to correct the overall power factor to unity.

P 12.7-29 A source with a complex internal impedance is connected to a load impedance $\mathbf{Z}_L$ as shown in Figure P 12.7-29. The load voltage is 100 V rms and absorbs 1 kW with a power factor of 0.8 lagging.

(a) Determine the source voltage $v_s(t)$.
(b) Determine the type and value of the element to be inserted in parallel with the load so that maximum power is transferred to the load. The source frequency is 200 rad/s.

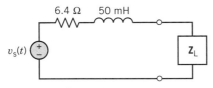

Figure P 12.7-29

P 12.7-30 A circuit is shown in Figure P 12.7-30 with an unknown impedance $\mathbf{Z}$. However, it is known that $v(t) = 100 \cos(100t + 20°)$ V and $i(t) = 25 \cos(100t - 10°)$ A. (a) Find $\mathbf{Z}$. (b) Find the power absorbed by the impedance. (c) Determine the type of element and its magnitude that should be placed across the impedance $\mathbf{Z}$ (connected to terminals a–b) so that the voltage $v(t)$ and the current entering the parallel elements are in phase.

Figure P 12.7-30

Section 12-8 The Transformer

P 12.8-1 Two magnetically coupled coils are connected as shown in Figure P 12.8-1. Using a source v_s connected to terminals a–b, show that an equivalent inductance at terminals a–b is $L_{ab} = L_1 + L_2 + 2M$.

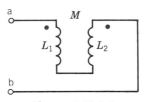

Figure P 12.8-1

P 12.8-2 Two magnetically coupled coils are shown connected in Figure P 12.8-2. Find the equivalent inductance L_{ab}.

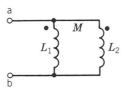

Figure P 12.8-2

P 12.8-3 Find the total energy stored in the circuit shown in Figure P 12.8-3 at $t = 0$ if the secondary winding is (a) open-circuited, (b) short-circuited, (c) connected to the terminals of a 7-Ω resistor.
Answer: (a) 15 J (b) 0 J (c) 5 J

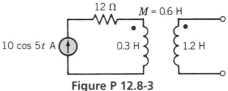

Figure P 12.8-3

P 12.8-4 A circuit with a mutual inductance is shown in Figure P 12.8-4. Find the ratio $\mathbf{V}_2/\mathbf{V}_1$ when $\omega = 5000$.

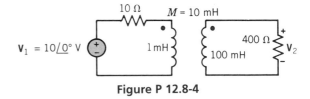

Figure P 12.8-4

P 12.8-5 The two windings of a transformer are connected in series and the measured inductance is 0.4 H. With the reversal of the connections of one coil, the measured inductance is 0.8 H. Determine the mutual inductance, M, of the transformer.

P 12.8-6 The source voltage of the circuit shown in Figure P 12.8-6 is $v_s = 141.4 \cos 100t$ V. Determine $i_1(t)$ and $i_2(t)$.

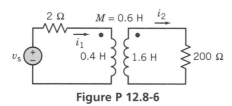

Figure P 12.8-6

P 12.8-7 A circuit with three mutual inductances is shown in Figure P 12.8-7. When $v_s = 10 \sin 2t$ V, $M_1 = 2$ H, and $M_2 = M_3 = 1$ H, determine the capacitor voltage $v(t)$.

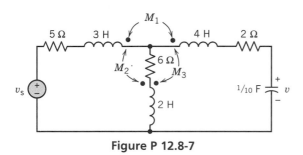

Figure P 12.8-7

P 12.8-8 Determine the currents i_1 and i_2 for the circuit shown in Figure P 12.8-8 when $v_s = 12 \sin 10t$ V and $i_s = 6 \cos(10t + 45°)$ A.

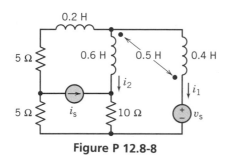

Figure P 12.8-8

P 12.8-9 A magnetic amplifier can be used to deliver large output power to a load R_L. Determine $i_o(t)$ for the circuit of Figure P 12.8-9 when $v_s = 2 \cos 500t$ V and $R_L = 3$ Ω. Determine the output power to the load.

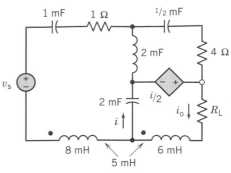

Figure P 12.8-9 Magnetic amplifier circuit.

P 12.8-10 Find the frequency ω for which the reflected impedance in Figure 12-23 is purely real if $L_2 = 40$ mH and Z_2 is a 10 kΩ resistor in series with a 100-μF capacitor.

P 12.8-11 Determine $v(t)$ for the circuit of Figure P 12.8-11 when $v_s = 10 \cos 30t$ V.

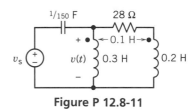

Figure P 12.8-11

P 12.8-12 Find the input impedance, **Z**, of the circuit of Figure P 12.8-12 when $\omega = 1000$ rad/s.

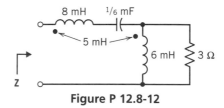

Figure P 12.8-12

Section 12-9 The Ideal Transformer

P 12.9-1 Find $\mathbf{V}_1$, $\mathbf{V}_2$, $\mathbf{I}_1$, and $\mathbf{I}_2$ for the circuit of Figure P 12.9-1, when $n = 5$.

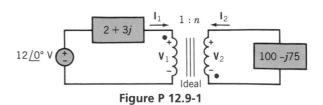

Figure P 12.9-1

P 12.9-2 A circuit with a transformer is shown in Figure P 12.9-2. (a) Determine the turns ratio. (b) Determine the value of R_{ab}. (c) Determine the current supplied by the voltage source.

Answer: (a) $n = 5$
        (b) $R_{ab} = 400 \ \Omega$

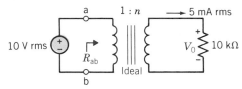

Figure P 12.9-2

P 12.9-3 Find the voltage $\mathbf{V}_c$ in the circuit shown in Figure P 12.9-3. Assume an ideal transformer. The turns ratio is $n = 1/3$.

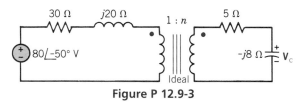

Figure P 12.9-3

P 12.9-4 An ideal transformer is connected in the circuit shown in Figure P 12.9-4, where $v_s = 50 \cos 1000t$ V and $n = N_2/N_1 = 5$. Calculate $\mathbf{V}_1$ and $\mathbf{V}_2$.

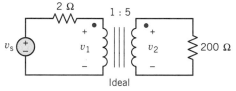

Figure P 12.9-4

P 12.9-5 The circuit of Figure P 12.9-5 is operating at 10^5 rad/s. Determine the inductance L and the turns ratio n to achieve maximum power transfer to the load.

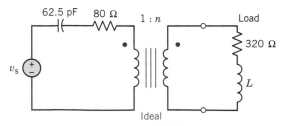

Figure P 12.9-5

P 12.9-6 Find the Thévenin equivalent at terminals a–b for the circuit of Figure P 12.9-6 when $v = 16 \cos 3t$ V.

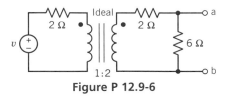

Figure P 12.9-6

P 12.9-7 Find the input impedance $\mathbf{Z}$ for the circuit of Figure P 12.9-7.

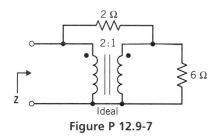

Figure P 12.9-7

P 12.9-8 A circuit is arranged as shown in Figure 12-24 where $\mathbf{V}_s = 100\underline{/0^\circ}$, $\mathbf{Z}_s = 10 \ \Omega$, and $\mathbf{Z}_2 = 1 \ \text{k}\Omega$. Find the turns ratio, n, so that maximum power transfer is achieved, and then find the power delivered to $\mathbf{Z}_2$.

Answer: $n = 10$, $P = 125$ W

P 12.9-9 Determine the node voltages for the circuit shown in Figure P 12.9-9 when $n = 4$.

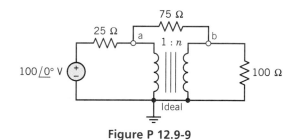

Figure P 12.9-9

P 12.9-10 A circuit with two ideal transformers is shown in Figure P 12.9-10, with $L = 2$ mH and $C = 100 \ \mu$F operating at $\omega = 1$ krad/s. Select the turns ratio of each transformer in order to achieve maximum power to R_L. For your design, determine the power delivered to R_L when $v_s = 28 \cos \omega t$ V.

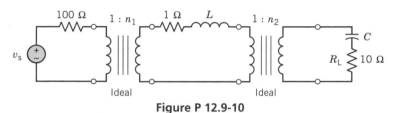

Figure P 12.9-10

P 12.9-11 In less-developed regions in mountainous areas, small hydroelectric generators are used to serve several residences (Mackay, 1990). Assume each house uses an electric range and an electric refrigerator, as shown in Figure P 12.9-11. The generator is represented as $\mathbf{V}_s$ operating at 60 Hz and $\mathbf{V}_2 = 230\underline{/0°}$ V. Calculate the power consumed by six homes connected to the hydro-electric generator when $n = 5$.

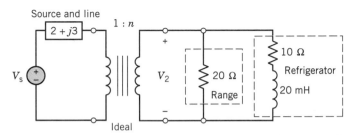

Figure P 12.9-11 Hydroelectric generator and the load for one residence.

P 12.9-12 An amplifier circuit with a transformer coupled output is shown in Figure P 12.9-12. It is desired to transfer maximum power to the output resistor R, which is equal to 100 Ω. Select the transformer turns ratio n and the compensator impedance $\mathbf{Z}$. Determine the value of the maximum power output when $v_s = 20 \cos 100t$ V.

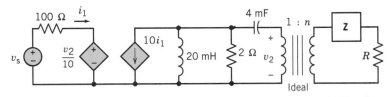

Figure P 12.9-12 Amplifier circuit with a transformer coupled load.

PSpice PROBLEMS

SP 12-1 Calculate $\mathbf{V}_1$ and $\mathbf{V}_2$ for Problem 12.9-4 using a PSpice program. Use appropriately large L_1 and L_2 to approximate an ideal transformer.

SP 12-2 Determine $i_1(t)$ and $i_2(t)$ of Problem 12.8-6 using a PSpice program.

SP 12-3 Determine $v(t)$ for Problem 12.8-7 using a PSpice program.

SP 12-4 For the circuit shown in Figure SP 12-4, determine the voltage, $v(t)$, when $v_s = 200 \sin(\omega t + 45°)$ V, $\omega = 10^4$ rad/s, and $R = 100 \Omega$.

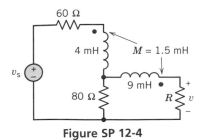

Figure SP 12-4

SP 12-5 For the circuit shown in Figure SP 12-5, determine $i(t)$ using a PSpice program. The two sources operate at $\omega = 500$ rad/s, $v_1 = 100 \cos \omega t$ V, and $v_2 = 100 \cos(\omega t - 90°)$ V.

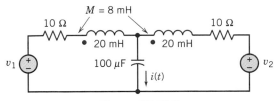

Figure SP 12-5

SP 12-6 Determine the ratio $\mathbf{V}_2/\mathbf{V}_1$ for Problem 12.8-6.

SP 12-7 The circuit with an iron core and with $k = 1$ is shown in Figure SP 12-7.

(a) Determine the ratio $\mathbf{V}_2/\mathbf{V}_1$ for $\omega = 1000$ rad/s analytically, by assuming that the transformer is ideal.
(b) Determine the actual ratio using PSpice and including the 2-Ω resistance.

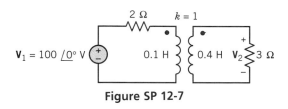

Figure SP 12-7

SP 12-8 Solve Exercise 12-20 using PSpice.

VERIFICATION PROBLEMS

VP 12-1 A laboratory report states that the average power is 214.6 W and the reactive power is 1,048 VAR delivered to the load shown in the circuit of Figure VP 12-1. Verify these values when $v_s = 115\sqrt{2} \cos(120\pi t)$ V.

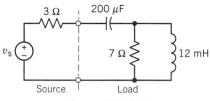

Figure VP 12-1

VP 12-2 A source and two loads are shown in Figure VP 12-2. One calculation reports that the power delivered to the total load is 5087 W. Check this report and calculate the overall power factor of the total load.

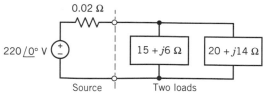

Figure VP 12-2

VP 12-3 A computer calculation states that the rms value of $v(t)$ of Figure VP 12-3 is 8.366 V. Verify this result.

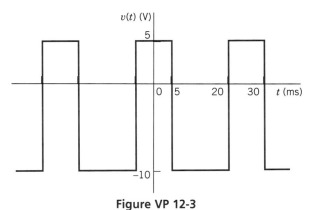

Figure VP 12-3

VP 12-4 A student report states that $\mathbf{Z}_{in} = j30$ Ω for the circuit of Figure VP 12-4. Verify this result.

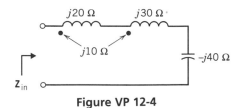

Figure VP 12-4

DESIGN PROBLEMS

DP 12-1 It is desired that the input impedance of the series RLC circuit shown in Figure DP 12-1 is $\mathbf{Z}_{in} = Z\underline{/\theta}$ where $Z = 7.21$ and $\theta = 33.7°$. Select the inductance L when the frequency of the input source is 4 rad/s.

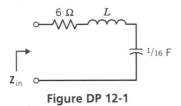

Figure DP 12-1

DP 12-2 A capacitor is in series with a 750-Ω resistor, as shown in Figure DP 12-2. The input is $v_s = 240 \cos \omega t$ V where $\omega = 2\pi(400)$. It is desired to limit the magnitude of the current to 0.2 A. Determine the required capacitance C.

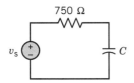

Figure DP 12-2

DP 12-3 The input admittance $\mathbf{Y}_{in} = Y\underline{/\theta}$ of the circuit shown in Figure DP 12-3 is required to be $2.8 < Y < 2.9$ when the source frequency is 50 krad/s. Determine the required inductance L and the resulting angle θ.

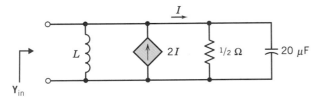

Figure DP 12-3

DP 12-4 Select the tuning capacitor C so that the input impedance of the circuit shown in Figure DP 12-4 is $\mathbf{Z}_{in} = Z\underline{/\theta}$ where $4.2 < Z < 4.6$. The frequency of operation is $\omega = 100$ rad/s.

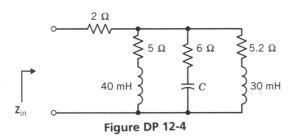

Figure DP 12-4

DP 12-5 Select the turns ratio n necessary to provide maximum power to the resistor R of the circuit shown in Figure DP 12-5. Assume an ideal transformer. Select n when $R = 4$ and 8 Ω.

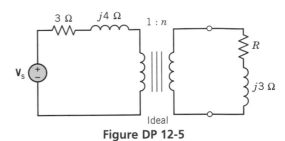

Figure DP 12-5

DP 12-6 A radio receiver operates at $\omega = 10^6$ rad/s with a load of a capacitor and resistor as shown in Figure DP 12-6. Select the cable R and L in order to deliver maximum power to the load. Determine the power delivered to the load when $v_s = V_1 \cos(\omega t + 30°)$ V and $V_1 = 0.045$ V.

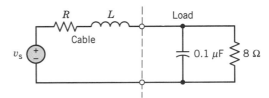

Figure DP 12-6 Radio receiver circuit.

DP 12-7 An amplifier in a short-wave radio operates at 100 kHz. The load $\mathbf{Z}_2$ is connected to a load through an ideal transformer, as shown in Figure 12-25. The load is a series connection 10-Ω resistance and 10-μH inductance. The $\mathbf{Z}_s$ consists of a 1-Ω resistance and a 1-μH inductance.

(a) Select an integer n in order to maximize the energy delivered to the load. Calculate $\mathbf{I}_2$ and the energy to the load.

(b) Add a capacitance C in series with Z_2 in order to improve the energy delivered to the load.

DP 12-8 A new electronic lamp (E-lamp) has been developed that uses a radio-frequency sinusoidal oscillator and a coil to transmit energy to a surrounding cloud of mercury gas as shown in Figure DP 12-8a. The mercury gas emits ultraviolet light which is transmitted to the phosphor coating which, in turn, emits visible light. A circuit model of the E-lamp is shown in Figure DP 12-8b. The capacitance C and the resistance R are dependent upon the lamp spacing design and the type of phosphor. Select R and C so that maximum power is delivered to R, which relates to the phosphor coating (Adler, 1992). The circuit operates at $\omega_0 = 10^7$ rad/s.

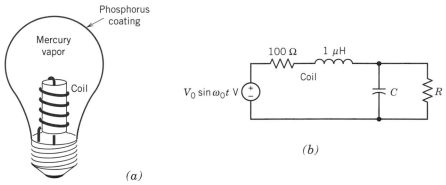

(a)

(b)

Figure DP 12-8 Electronic lamp.

CHAPTER 13

FREQUENCY RESPONSE

PREVIEW

Modern communication systems use devices called filters to separate electric signals on the basis of their frequency content. Therefore, it is important to describe the frequency-dependent relationships, both amplitude and phase, between the input sinusoidal signal and the output sinusoidal signal.

In this chapter we describe the frequency response of selected circuits that exhibit desirable characteristics and provide suitable filter characteristics. We discuss the parallel resonant circuit and the series resonant circuit, which are important circuits with many applications. We also describe the development of the frequency response of the logarithmic gain and phase angle values plotted on a log-frequency base for a function $H(j\omega)$. This plot is called a Bode diagram and is widely used for circuit analysis.

13-1 DESIGN CHALLENGE

BANDPASS FILTER

An instrument uses a bandpass filter and amplifier to amplify the input signal and reduce the effects of spurious noise. It is desired to obtain a bandpass filter with a gain of 200 at the resonant frequency of 500 Hz and a bandwidth of 50 Hz. Since this is a low-frequency filter and we desire a substantial gain, the design is going to be implemented using an op amp circuit with only resistors and capacitors.

Define the Situation

1 A bandpass filter is an example of a resonant circuit.
2 We can use an op amp and resistors and capacitors to obtain a resonant circuit.
3 Assume that the op amp is ideal.

The Goal

Design an op amp circuit that provides a resonant circuit frequency response.

The statement of this design problem uses terms like "bandwidth" and "resonant frequency." These terms are used to describe properties of the "frequency response" of a circuit. This chapter introduces the concept of frequency response and shows how a frequency response can be used to represent a linear circuit. It is common practice for engineers to specify a circuit by describing its frequency response.

We need to know more about frequency responses before we can design this bandpass filter. We will return to this design problem at the end of this chapter.

13-2 ‖ ELECTRONIC COMMUNICATION SYSTEMS

Our modern existence depends on communication systems such as television, radio, and the telephone. In this century, the telegraph emerged as an important communication system and then declined as competing systems came into existence. The rise and fall of the telegraph are shown in Figure 13-1.

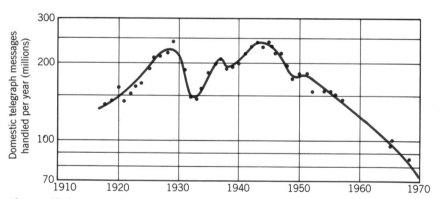

Figure 13-1 Rise and fall of the telegraph in the United States.

The five inventions that have done the most to change the twentieth century are, in order of occurrence, the telegraph, the telephone, radio, television, and the space satellite. No sooner had Bell invented the telephone than people wanted it. By 1892, 240,000 telephones were in use in the United States. By 1956, America's 50 million telephones accounted for more than half of the world total. In America nearly 3.5 percent of the total gross national product is spent on telecommunication products and services. As shown in Figure 13-2, the number of telephone lines in a nation increases as the wealth of its inhabitants increases.

By 1950, 90 percent of U.S. households had a radio, as shown in Figure 13-3. With the introduction in 1972 of satellites for relaying telephone and television signals, the real cost of telecommunications continued the decline that has led to a massive use of telephones and wide distribution of television programs.

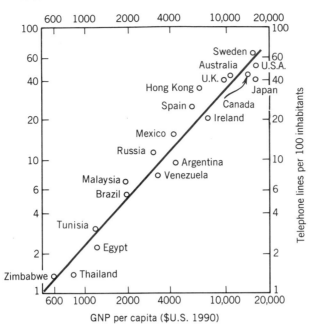

Figure 13-2 The number of telephone lines per 100 inhabitants rises as the GNP per capita increases.

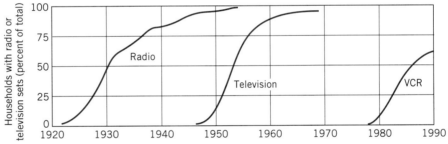

Figure 13-3 Ninety percent of U.S. households had a radio by 1950 and a television set by 1965. Sixty percent of U.S. households had a videocassette recorder by 1990.

An international phone call from Europe to America will most probably be beamed 23,000 miles into space and bounced back to earth from one of 16 satellites, each of which is the size of a car. There are currently about 120,000 transatlantic satellite voice channels available, and cable provides another 10,000 channels.

As the use of radio and the telephone grew in the 1920s, many engineers designed circuits that operated at higher and higher frequencies. Coaxial cable was proposed as a telephone transmission line to carry signals with frequencies exceeding 10 MHz. By 1921 many radio engineers were transmitting signals at 3 MHz to 5 MHz. The late 1930s saw the advent of FM (frequency modulation) radio transmitted signals above 30 MHz.

Thus, it became clear that the analysis of circuits at higher frequencies required a means of portraying the response of a circuit at selected frequencies. Furthermore, it became important to design circuits that could select a signal at one frequency while rejecting other undesired frequencies. G. A. Campbell was issued the first patent on filters in 1917. An *electric filter* is a circuit for separating electric signals on the basis of their frequency. In this chapter we explore the response of a circuit to different frequencies.

13-3 ‖ FREQUENCY RESPONSE

The *frequency response* of a circuit is the frequency-dependent relationship in both magnitude and phase between a steady-state sinusoidal input and a steady-state sinusoidal output signal. The frequency response of a circuit is an important tool for the circuit analyst since it provides information regarding the impact of the circuit on sinusoids of selected frequencies. For example, let us assume that the input signal of a circuit consists of sinusoids in the range 20 Hz to 5000 Hz. These are the frequencies in the range of the typical human voice. If, for this range of frequencies, the circuit can provide an output that is twice the input magnitude, and if the phase shift between input and output is proportional to frequency, we will have an amplified and faithful reproduction of the original signal. On the other hand, if the magnitude ratio and phase shift are different for 200 Hz and 2000 Hz, we can expect the output signal to be distorted. A filter circuit is often designed to achieve the properties of discrimination between one frequency and another. For example, a filter circuit in your AM radio is able, when properly tuned, to select one station operating at 800 kHz while rejecting the station broadcasting at 700 kHz.

In 1938 William Hewlett and David Packard built an audio oscillator, as shown in Figure 13-4. This oscillator provided a sinusoidal steady-state signal in the frequency range 20 Hz to 20 kHz, which is the range of human hearing.

Figure 13-4 Hewlett–Packard's first product, the model 200A audio oscillator (preproduction version). William Hewlett and David Packard built an audio oscillator in 1938, from which the famous firm grew. Courtesy of Hewlett–Packard Company.

13-4 ‖ FREQUENCY RESPONSE OF *RL* AND *RC* CIRCUITS

A *first-order* circuit contains only one irreducible energy storage element, either a capacitance or an inductance.

An op amp circuit with a storage element is a frequency-dependent circuit. The effect of the circuit on its input signal depends on the frequency of that input. The network function, frequency response, and Bode plot are tools that are used to represent, understand, and design frequency-dependent networks.

Consider the op amp circuit shown in Figure 13-5 when the input is

$$v_{\text{in}}(t) = 1 \cos 6283t \text{ V}$$

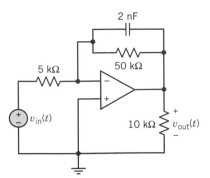

Figure 13-5 An op amp circuit.

Then the steady-state response can be found to be

$$v_{out}(t) = 8.5 \cos(6283t + 148°) \text{ V}$$

The **gain** of this circuit at the frequency $\omega = 6283$ rad/s is

$$\text{gain} = \frac{\text{output amplitude}}{\text{input amplitude}} = \frac{8.5}{1} = 8.5$$

This gain is unitless because both amplitudes have units of volts. Since the gain is greater than 1, the output sinusoid is larger than the input sinusoid. This circuit is said to **amplify** its input. When the gain of a circuit is less than 1, the output sinusoid is smaller than the input sinusoid. This circuit is said to **attenuate** its input.

The **phase shift** of this circuit is

$$\text{phase shift} = \text{output phase} - \text{input phase} = 148° - 0° = 148°$$

The phase shift determines the amount of time the output is advanced or delayed with respect to the input. Notice that

$$B \cos(\omega t + \theta) = B \cos\left(\omega\left(t + \frac{\theta}{\omega}\right)\right) = B \cos(\omega(t + t_0))$$

where θ is the phase angle in **radians** and $t_0 = \theta/\omega$. The positive peaks of $B \cos(\omega t + \theta)$ occur when

$$\omega t + \theta = n(2\pi)$$

and solving for t we have

$$t = \frac{n(2\pi)}{\omega} - t_0 = nT - t_0$$

where n is any integer and T is the period of the sinusoid.

The positive peaks of $A \cos \omega t$ occur at $t = \dfrac{n(2\pi)}{\omega}$

and the positive peaks of $B \cos(\omega t + \theta)$ occur at $t = \dfrac{n(2\pi)}{\omega} - t_0$

A phase shift of θ rad is seen to shift the output sinusoid by t_0 seconds. When the frequency is 6283 rad/s, a phase shift of 148 deg or 2.58 rad causes a shift in time of

$$t_0 = \frac{\theta}{\omega} = \frac{2.58 \text{ rad}}{6283 \text{ rad/s}} = 410 \ \mu s$$

In Figure 13-6, the positive peaks of the input sinusoid occur at 0 ms, 1 ms, 2 ms, 3 ms, ... Positive peaks of the output sinusoid occur at 0.59 ms, 1.59 ms, 2.59 ms, 3.59 ms, ... Peaks of the output sinusoid occur 410 μs **before** the next peak of the input sinusoid. The output is **advanced** by 410 μs with respect to the input.

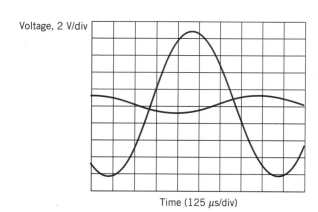

Voltage, 2 V/div

Time (125 μs/div)

Figure 13-6 Input and output sinusoids for the op amp circuit of Figure 13-5.

Notice that

$$v_{out}(t) = 8.5 \cos(6283t + 148°) = 8.5 \cos(6283t - 212°)$$

since a phase shift of 360 deg does not change the sinusoid. A phase shift of -212 deg or -3.70 rad causes a shift in time of

$$t_0 = \frac{-3.70 \text{ rad}}{6283 \text{ rad/s}} = -590 \ \mu s$$

Peaks of the output sinusoid occur 590 μs **after** the next peak of the input sinusoid. The output is **delayed** by 590 μs with respect to the input.

A phase shift that advances the output is called a **phase lead.** A phase shift that delays the output is called a **phase lag.**

At the frequency $\omega = 6283$ rad/s this circuit amplifies its input by a factor of 8.5 and advances it by 410 μs or, equivalently, delays it by 590 μs. The circuit of Figure 13-5 has a phase lead of 148 deg or, equivalently, a phase lag of 212 deg.

Suppose the voltages $v_{in}(t)$ and $v_{out}(t)$ are measured using an oscilloscope. Figure 13-6 shows the waveforms that would be displayed on the screen of the oscilloscope. Notice that the scales are shown but the axes are not. It is customary to take the angle of the input signal to be 0 deg, i.e.,

$$v_{in}(t) = A \cos \omega t$$

Then

$$v_{out} = B \cos(\omega t + \theta)$$

so

$$\text{gain} = \frac{B}{A}$$

$$\text{phase shift} = \theta$$

Now let us consider a circuit when the frequency of the input is changed. When the input is

$$v_{in}(t) = 1 \cos 3141.6t$$

the steady-state response of this circuit can be found to be

$$v_{out}(t) = 9.45 \cos(3141.6t + 163°)$$

The gain and phase shift of this circuit at the frequency $\omega = 3141.6$ rad/s are

$$\text{gain} = \frac{\text{output amplitude}}{\text{input amplitude}} = \frac{9.54}{1} = 9.54$$

and

$$\text{phase shift} = \text{output phase} - \text{input phase} = 163° - 0° = 163°$$

Changing the frequency of the input has changed the gain and phase shift of this circuit. Apparently, the gain and the phase shift of this circuit are functions of the frequency of the input. Table 13-1 shows the values of the gain and phase shift corresponding to several choices of the input frequency.

Table 13-1
Frequency Response Data for a Circuit

f (Hz)	ω (rad/s)	Gain	Phase Shift
100	628.3	9.98	176°
500	3,141.6	9.54	163°
1,000	6,283	8.50	148°
5,000	31,416	3.03	108°
10,000	62,830	1.57	99°

The **frequency response** is the steady-state response of a circuit to a sinusoidal input signal as the frequency of the sinusoid is varied. The frequency-dependent relation, in both gain and phase shift, is between an input sinusoidal steady-state signal and an output sinusoidal signal.

The frequency response can be represented by the ratio of the output response $\mathbf{Y}(j\omega)$ to the sinusoidal input signal $\mathbf{X}(j\omega)$ as

$$\frac{\mathbf{Y}(j\omega)}{\mathbf{X}(j\omega)} = \mathbf{H}(j\omega)$$

We will call the ratio $\mathbf{H}(j\omega)$ a voltage ratio, a current ratio, or a *network function*.

A table, such as Table 13-1, is an effective but clumsy way to represent the dependence of the gain and phase shift on the frequency. A plot of gain and/or phase would be helpful.

Such plots are called frequency response plots. **Network functions** provide a way to obtain equations that describe the frequency dependence of the gain and/or phase shift.

We note that

$$| \mathbf{H}(\omega) | = \frac{| \mathbf{V}_{out}(\omega) |}{| \mathbf{V}_{in}(\omega) |} = \text{gain}$$

$$\underline{/\mathbf{H}(\omega)} = \underline{/\mathbf{V}_{out}(\omega)} - \underline{/\mathbf{V}_{in}(\omega)} = \text{phase shift}$$

$$| \mathbf{V}_{out}(\omega) | = | \mathbf{H}(\omega) | \, | \mathbf{V}_{in}(\omega) |$$

$$\underline{/\mathbf{V}_{out}(\omega)} = \underline{/\mathbf{H}(\omega)} + \underline{/\mathbf{V}_{in}(\omega)}$$

As a consequence, when the input is represented as

$$v_{in}(t) = A \cos(\omega_0 t + \theta)$$

the output can be represented as

$$v_{out}(t) = A | \mathbf{H}(\omega_0) | \cos(\omega_0 t + \theta + \underline{/\mathbf{H}(\omega_0)})$$

In these equations ω_0 indicates a specific value of the variable ω.

Let us consider the *RL* circuit shown in Figure 13-7 when $v_s = V_m \cos \omega t$. Using phasors,

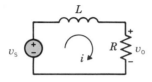

L

v_s i R v_0

Figure 13-7 The *RL* circuit.

we may show that

$$\mathbf{I} = \frac{\mathbf{V}_s}{\mathbf{Z}} \tag{13-1}$$

where

$$\mathbf{Z} = R + j\omega L$$
$$= Z\underline{/\theta}$$

We will seek the ratio of $\mathbf{V}_0/\mathbf{V}_s$ when the output voltage is across the resistor. Then

$$\mathbf{V}_o = \mathbf{I}R$$
$$= \frac{V_m R}{Z} \underline{/-\theta} \tag{13-2}$$

and

$$\frac{\mathbf{V}_o}{\mathbf{V}_s} = \frac{R}{Z} \underline{/-\theta} \tag{13-3}$$

Therefore, we have

$$v_o = \frac{RV_m}{Z} \cos(\omega t - \theta) \tag{13-4}$$

where

$$Z = \sqrt{R^2 + (\omega L)^2}$$

$$\theta = \tan^{-1} \frac{\omega L}{R}$$

We can determine and plot the ratio V_o/V_m and the phase (θ) for a range of ω of interest.

The frequency response portrays the magnitude ratio, V_o/V_m, and the phase shift as ω increases. In general, since the source will be taken as $v_s = V_m \cos\omega t$ with a frequency ω and a magnitude V_m, we will seek to determine the frequency response of the voltage ratio

$$\frac{\mathbf{V_o}}{\mathbf{V_s}} = \mathbf{H}(j\omega) \tag{13-5}$$

where $\mathbf{H}$ is a quantity that is a function of ω.

For the RL circuit we have

$$\mathbf{H} = H\underline{/\phi}$$

where H is the magnitude of $\mathbf{H}$ and ϕ is the phase shift.

Of course, it is also possible to find the frequency response of the ratio

$$\frac{\mathbf{V}}{\mathbf{I}} = \mathbf{Z}(j\omega)$$

$$= Z\underline{/\theta}$$

In this case, we use the magnitude of the impedance $\mathbf{Z}$ and the phase of the impedance, which provide the frequency response. Thus for the RL circuit of Figure 13-7,

$$Z(\omega) = \sqrt{R^2 + (\omega L)^2}$$

and

$$\theta = \tan^{-1}\frac{\omega L}{R}$$

Let us move toward obtaining a plot of the frequency response of the ratio $\mathbf{V_o}/\mathbf{V_s}$ for the RL circuit shown in Figure 13-7, where we found in Eq. 13-3 that the ratio is

$$\frac{\mathbf{V_o}}{\mathbf{V_s}} = \mathbf{H}(j\omega)$$

$$= \frac{R}{Z}\underline{/\phi}$$

Therefore, the magnitude H is

$$H = \frac{R}{[R^2 + (\omega L)^2]^{1/2}} \tag{13-6}$$

and the phase is

$$\phi = -\tan^{-1}\frac{\omega L}{R} \tag{13-7}$$

Rearranging Eq. 13-6, we obtain

$$H = \frac{1}{[1 + (\omega L/R)^2]^{1/2}}$$

$$= \frac{1}{[1 + (\omega\tau)^2]^{1/2}}$$

$$= \frac{1}{[1 + (\omega/\omega_0)^2]^{1/2}} \tag{13-8}$$

where $\tau = L/R$ and $\omega_0 = R/L$. Similarly, the phase is

$$\phi = -\tan^{-1}\omega\tau$$
$$= -\tan^{-1}(\omega/\omega_0) \tag{13-9}$$

At this point, it will be worthwhile to divert our attention briefly to the *RC* circuit shown in

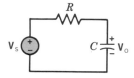

Figure 13-8 An *RC* circuit.

Figure 13-8 and determine $\mathbf{V}_o/\mathbf{V}_s$. Using the voltage divider principle, one finds

$$\frac{\mathbf{V}_o}{\mathbf{V}_s} = \frac{\mathbf{Z}_c}{R + \mathbf{Z}_c}$$

$$= \frac{(1/j\omega C)}{R + (1/j\omega C)}$$

$$= \frac{1}{j\omega RC + 1}$$

$$= \frac{1}{j\omega/\omega_0 + 1} \tag{13-10}$$

where $\omega_0 = 1/RC$. Therefore,

$$H = \frac{1}{[1 + (\omega/\omega_0)^2]^{1/2}} \tag{13-11}$$

and

$$\phi = -\tan^{-1}\omega/\omega_0 \tag{13-12}$$

Note that Eq. 13-11 is of the same form as Eq. 13-8 and that Eq. 13-9 is of the same form as Eq. 13-12. Thus, we have the first-order frequency response in the form

$$H = \frac{1}{\sqrt{1 + (\omega/\omega_0)^2}}$$

and

$$\phi = -\tan^{-1}\omega/\omega_0$$

We wish to plot the frequency response for the range of frequency from $\omega = 0$ to $\omega = \infty$. When $\omega = 0$, $H = 1$ and $\phi = 0°$. When $\omega = \infty$, $H = 0$ and $\phi = -90°$. These two values are summarized in Table 13-2.

Table 13-2
Frequency Response of a First-Order Circuit

Frequency, ω	0	$\omega_0/2$	ω_0	$2\omega_0$	$5\omega_0/2$	∞
Magnitude, H	1	0.89	$1/\sqrt{2}$	0.45	0.37	0
Phase, ϕ	0°	$-26.6°$	$-45°$	$-63.4°$	$-68.2°$	$-90°$

When $\omega = \omega_0$, we have $H = 1/\sqrt{2}$ and $\phi = -\tan^{-1} 1 = -45°$. The values of H and ϕ are summarized in Table 13-2 for selected values of ω. The plot of H and ϕ for $0 \leq \omega \leq 5\omega_0/2$ is shown in Figure 13-9.

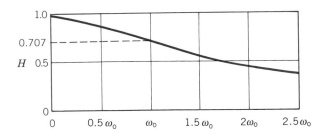

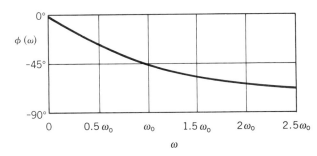

Figure 13-9 Frequency response for Eqs. 13-8 and 13-9, which represent a first-order RL or RC circuit.

When $\omega = \omega_0$, $H = 1/\sqrt{2}$ as shown in Figure 13-9a. The square of the output amplitude is reduced by one-half. For this reason ω_0 is called the *half-power frequency*.

Let us consider the op amp circuit of Figure 13-10. We seek to find the voltage ratio

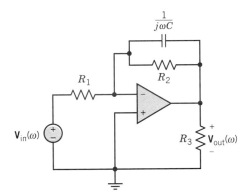

Figure 13-10 The frequency domain representation of the op amp circuit of Figure 13-5.

(network function) $\mathbf{H}(j\omega) = \mathbf{V}_{out}/\mathbf{V}_{in}$. Write the node equation at node a and assume an ideal op amp. Then we have

$$\frac{\mathbf{V}_{in}(\omega)}{R_1} + \frac{\mathbf{V}_{out}(\omega)}{R_2} + j\omega C \mathbf{V}_{out}(\omega) = 0$$

This implies

$$\mathbf{H}(\omega) = \frac{\mathbf{V}_{out}(\omega)}{\mathbf{V}_{in}(\omega)} = \frac{-R_2}{R_1 + j\omega C R_1 R_2}$$

The gain of this circuit is

$$\text{gain} = |\mathbf{H}(\omega)| = H = \frac{R_2/R_1}{\sqrt{1 + \omega^2 C^2 R_2^2}}$$

The phase shift of this circuit is

$$\text{phase shift} = \underline{/\mathbf{H}(\omega)} = 180° - \tan^{-1}(\omega C R_2)$$

When $R_1 = 5$ kΩ, $R_2 = 50$ kΩ, and $C = 2$ nF,

$$\mathbf{H}(\omega) = \frac{-10}{1 + (j\omega/10{,}000)}$$

$$\text{gain} = |\mathbf{H}(\omega)| = \frac{10}{\sqrt{1 + (\omega^2/10^8)}}$$

$$\text{phase shift} = \underline{/\mathbf{H}(\omega)} = 180° - \tan^{-1}(\omega/10{,}000)$$

13-5 BODE PLOTS

It is common to use logarithmic plots instead of linear plots. The logarithmic plots are called *Bode plots* in honor of H. W. Bode, who used them extensively in his work with amplifiers at Bell Telephone Laboratories in the 1930s and 1940s. A Bode plot is a plot of log-gain and phase angle values using a log-frequency horizontal axis. The plotting of the frequency response can be systematized and simplified by use of logarithmic plots. The use of semilog paper eliminates the need to take logarithms of very many numbers and also expands the range of frequencies portrayed on the horizontal axis.

The ratio **H** can be written as

$$\mathbf{H} = H\underline{/\phi}$$
$$= He^{j\phi} \tag{13-13}$$

The natural logarithm of Eq. 13-13 is

$$\ln \mathbf{H} = \ln H + j\phi \tag{13-14}$$

where $\ln H$ is the magnitude in nepers. The logarithm of the magnitude is normally expressed in terms of the logarithm to the base 10, so we use

$$\text{logarithmic gain} = 20 \log_{10} H \tag{13-15}$$

and the unit is decibel (dB). A decibel conversion table is given in Table 13-3.

The **Bode plot** is a chart of gain in decibels and phase in degrees versus the logarithm of frequency.

The *logarithmic magnitude* in dB and the angle $\phi(\omega)$ are plotted versus frequency, ω. For the Bode diagram, we use a logarithmic scale for ω so that we may show the frequency of response for a wide range of ω. The primary advantage of the logarithmic plot is the conversion of multiplicative factors into additive factors. Let us obtain the frequency

Table 13-3
A Decibel Conversion Table

Magnitude, H	$20 \log H$ (dB)
0.1	-20.00
0.2	-13.98
0.4	-7.96
0.6	-4.44
1.0	0.0
1.2	1.58
1.4	2.92
1.6	4.08
2.0	6.02
3.0	9.54
4.0	12.04
5.0	13.98
6.0	15.56
7.0	16.90
10.0	20.00
100.0	40.00

response in Bode form of the gain ratio

$$\mathbf{H} = \frac{1}{(j\omega/\omega_0) + 1}$$
$$= H\underline{/\phi} \tag{13-16}$$

where

$$H = \frac{1}{\sqrt{1 + (\omega/\omega_0)^2}}$$

and

$$\phi = -\tan^{-1} \omega/\omega_0$$

The logarithmic gain is

$$20 \log H = 20 \log \frac{1}{\sqrt{1 + (\omega/\omega_0)^2}}$$
$$= -10 \log[1 + (\omega/\omega_0)^2] \tag{13-17}$$

For small frequencies, that is, $\omega \ll \omega_0$, the logarithmic gain is

$$20 \log H \cong -10 \log 1$$
$$= 0 \text{ dB}$$

For very large frequencies, that is, $\omega \gg \omega_0$, we have

$$20 \log H = -20 \log \omega/\omega_0 \tag{13-18}$$

At $\omega = \omega_0$, the half-power frequency, we have

$$20 \log H = 20 \log(1/\sqrt{2})$$
$$= -10 \log 2$$
$$= -3.01 \text{ dB}$$

Table 13-4

Logarithmic Magnitude and Phase for the *RL* or *RC* Circuit of Eq. 13-16

ω/ω_0	0.10	0.50	0.764	1.0	1.31	2	5	10
$20\log\dfrac{1}{\sqrt{1+(\omega/\omega_0)^2}}$, dB	-0.04	-1.0	-2.0	-3.0	-4.3	-7.0	-14.2	-20.0
$\phi(\omega)$, deg	$-5.7°$	$-26.6°$	$-37.4°$	$-45.0°$	$-52.6°$	$-63.4°$	$-78.7°$	$-84.3°$

The logarithmic magnitude for the circuit is summarized in Table 13-4. The Bode plot is shown in Figure 13-11 for two decades of ω.

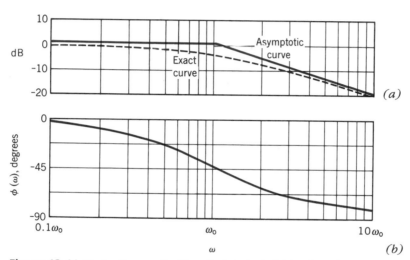

Figure 13-11 Bode diagram for $\mathbf{H} = (1 + j\omega/\omega_0)^{-1}$. The dotted curve is the exact curve for the magnitude. The solid curve for the magnitude is an asymptotic approximation.

The magnitude characteristic does not exhibit a sharp break. Nevertheless, we designate the frequency at which the magnitude is $1/\sqrt{2}$ times the magnitude at $\omega = 0$ as a special frequency. On the Bode diagram, the magnitude drop of $1/\sqrt{2}$ results in a logarithmic drop of approximately -3 dB at $\omega = \omega_0$. The frequency $\omega = \omega_0$ is often called the *break frequency* or *corner frequency*.

An interval between two frequencies with a ratio equal to 10 is called a decade, so the range of frequencies from ω_1 to ω_2 where $\omega_2 = 10\omega_1$ is called a decade. The difference between the logarithmic gains over a decade of frequency for $\omega \gg 1/\tau$ using Eq. 13-18 is

$$20\log H(\omega_1) - 20\log H(\omega_2) = -20\log\omega_1\tau - (-20\log\omega_2\tau)$$

$$= -20\log\frac{\omega_1\tau}{\omega_2\tau}$$

$$= -20\log\left(\frac{1}{10}\right)$$

$$= +20\text{ dB}$$

Thus, the slope of the asymptotic line for this first-order circuit when $\omega \gg 1/\tau$ is -20 dB/decade, as shown in Figure 13-11. The asymptotic curve intersects the 0-dB line at $\omega = \omega_0 = 1/\tau$, the break frequency.

Of course, **H** may take on forms other than that of Eq. 13-16. For example, consider the circuit shown in Figure 13-12. The voltage gain is

$$\mathbf{H} = \frac{\mathbf{V}_o}{\mathbf{V}_s} = \frac{R + j\omega L}{R_s + R + j\omega L} = \frac{R(1 + j\omega\tau_1)}{(R_s + R)(1 + j\omega\tau_2)}$$

$$= \frac{R(1 + j\omega/\omega_1)}{(R_s + R)(1 + j\omega/\omega_2)} \tag{13-19}$$

where $\omega_1 = 1/\tau_1 = R/L$ and $\omega_2 = 1/\tau_2 = (R_s + R)/L$. Here we use the break points written as ω_1 and ω_2 as most engineers prefer.

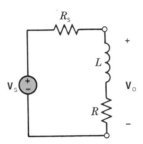

Figure 13-12 Source voltage delivering power to a load impedance consisting of L and R.

We have a magnitude ratio of

$$H = k \frac{[1 + (\omega/\omega_1)^2]^{1/2}}{[1 + (\omega/\omega_2)^2]^{1/2}} \tag{13-20}$$

where $k = R/(R_s + R)$.

The phase of **H** is

$$\phi = \tan^{-1} \omega/\omega_1 - \tan^{-1} \omega/\omega_2 \tag{13-21}$$

In this case, the Bode diagram is as shown in Figure 13-13, where, for convenience, we

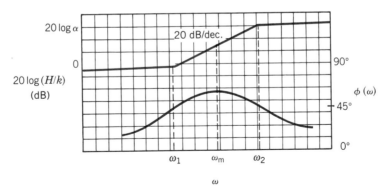

Figure 13-13 Bode diagram for the circuit of Figure 13-12. The asymptotic approximation for the magnitude curve is shown where $\alpha = \omega_2/\omega_1$.

choose to plot the magnitude of H/k as shown in the figure. Note that for lower frequencies, the phase increases and reaches a maximum at $\omega = \omega_m = \sqrt{\omega_1\omega_2}$. The asymptotic magnitude reaches a maximum value at $\omega = \omega_2$, where it attains a value of $20 \log(\omega_2/\omega_1)$. When $\omega_1 < \omega < \omega_2$ the slope of the asymptotic approximation for the magnitude curve is 20 dB/decade.

Example 13-1

Draw the Bode diagram for the ratio $\mathbf{V}_o/\mathbf{V}_s$ for the circuit of Figure 13-14.

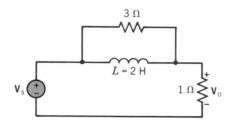

Figure 13-14 Source voltage delivering power to a 1-Ω resistor.

Solution

$$\frac{\mathbf{V}_o}{\mathbf{V}_s} = \mathbf{H} = \frac{1}{1 + (j\omega L)3/(3 + j\omega L)}$$

$$= \frac{1 + j(2\omega/3)}{1 + j(8\omega/3)}$$

$$= \frac{1 + j(\omega/\omega_1)}{1 + j(\omega/\omega_2)}$$

$$= \frac{H_n \underline{/\theta_n}}{H_d \underline{/\theta_d}} \tag{13-22}$$

where H_n is the magnitude of the numerator of $\mathbf{H}$, θ_n is the phase of the numerator of $\mathbf{H}$, H_d is the magnitude of the denominator, and θ_d is the phase of the denominator of $\mathbf{H}$. Then the phase of $\mathbf{H}$ is

$$\phi = \theta_n - \theta_d$$

$$= \tan^{-1} \frac{2\omega}{3} - \tan^{-1} \frac{8\omega}{3} \tag{13-23}$$

The logarithmic gain of Eq. 13-22 is

$$20 \log H = 20 \log \frac{H_n}{H_d} = 20 \log H_n - 20 \log H_d \tag{13-24}$$

Therefore, we see that one way to plot the logarithmic magnitude is to plot the logarithmic numerator magnitude and the logarithmic denominator magnitude of $\mathbf{H}$ and then subtract as required in Eq. 13-24. We will use the asymptotic approximation for the numerator and denominator to obtain the total decibel magnitude. The asymptotic approximations for 20 log H_n and 20 log H_d are shown in Figure 13-15a. The break point for the numerator occurs when $\omega_1 = 3/2$, and the break point for the denominator occurs when $\omega_2 = 3/8$. Completing the subtraction of Eq. 13-24, we obtain 20 log H as shown in Figure 13-15b. Finally, we calculate the phase ϕ, using Eq. 13-23, and tabulate the results as listed in Table 13-5. Then we may plot the phase as shown in Figure 13-15c.

Table 13-5

The Phase for Example 13-1

ω (rad/s)	0.1	0.375	0.75	1.0	1.5	5.0	10.0
ϕ, degrees	-11.1	-31.0	-36.9	-35.8	-31.0	-12.4	-6.4

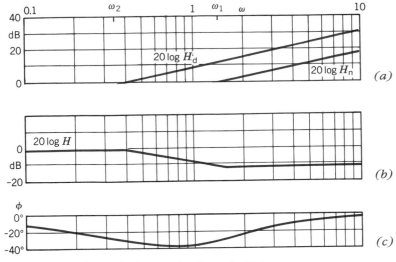

Figure 13-15 The Bode diagram for Example 13-1.

Example 13-2

An operational amplifier circuit is shown in Figure 13-16. Find the ratio $\mathbf{V}_o/\mathbf{V}_s$ and sketch the Bode plot when $R_2 = 10R_1$ and $R_2C = 0.1$.

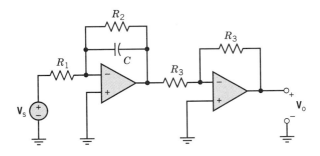

Figure 13-16 Operational amplifier circuit for Example 13-2.

Solution

Assuming an ideal model for the operational amplifier, we have (see Section 11-14)

$$\frac{\mathbf{V}_o}{\mathbf{V}_s} = \mathbf{H} = \frac{\mathbf{Z}_2}{\mathbf{Z}_1}$$

Since

$$\mathbf{Z}_2 = \frac{\mathbf{Z}_c R_2}{\mathbf{Z}_c + R_2} = \frac{R_2}{1 + j\omega R_2 C}$$

and

$$\mathbf{Z}_1 = R_1$$

we have

$$\mathbf{H} = \frac{R_2}{R_1} \frac{1}{1 + j\omega/\omega_0}$$

where $\omega_0 = 1/R_2C = 10$ and $R_2/R_1 = 10$.

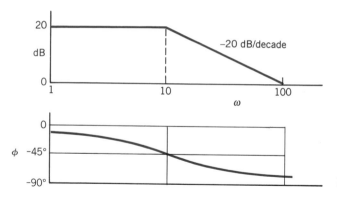

Figure 13-17 Bode plot for Example 13-2.

The Bode plot for this **H** is sketched in Figure 13-17 using the asymptotic approximation for the logarithmic magnitude.

EXERCISE 13-1

By how many decibels does $|\mathbf{V}_o/\mathbf{V}_s|$ lie above the reference level of zero when $|\mathbf{V}_o/\mathbf{V}_s|$ (a) equals 2, (b) equals 0.5?
Answer: (a) $+6.02$ dB   (b) -6.02 dB

EXERCISE 13-2

In a certain frequency range, the magnitude is $H = 1/\omega^2$. What is the slope of the Bode plot in this range, expressed in decibels per decade?
Answer: -40 dB/decade

EXERCISE 13-3

The magnitude ratio of $|\mathbf{V}_o/\mathbf{V}_s| = H$ depends on frequency as

$$H = \frac{A\omega}{(B + C\omega^2)^{1/2}}$$

Find (a) the break frequency, (b) the slope of the asymptotic line for ω above the break frequency in decibels per decade, and (c) the slope of the Bode plot below the break frequency.
Answer: (a) $\omega = \sqrt{B/C}$
　　　　 (b) zero
　　　　 (c) 20 dB/decade

EXERCISE 13-4

A first-order circuit is shown in Figure E 13-4. Determine the ratio $\mathbf{V}_o/\mathbf{V}_s$ and sketch the Bode diagram when $RC = 0.1$ and $R_1/R_2 = 3$.

Answer: $\mathbf{H} = \left(1 + \dfrac{R_1}{R_2}\right)\dfrac{1}{1 + j\omega RC}$

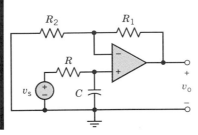

Figure E 13-4

EXERCISE 13-5

(a)   Draw the Bode diagram of the ratio $\mathbf{V_o}/\mathbf{V_s}$ for the circuit of Figure E 13-5.
(b)   Determine $v_o(t)$ when $v_s = 10 \cos 20t$ V.
Answer: (b) $v_o = 4.18 \cos(20t - 24.3°)$ V

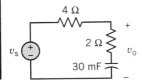

Figure E 13-5

13-6   PARALLEL RESONANT CIRCUITS

Many applications call for narrowband filters that pass the sinusoidal steady-state signal at one frequency and tend to reject sinusoidal signals at other frequencies.

The current gain $\mathbf{I_o}/\mathbf{I_s}$ of the parallel circuit shown in Figure 13-18 is

$$\mathbf{H} = \frac{\mathbf{I_o}}{\mathbf{I_s}} = \frac{1}{R\mathbf{Y}} \tag{13-25}$$

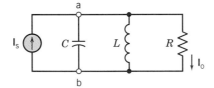

Figure 13-18 Parallel resonant circuit.

where $\mathbf{Y}$ is the admittance of the three parallel elements. Therefore,

$$\mathbf{Y} = \mathbf{Y_R} + \mathbf{Y_C} + \mathbf{Y_L}$$
$$= \frac{1}{R} + j\omega C + \frac{1}{j\omega L}$$
$$= \frac{1}{R} + j\left(\omega C - \frac{1}{\omega L}\right) \tag{13-26}$$

Substituting Eq. 13-26 into Eq. 13-25, we have

$$\mathbf{H} = \frac{1}{1 + j(\omega C - 1/\omega L)R} \tag{13-27}$$

The imaginary term is equal to zero at the *resonant frequency* when $\omega C = 1/\omega L$. The *resonant frequency* of a parallel resonant circuit is defined as the frequency ω_0 when the admittance $\mathbf{Y}$ is nonreactive. The voltage v_{ab} and the current i_s are in phase when $\mathbf{Y}$ is nonreactive. Resonance occurs when, for Eq. 13-26, we have $\omega C = 1/\omega L$. Therefore, the resonant frequency is

$$\omega_0 = \frac{1}{\sqrt{LC}} \tag{13-28}$$

A **resonant circuit** is a combination of frequency-sensitive elements connected to provide a frequency-selective response.

We also denote another parameter as the *quality factor Q* for a parallel resonant circuit as

$$Q = \omega_0 CR = \frac{R}{\omega_0 L} \tag{13-29}$$

where Q is a dimensionless ratio. In essence, Q is a measure of the energy storage property of a circuit in relation to its energy dissipation property. The definition of Q is

$$Q = 2\pi \frac{\text{maximum energy stored}}{\text{energy dissipated per cycle}}$$

which can be shown, with some effort, to yield Eq. 13-29.

Multiply $Q = \omega_0 CR$ from Eq. 13-29 by ω/ω_0 to obtain

$$\frac{\omega}{\omega_0} Q = \omega CR \tag{13-30}$$

Then, similarly, multiply $Q = R/\omega_0 L$ from Eq. 13-29 by ω_0/ω to obtain

$$\frac{\omega_0}{\omega} Q = \frac{R}{\omega L} \tag{13-31}$$

Substitute Eqs. 13-30 and 13-31 into Eq. 13-27 to obtain $\mathbf{H}$ in terms of Q and ω_0 as

$$\mathbf{H} = \frac{1}{1 + jQ(\omega/\omega_0 - \omega_0/\omega)} \tag{13-32}$$

Therefore, the magnitude is

$$H = \frac{1}{[1 + Q^2(\omega/\omega_0 - \omega_0/\omega)^2]^{1/2}} \tag{13-33}$$

and the phase shift is

$$\phi = -\tan^{-1} Q \left(\frac{\omega}{\omega_0} - \frac{\omega_0}{\omega} \right) \tag{13-34}$$

First, let us determine the phase and magnitude of H at selected frequencies and record them in Table 13-6. The magnitude H is zero when $\omega = 0$ and $\omega = \infty$. Clearly, the magnitude is $H = 1$ at resonance (ω_0). The phase is $+90°$ at $\omega = 0$ and $-90°$ at $\omega = \infty$. Now there are two frequencies, ω_1 and ω_2, that yield $H = 1/\sqrt{2}$. Examining Eq. 13-33, we

Table 13-6
Magnitude and Phase of **H** at Selected Frequencies

ω	0	ω_1	ω_0	ω_2	∞
H	0	$1/\sqrt{2}$	1	$1/\sqrt{2}$	0
ϕ	90°	45°	0°	$-45°$	$-90°$

note that $H = 1/\sqrt{2}$ occurs when

$$Q^2 \left(\frac{\omega}{\omega_0} - \frac{\omega_0}{\omega} \right)^2 = 1 \tag{13-35}$$

Equation 13-35 can be rearranged as a fourth-order equation in terms of ω. Solving for the solutions of interest, we have

$$\omega_1 = \omega_0 \sqrt{1 + \left(\frac{1}{2Q} \right)^2} - \frac{\omega_0}{2Q} \tag{13-36}$$

and

$$\omega_2 = \omega_0 \sqrt{1 + \left(\frac{1}{2Q} \right)^2} + \frac{\omega_0}{2Q} \tag{13-37}$$

The circuit *bandwidth BW* is defined as the range of frequencies that lie between the two frequencies where the magnitude of the gain is $1/\sqrt{2}$.

The **bandwidth** of a frequency-selective circuit is the frequency range between the points where the magnitude of the gain drops to $1/\sqrt{2}$ times the maximum value. Therefore,

$$BW = \omega_2 - \omega_1$$
$$= \frac{\omega_0}{Q} \tag{13-38}$$

A circuit with a high Q will have a narrow bandwidth. For example, if $Q = 100$ and $\omega_0 = 100$ krad/s, then $BW = 1$ krad/s.

Since we are normally interested in cases where $Q > 10$, then $(1/2Q)^2 \ll 1$ and Eqs. 13-36 and 13-37 reduce to

$$\omega_1 \cong \omega_0 - \frac{BW}{2} \tag{13-39}$$

and

$$\omega_2 \cong \omega_0 + \frac{BW}{2} \tag{13-40}$$

When $Q > 10$, the magnitude curve has approximate arithmetic symmetry around ω_0 as shown in Figure 13-19. Regardless of Q, the response is symmetrical on a log frequency scale.

The frequency response of a resonant circuit is shown in Figure 13-20. This diagram is for small deviations of frequency from ω_0 and for relatively high Q, where δ is

$$\delta = \frac{\omega - \omega_0}{\omega_0} \tag{13-41}$$

The curves of Figure 13-20 are derived as follows.

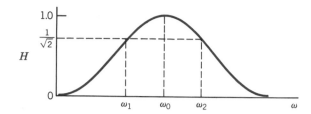

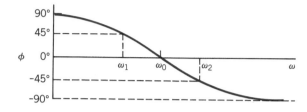

Figure 13-19 Magnitude and phase of the parallel resonant circuit.

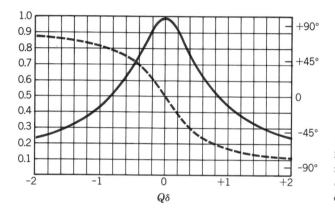

Figure 13-20 Resonance curve for the parallel and series resonant circuit for $\delta \ll 1$, where $\delta = (\omega - \omega_0)/\omega_0$. The phase curve is dashed.

Repeating Eq. 13-32, we have

$$\mathbf{H} = \frac{1}{1 + jQ(\omega/\omega_0 - \omega_0/\omega)} \tag{13-42}$$

and define

$$\delta = \frac{\omega - \omega_0}{\omega_0} = \frac{\omega}{\omega_0} - 1 \tag{13-43}$$

where δ represents the proportional amount we deviate in frequency from ω_0. Then consider the denominator factor $(\omega/\omega_0 - \omega_0/\omega)$ of Eq. 13-42. If we substitute Eq. 13-43, we have

$$\frac{\omega}{\omega_0} - \frac{\omega_0}{\omega} = (\delta + 1) - \left(\frac{1}{\delta + 1}\right)$$

$$= \frac{(\delta + 1)^2 - 1}{\delta + 1}$$

$$= \frac{\delta^2 + 2\delta}{\delta + 1} \tag{13-44}$$

Using $\delta \ll 1$ for small deviations from ω_0, we have

$$\frac{\omega}{\omega_0} - \frac{\omega_0}{\omega} \cong 2\delta \qquad (13\text{-}45)$$

Substituting Eq. 13-45 into Eq. 13-42, we have

$$\mathbf{H} = \frac{1}{1 + j2Q\delta} \qquad (13\text{-}46)$$

which is an approximation that holds as long as $\delta \ll 1$. The resonance curve for this approximation is given in Figure 13-20.

EXERCISE 13-6

(a) A parallel resonant circuit is resonant at 800 kHz. Assuming the input signal has an amplitude of 1, determine the response at 850 kHz when $Q = 100$. (b) Also calculate the bandwidth.
Answer: (a) $H = 0.08$, $\phi = -85.4°$ (b) $BW = 8$ kHz

EXERCISE 13-7

Consider the circuit of Exercise 13-6 with an input signal at 810 kHz. Using Figure 13-20, find the magnitude and phase at 810 kHz. Check your results by calculating them.
Answer: $H = 0.37$, $\phi = -68°$

EXERCISE 13-8

For the *RLC* parallel resonant circuit when $R = 8$ kΩ, $L = 40$ mH, and $C = 0.25$ μF, find (a) Q and (b) bandwidth.
Answer: (a) $Q = 20$ (b) $BW = 500$ rad/s

EXERCISE 13-9

A high-frequency RLC parallel resonant circuit is required to operate at $\omega_0 = 10$ Mrad/s with a bandwidth of 200 krad/s. Determine the required Q and L when $C = 10$ pF.
Answer: $Q = 50$, $L = 1$ mH

13-7 SERIES RESONANCE

The series resonant circuit is shown in Figure 13-21. The voltage ratio of interest is

$$\mathbf{H} = \frac{\mathbf{V}_o}{\mathbf{V}_s} = \frac{R}{R + j\omega L + 1/j\omega C}$$

$$= \frac{1}{1 + j(\omega L/R - 1/R\omega C)} \qquad (13\text{-}47)$$

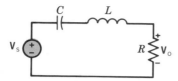

Figure 13-21 Series resonant circuit.

Again, we see that the ratio is without an imaginary term when $\omega L = 1/\omega C$ and the resonant frequency is

$$\omega_0^2 = \frac{1}{LC} \tag{13-48}$$

The resonant frequency in a series RLC circuit is defined as the frequency when the total impedance becomes real (nonreactive). Also, we define the quality factor Q of the series resonant circuit as

$$Q = \frac{\omega_0 L}{R} = \frac{1}{\omega_0 RC} \tag{13-49}$$

As before, the bandwidth of the circuit is

$$BW = \frac{\omega_0}{Q} \tag{13-50}$$

Multiply $Q = \omega_0 L/R$ from Eq. 13-49 by ω/ω_0 to obtain

$$\frac{\omega}{\omega_0} Q = \frac{\omega L}{R} \tag{13-51}$$

Similarly, multiply $Q = 1/\omega_0 RC$ by ω_0/ω to obtain

$$\frac{\omega_0}{\omega} Q = \frac{1}{\omega RC} \tag{13-52}$$

Substitute Eqs. 13-51 and 13-52 into Eq. 13-47 to obtain

$$\mathbf{H} = \frac{1}{1 + jQ(\omega/\omega_0 - \omega_0/\omega)} \tag{13-53}$$

which is equal to Eq. 13-32 for the parallel resonant circuit. Note, however, that the quality factor is defined differently for the series circuit than it was for the parallel resonant circuit. However, the other relationships for bandwidth and ω_2, ω_1, and δ remain consistent for the two circuits. The factors for each circuit are recorded in Table 13-7. As you can see, the only two differences between the parallel resonant and series resonant circuits are the definition of $\mathbf{H}$ and the definition of the quality factor Q and, of course, the layout (form) of the circuit. Thus, we can proceed with all the aids developed in Section 13-6 and summarized in Table 13-7 to find the appropriate parameters for a desired frequency response.

It will be worthwhile at this point to show how the quality factor is obtained. The quality factor is defined as

$$Q = 2\pi \frac{\text{maximum energy stored}}{\text{energy dissipated per cycle}} \tag{13-54}$$

For a circuit at resonance, the total stored energy is a constant. At an instant when the voltage across the capacitance is zero, the current in the inductance is maximum and all the stored energy is in the inductance. At an instant when the capacitor voltage is maxi-

Table 13-7

Characteristics of Resonant Circuits

Factor	Parallel Resonant Circuit	Series Resonant Circuit
Gain ratio, $\mathbf{H}$	$\dfrac{\mathbf{I}_o}{\mathbf{I}_s}$	$\dfrac{\mathbf{V}_o}{\mathbf{V}_s}$
Quality factor, Q	$\omega_0 CR = \dfrac{R}{\omega_0 L}$	$\dfrac{\omega_0 L}{R} = \dfrac{1}{\omega_0 RC}$
$\mathbf{H}$ in terms of Q and ω_0	$\dfrac{1}{1 + jQ(\omega/\omega_0 - \omega_0/\omega)}$	$\dfrac{1}{1 + jQ(\omega/\omega_0 - \omega_0/\omega)}$
Resonant frequency, ω_0	$\dfrac{1}{\sqrt{LC}}$	$\dfrac{1}{\sqrt{LC}}$
Bandwidth, BW	$\dfrac{\omega_0}{Q}$	$\dfrac{\omega_0}{Q}$
ω_2, ω_1	$\omega_0 \sqrt{1 + \left(\dfrac{1}{2Q}\right)^2} \pm \dfrac{\omega_0}{2Q}$	$\omega_0 \sqrt{1 + \left(\dfrac{1}{2Q}\right)^2} \pm \dfrac{\omega_0}{2Q}$
ω_2, ω_1 for $Q > 10$	$\omega_0 \pm \dfrac{BW}{2}$	$\omega_0 \pm \dfrac{BW}{2}$
$\mathbf{H}$ for small deviation, $\delta = \dfrac{\omega - \omega_0}{\omega_0}$	$\dfrac{1}{1 + j2Q\delta}$	$\dfrac{1}{1 + j2Q\delta}$

mum, the inductor current is zero and all the stored energy is in the capacitor. For the series circuit, the current is $i = I_m \cos \omega t$ and the maximum stored energy is then

$$w = \tfrac{1}{2}Li^2 = \tfrac{1}{2}LI_m^2$$

The energy dissipated per cycle is the average power divided by the frequency, f_0. Then for the series resonant circuit we obtain

$$Q_s = 2\pi \frac{LI_m^2}{I_m^2 R/f_0} = 2\pi f_0 \frac{L}{R} = \frac{\omega_0 L}{R} \tag{13-55}$$

Example 13-3

A series resonant circuit has $R = 2\ \Omega$, $L = 1\ \text{mH}$, and $C = 0.1\ \mu\text{F}$. Find ω_0, BW, and Q, and determine the response of the circuit at $\omega = 1.02\omega_0$.

Solution

First, we determine the resonant frequency as

$$\omega_0 = \frac{1}{\sqrt{LC}}$$

$$= \frac{1}{[(10^{-3})10^{-7}]^{1/2}}$$

$$= 10^5\ \text{rad/s}$$

Then the quality factor is

$$Q = \frac{\omega_0 L}{R}$$

$$= \frac{(10^5)10^{-3}}{2}$$

$$= 50$$

Therefore, the bandwidth is

$$BW = \frac{\omega_0}{Q} = \frac{10^5}{50} = 2 \text{ krad/s}$$

$$= 318 \text{ Hz}$$

We also wish to determine the response when $\omega = 1.02\omega_0$ or

$$\delta = \frac{\omega - \omega_0}{\omega_0} = 0.02$$

We can use Figure 13-20 to find the response, since $\delta \ll 1$. Since $Q = 50$, we have $Q\delta = 50(0.02) = 1$. Examining Figure 13-20, we have

$$H = \left| \frac{\mathbf{V}_o}{\mathbf{V}_s} \right| = 0.45$$

and

$$\phi = -63°$$

EXERCISE 13-10

A series resonant circuit has $L = 1$ mH and $C = 10$ μF. Find the required Q and R when it is desired that the bandwidth be 15.9 Hz.
Answer: $Q = 100$, $R = 0.1$ Ω

EXERCISE 13-11

A series resonant circuit has an inductor $L = 10$ mH. (a) Select C and R so that $\omega_0 = 10^6$ rad/s and the bandwidth is $BW = 10^3$ rad/s. (b) Find the response $\mathbf{H}$ of this circuit for a signal at $\omega = 1.05 \times 10^6$ rad/s.
Answers: (a) $C = 100$ pF, $R = 10$ Ω

(b) $\mathbf{H} = \dfrac{1}{1 + j97.6}$

EXERCISE 13-12

The input voltage for the circuit of Figure E 13-12 is $v_s = 10 \cos \omega t$ V where $\omega = 5 \times 10^3$ rad/s. Determine the circuit bandwidth and the output voltage $v(t)$ using Figure 13-20.

Answer: $BW = 159$ Hz
$$v = 7.07 \cos(\omega t - 45°) \text{ V}$$

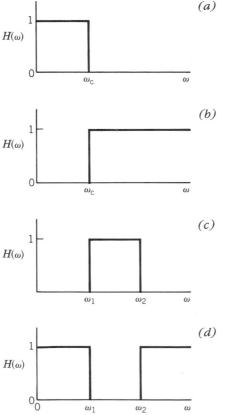

Figure E 13-12

13-8 | FILTER CIRCUITS

As we have determined in the preceding sections, the frequency response of individual circuits provides selective characteristics, passing some frequency components unaffected, while relatively attenuating other frequency components. An electrical *filter* is a circuit that is designed to introduce magnitude gain and/or loss over a prescribed range of frequencies.

By shaping $\mathbf{H}(j\omega)$ over a band of frequencies, one can discriminate sinusoidal waves according to their frequencies. In this section we consider only the shaping of the magnitude gain, H. The idealized magnitude characteristics of four widely used filters are shown in Figure 13-22.

Figure 13-22 Ideal magnitude characteristic of four filters: *(a)* low pass, *(b)* high pass, *(c)* bandpass, and *(d)* bandstop.

A **filter circuit** introduces a frequency-selective magnitude in order to pass signals of desired frequencies and reject others.

The low-pass filter shown in Figure 13-22a will ideally pass all frequencies up to ω_c and perfectly reject all frequencies above ω_c, where ω_c is called the cutoff frequency. We note that the op amp circuit of Figure 13-5 as well as the RL circuit of Figure 13-7 and the RC circuit of Figure 13-8 have a magnitude characteristic that is an approximation of the ideal response. Therefore, they are called low-pass filter circuits. The cutoff frequency ω_c is equal to $\omega = \omega_c = 1/\tau$, the break frequency of the RL or RC circuit.

A **low-pass filter** tends to pass signals at frequencies up to the cutoff frequency ω_c and reject signals of frequencies above ω_c.

The high-pass characteristic of Figure 13-22b has a cutoff frequency ω_c equal to the break or corner frequency $\omega = 1/\tau$ and passes frequencies above ω_c.

The parallel resonant circuit of Figure 13-18 and the series resonant circuit of Figure 13-21 provide a bandpass characteristic that is an approximate form of the idealized response of Figure 13-22c.

The ideal bandstop filter of Figure 13-22d may be approximated by a circuit using two energy storage elements as shown in Figure 13-23a. This filter uses only passive elements and is called a passive filter.

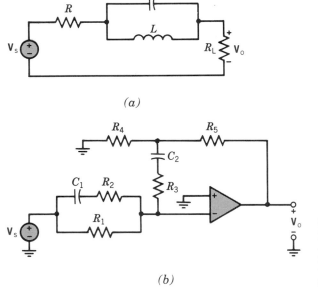

(a)

(b)

Figure 13-23 (a) A bandstop filter, also called a notch filter. $\mathbf{H} = \mathbf{V}_o/\mathbf{V}_s$. (b) Operational amplifier circuit that provides a gain $\mathbf{H} = \mathbf{V}_o/\mathbf{V}_s$ with a bandstop characteristic.

When a frequency-selective filter includes amplifiers as well as passive circuit elements, it is called an active filter.

The circuit of Figure 13-22a may require the use of large, heavy and expensive inductors. Often a bandstop filter is implemented using resistors, capacitors and op amps. Figure 13-23b shows an example of such a band-stop filter.

Example 13-4

Design a bandstop filter of the form of Figure 13-23a to reject a 60-Hz sinusoid while passing other frequencies. The bandwidth of the rejection desired is 30 Hz and $R = R_L = 600\ \Omega$.

Solution

First, we find the ratio $\mathbf{H} = \mathbf{V}_o/\mathbf{V}_s$. Using $\mathbf{Z}$ as the impedance of the parallel capacitor and inductor, we have

$$\mathbf{H} = \frac{R_L}{R + \mathbf{Z} + R_L} \tag{13-56}$$

and

$$\mathbf{Z} = \frac{j\omega L(1/j\omega C)}{j\omega L + 1/j\omega C} = \frac{Ls(1/Cs)}{Ls + 1/Cs} = \frac{Ls}{LCs^2 + 1} \tag{13-57}$$

where $s = j\omega$ for notational ease. Substituting Eq. 13-57 into Eq. 13-56, we have

$$\mathbf{H} = \frac{R_L(s^2 + \omega_0^2)}{(R + R_L)(s^2 + \omega_b s + \omega_0^2)}$$

where

$$\omega_b = \frac{1}{(R + R_L)C}$$

and ω_0 is the resonant frequency so that $\omega_0^2 = 1/LC$. Divide the top and bottom by s^2 to obtain

$$\mathbf{H} = \frac{R_L[1 + (\omega_0/s)^2]}{(R + R_L)[1 + \omega_b/s + (\omega_0/s)^2]}$$

Now let $s = j\omega$ to obtain

$$\mathbf{H} = \frac{R_L[1 - (\omega_0/\omega)^2]}{(R + R_L)[1 - j(\omega_b/\omega) - (\omega_0/\omega)^2]} \tag{13-58}$$

At the rejection frequency $\omega = \omega_0$, the gain magnitude is $H = 0$. Thus, the 60-Hz sinusoid is completely rejected if we set $\omega_0 = 2\pi(60) = 377$.

When $\omega \ll \omega_0$, we have

$$H \cong \frac{R_L}{R + R_L}$$

Similarly, when $\omega \gg \omega_0$, we have

$$H \cong \frac{R_L}{R + R_L}$$

A plot of H/H_0 where $H_0 = R_L/(R + R_L)$ is shown in Figure 13-24. With a fair amount of algebraic manipulation, you can show that the bandwidth $BW = \omega_2 - \omega_1$ is equal to ω_b.

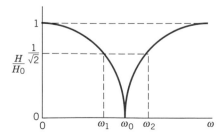

Figure 13-24 Magnitude characteristic for the band rejection circuit of Figure 13-23a.

Then, since it is required that $BW = \omega_b = 2\pi(30)$, we have

$$\omega_b = \frac{1}{(R + R_L)C}$$

or

$$188.5 = \frac{1}{(1200)C}$$

Therefore, $C = 4.4\ \mu F$. Since $\omega_0 = 377$ and $\omega_0^2 = 1/LC$, we find that

$$L = \frac{1}{\omega_0^2 C}$$

$$= 1.6\ H$$

EXERCISE 13-13

Design a bandstop filter using the circuit of Figure 13-23a to reject a 400-Hz sinusoid while passing other frequencies. The bandwidth of the rejection desired is 80 Hz and $R = 100\ \Omega$ and $R_L = 400\ \Omega$.
Answer: $C = 4\ \mu F$, $L = 40\ mH$

13-9 ‖ POLE–ZERO PLOTS AND THE s-PLANE

We may readily represent a network function using the operator s so that $s = d/dt$ and $1/s = \int dt$. In practice this means we replace $j\omega$ by s. Instead of writing $j\omega L$ we use sL and for $1/j\omega C$ use $1/sC$. Then, we may obtain $H(s) = V_2(s)/V_1(s)$ for a filter circuit as described in Section 13-8. In general, we may write

$$H(s) = \frac{N(s)}{D(s)}$$

$$= \frac{K(s - z_1)(s - z_2)\cdots(s - z_m)}{(s - p_1)(s - p_2)\cdots(s - p_n)}$$

where the roots of the numerator polynomial, $z_1, z_2, \ldots, z_m$ are called the zeros of $H(s)$ and the roots of the denominator polynomial, $p_1, p_2, \ldots, p_n$ are called the poles of $H(s)$. If we assume that the excitation is a complex exponential, we have $s = \sigma + j\omega$ and the frequency response is obtained when $s = j\omega$, a special case. All transfer functions $H(s)$ can be written as a ratio of two polynomials in s with real coefficients.

A **zero** is a particular value of s for which the magnitude of $H(s)$ goes to zero.

A **pole** is a particular value of s for which the magnitude of $H(s)$ goes to infinity.

Complex frequency is defined as $s = \sigma + j\omega$ where $\sigma = \mathrm{Re}(s)$ and $\omega = \mathrm{Im}(s)$. The domain of all possible values of s is called the complex-frequency plane or s-plane.

Consider a specific gain ratio $H(s)$ as follows:

$$H(s) = \frac{s - z_1}{(s - p_1)(s - p_2)}$$

$$= \frac{s + 2}{(s + 1 + j)(s + 1 - j)}$$

The poles and the zero are shown in the s-plane in Figure 13-25 using an $\times$ to designate a pole and a circle to designate a zero.

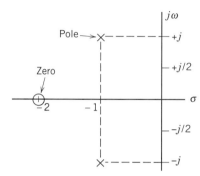

Figure 13-25 Two poles and a zero on the s-plane.

The shape of a circuit's frequency response is critically influenced by the poles and zeros of the gain ratio $H(s)$. The poles are the roots of the characteristic equation $D(s) = 0$. A pole is a particular value of s for which $H(s)$ approaches infinity. The poles of $H(s)$ are $p_1, p_2, \ldots, p_n$.

Let us consider the series resonant circuit of Figure 13-21 and obtain the gain ratio $H(s)$. The differential equation from Kirchhoff's voltage law is

$$\frac{1}{C} \int i \, dt + L \frac{di}{dt} + Ri = v_s(t)$$

Using the operator s, we have

$$\left(\frac{1}{Cs} + Ls + R \right) i = v_s$$

and the output is $v_o = Ri$. Therefore,

$$\frac{V_o}{V_s} = H(s) = \frac{R}{(1/Cs) + Ls + R}$$

$$= \frac{RCs}{LCs^2 + RCs + 1}$$

$$= \frac{(R/L)s}{s^2 + (R/L)s + 1/LC}$$

$$= \frac{(\omega_o/Q)s}{s^2 + (\omega_o/Q)s + \omega_o^2}$$

where Q = quality factor and ω_o = resonant frequency. As an example, consider the series resonant circuit with $\omega_o^2 = 401$, $\omega_0 \approx 20$, and $Q = 10$. Then, we have

$$H(s) = \frac{2s}{s^2 + 2s + 401}$$

$$= \frac{2s}{(s + 1 - j20)(s + 1 + j20)}$$

As shown in Figure 13-26, the zero is at the origin and the two complex poles are at $s = -1 \pm j20$. $H(s)$ may be written as

$$H(s) = \frac{2(s - z_1)}{(s - p_1)(s - p_1^*)}$$

where z_1 is the zero and p_1 and p_1^* are the poles. In this case, $z_1 = 0$ and $p_1 = -1 + j20$. Then, at $s = j\omega$ we obtain

$$H(j\omega) = \frac{2(j\omega - z_1)}{(j\omega - p_1)(j\omega - p_1^*)}$$

The phasor from p_1 to $j\omega$ is shown in color in Figure 13-26 with a magnitude M_1 at an angle

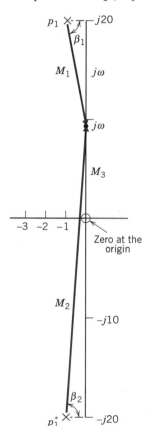

Figure 13-26 The poles and zero for $H(s)$ and the phasors at $j\omega$.

β_1. $H(j\omega)$ can be rewritten as

$$H(j\omega) = \frac{M_3 \underline{/90°}}{M_1 \underline{/\beta_1}\, M_2 \underline{/\beta_2}}$$

$$= \frac{M_3}{M_1 M_2} \underline{/90° - \beta_1 - \beta_2}$$

where $M_3 = 2\omega$. Note that at $j\omega = j20$, M_1 attains a minimum value and the overall magnitude $|H|$ attains a maximum value at or near resonance ($\omega_o = \sqrt{401} = 20.025$).

Thus, for any point $j\omega$ we simply draw phasors from all the poles and zeros to $j\omega$, measure their lengths and angles, and calculate the amplitude and phase. Quite often, we need only a few points to sketch the frequency response using the above equation.

EXERCISE 13-14

The series resonance circuit of Figure 13-21 has $\omega_0^2 = 6404$ and $Q = 20$. Determine $H(s)$ and show the poles and zeros on the s-plane.

Answer: $\quad H(s) = \dfrac{4s}{s^2 + 4s + 6404} = \dfrac{4s}{(s + 2 + j80)(s + 2 - j80)}$

13-10 FREQUENCY RESPONSE OF A MULTIFACTOR H($j\omega$)

A gain ratio $\mathbf{H}(j\omega)$ can be quite intricate and can include many terms. The primary advantage of the logarithmic plot is the conversion of multiplicative factors such as $(j\omega/\omega_0 + 1)$ into additive factors $20\log(j\omega/\omega_0 + 1)$ by virtue of the definition of logarithmic gain. This can readily be ascertained by considering a generalized gain function as

$$H(s) = \frac{K_b \, \Pi_{i=1}^{D} (1 + s/\omega_i)}{s^N \Pi_{m=1}^{M} (1 + s/\omega_m) \Pi_{k=1}^{R} [1 + (2\zeta_k s/\omega_{0k}) + (s/\omega_{0k})^2]} \qquad (13\text{-}59)$$

where the notation $\Pi_{i=1}^{D}$ means the product of D terms.

The roots of the numerator polynomial are called the *zeros* since the magnitude of the numerator will equal zero when $s = -\omega_1$. The roots of the denominator polynomial equation are called the *poles* since the magnitude of the function, $\mathbf{H}(s)$, will become infinite when $s = -\omega_m$ for the term $(1 + s/\omega_m)$.

If we let $s = j\omega$ in order to obtain the frequency response, Eq. 13-59 becomes

$$\mathbf{H}(j\omega) = \frac{K_b \, \Pi_{i=1}^{D} (1 + j\omega/\omega_i)}{(j\omega)^N \Pi_{m=1}^{M} (1 + j\omega/\omega_m) \Pi_{k=1}^{R} [1 + (2\zeta_k/\omega_{0k})j\omega + (j\omega/\omega_{0k})^2]}$$

This gain function includes D numerator factors, N factors of the form $(j\omega)$ in the denominator, M denominator factors $(1 + j\omega/\omega_m)$, and R second-order denominator factors. Obtaining the plot of such a function would be a formidable task indeed. However, the logarithmic magnitude of $\mathbf{H}(j\omega)$ is

$$20\log|\mathbf{H}(\omega)| = 20\log K_b + 20\sum_{i=1}^{D}\log|1 + j\omega/\omega_i|$$

$$- 20\log|(j\omega)^N| - 20\sum_{m=1}^{M}\log|1 + j\omega/\omega_m|$$

$$- 20\sum_{k=1}^{R}\log\left|1 + \left(\frac{2\zeta_k}{\omega_{0k}}\right)j\omega + \left(\frac{j\omega}{\omega_{0k}}\right)^2\right| \qquad (13\text{-}60)$$

and the Bode diagram can be obtained by adding the plot due to each individual factor.

Furthermore, the separate phase angle plot is obtained as

$$\phi(\omega) = + \sum_{i=1}^{D} \tan^{-1}(\omega/\omega_i) - N(90°) - \sum_{m=1}^{M} \tan^{-1}(\omega/\omega_m)$$
$$- \sum_{k=1}^{R} \tan^{-1}\left(\frac{2\zeta_k \omega_{0k}\, \omega}{\omega_{0k}^2 - \omega^2}\right) \qquad (13\text{-}61)$$

which is simply the summation of the phase angles due to individual factors of the gain function.

Therefore, the four different kinds of factors that may occur in a generalized gain function are

1 Constant gain K_b;
2 Poles (or zeros) at the origin of the s-plane;
3 Poles or zeros of the first-order factor, $(s/\omega_i + 1)$;
4 Complex conjugate poles (or zeros) of the quadratic factor,
      $[1 + (2\zeta/\omega_0)j + (j\omega/\omega_0)^2]$.

We can determine the logarithmic magnitude plot and phase angle for these four factors and then utilize them to obtain a Bode diagram for any general form of a gain function. Typically, the curves for each factor are obtained and then added together graphically to obtain the curves for the complete gain function. Furthermore, this procedure can be simplified by using the asymptotic approximation to these curves and obtaining the actual curves only at specific important frequencies.

Constant gain K_b. The logarithmic gain is

$$20 \log K_b = \text{constant in dB}$$

and the phase angle is zero. The gain curve is simply a horizontal line on the Bode diagram as shown in Figure 13-27.

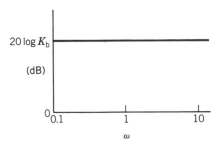

Figure 13-27 The Bode diagram for a constant gain K_b for two decades of frequency.

Poles or zeros at the origin. A term of the form $(j\omega)^{-1}$ has a logarithmic magnitude for a pole as

$$20 \log \left| \frac{1}{j\omega} \right| = -20 \log \omega \text{ dB}$$

and a phase angle $\phi(\omega) = -90°$. The slope of the magnitude curve is -20 dB/decade for a pole. Similarly, for a multiple term of the form $(j\omega)^{-N}$ (corresponding to N number of poles at the origin), we have

$$20 \log \left| \frac{1}{(j\omega)^N} \right| = -20 N \log \omega$$

and the phase is $\phi(\omega) = -90N°$. In this case, the slope due to the multiple pole is $-20N$

dB/decade. For a term of the form $(j\omega)$, we have a logarithmic magnitude for a zero at the origin as

$$20 \log | j\omega | = +20 \log \omega$$

where the slope is $+20$ dB/decade and the phase angle is $+90°$. The Bode diagram of the magnitude and phase angle of $(j\omega)^{\pm N}$ is shown in Figure 13-28 for $N = 1$ and $N = 2$.

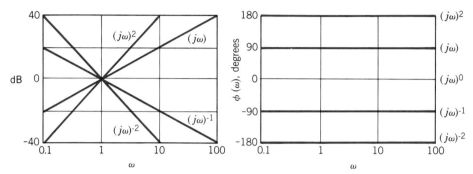

Figure 13-28 Bode diagram for $(j\omega)^{\pm N}$ for three decades of frequency.

Real and nonzero poles or zeros. The factor $(1 + j\omega/\omega_1)^{-1}$ was considered previously and we found that

$$20 \log \left| \frac{1}{1 + j\omega/\omega_1} \right| = -10 \log(1 + \omega^2/\omega_1^2)$$

The asymptotic curve for $\omega \ll \omega_1$ is $20 \log 1 = 0$ dB; and the asymptotic curve for $\omega \gg \omega_1$ is $-20 \log \omega/\omega_1$, which has a slope of -20 dB/decade. The intersection of the two asymptotes occurs when

$$20 \log 1 = 0 \text{ dB} = -20 \log \omega/\omega_1$$

or when $\omega = \omega_1$, which is the *break frequency.* The actual logarithmic gain when $\omega = \omega_1$ is -3 dB for this factor. The phase angle is $\phi(\omega) = -\tan^{-1} \omega/\omega_1$ for the denominator factor. The Bode diagram of a factor $(1 + j\omega/\omega_1)^{-1}$ was shown previously in Figure 13-11.

The Bode diagram of a zero factor $(1 + j\omega/\omega_2)$ is obtained in the same manner as that of the pole. However, the slope is positive at $+20$ dB/decade for $\omega > \omega_2$ and the phase angle is $\phi(\omega) = +\tan^{-1} \omega/\omega_2$.

Complex conjugate poles or zeros $[1 + (2\zeta\omega/\omega_0)j + (j\omega/\omega_0)^2]$. The quadratic factor for a pair of complex conjugate poles can be written in normalized form as

$$(1 + j2\zeta u - u^2)^{-1} \tag{13-62}$$

where $u = \omega/\omega_0$ and ζ is the dimensionless damping ratio.

This equation may also be written in the form

$$\left[1 + j \left(\frac{u}{Q} \right) - u^2 \right]^{-1}$$

where $Q = 1/2\zeta$ and Q is the familiar quality factor discussed in Section 13-6. Consider a gain ratio, $H(s)$, where

$$\mathbf{H}(s) = \frac{\omega_0^2}{s^2 + as + \omega_0^2} = \frac{\omega_0^2}{s^2 + 2\zeta\omega_0 s + \omega_0^2}$$

The roots of the characteristic equation of the denominator are

$$s_{1,2} = -\frac{a}{2} \pm \sqrt{\left(\frac{a}{2}\right)^2 - \omega_0^2}$$

The roots are equal and real when the square root term is zero or when $a = 2\omega_0$. This condition is obtained when $\zeta = 1$, which is called critical damping. In general, we note that the damping ratio ζ is

$$\zeta = \frac{a}{2\omega_0}$$

It can then be shown that $\zeta = 1/2Q$ for a resonant circuit.

Then the logarithmic magnitude is

$$20 \log |\mathbf{H}(\omega)| = -10 \log[(1 - u^2)^2 + 4\zeta^2 u^2]$$

and the phase angle is

$$\phi(\omega) = -\tan^{-1}\left(\frac{2\zeta u}{1 - u^2}\right)$$

When $u \ll 1$, the magnitude is

$$20 \log H = -10 \log 1 = 0 \text{ dB}$$

and the phase angle approaches $0°$. When $u \gg 1$, the logarithmic magnitude approaches

$$20 \log H = -10 \log u^4 = -40 \log u$$

which results in a curve with a slope of -40 dB/decade. The phase angle, when $u \gg 1$, approaches $-180°$. The magnitude asymptotes meet at the 0-dB line when $u = \omega/\omega_0 = 1$. However, the difference between the actual magnitude curve and the asymptotic approximation is a function of the damping ratio (or the quality factor Q). The Bode diagram of a quadratic factor due to a pair of complex conjugate poles is shown in Figure 13-29. The maximum value of the frequency response, $M_{p\omega}$, occurs at the *peak frequency* ω_p. When the damping ratio approaches zero, ω_p approaches ω_0, the resonant frequency. The peak frequency is determined by taking the derivative of the magnitude of Eq. 13-62 with respect to the normalized frequency u and setting it equal to zero. The peak frequency is represented by the relation

$$\omega_p = \omega_0 \sqrt{1 - 2\zeta^2} \qquad \zeta < 0.707 \qquad (13\text{-}63)$$

and the maximum value of the magnitude $|H(\omega)|$ is

$$M_{p\omega} = |H(\omega_p)| = (2\zeta\sqrt{1 - \zeta^2})^{-1} \qquad \zeta < 0.707 \qquad (13\text{-}64)$$

for a pair of complex poles. The maximum value of the frequency response, $M_{p\omega}$, and the peak frequency ω_p are shown as a function of the damping ratio ζ for a pair of complex poles in Figure 13-30. When $\zeta < 0.1$ or $Q > 5$, we may assume that $\omega_p = \omega_0$ and the peak occurs at the resonant frequency. Note that when $\omega = \omega_0$, $H = 1/2\zeta$ and

$$20 \log H = -20 \log 2\zeta \text{ dB}$$

Thus, $20 \log H = 0$ dB for $\zeta = 0.5$ and $+6$ dB for $\zeta = 0.25$.

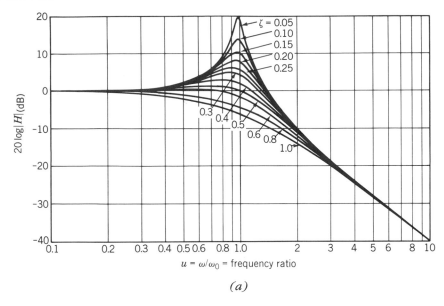

(a)

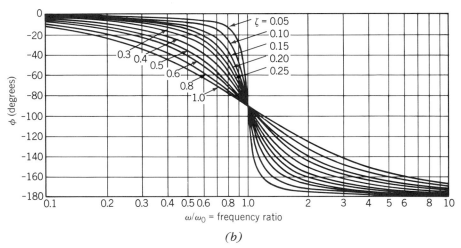

(b)

Figure 13-29 Bode diagram of $\mathbf{H}(j\omega) = [1 + (2\zeta/\omega_0)j\omega + (j\omega/\omega_0)^2]^{-1}$ for two decades of frequency.

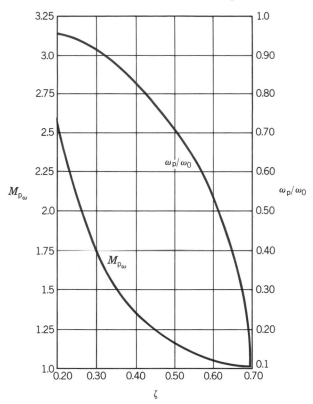

Figure 13-30 Maximum of the frequency response $M_{p\omega}$ and the peak frequency ω_p versus ζ for a pair of complex conjugate poles.

Example 13-5

Plot the Bode diagram of

$$\mathbf{H}(s) = \frac{5(1 + 0.1s)}{s(1 + 0.5s)[1 + 0.6(s/50) + (s/50)^2]}$$

Solution

The factors, in order of their occurrence as frequency ω increases, are

1 A constant gain $K = 5$;
2 A pole of the form, s, at the origin;
3 A pole at $s = -2$, from $(1 + 0.5s)$;
4 A zero at $s = -10$, from $(1 + 0.1s)$;
5 A pair of complex poles with $\omega_0 = 50$, from $[1 + 0.6(s/50) + (s/50)^2]$.

First, we plot the decibel magnitude curve for each factor.

1 The constant gain is $20 \log 5 = 14$ dB and is shown as curve 1 in Figure 13-31.
2 The decibel magnitude of the term of the form $(j\omega)$ is $20 \log(1/\omega) = -20 \log \omega$. The curve for this factor has a slope of -20 dB per decade and intersects the 0-dB line when $\omega = 1$. This factor is shown as curve 2 in Figure 13-31.
3 The asymptotic approximation of the decibel magnitude of the pole at $s = -2$ has a slope of -20 dB/decade beyond the break frequency at $\omega = 2$. The asymptotic magnitude below the break frequency is 0 dB, as shown for factor 3 in Figure 13-31 (see curve 3).

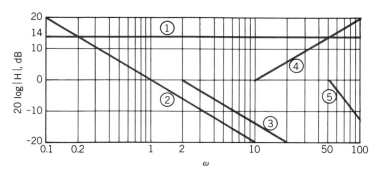

Figure 13-31 Magnitude asymptotes of poles and zeros for Example 13-5.

4 The asymptotic magnitude for the zero at $s = -10$ has a slope of $+20$ dB/decade beyond the break frequency at $\omega = 10$. Below the break frequency, the decibel magnitude curve is 0 dB, as shown for factor 4 in Figure 13-31.

5 The asymptotic approximation for the fifth factor has a slope of -40 dB/decade (see Figure 13-29) for $\omega > \omega_0$ and the break frequency is $\omega = \omega_0$. When $\omega < \omega_0$, the asymptotic approximation is 0 dB. This is shown as factor 5 in Figure 13-31.

The total asymptotic approximation can be obtained by adding the asymptotes due to each factor, thus yielding the solid line shown in Figure 13-32a. The exact curve is

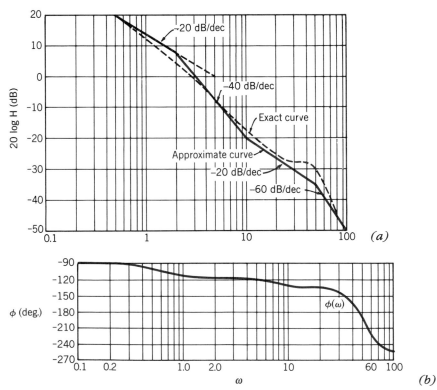

Figure 13-32 (a) Asymptotic curve and exact decibel curve and (b) phase characteristic for $\mathbf{H}(j\omega)$ of Example 13-5.

obtained by calculating the exact magnitude in decibels at selected frequencies. The approximation for factor 5 will differ from the exact at $\omega = \omega_0 = 50$ since $\zeta = 0.3$ and the actual curve will exhibit a peak (see Figure 13-29).

For Example 13-5 we used the *additive* property of 20 log |H| and plotted the magnitude asymptotes of the poles and zeros as illustrated in Figure 13-31. We then graphically added all the factors to obtain the asymptotic curve of Figure 13-32. An alternative method is to develop the total magnitude plot by starting at low frequencies and sequentially adding the effect of each pole and zero as ω increases. We call this the *sequential* method. Let us illustrate this method with $H(s)$ of Example 13-5.

Example 13-6

Plot the Bode diagram of

$$\mathbf{H}(s) = \frac{5(1 + 0.1s)}{s(1 + 0.5s)[1 + 0.6(s/50) + (s/50)^2]}$$

Solution

Rewriting $\mathbf{H}(s)$ with $s = j\omega$ we obtain

$$\mathbf{H}(j\omega) = \frac{5(1 + 0.1j\omega)}{j\omega(1 + 0.5j\omega)\left[1 + 0.6\left(\dfrac{j\omega}{50}\right) - \left(\dfrac{\omega}{50}\right)^2\right]}$$

Notice that when $\omega < 0.1$, then

$$\mathbf{H}(j\omega) \simeq \frac{5(1)}{j\omega(1)[1]}$$

since all the terms in the brackets and parentheses are essentially equal to 1 because the terms involving ω have become negligibly small.

Starting at very low frequencies we have 20 log |H| $= 20\log\left(\dfrac{5}{\omega}\right)$, which has a slope of -20 dB/decade and 20 log |H| $= 20$ dB at $\omega = 0.5$ rad/s. Accounting for the gain 5 and the low-frequency pole, we plot this factor as shown in Figure 13-33. Then the next factor

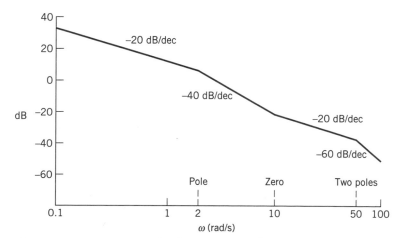

Figure 13-33 Asymptotic plot for Example 13-6 using the sequential method.

is the pole at $s = -2$, which introduces an asymptotic gain of -20 dB/decade. Thus, the total asymptotic slope is -40 dB/decade for $2 < \omega < 10$. A zero occurs at $s = -10$, which adds a positive slope of $+20$ dB/decade for $10 < \omega < 50$. The cumulative asymptotic slope for $10 < \omega < 50$ is -20 dB/decade. The complex poles occur with $\omega_o = 50$, adding a factor of -40 dB/decade, resulting in a cumulative slope for $\omega > 50$ of -60 dB/decade as shown in Figure 13-33.

When ω is very low, the phase is $-90°$. The total phase ϕ may be calculated from the total of all phase shifts, which is

$$\phi = \phi_2 + \phi_3 + \phi_4 + \phi_5$$

$$= -90° - \tan^{-1} 0.5\omega + \tan^{-1} 0.1\omega - \tan^{-1} \frac{2\zeta u}{1 - u^2}$$

where ϕ_n is the phase of the nth factor and $u = \omega/\omega_0 = \omega/50$.

Calculating ϕ at several frequencies, we summarize the results in Table 13-8. Using these data and other calculated points, we may obtain the phase curve as shown in Figure 13-32b.

Table 13-8

ω (rad/s)	0.1	1	10	50	100
ϕ (deg)	-91	-112	-125	-189	-250

We have introduced a useful technique for obtaining a Bode plot by adding the effect of each factor. In practice, computer programs such as PSpice are generally available to calculate and plot Bode diagrams.

Table 13-9
The Sequential Method of Drawing the Asymptotic
Bode Magnitude Plot

1. The low-frequency response is obtained when $\omega <$ all the real poles and zeros and any complex poles. Then
   $$20 \log | H(\omega) | = 20 \log(K/\omega^N)$$
   is the low-frequency asymptote of the magnitude. The phase is $\phi = -90°N$.
2. Increase ω and note that the slope of $20 \log | H |$
   (a) decreases by 20 dB/decade as ω increases past a pole p_j,
   (b) increases by 20 dB/decade as ω increases past a zero z_i.

The general rules for the sequential method as summarized in Table 13-9 for the form

$$\mathbf{H}(j\omega) = \frac{K \prod_{i=1}^{D} \left(1 + \dfrac{j\omega}{z_i}\right)}{j\omega^N \prod_{j=1}^{M} \left(1 + \dfrac{j\omega}{p_j}\right)}$$

where $u = \omega/\omega_o$, z_i = zeros, and p_j = poles.

EXERCISE 13-15

Draw the Bode diagram for the $\mathbf{H}(j\omega)$ of Example 13-5 when the constant gain is increased from 5 to 10 and the zero occurs at $\omega = 1$ instead of $\omega = 10$. The other poles and zeros remain as in Example 13-5. Therefore, we have

$$\mathbf{H}(j\omega) = \frac{10(1 + j\omega)}{j\omega(1 + j0.5\omega)(1 + j0.6(\omega/50) + (j\omega/50)^2)}$$

13-11 FREQUENCY RESPONSE OF OP AMP CIRCUITS

The gain of an op amp is not infinite; rather, it is finite and decreases with frequency. The gain A is a function of ω as

$$A(j\omega) = \frac{A_o}{1 + j\omega/\omega_1}$$

where A_o is the dc gain and ω_1 is the break frequency. The dc gain is normally greater than 10^4 and ω_1 is less than 100. A circuit model of a frequency-dependent nonideal op amp is shown in Figure 13-34. The break frequency is $\omega_1 = 1/RC$.

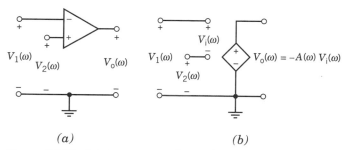

(a)   (b)

Figure 13-34 *(a)* An operational amplifier and *(b)* a frequency-dependent model of an operational amplifier.

Let us consider an example of an op-amp circuit incorporating a frequency-dependent op amp.

Example 13-7

A noninverting amplifier incorporates a frequency-dependent op amp. (a) Determine and plot the Bode diagram for $|A(j\omega)|$ when $A_o = 10^5$ and $\omega_1 = 10$ rad/s. (b) Determine and plot the Bode diagram for the noninverting amplifier of Figure 13-35 when $R_2 = 100$ kΩ and $R_1 = 10$ kΩ using the op amp of part (a).

Solution

The Bode plot of $20 \log |A(\omega)|$ is shown in Figure 13-36. Note that the magnitude is equal to 1 (0 dB) at $\omega = 10^6$ rad/s. The gain bandwidth is equal to 10^6 rad/s. When $R_2 = 100$ kΩ

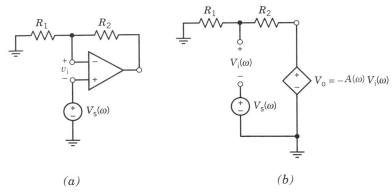

(a)   *(b)*

Figure 13-35 *(a)* A noninverting amplifier and *(b)* an equivalent circuit incorporating the frequency-dependent model of the operational amplifier.

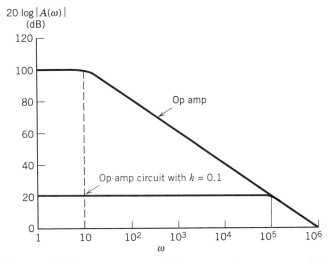

Figure 13-36 Bode magnitude diagram of the op amp and the noninverting op amp circuit (in color).

and $R_1 = 10$ kΩ, the gain ratio for the op amp circuit is

$$\frac{\mathbf{V}_o}{\mathbf{V}_s} = \frac{A(\omega)}{1 + kA(\omega)}$$

where $k = R_1/(R_1 + R_2)$ and A is a function of ω. Then we have

$$\frac{\mathbf{V}_o}{\mathbf{V}_s} = \frac{A_o/(1 + j\omega/\omega_1)}{1 + kA_o/(1 + j\omega/\omega_1)}$$

$$= \frac{A_o}{1 + j\omega/\omega_1 + kA_o}$$

$$= \frac{A_c}{(1 + j\omega/A_2\omega_1)}$$

where A_c is the total circuit gain defined as $A_c = A_o/(1 + kA_o)$ at dc and $A_2 = 1 + kA_o$.

For this example, $k = 1/10$ and $A_0 = 10^5$ and thus we have $A_c = 10$, $A_2 = 10^4$, and $\omega_1 A_2 = 10^5$. Therefore,

$$\frac{\mathbf{V}_o}{\mathbf{V}_s} = \frac{10}{1 + j10^{-5}\omega}$$

This circuit has a magnitude curve as shown in color in Figure 13-36. Note that the noninverting op amp has a low-frequency gain of 20 dB and a break frequency of 10^5 rad/s. The gain-bandwidth product remains 10^6 rad/s.

13-12 | DISTORTION IN FILTER CIRCUITS

Consider a circuit with a voltage gain $\mathbf{H} = \mathbf{V}_o/\mathbf{V}_s$ and an input as follows:

$$v_s = A \cos \omega_1 t + B \cos \omega_2 t$$

Let $K_1 = |H(\omega_1)|$, $K_2 = |H(\omega_2)|$, $\phi_1 = \underline{/\mathbf{H}(\omega_1)}$, and $\phi_2 = \underline{/\mathbf{H}(\omega_2)}$, which represents the magnitude and phase of the gain ratio $\mathbf{H}$. The output of the circuit is

$$v_o = AK_1 \cos(\omega_1 t + \phi_1) + BK_2 \cos(\omega_2 t + \phi_2)$$

If we desire a filter that has an output proportional to the input and a faithful reproduction of the input, we seek $K_1 = K_2 = K$ or $|H(\omega)| = K$. Then the gain is constant over the frequency range of interest. Furthermore, for a faithful reproduction we require $\phi(\omega) = m\pi - \omega T$ where T is a constant and m is an even integer.

A filter provides *amplitude distortion* when $K_1 \neq K_2$ and $|H(\omega)|$ is not constant. Furthermore, a filter provides *phase distortion* when $\phi(\omega) \neq m\pi - \omega T$. For example, in TV applications phase distortion can result in a smeared picture (Dorf, 1993). For the distortionless case (perfect fidelity) we require $|H(\omega)| = K$ and $\phi = m\pi - \omega T$. At a given frequency the output is

$$\begin{aligned} \text{Output} &= KA \cos(\omega t + m\pi - \omega T) \\ &= KA \cos(\omega(t - T) + m\pi) \\ &= KA \cos \omega(t - T) \end{aligned}$$

Then for the two distinct frequencies

$$\begin{aligned} v_0(t) &= K[A \cos \omega_1(t - T) + B \cos \omega_2(t - T)] \\ &= Kv_s(t - T) \end{aligned}$$

Thus, the distortionless filter provides an output with a constant gain K and a time delay T. Therefore, we desire a low-pass filter to have the magnitude characteristic as shown in Figure 13-22a and a phase characteristic of

$$\frac{d\phi(\omega)}{dt} = -T$$

As an example, consider the low-pass filter

$$H(s) = \frac{\omega_0^2}{s^2 + 2\zeta\omega_0 s + \omega_0^2}$$

with $\zeta = 0.5$. The Bode diagram for this gain ratio is shown in Figure 13-29. The gain is close to a constant equal to 1.0 for $\omega < 1.2\omega_0$. The phase change for $0.5 \leq \omega/\omega_0 \leq 2.0$ is approximately linear. The phase for a range of frequencies is recorded in Table 13-10 and shows a relatively linear slope $\Delta\phi \approx -42°/\Delta u$ where $u = \omega/\omega_0$.

Table 13-10

ϕ, phase (deg)	$-35°$	$-45°$	$-90°$	$-110°$	$-140°$
Frequency, ω/ω_0	0.5	0.6	1.0	1.2	2.0

13-13 ∥ FREQUENCY RESPONSE USING PSpice

The frequency response of a circuit can be readily obtained using the .AC statement. The .AC statement is written as

```
.AC < sweep type > < n > < start freq > < end freq >
```

The sweep type may be linear (LIN), octave (OCT), or decade (DEC). We will first use the LIN sweep in order to obtain a plot of the circuit response of the circuit shown in Figure 13-37. Using the approach of Section 13-6, which determines the resonant frequency

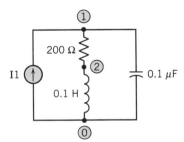

Figure 13-37 An *RLC* resonant circuit.

when the imaginary part of **H** is zero, we find that

$$\omega_0^2 = \left[\frac{1}{LC} - \left(\frac{R}{L}\right)^2 \right]$$

For the values of Figure 13-37, we obtain $\omega_0 = 9798$ rad/s or $f_0 = 1559.4$ Hz. Then, we will seek to plot the response between 100 Hz and 10 kHz. We use the PSpice program shown in Figure 13-38 to plot the magnitude of the voltage across the capacitor. The resulting frequency response is shown in Figure 13-39, where the plot is obtained for two decades.

```
RESONANT CIRCUIT
I1 0 1 AC 0.1
R 1 2 200
L 2 0 0.1
C 1 0 0.1U
.AC LIN 100   1E2 1E4
.PLOT AC VM(1,0)
.PROBE
.END
```

Figure 13-38 The PSpice program used to obtain a plot of the circuit response with I1 = 0.1 A.

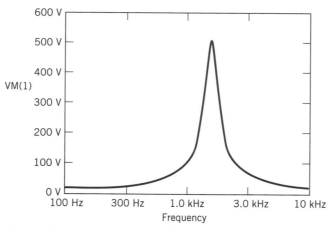

Figure 13-39 The magnitude of the capacitor voltage for two decades of frequency (frequency is a logarithmic scale).

We can alter the PSpice program by requesting a plot of decibel and phase by calling for VDB(1,0) and VP(1,0). The resulting plots are shown in Figure 13-40*a* and Figure 13-40*b*.

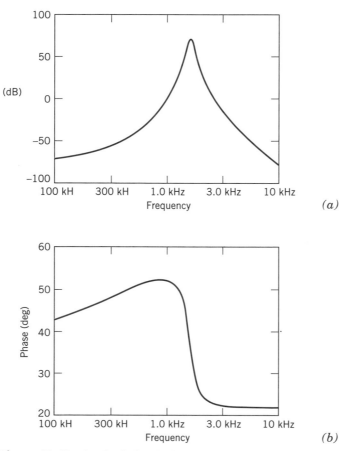

(a)

(b)

Figure 13-40 The decibel and phase versus frequency of the *RLC* circuit yielding the Bode diagram for the capacitor voltage.

If we wish to identify the resonant frequency f_0 and the bandwidth BW, we need to rewrite the PSpice program to print VDB(1,0) over a narrow frequency range so that we can accurately determine the values. If we choose a linear frequency sweep of 1.3 kHz to 1.9 kHz, we obtain a printed output as shown in Figure 13-41. By inspecting the response

```
    FREQ        VDB(1,0)

 (*)----------    4.0000E+01    4.5000E+01    5.0000E+01    5.5000E+01    6.0000E+01

                 - - - - - - - - - - - - - - - - - - - - - - - - - - - - - - - - - -
   1.300E+03    4.711E+01.                    .        *          .             .             .
   1.315E+03    4.753E+01 .                    .         *         .             .             .
   1.331E+03    4.795E+01 .                    .          *        .             .             .
   1.346E+03    4.838E+01 .                    .           *       .             .             .
   1.362E+03    4.882E+01 .                    .            *       .             .             .
   1.377E+03    4.927E+01 .                    .             * .     .             .             .
   1.392E+03    4.973E+01 .                    .              *.     .             .             .
   1.408E+03    5.019E+01 .                    .               *     .             .             .
   1.423E+03    5.066E+01 .                    .               . *    .             .             .
   1.438E+03    5.112E+01 .                    .               .  *   .             .             .
   1.454E+03    5.158E+01 .                    .               .   *  .             .             .
   1.469E+03    5.203E+01 .                    .               .    * .             .             .
   1.485E+03    5.246E+01 .                    .               .     *.             .             .
   1.500E+03    5.286E+01 .                    .               .      *             .             .
   1.515E+03    5.323E+01 .                    .               .      .*            .             .
   1.531E+03    5.355E+01 .                    .               .      . *           .             .
   1.546E+03    5.381E+01 .                    .               .      .  *          .             .
   1.562E+03    5.400E+01 .                    .               .      .  *          .             .
   1.577E+03    5.421E+01 .                    .               .      .   *         .             .
   1.592E+03    5.415E+01 .                    .               .      .  *          .             .
   1.608E+03    5.410E+01 .                    .               .      .  *          .             .
   1.623E+03    5.398E+01 .                    .               .      . *           .             .
   1.638E+03    5.379E+01 .                    .               .      .*            .             .
   1.654E+03    5.354E+01 .                    .               .     *.             .             .
   1.669E+03    5.325E+01 .                    .               .    * .             .             .
   1.685E+03    5.292E+01 .                    .               .   *  .             .             .
   1.700E+03    5.256E+01 .                    .               .  *   .             .             .
   1.715E+03    5.219E+01 .                    .               . *    .             .             .
   1.731E+03    5.181E+01 .                    .               .*     .             .             .
   1.746E+03    5.142E+01 .                    .              *.      .             .             .
   1.762E+03    5.104E+01 .                    .             * .      .             .             .
   1.777E+03    5.066E+01 .                    .            *       .             .             .
   1.792E+03    5.028E+01 .                    .          .*        .             .             .
   1.808E+03    4.991E+01 .                    .          *         .             .             .
   1.823E+03    4.955E+01 .                    .        *.          .             .             .
   1.838E+03    4.920E+01 .                    .       *           .             .             .
   1.854E+03    4.886E+01 .                    .      *            .             .             .
   1.869E+03    4.853E+01 .                    .     *             .             .             .
   1.885E+03    4.821E+01 .                    .    *              .             .             .
   1.900E+03    4.790E+01 .                    .    *              .             .             .
                 - - - - - - - - - - - - - - - - - - - - - - - - - - - - - - - - - -
```

Figure 13-41 The response of the RLC circuit for the frequency range 1.3 kHz to 1.9 kHz.

of the RLC circuit, we find that VDB(1,0) has a maximum of 54.21 dB at $f_0 = 1{,}577$ Hz. The two frequencies where the response is down 3 dB are 1440 Hz and 1754 Hz. Thus, the bandwidth is

$$BW = 1754 - 1440 = 314 \text{ Hz}$$

Now, let us consider a two-stage op amp circuit as shown in Figure 13-42. We assume an ideal op amp with $R_i = \infty$, $R_o = 0$, and $A = 10^6$. The PSpice program is shown in Figure

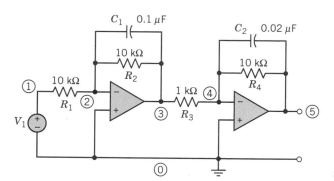

Figure 13-42 A two-stage op amp circuit.

13-43. The two op amps are represented by E1 and E2, and the goal is to determine the frequency response and step response of the circuit. We use V1 as an ac source with a magnitude of 1 for the AC calculation statement. In addition, we specify V1 as a step input using a PULSE statement that starts at $t = 0$ and jumps to 5 V after a short TSTEP as required for determining the transient response using the TRAN statement. The .TRAN statement calls for 40 points of the response to be calculated over a total period of 8 ms.

```
OP AMP CIRCUIT
V1 1 0 AC 1 PULSE(0 5)
R1 1 2 10K
R2 2 3 10K
C1 2 3 0.1U
R3 3 4 1K
R4 4 5 10K
C2 4 5 .02U
E1 3 0 0 2 1MEG
E2 5 0 0 4 1MEG
.TRAN 0.2M 8M
.AC DEC 40 15.915 1591.55
.PLOT TRAN V(5)
.PLOT AC V(5)
.PROBE
.END
```

Figure 13-43 PSpice program for the two-stage op amp circuit.

The .AC statement calls for 40 points to be plotted for the two decades from 15.92 Hz to 1592 Hz. The transient response to a step input is shown in Figure 13-44. The frequency response data for the two decades specified are listed in Figure 13-45. The Bode diagram provided by the PROBE statement is shown in Figure 13-46 for three decades. This three-decade chart was obtained by the PROBE format, which sets up the required decades to accommodate the plot from 15.9 Hz to 1591 Hz.

```
 ****        TRANSIENT ANALYSIS

 ***********************************************************************

   TIME         V(5)

 (*)----------    0.0000E+00   2.0000E+01   4.0000E+01   6.0000E+01   8.0000E+01

                ----------------------------------------------------------------
   0.000E+00   0.000E+00  *           .            .            .            .
   2.000E-04   1.271E+00  .*          .            .            .            .
   4.000E-04   6.534E+00  .      *    .            .            .            .
   6.000E-04   1.301E+01  .          *.            .            .            .
   8.000E-04   1.921E+01  .            *.          .            .            .
   1.000E-03   2.462E+01  .            .  *        .            .            .
   1.200E-03   2.920E+01  .            .      *    .            .            .
   1.400E-03   3.299E+01  .            .        *  .            .            .
   1.600E-03   3.604E+01  .            .          *.            .            .
   1.800E-03   3.857E+01  .            .           .*           .            .
   2.000E-03   4.066E+01  .            .           .  *         .            .
   2.200E-03   4.237E+01  .            .           .     *      .            .
   2.400E-03   4.374E+01  .            .           .     *      .            .
   2.600E-03   4.488E+01  .            .           .      *     .            .
   2.800E-03   4.581E+01  .            .           .       *    .            .
   3.000E-03   4.658E+01  .            .           .       *    .            .
   3.200E-03   4.719E+01  .            .           .        *   .            .
   3.400E-03   4.7770+01  .            .           .        *   .            .
   3.600E-03   4.812E+01  .            .           .        *   .            .
   3.800E-03   4.847E+01  .            .           .         *  .            .
   4.000E-03   4.874E+01  .            .           .         *  .            .
   4.200E-03   4.897E+01  .            .           .         *  .            .
   4.400E-03   4.916E+01  .            .           .         *  .            .
   4.600E-03   4.931E+01  .            .           .         *  .            .
   4.800E-03   4.944E+01  .            .           .         *  .            .
   5.000E-03   4.954E+01  .            .           .         *  .            .
   5.200E-03   4.962E+01  .            .           .         *  .            .
   5.400E-03   4.969E+01  .            .           .         *  .            .
   5.600E-03   4.975E+01  .            .           .         *  .            .
   5.800E-03   4.979E+01  .            .           .         *  .            .
   6.000E-03   4.983E+01  .            .           .         *  .            .
   6.200E-03   4.986E+01  .            .           .         *  .            .
   6.400E-03   4.989E+01  .            .           .         *  .            .
   6.600E-03   4.991E+01  .            .           .         *  .            .
   6.800E-03   4.992E+01  .            .           .         *  .            .
   7.000E-03   4.994E+01  .            .           .         *  .            .
   7.200E-03   4.995E+01  .            .           .         *  .            .
   7.400E-03   4.996E+01  .            .           .         *  .            .
   7.600E-03   4.997E+01  .            .           .         *  .            .
   7.800E-03   4.997E+01  .            .           .         *  .            .
   8.000E-03   4.998E+01  .            .           .        *   .            .
                ----------------------------------------------------------------
```

Figure 13-44 The response to a step input of 5 V for the two-stage op amp circuit.

```
*FREQUENCY RESPONSE

****        AC ANALYSIS

************************

FREQ          V(5)

1.592E+01   9.948E+00  .        1.686E+02   6.716E+00  .
1.686E+01   9.942E+00  .        1.786E+02   6.492E+00  .
1.786E+01   9.935E+00  .        1.892E+02   6.264E+00  .
1.892E+01   9.927E+00  .        2.004E+02   6.032E+00  .
2.004E+01   9.918E+00  .        2.122E+02   5.797E+00  .
2.122E+01   9.909E+00  .        2.248E+02   5.561E+00  .
2.248E+01   9.898E+00  .        2.381E+02   5.323E+00  .
2.381E+01   9.885E+00  .        2.522E+02   5.087E+00  .
2.522E+01   9.872E+00  .        2.672E+02   4.851E+00  .
2.672E+01   9.856E+00  .        2.830E+02   4.618E+00  .
2.830E+01   9.839E+00  .        2.998E+02   4.388E+00  .
2.998E+01   9.820E+00  .        3.175E+02   4.162E+00  .
3.175E+01   9.799E+00  .        3.364E+02   3.940E+00  .
3.364E+01   9.775E+00  .        3.563E+02   3.722E+00  .
3.563E+01   9.749E+00  .        3.774E+02   3.511E+00  .
3.774E+01   9.719E+00  .        3.998E+02   3.305E+00  .
3.998E+01   9.686E+00  .        4.235E+02   3.106E+00  .
4.235E+01   9.650E+00  .        4.485E+02   2.913E+00  .
4.485E+01   9.610E+00  .        4.751E+02   2.727E+00  .
4.751E+01   9.565E+00  .        5.033E+02   2.548E+00  .
5.033E+01   9.516E+00  .        5.331E+02   2.377E+00  .
5.331E+01   9.461E+00  .        5.647E+02   2.212E+00  .
5.647E+01   9.401E+00  .        5.981E+02   2.055E+00  .
5.981E+01   9.334E+00  .        6.336E+02   1.906E+00  .
6.336E+01   9.261E+00  .        6.711E+02   1.764E+00  .
6.711E+01   9.182E+00  .        7.109E+02   1.629E+00  .
7.109E+01   9.094E+00  .        7.530E+02   1.502E+00  .
7.530E+01   8.999E+00  .        7.976E+02   1.382E+00  .
7.976E+01   8.895E+00  .        8.449E+02   1.269E+00  .
8.449E+01   8.783E+00  .        8.950E+02   1.163E+00  .
8.950E+01   8.662E+00  .        9.480E+02   1.064E+00  .
9.480E+01   8.531E+00  .        1.004E+03   9.723E+01  .
1.004E+02   8.391E+00  .        1.064E+03   8.865E+01  .
1.064E+02   8.241E+00  .        1.127E+03   8.069E+01  .
1.127E+02   8.081E+00  .        1.193E+03   7.333E+01  .
1.193E+02   7.912E+00  .        1.264E+03   6.654E+01  .
1.264E+02   7.733E+00  .        1.339E+03   6.029E+01  .
1.339E+02   7.546E+00  .        1.418E+03   5.456E+01  .
1.418E+02   7.350E+00  .        1.502E+03   4.930E+01  .
1.502E+02   7.145E+00  .        1.592E+03   4.450E+01  .
1.592E+02   6.934E+00  .
```

Figure 13-45 The magnitude of the output voltage versus frequency for the two-stage op amp circuit.

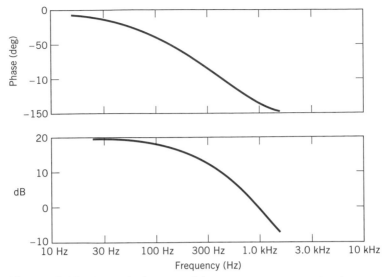

Figure 13-46 The Bode diagram of the two-stage op amp circuit.

EXERCISE 13-16

Obtain the Bode diagram for Example 13-2 using PSpice.

13-14 THE ALL-PASS FILTER AND FILTER NETWORK FUNCTIONS

In the previous examples, the poles and zeros of the gain ratio $H(s)$ (also called a transfer function) have been restricted to the left-hand s-plane. However, a system may have zeros located in the right-hand s-plane. Gain ratios with zeros in the right-hand s-plane are classified as nonminimum phase-shift transfer functions (gain ratios). If the zeros of a transfer function are all reflected about the $j\omega$-axis, there is no change in the shape of the magnitude plot of the transfer function (gain ratio), and the only difference is in the phase-shift characteristics. If the phase characteristics of the two system functions are compared, it can be readily shown that the net phase shift over the frequency range from zero to infinity is less for the system with all its zeros in the left-hand s-plane. Thus, the transfer function $H_1(s)$, with all its zeros in the left-hand s-plane, is called a minimum phase function. The transfer function $H_2(s)$, with $|H_2(\omega)| = |H_1(\omega)|$ and all the zeros of $H_1(s)$ are reflected about the $j\omega$-axis into the right-hand s-plane to form the zeros of $H_2(s)$, is called a nonminimum phase transfer function. Reflection of any zero or pair of zeros into the right half-plane results in a *nonminimum phase shift* transfer function.

The two pole–zero patterns shown in Figures 13-47a and 13-47b have the same amplitude characteristics, as can be deduced from the phasor lengths. However, the phase characteristics are different for Figures 13-47a and 13-47b. The minimum phase characteristic of Figure 13-47a and the nonminimum phase characteristic of Figure 13-47b

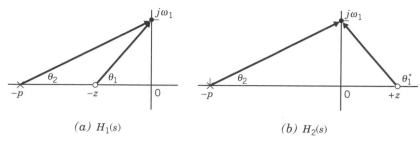

(a) $H_1(s)$ *(b)* $H_2(s)$

Figure 13-47 Pole-zero patterns giving the same amplitude response and different phase characteristics.

are shown in Figure 13-48. Clearly, the phase shift of

$$H_1(s) = \frac{s + z}{s + p}$$

ranges over less than 80°. The meaning of the term minimum phase is illustrated by Figure 13-48. The range of phase shift of a minimum phase gain ratio is the least possible or minimum corresponding to a given amplitude curve, whereas the range of the nonminimum phase curve is greater than the minimum possible for the given amplitude curve.

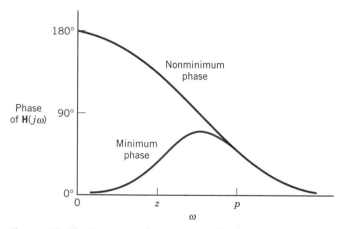

Figure 13-48 The phase characteristics for the minimum phase and nonminimum phase transfer functions.

A particularly interesting nonminimum phase network is the *all-pass network,* which can be realized with the op amp circuit shown in Figure 13-49. This circuit can be used to provide a specified time delay over a region of frequencies. Note that we have placed the positive terminal of V_1 to ground to account for the inversion due to the op amp. Then, for an ideal op amp, the transfer function is

$$H(s) = \frac{V_o(s)}{V_1(s)} = \frac{s - b}{s + b}$$

where $b = 1/R_2C$. Therefore, the magnitude of $H(j\omega)$ is

$$| H(\omega) | = 1$$

for all ω. This circuit passes all frequencies equally. The phase is 180° at $\omega = 0$, 90° at $\omega = b$, and 0° at $\omega = \infty$.

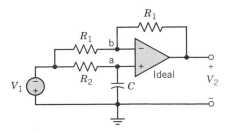

Figure 13-49 An op amp circuit that acts as an all-pass circuit.

In general, every *RLC* network is a filter and has a filter network function, often called a transfer function, $H(s)$. A lowpass filter passes low frequencies and rejects high frequencies with a -3-dB frequency called the cutoff frequency ω_c. A highpass filter passes high frequencies and rejects low frequencies with a cutoff frequency ω_c. A bandpass filter passes a band of frequencies with a center frequency ω_0 and a bandwidth ω_B. The band stop filter rejects a frequency at ω_0 and has a bandwidth ω_B. An all-pass filter has $|H(\omega)| = 1$ for all frequencies. The network function $H(s)$ is summarized for all first- and second-order filters in Table 13-11.

Table 13-11
Network Functions of Filters

Type	Order of Filter Function	Network Function $H(s)$
Lowpass	First order	$H(s) = \dfrac{\omega_c}{s + \omega_c}$
	Second order	$H(s) = \dfrac{\omega_c^2}{s^2 + 2\zeta\omega_c s + \omega_c^2}$
Highpass	First order	$H(s) = \dfrac{s}{s + \omega_c}$
	Second order	$H(s) = \dfrac{s^2}{s^2 + 2\zeta\omega_c s + \omega_c^2}$
Bandpass ω_B = bandwidth ω_0 = center frequency	Second order	$H(s) = \dfrac{s\omega_B}{s^2 + \omega_B s + \omega_0^2}$
Bandstop ω_B = bandwidth ω_0 = center (reject) frequency	Second order	$H(s) = \dfrac{s^2 + \omega_0^2}{s^2 + \omega_B s + \omega_0^2}$
All-pass filter	First order	$H(s) = \dfrac{s - b}{s + b}$
	Second order	$H(s) = \dfrac{(s - b)^2}{(s + b)^2}$

13-15 | VERIFICATION EXAMPLES

Example 13V-1

Figure 13V-1 shows the frequency response of a bandpass filter using PSpice. Such a filter can be represented by

$$\frac{V_0(j\omega)}{V_{in}(j\omega)} = H(\omega)$$

$$= \frac{H_0}{1 + jQ\left(\dfrac{\omega}{\omega_0} - \dfrac{\omega_0}{\omega}\right)}$$

where $V_{in}(j\omega)$ and $V_0(j\omega)$ are the input and output of the filter. This filter was designed to satisfy the specifications

$$\omega_0 = 2\pi 1000, \qquad Q = 10, \qquad H_0 = 100$$

Determine if the specifications are satisfied.

Solution

The frequency response was obtained by analyzing the filter using PSpice. The vertical axis of Fig. 13V-1 gives the magnitude of $\mathbf{H}(j\omega)$ in decibels. The horizontal axis gives the frequency in hertz. Three points on the frequency response have been labeled, giving the frequency and magnitude at each point. We want to use this information from the frequency response to check the filter to see if it has the correct values of ω_0, Q, and H_0.

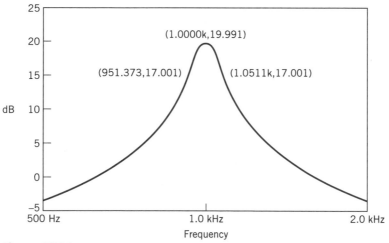

Figure 13V-1 A bandpass frequency response.

The three labeled points on the frequency response have been carefully selected. One of these labels indicates that the magnitude of $H(\omega)$ and frequency at the peak of the frequency response are 20 dB and 1000 Hz. This peak occurs at the resonant frequency, so

$$\omega_0 = 2\pi 1000$$

The magnitude at the resonant frequency is H_0, so

$$20 \log_{10} H_0 = 20$$

or

$$H_0 = 100$$

The other two labeled points were chosen so that the magnitudes are 3 dB less than the magnitude at the peak. The frequencies at these points are 951 Hz and 1051 Hz. The difference of these two frequencies is the bandwidth, BW, of the frequency response. Finally, Q is calculated from the resonant frequency, ω_0 and the bandwidth, BW

$$Q = \frac{\omega_0}{BW}$$

$$= \frac{2\pi 1000}{2\pi(1051 - 951)}$$

$$= 10$$

In this example, three points on the frequency response were used to verify that the bandpass filter satisfied the specifications for its resonant frequency, gain, and quality factor.

Example 13V-2

An old lab report from a couple of years ago includes the following data about a particular circuit:

1 The magnitude and phase frequency responses shown in Figure 13V-2;
2 When the input to the circuit was

$$v_{in} = 4 \cos(2\pi 1200)$$

the steady-state response was

$$v_{out} = 6.25 \cos(2\pi 1200 + 110°).$$

Are these data consistent?

Solution

There are three things that need to be checked: the frequencies, the amplitudes, and the phase angles. The frequencies of both sinusoids are the same, which is good because the circuit must be linear if it is to be represented by a frequency response and the steady-state response of a linear circuit to a sinusoidal input is a sinusoid at the same frequency as the input. The frequency of the input and output sinusoids is

$$\omega = 2 \cdot \pi \cdot 1200 \text{ rad/s}$$

or

$$f = 1200 \text{ Hz}$$

Fortunately the gain and phase shift at 1200 Hz have been labeled on the frequency response plots shown in Figure 13V-2. The gain at 1200 Hz is labeled as 3.9 dB, which means that

$$\frac{|V_{out}|}{|V_{in}|} = 3.9 \text{ dB}$$

$$= 1.57$$

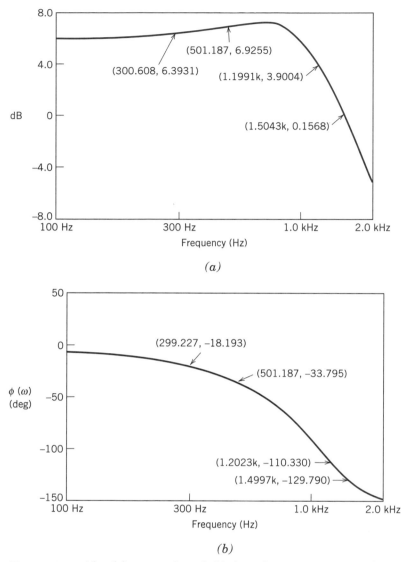

Figure 13V-2 The (a) magnitude and (b) phase frequency response of the circuit.

where V_{in} and V_{out} are the phasors corresponding to $v_{in}(t)$ and $v_{out}(t)$. Let us check this against the data about the input and output sinusoids. Since the magnitudes of the phasors are equal to the amplitudes of the corresponding sinusoids,

$$\frac{|V_{out}|}{|V_{in}|} = \frac{6.25}{4}$$

$$= 1.56$$

This is very good agreement for experimental work.

Next consider the phase shift. The frequency response indicates that the phase shift at 1200 Hz is $-100°$, which means

$$\underline{/V_{out}} - \underline{/V_{in}} = -110°$$

Let us check this against the data about the input and output sinusoids. Since the angles of the phasors are equal to the phase angles of the corresponding sinusoids,

$$\underline{/V_{out}} - \underline{/V_{in}} = 110° - 0°$$
$$= 110°$$

The signs of the phase angles do not match. At a frequency of 1200 Hz a phase angle of 110° indicates that the peaks of the output sinusoid will follow the peaks of the input sinusoid by

$$t_0 = \frac{110°}{360°} \cdot \frac{1}{1200}$$
$$= 2.55 \text{ ms}$$

while a phase angle of −110° indicates that the peaks of the output sinusoid will precede the peaks of the input sinusoid by 2.55 ms.

We have found an error in the old lab report and proposed an explanation for the error.

13-16 DESIGN CHALLENGE SOLUTION

BANDPASS FILTER

An instrument uses a bandpass filter and amplifier to amplify the input signal and reduce the effects of spurious noise. It is desired to obtain a bandpass filter with a gain of 200 at the resonant frequency of 500 Hz and a bandwidth of 50 Hz. Since this is a low-frequency filter and we desire a substantial gain, the design is going to be implemented using an op amp circuit with only resistors and capacitors.

Define the Situation

1 A bandpass filter is an example of a resonant circuit.
2 We can use an op amp and resistors and capacitors to obtain a resonant circuit.
3 Assume that the op amp is ideal.

The Goal

Design an op amp circuit that provides a resonant circuit frequency response.

Generate a Plan

1 Select an op amp circuit that will provide a second-order bandpass filter characteristic.
2 Determine the gain function $H(j\omega) = V_0(j\omega)/V(j\omega)$ of the circuit.
3 Select the capacitor and resistor values to achieve the desired gain and bandwidth.
4 Plot the frequency response for the actual op amp.

Take Action

We need to obtain a resonant circuit with the gain characteristic of Eq. 13-62 that exhibits a resonance for high Q (low ζ) as shown in Figure 13-29. We will use only capacitors and resistors and an op amp (assumed ideal).

A common op-amp bandpass circuit is shown in Figure 13D-1, where we have assumed that both capacitors have a value of C (for algebraic convenience). If the op amp is ideal,

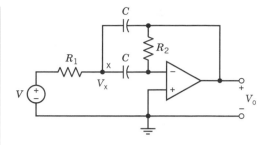

Figure 13D-1 A bandpass filter.

the current into the minus terminal of the op amp is zero. Then, we have

$$\frac{V_x}{Z_c} = \frac{-V_o}{R_2}$$

Therefore

$$V_x = \frac{-V_o}{R_2 C s} \tag{13D-1}$$

where

$$Z_c = \frac{1}{j\omega C} \quad \text{and} \quad s = j\omega$$

Using Kirchhoff's current law at node x, we obtain

$$\frac{V - V_x}{R_1} = \frac{V_x}{Z_c} + \frac{V_x - V_o}{Z_c} \tag{13D-2}$$

Substitute Eq. 13D-1 into Eq. 13D-2 and rearrange to obtain

$$\frac{V_o}{V} = \frac{-j\omega R_2 C}{1 + j\omega 2 R_1 C - \omega^2 C^2 R_1 R_2} \tag{13D-3}$$

Let

$$\omega_0^2 = \frac{1}{R_1 R_2 C^2}, \quad Q = \frac{1}{2}\left(\frac{R_2}{R_1}\right)^{1/2}$$

and then rewrite Eq. 13D-3 as

$$H(j\omega) = \frac{-j2Qu}{1 + j(u/Q) - u^2}$$

where $u = \omega/\omega_0$, as in Eq. 13-62, and $Q = 1/2\zeta$. This resonant response can be exhibited by letting $u = \omega_0/\omega = 1$ and $Q = \omega_0/BW = 500/50 = 10$. Then, we have

$$H(j\omega_0) = \frac{-j20}{1 + 0.1j - 1}$$
$$= -200$$

which is the gain we require at resonance. From Eq. 13-38 we note that the bandwidth BW is

$$BW = \frac{\omega_0}{Q}$$

Thus, if $Q = 10$, we have $BW = \omega_0/10$, as required for this design.

In order to select the components, we will try $C = 10$ nF and use $Q = 10$. Note that

$$\omega_0 Q = \frac{1}{2R_1 C}$$

Therefore, we require

$$R_1 = \frac{1}{2\omega_0 Q C}$$
$$= 1.59 \text{ k}\Omega$$

Furthermore,

$$BW = \frac{\omega_0}{Q} = \frac{2}{R_2 C}$$

Therefore, we have

$$R_2 = \frac{2}{BW \cdot C}$$
$$= 636.6 \text{ k}\Omega$$

A sketch of the frequency response of the resonant circuit is shown in Figure 13D-2.

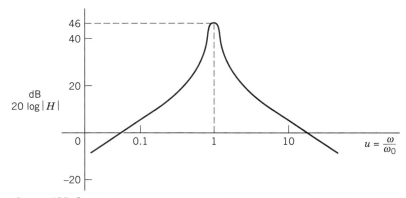

Figure 13D-2 The frequency response of the op amp circuit of Figure 13D-1.

The frequency response of the actual circuit and op amp is then shown to be essentially equal to the response of the ideal op amp circuit.

13-17 ‖ DESIGN EXAMPLE—TWO FILTER SPLITTER

Problem

An oscilloscope has two graphical traces available, and a signal is $v_s = 1 \cos \omega_1 t + 1 \cos \omega_2 t$ where $f_1 = 1$ kHz and $f_2 = 5$ kHz. It is required to design a two-filter circuit to deliver the 1-kHz signal to the oscilloscope trace circuit 1 represented by R_1 while rejecting the 5-kHz signal. Similarly, the second filter must deliver the 5-kHz signal to the second trace circuit represented by R_2 while rejecting the 1-kHz signal. This signal splitter circuit, as shown in Figure 13-50, incorporates resonant circuits utilizing two

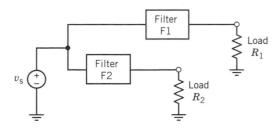

Figure 13-50 Two-filter circuit.

capacitors, an unspecified inductor, and a 10-mH inductor, for each filter. Assume that the inductors are ideal (coil resistance is zero) and then check the results taking into account the actual value of the resistance of an inductor coil that is $R = 1\ \Omega$.

Define the Situation

1 An inductor is represented by a series R and L.
2 Two capacitors and two inductors are available for each filter. One inductor is 10 mH.

The Goal

Design one filter to pass a signal at one frequency f_1 and reject the signal at the other frequency f_2. Similarly design a second filter to pass the signal at frequency f_2 and reject the signal at frequency f_1.

Generate a Plan

1 Use ideal components for resonant circuits.
2 Develop a filter circuit using two resonant circuits.
3 Use one resonant circuit to reject the unwanted signal and the second resonant circuit to pass the desired signal.
4 Proceed for filter F1.
5 Proceed for filter F2.

Take Action

We will proceed to design a filter F1 to pass the 1-kHz signal and reject the 5-kHz signal. The load resistor of F1 is R_1. We can consider a voltage divider circuit so that the parallel resonant circuit **Z** consisting of an inductor and a capacitor, as shown in Figure 13-51, will

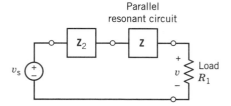

Figure 13-51 A voltage divider circuit to be used to implement the filters F1 and F2.

resonate at 5 kHz. The output of the voltage divider (ignore Z_2 at first) is

$$\mathbf{V} = \frac{R_1 \mathbf{V}_s}{R_1 + \mathbf{Z}}$$

The filter F1 consisting of an inductor and two capacitors is shown in Figure 13-52. For an

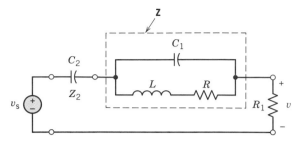

Figure 13-52 A multiresonant filter circuit F1 for passing a signal at 1 kHz and rejecting the signal at 5 kHz.

ideal inductor ($R = 0$), the impedance **Z** at its resonant frequency is infinite and the output voltage v across R_1 is zero. Therefore, the parallel resonance is to occur at $\omega_1 = 2\pi(5 \times 10^3)$ and $\omega_1^2 = 1/LC_1$. Using the inductor $L = 10$ mH, we have

$$C_1 = \frac{1}{L\omega_1^2} = \frac{1}{10^{-2}(\pi \times 10^4)^2}$$
$$= 0.101 \times 10^{-6}$$
$$= 0.101 \ \mu F$$

It is desired to pass the 1-kHz signal unimpeded to the load resistor. The impedance of the parallel circuit **Z** at 1 kHz can be canceled by a series impedance Z_2 obtained by a capacitor C_2. At 1 kHz the impedance **Z**, ignoring R, is

$$\mathbf{Z} = \frac{(j\omega L)(-j/\omega C)}{j\omega L - j1/\omega C}$$
$$= \frac{(j62.8)(-j1571)}{j(62.8 - 1571)}$$
$$= j65.5 \ \Omega$$

The series impedance $\mathbf{Z}_2 + \mathbf{Z}$ is equal to zero when $\mathbf{Z}_2 = -j65.5$. Therefore, we need a capacitor C_2 to obtain the series resonance condition. Then,

$$C_2 = \frac{1}{(2\pi \times 10^3)(65.5)} = 2.42 \ \mu F$$

Using PSpice, we plot $| V |$ as a function of frequency, as shown in Figure 13-53, with

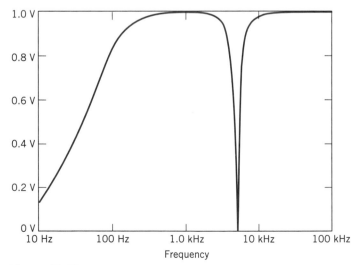

Figure 13-53 The magnitude of v with $R = 1 \ \Omega$ and $R_1 = 1 \ k\Omega$ for filter F1.

$R = 1 \ \Omega$ (the actual condition). The input signal has a magnitude of 1, and we expect $| V | = 1$ at 1 kHz and $| V | = 0$ at 5 kHz for the ideal case. Note that the actual voltage is very close to the ideal case.

In order to design filter F2, we select the L and C of the parallel resonant circuit so that the resonance occurs at 1 kHz. We will use $L = 10$ mH as before. Then, we find that the required capacitor is $C = 2.53 \ \mu F$ and the rejection of the 1-kHz signal is achieved (see Figure 13-54). The impedance Z at 5 kHz is

$$Z = -j13.09$$

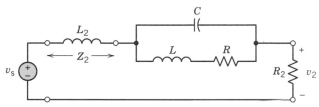

Figure 13-54 A multiresonant filter F2 for passing a signal at 5 kHz and rejecting a signal at 1 kHz.

Therefore, in order to pass the signal at 5 kHz with no loss, we set Z_2 of Figure 13-51 to $Z_2 = +j13.09$. This requires an inductance for the impedance Z_2 and

$$\omega L_2 = 13.09$$

or

$$L_2 = \frac{13.09}{2\pi \times 5 \times 10^3}$$
$$= 0.42 \times 10^{-3}$$
$$= 0.42 \text{ mH}$$

Then the filter F2 is as shown in Figure 13-54.

SUMMARY

Modern communication systems extensively use electric filter circuits to separate electric signals on the basis of their frequency. Therefore, it is important to describe the frequency-dependent relation, for both gain and phase, between an input sinusoidal steady-state signal and an output sinusoidal signal at steady state. This relationship is called the frequency response of a circuit.

We have described the frequency response relationship for a first-order RL or RC circuit as well as for an RLC circuit. The gain ratio, impedance, or admittance is used to describe the frequency response of a circuit.

The parallel RLC circuit exhibits a resonant response as a function of frequency. A resonant condition of a parallel RLC circuit is exhibited at a frequency where the admittance of the circuit is nonreactive. The resonant condition of a series RLC circuit is exhibited when the impedance of the circuit is nonreactive.

We found that it is convenient to use the Bode plot to display the frequency response of a circuit. A Bode plot is a plot of log-gain and phase angle values on a log-frequency base for the selected impedance, admittance, or gain ratio. An approximate Bode diagram can readily be obtained using an asymptotic approximation for the log-magnitude portion of the plot. This approach uses a change in slope of the asymptotic curve that occurs at each break frequency indicated by a pole or a zero of $\mathbf{H}(j\omega)$.

The concept of the bandwidth of a resonant circuit is used to describe the range of frequencies for which the magnitude of the gain is $\geq 1/\sqrt{2}$ for H/H_0, where H_0 is the magnitude of H at resonance. For a parallel or series resonant circuit $H_0 = 1$.

An electrical filter circuit is designed to introduce magnitude gain and/or loss over a designated range of frequencies. There are in general four types of ideal filters: low-pass, high-pass, bandpass, and band rejection filters. These ideal filters can only be approximated by actual circuits containing a finite number of energy storage elements. We can use *RLC* circuits as well as operational amplifier circuits to realize the desired filter form.

The analysis of filter circuits using frequency response plots enables the engineer to determine the quality and characteristics of the circuit.

TERMS AND CONCEPTS

Active Filter   A frequency-selective circuit including amplifiers as well as passive elements.

Asymptotic Curve   Plot of the logarithmic gain of a Bode diagram with a straight line representing each portion of the total curve.

Bandpass Filter   Circuit that will ideally pass unimpeded all frequencies in a selected range of frequencies and reject all frequencies outside the range.

Bandstop Filter   Circuit that will ideally reject all frequencies in a selected range of frequencies and pass unimpeded all frequencies outside the range.

Bandwidth (BW)   Range of frequencies that lie between the two frequencies where the magnitude of the gain ratio is $1/\sqrt{2}$ for H/H_0, where H_0 is the magnitude of H at resonance. For a parallel or series resonant circuit $H_0 = 1$.

Bode Plot   Plot of log-gain and phase angle values on a log-frequency base for a function $\mathbf{H}(j\omega)$.

Break Frequency   The point where the asymptotic curve for its logarithmic gain exhibits a sharp change in slope.

Corner Frequency   See Break Frequency.

Damping Ratio, ζ   Measure of damping for a characteristic equation with complex roots.

Decibels   Units of the logarithmic gain.

Electric Filter   See Filter.

Filter   Circuit designed to provide a magnitude gain and/or loss over a prescribed range of frequencies.

First-Order Circuit   Circuit that contains only one energy storage element, either capacitance or inductance.

Frequency Response   Frequency-dependent relation, in both gain and phase, between an input sinusoidal steady-state signal and an output sinusoidal signal.

Gain Ratio   Ratio of a selected output signal to a designated input signal, written as $\mathbf{H} = \mathbf{X}_o/\mathbf{X}_{in}$ where $\mathbf{X}$ may be a phasor voltage or current.

High-Pass Filter   Filter that will ideally pass all frequencies above the cutoff frequency ω_c and reject all frequencies below the cutoff frequency.

Logarithmic Gain $20 \log_{10} H$ with units of decibels (dB).

Low-Pass Filter   Filter that will ideally pass all frequencies up to the cutoff frequency ω_c and perfectly reject all frequencies above ω_c.

Parallel Resonant Circuit   Circuit with a resistor, capacitor, and inductor in parallel.

Peak Frequency, ω_p Frequency at which a peak magnitude occurs for the pair of complex poles.

Poles Roots of the denominator polynomial of the gain ratio $\mathbf{H}(s)$.

Quality Factor, Q Measure of the energy storage property in relation to the energy dissipation property of a circuit. For a parallel resonant circuit $Q = R/\omega_0 L = \omega_0 RC$. A measure of the frequency selectivity of a filter.

Resonance Condition in a circuit, occurring at the resonant frequency, when $\mathbf{H}(j\omega)$ becomes a real number (nonreactive).

Resonant Frequency Frequency ω_0 at which a gain ratio $\mathbf{H}(j\omega)$ is nonreactive. For a parallel resonant circuit the resonant frequency occurs when its admittance is nonreactive. $\omega_0 = 1/\sqrt{LC}$ for both the parallel and the series RLC circuits.

Series Resonant Circuit Circuit with a series connection of a resistor, capacitor, and inductor.

Zeros Roots of the numerator polynomial of the gain ratio $\mathbf{H}(s)$.

REFERENCES Chapter 13

Bak, David J. "Stethoscope Allows Electronic Amplification," *Design News,* December 15, 1986, pp. 50–51.

Brown, S. F. "Predicting Earthquakes," *Popular Science,* June 1989, pp. 124–125.

Dorf, Richard C. *Electrical Engineering Handbook,* CRC Press, Boca Raton, FL, 1993.

Lewis, Raymond. "A Compensated Accelerometer," *IEEE Transactions on Vehicular Technology,* August 1988, pp. 174–178.

Loeb, Gerald E. "The Functional Replacement of the Ear," *Scientific American,* February 1985, pp. 104–108.

PROBLEMS

Section 13-4 Frequency Response of *RL* and *RC* Circuits

P 13.4-1 Determine the ratio $\mathbf{H} = \mathbf{V}_o/\mathbf{V}_s$ for the circuit of Figure P 13.4-1. Determine $|\mathbf{H}|$ and $\phi(\omega)$ and plot the magnitude and phase on a linear scale for $|\mathbf{H}|$ and ϕ versus a logarithmic scale for ω.

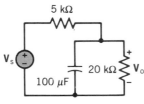

Figure P 13.4-1

P 13.4-2 Consider the circuit shown in Figure P 13.4-2, where $\mathbf{V}_i$ and $\mathbf{V}_o$ are phasor voltages. (a) Find expressions for $\mathbf{V}_o/\mathbf{V}_i$ and $|\mathbf{V}_o/\mathbf{V}_i|$, both as a function of the radian frequency ω. (b) Sketch $|\mathbf{V}_o/\mathbf{V}_i|$ versus ω for $\omega > 0$. Does your sketch give the correct limiting results for $\omega \rightarrow 0$ and $\omega \rightarrow \infty$? Briefly justify your answers. (c) Sketch the phase of $\mathbf{V}_o/\mathbf{V}_i$ for $\omega > 0$.

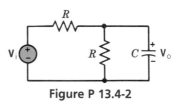

Figure P 13.4-2

P 13.4-3 Determine the ratio $\mathbf{H} = \mathbf{V}_o/\mathbf{V}_s$ for the circuit of Figure P 13.4-3. Sketch $|\mathbf{H}|$ and $\phi(\omega)$.

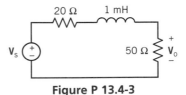

Figure P 13.4-3

Section 13-5 Bode Plots

P 13.5-1 A linear model of an operational amplifier that includes the effect of the output capacitance is shown in Figure P 13.5-1. Determine $\mathbf{V}_o/\mathbf{V}_s$ and the bandwidth of the operational amplifier when $C = 10$ nF and $A = 10$. Draw the Bode diagram for $10^2 < \omega < 10^8$. Determine the gain bandwidth product, which is equal to the gain at low frequencies times the radian frequency where $|\mathbf{H}| = (1/\sqrt{2})H_{max}$.
Answer: Gain bandwidth $= 10^9$ rad/s

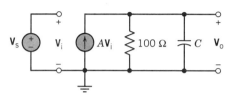

Figure P 13.5-1 A linear model of the 741 operational amplifier that includes the effect of the output capacitance C.

P 13.5-2 Consider the circuit shown in Figure P 13.5-2a, where the two-terminal circuit contained within the box is known to consist of two linear circuit elements. The magnitude plot for the Bode diagram of $|\mathbf{Z}|$ is shown in Figure P 13.5-2b. Find the two-element circuit and specify the numerical value of each element.

(a)

(b)

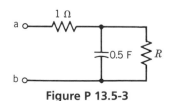

Figure P 13.5-2

P 13.5-3 The circuit shown in Figure P 13.5-3 has a device that can be represented by a negative resistor so that $R = -20$ kΩ. (a) Find the admittance at terminals a–b. (b) Plot the Bode diagram for the admittance of the circuit.

Figure P 13.5-3

P 13.5-4 Tissue electrodes are used by physicians to form the interface that conducts current to the target tissue of the human body. The electrode in tissue can be modeled by the RC circuit shown in Figure P 13.5-4. The value of each element depends on the electrode material and physical construction as well as the character of the tissue being probed. Find the Bode diagram for $\mathbf{V}_o/\mathbf{V}_s = \mathbf{H}(j\omega)$ when $R_1 = 1$ kΩ, $C = 1$ μF, and the tissue resistance is $R_t = 5$ kΩ.

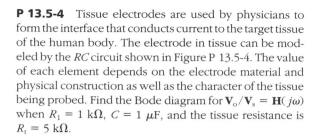

Figure P 13.5-4

P 13.5-5 Draw the Bode diagram for the ratio $\mathbf{V}_o/\mathbf{V}_s$ for the circuit of Figure P 13.5-5.

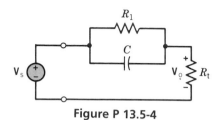

Figure P 13.5-5

P 13.5-6 A circuit has a voltage ratio

$$\mathbf{H}(j\omega) = \frac{K(1 + j\omega/z)}{j\omega}$$

(a) Find the high and low-frequency asymptotes of the magnitude Bode plot.
(b) The high- and low-frequency asymptotes comprise the magnitude Bode plot. Over what ranges of frequencies is the asymptotic magnitude Bode plot of $\mathbf{H}(j\omega)$ within 1% of the actual value of $\mathbf{H}(j\omega)$ in decibels?

Section 13-6 Parallel Resonant Circuits

P 13.6-1 For a parallel *RLC* circuit with $R = 10$ kΩ, $L = 1/120$ H, and $C = 1/30$ μF, find ω_0, Q, ω_1, and the bandwidth *BW*.

Answer: $\omega_0 = 60$ krad/s
$Q = 20$
$\omega_1 = 58.519$ krad/s
$BW = 3$ krad/s

P 13.6-2 A parallel resonant *RLC* circuit is driven by a current source $I_s = 0.2 \cos \omega t$ mA and shows a maximum response of 8 V at $\omega = 2500$ rad/s and 4 V at 2200 rad/s. Find R, L, and C.

Answer: $R = 40$ kΩ
$L = 14.5$ mH
$C = 11$ μF

P 13.6-3 For the *RLC* parallel resonant circuit, show that

$$Q = \omega_0 RC$$

by using the definition of Q expressed in terms of energy stored and dissipated.

P 13.6-4 An R, L, C parallel resonant circuit may be represented by the equation

$$\mathbf{H} = \frac{1}{R\mathbf{Y}} = \frac{(\omega_0/Q)s}{s^2 + (\omega_0/Q)s + \omega_0^2}$$

where $s = j\omega$. One can use a computer program to obtain a plot of H versus ω as shown in Figure P 13.6-4 for $Q = 10$, 20, and 50. Use a computer program to plot H versus ω for $Q = 5$, 30, and 100.

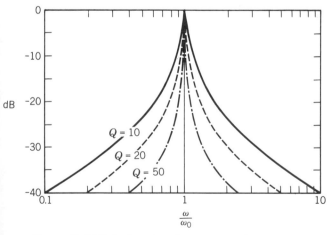

Figure P 13.6-4 The plot of **H** versus ω for $Q = 10$, 20, and 50.

P 13.6-5 The circuit shown in Figure P 13.6-5 represents a capacitor, coil, and resistor in parallel. Calculate the resonant frequency, bandwidth, and Q for the circuit.

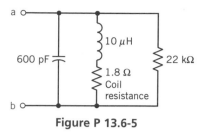

Figure P 13.6-5

P 13.6-6 Consider the simple model of an electric power system as shown in Figure P 13.6-6. The inductance, $L = 0.25$ H, represents the power line and transformer. The customer's load is $R_L = 100$ Ω and the customer adds $C = 25$ μF to increase the magnitude of V_0. The source is $V_s = 1000 \cos 400t$ V, and it is desired that $|\mathbf{V}_o|$ also be 1000 V. (a) Find $|\mathbf{V}_o|$ for $R_L = 100$ Ω. (b) When the customer leaves for the night, he turns off much of his load, making $R_L = 1$ kΩ, at which point sparks and smoke begin to appear in the equipment still connected to the power line. The customer calls you in as a consultant. Why did the sparks appear when $R_L = 1$ kΩ? (c) If 100 $\Omega \leq R_L \leq 1000$ Ω, obtain a formula that will determine the appropriate C so that $|\mathbf{V}_o|$ is always 1000 V.

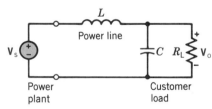

Figure P 13.6-6 Model of an electric power system.

Section 13-7 Series Resonance

P 13.7-1 (a) Consider the series resonant circuit shown in Figure P 13.7-1 and find $\mathbf{H}(j\omega) = \mathbf{V}_o/\mathbf{V}_s$. (b) Sketch the Bode diagram for the circuit, assuming that $2 < Q < 10$. (c) Find the expression for the resonant frequency ω_0. (d) Find the peak frequency ω_p if $Q = 5$ and $\omega_0 = 1000$.

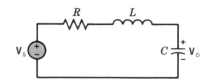

Figure P 13.7-1 A series resonant circuit.

P 13.7-2 A series resonant RLC circuit has $L = 10$ mH, $C = 0.01$ μF, and $R = 100$ Ω. Determine ω_0, Q, BW, and the magnitude of the response at $\omega = 1.1 \times 10^5$ rad/s.
Answer: $\omega_0 = 10^5$, $Q = 10$, $BW = 10^4$, $H = 0.44$

P 13.7-3 A quartz crystal exhibits the property that when mechanical stress is applied across its faces, a potential difference develops across opposite faces. When an alternating voltage is applied, mechanical vibrations occur and electromechanical resonance is exhibited. A crystal can be represented by a series RLC circuit. A specific crystal has a model with $L = 1$ mH, $C = 10$ μF, and $R = 1$ Ω. Find ω_0, Q, and the bandwidth.
Answer: $\omega_0 = 10^4$ rad/s, $Q = 10$, $BW = 10^3$ rad/s

P 13.7-4 The resistance R and capacitance C of the load can be varied as shown in Figure P 13.7-4. The sinusoidal source operates at 1 MHz and it is desired to deliver maximum power to the load. Select R and C for maximum power transfer and calculate the Q of the total circuit.

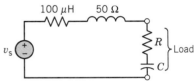

Figure P 13.7-4

P 13.7-5 Determine (a) the resonant frequency of the circuit shown in Figure P 13.7-5 and select C and L to obtain $\omega_0 = 100$ rad/s when $R_1 = R_2 = 1$ Ω.

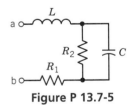

Figure P 13.7-5

P 13.7-6 For the circuit shown in Figure P 13.7-6, (a) derive an expression for the magnitude response $|\mathbf{Z}_{in}|$ versus ω; (b) sketch $|\mathbf{Z}_{in}|$ versus ω; (c) find $|\mathbf{Z}_{in}|$ at $\omega = 1/\sqrt{LC}$.

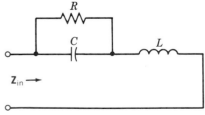

Figure P 13.7-6

P 13.7-7 A simple voltage divider with two resistors does not work as we expect at higher frequencies because of a parasitic capacitance shunting R_2. To remedy the capacitance problem, another capacitance C_1 is added

in parallel with R_1 as shown in Figure P 13.7-7. (a) Determine the relationship for $R_1 C_1$ and $R_2 C_2$ so that the output voltage is $\mathbf{V}_o = k\mathbf{V}_s$ and is not a function of frequency. (b) Plot the gain magnitude curve V_o/kV_s (dB) for $R_1 = R_2 = 10$ kΩ, $C_2 = 0.1$ μF, and three values of C_1: 1 μF, 0.1 μF, and 0.05 μF.
Answer: (a) $R_1 C_1 = R_2 C_2$

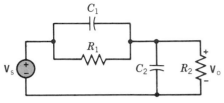

Figure P 13.7-7

P 13.7-8 A circuit is represented by

$$\mathbf{H}(s) = \frac{0.1(1 + j\omega)}{1 + j\omega\tau}$$

where $s = j\omega$ and $\tau = 0.1$. Plot the Bode diagram.

P 13.7-9 A circuit $\mathbf{Z}(j\omega)$ has the Bode magnitude diagram shown in Figure P 13.7-9. (a) Find $\mathbf{Z}(j\omega)$. (b) What is the magnitude of $\mathbf{Z}(j\omega)$ (in decibels) as ω approaches infinity?

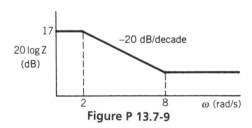

Figure P 13.7-9

Section 13-8 Filter Circuits

P 13.8-1 For the circuit shown in Figure P 13.8-1 determine (a) the type of filter and (b) the magnitude and phase of the frequency response when $R = 1$ kΩ, $C = 0.1$ μF, and $L = 1$ mH.
Answer: (a) Lowpass,
(b) $H = (1 + 3.95 \times 10^{-11} f^2)^{-1/2}$,
$\phi = -\tan^{-1} 2\pi f \times 10^{-6}$

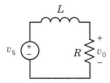

Figure P 13.8-1

P 13.8-2 Repeat P 13.8-1 for the circuit of Figure P 13.8-2.

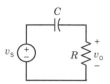

Figure P 13.8-2

P 13.8-3 An audio amplifier has a gain

$$\frac{\mathbf{V}_o}{\mathbf{V}_s} = \frac{1000a^2s}{(s + a)(s + 1000a)}$$

where $s = j\omega$. The goal is to pass unimpeded all the frequencies between 20 Hz and 20 kHz. Select the parameter a and draw the Bode diagram of this amplifier for $\omega = 1$ to $\omega = 10^8$, using the asymptotic approximation for the decibel magnitude and an approximate curve for the phase.

P 13.8-4 A bandstop filter can be used to eliminate unwanted signals. One common application is the removal of interference with television channel 6 due to nearby FM radio transmitters. In this problem, we wish to design a bandstop filter to eliminate the signal from an FM station broadcasting at 90 MHz.

The circuit for the receiving antenna for the TV set is represented by $\mathbf{V}_s$ and R_s, where $R_s = 300\ \Omega$. The impedance of the television set is $R_L = 300\ \Omega$, as shown in Figure P 13.8-4. Design a filter $\mathbf{V}_o/\mathbf{V}_s$ so that the 90-MHz signal is removed without affecting the performance due to channel 6, which operates in the band 82 MHz to 88 MHz. Use only an R, L, and C network, since amplifiers that work at 90 MHz would be quite expensive.

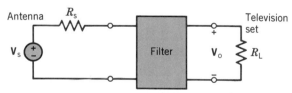

Figure P 13.8-4 A bandstop filter for elimination of television set interference.

P 13.8-5 Determine the value of R required for the circuit of Figure P 13.8-5 so that the maximum value of $H = 50$. For this value calculate the center frequency of the filter and the bandwidth.
Answer: $R = 0.174\ \Omega$, $BW = 1.11$ kHz

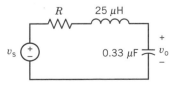

Figure P 13.8-5

P 13.8-6 A filter circuit is shown in Figure P 13.8-6. Determine the type of filter and the center frequency for

$$\mathbf{Z} = \frac{\mathbf{V}_o}{\mathbf{I}_s}.$$

Answer: bandstop filter, $f_0 = 2.32$ kHz

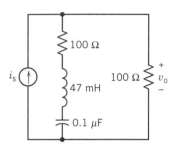

Figure P 13.8-6

P 13.8-7 A filter circuit is shown in Figure P 13.8-7. Determine the voltage ratio $\mathbf{H} = \mathbf{V}_o/\mathbf{V}_s$ for this circuit. Describe the type of filter.

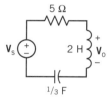

Figure P 13.8-7

P 13.8-8 A bandpass amplifier has a frequency response for the magnitude ratio H, shown in Figure P 13.8-8. Find the bandwidth and the magnitude of H at the resonant frequency. Determine the Q of the circuit.
Answer: $BW = 0.2$ MHz
$H = 3.16$
$Q = 50$

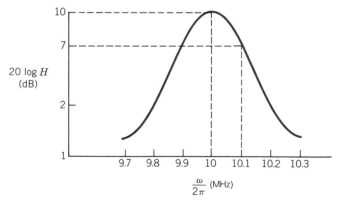

Figure P 13.8-8 A bandpass amplifier.

P 13.8-9 The filter circuit shown in Figure P 13.8-9 has $\mathbf{H}(j\omega) = \mathbf{V}_o/\mathbf{V}_s$. The circuit is to pass a 20-kHz signal while rejecting a 30-kHz signal.

(a) Select an appropriate value of L.
(b) Choose an appropriate inductive or capacitance element and its value for the parallel admittance $\mathbf{Y}_1$. Assume that $R_L > 100\ \Omega$.

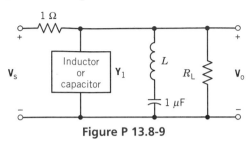

Figure P 13.8-9

P 13.8-10 Repeat Problem 13.8-9 when the filter is to reject the 30-kHz signal (as in P 13.8-9) but is to pass a 40-kHz signal.

P 13.8-11 A circuit used to reject two unwanted signal frequencies is shown in Figure P 13.8-11. Sketch the magnitude of the frequency response of the circuit and determine the two frequencies rejected by the circuit.

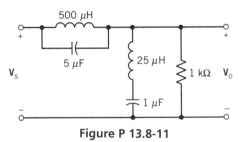

Figure P 13.8-11

P 13.8-12 The circuit shown in Figure P 13.8-12 has an input signal

$$v_s = 10 \sin \omega_1 t + 10 \sin \omega_2 t\ \text{V}$$

where $\omega_1 = 60{,}000\pi$ and $\omega_2 = 120{,}000\pi$.

(a) Determine the value of C and the type and component value of the element of $\mathbf{Z}_1$ so that the circuit rejects the 30-kHz signal and transmits the 60-kHz signal.
(b) For the values in part (a), find the output signal.
(c) Obtain the frequency response of the circuit.

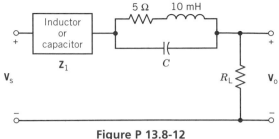

Figure P 13.8-12

P 13.8-13 A filter circuit is shown in Figure P 13.8-13. This filter circuit is called a double-tuned circuit since it may reject a signal at one frequency and pass a signal at another frequency. Assume that the inductor $L_S = 100$ mH and that the unspecified element is a capacitor or an inductor.

(a) Select the unknown element and design a double-tuned filter to reject a signal at 50 kHz while passing a signal at 100 kHz. Determine the necessary component values.
(b) Repeat (a) if the rejection frequency is 100 kHz and the pass frequency is 50 kHz.

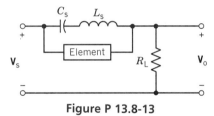

Figure P 13.8-13

Section 13-9 Pole–Zero Plots and the s-Plane

P 13.9-1 A circuit has $H(s)$ as

$$H(s) = \frac{10s + 400}{s^3 + 10s^2 + 100s}$$

(a) Determine the poles and zeros and plot them on the s-plane.
(b) Sketch the magnitude portion of the Bode diagram.

P 13.9-2 A circuit has $H(s)$ as

$$H(s) = \frac{50s}{s^2 + 50s + 10^6}$$

(a) Determine the poles and zeros and plot them on the s-plane.
(b) Sketch the magnitude portion of the Bode diagram.

P 13.9-3 A filter with

$$H(s) = \frac{1}{s + 1}$$

and

$$|H(\omega)| = \frac{1}{(\omega^2 + 1)^{1/2}}$$

is called a Butterworth filter. The general nth-order Butterworth filter has

$$|H(\omega)| = \frac{1}{(\omega^{2n} + 1)^{1/2}}$$

where $\omega = 1$ is the nominal cutoff frequency for the filter. (a) Determine the second- and fourth-order Butterworth filters and plot $20 \log |H|$ for $n = 1, 2,$ and 3. Describe the improvement achieved by higher n filters by determining $|H(\omega)|$ at $\omega = 2$. (b) Obtain $H(s)$ for $n = 2$ and 3 and plot the pole–zero diagram on the s-plane.

Section 13-10   Frequency Response of a Multifactor $H(j\omega)$

P 13.10-1   A low-pass filter has a gain function

$$H(s) = \frac{100}{s^2 + 10s + 100}$$

Plot $20 \log| H(j\omega)|$ and determine the bandwidth of the filter.

P 13.10-2   A high-pass filter has a gain function

$$H(s) = \frac{10s^2}{s^2 + 1000s + 10^8}$$

Plot $20 \log| H(j\omega)|$ and determine the maximum value of $20 \log| H(\omega)|$ and the frequency at which it occurs.

P 13.10-3   The cochlear implant is intended for patients with deafness due to malfunction of the sensory cells of the cochlea in the inner ear [Loeb]. These devices use a microphone for picking up the sound and a processor for converting to electrical signals, and they transmit these signals to the nervous system. A cochlear implant relies on the fact that many of the auditory nerve fibers remain intact in patients with this form of hearing loss. The overall transmission from microphone to nerve cells is represented by the gain function

$$H(j\omega) = \frac{10(j\omega/50 + 1)}{(j\omega/2 + 1)(j\omega/20 + 1)(j\omega/80 + 1)}$$

Plot the Bode diagram for $H(j\omega)$ for $1 \le \omega \le 100$.

P 13.10-4   The magnitude and phase of a circuit are measured and the data are shown in Table P 13.10-4. Determine an estimate of $H(j\omega)$ when the peak magnitude occurs at $\omega = 4.1$.

P 13.10-5   The amplitude gain of a circuit is shown in Figure P 13.10-5 for $1 \le \omega \le 10^5$ rad/s. Find $H(j\omega)$ by estimating the asymptotic approximation for the Bode diagram of the circuit. The peak value of $20 \log H$ is 43 dB at $\omega = 450$. You may assume that this circuit's gain function $H(j\omega)$ includes only first-order factors of the form $(1 + j\omega\tau)$.

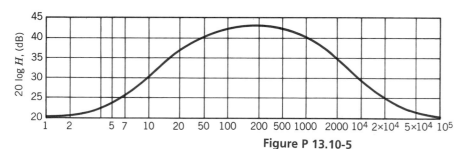

Figure P 13.10-5

P 13.10-6   In the medical world the stethoscope remains the most practical instrument available for listening for sounds arising within the heart and lungs. A recently introduced electronic stethoscope uses 10 bandpass amplifiers to let the physician concentrate on specific sounds [Bak]. The gain characteristic of one bandpass amplifier is represented by

$$H = \frac{(j\omega/100 + 1)}{[1 + ju/10 + (ju)^2](j\omega/5000 + 1)}$$

where $u = \omega/\omega_0$ and $\omega_0 = 1000$. (a) Draw the exact Bode diagram of this filter for $100 \le \omega \le 10^4$. (b) Find the

Table P 13.10-4

ω	0.1	1	2	3	4	4.1	5	8	100
$|H|$	0.01	0.12	0.29	0.60	1.0	2.2	0.78	0.32	0.02
ϕ	90°	83°	73°	53°	7°	0°	−39°	−71°	−89°

bandpass bandwidth *BW*. (c) Determine the *Q* of this circuit. (d) Find the resulting gain at ω_0.

P 13.10-7 Determine $\mathbf{H}(j\omega)$ from the asymptotic Bode diagram in Figure P 13.10-7.

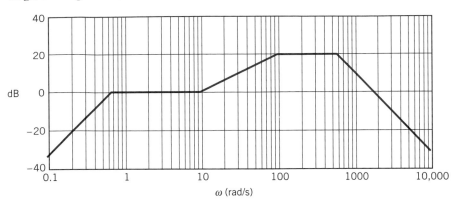

Figure P 13.10-7

P 13.10-8 The frequency response of an unknown circuit is measured and drawn with the asymptotes for 20 log *H* as shown in Figure P 13.10-8. The phase is also shown. Determine the equation for $\mathbf{H}(j\omega)$, indicating the corner frequencies.

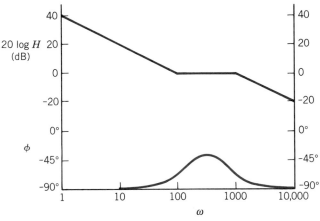

Figure P 13.10-8 The frequency response of an unknown circuit.

P 13.10-9 (a) For the circuit of Figure P 13.10-9*a* derive an expression for the magnitude of the gain function $|\mathbf{H}_1(j\omega)| = |\mathbf{V}_1/\mathbf{V}_s|$ and sketch it as a function of frequency ω. (Include values $\omega = 0$, $\omega = \omega_1$, $\omega = \infty$.) Assume the break frequency is ω_1. (b) For the circuit of Figure P 13.10-9*b* derive an expression for the magnitude of the gain function $|\mathbf{H}_2(j\omega)| = |\mathbf{V}_2/\mathbf{V}_1|$ and sketch it versus frequency ω on the same graph as $|\mathbf{H}_1(\omega)|$. (Assume $\omega_2 < \omega_1$ and include value of $\omega = 0$, $\omega = \omega_2$, $\omega = \infty$.) (c) Each of the above filters is known as a first-order filter. Create a second-order filter by obtaining

an expression for $|\mathbf{H}(\omega)| = |\mathbf{H}_1(\omega)\,\mathbf{H}_2(\omega)|$. Draw a rough sketch of $|\mathbf{H}(\omega)|$ versus ω. Use the same graph page as above. Do not use your expression for $|\mathbf{H}(\omega)|$, but instead think about the plots from parts a and b and why the plot of $|\mathbf{H}(\omega)|$ versus ω for this second-order filter has a point of inflection. (d) Can you think of any reason why the circuit of Figure P 13.10-9*c* will not perform in the same way as the circuit described by $\mathbf{H}(\omega)$ in part (c)? Hint: Look at the impedance to the right of the capacitor.

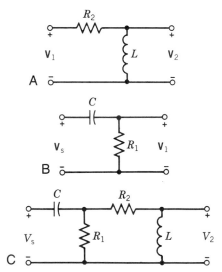

Figure P 13.10-9 *(a)* Circuit for $\mathbf{H}_1$. *(b)* Circuit for $\mathbf{H}_2$. *(c)* Circuit for $\mathbf{H}$.

P 13.10-10 Two cascade amplifiers are used to obtain an output voltage as shown in Figure P 13.10-10. Plot the magnitude portion of the Bode diagram of the overall gain $\mathbf{V}_o/\mathbf{V}_s$. Each amplifier is of the form

$$H(s) = \frac{As}{(1 + s/\omega_L)(1 + s/\omega_h)}$$

where $s = j\omega$.

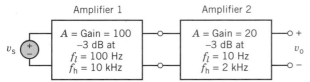

Figure P 13.10-10 Two cascaded amplifiers.

P 13.10-11 A circuit called a bridged T network is shown in Figure P 13.10-11. The relationship for $\mathbf{V}_o/\mathbf{V}_s$ is

$$\mathbf{H} = \frac{(j\omega/\omega_0)^2 + (j\omega/Q\omega_0) + 1}{(j\omega/\omega_1 + 1)(j\omega/\omega_2 + 1)}$$

A certain circuit is designed so that $\omega_0 = 2500$, $Q = 10/3$, $\omega_1 = 370$, and $\omega_2 = 20,000$. (a) Plot the Bode diagram of this circuit for $\omega = 10$ to $\omega = 10^5$. (b) Describe the type of filter. (c) Find the band of frequencies for the circuit that results in an attenuation (gain reduction) of 20 dB or greater.

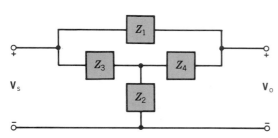

Figure P 13.10-11 A bridged T circuit.

P 13.10-12 Determine $H(s)$ for the circuit of Figure P 13.10-12 and sketch the Bode diagram.

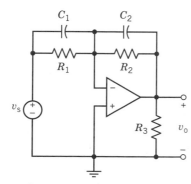

Figure P 13.10-12

P 13.10-13 Powerful medium-wave or long-wave transmitters operating not far from an active short-wave antenna can cause serious interference in the reception of short-wave signals. The interfering signals can be suppressed effectively by a critically damped high-pass filter (Becker, 1992).

The goal is to pass frequencies above 4 MHz unhindered, while reducing the effect of low-frequency signals. The circuit is shown in Figure P 13.10-13.

(a) Determine $V_0(s)/V_1(s)$.
(b) Plot the Bode diagram of the transfer function.

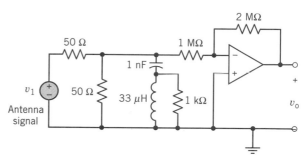

Figure P 13.10-13 High-pass filter.

P 13.10-14 An advanced circuit has three sources with different frequencies, as shown in Figure P 13.10-14. Determine the voltage $v(t)$ and the current $i(t)$ when $i_1 = 12 \cos 400t$ A, $i_2 = 10$ A, and $v_1 = 80 \sin 1000t$ V. Note the existence of three resonant circuits within the overall circuit.

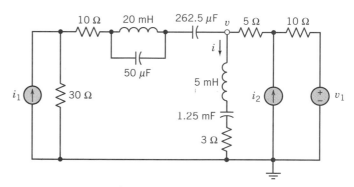

Figure P 13.10-14

Section 13-11 Frequency Response of Op Amp Circuits

P 13.11-1 Consider the circuit shown in Figure P 13.11-1. (a) Find the gain $\mathbf{H}(j\omega) = \mathbf{V}_o/\mathbf{V}_s$. (b) Plot the Bode diagram for $\mathbf{H}(j\omega)$ for $R_1 = R_2 = 20$ Ω, $C = 2.4$ μF, and $L = 0.25$ mH.

Answer: $\mathbf{H}(j\omega) = LC\left[(j\omega)^2 + j\frac{R_2}{L}\omega + \frac{1}{LC}\right]$

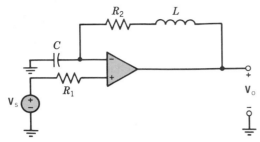

Figure P 13.11-1 An *RLC* operational amplifier circuit.

P 13.11-2 An operational amplifier can be accurately described by its frequency response with the equation

$$\mathbf{H}(j\omega) = \frac{\mathbf{V}_o}{\mathbf{V}_s} = \frac{K\omega_1}{j\omega + \omega_1} = \frac{K}{j\omega\tau + 1}$$

where $K\omega_1$ is called the gain–bandwidth product. The gain–bandwidth product of a specific amplifier is 20 MHz and $K = 10^5$. Sketch the magnitude (dB) portion of the Bode diagram.

P 13.11-3 For the circuit shown in Figure P 13.11-3, (a) show that $\mathbf{V}_o(\omega)/\mathbf{V}_s(\omega)$ is given by

$$\frac{\mathbf{V}_o(\omega)}{\mathbf{V}_s(\omega)} = \frac{1 + j\omega R_1 C}{1 + j\omega RC} \cdot \frac{R}{R_1}$$

where R is the equivalent resistance for R_1 and R_2 in parallel. (b) Sketch $|\mathbf{V}_o(\omega)/\mathbf{V}_s(\omega)|$ and determine the filter type of the above circuit. (c) Find the frequency ω_1 when the gain ratio 20 log H is 3 dB above its low-frequency value. By choosing reasonable values for R_1, and C, design a filter with $\omega_1 = 10^4$ rad/s.

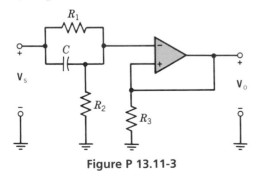

Figure P 13.11-3

P 13.11-4 A bandpass filter can be obtained from the circuit shown in Figure P 13.11-4. (a) Find $\mathbf{V}_o/\mathbf{V}_s(j\omega)$. (b) Find the bandwidth BW and resonant frequency ω_o. Note the sign of the source. Assume an ideal operational amplifier.

Answer: $BW = \dfrac{C_1 + C_2}{R_2 C_1 C_2}$

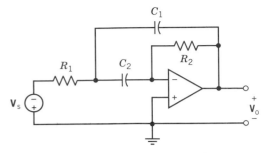

Figure P 13.11-4 A bandpass filter.

P 13.11-5 Find the gain function $\mathbf{H}(j\omega)$ for the circuit of Figure P 13.11-5. Sketch the Bode diagram for the circuit.

Answer: $\mathbf{H} = \dfrac{1 - j(\omega/\omega_0)}{1 + j(\omega/\omega_0)}, \quad \omega_0 = \dfrac{1}{RC}$

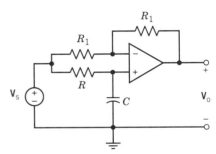

Figure P 13.11-5

P 13.11-6 A bandpass filter can be achieved using the circuit of Figure P 13.11-6. The advantage of this circuit is that we are not required to use inductors. Find (a) the magnitude of $\mathbf{H} = \mathbf{V}_o/\mathbf{V}_s$, (b) the low- and high-frequency cutoff frequencies ω_1 and ω_2, and (c) the peak gain in the passband.

Answer: (b) $\omega_1 = \dfrac{1}{R_1 C_1}$

(c) midband gain $= \dfrac{R_2}{R_1}$

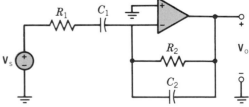

Figure P 13.11-6 A bandpass filter.

P 13.11-7 An amplifier uses two identical operational amplifier circuits as shown in Figure P 13.11-7. It is desired to set the cutoff or break frequency at $\omega_c = 1000$ for this low-pass amplifier with a gain $\mathbf{H} = \mathbf{V}_o/\mathbf{V}_s$. Also, it is desired that the low-frequency gain of $\mathbf{H} = 0$ dB. (a) Find the required R_1, R_2, and C. (b) Find the rejection in decibels for a signal at $\omega = 10{,}000$.

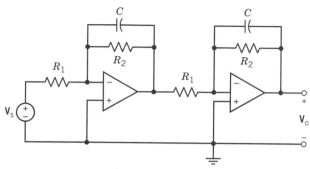

Figure P 13.11-7

P 13.11-8 An operational amplifier circuit is shown in Figure P 13.11-8, where $R_2 = 5\ \text{k}\Omega$ and $C = 0.02\ \mu\text{F}$. (a) What is the gain of the circuit, $\mathbf{V_o}/\mathbf{V_s}$, for $\omega = 0$? (b) Find the expression for $\mathbf{V_o}/\mathbf{V_s}(j\omega)$ and sketch the Bode diagram using asymptotic approximations. (c) At what frequency does $|\mathbf{V_o}|$ fall to $1/\sqrt{2}$ of its low-frequency value?

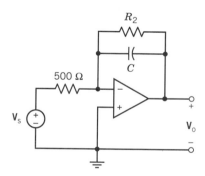

Figure P 13.11-8

P 13.11-9 A unity gain, low-pass filter is obtained from the operational amplifier circuit shown in Figure P 13.11-9. Determine the gain function $\mathbf{V_o}/\mathbf{V_s} = \mathbf{H}$ and sketch the Bode diagram.

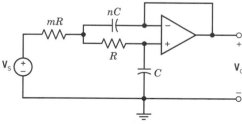

Figure P 13.11-9

P 13.11-10 Calculate the high and low cutoff frequencies for the bandpass filter shown in Figure P 13.11-10

when $R_1 = R_2 = 20\ \text{k}\Omega$, $C_1 = 0.2\ \mu\text{F}$, and $C_2 = 1\ \text{nF}$. Determine the circuit bandwidth. Assume ideal op amps.

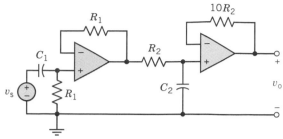

Figure P 13.11-10

P 13.11-11 The audio range can be considered to be between $\omega_1 = 100\ \text{rad/s}$ and $\omega_2 = 10^5\ \text{rad/s}$. Using operational amplifiers and resistors and capacitors, design a bandpass filter by using a low-pass filter cascaded with a high-pass filter to obtain

$$\mathbf{H}(j\omega) = \frac{10(j\omega/\omega_1 + 1)}{(j\omega/\omega_2 + 1)}$$

P 13.11-12 Determine the voltage gain and the cutoff frequency for the high-pass filter shown in Figure P 13.11-12. Draw the magnitude portion of the Bode diagram for (a) $C = 50\ \text{nF}$ and $R = 4\ \text{k}\Omega$ and (b) $C = 100\ \text{nF}$ and $R = 2\ \text{k}\Omega$. Assume an ideal op amp.

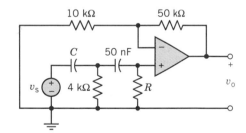

Figure P 13.11-12

P 13.11-13 Determine the gain ratio $H(s)$ for the op amp circuit shown in Figure P 13.11-13 and plot the Bode diagram. Assume ideal op amps.

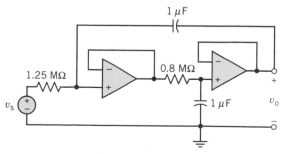

Figure P 13.11-13

P 13.11-14 An acoustic sensor operates in the range of 5 kHz and is represented by v_s in Figure P 13.11-14. It is specified that the bandpass filter shown in the figure pass an input signal within the specified bandwidth. Determine the bandwidth and center frequency of the circuit when the op amp is assumed ideal.

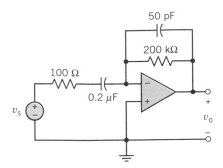

Figure P 13.11-14

P 13.11-15 A bandpass filter can be achieved by using two operational amplifiers in the circuit shown in Figure P 13.11-15. (a) Assume the operational amplifiers are ideal and find $\mathbf{H} = \mathbf{V}_o/\mathbf{V}_s$. (b) Plot the Bode diagram for this circuit when $R_1 = 1\,\text{k}\Omega$, $R_2 = 100\,\Omega$, $C_1 = 1\,\mu\text{F}$, and $C_2 = 0.1\,\mu\text{F}$. (c) Find ω_0 and Q for the circuit. (d) Find the bandwidth of the circuit.

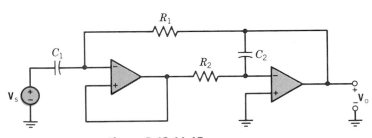

Figure P 13.11-15

P 13.11-16 An op amp circuit is shown in Figure P 13.11-16. Determine $H(s)$ when $a > 10$ and the op amp is ideal. Select a, R, and C so that the resonant gain is 201 and the resonant frequency is 10^5 rad/s. Draw the resulting Bode diagram.

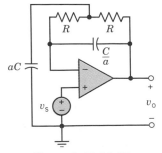

Figure P 13.11-16

P 13.11-17 A low-pass filter is shown in Figure P 13.11-17.

(a) Assume an ideal op amp and find the voltage ratio $\mathbf{V}_o/\mathbf{V}_s$.
(b) Sketch the Bode plot and find the cutoff frequency when $C = 20\,\text{nF}$, $R = 1.2\,\text{k}\Omega$, and $R_1/R_2 = 1$.

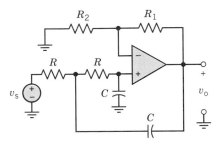

Figure P 13.11-17 Low-pass filter.

P 13.11-18 The input to the circuit of Figure P 13.11-18 is

$$v_s = 50 + 30\cos(500t + 115°) - 20\cos(2500t + 30°)\ \text{mV}$$

Find the steady-state output v_o for (a) $C = 0.1\,\mu\text{F}$ and (b) $C = 0.01\,\mu\text{F}$. Assume an ideal op amp.

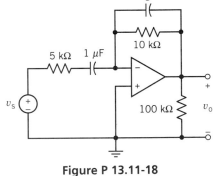

Figure P 13.11-18

Section 13-14 The All-Pass Filter

P 13.14-1 (a) Find $\mathbf{H}(j\omega) = \mathbf{V}_o/\mathbf{V}_s$ of the phase shifter circuit shown in Figure P 13.14-1. (b) Sketch the Bode diagram of this circuit.

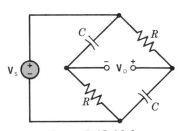

Figure P 13.14-1

P 13.14-2 Plot the magnitude and phase of $\mathbf{H}(j\omega)$ on a Bode diagram for the circuit shown in Figure P 13.14-2 where $\mathbf{H}(j\omega) = \mathbf{V}_o/\mathbf{V}_1$.

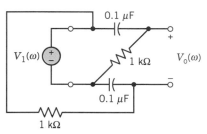

Figure P 13.14-2

P 13.14-3 Sketch the magnitude $|\mathbf{H}(\omega)|$ and the phase versus frequency for the circuit in Figure P 13.14-3 when $\mathbf{H}(j\omega) = \mathbf{V}_o/\mathbf{V}_1$, $C = 0.1~\mu F$, and $R = 1~k\Omega$.

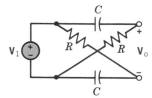

Figure P 13.14-3

PSpice PROBLEMS

SP 13-1 Obtain the Bode diagram for Problem 13.7-1.

SP 13-2 Obtain the Bode diagram for Problem 13.11-1.

SP 13-3 Obtain the Bode diagram for Problem 13.10-3.

SP 13-4 Obtain a PSpice program for obtaining the frequency response of filter F1 for the circuit of Figure 13-50. Verify the response and compare it with the result shown in Figure 13-51.

SP 13-5 Obtain the Bode diagram for Problem 13.14-2 using PSpice.

SP 13-6 The frequency response of an actual op amp is dependent on a gain $A(j\omega)$, which is a function of frequency. One model of a frequency-dependent op amp circuit is shown in Figure SP 13-6, where R_1 and R_2 are used to establish an inverting amplifier with $R_2 = 10R_1$.

Assume that the op amp has $R_i = 1~M\Omega$, $R_o = 100~\Omega$, and $A = 10^6$ with R and C establishing the bandwidth of the op amp itself. The op amp has a gain–bandwidth product equal to $Af = 10^9$ where f is the 3-dB bandwidth in hertz. Select the appropriate R and C to fit the model, and use PSpice to obtain the Bode diagram for the circuit.

SP 13-7 Use PSpice to obtain the frequency response of the crossover network of Design Problem 13-1 when all of the speaker impedances are 8 Ω, $L_1 = 2.5~mH$, $L_2 = 364~\mu H$, $C_2 = 13.76~\mu F$, and $C_3 = 5~\mu F$. Verify that the three speakers handle different parts of the audio-system bandwidth.

SP 13-8 Obtain the Bode diagram for $\mathbf{V}/\mathbf{I}$ of the circuit shown in Figure SP 13-8. Determine the bandwidth of the circuit.

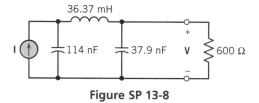

Figure SP 13-8

SP 13-9 The model of an amplifier for a phonograph stereo is shown in Figure SP 13-9. Plot the dB magnitude portion of the Bode diagram for $\mathbf{V}_o/\mathbf{V}_s$ for 20 Hz to 20 kHz.

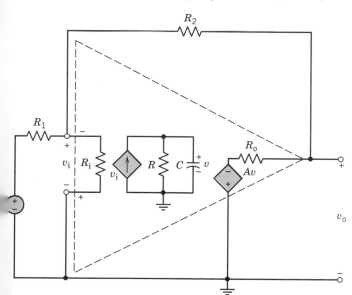

Figure SP 13-6 Bandpass filter.

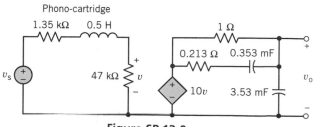

Figure SP 13-9

SP 13-10 A low-pass filter is shown in Figure SP 13-10. The output of a two-stage filter is v_1 and the output of a three-stage filter is v_2. Plot the Bode diagram of $\mathbf{V}_1/\mathbf{V}_s$ and $\mathbf{V}_2/\mathbf{V}_s$ and compare the results when $L = 10$ mH and $C = 1\ \mu$F.

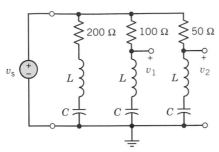

Figure SP 13-10

SP 13-11 Use PSpice to verify the design obtained for Design Problem 13-2 by plotting the dB versus frequency for 10 kHz $< f <$ 1 MHz.

SP 13-12 An acoustic sensor operates in the range of 5 kHz to 25 kHz and is represented in Figure SP 13-12 by v_s. It is specified that the bandpass filter shown in the figure pass the signal in the frequency range within 3 dB of the center frequency gain. Determine the bandwidth and center frequency of the circuit when the op amp has $R_i = 500$ kΩ, $R_o = 1$ kΩ, and $A = 10^6$.

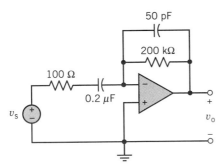

Figure SP 13-12

SP 13-13 A circuit with a transformer is shown in Figure SP 13-13. The input current i is a 1-mA sinusoidal signal. Determine and plot the magnitude of the output signal, v, for a frequency range between 0.94 MHz and 1.06 MHz. This circuit is designed to pass a signal between 0.98 MHz and 1.02 MHz and to severely reject signals at (1.00 ± 0.05) MHz. Discuss the performance of the circuit.

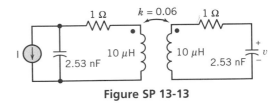

Figure SP 13-13

SP 13-14 A series connection consists of a resistance of 33 kΩ, a capacitance of 4.7 nF and a voltage source $v_s = 1\sin \omega t$. Use PSpice to obtain the Bode plot for 10 Hz to 100 kHz.

SP 13-15 Frequently, audio systems contain two or more loudspeakers that are intended to handle different parts of the audio-frequency spectrum. In a three-way setup, one speaker, called a woofer, handles low frequencies. A second, the tweeter, handles high frequencies, while a third, the midrange, handles the middle range of the audio spectrum.

A three-way filter, called a crossover network, is used to split the audio signal into the three bands of frequencies suitable for each speaker. There are many and varied designs. A simple one is based on series LR, CR, and resonant RLC circuits as shown in Figure SP 13-15. All speaker impedances are assumed resistive. The conditions are (1) woofer, at the crossover frequency: $X_{L1} = R_w$; (2) tweeter, at the crossover frequency $X_{C3} = R_T$; (3) midrange, components C_2, L_2, and R_{MR} form a series resonant circuit with upper and lower cutoff frequencies f_u and f_L, respectively. The resonant frequency $= (f_u f_L)^{1/2}$.

When all the speaker resistances are 8 Ω, determine the frequency response and determine the cutoff frequencies. Plot the Bode diagram for the three speakers. Determine the bandwidth of the midrange speaker section.

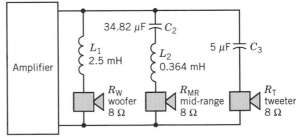

Figure SP 13-15 Three-way filter for a speaker system.

VERIFICATION PROBLEMS

VP 13-1 A laboratory report describes the circuit of Figure VP 13-1 as a low-pass filter with a corner frequency (cutoff frequency) of 10^6 rad/s. Verify these results.

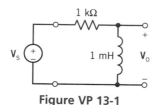

Figure VP 13-1

VP 13-2 A parallel resonant circuit (see Figure 13-16) has $Q = 70$ and a resonant frequency $\omega_0 = 10,000$ rad/s. A student report states that the bandwidth of this circuit is 71.43 rad/s. Verify this result.

VP 13-3 A series resonant circuit (see Figure 13-19) has $L = 1$ mH, $C = 10\ \mu F$, and $R = 0.5\ \Omega$. A software program report states that the resonant frequency is $f_0 = 1.59$ kHz and the bandwidth is $BW = 79.6$ Hz. Are these results correct?

VP 13-4 A series resonant circuit (see Figure 13-19) has $L = 10$ mH, $C = 1\ \mu F$, and $R = 2.5\ \Omega$. A student report states that the bandwidth is 40 Hz. Is this an accurate result?

VP 13-5 A series resonant circuit is shown in Figure 13-19 with an output $v_o = R_i$. A report states that the bandwidth of the circuit is $BW = 159$ Hz when $L = 1$ mH, $C = 0.1\ \mu F$, and $R = 1\ \Omega$. Check the accuracy of this report.

DESIGN PROBLEMS

DP 13-1 Design a circuit that has a low-frequency gain of 2, a high-frequency gain of 5, and makes the transition of $H = 2$ to $H = 5$ between the frequencies of 1 kHz and 10 kHz.

DP 13-2 Determine L and C for the circuit of Figure DP 13-2 in order to obtain a low-pass filter with a magnitude of -3 dB at 100 kHz.

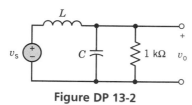

Figure DP 13-2

DP 13-3 British Rail has constructed an instrumented rail car that can be pulled over its tracks at speeds up to 180 km/hr and will measure the track-grade geometry. Using such a rail car, British Rail can monitor and track gradual degradation of the rail grade, especially the banking of curves, and permit preventive maintenance to be scheduled as needed well in advance of track-grade failure.

The instrumented rail car has numerous sensors, such as angular-rate sensors (devices that output a signal proportional to rate of rotation) and accelerometers (devices that output a signal proportional to acceleration), whose signals are filtered and combined in a fashion to create a composite sensor called a "compensated accelerometer" (Lewis, 1988). A component of this composite sensor signal is obtained by integrating and high-pass filtering an accelerometer signal. A first-order low-pass filter will ap-

proximate an integrator at frequencies well above the break frequency. This can be seen by computing the phase shift of the filter-transfer function at various frequencies. At sufficiently high frequencies, the phase shift will approach 90°, the phase characteristic of an integrator.

A circuit has been proposed to filter the accelerometer signal, as shown in Figure DP 13-3. The circuit is comprised of three sections, labeled A, B, and C. For each section, find an expression for and name the function performed by that section. Then find an expression for the gain function of the entire circuit, $\mathbf{V}_o/\mathbf{V}_s$. For the component values, evaluate the magnitude and phase of the circuit response at 0.01, 0.02, 0.05, 0.1, 0.2, 0.5, 1.0, 2.0, 5.0, and 10.0 Hz. Draw a Bode diagram. At what frequency is the phase response approximately equal to 0°? What is the significance of this frequency?

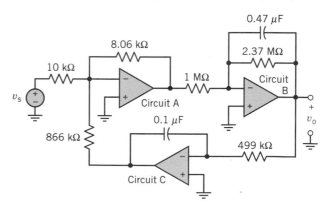

Figure DP 13-3

DP 13-4 Design a bandpass filter with a center frequency of 100 kHz and a bandwidth of 10 kHz using the circuit shown in Figure DP 13-4. Assume that $C = 100$ pF and find R and R_3 when the nonideal op amp has $R_i = 100$ kΩ, $R_o = 100$ Ω, and $A = 10^5$. Use PSpice to verify the design.

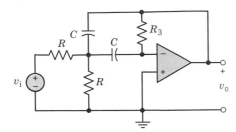

Figure DP 13-4

DP 13-5 Strain-sensing instruments can be used to measure orientation and magnitude of strains running in more than one direction. The search for a way to predict earthquakes focuses on identifying precursors, or changes, in the ground that reliably warn of an impending event. Because so few earthquakes have occurred precisely at instrumented locations, it has been a slow and frustrating quest. Laboratory studies show that before rock actually ruptures—precipitating an earthquake—its rate of internal strain increases. The material starts to fail before it actually breaks. This prelude to outright fracture is called "tertiary creep" (Brown, 1989).

The frequency of strain signals varies from 0.1 to 100 rad/s. A bandpass filter is used to pass these frequencies, so that

$$H(s) = \frac{Ks}{(s + 1)\left(\dfrac{s}{\omega_1} + 1\right)}$$

where $s = j\omega$. Plot the Bode diagram and select ω_1 and K so that the maximum gain is 20 dB at $\omega = 3.16$ rad/s.

DP 13-6 A communication transmitter requires a bandpass filter to eliminate low-frequency noise from nearby traffic. Measurements indicate that the range of traffic rumble is $2 < \omega < 12$ rad/s. A designer proposes a filter as

$$H(s) = \frac{(1 + s/\omega_1)^2(1 + s/\omega_3)}{(1 + s/\omega_2)^3}$$

where $s = j\omega$.

It is desired that signals with $\omega > 100$ rad/s pass with less than 3 dB loss while the traffic rumble be reduced by 46 dB or more. Select ω_1, ω_2, and ω_3 and plot the Bode diagram.

DP 13-7 It is desired to obtain a low-pass filter using an op amp configuration of the form shown in Figure 11-44. The goal is to achieve an overall gain of 20 dB and a bandwidth of 10^4 rad/s. Thus, the de-

sired overall response is

$$H(s) = \frac{10}{1 + (s/\omega_1)^n}$$

where $s = j\omega$, $\omega_1 = 10^4$, and $n = $ number of op amp cascaded circuit stages. Design the op amp circuit required for $n = 1$, $n = 2$, and $n = 3$ stages. Plot the magnitude plot (in decibels) for the Bode diagram and determine the gain at $\omega = 4 \times 10^4$ rad/s. The goal is to have the three-stage circuit more closely approximate the ideal filter characteristic. Discuss the results of your design.

DP 13-8 A communication transmitter requires a bandstop filter to eliminate low-frequency noise from nearby auto traffic. Measurements indicate that the range of traffic rumble is 2 rad/s $< \omega <$ 12 rad/s. A designer proposes a filter as

$$H(s) = \frac{(1 + s/\omega_1)^2(1 + s/\omega_3)^2}{(1 + s/\omega_2)^2(1 + s/\omega_4)^2}$$

where $s = j\omega$. It is desired that signals above 130 rad/s pass with less than 4 dB loss while the traffic rumble be reduced by 35 dB or more. Select ω_1, ω_2, ω_3, and ω_4 and plot the Bode diagram.

DP 13-9 An op-amp low-pass filter is shown in Figure DP 13-9. Find $H(s)$ and select R_1, R_2 and C so that $\zeta = 0.5$ and $\omega_o = 1000$ rad/s (see Figure 13-27).

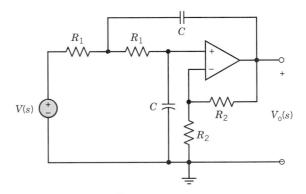

Figure DP 13-9

DP 13-10 For the circuit of Figure DP 13-10, select R_1 and R_2 so that the gain at high frequencies is 10 and the phase shift is 195° at $\omega = 1000$ rad/s. Determine the gain at $\omega = 10$ rad/s. Describe the type of filter.

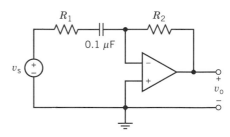

Figure DP 13-10

CHAPTER 14

THE LAPLACE TRANSFORM

PSpice Problems

Verification Problems

Design Problems

PREVIEW

In this chapter we introduce a very powerful tool for the analysis of circuits. The Laplace transform enables the circuit analyst to convert the set of differential equations describing a circuit to the complex frequency domain, where they become a set of linear algebraic equations. Then, using straightforward algebraic manipulation, one may solve for the variables of interest. Finally, one uses the inverse transform to go back to the time domain and express the desired response in terms of time. This is a powerful tool, indeed!

Also, we introduce the concept of impedance in the complex frequency domain. Thus, we may again use the impedance method to analyze circuits using techniques such as Thévenin's theorem and source transformations as described in earlier chapters.

14-1 DESIGN CHALLENGE

SPACE SHUTTLE CARGO DOOR

The U.S. space shuttle Atlantis docked with Russia's Mir space station on June 28, 1995 as shown in the cover photo of this book. The electromagnet for opening a cargo door on the NASA space shuttle requires 0.1 A before activating. The electromagnetic coil is represented by L as shown in Figure 14D-1. The activating current is designated $i_1(t)$. The time period required for i_1 to reach 0.1 A is specified as less than 3 seconds. Select a suitable value of L.

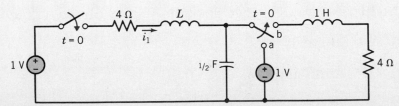

Figure 14D-1 The control circuit for a cargo door on the NASA space shuttle.

Define the Situation, State the Assumptions, and Develop a Model

1 The two switches are thrown at $t = 0$ and the movement of the second switch from terminal a to terminal b occurs instantaneously.
2 The switches prior to $t = 0$ were in position for a long time.

The Goal

Determine a value of L so that the time period for the current $i_1(t)$ to attain a value of 0.1 A is less than 3 seconds.

The circuit shown in Figure 14D-1 is a third-order circuit because it contains three energy storage devices: a capacitor and two inductors. We will need to calculate the transient response of this third-order circuit in order to design the circuit. In this chapter we will see that the Laplace Transform can be used to calculate this transient response. We will return to this design problem at the end of the chapter.

14-2 | COMMUNICATIONS AND AUTOMATION

The use of electric circuits for communications systems has grown over the past century. As society's ability to categorize, store, and transmit data increased, the concept of an information society became prevalent. An *information society* is one in which there is an abundance (some say an excess) in quantity and quality of information with all the necessary facilities for its distribution. Electrical communication systems facilitate the quick, efficient distribution and conversion of this information at a price that almost anyone can afford.

The balance of employment and products in the United States has been shifting toward information-related activities for more than 40 years. In 1988, according to some observers, about one-half of the U.S. worker population was employed in one form or another of information work, in contrast to 3 percent in agriculture and 22 percent in manufacturing.

Researchers have defined an *information ratio,* which is the percentage of household expenditures for various kinds of information-related activities, such as the time spent with reading, television, radio, and the like (Dordick, 1986). The information ratio for selected countries is shown in Figure 14-1. Note how all nations are moving up the trend line as they increase their information ratio and their per capita income.

The information technologies, communications, and computers have reached the office, home, and factory. Industrialization is dependent on the use of information to control the production process in the factory. In 1908 Henry Ford developed his assembly line, which was to turn out millions of Model T's (Dorf, 1974). The term automation first became popular in the automobile industry. *Automation* is the automatic operation and control of a process, device, or system. For example, a factory may use an automatically controlled machine to produce a product within specified tolerances.

A large impetus to the theory and practice of automation, also often called *automatic control,* occurred during World War II when it became necessary to design and construct automatic airplane pilots, gun control devices, and positioning systems for radar antennas. These new techniques moved to the factory after 1950.

It has been to the advantage of the producer to automate in order to reduce costs and increase quality. Furthermore, many conditions require remotely controlled devices. For example, the remotely controlled device shown in Figure 14-2 is used to enter and to move sensors about radiation hazard areas in nuclear power plants.

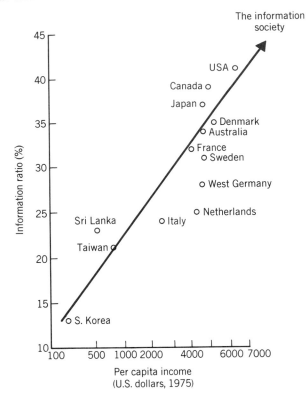

Figure 14-1 Information ratio (or index) for selected countries. As nations industrialize, they move toward becoming information societies. Adapted from Dordick, p. 14.

With the advent of the requirement to control complex devices such as the Surveyor vehicle or a robot, it became necessary to describe the characteristics and behavior of complex communication and control circuits and systems as they experience myriad excitation inputs.

In our preceding studies, we considered the analysis of circuits that experience one or more switch changes and have constant sources. Later, we considered the analysis of first- and second-order circuits described by differential equations. Then we considered the steady-state response of circuits with sinusoidal sources. In this chapter we consider a method of analysis that is particularly useful for the analysis of circuits incorporating independent sources that take on many forms, which may include a sinusoid, and exponential, or a step function.

Figure 14-2 The Surveyor robot assumes a role in nuclear power plants where limitations or hazards exist for humans. The Surveyor can climb steps and carry sensors into radiation areas. Surveyor is an untethered, remote-controlled surveillance system. Courtesy of Electric Power Research Institute.

14-3 ‖ LAPLACE TRANSFORM

As we have seen in earlier chapters, it is useful to *transform* the equations describing a circuit from the time domain into the frequency domain, then perform an analysis, and finally transform the problem's solution back to the time domain. You will recall that in Chapter 11 we defined the phasor as a mathematical transformation used to simplify the solution of the circuit response to a sinusoidal source at steady-state. The phasor transformation resulted in the problem being transformed to one of algebraic manipulation of complex numbers.

A **transform** is a change in the mathematical description of a physical variable to facilitate computation.

The transform method is summarized in Figure 14-3.

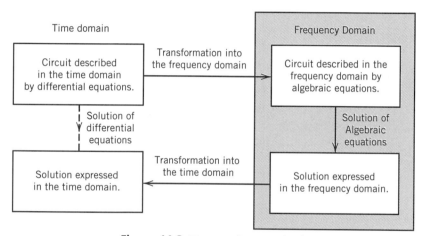

Figure 14-3 The transform method.

We are interested in using the transform that will take the differential equations describing a circuit into the complex frequency domain, where we may readily perform the required algebraic manipulations to determine the desired response. Then we may use an inverse transform to obtain the response in the time domain.

Pierre-Simon Laplace, who is shown in Figure 14-4, is credited with a transform that bears his name. The *Laplace transform* is defined as

$$\mathcal{L}[f(t)] = \int_0^\infty f(t)e^{-st}\,dt \tag{14-1}$$

and

$$\mathcal{L}[f(t)] = F(s) \tag{14-2}$$

so $\mathcal{L}$ implies the Laplace transform. Here the complex frequency is $s = \sigma + j\omega$. We only treat the one-sided (unilateral) transform for $t \geq 0$.

The notation implies that once the integral has been evaluated, $f(t)$, which is a time domain function, is transformed to $F(s)$, which is a frequency domain function.

We should stop and ask under what conditions the integral of Eq. 14-1 converges to a

Figure 14-4 Pierre-Simon Laplace (1749–1827) is credited with the transform that bears his name. Courtesy of Burndy Library.

finite value. It can be shown that the integral converges when

$$\int_0^\infty |f(t)| \, e^{-\sigma_1 t} \, dt < \infty$$

for some real positive σ_1. If the magnitude of $f(t)$ is $|f(t)| < Me^{\alpha t}$ for all positive t, the integral will converge for $\sigma_1 > \alpha$. The region of convergence is therefore given by $\infty > \sigma_1 > \alpha$, and σ_1 is known as the abscissa of absolute convergence. Functions of time, $f(t)$, that are physically possible always have a Laplace transform. Also, we will assume that $f(t) = 0$ for $t < 0$. We assume that the initial conditions for the circuit account for the circuit's behavior prior to $t = 0$. We will consider $f(t)$ to be continuous after $t = 0$. However, we can permit a discontinuity at $t = 0$ and assume the transform occurs for $t > 0$. Thus, we permit a discontinuity between $t = 0^-$ and $t = 0^+$.

If we have a transformation $\mathcal{L}[f(t)]$, then we must have an inverse transformation $\mathcal{L}^{-1}[F(s)] = f(t)$. If we develop a series of transformations using Eq. 14-1, we may obtain a table of *transform pairs* as listed in Table 14-1.

Table 14-1
Important Laplace Transform Pairs

$f(t), t \geq 0$	$F(s) = \mathcal{L}\{f(t)u(t)\}$
Step function, $u(t)$	$\dfrac{1}{s}$
e^{-at}	$\dfrac{1}{s+a}$
$\sin \omega t$	$\dfrac{\omega}{s^2 + \omega^2}$
$\cos \omega t$	$\dfrac{s}{s^2 + \omega^2}$
$e^{-at}f(t)$	$F(s+a)$
t^n	$\dfrac{n!}{s^{n+1}}$
$f^{(k)}(t) = \dfrac{d^k f(t)}{dt^k}$	$s^k F(s) - s^{k-1}f(0^+) - s^{k-2}f'(0^+) - \cdots - f^{(k-1)}(0^+)$

This table can be constructed by using Eq. 14-1. Again, note that we only consider the Laplace transform for signals with nonzero values only for $t \geq 0$. We have $f(t)u(t) = 0$ for $t < 0$. For example, consider $f(t) = e^{-at}u(t)$, where $a > 0$ and $u(t)$ is the unit step function. Then we have

$$\mathcal{L}[e^{-at}u(t)] = F(s) = \int_0^\infty e^{-at}e^{-st} \, dt$$

$$= \frac{-e^{-(s+a)t}}{s+a} \bigg|_0^\infty$$

$$= \frac{1}{s+a} \tag{14-3}$$

since the value at the upper limit is zero because the convergence condition is satisfied by $\sigma > -a$.

Now let us find the transform of the familiar unit step function $u(t)$. We have

$$\mathcal{L}[u(t)] = F(s) = \int_0^\infty e^{-st} \, dt$$

$$= \frac{1}{s}$$

where convergence is satisfied by $\sigma > 0$.

Let us obtain the transform of the first derivative of $f(t)$, denoted as $f'(t)$. We have

$$\mathcal{L}[f'(t)] = \mathcal{L}\left[\frac{df}{dt}\right] = \int_0^\infty f'(t)e^{-st} \, dt$$

Integrating by parts, we have

$$\mathcal{L}[f'(t)] = s\int_0^\infty f(t)e^{-st} \, dt + f(t)e^{-st} \bigg|_0^\infty$$

Again, we assume that the integrated term vanishes at the upper limit, so that

$$\mathcal{L}[f'(t)] = sF(s) - f(0) \tag{14-4}$$

Thus, the Laplace transform of the derivative of a function $f(t)$ is s times the Laplace transform of the function minus the initial condition.

For example, if $f(t) = e^{-t}$, we have

$$\mathcal{L}[f'(t)] = s\left(\frac{1}{s+1}\right) - f(0)$$

where $1/(s+1)$ is the Laplace transform of $f(t) = e^{-t}$, $t \geq 0$.

Another important property of the Laplace transform is linearity. That is, if $F_1(s)$ and $F_2(s)$ are the Laplace transforms of the time functions $f_1(t)$ and $f_2(t)$, respectively, then

$$\mathcal{L}[a_1 f_1(t) + a_2 f_2(t)] = a_1 F_1(s) + a_2 F_2(s) \tag{14-5}$$

for arbitrary constants a_1 and a_2.

Uniqueness is a fundamental property of the Laplace transform. Thus, if two time functions $f_1(t)$ and $f_2(t)$ have the same transform $F(s)$, then $f_1 = f_2$. Therefore, for $f(t)$ a unique function $F(s)$ exists. Conversely, given a Laplace transform $F(s)$, there is a unique time function $f(t)$. This property is written as

$$f(t) = \mathcal{L}^{-1}[F(s)]$$

meaning that $f(t)$ is the *inverse Laplace transform* of $F(s)$ and exists for $t \geq 0$.

The inverse Laplace transform is defined by the complex inversion integral

$$f(t) = \frac{1}{2\pi j} \int_{\alpha - j\infty}^{\alpha + j\infty} F(s)e^{st}ds \tag{14-6}$$

The uniqueness of the Laplace transform enables us to avoid this complex integration. Instead, we will build Laplace transform tables, such as Tables 14-1 and 14-3, by calculating the Laplace transform of several key functions. When we want to find the inverse Laplace transform, we will look in these tables. For example, suppose we want to find the inverse Laplace transform of

$$F(s) = \frac{2}{s}$$

Unfortunately $2/s$ is not in Table 14-1 or 14-3. Linearity tells us that

$$\mathcal{L}\left\{\frac{2}{s}\right\} = 2\mathcal{L}\left\{\frac{1}{s}\right\}$$

Further, $1/s$ is in Table 14-1. Thus, we find

$$f(t) = 2u(t)$$

Notice that we needed both a Laplace transform pair, $2[u(t)] = 1/s$, and a Laplace transform property, linearity. The success of this method depends on having a supply of Laplace transform pairs (Tables 14-1 and 14-3) and Laplace transform properties (Table 14-2).

Table 14-2
Laplace Transform Properties

Property	Relationship
Linearity	$\mathcal{L}[a_1 f_1(t) + a_2 f_2(t)] = a_1 F_1(s) + a_2 F_2(s)$
Time delay	$\mathcal{L}[f(t - \tau)u(t - \tau)] = e^{-s\tau}F(s)$
Impulse	$\mathcal{L}[\delta(t)] = 1$
Frequency shift	$\mathcal{L}[e^{-at}f(t)] = F(s + a)$
Product of time and a function	$\mathcal{L}[tf(t)] = \dfrac{-dF(s)}{ds}$
	where $F(s) = \mathcal{L}[f(t)]$

Example 14-1
Find the Laplace transform of $\sin \omega t$.

Solution
Since

$$\sin \omega t = \frac{1}{2j}(e^{j\omega t} - e^{-j\omega t})$$

and since we know $\mathcal{L}(e^{at}) = 1/(s - a)$, we then have

$$\mathcal{L}(\sin \omega t) = \frac{1}{2j}\left(\frac{1}{s - j\omega} - \frac{1}{s + j\omega}\right)$$

$$= \frac{(s + j\omega) - (s - j\omega)}{2j(s - j\omega)(s + j\omega)}$$

$$= \frac{\omega}{s^2 + \omega^2}$$

EXERCISE 14-1

Find the Laplace transform of $f(t) = \cos \omega t$.

Answer: $F(s) = \dfrac{s}{s^2 + \omega^2}$

EXERCISE 14-2

Using the linearity property, find the Laplace transform of $f(t) = e^{-2t} + \sin t$.

Answer: $F(s) = \dfrac{s^2 + s + 3}{(s + 2)(s^2 + 1)}$

14-4 ‖ IMPULSE FUNCTION AND TIME SHIFT PROPERTY

Let us consider a special function called the *impulse function,* denoted as $\delta(t)$. First, we consider the pulse centered at $t = 0$ as shown in Figure 14-5. We have

$$f(t) = \frac{1}{a} \qquad -\frac{a}{2} < t < \frac{a}{2}$$

$$= 0 \qquad \text{all other } t \tag{14-7}$$

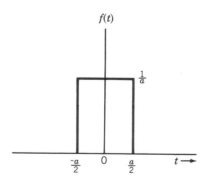

Figure 14-5 Pulse of width a and of height $1/a$ centered at $t = 0$.

where a is very small. Note that the area under the pulse remains equal to 1 regardless of the value of a. If a becomes progressively smaller, the area remains equal to 1, but we have a large pulse height with a very narrow width.

An **impulse** is a pulse of infinite amplitude for an infinitesimal time whose area $\int_{-\infty}^{\infty} f(t)\, dt$ is finite.

The *impulse function* is defined as

$$\delta(t) = 0 \qquad \text{for } t \neq 0$$

and

$$\int_{-\infty}^{\infty} \delta(t) \, dt = 1 \tag{14-8}$$

Therefore, we have $f(t)\delta(t) = f(0)\delta(t)$ since $\delta(t) = 0$ for $t \neq 0$.

Let us determine the Laplace transform of the impulse function. By definition, we have

$$\mathcal{L}[\delta(t)] = \int_{0^-}^{\infty} e^{-st}\delta(t) \, dt \tag{14-9}$$

where the lower limit is 0^- since we have an infinite discontinuity at $t = 0$. Since $\delta(t) = 0$ for $t \neq 0$, the integral of Eq. 14-9 is evaluated between 0^- and 0^+ to obtain

$$\mathcal{L}[\delta(t)] = e^{-st}\Big|_{\text{at } t=0}$$
$$= 1$$

Now let us consider the transform of a time function shifted τ seconds in time. If we have $f(t)$ as shown in Figure 14-6a and we wish to shift (delay) it to τ seconds later, we may write it as

$$f(t - \tau)u(t - \tau)$$

where $u(t - \tau)$ is the unit step function, which is zero for $t < \tau$ and 1 for $t > \tau$. The delayed function is shown in Figure 14-6b.

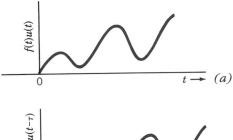

Figure 14-6 *(a)* A function of time $f(t)$ and *(b)* the function delayed by τ s.

To obtain the transform of the time-shifted function, we use the definition of the transform to obtain

$$\mathcal{L}[f(t - \tau) \, u(t - \tau)] = \int_{0}^{\infty} f(t - \tau)u(t - \tau)e^{-st} \, dt$$
$$= \int_{\tau}^{\infty} f(t - \tau)e^{-st} \, dt$$

Now let $t - \tau = x$ to obtain

$$\mathcal{L}[f(t - \tau)\, u(t - \tau)] = \int_0^\infty f(x)e^{-s(\tau + x)}\, dx$$

$$= e^{-s\tau} \int_0^\infty f(x)e^{-sx}\, dx$$

$$= e^{-s\tau}F(s) \tag{14-10}$$

Thus, for example, if we have a step function of amplitude A delayed by 2 seconds, we have the Laplace transform

$$\mathcal{L}[Au(t - 2)] = e^{-2s}\left(\frac{A}{s}\right)$$

since $\mathcal{L}[Au(t)] = A/s$.

Another property can be obtained from the Laplace transform of $e^{-at}f(t)$ as follows:

$$\mathcal{L}[e^{-at}f(t)] = \int_0^\infty e^{-at}f(t)e^{-st}\, dt$$

$$= \int_0^\infty f(t)e^{-(s + a)}\, dt$$

$$= F(s + a) \tag{14-11}$$

This property is called the frequency shift property. The frequency shift property and the properties of the impulse, the time delay, and linearity are summarized in Table 14-2.

We may use the frequency shift property of Eq. 14-11 to obtain additional transform relationships. For example, consider

$$\mathcal{L}(e^{-at} \sin \omega t) = F(s + a)$$

Since

$$\mathcal{L}(\sin \omega t) = \frac{\omega}{s^2 + \omega^2}$$

we have

$$\mathcal{L}(e^{-at} \sin \omega t) = \frac{\omega}{(s + a)^2 + \omega^2} \tag{14-12}$$

Example 14-2

Find the Laplace transform of $g(t) = e^{-4t}u(t - 3)$, which is shown in Figure 14-7.

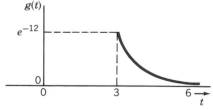

Figure 14-7 Delayed function $g(t) = e^{-4t}u(t - 3)$.

Solution
First, find the transform of $u(t - 3)$. Using the time delay property, we have

$$F(s) = \mathcal{L}[u(t - 3)] = e^{-3s}\left(\frac{1}{s}\right)$$

Then, using the frequency shift property, we have

$$
\begin{aligned}
G(s) &= \mathcal{L}[e^{-4t}u(t - 3)] \\
&= \mathcal{L}[e^{-4t}f(t)] \\
&= F(s + 4)
\end{aligned}
$$

Therefore, we have

$$G(s) = e^{-3(s+4)}\frac{1}{s + 4}$$

$$= e^{-12}\frac{e^{-3s}}{s + 4}$$

Another property of great utility arises for the product of time t and $f(t)$. Start with the derivative of $F(s)$ with respect to s, where $F(s) = \mathcal{L}[f(t)]$. Then we have

$$
\begin{aligned}
\frac{dF(s)}{ds} &= \int_0^\infty \frac{d}{ds}f(t)e^{-st}\,dt \\
&= \int_0^\infty -tf(t)e^{-st}\,dt
\end{aligned}
$$

Consequently,

$$\mathcal{L}[tf(t)] = -\frac{dF(s)}{ds} \qquad (14\text{-}13)$$

where

$$F(s) = \mathcal{L}[f(t)]$$

This property is also included in Table 14-2.

For example, let us find the Laplace transform of $tu(t)$. We have

$$\mathcal{L}[tu(t)] = -\frac{dF(s)}{ds}$$

$$= -\frac{d}{ds}\left(\frac{1}{s}\right)$$

where

$$F(s) = \mathcal{L}[u(t)] = 1/s$$

Completing the derivative,

$$\mathcal{L}[tu(t)] = \frac{1}{s^2}$$

These additional transform pairs are summarized in Table 14-3.

Table 14-3

Additional Transform Pairs

$f(t),\ t \geq 0$	$F(s)$
te^{-at}	$\dfrac{1}{(s + a)^2}$
t^n	$\dfrac{n!}{s^{n+1}}$
$e^{-at}\sin \omega t$	$\dfrac{\omega}{(s + a)^2 + \omega^2}$
$e^{-at}\cos \omega t$	$\dfrac{s + a}{(s + a)^2 + \omega^2}$
$e^{-bt}t^n$	$\dfrac{n!}{(s + b)^{n+1}}$

EXERCISE 14-3

Using the linearity property and the frequency shift property, find $\mathcal{L}[2u(t) + 3e^{-4t}u(t)] = P(s)$.

Answer: $P(s) = \dfrac{2}{s} + \dfrac{3}{s + 4}$

EXERCISE 14-4

Using the time shift property, find $F(s) = \mathcal{L}[\sin(t - \tau)u(t - \tau)]$ where $\tau = 2$.

Answer: $F(s) = \dfrac{e^{-2s}}{s^2 + 1}$

EXERCISE 14-5

Find $\mathcal{L}(te^{-t}) = F(s)$.

Answer: $F(s) = \dfrac{1}{(s + 1)^2}$

EXERCISE 14-6

Determine the Laplace transform of $f(t)$ shown in Figure E 14-6.

Answer: $F(s) = \dfrac{(5 - 6s)e^{-4.2s} + 15s - 5}{3s^2}$

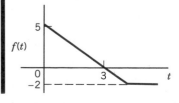

Figure E 14-6

EXERCISE 14-7

Use the Laplace transform to obtain the transform of the signal $f(t)$ shown in Figure E 14-7.

Answer: $F(s) = \dfrac{3(1 - e^{-2s})}{s}$

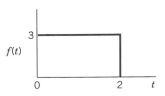

Figure E 14-7

EXERCISE 14-8

Determine the Laplace transform of $f(t)$ shown in Figure E 14-8.

Answer: $F(s) = \dfrac{5}{2s^2}(1 - e^{-2s} - 2se^{-2s})$

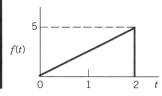

Figure E 14-8

14-5 INVERSE LAPLACE TRANSFORM

We have built up a set of transform pairs in Table 14-1 and Table 14-3. We can use these relationships between $F(s)$ and $f(t)$ to find

$$f(t) = \mathcal{L}^{-1}[F(s)]$$

where the symbol $\mathcal{L}^{-1}$ denotes the inverse transform.

Suppose we have

$$F(s) = \frac{2}{s} + \frac{3}{s + 2} + \frac{9}{s^2 + 9}$$
$$= F_1(s) + F_2(s) + F_3(s) \qquad (14\text{-}14)$$

We proceed to find the inverse transform for each term and obtain

$$f(t) = \mathcal{L}^{-1}[F(s)] = 2 + 3e^{-2t} + 3 \sin 3t \qquad t \geq 0$$

Hence if we wish to obtain the inverse transform of a general form of $F(s)$, we attempt to obtain a *partial fraction expansion* of $F(s)$ to get a series of terms similar to Eq. 14-14 so that we may find the inverse transform of each term.

For example, consider the case where we wish to find the expansion of $F(s)$ as follows:

$$F(s) = \frac{s + 3}{(s + 1)(s + 2)} = \frac{A}{(s + 1)} + \frac{B}{(s + 2)}$$
$$= F_1(s) + F_2(s) \qquad (14\text{-}15)$$

We may evaluate A by multiplying through by the denominator of $F_1(s)$ and setting s equal to -1. Following this procedure, we multiply by $s + 1$ to obtain

$$(s + 1)F(s) = \frac{s + 3}{s + 2} = A + \frac{B(s + 1)}{s + 2}$$

Then we let $s = -1$ to obtain

$$\frac{s + 3}{s + 2}\bigg|_{s=-1} = A + \frac{B(s + 1)}{s + 2}\bigg|_{s=-1}$$

or

$$\frac{-1 + 3}{-1 + 2} = A + 0$$

Consequently, $A = 2$.

Following the same procedure for $F_2(s)$, we have

$$\frac{s + 3}{s + 1}\bigg|_{s=-2} = B$$

or

$$B = -1$$

Therefore, substituting A and B into Eq. 14-15, we have

$$F(s) = \frac{2}{s + 1} + \frac{-1}{s + 2}$$

Taking the inverse Laplace transform of each term, we have

$$f(t) = 2e^{-t} - e^{-2t}, \qquad t \geq 0$$

$F(s)$ of Eq. 14-14 has a numerator of order 1 and a denominator of order 2. In general, we write

$$F(s) = \frac{N(s)}{D(s)}$$

$$= \frac{b_m s^m + b_{m-1} s^{m-1} + \cdots + b_1 s + b_0}{a_n s^n + a_{n-1} s^{n-1} + \cdots + a_1 s + a_0} \qquad (14\text{-}16)$$

where the coefficients of the polynomials are real numbers. The function $F(s)$ is said to be a rational function of s since it is the ratio of two polynomials in s. The roots of the denominator polynomial $D(s)$ are the roots of the equation $D(s) = 0$ and are called the *poles* of $F(s)$. The roots of the numerator polynomial are called the *zeros* of $F(s)$. Normally, we have the case $n > m$. For the case $n = m$, we first perform long division to obtain

$$F(s) = (b_m/a_n) + G(s)$$

Then we proceed to find $f(t)$ by inverse transformation of all the terms after obtaining the partial fraction of $G(s)$.

As an example of the special case $m = n$ for Eq. 14-16, consider

$$F(s) = \frac{s}{s + 1}$$

Use long division to obtain

$$F(s) = 1 + \frac{-1}{s + 1}$$

Then we obtain the inverse transform of each term to yield

$$f(t) = \delta(t) - 1e^{-t}, \qquad t \geq 0$$

where $\delta(t)$ is the impulse function.

If we have repeated poles, we use a partial fraction expansion that includes all powers of the term $(s + p)$ up to the power of the repeated poles of $F(s)$. For example, if

$$F(s) = \frac{4}{(s + 1)^2(s + 2)}$$

we may arrange the partial fraction expansion as

$$F(s) = \frac{4}{(s + 1)^2(s + 2)} = \frac{A}{(s + 1)^2} + \frac{B}{s + 1} + \frac{C}{s + 2} \qquad (14\text{-}17)$$

First, we may evaluate C by multiplying through by $s + 2$ and letting $s = -2$. Then we have $C = 4$. To obtain A, we multiply through by $(s + 1)^2$ and let $s = -1$, obtaining

$$\left. \frac{4}{(s + 2)} \right|_{s = -1} = A$$

or $A = 4$.

To find B, since we cannot use $s = -1$ again, we multiply both sides of Eq. 14-17 by the denominator polynomial, $D(s)$, to obtain

$$4 = A(s + 2) + B(s + 1)(s + 2) + C(s + 1)^2$$
$$= 4(s + 2) + B(s + 1)(s + 2) + 4(s + 1)^2$$

Equating coefficients of s^2 yields

$$B + 4 = 0$$

or $B = -4$. Then

$$F(s) = \frac{4}{(s + 1)^2} - \frac{4}{s + 1} + \frac{4}{s + 2}$$

Recall from Table 14-3 that

$$\mathcal{L}^{-1}\left[\frac{a}{(s + b)^2} \right] = ate^{-bt}$$

Therefore, we have

$$f(t) = 4(te^{-t} - e^{-t} + e^{-2t}) \qquad t \geq 0$$

Let us consider another approach. In the case of multiple roots, where a root r_i is repeated m times, the partial fraction expansion must include

$$\frac{b_1}{(s - r_i)} + \cdots + \frac{b_l}{(s - r_i)^l} + \cdots + \frac{b_m}{(s - r_i)^m}$$

It can be shown that the coefficient b_{m-k} is

$$b_{m-k} = \frac{1}{k!} \frac{d^k}{ds^k} [(s - r_i)^m F(s)] \Big|_{s=r_i}$$

where $k = 0, 1, \ldots, m - 1$.

Therefore, to evaluate B of Eq. 14-17, we set $k = 1$ and evaluate

$$\begin{aligned}
B = b_1 &= \frac{d}{ds} (s - r_i)^2 F(s) \Big|_{s=r_i} \\
&= \frac{d}{ds} (s + 1)^2 F(s) \Big|_{s=-1} \\
&= \frac{d}{ds} \frac{4}{s + 2} \Big|_{s=-1} \\
&= \frac{-4}{(s + 2)^2} \Big|_{s=-1} = -4
\end{aligned}$$

Often we will encounter an $F(s)$ that has two complex poles, as follows:

$$F(s) = \frac{N(s)}{(s + a + j\omega)(s + a - j\omega)} \tag{14-18}$$

where $N(s)$ is the unspecified numerator polynomial. The $F(s)$ may also be written as

$$F(s) = \frac{N(s)}{(s + a)^2 + \omega^2} = \frac{N(s)}{D(s)} \tag{14-19}$$

If $N(s) = s + a$, we know from Table 14-3 that

$$f(t) = e^{-at} \cos \omega t$$

If $N(s) = d$, we find from Table 14-3 that

$$f(t) = \frac{d}{\omega} e^{-at} \sin \omega t \geq 0$$

Example 14-3
Find $f(t)$ when

$$F(s) = \frac{10}{(s + 2)(s^2 + 6s + 10)}$$

Solution
The roots of the quadratic $(s^2 + 6s + 10)$ are complex, and we may write $F(s)$ as

$$F(s) = \frac{10}{(s + 2)[(s + 3)^2 + 1]} = \frac{10}{(s + 2)(s + 3 - j)(s + 3 + j)}$$

Using a partial fraction expansion, we have

$$F(s) = \frac{10}{(s + 2)(s^2 + 6s + 10)} = \frac{A}{s + 2} + \frac{B}{s - r_1} + \frac{B^*}{s - r_1^*} \tag{14-20}$$

where $r_1 = -3 + j$ and r_1^* is the conjugate of r_1. It is easy to show that we obtain B^* for the partial factor $B^*/(s - r_1^*)$. Using the partial fraction method, we find

$$A = 5 \qquad B = -\frac{5}{2} + j\frac{5}{2} \qquad B^* = -\frac{5}{2} - j\frac{5}{2}$$

Combining the two complex terms, we have

$$F(s) = \frac{5}{s + 2} + \frac{B}{s - r_1} + \frac{B^*}{s - r_1^*}$$

$$= \frac{5}{s + 2} + \frac{-5s - 20}{(s + 3)^2 + 1}$$

Rearranging the second term into two terms for which the inverse will yield a cosine and a sine function, we have

$$F(s) = \frac{5}{s + 2} - \frac{5(s + 3)}{(s + 3)^2 + 1} - \frac{5}{(s + 3)^2 + 1}$$

Using Table 14-3, we obtain

$$f(t) = 5e^{-2t} - 5e^{-3t} \cos t - 5e^{-3t} \sin t$$

$$= 5e^{-2t} - 5e^{-3t}(\cos t + \sin t), \qquad t \geq 0$$

Since the function $F(s) = \omega_n^2/(s^2 + 2\zeta\omega_n s + \omega_n^2)$ occurs frequently where the roots of the denominator are complex, we seek the inverse transform $f(t)$.

When we have

$$F(s) = \frac{\omega_n^2}{s^2 + 2\zeta\omega_n s + \omega_n^2}$$

we obtain

$$f(t) = \frac{\omega_n}{b} e^{-\zeta\omega_n t} \sin \omega_n bt$$

where $\zeta < 1$ and $b = \sqrt{1 - \zeta^2}$.

Table 14-4

Transforms of F(s) with Complex Poles

$f(t),\ t \geq 0$	$F(s)$
1. $\dfrac{\omega_n}{b} e^{-\zeta\omega_n t} \sin \omega_n bt$ where $b = \sqrt{1 - \zeta^2}$	$\dfrac{\omega_n^2}{s^2 + 2\zeta\omega_n s + \omega_n^2}$
2. $e^{-at}(2c \cos \omega t - 2d \sin \omega t)$	$\dfrac{c + jd}{s + a - j\omega} + \dfrac{c - jd}{s + a + j\omega}$
3. $2me^{-at} \cos(\omega t + \theta)$ where $m = \sqrt{c^2 + d^2}$ and $\theta = \tan^{-1}(d/c)$	$\dfrac{me^{j\theta}}{s + a - j\omega} + \dfrac{me^{-j\theta}}{s + a + j\omega}$
4. $\dfrac{1}{\omega}(a^2 + \omega^2)^{1/2} e^{-at} \sin(\omega t + \phi)$ $\phi = \tan^{-1}(\omega/-a)$	$\dfrac{s}{(s + a)^2 + \omega^2}$

Another useful relationship, which you are invited to prove in Exercise 14-9, is for

$$F(s) = \frac{c + jd}{s + a - j\omega} + \frac{c - jd}{s + a + j\omega}$$

$$= \frac{2[cs + (ca - \omega d)]}{(s + a)^2 + \omega^2} \tag{14-21}$$

Then

$$f(t) = 2me^{-at}\cos(\omega t + \theta) \qquad t \geq 0$$

where

$$m = \sqrt{c^2 + d^2} \quad \text{and} \quad \theta = \tan^{-1}\left(\frac{d}{c}\right)$$

Four forms of $F(s)$ with complex poles are shown in Table 14-4 along with their inverse transforms.

Recall from Chapter 13 that a function $F(s)$ in the form of Eq. 14-21 has two complex poles that may be shown on the s-plane. Figure 14-8 shows a three-dimensional plot of $|F(s)|$ for the left-hand s-plane when $\omega_n = 5$ and $\zeta = 0.1$. Then, we have

$$F(s) = \frac{25}{s^2 + s + 25}$$

$$= \frac{25}{(s - p_1)(s - p_1^*)}$$

where $p_1 = -0.5 + j4.97$. Note the resonant frequency response at $s = +j\omega \cong j5$.

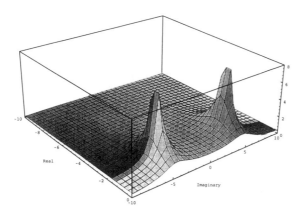

Figure 14-8 A three-dimensional plot of $|F(s)|$ on the left-hand s-plane. Note that the plot caps the magnitude at $s = p_1$ at 8 when it actually goes to infinity at that point. Courtesy of Mark A. Yoder of Rose–Hulman Institute.

EXERCISE 14-9

Prove that

$$\mathcal{L}^{-1}\left[\frac{cs + (ca - \omega d)}{(s + a)^2 + \omega^2}\right]$$

(for Eq. 14-21) is $f(t) = me^{-at}\cos(\omega t + \theta)$ where $m = \sqrt{c^2 + d^2}$ and $\theta = \tan^{-1}(d/c)$.

EXERCISE 14-10

Find the inverse transform of $F(s)$ expressing $f(t)$ in cosine and angle forms.

(a) $F(s) = \dfrac{8s - 3}{s^2 + 4s + 13}$ (b) $F(s) = \dfrac{3e^{-s}}{s^2 + 2s + 17}$

Answer: (a) $f(t) = 10.2e^{-2t}\cos(3t + 38.4°),\ t \geq 0$

(b) $f(t) = \frac{3}{4}e^{-(t-1)}\sin[4(t - 1)],\ t \geq 1$

EXERCISE 14-11

Find the inverse transform of $F(s)$.

(a) $F(s) = \dfrac{s^2 - 5}{s(s + 1)^2}$ (b) $F(s) = \dfrac{4s^2}{(s + 3)^3}$

Answer: (a) $f(t) = -5 + 6e^{-t} + 4te^{-t},\ t \geq 0$

(b) $f(t) = 4e^{-3t} - 24te^{-3t} + 18t^2e^{-3t},\ t \geq 0$

14-6 | INITIAL AND FINAL VALUE THEOREMS

The initial value of a function $f(t)$ is the value at $t = 0$, provided that $f(t)$ is continuous at $t = 0$. If $f(t)$ is discontinuous at $t = 0$, then the initial value is the limit as $t \rightarrow 0^+$, where t approaches $t = 0$ from positive time.

A function's *initial value* may be found from

$$f(0) = \lim_{s \to \infty} sF(s) \tag{14-22}$$

To prove the relationship, substitute the definition of $F(s)$ on the right side of Eq. 14-22 to obtain

$$\lim_{s \to \infty} sF(s) = \lim_{s \to \infty} s \int_0^\infty f(t)e^{-st}\, dt$$

$$= \lim_{s \to \infty} \int_0^\infty [f(t)se^{-st}]\, dt$$

where lim is used for the notation "limit." The integrand within the brackets is very small except near $t = 0$. Therefore,

$$\lim_{s \to \infty} sF(s) \cong \lim_{s \to \infty} f(0) \int_0^\infty se^{-st}\, dt$$

$$= f(0) \lim_{s \to \infty} \left[\frac{-se^{-st}}{s} \Big|_0^\infty \right]$$

$$= f(0)$$

The initial value theorem is easy to apply for a specific $F(s)$. Consider the case where

$$F(s) = \frac{-2s^3 + 7s^2 + 2s + 9}{3s^4 + 3s^3 + 2s^2 + 6}$$

Hence

$$f(0) = \lim_{s \to \infty} sF(s) = \lim_{s \to \infty} s \left(\frac{-2s^3}{3s^4} \right) = \frac{-2}{3}$$

If we have

$$F(s) = \frac{\omega_0}{s^2 + as + \omega_0^2}$$

then

$$f(0) = \lim_{s \to \infty} sF(s) = \lim_{s \to \infty} s \frac{\omega_0}{s^2 + as + \omega_0^2}$$
$$= 0$$

The *final value* of a function $f(t)$ is $\lim_{t \to \infty} f(t)$ where

$$\lim_{t \to \infty} f(t) = \lim_{s \to 0} sF(s)$$

To prove the relationship, first note that by the derivative property

$$\mathcal{L} \left(\frac{df}{dt} \right) = sF(s) - f(0) \tag{14.23}$$

and

$$\mathcal{L} \left(\frac{df}{dt} \right) = \int_0^\infty \left(\frac{df}{dt} \right) e^{-st} \, dt \tag{14-24}$$

Equating Eqs. 14-23 and 14-24, we have

$$\int_0^\infty \frac{df}{dt} e^{-st} \, dt = sF(s) - f(0) \tag{14-25}$$

and we take the limit as $s \to 0$ for both sides of Eq. 14-25 to obtain

$$\lim_{s \to 0} \int_0^\infty \frac{df}{dt} e^{-st} \, dt = \lim_{s \to 0} [sF(s) - f(0)] \tag{14-26}$$

We assume that $f(t)$ has a limit as $t \to \infty$ and use integration by parts for the integral of the left-hand side of Eq. 14-26 to obtain

$$\int_0^\infty \frac{df}{dt} e^{-st} \, dt = f(\infty) - f(0) \tag{14-27}$$

Therefore,

$$\lim_{s \to 0} \int_0^\infty \frac{df}{dt} e^{-st} \, dt = \lim_{s \to 0} [f(\infty) - f(0)] \tag{14-28}$$

Equating (14-26) and (14-28), we have

$$\lim_{s \to 0} [f(\infty) - f(0)] = \lim_{s \to 0} [sF(s) - f(0)]$$

Evaluating the limit we obtain

$$f(\infty) = \lim_{s \to 0} sF(s) \tag{14-29}$$

and this relation is called the *final value theorem*. This theorem may be applied if, and only if, all the poles of $F(s)$ have a real part that is negative.

The final value theorem is a useful property since we may determine $f(\infty)$ directly from $F(s)$. However, one must be careful in using the final value theorem since the function may not have a final value as $t \to \infty$. For example, consider $f(t) = \sin \omega t$ where $F(s) = \omega/(s^2 + \omega^2)$. Now we know that $\lim_{t \to \infty} \sin \omega t$ does not exist. However, if we unwittingly used the final value theorem, we would obtain

$$\lim_{s \to 0} sF(s) = \lim_{s \to 0} \frac{s\omega}{s^2 + \omega^2} = 0$$

while the actual function $f(t)$ does not have a limiting value as $t \to \infty$. We should not use the final value theorem in this case since the poles of $F(s)$ do not have a real part that is negative.

Consider the function

$$F(s) = \frac{N(s)}{D(s)}$$

$$= \frac{N(s)}{(s - p_1)(s - p_2) \cdots (s - p_n)}$$

where p_i are the poles of $F(s)$. Since we take $\lim_{s \to 0} sF(s)$ for the final value, we require that the poles of $F(s)$ satisfy the requirement that $\mathrm{Re}\{p_i\} < 0$, except that one pole may be $p_j = 0$ since it cancels with the multiplicative factor s. Thus, we require that

$$\mathrm{Re}\{p_i\} < 0 \qquad \text{for all } i$$

except that p_i may be zero for one value of i.

Look again at

$$F(s) = \frac{\omega}{s^2 + \omega^2}$$

when $f(t) = \sin \omega t$. Then we note that

$$F(s) = \frac{\omega}{(s + j\omega)(s - j\omega)}$$

and this function fails the requirement that $\mathrm{Re}\{p_i\} < 0$ for all but one p_i. Thus, it does not have a final value.

Consider the function $F(s) = (s^2 + 4)/(s^3 + 3s^2 + 2s)$. We have

$$F(s) = \frac{s^2 + 4}{s(s + 1)(s + 2)}$$

Since two of the poles are $p_i < 0$ and one pole is at $p = 0$, we may proceed to find $\lim_{t \to \infty} f(t)$ as

$$f(\infty) = \lim_{s \to 0} sF(s)$$

$$= \lim_{s \to 0} \frac{s^2 + 4}{(s + 1)(s + 2)}$$

$$= \frac{4}{2}$$

$$= 2$$

EXERCISE 14-12

Find the initial and final values of $f(t)$ when

(a) $F(s) = \dfrac{6s + 5}{s^2 + 2s + 1}$ (b) $F(s) = \dfrac{6}{s^2 - 2s + 1}$

Answer: (a) $f(0) = 6, f(\infty) = 0$

(b) $f(0) = 0, f(\infty)$, no final value

14-7 SOLUTION OF DIFFERENTIAL EQUATIONS DESCRIBING A CIRCUIT

We may solve a set of differential equations describing a circuit using the Laplace transform of the variable, $x(t)$, and its derivatives.

Consider the circuit of Figure 14-9 when it is known that the initial value of the inductor current is $i(0) = I_0$. We wish to determine $i(t)$, the response of the circuit excited by a source $v(t)$.

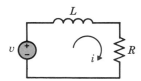

Figure 14-9 An RL circuit with an initial current $i(0) = I_0$.

The differential equation describing the circuit is

$$v = L\frac{di}{dt} + Ri \qquad (14\text{-}30)$$

where $i(0) = I_0$. Recall from Table 14-1 that

$$\mathcal{L}\left(\frac{di}{dt}\right) = sI(s) - I(0)$$

where $I(s) = \mathcal{L}[i(t)]$. Also, we will use $V(s) = \mathcal{L}[v(t)]$. Then Eq. 14-30 becomes

$$V(s) = L[sI(s) - i(0)] + RI(s)$$

Let $i(0) = I_0$ and rearrange, obtaining

$$(R + Ls)I(s) = V(s) + LI_0$$

Then, since we wish to determine $i(t)$, we solve for $I(s)$ to obtain

$$I(s) = \frac{V(s) + LI_0}{R + Ls}$$

If, for example, $v(t) = M$ for $t \geq 0$, we have a step function and $V(s) = M/s$. Then $I(s)$ is

$$I(s) = \frac{M/s + LI_0}{R + Ls}$$

If $R = 1\ \Omega$, $L = \frac{1}{2}$ H, $I_0 = 1$ A, and $M = 2$ V, we have

$$I(s) = \frac{2/s + \frac{1}{2}}{1 + \frac{1}{2}s}$$

$$= \frac{s + 4}{s(s + 2)} \qquad (14\text{-}31)$$

To find $i(t)$ we perform the inverse transform by using the partial fraction expansion of Eq. 14-31 as follows:

$$I(s) = \frac{A}{s} + \frac{B}{s + 2}$$

Then we find that $A = 2$ and $B = -1$. Therefore,

$$i(t) = \mathcal{L}^{-1}\{I(s)\} = 2 - e^{-2t} \qquad t \geq 0 \qquad (14\text{-}32)$$

Let us use the initial and final value theorems to show how we may obtain $i(0)$ and $i(\infty)$ from $I(s)$. We find the initial value $i(0)$ from

$$i(0) = \lim_{s \to \infty} sI(s) = 1$$

which checks with the initial value given as $I_0 = 1$. The final value $I(\infty)$ is found from

$$i(\infty) = \lim_{s \to 0} sI(s) = 2$$

which checks with the final value evaluated from Eq. 14-32.

The general method of obtaining the solution of one or more differential equations describing a circuit is summarized in Table 14-5. It will, in a majority of cases, be most effective in identifying the variables of interest as the capacitor voltages and the inductor currents, since we will normally know the initial conditions of these variables. Recall that in Chapter 10 we defined these variables as the state variables.

Example 14-4

Find $i(t)$ and $v_c(t)$ for the circuit shown in Figure 14-10 when $v_c(0) = 10$ V and $i(0) = 0$. The input source is $v_1 = 15u(t)$ V. Choose R so that the roots of the characteristic equation are real.

Table 14-5
Laplace Transform Method for Solving a Set of Differential Equations

---

1. Identify the circuit variables such as the inductor currents and the capacitor voltages.
2. Write the differential equations describing the circuit and identify the initial conditions of the circuit variables.
3. Obtain the Laplace transform of all the terms of the differential equation.
4. Using Cramer's rule or a similar method, solve for one or more of the unknown variables, obtaining the solution in the s domain.
5. Obtain the inverse transform of the unknown variables and thus the solution in the time domain.

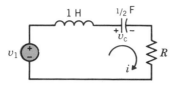

Figure 14-10 Circuit of Example 14-4.

Solution

We will identify the two variables of interest as $v_c(t)$ and $i(t)$. Then the KVL equation around the loop is

$$L\frac{di}{dt} + v_c + Ri = v_1 \tag{14-33}$$

The differential equation describing the variable v_c is

$$C\frac{dv_c}{dt} = i \tag{14-34}$$

These two differential equations are sufficient to solve for the two unknown variables. The Laplace transform of Eq. 14-33 is

$$L[sI(s) - i(0)] + V_c(s) + RI(s) = V_1(s) \tag{14-35}$$

The Laplace transform of Eq. 14-34 is

$$C[sV_c(s) - v_c(0)] = I(s) \tag{14-36}$$

Rearranging Eqs. 14-35 and 14-36 and noting that $i(0) = 0$, we have

$$(R + Ls)I(s) + V_c(s) = V_1(s)$$

and

$$-I(s) + CsV_c(s) = Cv_c(0)$$

Substituting the values for C, L, and $v_c(0)$, we obtain

$$(R + s)I(s) + V_c(s) = V_1(s) \tag{14-37}$$

and

$$-I(s) + \tfrac{1}{2}sV_c(s) = 5 \tag{14-38}$$

Since v_1 is a step of magnitude of 15 at $t = 0$, we have $V_1(s) = 15/s$. Substituting $V_1(s)$ into Eq. 14-37 and solving for $I(s)$ using Cramer's rule, we have

$$\begin{aligned}
I(s) &= \frac{V_1(s)(\tfrac{1}{2}s) - 5}{(R + s)(\tfrac{1}{2}s) - (-1)} \\[2mm]
&= \frac{(15/s)(s) - 5(2)}{(R + s)(s) + 2} \\[2mm]
&= \frac{5}{s^2 + Rs + 2} \tag{14-39}
\end{aligned}$$

The inverse transform of $I(s)$ will depend on the value of R. The two roots of the denominator are equal when $R = 2\sqrt{2}$, and the roots will be real but unequal when $R > 2\sqrt{2}$. When $R < 2\sqrt{2}$ the roots are complex.

Assuming the value of $R = 3$ we have

$$I(s) = \frac{5}{s^2 + 3s + 2}$$

$$= \frac{5}{(s + 1)(s + 2)}$$

Using a partial fraction expansion, we have

$$I(s) = \frac{A}{s + 1} + \frac{B}{s + 2}$$

where we find that $A = 5$ and $B = -5$. Therefore, using Table 14-1, we find that

$$i(t) = 5e^{-t} - 5e^{-2t} \text{ A}, \qquad t \geq 0$$

Note that $I(0) = 0$ and $I(\infty) = 0$ as required.

Solving Eqs. 14-37 and 14-38 for $V_c(s)$, we have

$$V_c(s) = \frac{(R + s)10 + 2V_1(s)}{(R + s)s + 2}$$

$$= \frac{10(R + s) + 2(15/s)}{s^2 + Rs + 2}$$

$$= \frac{10s^2 + 10Rs + 30}{s(s^2 + Rs + 2)}$$

Therefore, when $R = 3$ we have

$$V_c(s) = \frac{10(s^2 + 3s + 3)}{s(s + 1)(s + 2)}$$

Using partial fraction expansion, we have

$$V_c(s) = \frac{A}{s} + \frac{B}{s + 1} + \frac{C}{s + 2}$$

where $A = 15$, $B = -10$, and $C = 5$. Therefore, the capacitor voltage is

$$v_c(t) = 15 - 10e^{-t} + 5e^{-2t} \text{ V}, \qquad t \geq 0$$

Note that $v_c(0) = 10$ V and $v_c(\infty) = 15$ V as required.

Example 14-5
Find $v_c(t)$ for $t \geq 0$ for the circuit of Figure 14-11.

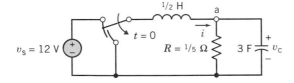

Figure 14-11 Circuit of Example 14-5.

Solution
First, we identify the two variables as i and v_c. Determine the initial conditions by considering the circuit with the constant 12-V source connected for a long time. Then, replacing the inductor with a short circuit and the capacitor with an open circuit, as shown in Figure 14-12, it is clear that $i(0) = 60$ A and $v_c(0) = 12$ V.

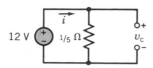

Figure 14-12 Circuit of Figure 14-11 (Example 14-5) at $t = 0^-$.

We require two first-order differential equations in terms of v_c and i for $t \geq 0$. The KVL equation for the mesh with i of Figure 14-11 is

$$L\frac{di}{dt} + v_c = 0 \qquad (14\text{-}40)$$

The equation for the capacitor current i_c at node a is

$$i_c + \frac{v_c}{R} - i = 0$$

Since $i_c = C\, dv_c/dt$, we have

$$C\frac{dv_c}{dt} + \frac{v_c}{R} - i = 0 \qquad (14\text{-}41)$$

Equations 14-40 and 14-41 describe the system completely in terms of the state variables v_c and i. Taking the Laplace transform of Eqs. 14-40 and 14-41, we obtain

$$L[sI(s) - I(0)] + V_c(s) = 0$$

and

$$C[sV_c(s) - v_c(0)] + \frac{V_c(s)}{R} - I(s) = 0$$

Substituting the values for L, C, R, $i(0)$, and $v_c(0)$, we have

$$\tfrac{1}{2}[sI(s) - 60] + V_c(s) = 0$$

and

$$3[sV_c(s) - 12] + 5V_c(s) - I(s) = 0$$

Rearranging in a form more suitable for Cramer's rule, we obtain

$$sI(s) + 2V_c(s) = 60 \qquad (14\text{-}42)$$

$$-I(s) + (3s + 5)V_c(s) = 36 \qquad (14\text{-}43)$$

Solving for $V_c(s)$, we have

$$V_c(s) = \frac{36s + 60}{s(3s + 5) + 2}$$

$$= \frac{36s + 60}{3s^2 + 5s + 2}$$

$$= \frac{36s + 60}{3(s + \tfrac{2}{3})(s + 1)}$$

Using a partial fraction expansion, we have

$$V_c(s) = \frac{A}{s + \tfrac{2}{3}} + \frac{B}{s + 1}$$

where $A = 36$ and $B = -24$. Then obtaining the inverse transform, we have

$$v_c(t) = 36e^{-2t/3} - 24e^{-t} \text{ V} \qquad t \geq 0 \qquad (14\text{-}44)$$

EXERCISE 14-13

Find $i(t)$ for the circuit of Figure E 14-13 when $i_1(t) = 7e^{-6t}$ A for $t \geq 0$ and $i(0) = 0$.

Answer: $\quad i(t) = -\dfrac{35}{16} e^{-6t} + \dfrac{35}{16} e^{-2t}$ A, $t \geq 0$

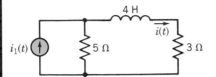

Figure E 14-13

EXERCISE 14-14

Find $v_2(t)$ for the circuit of Figure E 14-14 for $t \geq 0$. Hint: Write the node equations at a and b in terms of v_1 and v_2. The initial conditions are $v_1(0) = 10$ V and $v_2(0) = 25$ V. The source is $v_s = 50 \cos 2t \, u(t)$ V.

Answer: $\quad v_2(t) = \frac{23}{3} e^{-t} + \frac{16}{3} e^{-4t} + 12 \cos 2t + 12 \sin 2t$ V, $t \geq 0$

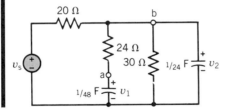

Figure E 14-14

EXERCISE 14-15

Determine $f(t)$ for a differential equation

$$\frac{d^2f}{dt^2} + 5\frac{df}{dt} + 6f = 10e^{-3t} \qquad t \geq 0$$

when $f(0) = 2$ and $\left.\dfrac{df}{dt}\right|_{t=0} = 0$.

Answer: $\quad f(t) = -10te^{-3t} - 14e^{-3t} + 16e^{-2t}$, $t \geq 0$

14-8 TRANSFER FUNCTION AND IMPEDANCE

The *transfer function* of a circuit is defined as the ratio of the response $Y(s)$ of the circuit to an excitation $X(s)$ expressed in the complex frequency domain. The transfer function, denoted by $H(s)$, is then expressed as

$$H(s) = \frac{Y(s)}{X(s)} \tag{14-45}$$

This ratio is obtained assuming all initial conditions are equal to zero.

Consider the RL circuit shown in Figure 14-13 and let us find the transfer function

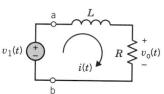

Figure 14-13 An RL circuit with the variables expressed in the time domain.

$H(s) = V_0(s)/V_1(s)$. Setting the initial conditions to zero, we write the KVL for the mesh as

$$Ri + L\frac{di}{dt} = v_1 \tag{14-46}$$

The Laplace transform of Eq. 14-46 yields

$$RI(s) + LsI(s) = V_1(s) \tag{14-47}$$

Since $v_0 = Ri$, we have $V_0(s) = RI(s)$ in the s-domain. Then, solving for $I(s)$ from Eq. 14-47, we have

$$I(s) = \frac{V_1(s)}{R + Ls}$$

and therefore

$$V_0(s) = RI(s)$$
$$= \frac{RV_1(s)}{R + Ls}$$

Then the transfer function is

$$H(s) = \frac{V_0(s)}{V_1(s)} = \frac{R}{R + Ls} \tag{14-48}$$

At this point, it is worthwhile to introduce the concept of impedance expressed in the complex frequency domain. Recall that earlier we introduced the concept of impedance in the frequency domain and expressed impedance as

$$\mathbf{Z}(j\omega) = \frac{\mathbf{V}(j\omega)}{\mathbf{I}(j\omega)}$$

where $\mathbf{V}(j\omega)$ and $\mathbf{I}(j\omega)$ are phasor quantities.

We define the *impedance* of an element in the complex frequency domain, s, as

$$Z(s) = \frac{V(s)}{I(s)} \tag{14-49}$$

where the element is assumed to have no initial stored energy. Similarly we define the *admittance* of an element as

$$Y(s) = \frac{I(s)}{V(s)}$$

If we have a resistor R, the impedance is R since the voltage–current ratio expressed in the s-domain is R. We can show this by writing the time domain equation for the voltage–current relationship of a resistor, which is

$$v(t) = Ri(t)$$

Taking the Laplace transform of both sides, we have

$$V(s) = RI(s)$$

and, therefore,

$$Z_R(s) = \frac{V(s)}{I(s)} = R \qquad (14\text{-}50)$$

Now consider the voltage–current relationship for an inductor with a zero initial condition. We have

$$v(t) = L\frac{di}{dt}$$

Taking the Laplace transform and recalling that the initial condition is zero, we have

$$V(s) = LsI(s)$$

Therefore, the impedance of the inductor is

$$Z_L(s) = \frac{V(s)}{I(s)} = Ls \qquad (14\text{-}51)$$

Finally, the time domain voltage–current relationship for a capacitor is

$$i = C\frac{dv}{dt}$$

Taking the Laplace transform, we have (no initial stored energy)

$$I(s) = CsV(s)$$

Then the impedance is

$$Z_c(s) = \frac{V(s)}{I(s)} = \frac{1}{Cs} \qquad (14\text{-}52)$$

The impedances of the three elements, R, L, and C, are summarized in Table 14-6.

Table 14-6
Impedance of an Element Expressed in the Laplace Form*

Element	Time Domain Relationship	s-Domain Relationship	Impedance $Z(s) = \dfrac{V(s)}{I(s)}$
Resistor, R	$v = iR$	$V(s) = RI(s)$	R
Inductor, L	$v = L\dfrac{di}{dt}$	$V(s) = LsI(s)$	Ls
Capacitor, C	$i = C\dfrac{dv}{dt}$	$I(s) = CsV(s)$	$\dfrac{1}{Cs}$

* Stored energy at $t = 0^-$ is assumed to be zero.

We can use the impedance in the Laplace form for determining a transfer function or a Thévenin equivalent impedance of a circuit, since the transfer function assumes zero initial conditions. Now referring back to the RL network of Figure 14-13, since we wish to determine the transfer function, $H(s) = V_0(s)/V_1(s)$, we may redraw the circuit in terms of variables of s and the impedance of each element as shown in Figure 14-14. Then, using KVL, we have the mesh equation

$$RI(s) + (Ls)I(s) = V_1(s)$$

and

$$V_0(s) = I(s)R$$

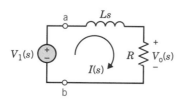

Figure 14-14 The RL circuit of Figure 14-13 with the variables expressed in the s-domain.

Therefore,

$$H(s) = \frac{V_0(s)}{V_1(s)} = \frac{R}{R + Ls}$$

$$= \frac{1}{\tau s + 1}$$

where $\tau = L/R$.

If we wish to determine the Thévenin equivalent impedance $Z_t(s)$ at terminals a−b of the circuit shown in Figure 14-14, we seek

$$Z_t(s) = \frac{V_1(s)}{I(s)}$$

The mesh equation is

$$RI(s) + LsI(s) = V_1(s)$$

Therefore, we obtain

$$Z_t(s) = \frac{V_1(s)}{I(s)} = \frac{1}{R + Ls}$$

Example 14-6

The circuit of Figure 14-15 represents a linear model of a low-frequency amplifier. (a) Redraw the circuit model in terms of admittances with units of siemens. (b) Using this model, find $H(s) = V_0(s)/V_1(s)$. (c) Determine $v_0(t)$ when $v_1(t) = 1u(t)$ V given that there is no initial stored energy. Assume that $R_1 = 0.1\ \Omega$, $R_2 = 0.5\ \Omega$, $C_1 = 100$ F, $C_2 = 3$ F, and $R_3 = 1\ \Omega$.

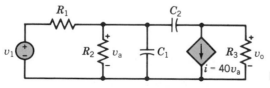

Figure 14-15 Circuit model of Example 14-6.

Solution

To redraw the circuit, note that for a resistor the admittance is to be stated in terms of siemens. For example, for R_1

$$G_1 = \frac{1}{R_1}$$
$$= 10 \text{ S}$$

For a capacitor the admittance is $Y = Cs$. Then for C_1

$$Y = C_1 s$$
$$= 100s \text{ S}$$

The circuit with all admittances is shown in Figure 14-16. At node a we have the node voltage V_a and the KCL equation is

$$(V_a - V_1)10 + 100sV_a + 2V_a + (V_a - V_0)3s = 0 \tag{14-53}$$

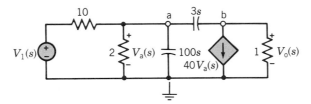

Figure 14-16 Circuit model of Example 14-6 with admittances for each element in siemens and current in amperes.

At node b the node voltage is V_0 and the KCL equation is

$$(V_0 - V_a)3s + 1V_0 + 40V_a = 0 \tag{14-54}$$

Rearranging the equations, we have

$$(12 + 103s)V_a + (-3s)V_0 = 10V_1(s) \tag{14-55}$$

and

$$(40 - 3s)V_a + (1 + 3s)V_0 = 0 \tag{14-56}$$

Using Cramer's rule to solve for $V_0(s)$, we have

$$V_0 = \frac{10(3s - 40)}{300s^2 + 259s + 12} V_1(s)$$

$$= \frac{0.1(s - 13.3)}{(s + 0.049)(s + 0.81)} V_1(s)$$

Recall that $v_1 = 1\,u(t)$ so that $V_1(s) = 1/s$; hence we have

$$V_0(s) = \frac{0.1(s - 13.3)}{s(s + 0.049)(s + 0.81)}$$

$$= \frac{-33.3}{s} + \frac{35.5}{s + 0.049} - \frac{2.26}{s + 0.81} \tag{14-57}$$

Then, taking the inverse transform,

$$v_0(t) = -33.3 + 35.5e^{-0.049t} - 2.26e^{-0.81t} \text{ V} \qquad t \geq 0$$

The two time constants are $\tau_1 = 1/0.049 = 20.4$ s and $\tau_2 = 1/0.81 = 1.23$ s. The transient response will be over and steady state will be achieved within 5 times the slower time constant. Therefore, the steady-state condition will be achieved by 100 s.

Example 14-7

Find the transfer function $V_0(s)/V_1(s)$ for the circuit of Figure 14-17a.

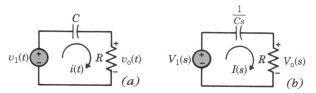

Figure 14-17 An RC circuit expressed (a) in the time domain and (b) in the s-domain in terms of impedances.

Solution

The first step is to convert the circuit to s-domain variables and impedances as shown in Figure 14-17b. Using the impedances and the voltage divider principle, we have

$$\frac{V_0(s)}{V_1(s)} = \frac{R}{R + Z_c}$$

$$= \frac{R}{R + 1/Cs}$$

$$= \frac{RCs}{RCs + 1}$$

$$= \frac{\tau s}{\tau s + 1}$$

where $\tau = RC$.

With the concept of impedance in the s-domain, we note that the principle of superposition and the Thévenin and Norton theorems will apply as they did earlier for circuits with resistors and for impedances defined in terms of phasor currents and phasor voltages.

For example, consider the circuit shown in Figure 14-18a represented in the time domain. The Laplace transform of the circuit variables and the impedance of each element is shown in Figure 14-18b. Using the concept of parallel impedances, we note that the impedance Z_p of the two parallel elements is

$$Z_p = \frac{RZ_c}{R + Z_c}$$

$$= \frac{R(1/Cs)}{R + 1/Cs}$$

$$= \frac{R}{RCs + 1}$$

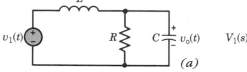
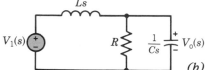

Figure 14-18 An RLC circuit expressed (a) in the time domain and (b) in the s-domain in terms of impedances.

Then, using the voltage divider principle, we find $V_0(s)$ as

$$V_0(s) = \frac{Z_p}{Z_p + Ls} V_1(s)$$

or

$$H(s) = \frac{V_0(s)}{V_1(s)} = \frac{Z_p}{Z_p + Ls}$$

$$= \frac{R}{R + Ls(RCs + 1)}$$

$$= \frac{R}{RLCs^2 + Ls + R}$$

$$= \frac{1/LC}{s^2 + (1/RC)s + 1/LC} \qquad (14\text{-}58)$$

When $v_1(t) = \delta(t)$, the unit impulse, $V_1(s) = 1$. Then we have

$$V_0(s) = H(s)$$

For this case $v_0(t)$ is called the impulse response and is denoted as $h(t) = \mathcal{L}^{-1}[H(s)]$. The *impulse response $h(t)$* is defined as the output response of a circuit when the input is the unit impulse $\delta(t)$ with no initial stored energy in the circuit at $t = 0$.

Consider the circuit of Figure 14-18 with the transfer function of Eq. 14-58. For example, when $1/LC = 2$ and $1/RC = 3$, we have

$$H(s) = \frac{2}{s^2 + 3s + 2}$$

$$= \frac{2}{(s + 1)(s + 2)}$$

Then the impulse response is

$$h(t) = \mathcal{L}^{-1}[H(s)]$$
$$= 2(e^{-t} - e^{-2t}) \qquad t \geq 0$$

EXERCISE 14-16

Determine $i(t)$ when $i(0^-) = 3$ A for the circuit shown in Figure E 14-16.
Answer: $i = 2 + 1e^{-2t}$ A, $t \geq 0$

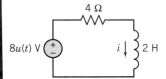

Figure E 14-16

EXERCISE 14-17

Determine the impedance Z_{ab} at terminals a–b of the circuit of Figure E 14-17.
Answer: $Z_{ab} = \dfrac{12s}{3s^2 + 4s + 6}$

Figure E 14-17

14-9 CIRCUIT ANALYSIS USING IMPEDANCE AND INITIAL CONDITIONS

The time response of a circuit can be determined from a circuit using Laplace transform models that include impedances and the initial conditions of energy storage elements. We extend Table 14-6 to include the effect of initial conditions. The v–i relationship for the resistor is Ohm's Law as:

$$v(t) = i(t)R$$

Therefore, the Laplace transform domain relationship for a resistor R is

$$V(s) = I(s)R$$

as summarized in Table 14-7.

A capacitor is represented by its time domain equation

$$v_c(t) = \frac{1}{C}\int_0^t i_c\, dt + v_c(0) \tag{14-59}$$

The Laplace transform of Eq. 14-59 is

$$V_c(s) = \frac{I_c(s)}{Cs} + \frac{v_c(0)}{s} \tag{14-60}$$

The model of the Laplace transform relationship for the capacitor is summarized in Table 14-7.

Table 14-7
Impedance Relationships with Initial Conditions

Element	Circuit	Time Domain	s-Domain Relationship	Impedance Model
Resistor		$v = iR$	$V(s) = I(s)R$	
Capacitor		$v_c = \dfrac{1}{C}\displaystyle\int_0^t i_c\, dt + v_c(0)$	$V_c(s) = \dfrac{1}{Cs} I_c(s) + \dfrac{v_c(0)}{s}$	
Inductor		$v_L = L\dfrac{di_i}{dt}$	$V_L(s) = Ls I_L(s) - Li_L(0)$	

An inductor is represented by its time domain equation

$$v_L(t) = L\frac{di}{dt} \tag{14-61}$$

The Laplace transform of Eq. 14-61 is

$$V_L(s) = LsI(s) - Li(0) \tag{14-62}$$

The model of the Laplace transform relationship for the inductor in Table 14-7.

Let us use these relationships to solve for the current in the inductor, i_2, for the circuit shown in Figure 14-19a. The Laplace transform model of the circuit is shown in Figure 14-19b when the initial conditions are $v_c(0) = 8$ V and $i_L(0) = 4$ A.

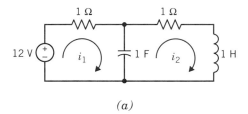

(a)

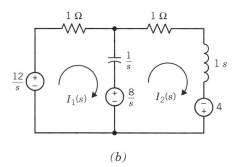

(b)

Figure 14-19 *(a)* Circuit with mesh currents. *(b)* Laplace transform model of circuit.

The mesh current equations are

$$\left(1 + \frac{1}{s}\right) I_1(s) - \frac{1}{s} I_2(s) = \frac{12}{s} - \frac{8}{s}$$

and

$$-\frac{1}{s} I_1(s) + \left(1 + s + \frac{1}{s}\right) I_2(s) = 4 + \frac{8}{s}$$

Using Cramer's rule for $I_2(s)$ we obtain

$$I_2(s) = \frac{4(s^2 + 3s + 3)}{s(s^2 + 2s + 2)}$$

The convenient partial fraction expansion is

$$\frac{I_2(s)}{4} = \frac{A}{s} + \frac{Bs + D}{s^2 + 2s + 2} = \frac{s^2 + 3s + 3}{s(s^2 + 2s + 2)}$$

Then, we determine that $A = 1.5$, $B = -0.5$, and $D = 0$. Then, we can state

$$\frac{I_2(s)}{4} = \frac{1.5}{s} + \frac{-0.5s}{(s + 1)^2 + 1}$$

Using the inverse Laplace transform Tables 14-1 and 14-4 we obtain

$$i_2(t) = \{6 + 2\sqrt{2}e^{-t}\sin(t - 45°)\}u(t) \text{ A}$$

Checking the initial value of i_2 we get $i_2(0) = 4$ A, which verifies the correct value. The final value is $i_2(\infty) = 6$ A.

14-10 ‖ CONVOLUTION THEOREM

The *convolution* is defined for an output $y(t)$ as the integral

$$y(t) = \mathcal{L}^{-1}[H(s)F(s)] = \int_0^t h(\tau)f(t - \tau)\, d\tau \tag{14-63}$$

That is, the product of a transfer function $H(s)$ times a signal $F(s)$ in the s-domain is equivalent to the Laplace transform of an integral given on the right-hand side of, Eq. 14-63.

To prove the convolution integral, consider the product $H(s)F(s)$, where

$$H(s) = \int_0^\infty h(\tau)e^{-s\tau}\, d\tau$$

The product is

$$H(s)F(s) = \int_0^\infty h(\tau)e^{-s\tau}F(s)\, d\tau \tag{14-64}$$

However, recall that the time shift property is

$$\mathcal{L}[f(t - \tau)] = e^{-s\tau}F(s) \tag{14-65}$$

Therefore, substituting Eq. 14-65 into Eq. 14-64, we have

$$H(s)F(s) = \int_0^\infty h(\tau)\mathcal{L}[f(t - \tau)]\, d\tau$$

$$= \int_0^\infty h(\tau)\left[\int_0^\infty f(t - \tau)e^{-st}\, dt\right] d\tau$$

Rearranging the order of the integrals, we have

$$H(s)F(s) = \int_0^\infty e^{-st}\left[\int_0^t h(\tau)f(t - \tau)\, d\tau\right] dt$$

$$= \mathcal{L}\left[\int_0^t h(\tau)f(t - \tau)\, d\tau\right] \tag{14-66}$$

Usually the convolution is denoted by an asterisk so that when we seek the convolution of $h(t)$ with $f(t)$ we write

$$\mathcal{L}^{-1}[H(s)F(s)] = h(t) * f(t) = \int_0^t h(\tau)f(t - \tau)\, d\tau \tag{14-67}$$

The **convolution** of two functions of time, $f(t)$ and $h(t)$, is described as $f(t) * h(t)$ and is identical to the inverse transform of $F(s)H(s)$.

For example, find the convolution of $h(t) = e^{-t}$ and $f(t) = e^{-2t}$:

$$h(t) * f(t) = \mathcal{L}^{-1}[H(s)F(s)]$$

$$= \mathcal{L}^{-1}\left[\left(\frac{1}{s+1}\right)\left(\frac{1}{s+2}\right)\right]$$

$$= \mathcal{L}^{-1}\left[\frac{1}{s+1} + \frac{-1}{s+2}\right]$$

$$= e^{-t} - e^{-2t} \qquad t \geq 0$$

Example 14-8

Determine the output $v(t)$ for the circuit shown in Figure 14-20 using the convolution integral and the Laplace transform. The input voltage is $v_1(t) = 2e^{-100t}u(t)$ V.

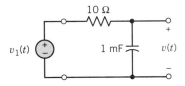

Figure 14-20 An RC circuit.

Solution

Let us first determine the output using the convolution integral. The transfer function of the circuit is

$$H(s) = \frac{1}{RCs+1}$$

$$= \frac{100}{s+100}$$

Therefore, the impulse response is

$$h(t) = 100e^{-100t}u(t)$$

Using Eq. 14-67, we have

$$v(t) = h(t) * v_1(t) = 200\int_0^t e^{-100\tau}u(\tau)e^{-100(t-\tau)}u(t-\tau)\,d\tau$$

$$= 200e^{-100t}\int_0^t d\tau$$

$$= 200te^{-100t} \qquad t > 0$$

We now proceed to determine $v(t)$ using $V_1(s)$ and $H(s)$. Using the Laplace transform, we find $V_1(s)$ as

$$V_1(s) = \mathcal{L}[2e^{-100t}u(t)]$$

$$= \frac{2}{s+100}$$

Then, the output is

$$V(s) = H(s)V_1(s)$$

$$= \left(\frac{100}{s+100}\right)\left(\frac{2}{s+100}\right)$$

$$= \frac{200}{(s+100)^2}$$

Using the first entry in Table 14-3, we obtain

$$v(t) = 200te^{-100t} \qquad t > 0$$

EXERCISE 14-17

(a) Find $H(s) = V_c(s)/V_1(s)$ for the circuit of Figure E 14-17. (b) Find the impulse response $h(t)$ of this circuit.

Answer: (a) $H(s) = \dfrac{1}{s + 1.25}$

(b) $h(t) = e^{-1.25t} \qquad t > 0$

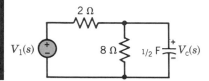

Figure E 14-17

EXERCISE 14-18

Find the convolution of $h(t) = e^{-2t}$ and $f(t) = u(t)$.

14-11 STABILITY

A circuit is said to be *stable* when the response to a bounded input signal is a bounded output signal. Consider a circuit represented by the transfer function $H(s)$. The impulse response is

$$h(t) = \mathcal{L}^{-1}\{H(s)\}$$

Thus, we require for stability that

$$\lim_{t \to \infty} |h(t)| = \text{finite}$$

If $H(s)$ is written in terms of its poles, we have

$$H(s) = \frac{N(s)}{(s - p_1)(s - p_2) \cdots (s - p_N)}$$

and, therefore,

$$h(t) = \sum_{i=1}^{N} A_i e^{p_i t}\, u(t)$$

Thus we require, for a stable circuit, that all the poles of $H(s)$ lie in the left-hand s-plane. A simple pole in the right-hand plane where $\sigma_1 > 0$ can be represented as

$$H(s) = \frac{K}{s - \sigma_1}$$

Then, $h(t) = Ke^{\sigma_1 t}$ and the response to an impulse is unbounded if $\sigma_1 > 0$.

Example 14-9

Determine $H(s) = V_3(s)/V(s)$ for the op amp circuit shown in Figure 14-21 and determine if the circuit is stable. Assume ideal op amps and the input signal is $v(t)$.

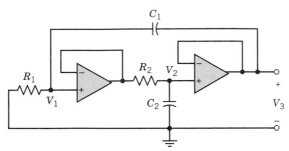

Figure 14-21 A two-op amp circuit with a feedback capacitor.

Solution

The node equations at nodes 1 and 2, using the notation $V_n = V_n(s)$, are

1   $\dfrac{V_1 - V}{R_1} + (V_1 - V_3)sC_1 = 0$

2   $\dfrac{V_2 - V_1}{R_2} + V_2 sC_2 = 0$

Since, for an ideal op amp, $V_3 = V_2$ due to the buffer amplifier stage, we rewrite the two node equations as

1   $\left(\dfrac{1}{R_1} + sC_1\right)V_1 + (-sC_1)V_2 = \dfrac{V}{R_1}$

2   $\left(-\dfrac{1}{R_2}\right)V_1 + \left(\dfrac{1}{R_2} + sC_2\right)V_2 = 0$

Using Cramer's rule, we solve for V_2 in terms of V as

$$\frac{V_2(s)}{V(s)} = \frac{1}{R_1 R_2 C_1 C_2 s^2 + R_2 C_2 s + 1}$$

$$= \frac{\omega_0^2}{s^2 + (1/RC)s + \omega_0^2}$$

and $\omega_0^2 = 1/R_1 R_2 C_1 C_2$. This circuit is stable, since the two roots of the characteristic equation always lie in the left-hand s-plane for all positive values of R and C.

Example 14-10

Determine $H(s) = V_3(s)/V(s)$ for the op amp circuit shown in Figure 14-22 and determine if the circuit is stable. Assume ideal op amps.

Solution

The circuit consists of two ideal integrators in cascade (see Section 7-10). Therefore, we have

$$\frac{V_3(s)}{V(s)} = \frac{1}{(RC)^2 s^2}$$

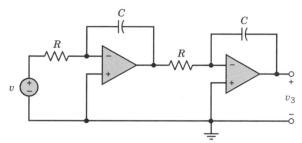

Figure 14-22 A two-stage op amp circuit.

The two poles of the circuit lie at the origin of the s-plane. If $V(s) = 1$, which represents an impulse input, we have

$$H(s) = V_3(s) = \frac{1}{(RC)^2 s^2}$$

Using Table 14-1, we find the impulse response as

$$h(t) = \frac{t}{(RC)^2}$$

which is unbounded since $h(t) \to \infty$ as $t \to \infty$. Thus, this circuit is said to be unstable.

A stable circuit has a steady-state response while an unstable circuit does not. Thus, frequency response analysis and phasor analysis may only be applied to stable circuits.

The stability of a circuit can be determined from the circuit's frequency response examining both phase and magnitude. Stability cannot be determined from the magnitude of the frequency response alone. Indeed, both $H_1(s) = 1/(s + 1)$ and $H_2(s) = 1/(s - 1)$ have the same magnitude, but $H_1(s)$ is stable and $H_2(s)$ is unstable.

The transfer function of a second-order circuit can be written as

$$H(s) = \frac{N(s)}{s^2 + a_1 s + a_0}$$

$$= \frac{N(s)}{s^2 + 2\zeta\omega_0 s + \omega_0^2} = \frac{N(s)}{s^2 + \omega_0 s/Q + \omega_0^2}$$

When $\zeta = 0$ or $Q = \infty$ we have

$$H(s) = \frac{N(s)}{s^2 + \omega_0^2} = \frac{N(s)}{(s + j\omega_0)(s - j\omega_0)}$$

and this circuit is called an *oscillator* since the response is a sinusoidal signal. For example, the impulse response of

$$H(s) = \frac{100}{s^2 + 100}$$

is $h(t) = 10 \sin 10t$ for $t > 0$. Thus, the response to a unit impulse is a sinusoidal oscillator.

EXERCISE 14-19

The transfer function of a second-order circuit is

$$H(s) = \frac{25}{s^2 + (8 - k)s + 100}$$

Determine the required range of k for stable operation when k is $k \geq 0$.
Answer: $0 \leq k < 8$

EXERCISE 14-20

For the transfer function of Exercise 14-19 determine the required value of k so that the circuit is an oscillator.
Answer: $k = 8$

14-12 VERIFICATION EXAMPLES

Example 14V-1

A circuit is specified to have a transfer function of

$$H(s) = \frac{V_0(s)}{V_1(s)} = \frac{25}{s^2 + 10s + 125} \tag{14V-1}$$

and a step response of

$$v_0(t) = 0.1(2 - e^{-5t}(3 \cos 10t + 2 \sin 10t))u(t) \tag{14V-2}$$

Are these specifications consistent? Let us verify their consistency.

Solution

If the specifications are consistent, then the unit step response and the transfer function will be related by

$$\mathcal{L}[v_0(t)] = H(s) \frac{1}{s} \tag{14V-3}$$

where $V_1(s) = 1/s$.

This equation can be verified either by calculating the Laplace transform of $v_0(t)$ or by calculating the inverse Laplace transform of $H(s)/s$. Both of these calculations involve a bit of algebra. The final and initial value theorems provide a quicker, though less conclusive, check. (If either the final or initial value theorem is not satisfied, then we know that the step response is not consistent with the transfer function. The step response could be inconsistent with the transfer function even if both the final and initial value theorems are satisfied.) Let us see what the final and initial value theorems tell us.

The final value theorem requires that

$$v_0(\infty) = \lim_{s \to 0} s \left[H(s) \frac{1}{s} \right] \tag{14V-4}$$

From Eq. 14V-1, we substitute $H(s)$ obtaining

$$\lim_{s \to 0} s \left[\frac{25}{s^2 + 10s + 125} \cdot \frac{1}{s} \right] = \lim_{s \to 0} \left[\frac{25}{s^2 + 10s + 125} \right]$$

$$= \frac{25}{125} \tag{14V-5}$$

$$= 0.2$$

From Eq. 14V-2, we evaluate at $t = \infty$ obtaining

$$v_0(\infty) = 0.1 (2 - e^{-\infty}(2 \cos \infty + \sin \infty))$$
$$= 0.1(2 - 0) \tag{14V-6}$$
$$= 0.2$$

so the final value theorem is satisfied.

Next, the initial value theorem requires that

$$v_0(0) = \lim_{s \to \infty} s \left[H(s) \frac{1}{s} \right] \tag{14V-7}$$

From Eq. 14V-1, we substitute $H(s)$ obtaining

$$\lim_{s \to \infty} s \left[\frac{25}{s^2 + 10s + 125} \cdot \frac{1}{s} \right] = \lim_{s \to \infty} \frac{25/s^2}{1 + 10/s + 125/s^2} \tag{14V-8}$$

$$= \frac{0}{1}$$

$$= 0$$

From Eq. 14V-2, we evaluate at $t = 0$ to obtain

$$v_0(0) = 0.1(2 - e^{-0}(3 \cos 0 + 2 \sin 0))$$
$$= 0.1(2 - 1(3 + 0)) \tag{14V-9}$$
$$= -0.1$$

The initial value theorem is not satisfied, so the step response is not consistent with the transfer function.

Example 14V-2

A circuit is specified to have a transfer function of

$$H(s) = \frac{V_0(s)}{V_1(s)} = \frac{25}{s^2 + 10s + 125} \tag{14V-10}$$

and a unit step response of

$$v_0(t) = 0.1(2 - e^{-5t}(2 \cos 10t + 3 \sin 10t))u(t) \tag{14V-11}$$

Are these specifications consistent? (This step response is a slightly modified version of the step response considered in the previous example.)

Solution

The reader is invited to verify that both the final and initial value theorems are satisfied. This suggests, but does not guarantee, that the transfer function and step response are consistent. To guarantee consistency it is necessary to verify that

$$\mathcal{L}[v_0(t)] = H(s) \frac{1}{s} \tag{14V-12}$$

either by calculating the Laplace transform of $v_0(t)$ or by calculating the inverse Laplace transform of $H(s)/s$. Recall the input is a unit step, so $V_1(s) = 1/s$. We will calculate the Laplace transform of $v_0(t)$ as follows:

$$\mathcal{L}[0.1(2 - e^{-5t}(2 \cos 10t + 3 \sin 10t))u(t)]$$

$$= 0.1 \left[\frac{2}{s} - 2\frac{(s + 5)}{(s + 5)^2 + 10^2} + 3\frac{10}{(s + 5)^2 + 10^2} \right]$$

$$= 0.1 \left[\frac{2}{s} + \frac{-2s + 20}{s^2 + 10s + 125} \right]$$

$$= \frac{2s + 25}{s(s^2 + 10s + 125)}$$

Since this is not equal to $H(s)/s$, Eq. 14 V-12 is not satisfied. The step response is not consistent with the transfer function even though the initial and final values of $v_0(t)$ are consistent.

EXERCISE 14V-1

A circuit is specified to have a transfer function of

$$H(s) = \frac{25}{s^2 + 10s + 125}$$

and a unit step response of

$$v_0(t) = 0.1(2 - e^{-5t}(2 \cos 10t + \sin 10t))u(t)$$

Verify that these specifications are consistent.

14-13 DESIGN CHALLENGE SOLUTION

SPACE SHUTTLE CARGO DOOR

Problem

The U.S. space shuttle Atlantis docked with Russia's Mir Space station on June 28, 1995 as shown in the cover photo of this book. The electromagnet for opening a cargo door on the NASA space shuttle requires 0.1 A before activating. The electromagnetic coil is represented by L as shown in Figure 14D-1. The activating current is designated $i_1(t)$. The time period required for i_1 to reach 0.1 A is specified as less than 3 seconds. Select a suitable value of L.

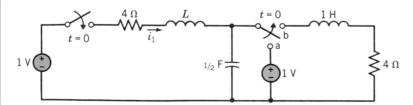

Figure 14D-1 The control circuit for a cargo door on the NASA space shuttle.

Define the Situation, State the Assumptions, and Develop a Model

1 The two switches are thrown at $t = 0$ and the movement of the second switch from terminal a to terminal b occurs instantaneously.
2 The switches prior to $t = 0$ were in position for a long time.

The Goal

Determine a value of L so that the time period for the current $i_1(t)$ to attain a value of 0.1 A is less than 3 seconds.

Generate a Plan

1 Determine the initial conditions for the two inductor currents and the capacitor voltage.
2 Designate two mesh currents and write the two mesh KVL equations using the Laplace transform of the variables and the impedance of each element.
3 Select a trial value of L and solve for $I_1(s)$.
4 Determine $i_1(t)$.
5 Sketch $i_1(t)$ and determine the time instant t_1 when $i_1(t_1) = 0.1$ A.
6 Check if $t_1 < 3$ seconds, and if not return to step 3 and select another value of L.

Prepare the Plan

Goal	Equation	Need	Information
Determine the initial conditions at $t = 0$.	$i(0) = i(0^-)$ $v_c(0) = v_c(0^-)$	Prepare a sketch of the circuit at $t = 0^-$. Find $i_1(0^-)$, $i_2(0^-)$, $v_c(0^-)$.	
Designate two mesh currents and write the mesh KVL equations.		$I_1(s)$, $I_2(s)$. The initial conditions $i_1(0)$, $i_2(0)$	
Solve for $I_1(s)$ and select L.			Cramer's rule
Determine $i_1(t)$.	$i_1(t) = \mathcal{L}^{-1}\{I_1(s)\}$		Use a partial fraction expansion.
Sketch $i_1(t)$ and find t_1.	$i_1(t_1) = 0.1$ A		

Take Action Using the Plan

First, the circuit with the switches in position at $t = 0^-$ is shown in Figure 14D-2. Clearly, the inductor currents are $i_1(0^-) = 0$ and $i_2(0^-) = 0$. Furthermore, we have

$$v_c(0) = 1 \text{ V}$$

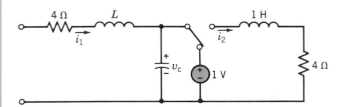

Figure 14D-2 The circuit of Figure 14D-1 at $t = 0^-$.

Second, redraw the circuit for $t > 0$ as shown in Figure 14D-3 and designate the two mesh currents i_1 and i_2 as shown.

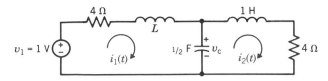

Figure 14D-3 The circuit of Figure 14D-1 for $t > 0$.

Recall that the impedance is Ls for an inductor and $1/Cs$ for a capacitor. We must account for the initial condition for the capacitor. Recall that the capacitor voltage may be written as

$$v_c(t) = v_c(0) + \frac{1}{C} \int_0^t i_c(\tau)\, d\tau$$

The Laplace transform of this equation is

$$V_c(s) = \frac{v_c(0)}{s} + \frac{1}{Cs} I_c(s)$$

where $I_c(s) = I_1(s) - I_2(s)$ in this case. We now may write the two KVL equations for the two meshes for $t \geq 0$ with $v_c(0) = 1$ V as

mesh 1: $-V_1(s) + (4 + Ls)I_1(s) + V_c(s) = 0$

mesh 2: $(4 + 1s)I_2(s) - V_c(s) = 0$

The Laplace transform of the input voltage is

$$V_1(s) = \frac{1}{s}$$

Also, note that for the capacitor, we have

$$V_c(s) = \frac{1}{s} + \frac{1}{Cs}(I_1(s) - I_2(s))$$

Substituting V_1 and V_c into the mesh equations, we have (when $C = 1/2$ F)

$$\left(4 + Ls + \frac{2}{s}\right) I_1(s) - \left(\frac{2}{s}\right) I_2(s) = 0$$

and

$$-\left(\frac{2}{s}\right) I_1(s) + \left(4 + s + \frac{2}{s}\right) I_2(s) = \frac{1}{s}$$

The third step requires the selection of the value of L and then solving for $I_1(s)$. Examine Figure 14D-3; the two meshes are symmetric when $L = 1$ H. Then, trying this value and using Cramer's rule, we solve for $I_1(s)$, obtaining

$$I_1(s) = \frac{\left(\dfrac{2}{s}\right)\dfrac{1}{s}}{\left(4 + s + \dfrac{2}{s}\right)^2 - \left(\dfrac{2}{s}\right)^2}$$

$$= \frac{2}{s(s^3 + 8s^2 + 20s + 16)}$$

Fourth, in order to determine $i_1(t)$, we will use a partial fraction expansion. Rearranging and factoring the denominator of $I_1(s)$, we determine that

$$I_1(s) = \frac{2}{s(s + 4)(s + 2)^2}$$

Hence, we have the partial fraction expansion

$$I_1(s) = \frac{A}{s} + \frac{B}{s + 4} + \frac{C}{(s + 2)^2} + \frac{D}{s + 2}$$

Then, we readily determine that $A = 1/8$, $B = -1/8$, and $C = -1/2$. In order to find D, we use the differentiation method of Section 14-5 to obtain

$$D = \frac{1}{(2 - 1)!} \frac{d}{ds} [(s + 2)^2 I_1(s)]_{s = -2}$$

$$= \frac{-2(2s + 4)}{s^4 + 8s^3 + 16s^2} \bigg|_{s = -2}$$

$$= 0$$

Therefore, using the inverse Laplace transform for each term, we obtain

$$i_1(t) = \frac{1}{8} - \frac{1}{8} e^{-4t} - \frac{1}{2} te^{-2t} \text{ A} \qquad t \geq 0$$

Finally, the sketch of $i_1(t)$ is shown in Figure 14D-4. It is clear that $i_1(t)$ has essentially reached a steady-state value of 0.125 A by $t = 4$ seconds.

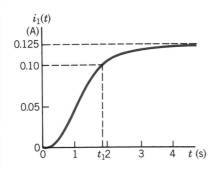

Figure 14D-4 The response of $i_1(t)$.

In order to find t_1 when

$$i_1(t_1) = 0.1 \text{ A}$$

we estimate that t_1 is approximately 2 seconds. After evaluating $i_1(t)$ for a few selected values of t near 2 seconds, we find that $t_1 = 1.8$ seconds. Therefore, the design requirements are satisfied for $L = 1$ H. Of course, other suitable values of L can be determined which will satisfy the design requirements.

TERMS AND CONCEPTS

Automatic Control   Use of a self-directing system to control a process to achieve a desired performance.

Automation   Automatic operation or control of a process, device, or system.

Complex Frequency $s = \sigma + j\omega$

Convolution Theorem   The convolution of $h(t)$ and $f(t)$ is equal to $\int_0^t h(\tau)f(t - \tau)\, d\tau$

Final Value Theorem   The value of $f(t)$ as t approaches infinity may be determined from $f(\infty) = \lim\limits_{s \to 0} sF(s)$.

Impedance in the s-domain   Ratio of the voltage $V(s)$ at a pair of terminals to the current $I(s)$ flowing into the positive voltage terminal, all expressed in the s-domain.

Impulse Function   A very short pulse such that its value is zero for $t \neq 0$ and the integral of the area of the pulse is 1.

Impulse Response   Output response of a circuit when the input is a unit impulse, with no initial stored energy in the circuit.

Information Ratio   Percentage of household expenditures for various kinds of information-related activities.

Information Society   A society in which there is an abundance in quantity and quality of information with all the necessary facilities for its distribution.

Initial Value Theorem   The initial value at time $t = 0$ may be determined from $f(0) = \lim\limits_{s \to \infty} sF(s)$.

Inverse Laplace Transform   Transform of $F(s)$ into the time domain to yield $f(t)$.

Laplace Transform   Transform of $f(t)$ into its s-domain form $F(s)$.

Partial Fraction Expansion   Expansion of $F(s)$ into a series of terms.

Poles   Roots of the denominator polynomial of the transfer function $H(s)$.

Transfer Function   Ratio of the response of a circuit to an excitation expressed as a function of complex frequency s. The initial conditions are assumed to be zero.

Transformation   Conversion of a set of equations from one domain to another, e.g., from the time domain to the complex frequency domain.

Transform Pairs   A function in the time domain, $f(t)$, and its Laplace transform $F(s)$.

Zeros   Roots of the numerator polynomial of the transfer function $H(s)$.

REFERENCES   Chapter 14

Agnew, J. "Simulating Audio Transducers with PSpice," *Electronic Design,* November 7, 1991, pp. 45–59.

Becker, J. "*RC* High-Pass Filter for Active Antennas," *Elektor Electronics,* February 1992, pp. 24–27.

Dordick, Herbert S. *Understanding Modern Telecommunications,* McGraw-Hill, New York, 1986.

Dorf, Richard C. *Technology, Society and Man,* Boyd and Fraser, San Francisco, 1974.

Dorf, Richard C. *Modern Control Systems,* 7th Edition, Addison-Wesley, Reading, Mass., 1995.

Garnett, G. H. "A High-Resolution, Multichannel Digital-to-Analog Converter," *Hewlett-Packard Journal,* February 1992, pp. 48–52.

Kullstam, Per A. "Heaviside's Operational Calculus," *IEEE Transactions on Education,* May 1991, pp. 155–166.

Nahin, Paul J. "Behind the Laplace Transform," *IEEE Spectrum,* March 1991, p. 60.

Nahin, Paul J. "Oliver Heaviside," *Scientific American,* June 1990, pp. 122–129.

Ruffell, J. "Switch-Mode Power Supply," *Elektor Electronics,* February 1992, pp. 62–66.

SUMMARY

With the advent of the increasing requirement to control complex circuits and systems, it is necessary to describe circuits and the excitation inputs and to analyze their time response. In this chapter, we showed that the use of the Laplace transform permits the analyst to transform a problem governed by differential equations to a problem that can be solved by linear algebraic equations.

Using the Laplace transform, we convert the description of a circuit from the time domain to the s-domain. Then, using algebra, we may solve for selected variables while readily accounting for the initial conditions of the energy storage elements. Then using the inverse Laplace transform, we are able to express the response in the time domain.

The Laplace transform enables us to represent several special time functions in the s-domain. For example, we may readily obtain the Laplace transform of an impulse function and a time-shifted function. Furthermore, the Laplace transform of the derivative of a time function is readily obtained and the initial conditions are easily taken into account. Thus, we may proceed to solve differential equations that describe complex circuits.

Given the function $F(s)$, we may use the initial and final value theorems, subject to some limitations, to find the initial and final values of the corresponding function of time, $f(t)$.

The concept of impedance in the s-domain is useful and permits us to represent a circuit by the ratio of the response to an excitation when all the initial conditions are set to zero. Thus, we may proceed to use all the previously developed theorems such as Thévenin's theorem, source transformations, and the maximum power theorem.

Finally, we noted the utility of the convolution theorem, which permits us to determine the inverse transform of $H(s)F(s)$ as the convolution of $h(t)$ and $f(t)$. One example of the response of a circuit to a repetitive square wave is shown in Figure 14-23.

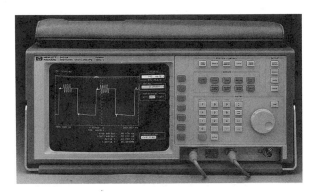

Figure 14-23 The HP 54510A oscilloscope displays the response of a circuit to a square wave signal. Note that each horizontal division is 0.1 μs. Courtesy of Hewlett-Packard Company.

PROBLEMS

Section 14-3   Laplace Transform

P 14.3-1 Find the Laplace transform, $F(s)$, when $f(t) = A \cos \omega t,\ t \ge 0$.

Answer: $F(s) = \dfrac{As}{s^2 + \omega^2}$

P 14.3-2 Find the Laplace transform, $F(s)$, when $f(t) = t,\ t \ge 0$.

P 14.3-3 Using the linearity property find the Laplace transform of $f(t) = e^{-3t} + t,\ t \ge 0$.

P 14.3-4 Using the linearity property find the Laplace transform of $f(t) = A(1 - e^{-bt})u(t)$.

Answer: $F(s) = \dfrac{Ab}{s(s + b)}$

Section 14-4   Impulse Function and Time Shift Property

P 14.4-1 Consider a pulse $f(t)$ defined by

$$f(t) = A \qquad 0 \le t \le T$$
$$= 0 \qquad \text{all other } t$$

Find $F(s)$.

Answer: $F(s) = \dfrac{A(1 - e^{-sT})}{s}$

P 14.4-2 Consider the pulse shown in Figure P 14.4-2, where the time function follows e^{at} for $0 < t < T$. Find $F(s)$ for the pulse.

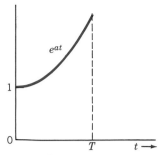

Figure P 14.4-2

P 14.4-3 Find the Laplace transform $F(s)$ for

(a) $f(t) = t^2 e^{-3t}, t \geq 0$;
(b) $f(t) = \delta(t - T), t \geq 0$;
(c) $f(t) = e^{-4t} \sin 5t, t \geq 0$.

P 14.4-4 Find the Laplace transform for $g(t) = e^{-t} u(t - 0.5)$.

P 14.4-5 Find the Laplace transform for $f(t) = \dfrac{-(t - T)}{T} u(t - T)$.

Answer: $F(s) = \dfrac{-1e^{-ST}}{Ts^2}$

Section 14-5   Inverse Laplace Transform

P 14.5-1 Find $f(t)$ when

$$F(s) = \frac{s + 3}{s^3 + 3s^2 + 6s + 4}$$

Answer: $f(t) = \frac{2}{3}e^{-t} - \frac{2}{3}e^{-t} \cos \sqrt{3}t + \dfrac{1}{\sqrt{3}} e^{-t} \times \sin \sqrt{3}t, t \geq 0$

P 14.5-2 Find $f(t)$ when

$$F(s) = \frac{s^2 - 2s + 1}{s^3 + 3s^2 + 4s + 2}$$

P 14.5-3 Find $f(t)$ when

$$F(s) = \frac{5s - 1}{s^3 - 3s - 2}$$

Answer: $f(t) = -e^{-t} + 2te^{-t} + e^{2t}, t \geq 0$

P 14.5-4 Find the inverse transform of

$$Y(s) = \frac{1}{s^3 + 3s^2 + 4s + 2}$$

Answer: $y(t) = e^{-t}(1 - \cos t), t \geq 0$

P 14.5-5 Find the inverse transform of

$$F(s) = \frac{2s + 6}{(s + 1)(s^2 + 2s + 5)}$$

P 14.5-6 Find the inverse transform of

$$F(s) = \frac{2s + 6}{s(s^2 + 3s + 2)}$$

Answer: $f(t) = [3 - 4e^{-t} + e^{-2t}] u(t)$

Section 14-6   Initial and Final Value Theorems

P 14.6-1 A function of time is represented by

$$F(s) = \frac{2s^2 - 3s + 4}{s^3 + 3s^2 + 2s}$$

(a) Find the initial value of $f(t)$ at $t = 0$.
(b) Find the value of $f(t)$ as t approaches infinity.

P 14.6-2 Find the initial and final values of $v(t)$ when

$$V(s) = \frac{(s + 16)}{s^2 + 4s + 12}$$

P 14.6-3 Find the initial and final values of $v(t)$ when

$$V(s) = \frac{(s + 10)}{(3s^3 + 2s^2 + 1s)}$$

P 14.6-4 Find the initial and final values of $f(t)$ when

$$F(s) = \frac{-2(s + 7)}{s^2 - 2s + 10}$$

Answer: initial value = -2, final value does not exist.

Section 14-7   Solution of Differential Equations Describing a Circuit

P 14.7-1 Find $i(t)$ for the circuit of Figure P 14.7-1 when $i(0) = 1$ A, $v(0) = 8$ V, and $v_1 = 2e^{-at}u(t)$ where $a = 2 \times 10^4$.

Answer: $i(t) = \frac{1}{15}(-10e^{-bt} + 3e^{-2bt} + 22e^{-4bt})$A, $t \geq 0$, $b = 10^4$.

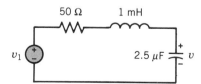

Figure P 14.7-1

P 14.7-2 All new homes are required to install a device called a ground fault circuit interrupter (GFCI) that will provide protection from shock. By monitoring the current going to and returning from a receptacle, a GFCI senses when normal flow is interrupted and switches off the power in 1/40 second. This is particularly important if you are holding an appliance shorted, through your body to ground. A circuit model of the GFCI acting to interrupt a short is shown in Figure P 14.7-2. Find the current flowing through the person, $i(t)$, for $t \geq 0$ when the short is initiated at $t = 0$. Assume $v = 160 \cos 400t$ and the capacitor is initially uncharged.

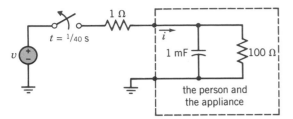

Figure P 14.7-2 Circuit model of person and appliance shorted to ground.

P 14.7-3 Using the Laplace transform, find $v_c(t)$ for $t > 0$ for the circuit shown in Figure P 14.7-3. The initial conditions are zero.
Answer: $v_c = -5e^{-2t} + 5(\cos 2t + \sin 2t)$

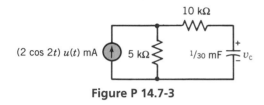

Figure P 14.7-3

P 14.7-4 Consider Figure P 14.7-4. Find $v_c(t)$ for $t > 0$ using the Laplace transform.

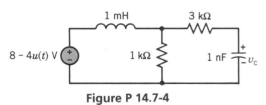

Figure P 14.7-4

P 14.7-5 Find $i_1(t)$ and $i_2(t)$ for $t > 0$ for the circuit of Figure P 14.7-5 using the Laplace transform.
Answer: $i_1 = (2.4e^{-bt} + 0.6e^{-6bt} + 3)$ mA
$i_2 = (1.2e^{-6bt} - 1.2e^{-bt})$ mA; $b = 5 \times 10^5$

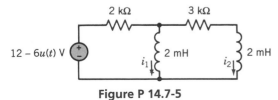

Figure P 14.7-5

P 14.7-6 Using Laplace transforms, find $v_c(t)$ for $t > 0$ for the circuit of Figure P 14.7-6 when (a) $C = 1/18$ F and (b) $C = 1/10$ F.

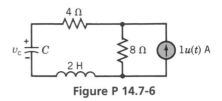

Figure P 14.7-6

P 14.7-7 The source for the circuit of Figure P 14.7-7 is $v_s = 40 \cos t$ V. Determine the current $I(s)$ when $i(0) = 6$ A and $di/dt(0) = 10$ A/s.

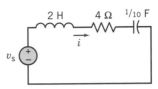

Figure P 14.7-7

P 14.7-8 Using the superposition principle, find $v_0(t)$ for the circuit of Figure P 14.7-8 when $v_1(t) = u(t)$ and $i(t) = e^{-2t}u(t)$ and there is no stored energy in the circuit at $t = 0$.

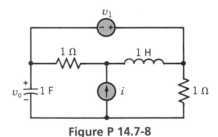

Figure P 14.7-8

P 14.7-9 A circuit with no stored energy at $t = 0^-$ is shown in Figure P 14.7-9. Determine and plot $v(t)$ for $t \geq 0$.

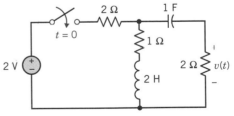

Figure P 14.7-9

P 14.7-10 Find $i(t)$ for the circuit of Figure P 14.7-10. Assume the switch has been open for a long time.
Answer: $i = -0.025e^{-200t} \sin 400t$ A, $t > 0$

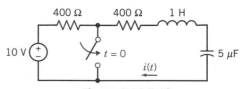

Figure P 14.7-10

Section 14-8 Transfer Function and Impedance
(a) *RLC* Circuits

P 14.8-1 The current source shown in Figure P 14.8-1 is $i(t) = tu(t)\ \mu$A. Find $v_0(t)$ when the initial value of v_0 is zero.

Answer: $v_0(t) = 1t - 10^{-3}(1 - e^{-10^3 t})$ mV, $t \geq 0$

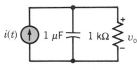

Figure P 14.8-1

P 14.8-2 For the circuit shown in Figure P 14.8-2, find the Thévenin equivalent of the circuit to the left of terminals a–b using the Laplace transform and impedances.

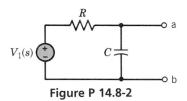

Figure P 14.8-2

P 14.8-3 For the circuit shown in Figure P 14.8-3, find the impedance of the circuit at input terminals a–b.

Answer: $Z_{in} = 1\Omega$

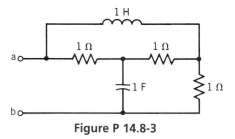

Figure P 14.8-3

P 14.8-4 Consider the circuit of Figure P 14.8-4, where the combination of R_2 and C_2 represents the input of an oscilloscope. The combination of R_1 and C_1 is added to the probe of the oscilloscope to shape the response $v_0(t)$ so that it will equal $v_1(t)$ as closely as possible. Find the necessary relationship for the resistors and capacitors so that $v_0 = av_1$ where a is a constant.

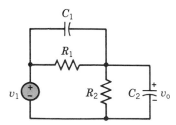

Figure P 14.8-4 Circuit for oscilloscope probe.

P 14.8-5 The impulse response of a circuit is

$$h(t) = \sqrt{2}e^{-t/\sqrt{2}} \sin \frac{t}{\sqrt{2}}, \qquad t \geq 0$$

(a) Find $H(s)$.
(b) Determine $|H(j\omega)|$.

Answer: (a) $H(s) = \dfrac{1}{s^2 + \sqrt{2}s + 1}$

P 14.8-6 A gyrator is an ideal element defined by the equations

$$i_1 = G_2 v_2$$

and

$$i_2 = -G_1 v_1$$

where the constant G_i is called the gyration conductance. The ideal model of the gyrator is shown in Figure P 14.8-6.

(a) Find $H(s) = V_2(s)/V_1(s)$ when an impedance $Z_2(s)$ is connected to the output terminals.
(b) Find $Z_{in}(s) = V_1(s)/I_1(s)$ when a load impedance Z_2 is connected to the output terminals.
(c) Find $Z_{in}(s)$ when Z_2 consists of a resistor R in parallel with a capacitor C and $G_1 = G_2 = 1$.

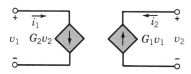

Figure P 14.8-6 Ideal model of gyrator.

P 14.8-7 Consider the circuit shown in Figure P 14.8-7. Show that by proper choice of L, the input impedance $Z = V_1(s)/I_1(s)$ can be made independent of s. What value of L satisfies this condition? What is the value of Z when it is independent of s?

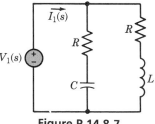

Figure P 14.8-7

P 14.8-8 The circuit shown in Figure P 14.8-8 has the initial conditions of $i_1(0) = -1$ A, $v_c(0) = -2$ V, and $i_2(0) = 3$ A. When $v_1 = 6e^{-t}u(t)$ V, determine $I_2(s) = \mathcal{L}\{i_2(t)\}$.

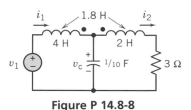

Figure P 14.8-8

P 14.8-9 A bridged-T circuit is often used as a filter and is shown in Figure P 14.8-9. Show that the transfer function of the circuit is

$$\frac{V_{out}(s)}{V_{in}(s)} = \frac{1 + (2R_1 + R_2)Cs + R_1R_2C^2s^2}{1 + 2R_1Cs + R_1R_2C^2s^2}$$

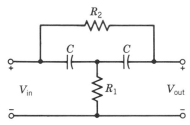

Figure P 14.8-9 Bridged-T circuit.

P 14.8-10 (a) Find $H(s) = V_0(s)/V_1(s)$ for the circuit of Figure P 14.8-10. (b) Determine $v_0(t)$ when the initial current in the inductors is zero. Note that the voltage v_0 appears across the series combination of the 150-Ω resistor and the 2-mH inductor.

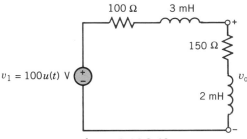

Figure P 14.8-10

P 14.8-11 A circuit is described by the transfer function

$$\frac{V_0}{V_1} = H(s) = \frac{9s + 18}{3s^3 + 18s^2 + 39s}$$

Find the response $v_0(t)$ when the input v_1 is a unit impulse at $t = 0$.

P 14.8-12 A circuit is shown in Figure P 14.8-12a, which has a response to a unit step, $v_1 = 1u(t)$ V, as shown in Figure P 14.8-12b. (a) Obtain the transfer function $V_0(s)/V_1(s)$. (b) Using the transfer function and the step response, calculate R. (c) Find the value of R that would lead to critical damping of the step response.

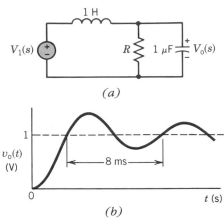

(a)

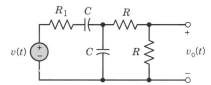

(b)

Figure P 14.8-12

P 14.8-13 An electric microphone and its associated circuit can be represented by the circuit shown in Figure P 14.8-13.

(a) Determine the transfer function $H(s) = V_0(s)/V(s)$.
(b) Sketch the magnitude portion of the Bode diagram for $H(j\omega)$ when $R_1 = 10\ \Omega$, $R = 50\ \Omega$, and $C = 100\ \mu$F (Agnew, 1991).

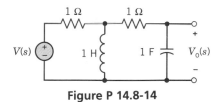

Figure P 14.8-13 Microphone circuit.

P 14.8-14 (a) Determine the transfer function $H(s) = V_0(s)/V(s)$ for the circuit shown in Figure P 14.8-14. (b) Obtain $H(j\omega)$ and determine the maximum of $|H(\omega)|$. Indicate the frequency, ω_{max}.

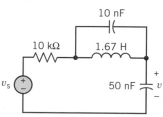

Figure P 14.8-14

P 14.8-15 Determine $v(t)$ for the circuit of Figure P 14.8-15 when $v_s = 1u(t)$ V.

Figure P 14.8-15

P 14.8-16 Modern electronic equipment generally needs one or more dc power supplies. Depending on the type of equipment, either a linear or a switch-mode stabilizer is used. Compared with the linear supply, the switch-mode power supply has some distinct advantages, including smaller size and higher efficiency for the same output power. They are used in television sets and computers. A circuit used to filter the high frequencies from the power supply output voltage is shown in Figure P 14.8-16 (Ruffel, 1992). Determine the transfer function $V_0(s)/V(s)$ and sketch the Bode diagram. The unwanted voltage, v, is a sinusoid at 15 kHz. Determine the attenuation at that frequency.

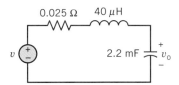

Figure P 14.8-16

P 14.8-17 Engineers had avoided inductance in long-distance circuits because it slows transmission. Heaviside proved that the addition of inductance to a circuit could enable it to transmit without distortion. George A. Campbell of the Bell Telephone Company designed the first practical inductance loading coils, in which the induced field of each winding of wire reinforced that of its neighbors so that the coil supplied proportionally more inductance than resistance. Each one of Campbell's 300 test

coils added 0.11 H and 12 Ω at regular intervals along 35 miles of telephone wire (Nahin, 1990). The loading coil balanced the effect of the leakage between the telephone wires represented by R and C in Figure P 14.8-17.
(a) Determine the transfer function $V_2(s)/V_1(s)$.
(b) When $C = 1$ mF and $R = (1/\sqrt{2})$ Ω determine $V_2(s)/V_1(s)$ and sketch the Bode diagram.

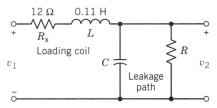

Figure P 14.8-17 Telephone and load coil circuit.

P 14.8-18 Determine $I(s)$ for the circuit of Figure P 14.8-18 when the initial conditions are zero.

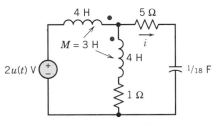

Figure P 14.8-18

P 14.8-19 Determine $v(t)$ for the circuit shown in Figure P 14.8-19 when $v_c(0) = -1$ V, $i(0) = 2$ A, $i_s = 4e^{-t}u(t)$ A, and $v_s = 3u(t)$ V.

Figure P 14.8-19

P 14.8-20 The motor circuit for driving the snorkel shown in Figure P 14.8-20a is shown in Figure P 14.8-20b. Find the motor current $I_2(s)$ when the initial conditions

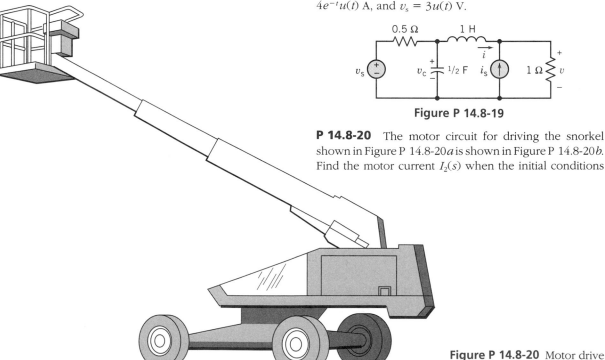

(a)

Figure P 14.8-20 Motor drive circuit for snorkel device.

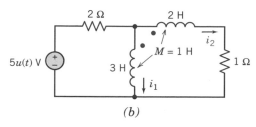

(b)

are $i_1(0^-) = 2$ A and $i_2(0^-) = 3$ A. Determine $i_2(t)$ and sketch it for 10 seconds. Does the motor current smoothly drive the snorkel?

Section 14-8 Transfer Function and Impedance
(b) Op Amp Circuits

P 14.8-21 Obtain $H(\omega) = V_2/V_1$ for the circuit shown in Figure P 14.8-21. Assume an ideal op amp. Sketch $|H(\omega)|$ for $\omega \geq 0$.

P 14.8-23 A high-pass filter using op amps is shown in Figure P 14.8-23. Determine $V_0(s)/V(s)$ and sketch the Bode diagram. Assume ideal op amps.

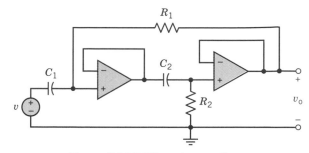

Figure P 14.8-23 High-pass filter.

P 14.8-24 A band-stop filter using op amps is shown in Figure P 14.8-24. Determine $V_0(s)/V(s)$ and sketch the Bode diagram. Assume ideal op amps.

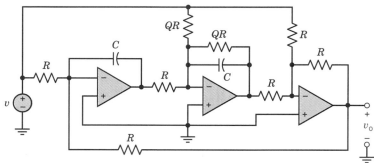

Figure P 14.8-24 Band-stop filter.

P 14.8-25 (a) Find $H(s) = V_0(s)/V_1(s)$ in terms of R_1, R_2, L, and C for the circuit in Figure P 14.8-25. (b) When $R_1 = R_2 = 1\ \Omega$, $C = 1$ F, and $L = 1$ H, find $H(s)$. Assume an ideal op amp.

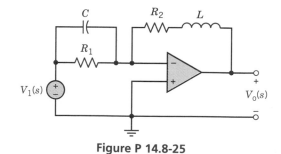

Figure P 14.8-25

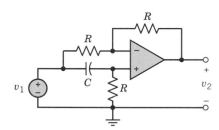

Figure P 14.8-21

P 14.8-22 An op amp circuit for a bandpass filter is shown in Figure P 14.8-22. Determine $V_0(s)/V(s)$ and sketch the Bode diagram. Assume ideal op amps.

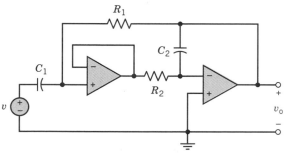

Figure P 14.8-22

P 14.8-26 A digital-to-analog converter (DAC) uses an op amp circuit shown in Figure P 14.8-26 (Garnett, 1992). The filter receives the pulse output from the DAC and produces the analog voltage. Although a number of different types of low-pass filters can be used, the one necessary requirement is that each of these filters present a load of no less than 100,000 Ω to the DAC channel output. If the filter input impedance at 39.06 kHz is lower than

100,000 Ω, DAC linearity will be degraded. Determine the transfer function of the filter, $V_0(s)/V(s)$. Select an appropriate op amp from Table 6-3 to achieve the input impedance requirement. Assume an ideal op amp.

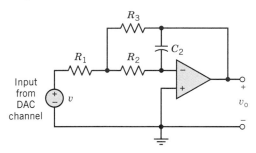

Figure P 14.8-26 Digital-analog converter filter.

P 14.8-27 The circuit shown in Figure P 14.8-27 can be used as a tone control for a stereo amplifier. Determine the transfer function $H(s)$. Assume an ideal op amp.

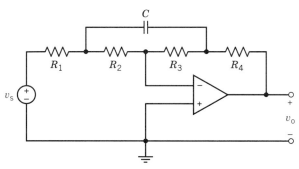

Figure P 14.8-27

Section 14-8 Transfer Function and Impedance
(c) Impulse Response

P 14.8-28 A circuit has a transfer function

$$H(s) = \frac{2}{s(s + 2)(s + 1)^2}$$

Find the impulse response $h(t)$. Use the final and initial value theorems to verify the values of $h(t)$ at $t = 0$ and $t = \infty$.
Answer: $h(t) = (1 - 2te^{-t} - e^{-2t})u(t)$

P 14.8-29 A circuit has a transfer function

$$H(s) = \frac{400}{s^2 + 400s + 2 \times 10^5}$$

Find the impulse response.
Answer: $h(t) = (e^{-200t} \sin 400t)\, u(t)$

P 14.8-30 A circuit has a transfer function

$$H(s) = \frac{4(s + 3)}{s^3 + 4s^2 + 4s}$$

Find the impulse response.

P 14.8-31 A series *RLC* circuit is shown in Figure P 14.8-31. Determine (a) the transfer function $H(s)$, (b) the impulse response, and (c) the step response for each set of parameter values given in the table below.

Table P 14.8-31

	L	C	R
a	2 H	0.025 F	18 Ω
b	2 H	0.025 F	8 Ω
c	1 H	0.391 F	4 Ω
d	2 H	0.125 F	8 Ω

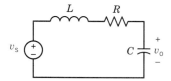

Figure P 14.8-31

Section 14-9 Circuit Analysis Using Impedance and Initial Conditions

P 14.9-1 Using Laplace transforms, find the response $v(t)$ for $t > 0$ for the circuit of Figure P 14.9-1 when $v_s = 6e^{-3t}u(t)$ V.
Answer: $v = \frac{44}{3}e^{-2t} + \frac{1}{3}e^{-5t} - 9e^{-3t}$ V

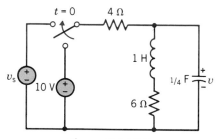

Figure P 14.9-1

P 14.9-2 Find $i(t)$ for $t > 0$ using Laplace transforms when $C = 1$ F and $i_s = e^{-t}u(t)$ A for the circuit of Figure P 14.9-2.

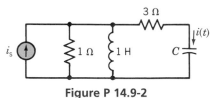

Figure P 14.9-2

P 14.9-3 Using the Laplace transform and state variables, find $i(t)$ for $t > 0$ when $i_1 = 0.1e^{-bt}$ A for the circuit of Figure P 14.9-3 when $b = 10^5$. Assume steady-state conditions at $t = 0^-$.

Answer: $i = \frac{1}{30}e^{-bt} + \frac{27}{40}e^{-6bt} - \frac{17}{24}e^{-10bt}$ A

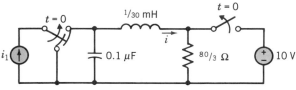

Figure P 14.9-3

P 14.9-4 Determine $i_2(t)$ for the circuit shown in Figure P 14.9-4. Assume the initial currents are zero.

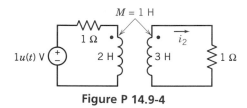

Figure P 14.9-4

P 14.9-5 Determine $v(t)$ for $t > 0$ for the circuit of Figure P 14.9-5 when $i_s = 5u(-t) + \delta(t)$ mA

Answer: $v(t) = 10e^{-5t}u(t)$ V

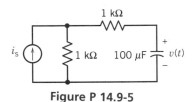

Figure P 14.9-5

P 14.9-6 Consider the circuit shown in Figure P 14.9-6 and $v_s = 1u(t)$ V. The initial conditions are zero, and it is required to find and plot $v_0(t)$. Use the state variables v and i and determine $v_0(t)$ for $t > 0$.

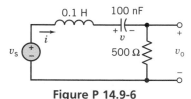

Figure P 14.9-6

P 14.9-7 Determine $v_0(t)$ when the capacitance has an initial voltage $v(0^-) = 5$ V, as shown in Figure P 14.9-7.

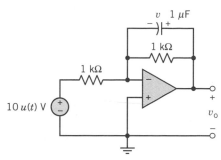

Figure P 14.9-7

Section 14-10   Convolution Theorem

P 14.10-1 Let $f(t)$ denote the 1-second pulse given by $f(t) = u(t) - u(t - 1)$. Determine the convolution $f(t) * f(t)$, which is the convolution of the pulse with itself. Draw $f * f$ versus time.

Answer: $f(t) * f(t) = tu(t) - 2(t - 1)u(t - 1) + (t - 2)u(t - 2)$

P 14.10-2 Consider a pulse of amplitude 2 and a duration of 2 second with its starting point at $t = 0$. (a) Find the convolution of this pulse with itself. (b) Draw the convolution $f(t) * f(t)$ versus time.

P 14.10-3 A circuit is shown in Figure P 14.10-3. Determine (a) the transfer function $V_2(s)/V_1(s)$ and (b) the response $v_2(t)$ when $v_1 = tu(t)$.

Answer: $v_2 = t - RC(1 - e^{-t/RC})$, $t \geq 0$

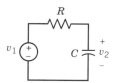

Figure P 14.10-3

P 14.10-4 Find the convolution of $h(t) = t$ and $f(t) = e^{-at}$ for $t > 0$ using the convolution integral and the inverse transform of $H(s)F(s)$.

Answer: $\dfrac{at - 1 + e^{-at}}{a^2}$, $t > 0$

Section 14-11   Stability

P 14.11-1 The transfer function of a circuit is

$$H(s) = \frac{k}{s + (3 - k)}$$

Determine the required range of k for stable operation.

P 14.11-2 The transfer function of a circuit is

$$H(s) = \frac{8}{s^2 + (6 - k)s + 8}$$

Determine the required range of k for stable operation.

P 14.11-3 The transfer function of a circuit is

$$H(s) = \frac{(s + 2)}{s^2 - 2s + 2}$$

Plot the poles and zeros on the s-plane and determine if the circuit is stable.

P 14.11-4 A circuit is shown in Figure P 14.11-4 with K unspecified. Determine the transfer function of the circuit and find the range of K so that the circuit is stable. Determine K so that the circuit acts as an oscillator.

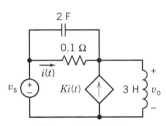

Figure P 14.11-4

P 14.11-5 Determine the transfer function $\dfrac{V_0(s)}{V_1(s)}$ of the circuit of Figure P 14.11-5. (a) Determine the value of K that causes the circuit to act as an oscillator. (b) Determine the impulse response of the circuit when $K = 1$, $R = 1$ kΩ, and $C = 0.5$ mF.

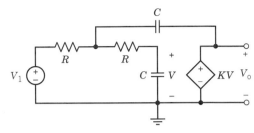

Figure P 14.11-5

PSpice PROBLEMS

SP 14-1 Plot the response $v_c(t)$ for Example 14.5.

SP 14-2 The circuit of Figure SP 14-2a has no stored energy at $t = 0$ and the input signal is a pulse, as shown in SP 14-2b. Determine $v(t)$ for a period of 10 seconds.

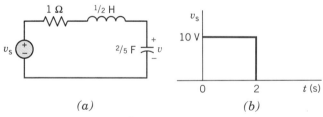

Figure SP 14-2

SP 14-3 Obtain $i(t)$ for the circuit of Figure 14.6-1 and verify the response obtained in Eq. 14-32 when $R = 1$ Ω, $L = 1/2$ H, $I_0 = 1$ A, and $M = 2$ V.

SP 14-4 Obtain the Bode diagram for the circuit of Problem 14.8-13 using PSpice.

SP 14-5 Plot $20 \log | H(\omega) |$ for Problem 14.8-14 using PSpice and determine the maximum of $| H(\omega) |$ and the frequency at which the maximum occurs.

SP 14-6 For Problem 14.8-12 determine R that leads to the step response shown in Figure P 14.8-12b. Using

PSpice plot $v_0(t)$ and determine the 8 ms period and the overshoot magnitude. Determine the value of R that will result in an 8% overshoot.

SP 14-7 Determine and plot the unit step and unit impulse response of the circuit of Figure SP 14-7 for each set of parameter values listed in the following table.

Table SP 14-7

	R	C	K
a	50 kΩ	2 μF	1
b	50 kΩ	2 μF	1.25
c	50 kΩ	2 μF	5.25

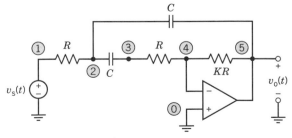

Figure SP 14-7

VERIFICATION PROBLEMS

VP 14-1 A laboratory circuit is represented by $Y(s) = H(s)X(s)$ where $h(t) = e^{-4t}$ and $x(t) = 8$ for $t > 0$. A student report states that $y(t) = 2(1 + e^{-4t})$, t > 0. Is this correct?

VP 14-2 A student has analyzed the circuit of Figure VP 14-2 and states that $v(t) = 5(1 - e^{-40t}) u(t)$ V. Verify this answer.

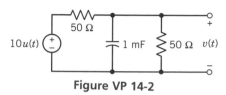

Figure VP 14-2

VP 14-3 The transfer function of a circuit is

$$H(s) = \frac{18}{s^2 + s - 2}$$

A student report states that the two poles are $s = -1$ and $s = -2$ and the circuit is stable. Is this report correct?

VP 14-4 An experimenter has built the circuit shown in Figure VP 14-4. She states that this circuit is stable. Can you prove otherwise? Assume an ideal op amp.

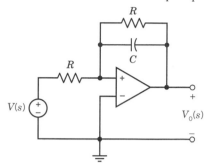

Figure VP 14-4

DESIGN PROBLEMS

DP 14-1 The circuit shown in Figure DP 14-1 represents an oscilloscope probe connected to an oscilloscope. Components C_1 and R_2 represent the input circuitry of the oscilloscope, and C_1 and R_1 represent the probe. Determine the transfer function $H(s) = V_0(s)/V(s)$. Determine the required relationship so that the natural response of the probe is zero. Determine the required relationship so that the step response is equal to the step input to within a gain constant. Is this achievement physically possible?

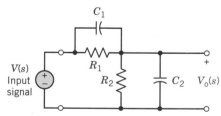

Figure DP 14-1 Oscilloscope probe circuit.

DP 14-2 A circuit with $v_1 = 1e^{-\alpha t}$ V is shown in Figure DP 14-2. Determine the required C so that the effect of the input v_1 does not appear at $v_o(t)$. Select C when $\alpha = 2$.

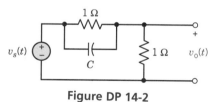

Figure DP 14-2

DP 14-3 A bicycle light is a useful accessory if you do a lot of riding at night. By lighting up the road and making you more visible to cars, it reduces the chances of an accident. While generator-powered incandescent lights are the most common type used on bikes, there are a number of reasons why fluorescent lights are more suitable. For one thing, fluorescent lights shine with a brighter light that fully covers the road, the rider, and the bike and really gets the attention of car drivers. Because they are shaped in narrow tubes that can mount alongside a bike's frame, fluorescent lights offer less wind resistance than a comparable headlight with a flat face. When used with a generator, a fluorescent light offers additional advantages over a conventional incandescent light, first, because it is more efficient—giving more light for the same pedaling effort—and, second, because it cannot be burned out by the overvoltage that a generator can produce when speeding down hills. Fluorescent lights also last longer than do incandescent bulbs, especially on a bicycle, where vibrations tend to weaken an incandescent bulb's filament.

A model of a fluorescent light for a bike is shown in Figure DP 14-3. Select L so that the bulb current rapidly rises to its steady-state value and only overshoots its final value by less than 10 percent.

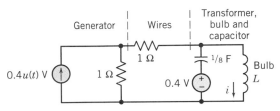

Figure DP 14-3 Fluorescent bicycle light circuit.

DP 14-4 A low-pass RLC circuit is shown in Figure DP 14-4.

(a) Determine the transfer function $V_2(s)/V_1(s)$.
(b) Plot the Bode diagram when $R = 25\ \Omega$, $L = 10$ mH, and $C = 1\ \mu$F for frequencies from 10 Hz to 10 kHz.
(c) Select R so that the magnitude portion of the Bode diagram does not vary more than ± 6 dB between 50 Hz and 2.8 kHz.

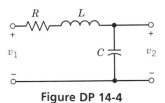

Figure DP 14-4

DP 14-5 A bandpass filter for a microphone is shown in Figure DP 14-5. The voice has a frequency range of 100 Hz to 3 kHz, and thus we need a bandwidth equal to this frequency range. Select R, L, and C and plot the Bode diagram.

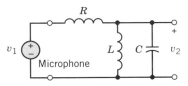

Figure DP 14-5 Bandpass filter.

DP 14-6 A bandstop filter is used to reduce the effects of a 60-Hz hum in a public auditorium speaker system. The bandstop filter is shown in Figure DP 14-6. It is desired to decrease the effect of the 60-Hz hum by, at least, 25 dB, and the bandwidth of the filter should be less than 10 Hz. Select suitable R, L, and C and plot the resulting Bode diagram for frequencies between 10 Hz and 100 Hz. The winding resistance of the coil is $R_1 = 0.02\ \Omega$ and $L \le 10$ H.

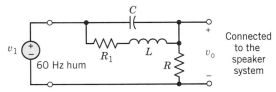

Figure DP 14-6 Band-stop filter.

DP 14-7 A TV cable company engineer has designed a notch (bandstop) filter shown in Figure DP 14-7. The engineer wants to reject channel 3, which operates at 59.6 MHz, with a rejection ratio for V_2/V_1 of at least 45 dB. Select a suitable C and determine the rejection ratio achieved.

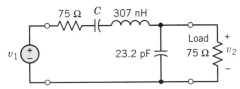

Figure DP 14-7 Band-stop filter.

CHAPTER 15

FOURIER SERIES

PREVIEW

Periodic waveforms arise in many practical circuits. We have, heretofore, developed skills in the analysis of circuits that have sinusoidal waveforms as input sources. Thus, in this chapter we will endeavor to represent a nonsinusoidal periodic waveform by a series of sinusoidal waveforms. Then we can use superposition to find the response of the circuit to each sinusoid in turn and thus, in the aggregate, determine the response to the irregular waveform.

The idea of describing waveforms as a series of sinusoidal functions is useful to today's engineer. We daily experience speech synthesizers and music synthesizers that generate speech sounds or musical sounds as a result of generating an appropriate series of sinusoidal signals that activate a stereo speaker. In this chapter we study the representation of a waveform by a series of sinusoidal signals of integer multiple frequencies.

15-1 DESIGN CHALLENGE

DC POWER SUPPLY

A laboratory power supply uses a nonlinear circuit called a rectifier to convert a sinusoidal voltage input into a dc voltage. The sinusoidal input

$$v_{ac}(t) = A \sin \omega_0 t$$

comes from the wall plug. In this example, $A = 160$ V and $\omega_0 = 377$ rad/s ($f_0 = 60$ Hz). Figure 15D-1 shows the structure of the power supply. The output of the rectifier is the absolute value of its input, that is,

$$v_s(t) = |A \sin \omega_0 t|$$

The purpose of the rectifier is to convert a signal that has an average value equal to zero into a signal that has an average value that is not zero. The average value of $v_s(t)$ will be used to produce the dc output voltage of the power supply.

The rectifier output is not a sinusoid, but it is a periodic signal. In this chapter we will see that periodic signals can be represented by Fourier series. The Fourier series of $v_s(t)$ will contain a constant, or dc, term and some sinusoidal terms. The purpose of the filter shown in Figure 15D-1 is to pass the dc term and attenuate the sinusoidal terms. The output of the filter, $v_0(t)$, will be a periodic signal and can be represented by a Fourier series. Because we are designing a dc power supply, the sinusoidal terms in the Fourier series of $v_0(t)$ are undesirable. The sum of these undesirable terms is called the ripple of $v_0(t)$.

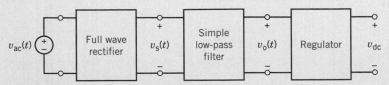

Figure 15D-1 Diagram of a power supply.

The challenge is to design a simple filter so that the dc term of $v_0(t)$ is at least 90 V and the size of the ripple is no larger than 5% of the size of the dc term.

We will return to this problem at the end of the chapter, after learning about Fourier series.

15-2 CHANNELS OF COMMUNICATION

Accuracy is the basic requirement in the transmission of signals. Signals in their original form are transmitted through a channel. For example, Morse code dot-dash transmission over the original Atlantic cable was distorted by the *RC* nature of the cable and its insulation material. Thus, the transmission rate was limited and the signal was distorted as it traveled over the cable.

The undersea cable transmits information slowly but uses only a narrow frequency channel. The telephone transmits information at an intermediate rate with moderate bandwidth requirements. Television transmits information at a high rate and requires a very wide frequency band. On long telephone lines, repeater amplifiers are spaced along the lines, regenerating the signal at points where it becomes weak. This allows transmission distance to be greatly extended.

Michael I. Pupin arrived in America in 1874 at age fifteen and spent the next five years preparing himself for admission to Columbia University. Pupin graduated with distinction in 1883 and after additional study at the University of Cambridge went to Berlin, where he worked under Helmholtz and G. Kirchhoff, receiving the doctorate in 1889. He then returned to Columbia to teach in the newly formed department of electrical engineering.

Professor Pupin discovered that the periodic insertion of inductance coils in telephone lines would improve their performance by reducing attenuation and distortion. For telephone signals, the line capacitance again limits the high frequencies to about 3000 Hz and reduces accuracy in understanding speech over the telephone. Pupin's loading coils balanced out the capacitive effects, thus extending the frequency range of the lines and improving the accuracy of transmission. For a time such lines were called "Pupinized."

15-3 THE FOURIER SERIES

A number of eighteenth-century mathematicians, including Euler and D. Bernoulli, were aware that a waveform $f(t)$ could be approximately represented by a finite weighted sum of harmonically related sinusoids. Baron Jean-Baptiste-Joseph Fourier proposed in 1807 that a periodic waveform could be broken down into an infinite series of simple sinusoids which, when added together, would construct the exact form of the original waveform. There are certain periodic functions for which an infinite series will exactly represent the function. There are also other certain functions for which the infinite series provides a good, but inexact, representation.

Let us consider the periodic function

$$f(t) = f(t + nT) \qquad n = \pm 1, \pm 2, \pm 3, \ldots$$

for every value of t, where T is the period. (The period is the smallest value of T that satisfies the equation above.)

The expression for a finite sum of harmonically related sinusoids called a *Fourier series* is

$$f(t) = a_0 + \sum_{n=1}^{N} a_n \cos n\omega_0 t + \sum_{n=1}^{N} b_n \sin n\omega_0 t$$

or

$$f(t) = C_0 + \sum_{n=1}^{N} C_n \cos(n\omega_0 t + \theta_n) \tag{15-1}$$

where $\omega_0 = 2\pi/T$ and a_0, a_n, and b_n (all real) are called the *Fourier trigonometric coefficients*. Using the alternative form of $f(t)$ given by the second equation, we have $C_n = (a_n^2 + b_n^2)^{1/2}$ and $\theta_n = -\tan^{-1} b_n/a_n$.

In general, it is easier to calculate a_n and b_n than it is to calculate the coefficients C_n and θ_n. We will see in Section 15-4 that this is particularly true when $f(t)$ is symmetric. On the other hand, the Fourier series involving C_n is more convenient for calculating the steady-state response of a linear circuit to a periodic input.

Examining Eq. 15-1, we find that most periodic waveforms can be described by this equation. For $n = 1$, one cycle covers T seconds and the corresponding sinusoid $C_1 \cos(\omega_0 t + \theta_1)$ is called the *fundamental*. For $n = k$, k cycles fall within T seconds and $C_k \cos(k\omega_0 t + \theta_k)$ is called the kth *harmonic term*. Similarly, ω_0 is called the fundamental frequency and $k\omega_0$ is called the kth harmonic frequency. The function $f(t)$ can be represented to any degree of accuracy by increasing the number of terms in its Fourier series. Certain functions can also be represented exactly by an infinite number of terms.

A **Fourier series** is an accurate representation of a periodic signal that consists of the sum of sinusoids at the fundamental and harmonic frequencies.

The nature of the waveform $f(t)$ depends on the amplitude and phase of every possible harmonic component, and we shall find it possible to generate waveforms that have extremely nonsinusoidal characteristics by an appropriate combination of sinusoidal functions. The waveforms $f(t)$ that can be described by Eq. 15-1 satisfy the following mathematical properties:

1 $f(t)$ is a single-valued function except at possibly a finite number of points.

2 The integral $\displaystyle\int_{t_0}^{t_0 + T} | f(t) | \, dt < \infty$ for any t_0.

3 $f(t)$ has a finite number of discontinuities within the period T.

4 $f(t)$ has a finite number of maxima and minima within the period T.

We shall consider $f(t)$ to represent a voltage or current waveform, and any voltage or current waveform that we can actually produce will certainly satisfy these conditions. We shall assume that the four conditions listed above are always satisfied.

Many electrical waveforms are periodic waveforms and can be described by a Fourier series. Once $f(t)$ is known, the Fourier coefficients may be calculated and the waveform is resolved into a dc term a_0 plus a sum of sinusoidal terms described by a_n and b_n. We will select N, the total number of terms, in order to achieve a faithful representation of $f(t)$ while accepting a certain degree of error.

Recall from calculus that sinusoids whose frequencies are integer multiples of some fundamental frequency $f_0 = 1/T$ form an orthogonal set of functions, that is,

$$\frac{2}{T} \int_0^T \sin \frac{2\pi nt}{T} \cos \frac{2\pi mt}{T} \, dt = 0 \qquad \text{all } n,m \tag{15-2}$$

and

$$\frac{2}{T} \int_0^T \sin \frac{2\pi nt}{T} \sin \frac{2\pi mt}{T} \, dt = \frac{2}{T} \int_0^T \cos \frac{2\pi nt}{T} \cos \frac{2\pi mt}{T} \, dt$$

$$= \begin{cases} 0 & n \neq m \\ 1 & n = m \neq 0 \end{cases} \tag{15-3}$$

The coefficients of Eq. 15-1 can be obtained from

$$a_0 = \frac{1}{T} \int_{t_0}^{T + t_0} f(t) \, dt \quad \text{the average over one period} \tag{15-4}$$

$$a_n = \frac{2}{T} \int_0^T f(t) \cos n\omega_0 t \, dt \qquad n > 0 \tag{15-5}$$

$$b_n = \frac{2}{T} \int_0^T f(t) \sin n\omega_0 t \, dt \qquad n > 0 \tag{15-6}$$

We now demonstrate how Eqs. 15-4 through 15-6 are obtained using the orthogonal relationships of Eqs. 15-2 and 15-3. For example, to obtain a_k we multiply Eq. 15-1 by $\cos k\omega_0 t$ and integrate both sides over one period of T, obtaining

$$\int_0^T f(t) \cos k\omega_0 t \, dt = \int_0^T a_0 \cos k\omega_0 t \, dt$$

$$+ \sum_{n=1}^N \int_0^T (a_n \cos n\omega_0 t + b_n \sin n\omega_0 t) \cos k\omega_0 t \, dt \tag{15-7}$$

Evaluating the right-hand side of Eq. 15-7, we find that the only nonzero term is for $n = k$, so that

$$\int_0^T f(t) \cos k\omega_0 t \, dt = a_k(T/2) \tag{15-8}$$

Therefore, we have derived Eq. 15-5 for all a_n. Following a similar approach and multiplying Eq. 15-1 by $\sin k\omega_0 t$, we can find Eq. 15-6. Equation 15-4 provides the dc term, which is the average value over the period T.

Example 15-1

Determine the Fourier series for the periodic rectangular wave shown in Figure 15-1. Plot the approximation of the waveform when $N = 7$.

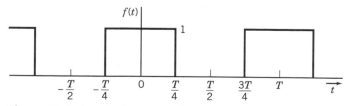

Figure 15-1 A periodic rectangular wave.

Solution

The fundamental radian frequency is $\omega_0 = 2\pi/T$. First, let us find a_0, the average value over the period T. Hence we obtain

$$a_0 = \frac{1}{T} \int_{-T/2}^{T/2} f(t) \; dt = \frac{1}{T} \int_{-T/4}^{T/4} 1 \; dt = \frac{1}{2}$$

We choose to integrate from $-T/2$ to $+T/2$ for convenience. Furthermore, the function is only nonzero from $-T/4$ to $+T/4$ for the selected period of integration. We call $f(t)$ of Figure 15-1 an *even function* since $f(t) = f(-t)$.

An **even function** $f(t)$ exhibits symmetry around the vertical axis at $t = 0$ so that $f(t) = f(-t)$.

Hence, Eq. 15-6 with $f(t) = 1$ for $-T/4 < t < T/4$ is

$$b_n = \frac{2}{T} \int_{-T/4}^{T/4} 1 \sin n\omega_0 t \; dt = 0$$

Therefore, all b_n from Eq. 15-6 will be zero. Thus, solving for a_n using Eq. 15-5, we obtain

$$a_n = \frac{2}{T} \int_{-T/4}^{T/4} 1 \cos n\omega_0 t \; dt$$

$$= \frac{2}{T\omega_0 n} \sin n\omega_0 t \Big|_{-T/4}^{T/4}$$

$$= \frac{1}{\pi n} \left[\sin\left(\frac{\pi n}{2}\right) - \sin\left(\frac{-\pi n}{2}\right) \right] \tag{15-9}$$

Equation 15-9 is equal to zero when $n = 2, 4, 6, \ldots$ and

$$a_n = \frac{2(-1)^q}{\pi n} \tag{15-10}$$

where $q = (n-1)/2$ and $n = 1, 3, 5, \ldots$. Thus the Fourier series is

$$f(t) = \frac{1}{2} + \sum_{n=1,\,\text{odd}}^{N} \frac{2(-1)^q}{\pi n} \cos n\omega_0 t$$

and $q = (n-1)/2$. Then, a_1 and a_3 are

$$a_1 = \frac{2}{\pi} \quad \text{and} \quad a_3 = \frac{-2}{3\pi}$$

Similarly,

$$a_5 = \frac{2}{5\pi} \quad \text{and} \quad a_7 = \frac{-2}{7\pi}$$

Thus, the a_n coefficients decrease as $1/n$ and the coefficient of the seventh harmonic is one-seventh of the value of the fundamental. The approximation of the function $f(t)$ when $N = 7$ is shown in Figure 15-2. Note that the error is readily discernible. Nevertheless, the approximation may be satisfactory in many cases. This approximation neglects the terms for $N \geq 9$ where $a_9 = 2/9\pi = 0.0707$. In order to attain a good approximation

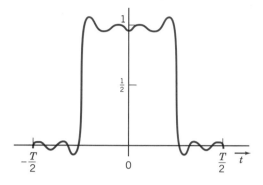

Figure 15-2 The approximation to the function $f(t)$ of Figure 15-1 for $N = 7$.

we may seek to include all terms whose magnitude exceeds a specified percentage of the magnitude of the fundamental coefficient a_1.

EXERCISE 15-1

Determine the Fourier series when $f(t) = K$, a constant.
Answer: $a_0 = K$, $a_n = b_n = 0$ for $n \geq 1$

EXERCISE 15-2

Determine the Fourier series when $f(t) = A \cos \omega_0 t$.
Answer: $a_0 = 0$, $a_1 = A$, $a_n = 0$ for $n > 1$, $b_n = 0$

15-4 Symmetry of the Function $f(t)$

Four types of symmetry can be readily recognized and then utilized to simplify the task of calculating the Fourier coefficients. They are

1 Even-function symmetry,
2 Odd-function symmetry,
3 Half-wave symmetry,
4 Quarter-wave symmetry.

A function is *even* when $f(t) = f(-t)$, and a function is *odd* when $f(t) = -f(-t)$. The function shown in Figure 15-1 is an even function. For even functions all $b_n = 0$ and

$$a_n = \frac{4}{T} \int_0^{T/2} f(t) \cos n\omega_0 t \, dt$$

For odd functions all $a_n = 0$ and

$$b_n = \frac{4}{T} \int_0^{T/2} f(t) \sin n\omega_0 t \, dt$$

An example of an odd function is $\sin \omega_0 t$. Another odd function is shown in Figure 15-3.

Half-wave symmetry for a function $f(t)$ is obtained when

$$f(t) = -f\left(t - \frac{T}{2}\right) \tag{15-11}$$

These half-wave symmetric waveforms have the property that the second half of each period looks like the first half turned upside down. The function shown in Figure 15-3 has half-wave symmetry. If a function has half-wave symmetry, then both a_n and b_n are zero for even values of n. We see that $a_0 = 0$ for half-wave symmetry since the average value of the function over one period is zero.

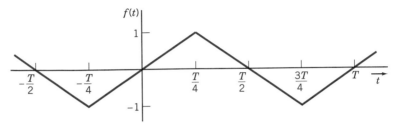

Figure 15-3 An example of an odd function with half-wave symmetry.

Quarter-wave symmetry is a term used to describe a function that has half-wave symmetry and, in addition, has symmetry about the midpoint of the positive and negative half-cycles. An example of an odd function with quarter-wave symmetry is shown in Figure 15-4. If a function is odd and quarter-wave, then $a_0 = 0$, $a_n = 0$ for all n, $b_n = 0$ for

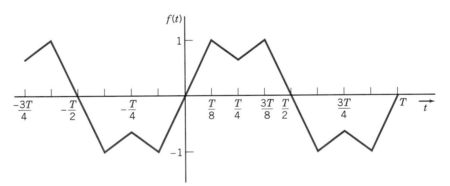

Figure 15-4 An odd function with quarter-wave symmetry.

even n, and for all odd n

$$b_n = \frac{8}{T}\int_0^{T/4} f(t) \sin n\omega_0 t \, dt$$

If a function is even and quarter-wave, then $a_0 = 0$, $b_n = 0$ for all n, $a_n = 0$ for even n, and for all odd n

$$a_n = \frac{8}{T}\int_0^{T/4} f(t) \cos n\omega_0 t \, dt$$

The calculation of the Fourier coefficients and the associated effects of symmetry of the waveform $f(t)$ are summarized in Table 15-1. Often the calculation of the Fourier series can be simplified by judicious selection of the origin ($t = 0$), since the analyst usually has the choice to arbitrarily select this point.

Table 15-1
Fourier Series and Symmetry

Symmetry	Fourier Coefficients
1. Odd function $f(t) = -f(-t)$	$a_n = 0$ for all n $b_n = \dfrac{4}{T}\displaystyle\int_0^{T/2} f(t) \sin n\omega_0 t \; dt$
2. Even function $f(t) = f(-t)$	$b_n = 0$ for all n $a_n = \dfrac{4}{T}\displaystyle\int_0^{T/2} f(t) \cos n\omega_0 t \; dt$
3. Half-wave symmetry $f(t) = -f\left(t - \dfrac{T}{2}\right)$	$a_0 = 0$ $a_n = 0$ for even n $b_n = 0$ for even n $a_n = \dfrac{4}{T}\displaystyle\int_0^{T/2} f(t) \cos n\omega_0 t \; dt$ for odd n $b_n = \dfrac{4}{T}\displaystyle\int_0^{T/2} f(t) \sin n\omega_0 t \; dt$ for odd n
4. Quarter-wave symmetry Half-wave symmetric and symmetric about the midpoints of the positive and negative half-cycles	A. Odd function: $\quad a_0 = 0,\; a_n = 0$ for all n $\qquad\qquad\qquad\quad b_n = 0$ for even n $\qquad\qquad b_n = \dfrac{8}{T}\displaystyle\int_0^{T/4} f(t) \sin n\omega_0 t \; dt$ for odd n B. Even function: $\quad a_0 = 0,\; b_n = 0$ for all n $\qquad\qquad\qquad\quad a_n = 0$ for even n $\qquad\qquad a_n = \dfrac{8}{T}\displaystyle\int_0^{T/4} f(t) \cos n\omega_0 t \; dt$ for odd n

Example 15-2

Determine the Fourier series for the triangular waveform $f(t)$ shown in Figure 15-5. Each increment of time on the horizontal grid is $\pi/8$ s, and the maximum and minimum values of $f(t)$ are 4 and -4, respectively. Determine how many terms are required (find N) in order to retain all terms whose magnitude exceeds 2% of the magnitude of the fundamental term.

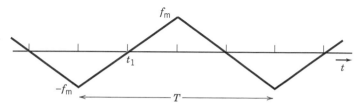

Figure 15-5 Waveform $f(t)$ for Example 15-2, where $f_m = 4$ and $T = \dfrac{\pi}{2}$ s.

Solution

We will attempt to obtain the most advantageous form of symmetry. If we choose t_1 as the origin, we obtain an odd function with half-wave symmetry. Since the function is then also symmetrical about the midpoint of the positive and negative half-cycles, we obtain quarter-wave symmetry.

We determine the fundamental frequency by noticing that $T = 4$ increments of time or $T = 4(\pi/8) = \pi/2$ s. Therefore,

$$\omega_0 = \frac{2\pi}{T} = 4 \text{ rad/s}$$

Using Table 15-1, we use entry 4A to obtain $a_0 = 0$, $a_n = 0$ for all n, $b_n = 0$ for even n, and for odd n

$$b_n = \frac{8}{T} \int_0^{T/4} f(t) \sin n\omega_0 t \, dt$$

For the interval $0 \leq t \leq T/4$ we have

$$f(t) = \frac{f_m}{T/4} t = \frac{4f_m}{T} t$$

and since $T = \pi/2$ and $f_m = $ maximum value $= 4$, we have

$$f(t) = \frac{4(4)}{\pi/2} t = \frac{32}{\pi} t \qquad 0 \leq t \leq T/4$$

Then, we obtain

$$b_n = \frac{8}{\pi/2} \left(\frac{32}{\pi}\right) \int_0^{T/4} t \sin n\omega_0 t \, dt$$

$$= \frac{512}{\pi^2} \left[\frac{\sin n\omega_0 t}{n^2 \omega_0^2} - \frac{t \cos n\omega_0 t}{n\omega_0}\right]_0^{T/4}$$

$$= \frac{32}{\pi^2 n^2} \sin \frac{n\pi}{2} \qquad \text{for odd } n$$

Therefore, the Fourier series is

$$f(t) = 3.24 \sum_{n=1}^{N} \frac{1}{n^2} \sin \frac{n\pi}{2} \sin n\omega_0 t \qquad \text{for odd } n$$

The first four terms of the Fourier series (up to and including $N = 7$) are

$$f(t) = 3.24(\sin 4t - \tfrac{1}{9} \sin 12t + \tfrac{1}{25} \sin 20t - \tfrac{1}{49} \sin 28t)$$

The next harmonic is for $n = 9$, and its magnitude is $3.24/81 = 0.04$, which is less than 2% of $a_1 = 3.24$. Therefore, we can attain the desired approximation by using the first four nonzero terms (including $N = 7$) of the series.

The Fourier series for selected functions $f(t)$ are provided in Table 15-2.

EXERCISE 15-3

Determine the Fourier series for the waveform $f(t)$ shown in Figure E 15-3. Each increment of time on the horizontal axis is $\pi/8$ s, and the maximum and minimum are $+1$ and -1, respectively.

Table 15-2
The Fourier Series for Selected Periodic Waveforms

Function	Fourier Series

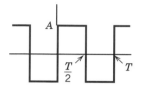

$$f(t) = \frac{4A}{\pi} \sum_{n=1}^{\infty} \frac{1}{2n-1} \sin(2n-1)\omega_0 t$$

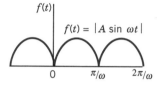

$$f(t) = \frac{2A}{\pi} - \frac{4A}{\pi} \sum_{n=1}^{\infty} \frac{1}{4n^2-1} \cos 2n\omega_0 t$$

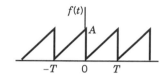

$$f(t) = \frac{A}{2} - \frac{A}{\pi} \sum_{n=1}^{\infty} \frac{\sin n\omega_0 t}{n}$$

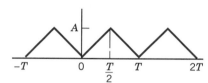

$$f(t) = \frac{A}{2} - \frac{4A}{\pi^2} \sum_{n=1}^{\infty} \frac{1}{(2n-1)^2} \cos(2n-1)\omega_0 t$$

Answer: $f(t) = \dfrac{4}{\pi} \sum_{n=1}^{N} \dfrac{1}{n} \sin n\omega_0 t; \qquad n$ odd, $\omega_0 = 4$ rad/s

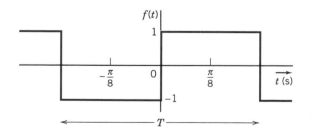

Figure E 15-3 The period $T = \dfrac{\pi}{2}$ s.

EXERCISE 15-4

Determine the Fourier series for the waveform $f(t)$ shown in Figure E 15-4. Each increment of time on the horizontal grid is $\pi/6$ s, and the maximum and minimum values of $f(t)$ are 2 and -2, respectively.

Answer: $f(t) = \dfrac{12}{\pi^2} \sum\limits_{n=1}^{N} \dfrac{1}{n^2} \sin(n\pi/3) \sin n\omega_0 t;$ n odd, $\omega_0 = 2$ rad/s

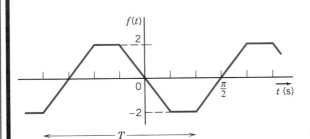

Figure E 15-4 The period $T = \pi$ s.

EXERCISE 15-5

For the periodic signal $f(t)$ shown in Figure E 15-5, determine if the Fourier series contains (a) sine and cosine terms and (b) even harmonics and (c) calculate the dc value.
Answer: (a) Yes, both sine and cosine terms; (b) no even harmonics; (c) $a_0 = 0$.

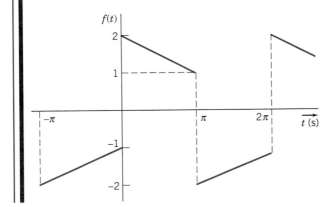

Figure E 15-5

15-5 EXPONENTIAL FORM OF THE FOURIER SERIES

The form of the Fourier series of Eq. 15-1 is often called the sine-cosine form or the rectangular form. An alternative is

$$f(t) = C_0 + \sum_{n=1}^{N} C_n \cos(n\omega_0 t + \theta_n) \tag{15-12}$$

where $C_0 = a_0$ = average value of $f(t)$ and

$$C_n = (a_n^2 + b_n^2)^{1/2} \quad \text{and} \quad \theta_n = \tan^{-1}(b_n/a_n)$$

or

$$a_n = C_n \cos \theta_n \quad \text{and} \quad b_n = C_n \sin \theta_n$$

TOOLBOX 15-1

EULER'S IDENTITY

$$\cos \beta = \tfrac{1}{2}(e^{j\beta} + e^{-j\beta})$$

See Appendix E for further details.

Since the function $\cos(n\omega_0 t + \theta_n)$ may be written in *exponential form* using Euler's identity, we have, with $N = \infty$,

$$f(t) = C_0 + \sum_{\substack{n = -\infty \\ n \neq 0}}^{\infty} C_n e^{jn\omega_0 t}$$

$$= \sum_{-\infty}^{\infty} C_n e^{jn\omega_0 t} \tag{15-13}$$

where C_n are the *complex* (phasor) *coefficients* defined by

$$C_n = \frac{1}{T} \int_{t_0}^{t_0 + T} f(t) e^{-jn\omega_0 t} \, dt \tag{15-14}$$

In Chapter 12 we used $\mathbf{C}_n$ to represent a phasor, but henceforth, for ease, we drop the boldface notation and use C_n to represent the phasor. Furthermore, $C_n = C_{-n}^*$ so that the coefficients for negative n are the complex conjugates of C_n for positive n.

Example 15-3
Determine the complex Fourier series for the waveform shown in Figure 15-6.

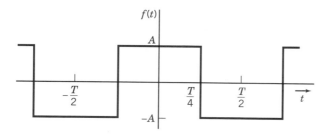

Figure 15-6 Square wave.

Solution
The average value of $f(t)$ is zero, so $C_0 = 0$. As shown in Figure 15-6, the function is even. Then, using Eq. 15-14, we select $t_0 = -T/2$ and define $jn\omega_0 = m$ to obtain

$$C_n = \frac{1}{T} \int_{-T/2}^{T/2} f(t) e^{-jn\omega_0 t} \, dt$$

$$= \frac{1}{T} \int_{-T/2}^{-T/4} -Ae^{-mt} \, dt + \frac{1}{T} \int_{-T/4}^{T/4} Ae^{-mt} \, dt + \frac{1}{T} \int_{T/4}^{T/2} -Ae^{-mt} \, dt$$

$$= \frac{A}{mT} \left(e^{-mt} \Big|_{-T/2}^{-T/4} - e^{-mt} \Big|_{-T/4}^{T/4} + e^{-mt} \Big|_{T/4}^{T/2} \right)$$

$$= \frac{A}{jn\omega_0 T} \left(2e^{jn\pi/2} - 2e^{-jn\pi/2} + e^{-jn\pi} - e^{jn\pi} \right)$$

$$= \frac{A}{2\pi n}\left(4\sin\frac{n\pi}{2} - 2\sin(n\pi)\right)$$

$$= \begin{cases} 0 & \text{for even } n \\ \dfrac{2A}{n\pi}\sin n\dfrac{\pi}{2} & \text{for odd } n \end{cases}$$

$$= A\frac{\sin x}{x}$$

where $x = n\pi/2$.

Since $f(t)$ is an even function, all C_n are real. We found $C_n = 0$ for n even. Furthermore, we have for $n = 1$

$$C_1 = A\frac{\sin \pi/2}{\pi/2} = \frac{2A}{\pi} = C_{-1}$$

For $n = 2$ we obtain

$$C_2 = C_{-2} = A\frac{\sin \pi}{\pi} = 0$$

and for $n = 3$

$$C_3 = A\frac{\sin(3\pi/2)}{3\pi/2} = \frac{-2A}{3\pi} = C_{-3}$$

The complex Fourier series is

$$f(t) = \cdots \frac{-2A}{3\pi}e^{-j3\omega_0 t} + \frac{2A}{\pi}e^{-j\omega_0 t} + \frac{2A}{\pi}e^{j\omega_0 t} + \frac{-2A}{3\pi}e^{j3\omega_0 t} + \cdots$$

$$= \frac{2A}{\pi}(e^{j\omega_0 t} + e^{-j\omega_0 t}) + \frac{-2A}{3\pi}(e^{j3\omega_0 t} + e^{-j3\omega_0 t}) + \cdots$$

$$= \frac{4A}{\pi}\cos \omega_0 t - \frac{4A}{3\pi}\cos 3\omega_0 t + \cdots$$

$$= \frac{4A}{\pi}\sum_{\substack{n=1 \\ n=\text{odd}}}^{\infty}\frac{(-1)^q}{n}\cos n\omega_0 t \qquad q = \frac{n-1}{2}$$

Note that the exponential series extends from $n = -\infty$ to $n = +\infty$. All voltages and currents of a circuit are real functions. Furthermore, for real $f(t)$ it follows that

$$C_n = C_{-n}$$

If it is known that $f(t)$ is real, only the positive-frequency part of the spectrum need be shown. If, in addition to being real, $f(t)$ is an odd function, then the coefficients C_n are all purely imaginary.

Example 15-4

Determine the complex Fourier coefficients for the square wave shown in Figure 15-7.

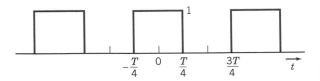

Figure 15-7 Square wave.

Solution

We note that the function is even, so we expect the coefficients to all be real. Using Eq. 15-14 and defining $m = jn\omega_0$, we have

$$C_n = \frac{1}{T} \int_{-T/4}^{T/4} 1 e^{-mt} \, dt$$

$$= \frac{1}{-mT} e^{-mt} \Big|_{-T/4}^{T/4}$$

$$= \frac{1}{-mT} (e^{-mT/4} - e^{+mT/4})$$

$$= \frac{1}{-jn2\pi} (e^{-jn\pi/2} - e^{+jn\pi/2})$$

$$= 0 \qquad n \text{ even}, \ n \neq 0$$

$$= (-1)^{(n-1)/2} \frac{1}{n\pi} \qquad n \text{ odd}$$

To find C_0 we determine the average value as

$$C_0 = \frac{1}{T} \int_0^T f(t) \, dt = \frac{1}{T} \int_{-T/4}^{T/4} 1 \, dt = \frac{1}{2}$$

Finally, we note that the conditions of symmetry will imply that if a function is half-wave symmetric, $C_n = 0$ for all even values of n and only odd harmonics exist (see Table 15-1). The Fourier coefficients for the complex Fourier series of selected waveforms are given in Table 15-3.

EXERCISE 15-6

Find the exponential Fourier coefficients for the function shown in Figure E 15-6.

Answer: $C_n = 0$ for even n, $C_n = \dfrac{2}{jn\pi}$ for odd n

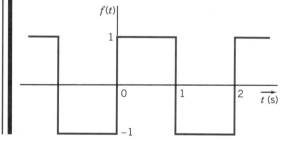

Figure E 15-6

EXERCISE 15-7

Determine the Fourier series in terms of sine and cosine terms for the waveform shown in Figure E 15-7.

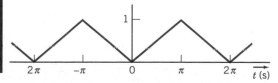

Figure E 15-7

Table 15-3
Complex Fourier Coefficients for Selected Waveforms

Waveform	Name of Waveform and Equation	Symmetry	C_n
1. 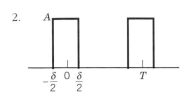	**Square wave** $$f(t) = \begin{cases} A, & \dfrac{-T}{4} < t < \dfrac{T}{4} \\ -A, & \dfrac{T}{4} \le t < \dfrac{3T}{4} \end{cases}$$	Even	$= A\dfrac{\sin n\pi/2}{n\pi/2},\ n$ odd $= 0,\ n = 0$ and n even
2. 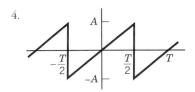	**Rectangular pulse** $$f(t) = A, \qquad \dfrac{-\delta}{2} < t < \dfrac{\delta}{2}$$	Even	$= A\dfrac{\delta}{T}\dfrac{\sin(n\pi\delta/T)}{(n\pi\delta/T)}$
3. 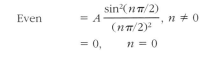	**Triangular wave**	Even	$= A\dfrac{\sin^2(n\pi/2)}{(n\pi/2)^2},\ n \ne 0$ $= 0, \qquad n = 0$
4. 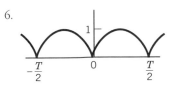	**Sawtooth wave** $$f(t) = 2At/T$$ $$\dfrac{-T}{2} < t < \dfrac{T}{2}$$	Odd	$= Aj(-1)^n/n\pi,\ n \ne 0$ $= 0, \qquad n = 0$
5. 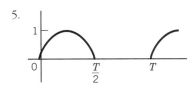	**Half-wave rectified sinusoid** $$f(t) = \begin{cases} \sin \omega_0 t, & 0 \le t < T/2 \\ 0, & -T/2 \le t < 0 \end{cases}$$		$= 1/\pi(1 - n^2),\ n$ even $= -j/4, \qquad n = \pm 1$ $= 0, \qquad$ otherwise
6.	**Full-wave rectified sinusoid** $$f(t) = \lvert \sin \omega_0 t \rvert$$	Even	$= 2/\pi(1 - n^2),\ n$ even $= 0, \qquad$ otherwise

EXERCISE 15-8

Determine the complex Fourier coefficients for the waveform shown in Figure E 15-8.

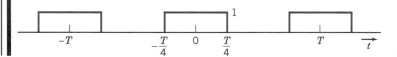

Figure E 15-8

15-6 | THE FOURIER SPECTRUM

If we plot the complex Fourier coefficients C_n as a function of angular frequency, we obtain a *Fourier spectrum*. Since C_n may be complex, we have

$$C_n = |C_n| \underline{/\theta_n} \qquad (15\text{-}15)$$

and we plot $|C_n|$ and $\underline{/\theta_n}$ as the *amplitude spectrum* and the *phase spectrum*, respectively. The Fourier spectrum exists only at the fundamental and harmonic frequencies and therefore is called a discrete or line spectrum. The amplitude spectrum appears on a graph as a series of equally spaced vertical lines with heights proportional to the amplitudes of the respective frequency components. Similarly, the phase spectrum appears as a series of equally spaced lines with heights proportional to the value of the phase at the appropriate frequency. The word "spectrum" was introduced into physics by Newton (1664) to describe the analysis of light by a prism into its different color components or frequency components.

Let us calculate the complex Fourier coefficients for the pulse of width δ repeated every T seconds, as shown in Figure 15-8. Then we will plot the *Fourier spectrum* for the function.

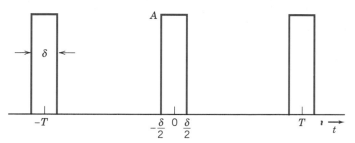

Figure 15-8 A periodic sequence of pulses each of width δ.

The **Fourier spectrum** is a graphical display of the amplitude and phase of the complex Fourier coefficients at the fundamental and harmonic frequencies.

The Fourier coefficients are

$$C_n = \frac{1}{T} \int_{-T/2}^{T/2} A e^{-jn\omega_0 t} \, dt \qquad (15\text{-}16)$$

For $n \neq 0$, we have

$$
\begin{aligned}
C_n &= \frac{A}{T} \int_{-\delta/2}^{\delta/2} e^{-jn\omega_0 t} \, dt \\
&= \frac{-A}{jn\omega_0 T} (e^{-jn\omega_0 \delta/2} - e^{jn\omega_0 \delta/2}) \\
&= \frac{2A}{n\omega_0 T} \sin\left(\frac{n\omega_0 \delta}{2}\right) \\
&= \frac{A\delta}{T} \frac{\sin(n\omega_0 \delta/2)}{(n\omega_0 \delta/2)} \\
&= \frac{A\delta}{T} \frac{\sin x}{x} \qquad\qquad\qquad (15\text{-}17)
\end{aligned}
$$

where $x = (n\omega_0 \delta/2)$ and $n \neq 0$. When $n = 0$, we have

$$
C_0 = \frac{1}{T} \int_{-\delta/2}^{\delta/2} A \, dt = \frac{A\delta}{T}
$$

One may show that $(\sin x)/x = 1$ for $x = 0$ by using L'Hôpital's rule. Also, $(\sin x)/x$ is zero whenever x is an integer multiple of π, that is,

$$
\frac{\sin(n\pi)}{n\pi} = 0 \qquad n = 1, 2, 3, \ldots
$$

We plot $|C_n|$ versus $\omega = n\omega_0$ in Figure 15-9a for n up to ± 15. Also, $|(\sin x)/x|$ is

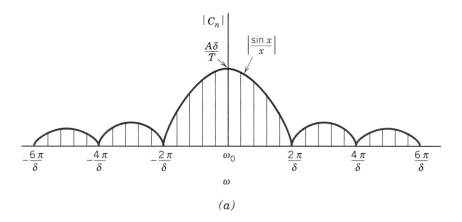

(a)

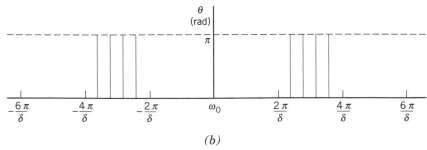

(b)

Figure 15-9 *(a)* The line spectra $|C_n|$ and the function $|(\sin x)/x|$ in color for pulse width $\delta = T/5$ for each pulse. *(b)* Phase spectra for the pulse width $\delta = T/5$.

shown in Figure 15-9a in color. Note that $|C_n| = 0$ when $n\omega_0\delta/2 = n\pi$, or

$$\frac{2\pi\delta}{2T} = \pi$$

This zero value occurs when $\delta/T = 1$ when $f(t) = A$ for all t. When $\delta = T/5$, as is the case for Figure 15-8, the zero value of $(\sin x)/x$ occurs at $\omega = 5\omega_0$, $\omega = 10\omega_0$, and so on. If δ/T is not an integer, the plot for $|C_n|$ still follows $|(\sin x)/x|$ but the spectral lines are not equal to zero for any n.

The phase spectrum is plotted in Figure 15-9b. The plot of phase spectrum is a description of the fact that between ω_0 and $2\pi/\delta$ the angle is zero and between $2\pi/\delta$ and $4\pi/\delta$ the angle is π radians.

It is useful to consider the case where we keep δ fixed and vary the period T. As T increases, two effects are noticed: (1) the amplitude of the spectrum decreases in proportion to $1/T$ and (2) the spacing between lines decreases in proportion to $2\pi/T$. As the period T increases, the spacing between components, $\Delta\omega = 2\pi/T$, becomes smaller and there are more and more frequency components in a given range of frequency. The amplitudes of these components decrease as T is increased, but the shape of the spectrum does not change. We conclude that the shape, or envelope, of the spectrum is dependent only on the pulse shape and not on the repetition period T.

It is also instructive to keep T fixed and vary δ (with the restriction that $\delta < T/2$). As δ increases, (1) the amplitude of the spectrum increases in proportion to δ and (2) the frequency content of the signal is compressed within an increasingly narrow range of frequencies. Thus there is an inverse relationship between pulse width in time and the frequency spread of the spectrum. A convenient measure of the frequency spread is the distance to the first zero crossing of the $(\sin x)/x$ function.

EXERCISE 15-9

Plot $|C_n|$ when $\delta = T/3$ and $\delta = T/8$ for the pulse of Figure 15-8, and compare the results with Figure 15-9.

15-7 THE TRUNCATED FOURIER SERIES

A practical calculation of the Fourier series requires that we truncate the series to a finite number of terms. Thus, the periodic function $f(t)$ is represented by a practical sum

$$f(t) \cong \sum_{n=-N}^{N} C_n e^{jn\omega_0 t} = S_N(t) \tag{15-18}$$

Nevertheless, it can be shown that the Fourier coefficients emanate from the minimization of the mean-square error. The error for N terms is

$$\epsilon(t) = f(t) - S_N(t) \tag{15-19}$$

If we use the mean-square error (MSE) defined as

$$\text{MSE} = \frac{1}{T}\int_0^T \epsilon^2(t)\,dt \tag{15-20}$$

then the minimum of the MSE is obtained when the coefficients, C_n, are the Fourier coefficients.

However, when we use a truncated series with $N < \infty$, we will usually experience an error. For example, if we use $N = 15$ terms to approximate a square wave, as shown in Figure 15-10, we find an oscillatory error. Furthermore, we observe an overshoot at each

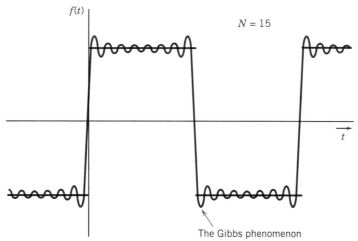

Figure 15-10 The Fourier series for $N = 15$ for a periodic square wave.

point of discontinuity where the function jumps from $-A$ to $+A$. It was shown by the nineteenth-century American mathematician Josiah W. Gibbs that the overshoot remains at about 10 percent at the edge, regardless of the number of terms included in the expansion. This property is known as the *Gibbs phenomenon*.

15-8 ║ CIRCUITS AND FOURIER SERIES

It is often desired to determine the response of a circuit excited by a periodic input signal $v_s(t)$. We can represent $v_s(t)$ by a Fourier series and then find the response of the circuit to the fundamental and each harmonic. Assuming the circuit is linear and the principle of superposition holds, we can consider that the total response is the sum of the response to the dc term, the fundamental, and each harmonic.

Consider the RC circuit shown in Figure 15-11 where we consider the output v_o and the input v_s. We can find V_o/V_s for each Fourier term by using phasor analysis.

Example 15-5
Find the steady-state response, $v_0(t)$, of the RC circuit shown in Figure 15-11a. The input, $v_s(t)$, is the square wave shown earlier in Figure 15-1. In this example, we will represent this square wave by the first four terms of its Fourier series. This Fourier series was calculated in Example 15-1. Assume that the parameters of the circuit are $R = 1\ \Omega$ and $C = 2$ F and that the period of the square wave is $T = \pi$ seconds.

Solution

In Example 15-1 we found that

$$v_s(t) \cong \frac{1}{2} + \sum_{\substack{n=1 \\ \text{odd}}}^{N} \frac{2(-1)^q}{n\pi} \cos n\omega_0 t$$

where $q = (n-1)/2$ and $n = 1, 2, 3, \ldots$.

The input $v_s(t)$ will be represented by the first four terms of this series. Since $\omega_o = 2$, we have,

$$v_s(t) = \frac{1}{2} + \frac{2}{\pi} \cos 2t - \frac{2}{3\pi} \cos 6t + \frac{2}{5\pi} \cos 10t$$

We will find the steady-state response, $v_o(t)$, using superposition. It is helpful to let $v_{sn}(t)$ denote the term of $v_s(t)$ corresponding to n. In this example, $v_s(t)$ has four terms, corresponding to $n = 0, 1, 3,$ and 5. Then

$$v_s(t) = v_{s0}(t) + v_{s1}(t) + v_{s3}(t) + v_{s5}(t)$$

where

$$v_{s0}(t) = \frac{1}{2} \qquad v_{s1}(t) = \frac{2}{\pi} \cos 2t$$

$$v_{s3}(t) = -\frac{2}{3\pi} \cos 6t \qquad \text{and} \qquad v_{s5}(t) = \frac{2}{5\pi} \cos 10t$$

Figure 15-11 illustrates the way that superposition is used in this example. First, since the series connection of the voltage sources with voltages $v_{s0}(t)$, $v_{s1}(t)$, $v_{s3}(t)$, and $v_{s5}(t)$ is equivalent to a single voltage source having voltage $v_s(t) = v_{s0}(t) + v_{s1}(t) + v_{s3}(t) + v_{s5}(t)$, the circuit shown in Figure 15-11b is equivalent to the circuit shown in Figure 15-11a.

Next, the principle of superposition is invoked to break the problem up into four simpler problems as shown in Figure 15-11c. Each circuit in Figure 15-11c is used to calculate the steady-state response to a single one of the voltage sources from Figure 15-11b. (When calculating the response to one voltage source, the other voltage sources are set to zero; that is, they are replaced by short circuits.) For example, the voltage $v_{o3}(t)$ is the steady-state response to $v_{s3}(t)$ alone. Superposition tells us that the response to all four voltage sources working together is the sum of the responses to the four voltage sources working separately, that is,

$$v_o(t) = v_{o0}(t) + v_{o1}(t) + v_{o3}(t) + v_{o5}(t)$$

The advantage of breaking the problem up into four simpler problems is that the input to each of the four circuits in Figure 15-11c is a sinusoid. The problem of finding the steady-state response to a periodic input has been reduced to the simpler problem of finding the steady-state response to a sinusoidal input. The steady-state response of a linear circuit to a sinusoidal input can be found using phasors. In Figure 15-11d the four circuits from Figure 15-11c have been redrawn using phasors and impedances. The impedance of the capacitor is

$$\mathbf{Z}_c = \frac{1}{jn\omega_0 C} \qquad \text{for } n = 0, 1, 3, 5$$

Each of the four circuits corresponds to a different value of n, so the impedance of the capacitor is different in each of the circuits. (The frequency of the input sinusoid is $n\omega_0$, so

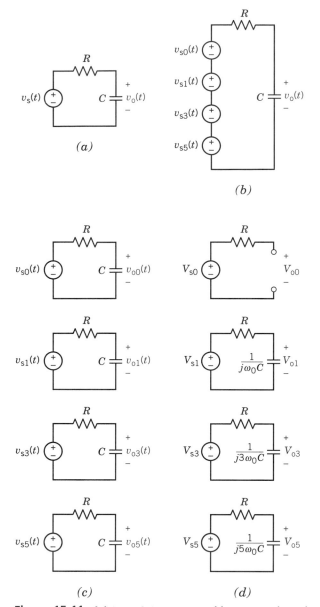

Figure 15-11 *(a)* An *RC* circuit excited by a periodic voltage $v_s(t)$. *(b)* An equivalent circuit. Each voltage source is a term of the Fourier series of $v_s(t)$. *(c)* Using superposition. Each input is a sinusoid. *(d)* Using phasors to find steady-state responses to the sinusoids.

each of the circuits corresponds to a different frequency.) Notice that when $n = 0$, $Z_c = \infty$, so the capacitor acts like an open circuit. The four circuits shown in Figure 15-11*d* are very similar. In each case voltage division can be used to write

$$\mathbf{V}_{on} = \frac{1/jn\omega_0 C}{R + 1/jn\omega_0 C}\mathbf{V}_{sn} \quad \text{for } n = 0, 1, 3, 5$$

where $\mathbf{V}_{sn}$ is the phasor corresponding to $v_{sn}(t)$ and $\mathbf{V}_{on}$ is the phasor corresponding to

$v_{on}(t)$. So

$$\mathbf{V}_{on} = \frac{\mathbf{V}_{sn}}{1 + jn\omega_0 CR} \quad \text{for } n = 0, 1, 3, 5$$

In this example, $\omega_0 CR = 4$ so

$$\mathbf{V}_{on} = \frac{\mathbf{V}_{sn}}{1 + j4n} \quad \text{for } n = 0, 1, 3, 5$$

Next, the steady-state response can be written as

$$v_{on}(t) = |V_{on}| \cos(n\omega_0 t + \underline{/V_{on}})$$

$$= \frac{|V_{sn}|}{\sqrt{1 + 16n^2}} \cos(n\omega_0 t + \underline{/V_{sn}} - \tan^{-1} 4n)$$

In this example,

$$|V_{s0}| = \frac{1}{2}$$

$$|V_{sn}| = \frac{2}{n\pi} \quad \text{for } n = 1, 3, 5$$

$$\underline{/V_{sn}} = 0 \quad \text{for } n = 0, 1, 3, 5$$

Therefore,

$$v_{o0}(t) = \frac{1}{2}$$

$$v_{on}(t) = \frac{2}{n\pi\sqrt{1 + 16n^2}} \cos(n2t - \tan^{-1} 4n) \quad \text{for } n = 1, 3, 5$$

Doing the arithmetic yields

$$v_{o0}(t) = \frac{1}{2}$$

$$v_{o1}(t) = 0.154 \cos(2t - 76°)$$

$$v_{o3}(t) = 0.018 \cos(6t - 85°)$$

$$v_{o5}(t) = 0.006 \cos(10t - 87°)$$

Finally, the steady-state response of the original circuit, $v_o(t)$, is found by adding up the partial responses

$$v_o(t) = \frac{1}{2} + 0.154 \cos(2t - 76°) + 0.018 \cos(6t - 85°) + 0.006 \cos(10t - 87°)$$

It is important to notice that superposition justifies adding the functions of time, $v_{o0}(t)$, $v_{o1}(t)$, $v_{o3}(t)$, and $v_{o5}(t)$ to get $v_o(t)$. The phasors $\mathbf{V}_{o0}$, $\mathbf{V}_{o1}$, $\mathbf{V}_{o3}$, and $\mathbf{V}_{o5}$ each correspond to a different frequency. A sum of these phasors has no meaning.

EXERCISE 15-10

Find the response of the circuit of Figure 15-11 when $R = 10$ kΩ, $C = 0.4$ mF, and v_s is the sawtooth wave considered in Example 15-2 (Figure 15-5). Include all terms that exceed 2% of the fundamental term.

Answer: $v_o(t) \approx 0.20 \sin(4t - 86°) - 0.008 \sin(12t - 89°)$ V

15-9 FOURIER SERIES OF A WAVEFORM USING PSpice

PSpice can be used to calculate the coefficients of the Fourier series of a periodic voltage or current. There are two steps to the procedure.

1 Simulate the circuit containing the voltage or current to obtain a transient response.
2 Use the .FOUR command to tell PSpice to compute the coefficients and to put them in the PSpice output file.

The response that PSpice calls the transient response is the response that we called the complete response in Chapter 9. In other words, the PSpice transient response includes both transient and steady-state parts. The transient part must die out before the PSpice can calculate the coefficients of the Fourier series. Indeed, there must be at least one full period of the steady-state response at the end of the PSpice transient response if the coefficients are to be calculated accurately. (Warning: PSpice will calculate coefficients even if this condition is not satisfied. The coefficients will not be accurate. PSpice will not issue a warning.)

The syntax of the .FOUR command is

.FOUR <fundamental frequency> <variable name> . . .

Example 15-6

Consider the circuit shown in Figure 15-12a. The input to this circuit is the period voltage, $v_{in}(t)$, shown in Figure 15-12b. The output of this circuit, $v_{out}(t)$, will also be a periodic

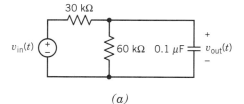

(a)

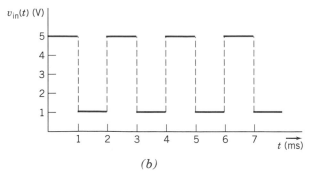

(b)

Figure 15-12 *(a)* An example circuit. *(b)* The periodic input voltage.

voltage. Both $v_{in}(t)$ and $v_{out}(t)$ will have the same fundamental frequency, 500 Hz. We will use PSpice to calculate the coefficients of the Fourier series of $v_{in}(t)$ and of $v_{out}(t)$. Figure 15-13 shows the PSpice input and output files. First, consider the input file. The command

.tran 10us 10ms uic

tells PSpice to run the simulation for 10 ms or 5 periods of $v_{in}(t)$. All transients will have died out before the final 2 ms. PSpice will calculate the Fourier series during the last full period of $v_{in}(t)$, i.e., during the last 2 ms of the simulation.

```
Illustration of the .FOUR command

Vin  1  0  pulse  ( 1  5  1ms  1us  1us  .999ms  2ms  )
R1   1  2  30k
R2   2  0  60k
C    2  0  0.10uF  ic=3

.tran  10us   10ms   uic
.four  500Hz   v(1)   v(2)

.end
```
(a)

```
FOURIER COMPONENTS OF TRANSIENT RESPONSE V(1)

DC COMPONENT =  3.000000E+00
```

HARMONIC NO	FREQUENCY (HZ)	FOURIER COMPONENT	NORMALIZED COMPONENT	PHASE (DEG)	NORMALIZED PHASE (DEG)
1	5.000E+02	2.547E+00	1.000E+00	1.791E+02	0.000E+00
2	1.000E+03	4.831E-09	1.897E-09	9.740E+01	-8.170E+01
3	1.500E+03	8.491E-01	3.334E-01	1.773E+02	-1.800E+00
4	2.000E+03	4.741E-09	1.862E-09	1.048E+02	-7.435E+01
5	2.500E+03	5.098E-01	2.002E-01	1.755E+02	-3.600E+00
6	3.000E+03	4.594E-09	1.804E-09	1.120E+02	-6.710E+01
7	3.500E+03	3.645E-01	1.431E-01	1.737E+02	-5.400E+00
8	4.000E+03	4.394E-09	1.726E-09	1.191E+02	-6.001E+01
9	4.500E+03	2.839E-01	1.115E-01	1.719E+02	-7.200E+00

```
    TOTAL HARMONIC DISTORTION =  4.291613E+01 PERCENT


FOURIER COMPONENTS OF TRANSIENT RESPONSE V(2)

DC COMPONENT =  2.007793E+00
```

HARMONIC NO	FREQUENCY (HZ)	FOURIER COMPONENT	NORMALIZED COMPONENT	PHASE (DEG)	NORMALIZED PHASE (DEG)
1	5.000E+02	2.668E-01	1.000E+00	9.818E+01	0.000E+00
2	1.000E+03	1.236E-03	4.633E-03	6.229E+00	-9.195E+01
3	1.500E+03	2.999E-02	1.124E-01	9.099E+01	-7.187E+00
4	2.000E+03	6.190E-04	2.320E-03	5.789E+00	-9.239E+01
5	2.500E+03	1.097E-02	4.111E-02	8.866E+01	-9.520E+00
6	3.000E+03	4.135E-04	1.550E-03	6.811E+00	-9.137E+01
7	3.500E+03	5.625E-03	2.119E-02	8.730E+01	-1.088E+01
8	4.000E+03	3.102E-04	1.163E-03	8.159E+00	-9.002E+01
9	4.500E+03	3.687E-03	1.382E-02	9.015E+01	-8.027E+00

```
    TOTAL HARMONIC DISTORTION =  1.224721E+01 PERCENT
```
(b)

Figure 15-13 PSpice *(a)* input and *(b)* output files for Example 15-6.

Notice that $v_{in}(t)$ is the node voltage at node 1 and that $v_{out}(t)$ is the node voltage at node 2 of the circuit. The command

.four 500Hz v(1) v(2)

tells PSpice that the fundamental frequency is 500 Hz and that we want the coefficients of the Fourier series of $v_{in}(t) = v(1)$ and of $v_{out}(t) = v(2)$.

Next, consider the PSpice output file shown in Figure 15-13. The coefficients of the Fourier series of $v_{in}(t)$ are tabulated under the heading "FOURIER COMPONENTS OF TRANSIENT RESPONSE V(1)." The Fourier series of $v_{in}(t)$ can be constructed from the coefficients in the Table shown in Figure 15-13 as

$$\begin{aligned} v_{in}(t) = 3 &+ 2.547 \cos(\pi 1000t + 179°) \\ &+ 0.8491 \cos(3\pi 1000t + 177°) \\ &+ 0.5098 \cos(5\pi 1000t + 175°) \\ &+ 0.364 \cos(7\pi 1000t + 174°) \\ &+ 0.2839 \cos(9\pi 1000t + 172°) \end{aligned}$$

Similarly, the Fourier series of $v_{out}(t)$ can be constructed from the coefficients in the table labeled "FOURIER COMPONENTS OF TRANSIENT RESPONSE V(2)."

$$\begin{aligned} v_{out}(t) = 2 &+ 0.2668 \cos(\pi 1000t + 98°) \\ &+ 0.02999 \cos(3\pi 1000t + 91°) \\ &+ 0.01097 \cos(5\pi 1000t + 89°) \\ &+ 0.005652 \cos(7\pi 1000t + 87°) \\ &+ 0.0003687 \cos(9\pi 1000t + 90°) \end{aligned}$$

15-10   VERIFICATION EXAMPLE

Figure 15V-1 shows the transfer characteristic of the saturation nonlinearity. Suppose that the input to this nonlinearity is

$$v_{in}(t) = A \sin \omega t$$

where $A > a$. Verify that the output of the nonlinearity will be a periodic function that can

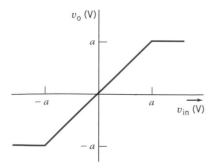

Figure 15V-1 The saturation nonlinearity.

be represented by the Fourier series

$$v_o(t) = b_1 \sin \omega t + \sum_{\substack{n=3 \\ \text{odd}}}^{N} b_n \sin n\omega t \qquad (15\text{V-}1)$$

where (Graham, 1971)

$$B = \sin^{-1}\left(\frac{a}{A}\right)$$

$$b_1 = \frac{2}{\pi} A \left[B + \frac{a}{A} \sqrt{1 - \left(\frac{a}{A}\right)^2} \right]$$

and

$$b_n = \frac{4A}{\pi(1 - n^2)} \left[\frac{a}{A} \frac{\cos(nB)}{n} - \sqrt{1 - \left(\frac{a}{A}\right)^2} \sin(nB) \right]$$

Solution

The output voltage, $v_o(t)$, will be a clipped sinusoid. We need to verify that Eq. 15V-1 does indeed represent a clipped sinusoid. A straightforward, but tedious, way to do this is to plot $v_o(t)$ versus t directly from Eq. 15V-1. Several computer programs, such as spreadsheets and equation solvers, are available to reduce the work required to produce this plot. Mathcad is one of these programs. In Figure 15V-2 Mathcad is used to plot $v_o(t)$ versus t. This plot verifies that the Fourier series in Eq. 15V-1 does indeed represent a clipped sinusoid.

15-11 DESIGN CHALLENGE SOLUTION

DC POWER SUPPLY

A laboratory power supply uses a nonlinear circuit called a rectifier to convert a sinusoidal voltage input into a dc voltage. The sinusoidal input

$$v_{ac}(t) = A \sin \omega_0 t$$

comes from the wall plug. In this example, $A = 160\,\text{V}$ and $\omega_0 = 377\,\text{rad/s}\,(f_0 = 60\,\text{Hz})$. Figure 15D-1 shows the structure of the power supply. The output of the rectifier is the absolute value of its input, that is,

$$v_s(t) = |\,A \sin \omega_0 t\,|$$

The purpose of the rectifier is to convert a signal that has an average value equal to zero into a signal that has an average value that is not zero. The average value of $v_s(t)$ will be used to produce the dc output voltage of the power supply.

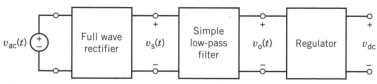

Figure 15D-1 Diagram of a power supply.

Plot a periodic signal from its coefficients.

Define n, the index for the summation:

$$N := 25 \qquad n := 3, 5, \ldots, N$$

Define any parameters that are used to make it easier to enter the coefficient of the Fourier series:

$$A := 12.5 \qquad a := 12 \qquad B := a \sin\left(\frac{a}{A}\right)$$

Enter the fundamental frequency:

$$\omega := 2 \cdot \pi \cdot 1000$$

Define an increment of time. Set up an index to run over two periods of the periodic signal:

$$T := \frac{2 \cdot \pi}{\omega} \qquad dt := \frac{T}{200} \qquad i := 1, 2, \ldots, 400 \qquad t_i := dt \cdot i$$

Enter the formulas for the coefficients of the Fourier series,

$$b_1 := \frac{2}{\pi} \cdot A \cdot \left[B + \frac{a}{A} \cdot \sqrt{1 - \left(\frac{a}{A}\right)^2}\right]$$

$$b_n := \frac{4 \cdot A}{\pi \cdot (1 - n^2)} \cdot \left[\frac{a}{A} \cdot \frac{\cos(n \cdot B)}{n} - \sqrt{1 - \left(\frac{a}{A}\right)^2} \cdot \sin(n \cdot B)\right]$$

Enter the Fourier series:

$$v(i) := b_1 \cdot \sin(\omega \cdot t_i) + \sum_{n=3}^{N} b_n \cdot \sin(n \cdot \omega \cdot t_i)$$

Plot the periodic signal:

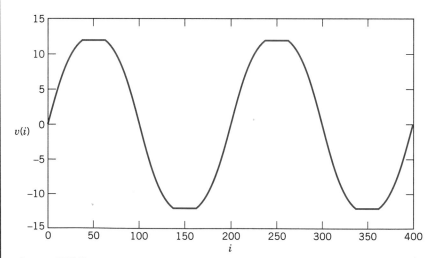

Figure 15V-2 Using Mathcad to verify the Fourier series of a clipped sinusoid.

The rectifier output is not a sinusoid, but it is a periodic signal. Periodic signals can be represented by Fourier series. The Fourier series of $v_s(t)$ will contain a constant, or dc, term and some sinusoidal terms. The purpose of the filter shown in Figure 15D-1 is to pass the dc term and attenuate the sinusoidal terms. The output of the filter, $v_o(t)$, will be a periodic signal and can be represented by a Fourier series. Because we are designing a dc power supply, the sinusoidal terms in the Fourier series of $v_o(t)$ are undesirable. The sum of these undesirable terms is called the ripple of $v_o(t)$.

The challenge is to design a simple filter so that the dc term of $v_o(t)$ is at least 90 V and the size of the ripple is no larger than 5% of the size of the dc term.

Define the Situation

1. From Table 15-2, the Fourier series of $v_s(t)$ is

$$v_s(t) = \frac{320}{\pi} - \sum_{n-1}^{N} \frac{640}{\pi(4n^2 - 1)} \cos(2 \cdot n \cdot 377 \cdot t)$$

Let $v_{sn}(t)$ denote the term of $v_s(t)$ corresponding to the integer n. Using this notation, the Fourier series of $v_s(t)$ can be written as

$$v_s(t) = v_{s0} + \sum_{n=1}^{N} v_{sn}(t)$$

2. Figure 15D-2 shows a simple filter. The resistance R_s models the output resistance of the rectifier. We have assumed that the input resistance of the regulator is large enough to be ignored. (The input resistance of the regulator will be in parallel with R and will probably be much larger than R. In this case, the equivalent resistance of the parallel combination will be approximately equal to R.)

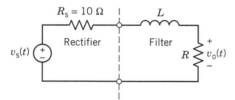

Figure 15D-2 A simple RL low-pass filter connected to the rectifier.

3. The filter output, $v_o(t)$, will also be a periodic signal and will be represented by the Fourier series

$$v_o(t) = v_{o0} + \sum_{n=1}^{N} v_{on}(t)$$

4. Most of the ripple in $v_o(t)$ will be due to $v_{o1}(t)$, the fundamental term of the Fourier series. The specification regarding the allowable ripple can be stated as

$$\text{amplitude of the ripple} \leq 0.05 \cdot \text{dc output}$$

Equivalently, we can state that we require

$$\max \left(\sum_{n=1}^{N} v_{on}(t) \right) \leq 0.05 \cdot v_{o0} \tag{15D-1}$$

For ease of calculation, we replace Eq. 15D-1 with the simpler condition

$$|v_{o1}(t)| \leq 0.04 \cdot v_{o0}$$

That is, the amplitude $v_{o1}(t)$ must be less than 4% of the dc term of the output (v_{o0} = dc term of the output).

State the Goal

Specify values of R and L so that

$$\text{dc output} = v_{o0} \geq 90$$

and

$$|v_{o1}(t)| \leq 0.04 \cdot v_{o0}$$

Generate a Plan

Use superposition to calculate the Fourier series of the filter output. First, the specification

$$\text{dc output} = v_{o0} \geq 90$$

can be used to determine the required value of R. Next, the specification

$$|v_{o1}(t)| \leq 0.04 \cdot v_{o0}$$

can be used to calculate L.

Take Action on the Plan

First, we will find the response to the dc term of $v_s(t)$. When the filter input is a constant and the circuit is at steady state, the inductor acts like a short circuit. Using voltage division

$$v_{o0} = \frac{R}{R + R_s} v_{s0}$$

$$= \frac{R}{R + 10} \cdot \frac{320}{\pi}$$

The specification that $v_{o0} \geq 90$ requires

$$90 \leq \frac{R}{R + 10} \cdot \frac{320}{\pi}$$

or

$$R \geq 75.9$$

Let us select

$$R = 80 \ \Omega$$

When $R = 80 \ \Omega$,

$$v_{o0} = 90.54 \text{ V}$$

Next, we find the steady-state response to a sinusoidal term, $v_{sn}(t)$. Phasors and impedances can be used to find this response. By voltage division

$$\mathbf{V}_{on} = \frac{R}{R + R_s + j2n\omega_0 L} \mathbf{V}_{sn}$$

We are particularly interested in $\mathbf{V}_{o1}$:

$$\mathbf{V}_{o1} = \frac{R}{R + R_s + j2\omega_0 L} \mathbf{V}_{s1}$$

$$= \frac{80}{90 + j754L} \cdot \frac{640}{\pi \cdot 3}$$

The amplitude of $v_{o1}(t)$ is equal to the magnitude of the phasor $\mathbf{V}_{o1}$. The specification on the amplitude of $v_{o1}(t)$ requires that

$$\frac{80}{\sqrt{90^2 + 754^2 L^2}} \cdot \frac{640}{\pi \cdot 3} \leq 0.04 \, v_{o1}$$

$$\leq 0.04 \cdot 90.54$$

That is,

$$L \geq 0.99 \text{ H}$$

Selecting

$$L = 1 \text{ H}$$

completes the design.

SUMMARY

Periodic waveforms arise in many circuits. For example, the form of the load current waveforms for selected loads is shown in Figure 15-14. While the load current for motors and incandescent lamps is of the same form as that of the source voltage, it is significantly altered for the power supplies, dimmers, and variable-speed drives as shown in Figure 15-14*b* and *c*. Electrical engineers have long been interested in developing the tools required to analyze circuits incorporating periodic waveforms.

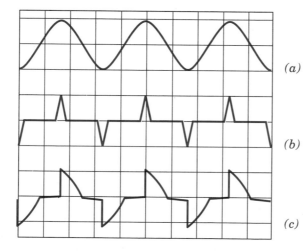

(a)

(b)

(c)

Figure 15-14 Load-current waveforms for *(a)* motors and incandescent lights, *(b)* switch-mode power supplies, and *(c)* dimmers and variable-speed drives. The vertical axis is current, and the horizontal axis is time. Source: *EPRI Journal*, August 1991.

The brilliant mathematician-engineer Jean-Baptiste-Joseph Fourier proposed in 1807 that a periodic waveform could be represented by a series consisting of cosine and sine terms with the appropriate coefficients. The integer multiple frequencies of the fundamental are called the harmonic frequencies (or harmonics).

With judicious use of symmetry of the waveforms, the Fourier coefficients may be readily calculated. The Fourier series may also be written as a series of terms of complex coefficients multiplying the complex exponential.

The line spectra consisting of the amplitude and phase of the complex coefficients of the Fourier series when plotted against frequency are useful for portraying the frequencies that represent a waveform.

The practical representation of a periodic waveform consists of a finite number of sinusoidal terms of the Fourier series. It is shown that while convergence occurs as n grows large, there always remains an error at the points of discontinuity of the waveform.

Finally, we demonstrated that if a periodic input waveform is represented by a Fourier series, then we can predict the response of a linear circuit by calculating the phasor response at each harmonic.

TERMS AND CONCEPTS

Complex Fourier Coefficients   Fourier coefficients for the exponential form of the Fourier series.

Even Function   A function with $f(t) = f(-t)$.

Fourier Trigonometric Coefficients   Coefficients of the sinusoidal terms that in sum represent a periodic function.

Fourier Series   Infinite sum of harmonically related sinusoids.

Fourier Spectrum   See Spectrum.

Fundamental Frequency   For a periodic function of period T, the frequency $\omega_0 = 2\pi/T$.

Gibbs Phenomenon   Regardless of the number of terms used, overshoot occurs at discontinuities when a waveform is represented by a Fourier series.

Harmonic Frequencies   Integer multiples of the fundamental frequency.

Odd Function   A function with $f(t) = -f(-t)$.

Period   Length of time, T, at which a waveform repeats itself.

Spectrum (Line)   A plot of the line spectra of amplitude and phase against frequency of the coefficients of the complex Fourier series.

REFERENCES   Chapter 15

Bracewell, Ronald N., "The Fourier Transform," *Scientific American,* June 1989, pp. 86–95.
Fourier, J.B.J., *Théorie Analytique de la Chaleur,* 1822, Dover Publishing, New York, 1985.
Graham, Dunstan, Analysis of Nonlinear Control Systems, Dover Publishing, New York, 1971.
Lamarre, Leslie, "Problems with Power Quality," *EPRI Journal,* August 1991, pp. 14–23.
Pickover, Clifford A., *Computers, Patterns, Chaos, and Beauty,* St. Martin's Press, New York, 1990, Chapter 2.

PROBLEMS

Section 15-3 The Fourier Series

P 15.3-1 Find the trigonometric Fourier series of Eq. 15-1 for a periodic function $f(t) = t^2$ over the period from $t = 0$ to $t = 2$.

P 15.3-2 Find the trigonometric Fourier series of Eq. 15-1 for the function of Figure P 15.3-2. The function is the positive portion of a cosine wave.

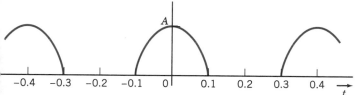

Figure P 15.3-2 Half-wave rectified cosine wave.

P 15.3-3 A "staircase" periodic waveform is described by its first cycle as

$$f(t) = \begin{cases} 1 & 0 < t < 0.25 \\ 2 & 0.25 < t < 0.5 \\ 0 & 0.5 < t < 1 \end{cases}$$

Find the Fourier series for this function.

P 15.3-4 Determine the Fourier series for the sawtooth function shown in Figure P 15.3-4.

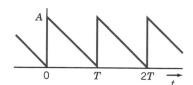

Figure P 15.3-4 Sawtooth wave.

P 15.3-5 Find the Fourier series for

$$f(t) = |A \sin \omega t|$$

P 15.3-6 Find the Fourier series for the periodic function $f(t) = t$ over the period from $t = 0$ to $t = 2$ s.

Section 15-4 Symmetry of the Function $f(t)$

P 15.4-1 A sawtooth wave shown in Figure P 15.4-1 can be represented by a Fourier series.

(a) Find the trigonometric series.
(b) Plot the series consisting of the fundamental and two harmonics, and compare it to the original waveform for $-\pi \le t \le \pi$.

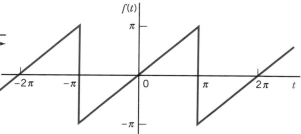

Figure P 15.4-1 Sawtooth wave.

P 15.4-2 Find the sine-cosine form of the Fourier series for the waveform of Figure P 15.4-2.

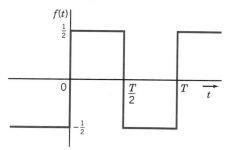

Figure P 15.4-2 Square waveform.

P 15.4-3 Determine the Fourier series for the waveform shown in Figure P 15.4-3. Calculate a_0, a_1, a_2, and a_3.

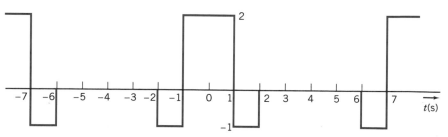

Figure P 15.4-3

P 15.4-4 Determine the Fourier series for

$$f(t) = |A \cos \omega t|$$

P 15.4-5 Determine the Fourier series for $f(t)$ shown in Figure P 15.4-5.
Answer: $a_n = a_0 = 0$; $b_n = 0$ for even n, $= 8/n^2\pi^2$ for $n = 1, 5, 9$, and $= -8/n^2\pi^2$ for $n = 3, 7, 11$

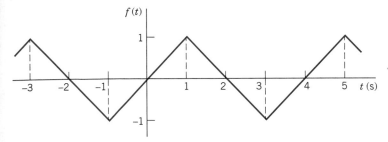

Figure P 15.4-5

P 15.4-6 Determine the Fourier series for the periodic signal shown in Figure P 15.4-6.
Answer:

$$f(t) = \frac{1}{2} + \frac{2}{\pi}\left(\sin t + \frac{1}{3}\sin 3t + \frac{1}{5}\sin 5t + \cdots\right)$$

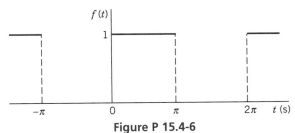

Figure P 15.4-6

Section 15-5 Exponential Form of the Fourier Series

P 15.5-1 Determine the exponential form of the Fourier series for the sawtooth wave shown in Figure P 15.4-1.

P 15.5-2 Determine the exponential form of the Fourier series for the waveform of Figure P 15.4-3.

P 15.5-3 Determine the exponential form of the Fourier series for the waveform of Figure P 15.5-3.

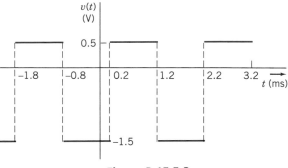

Figure P 15.5-3

P 15.5-4 Determine the Fourier series for the waveform of Figure P 15.5-4.

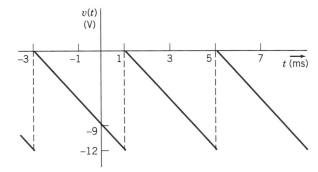

Figure P 15.5-4

P 15.5-5 Determine the exponential Fourier series of the periodic waveform shown in Figure P 15.5-5.
Answer: $C_n = \dfrac{1}{j2\pi n}$

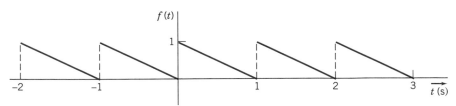

Figure P 15.5-5

P 15.5-6 A periodic function consists of rising and decaying exponentials of time constants of 0.2 s each and durations of 1 s each as shown in Figure P 15.5-6. Determine the exponential Fourier series for this function.

Answer: $C_n = \dfrac{5}{(j\pi n)(5 + j\pi n)}$, $n = 1, 3, 5$

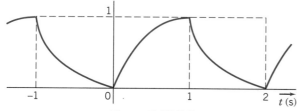

Figure P 15.5-6

Section 15-6 The Fourier Spectrum

P 15.6-1 Determine the cosine-sine Fourier series for the sawtooth waveform shown in Figure P 15.6-1. Draw the Fourier spectra for the first four terms including magnitude and phase.

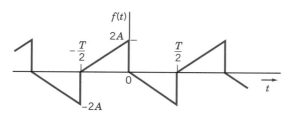

Figure P 15.6-1

P 15.6-2 The load current waveform of the variable-speed motor drive depicted in Figure 15-14c is shown in Figure P 15.6-2. The current waveform is a portion of $A \sin \omega_0 t$. Determine the Fourier series of this waveform, and draw the line spectra of $|C_n|$ for the first 10 terms.

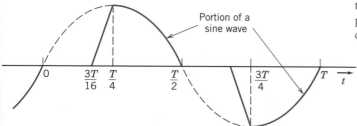

Figure P 15.6-2 The load current of a variable-speed drive.

P 15.6-3 Draw the Fourier spectra for the waveform shown in Figure P 15.5-3.

P 15.6-4 Draw the Fourier spectra for the waveform shown in Figure P 15.5-5.

P 15.6-5 Draw the Fourier spectra for the waveform shown in Figure P 15.5-6.

Section 15-8 Circuits and Fourier Series

P 15.8-1 A square wave, as shown in Figure P 15.8-1a, is applied to the *RL* circuit of Figure P 15.8-1b. Using the Fourier series representation, find the current $i(t)$.

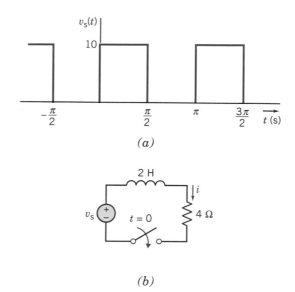

Figure P 15.8-1 (*a*) Square wave and (*b*) *RL* circuit.

P 15.8-2 In Section 15-2 we discussed the discovery by M.I. Pupin of the concept of telephone line balance. A length of cable can be represented by the *RC* circuit of Figure P 15.8-2, and the inductor *L* was inserted before being connected to another length of cable as shown. The voice sounds have significant harmonics up to 8000 Hz. Determine the required *L* when $R = 2 \, \Omega$, $C = 4 \, \text{mF}$, and the resistance R_L of the receiver is 100 Ω. The goal is to pass all frequencies up to 8000 Hz and reject frequency content above 8000 Hz for each length of cable.

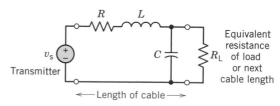

Figure P 15.8-2 A balanced cable according to Pupin.

P 15.8-3 Find the steady-state response for the output voltage, v_o, for the circuit of Figure P 15.8-3 when $v(t)$ is as described in P 15.5-4.

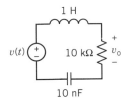

1 H

$v(t)$ 10 kΩ v_o +/−

10 nF

Figure P 15.8-3 ALC circuit.

P 15.8-4 Determine the value of the voltage $v_o(t)$ at $t = 4$ ms when v_{in} is shown in Figure P 15.8-4a and the circuit is shown in Figure P 15.8-4b.

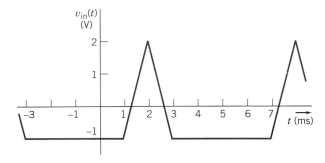

$v_{in}(t)$
(V)

(a)

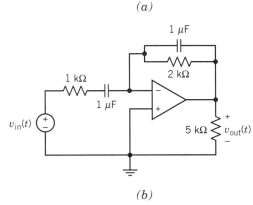

1 µF

1 kΩ

2 kΩ

1 µF

$v_{in}(t)$

5 kΩ $v_{out}(t)$

(b)

Figure P 15.8-4

PSpice PROBLEMS

SP 15-1 Use PSpice to determine the Fourier coefficients and the Fourier series for P 15.4-1.

SP 15-2 Use PSpice to determine the Fourier coefficients for P 15.3-4.

SP 15-3 Use PSpice to determine the Fourier coefficients for a voltage square waveform as shown in Figure 15-6 when $T = 1$ ms and $A = 40$ V.

SP 15-4 Use PSpice to determine the Fourier coefficients for $v(t)$ shown in Figure SP 15-4.

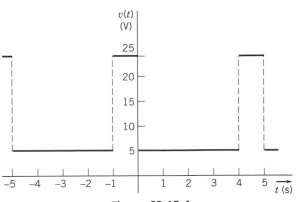

$v(t)$
(V)

Figure SP 15-4

SP 15-5 Use PSpice to determine the Fourier coefficients for $v(t)$ shown in Figure SP 15-5.

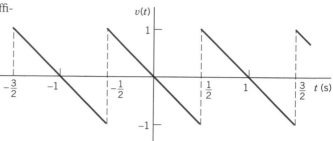

Figure SP 15-5

VERIFICATION PROBLEMS

VP 15-1 A computer printout states that the Fourier series for the signal $f(t)$ shown in Figure VP 15-1 is $a_0 = 2$, $a_1 = 1$, $a_n = 0$ for $n > 1$, and $b_n = 0$. Verify this result.

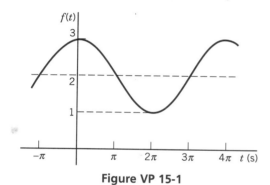

Figure VP 15-1

VP 15-2 A laboratory report states that the value of the dc term of the half-wave rectified voltage output is 255 V when the input to the rectifier is $400 \sin 300\ t$. Verify this result.

DESIGN PROBLEMS

DP 15-1 A periodic waveform shown in Figure DP 15-1a is the input signal of the circuit shown in Figure DP 15-1b. Select the capacitance C so that the magnitude of the third harmonic of $v_2(t)$ is less than 1.4 V and greater than 1.3 V. Write the equation describing the third harmonic of $v_2(t)$, for the value of C selected.

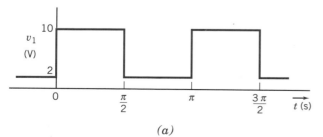

(a)

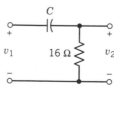

(b)

Figure DP 15-1

DP 15-2 A dc laboratory power supply uses a nonlinear circuit to convert a sinusoidal voltage obtained from the wall plug to a constant dc voltage. The wall plug voltage is $A \sin \omega_0 t$, where $f_0 = 60$ Hz and $A = 160$ V. The voltage is then rectified so that $v_s = |A \sin \omega_0 t|$. Using the filter circuit of Figure DP 15-2, determine the required inductance L so that the magnitude of each harmonic (ripple) is less than 4 percent of the dc component of the output voltage.

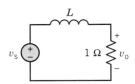

Figure DP 15-2 An *RL* circuit.

DP 15-3 A low-pass filter is shown in Figure DP 15-3. The input, v_s, is a half-wave rectified sinusoid with $\omega_0 = 800\pi$ (item 5 of Table 15-3). Select L and C so that the peak value of the first harmonic is $\frac{1}{20}$ of the dc component for the output, v_o.

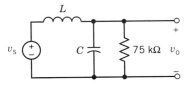

Figure DP 15-3 An RLC circuit.

CHAPTER 16

FOURIER TRANSFORM

PREVIEW

The Fourier series representation of a periodic time function $f(t)$ was studied in Chapter 15. In this chapter, we extend the concept to an integral transformation of periodic and nonperiodic functions $f(t)$ resulting in a frequency domain function called the Fourier transform.

We will develop an integral formula to obtain the frequency domain transform and the associated inverse formula to obtain the time function from the frequency domain function. Then we can use the frequency domain description of an input signal to a circuit along with the transfer function of the circuit to obtain the frequency domain description of an output variable. Finally, the time domain description of this output variable can be obtained by using the inverse Fourier transform.

16-1 DESIGN CHALLENGE

LOW-PASS FILTER CIRCUIT

Problem

An input signal $v(t)$ is used to control a valve. The dc component of $v(t)$ is required to actuate the valve. The higher frequency components of $v(t)$ need to be attenuated in order to reduce valve wear. A filter circuit is to be designed so that the magnitude of the output voltage at $\omega = 2$ rad/s is less than or equal to 1 percent of the magnitude of the output signal at $\omega = 0$. The input signal is $v(t) = 10e^{-t}u(t)$ V. A low-pass RLC filter circuit is proposed as shown in Figure 16D-1.

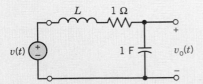

Figure 16D-1 RLC circuit.

Define the Situation, State the Assumptions, and Develop a Model

1 The *RLC* circuit is linear.
2 The input signal is a decaying exponential.

The Goal

Determine L to achieve

$$\frac{|V_0(2)|}{|V_0(0)|} \leq 0.01 \tag{16D-1}$$

We need to characterize $v(t)$ in the frequency domain. We proceed to develop this characterization and return to the problem at the end of the chapter.

16-2 FM RADIO

One problem that plagued radio from its earliest days was that of distortion of the transmitted signal. Any AM radio system is prone to static or distortion, and it was natural for radio engineers to try to reduce this distortion. Many tried and failed. Professor M. Pupin is reported to have summed it up with the words, "God gave men radio and the Devil made static."

Edwin H. Armstrong, who became interested in radio in high school, entered Columbia University in 1909 and studied electrical engineering under Michael Idvorsky Pupin, the inventor of the Pupin loading coil used in long-distance telegraphy and telephony (see Section 15-2). While still an undergraduate, Armstrong made the first of his many inventions. Later, in 1933, he obtained four patents dealing with frequency modulation (FM). Armstrong's approach was to step outside the existing state of the art. In his own words, "The invention of the FM system gave a reduction of interfering noises of hundreds of thousands of times. It did so by proceeding in exactly the opposite direction that mathematical theory had demonstrated one ought to go to reduce interference. It widened instead of narrowed the band." Many of the leading engineers of the day had considered FM impossible.

Nevertheless, the U.S. radio industry resisted the introduction of FM broadcasting. FM production was interrupted when the United States entered World War II, and the Federal Communications Commission dealt FM a setback in 1945 when it relegated it to a new frequency band, thus making more than 50 existing transmitters and half a million receivers obsolete. At the same time, FM came to be widely used in military and other mobile communications, radar, telemetering, and the audio portion of television; but widespread adoption of FM broadcasting came only after Armstrong's death in 1954.

16-3 THE TRANSFORM CONCEPT

A person's hearing is based on the ear automatically formulating a transform by converting sound into a frequency spectrum; the brain then converts this information into perceived sound. The *Fourier transform* is a function that describes the amplitude and phase of each sinusoid that corresponds to a specific frequency. Fourier originally derived an equation that described the conduction of heat in solid bodies using a sum of sinusoids. Fortunately, numerical methods are now available for computing Fourier transforms of functions of complex, irregular, nonperiodic waveforms.

The National Aeronautics and Space Administration improves the clarity and detail of pictures of celestial objects taken in space by means of Fourier analysis. Planetary probes and earth-orbiting satellites transmit images to the earth as a series of radio signals. A computer transforms these signals into the frequency domain by Fourier techniques. The computer than adjusts various components of each transform to enhance certain features and remove others—much as noise can be removed from the Fourier transforms of recorded music. Finally, the altered data are converted back to the time domain to reconstruct the image. This process can sharpen focus, filter out background fog, and change contrast.

The Fourier transform converts a signal $f(t)$ in the time domain to the frequency domain, where operations and processes can be applied to the transformed signal. Then the Fourier transform can be converted back to the time domain, resulting in an enhanced or improved signal $f_i(t)$.

A **transform** is a change in the mathematical description of a physical variable to facilitate computation.

16-4 ∥ THE FOURIER TRANSFORM

The Fourier transform is closely related to the Fourier series described in Chapter 15 and the Laplace transform described in Chapter 14. Recall from Chapter 15 that the periodic waveform $f(t)$ possesses a Fourier series with a line spectrum. As we increase the period T the fundamental frequency ω_0 becomes smaller since

$$\omega_0 = \frac{2\pi}{T}$$

The space between two spectra is $\Delta\omega = (n + 1)\omega_0 - n\omega_0 = \omega_0 = 2\pi/T$. Therefore, as T approaches infinity, $\Delta\omega$ approaches $d\omega$, an infinitesimal frequency increment. Furthermore, the number of frequencies in any given frequency interval increases as $\Delta\omega$ decreases. Thus, in the limit $n\omega_0$ approaches the continuous variable ω.

Consider the exponential Fourier series

$$f(t) = \sum_{n=-\infty}^{n=\infty} C_n e^{jn\omega_0 t} \qquad (16\text{-}1)$$

and

$$C_n = \frac{1}{T} \int_{-T/2}^{T/2} f(t) e^{-jn\omega_0 t} \, dt \qquad (16\text{-}2)$$

Multiplying Eq. 16-2 by T and letting T approach infinity, we have

$$C_n T = \int_{-\infty}^{\infty} f(t) e^{-j\omega t} \, dt \qquad (16\text{-}3)$$

Let $C_n T$ equal a new frequency function $F(j\omega)$ so that

$$F(j\omega) = \int_{-\infty}^{\infty} f(t)e^{-j\omega t}\, dt \tag{16-4}$$

where $F(j\omega)$ is the *Fourier transform* of $f(t)$. The inverse process is found from Eq. 16-1, where we let $C_n T = F(j\omega)$ so that

$$f(t) = \lim_{T \to 0} \sum_{n=-\infty}^{\infty} C_n T e^{jn\omega_0 t}\frac{1}{T}$$

$$= \lim_{T \to 0} \sum_{n=-\infty}^{\infty} F(j\omega)e^{jn\omega_0 t}\frac{\omega_0}{2\pi}$$

since $1/T = \omega_0/2\pi$. As $T \to 0$ the sum becomes an integral and the increment $\Delta\omega = \omega_0$ becomes $d\omega$. Then, we have

$$f(t) = \frac{1}{2\pi}\int_{-\infty}^{\infty} F(j\omega)e^{j\omega t}\, d\omega \tag{16-5}$$

Equation 16-5 is called the *inverse Fourier transform*. This pair of equations (Eqs. 16-4 and 16-5), called the Fourier transform pair, permits us to complete the Fourier transformation to the frequency domain and the inverse process to the time domain.

A given function of time $f(t)$ has a Fourier transform if the integral in Eq. 16-4 converges. The integral will converge if $f(t)$ is single-valued except at a finite number of points and $|f(t)|$ encloses a finite area over the range of integration. From a practical point of view, all pulses of finite duration in which we are interested have Fourier transforms. If $f(t)$ is different from zero over an infinite interval, then the convergence of the Fourier integral depends on the behavior of $f(t)$ as t approaches plus or minus infinity. A single-valued function that is nonzero over an infinite interval will have a Fourier transform if

$$\int_{-\infty}^{\infty} |f(t)|\, dt < \infty$$

and if the number of discontinuities in $f(t)$ is finite.

The Fourier transform pair is summarized in Table 16-1.

Example 16-1

Derive the Fourier transform of the aperiodic pulse shown in Figure 16-1.

Table 16-1
The Fourier Transform Pair

Equation	Name	Process
$F(j\omega) = \int_{-\infty}^{\infty} f(t)e^{-j\omega t}\, dt$	Transform	Time domain to frequency domain. Conversion of $f(t)$ into $F(j\omega)$
$f(t) = \dfrac{1}{2\pi}\int_{-\infty}^{\infty} F(j\omega)e^{j\omega t}\, d\omega$	Inverse transform	Frequency domain to time domain. Conversion of $F(j\omega)$ into $f(t)$

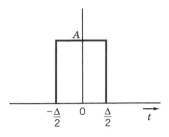

Figure 16-1 An aperiodic pulse.

Solution

Using the transform, we have

$$F(j\omega) = \int_{-\Delta/2}^{\Delta/2} Ae^{-j\omega t}\, dt$$

$$= \frac{A}{-j\omega}\, e^{-j\omega t}\, \bigg|_{-\Delta/2}^{\Delta/2}$$

$$= \frac{A}{-j\omega}\, (e^{-j\omega\,\Delta/2} - e^{j\omega\,\Delta/2})$$

$$= A\Delta\, \frac{\sin(\omega\Delta/2)}{\omega\Delta/2} \tag{16-6}$$

Thus the Fourier transform is of the form $(\sin x)/x$, where $x = \omega\Delta/2$ as shown in Figure 16-2. Note that $(\sin x)/x = 0$ when $x = \omega\Delta/2 = n\pi$ or $\omega = 2n\pi/\Delta$ as shown in Figure 16-2. We will denote $(\sin x)/x = Sa(x)$.

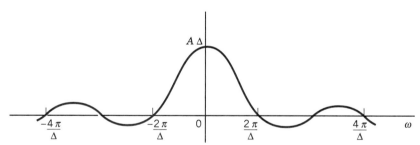

Figure 16-2 The Fourier transform for the rectangular aperiodic pulse is shown as a function of ω.

Let us consider the shifted version of the rectangular pulse of Figure 16-1 where $A = 1/\Delta$ and the width of the pulse approaches zero, $\Delta \to 0$, while the area of the rectangle remains equal to 1. Then we have the *unit impulse* $\delta(t - t_0)$ so that

$$\int_a^b \delta(t - t_0)\, dt = \begin{cases} 1 & a \le t_0 \le b \\ 0 & \text{otherwise} \end{cases} \tag{16-7}$$

We obtain the Fourier transform for a unit impulse at t_0 as

$$F(j\omega) = \int_{t_0-}^{t_0+} \delta(t - t_0)e^{-j\omega t}\, dt$$

$$= e^{-j\omega t_0} \tag{16-8}$$

When $t_0 = 0$ we have the special case

$$F(j\omega) = 1 \tag{16-9}$$

Thus, we note that $F(j\omega) = 1$ of a unit impulse located at the origin is constant and equal to 1 for all frequencies.

EXERCISE 16-1

Determine the Fourier transform of $f(t) = e^{-at}u(t)$, where $u(t)$ is the unit step function (see Section 9-5).

Answer: $F(j\omega) = \dfrac{1}{a + j\omega}$

16-5 FOURIER TRANSFORM PROPERTIES

We can derive some properties of the Fourier transform by writing $F(j\omega)$ in complex form as

$$F(j\omega) = X(\omega) + jY(\omega)$$

Alternatively, we have

$$F(j\omega) = |F(\omega)| e^{j\theta}$$

where $\theta = \tan^{-1}(Y/X)$. Note that we use $F(j\omega) = F(\omega)$ interchangeably. Furthermore,

$$F(-\omega) = F^*(\omega) \tag{16-10}$$

where $F^*(\omega)$ is the complex conjugate of $F(\omega)$.

If we have the Fourier transform of $f(t)$, we write

$$\mathscr{F}[f(t)] = F(\omega) \tag{16-11}$$

where the script $\mathscr{F}$ implies the Fourier transform. Then the inverse transform is written as

$$\mathscr{F}^{-1}[F(\omega)] = f(t) \tag{16-12}$$

Repeating the transformation equation, we have (Table 16-1)

$$F(\omega) = \int_{-\infty}^{\infty} f(t)e^{-j\omega t}\, dt \tag{16-13}$$

Then, if $\mathscr{F}[af_1(t)] = aF_1(\omega)$ and $\mathscr{F}[bf_2(t)] = bF_2(\omega)$, we have

$$\mathscr{F}[af_1 + bf_2] = \int_{-\infty}^{\infty} [af_1 + bf_2]e^{-j\omega t}\, dt$$

$$= \int_{-\infty}^{\infty} af_1 e^{-j\omega t}\, dt + \int_{-\infty}^{\infty} bf_2 e^{-j\omega t}\, dt$$

$$= aF_1(\omega) + bF_2(\omega) \tag{16-14}$$

This is known as the *linearity* property.

We now use the definition of the Fourier transform, Eq. 16-13, in the following examples to find several other properties.

Example 16-2

Find the Fourier transform of a time-shifted function $f(t - t_0)$.

Solution

$$\mathcal{F}[f(t - t_0)] = \int_{-\infty}^{\infty} f(t - t_0)e^{-j\omega t} \, dt$$

If we let $x = t - t_0$, we have

$$\mathcal{F}[f(t - t_0)] = \int_{-\infty}^{\infty} f(x)e^{-j\omega(x + t_0)} \, dx$$

$$= e^{-j\omega t_0}F(\omega) \tag{16-15}$$

where $F(\omega) = \mathcal{F}[f(t)]$.

Selected properties of the Fourier transform are summarized in Table 16-2. We can use these properties to derive Fourier transform pairs.

Table 16-2
Selected Properties of the Fourier Transform

Name of Property	Function of Time	Fourier Transform
1. Definition	$f(t)$	$F(\omega)$
2. Multiplication by constant	$Af(t)$	$AF(\omega)$
3. Linearity	$af_1 + bf_2$	$aF_1(\omega) + bF_2(\omega)$
4. Time shift	$f(t - t_0)$	$e^{-j\omega t_0}F(\omega)$
5. Time scaling	$f(at), a > 0$	$\dfrac{1}{a}F\left(\dfrac{\omega}{a}\right)$
6. Modulation	$e^{j\omega_0 t}f(t)$	$F(\omega - \omega_0)$
7. Differentiation	$\dfrac{d^n f(t)}{dt^n}$	$(j\omega)^n F(\omega)$
8. Convolution	$\displaystyle\int_{-\infty}^{\infty} f_1(x)f_2(t - x) \, dx$	$F_1(\omega)F_2(\omega)$
9. Time multiplication	$t^n f(t)$	$(j)^n \dfrac{d^n F(\omega)}{d\omega^n}$
10. Time reversal	$f(-t)$	$F(-\omega)$
11. Integration	$\displaystyle\int_{-\infty}^{t} f(\tau) \, d\tau$	$\dfrac{F(\omega)}{j\omega} + \pi F(0)\delta(\omega)$

EXERCISE 16-2

Find the Fourier transform of $f(at)$ for $a > 0$ when $F(\omega) = \mathcal{F}[f(t)]$.

Answer: $\mathcal{F}[f(at)] = \dfrac{1}{a}F\left(\dfrac{\omega}{a}\right)$

16-6 | TRANSFORM PAIRS

With the aid of the properties of the Fourier transform and the original defining equation, we can derive useful transform pairs and develop a table of these relationships. We have already derived the first three entries in Table 16-3, and we will add several more by using the properties of Table 16-2 and/or the original definition of the transformation.

Table 16-3

Fourier Transform Pairs

$f(t)$	Waveform	$F(\omega)$
1. Pulse $f_1(t) = Au\left(t + \dfrac{\Delta}{2}\right) - Au\left(t - \dfrac{\Delta}{2}\right)$	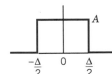	$A\,\Delta\,Sa\left(\dfrac{\omega\Delta}{2}\right)$
2. Impulse $\delta(t - t_0)$	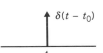	$e^{-j\omega t_0}$
3. Decaying exponential $Ae^{-at}u(t)$	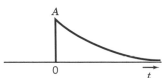	$\dfrac{A}{a + j\omega}$
4. Symmetric decaying exponential $e^{-a\lvert t\rvert}$	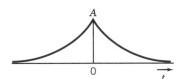	$\dfrac{2aA}{a^2 + \omega^2}$
5. Tone burst (gated cosine) $Af_1(t)\cos\omega_0 t$	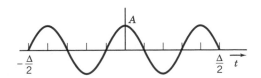	$\dfrac{A\,\Delta}{2}\,[Sa(\omega - \omega_0) + Sa(\omega + \omega_0)]$

(continued)

Table 16-3 *(continued)*
Fourier Transform Pairs

$f(t)$	Waveform	$F(\omega)$
6. Triangular pulse	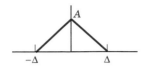	$A \Delta Sa^2 \left(\dfrac{\omega \Delta}{2} \right)$
7. $A Sa(bt) = A \dfrac{\sin bt}{bt}$	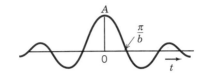	$\begin{cases} \dfrac{A\pi}{b} & \|\omega\| < b \\ 0 & \|\omega\| > b \end{cases}$
8. Constant dc $f(t) = A$	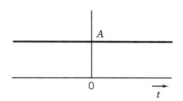	$2\pi A\, \delta(\omega)$
9. Cosine wave $A \cos \omega_0 t$	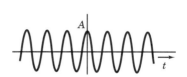	$\pi A\, [\delta(\omega + \omega_0) + \delta(\omega - \omega_0)]$
10. Signum $f(t) = \begin{cases} +1 & t > 0 \\ -1 & t < 0 \end{cases}$	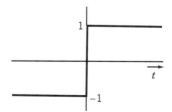	$\dfrac{2}{j\omega}$
11. Step input $Au(t)$	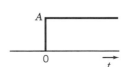	$A \left[\pi\delta(\omega) + \dfrac{1}{j\omega} \right]$

Note: $Sa(x) = (\sin x)/x$.

Example 16-3

Find the Fourier transform of $f(t) = Ae^{-a|t|}$, which is shown in Figure 16-3.

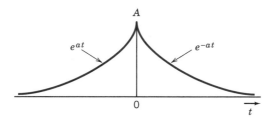

Figure 16-3 Waveform of Example 16-3.

Solution

We will break the function into two symmetric waveforms and use the linearity property. Then,

$$f(t) = f_1(t) + f_2(t)$$
$$= Ae^{-at}u(t) + Ae^{at}u(-t)$$

We have from entry 3 of Table 16-3

$$F_1(\omega) = \frac{A}{a + j\omega}$$

From property 10 of Table 16-2, we obtain

$$F_2(\omega) = F_1(-\omega)$$
$$= \frac{A}{a - j\omega}$$

Using the linearity property, we have

$$F(\omega) = F_1(\omega) + F_2(\omega)$$
$$= \frac{A}{a + j\omega} + \frac{A}{a - j\omega}$$
$$= \frac{2Aa}{a^2 + \omega^2} \tag{16-16}$$

This result is entry 4 in Table 16-3. Note that $F(\omega)$ is an even function.

Example 16-4

Find the Fourier transform of the gated cosine waveform $f(t) = f_1(t) \cos \omega_0 t$, where $f_1(t)$ is the rectangular pulse shown in Figure 16-1.

Solution

The Fourier transform of the rectangular pulse is entry 1 in Table 16-3 and is written as

$$F_1(\omega) = A \Delta (\sin x)/x$$

where $x = \omega\Delta/2$. The cosine function can be written as

$$\cos \omega_0 t = \frac{1}{2} (e^{j\omega_0 t} + e^{-j\omega_0 t})$$

Therefore,

$$f(t) = \frac{1}{2} f_1(t)e^{j\omega_0 t} + \frac{1}{2} f_1(t)e^{-j\omega_0 t}$$

Using the modulation property (entry 6) of Table 16-2, we obtain

$$F(\omega) = \tfrac{1}{2}F_1(\omega - \omega_0) + \tfrac{1}{2}F_1(\omega + \omega_0)$$

Therefore, using $F_1(\omega)$ from Eq. 16-6, we have

$$F(\omega) = \frac{A\Delta}{2} \frac{\sin[(\omega - \omega_0)\Delta/2]}{(\omega - \omega_0)\Delta/2} + \frac{A\Delta}{2} \frac{\sin[(\omega + \omega_0)\Delta/2]}{(\omega + \omega_0)\Delta/2} \qquad (16\text{-}17)$$

or, using $Sa(x) = (\sin x)/x$, we have

$$F(\omega) = \frac{A\Delta}{2} Sa\left[(\omega - \omega_0)\frac{\Delta}{2}\right] + \frac{A\Delta}{2} Sa\left[(\omega + \omega_0)\frac{\Delta}{2}\right]$$

EXERCISE 16-3

Show that the Fourier transform of a constant dc waveform $f(t) = A$ for $-\infty \le t \le \infty$ is $F(\omega) = 2\pi A\delta(\omega)$ by obtaining the inverse transform of $F(\omega)$.

16-7 ‖ THE SPECTRUM OF SIGNALS

The *spectrum,* also called the *spectral density,* of a signal $f(t)$ is its Fourier transform $F(\omega)$. We can plot $F(\omega)$ as a function of ω to show the spectrum. For example, for a rectangular pulse signal of Figure 16-1, we found that

$$F(\omega) = A\Delta Sa(\omega\Delta/2)$$

which is plotted in Figure 6-2. The spectrum of the rectangular pulse is real.

The Fourier transform of an impulse $\delta(t)$ is (item 2 of Table 16-3)

$$F(\omega) = 1$$

Thus, the spectrum of an impulse contains all frequencies, and a plot of the spectrum of the impulse is shown in Figure 16-4.

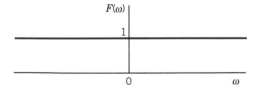

Figure 16-4 Spectrum of impulse $f(t) = \delta(t)$.

The Fourier transform of a constant dc signal of magnitude A is

$$F(\omega) = 2\pi A \delta(\omega)$$

which has a spectrum as shown in Figure 16-5. The integral of the impulse $\delta(\omega)$ has value unity at $\omega = 0$. The symbol for the impulse is a vertical line with an arrowhead.

Figure 16-5 Spectrum of constant dc signal of magnitude A. The symbol for an impulse is a vertical line with an arrowhead.

For completeness, let us examine a function that has a Fourier transform that is complex. When $f(t) = Ae^{-at}u(t)$,

$$F(\omega) = \frac{A}{a + j\omega}$$

In order to plot the spectrum, we calculate the magnitude and phase of $F(\omega)$ as

$$|F(\omega)| = \frac{A}{(a^2 + \omega^2)^{1/2}}$$

and

$$\phi(\omega) = -\tan^{-1}\omega/a$$

The Fourier spectrum is shown in Figure 16-6.

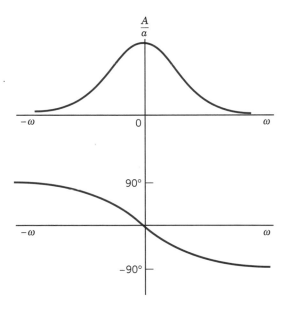

Figure 16-6 The Fourier spectrum for $f(t) = Ae^{-at}u(t)$.

The **Fourier spectrum** of a signal is a graph of the magnitude and phase of the Fourier transform of the signal.

EXERCISE 16-4

Calculate the Fourier transform and draw the Fourier spectrum for $f(t)$ shown in Figure E 16-4, where $f(t) = A\cos\omega_0 t$ for all t.
Answer: $F(\omega) = \pi A\delta(\omega + \omega_0) + \pi A\delta(\omega - \omega_0)$

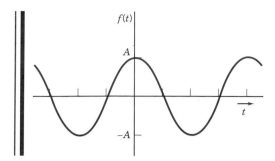

Figure E 16-4

16-8 ∥ THE ENERGY SPECTRUM

Consider the power dissipated by a 1-Ω resistor as

$$p = v^2(t)$$

and thus the energy is

$$\int_{-\infty}^{\infty} v^2(t)\ dt = \int_{-\infty}^{\infty} v(t)\mathcal{F}^{-1}[V(\omega)]\ dt$$

$$= \int_{-\infty}^{\infty} v(t)\ \frac{1}{2\pi}\int_{-\infty}^{\infty} V(\omega)e^{j\omega t}\ d\omega\ dt$$

We arrange the order of integration to obtain

$$\int_{-\infty}^{\infty} v^2(t)\ dt = \frac{1}{2\pi}\int_{-\infty}^{\infty} V(\omega)\int_{-\infty}^{\infty} v(t)e^{j\omega t}\ dt\ d\omega$$

$$= \frac{1}{2\pi}\int_{-\infty}^{\infty} V(\omega)V^*(\omega)\ d\omega$$

$$= \frac{1}{2\pi}\int_{-\infty}^{\infty} |V(\omega)|^2\ d\omega \tag{16-18}$$

Therefore, the energy absorbed by the 1-Ω resistor is equivalent to the integral of $|V(\omega)|^2$. Furthermore, we call the plot of $|V(\omega)|^2$ the *energy spectrum*, and Eq. 16-18 is called *Parseval's theorem*.

The energy spectrum, often called the energy density spectrum, can be shown by plotting $|V(\omega)|^2$ versus ω.

Speech and music contain information represented by $F(\omega)$ and can be recorded and plotted over time. Thus, if we plot $|F(\omega)|^2$ versus ω as well as time t, we have a three-dimensional plot as shown in Figures 16-7 and 16-8. Figure 16-8 shows the word "Hallelujah" from the Hallelujah Chorus, a piece from *The Messiah*, Handel's most successful and best-known oratorio, composed in the year 1741. The soprano, alto, tenor, and bass voices were each created electronically and subsequently combined using a multitrack recorder.

It is apparent that with the increasing availability and improvement of voice synthesis technology, singing synthesizers may herald the next revolution in music. Even the most proficient soprano or basso profundo is limited in the range, duration, and pitch of notes that can be generated because of the physical constraints of the vocal apparatus. A

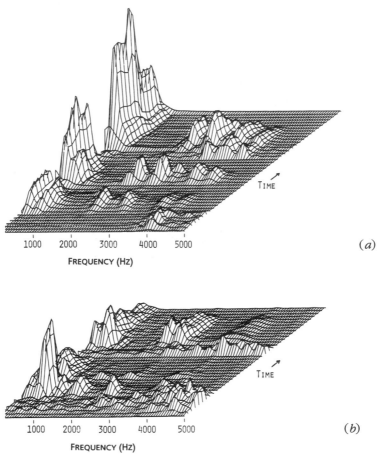

Figure 16-7 The energy spectrum for the word "seventeen" created by *(a)* a synthesizer and *(b)* a person.

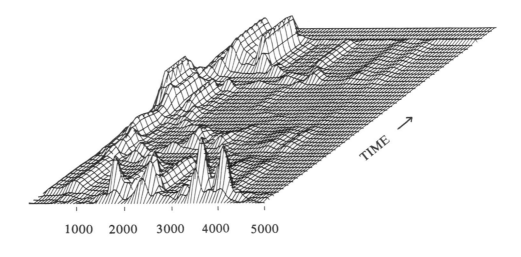

Figure 16-8 An electronically created synthetic voice singing "Hal-le-lu-jah!" is represented by its energy spectrum versus time.

synthesizer machine has no such limits, and future composers could create songs that only synthesizers can "sing."

Example 16-5

Find the total energy absorbed by a 1-Ω resistor having $v(t) = e^{-t}u(t)$ V using the time domain and the frequency domain.

Solution

In the time domain, we have

$$w = \int_{-\infty}^{\infty} v^2(t) \, dt = \int_{0}^{\infty} e^{-2t} \, dt = \frac{1}{2} \text{ J}$$

In the frequency domain, we have $V(\omega) = 1/(1 + j\omega)$ and hence

$$w = \frac{1}{2\pi} \int_{-\infty}^{\infty} |V(\omega)|^2 \, d\omega = \frac{1}{\pi} \int_{0}^{\infty} \frac{1}{1 + \omega^2} \, d\omega$$

$$= \frac{1}{\pi} \tan^{-1}\omega \Big|_{0}^{\infty} = \frac{1}{2} \text{ J}$$

EXERCISE 16-5

A function is $v(t) = 10e^{-2t}u(t)$ V. Determine the energy absorbed by a 1-Ω resistor when the voltage across the resistor is $v(t)$. Use both the time domain and the frequency domain methods.

Answer: 25 J

16-9 CONVOLUTION AND CIRCUIT RESPONSE

A circuit with an impulse response $h(t)$ and an input $f(t)$ has a response $y(t)$ that may be determined from the convolution integral. For the circuit shown in Figure 16-9, the convolution integral (see Section 14-9) is

$$y(t) = \int_{-\infty}^{\infty} h(x)f(t - x) \, dx$$

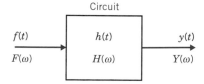

Circuit

$f(t)$ $h(t)$ $y(t)$

$F(\omega)$ $H(\omega)$ $Y(\omega)$

Figure 16-9 A linear circuit.

If we use the Fourier transform of the convolution integral, we have

$$\mathcal{F}[y(t)] = \int_{-\infty}^{\infty} \int_{-\infty}^{\infty} h(x)f(t - x) \, dx \, e^{-j\omega t} \, dt$$

$$= \int_{-\infty}^{\infty} h(x) \int_{-\infty}^{\infty} f(t - x)e^{-j\omega t} \, dt \, dx$$

Let $u = t - x$ to obtain

$$\mathcal{F}[y(t)] = \int_{-\infty}^{\infty} h(x) \int_{-\infty}^{\infty} f(u)e^{-j\omega(u+x)} \, du \, dx$$

$$= \int_{-\infty}^{\infty} h(x)e^{-j\omega x} \, dx \int_{-\infty}^{\infty} f(u)e^{-j\omega u} \, du$$

or

$$Y(\omega) = H(\omega)F(\omega) \qquad\qquad (16\text{-}19)$$

Thus, convolution in the time domain corresponds to multiplication in the frequency domain. When the input is an impulse, $f(t) = \delta(t)$, since $F(\omega) = 1$, we obtain the impulse response

$$Y(\omega) = H(\omega)$$

When the input is a sinusoid, the Fourier transform of the output is the steady-state response to that sinusoidal driving function.

In addition, the energy spectrum can be deduced using Eq. 16-19 so that

$$|Y(\omega)|^2 = |H(\omega)F(\omega)|^2$$
$$= |H(\omega)|^2 |F(\omega)|^2$$

This equation shows that the energy spectrum of the response of a linear system is the product of the energy spectrum of the input function and the squared amplitude of the circuit function. The phase characteristics of the network do not affect the energy density of the output.

Example 16-6

Find the response, $v_0(t)$, of the RL circuit shown in Figure 16-10 when $v(t) = 4e^{-2t}u(t)\,\mathrm{V}$. The initial condition is zero.

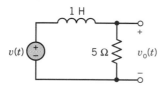

Figure 16-10 Circuit of Example 16-6.

Solution

Since $v(t) = 4e^{-2t}u(t)$, we obtain $V(\omega)$ as

$$V(\omega) = \frac{4}{2 + j\omega}$$

The circuit is represented by $H(\omega)$, and using the voltage divider principle, we have

$$H(\omega) = \frac{R}{R + j\omega L} = \frac{5}{5 + j\omega}$$

Then, we have

$$V_0(\omega) = H(\omega)V(\omega)$$

$$= \frac{20}{(5 + j\omega)(2 + j\omega)}$$

Expand using partial fractions to obtain[1]

$$V_0(\omega) = \frac{-20/3}{5 + j\omega} + \frac{20/3}{2 + j\omega}$$

Using the inverse transform for each term (entry 3 of Table 16-3), we have

$$v_0(t) = \frac{20}{3} (e^{-2t} - e^{-5t}) u(t) \text{ V}$$

The time domain responses obtained in this manner are responses of initially relaxed circuits. (No initial energy is stored.)

Example 16-7

Determine and plot the spectrum of the response $V_0(\omega)$ of the circuit of Figure 16-11 when $v = 10e^{-2t} u(t)$ V.

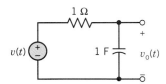

Figure 16-11 Circuit of Example 16-7.

Solution

The input signal $v(t)$ has a Fourier transform

$$V(\omega) = \frac{10}{2 + j\omega} = \frac{10}{(4 + \omega^2)^{1/2}} \underline{/- \tan^{-1} \omega/2}$$

The circuit transfer function is

$$H(j\omega) = \frac{1/j\omega C}{R + 1/j\omega C} = \frac{1}{1 + j\omega} = \frac{1}{(1 + \omega^2)^{1/2}} \underline{/- \tan^{-1} \omega}$$

Then the output is

$$V_0(\omega) = H(\omega) V(\omega)$$

$$= \frac{10}{(2 + j\omega)(1 + j\omega)}$$

Therefore

$$|V_0| = \frac{10}{[(4 + \omega^2)(1 + \omega^2)]^{1/2}}$$

and

$$\phi(\omega) = \underline{/V_0(\omega)} = - \tan^{-1} \frac{\omega}{2} - \tan^{-1} \omega$$

The calculated magnitude and phase for $V_0(\omega)$ are recorded in Table 16-4. For negative ω, $|V_0(\omega)| = |V_0(-\omega)|$ and

$$\phi(-\omega) = -\phi(\omega) \tag{16-20}$$

---

[1] See Section 14-5 for a review of partial fraction expansion.

Table 16-4

Fourier Response for Example 16-7

ω	0	1	2	3	5	∞
$\lvert V_0 \rvert$	5	3.16	1.58	0.88	0.36	0
$\phi(\omega)$	0°	−71.6°	−108.4°	−127.9°	−146.9°	−180°

Therefore, the Fourier spectrum of $V_0(\omega)$ is represented by the plot shown in Figure 16-12.

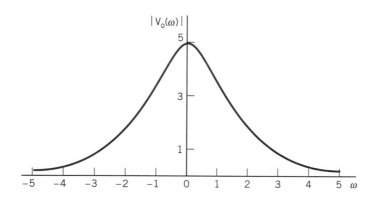

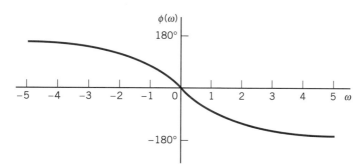

Figure 16-12 The amplitude and phase versus ω of the output voltage for Example 16-7.

EXERCISE 16-6

An ideal bandpass filter passes all frequencies between 24 rad/s and 48 rad/s without attenuation and completely rejects all frequencies outside this passband.

(a) Sketch $\lvert V_0 \rvert^2$ for the filter output voltage when the input voltage is

$$v(t) = 120e^{-24t}u(t) \text{ V}$$

(b) What percentage of the input signal energy is available in the signal at the output of the ideal filter?

Answer: (b) 20.5%

16-10 │ THE FOURIER TRANSFORM AND THE LAPLACE TRANSFORM

The tables of Laplace transforms, Table 14-1 and Table 14-3, developed in Chapter 14, can be used to obtain the Fourier transform of a function $f(t)$. Of course, the Fourier transform formally exists only when the Fourier integral, Eq. 16-4, converges. The Fourier integral will converge when all the poles of $F(s)$ lie in the left-hand s-plane and not on the $j\omega$ axis or at the origin.

If $f(t)$ is zero for $t \leq 0$ and $\int_0^\infty f(t)dt < \infty$, we can obtain the Fourier transform from the Laplace transform of $f(t)$ by replacing s by $j\omega$. Then

$$F(\omega) = F(s)\,|_{s=j\omega} \qquad (16\text{-}21)$$

where

$$F(s) = \mathcal{L}\{f(t)\}$$

For example, if (entry 3 of Table 16-3)

$$f(t) = Ae^{-at}u(t)$$

then, from Table 14-1,

$$F(s) = \frac{A}{s + a}$$

Therefore, with $s = j\omega$ we obtain the Fourier transform

$$F(\omega) = \frac{A}{a + j\omega}$$

If $f(t)$ is a real function with a nonzero value for negative time only, then we can reflect $f(t)$ to positive time, find the Laplace transform, and then find $F(\omega)$ by setting $s = -j\omega$. Therefore, when $f(t) = 0$ for $t \geq 0$ and $f(t)$ exists only for negative time, we have

$$F(\omega) = \mathcal{L}\{f(-t)\}\,|_{s=-j\omega} \qquad (16\text{-}22)$$

For example, consider the exponential function

$$\begin{aligned} f(t) &= 0 & t \geq 0 \\ &= e^{at} & t < 0 \end{aligned}$$

Then, reversing the time function, we have

$$f(-t) = e^{-at} \qquad t > 0$$

and, therefore,

$$F(s) = \frac{1}{s + a}$$

Hence, setting $s = -j\omega$, we obtain

$$F(\omega) = \frac{1}{a - j\omega}$$

Functions that are nonzero over all time can be divided into positive time and negative time functions. We then use Eqs. 16-21 and 16-22 to obtain the Fourier transform of each part. The Fourier transform of $f(t)$ is the sum of the Fourier transforms of the two parts.

For example, consider the function $f(t)$ with a nonzero value over all time where

$$f(t) = e^{-a|t|}$$

which is entry 4 in Table 16-3. The positive time portion of the function will be called $f^+(t)$, and the negative time portion will be called $f^-(t)$. Then,

$$f(t) = f^+(t) + f^-(t)$$

Hence

$$F(\omega) = \mathcal{L}\{f^+(t)\}_{s=j\omega} + \mathcal{L}\{f^-(-t)\}_{s=-j\omega} \tag{16-23}$$

In this case,

$$f^+(t) = e^{-at} \qquad t \ge 0$$

and

$$f^-(t) = e^{at} \qquad t \le 0$$

Note that $f^-(-t) = e^{-at}$. Then,

$$F^+(s) = \frac{1}{s+a} \quad \text{and} \quad F^-(s) = \frac{1}{s+a}$$

We obtain the total $F(\omega)$ as

$$F(\omega) = F^+(s)\,|_{s=j\omega} + F^-(s)\,|_{s=-j\omega}$$

$$= \frac{1}{a+j\omega} + \frac{1}{a-j\omega}$$

$$= \frac{2a}{\omega^2 + a^2}$$

The use of the Laplace transform to find the Fourier transform is summarized in Table 16-5. Remember that the method summarized cannot be used for $\sin \omega t$, $\cos \omega t$, or $u(t)$, since the poles of $F(s)$ lie on the $j\omega$ axis or at the origin.

Table 16-5
Obtaining the Fourier Transform Using the Laplace Transform

Case	Method
A. $f(t)$ nonzero for positive time only and $f(t) = 0$, $t < 0$	Step 1. $F(s) = \mathcal{L}\{f(t)\}$ 2. $F(\omega) = F(s)\,\|_{s=j\omega}$
B. $f(t)$ nonzero for negative time only and $f(t) = 0$, $t > 0$	Step 1. $F(s) = \mathcal{L}\{f(-t)\}$ 2. $F(\omega) = F(s)\,\|_{s=-j\omega}$
C. $f(t)$ nonzero over all time	Step 1. $f(t) = f^+(t) + f^-(t)$ 2. $F^+(s) = \mathcal{L}\{f^+(t)\}$ $\quad\; F^-(s) = \mathcal{L}\{f^-(-t)\}$ 3. $F(\omega) = F^+(s)\,\|_{s=j\omega} +$ $\quad\; F^-(s)\,\|_{s=-j\omega}$

Note: The poles of $F(s)$ must lie in the left-hand s-plane.

EXERCISE 16-7

Derive the Fourier transform for

$$f(t) = te^{-at} \qquad t \ge 0$$
$$= te^{at} \qquad t \le 0$$

Answer: $\dfrac{-j4a\omega}{(a^2 + \omega^2)^2}$

16-11 THE FOURIER TRANSFORM AND PSpice

Just as the Fourier transform is a special case of the Laplace transform, the Fourier integral (the function performing the transform) can be thought of as an extension of the Fourier series where, by extending the fundamental period to infinity, each harmonic component becomes infinitesimally close in "frequency." Thus, in the limit, the Fourier series becomes the Fourier integral. This means the Fourier transform converts a function of time to a function of frequency, and vice versa. The physical interpretation of the Fourier transform is the conversion of a time domain signal to the steady-state frequency content, or spectrum, that makes up the signal.

Probe, the graphical postprocessor for PSpice, has the capability of calculating the Fourier transform of the data sequence stored in the Probe data file. The Fourier transform in Probe is a *discrete Fourier transform* (DFT), where the Fourier integral has been replaced by a nearly equivalent summation formula applied to evenly spaced samples of the signal. Furthermore, the transform is accomplished by a special technique called a *fast Fourier transform* (FFT). This is a numerical procedure whereby, if the size of the data sequence is a power of two (such as 1024 or 4096), a much shorter sequence of calculations can be used to get the same results as the discrete Fourier transform. Even for modest data sets, the DFT is so time consuming that most computer applications use the FFT. Care must be exercised when interpreting the frequency spectrum produced by Probe. Probe scales this frequency spectrum. The scale factor depends on the duration of the time response used to calculate the frequency spectrum.

Consider a sine wave of 1000 Hz and peak voltage 0.60 V. We can use a voltage-controlled voltage source to calculate a transient analysis as shown in Figure 16-13. In this case, we have a simple polynomial $v = v_{10}^2$ for the input-output relation of the VCVS(E1). The polynomial uses the voltage (1,0) and the coefficient list for ascending powers of v_{10}. The transient response is run for 10 ms, and the t step is 50 μs.

```
SINE    WAVE
I1    0    1    sin  (0   0.60   1000)
R1    0    1    1
E1    2    0    POLY (1)    (1,0)    0   0   1
R2    2    0    1G
.TRAN    50U    10M    0    50U
.PROBE
.END
```

Figure 16-13 The Probe program used to calculate the Fourier transform of the transient output signal.

The general form of the polynomial voltage source is

E*name* <*output nodes*> POLY (*dim.*) <*control variables*> <*coeff.*>

where the number of nodes in the list of *control variables* must be twice the *dimension* specified. Not all values must be specified in the *coefficients* list; however, no coefficients may be skipped. This means that even if a coefficient has a value of zero, it must be entered in the list unless all of the remaining coefficients in the list are zero, in which case the list is ended.

The transient output signal of the VCVS is calculated at the t step intervals, and these samples are stored in Probe. We then plot V(1) versus frequency to display the input spectral line and plot V(2) versus frequency to display the output signal spectral line. The FFT of the input signal is a spectral line at 1000 Hz, and the FFT of the output of the VCVS is a spectral line at 2000 Hz. The height of the spectral line at 1000 Hz is 0.60, and the height of the output spectral line at 2000 Hz is 0.36.

16-12 | VERIFICATION EXAMPLE

Problem

An ideal low-pass filter has an input $v(t) = 4e^{-10t}u(t)$ V. The filter has a cutoff frequency of 10 rad/s. A laboratory report states that the ratio of energy at the filter output to the input energy is 0.75. Verify this result.

Solution

The Fourier transform is (see Table 16-3)

$$V(\omega) = \frac{4}{10 + j\omega}$$

The total energy of the input signal is

$$w_i = \int_0^\infty v^2 \, dt = \int_0^\infty 16e^{-20t} \, dt = \frac{16}{20} \text{ J}$$

The output signal is bandlimited so that the output signal energy is

$$w_0 = \frac{1}{2\pi} \int_0^{10} |V(\omega)|^2 \, d\omega = \frac{1}{\pi} \int_0^{10} \frac{16}{(100 + \omega^2)} \, d\omega$$

$$= \frac{16}{10\pi} \tan^{-1} \left(\frac{\omega}{10}\right) \Big|_0^{10}$$

$$= \frac{16}{40} \text{ J}$$

We obtain the ratio of output energy to input energy as

$$\frac{w_0}{w_i} = \frac{1}{2}$$

Thus, the result stated in the laboratory report is incorrect.

LOW-PASS FILTER CIRCUIT

Problem

An input signal $v(t)$ is used to control a valve. The dc component of $v(t)$ is required to actuate the valve. The higher frequency components of $v(t)$ need to be attenuated in order to reduce valve wear. A filter circuit is to be designed so that the magnitude of the output voltage at $\omega = 2$ rad/s is less than or equal to 1% of the magnitude of the output signal at $\omega = 0$. The input signal is $v(t) = 10e^{-t}u(t)$ V. A low-pass RLC filter circuit is proposed as shown in Figure 16D-1.

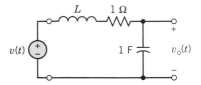

Figure 16D-1 RLC circuit.

Define the Situation, State the Assumptions, and Develop a Model

1 The RLC circuit is linear.
2 The input signal is a decaying exponential.

The Goal

Determine L to achieve

$$\frac{|V_0(2)|}{|V_0(0)|} \leq 0.01 \qquad\qquad (16\text{D-1})$$

Generate a Plan

1 Determine $V(\omega)$ and $H(\omega)$.
2 Calculate $V_0(\omega)$ and $|V_0(\omega)|$.
3 Using Eq. 16-24 as the requirement, evaluate the ratio and determine the required L.

Take Action Using the Plan

The Fourier transform of the input is

$$V(\omega) = \frac{10}{1 + j\omega}$$

The transfer function of the linear circuit is obtained using impedances as

$$\frac{V_0(\omega)}{V(\omega)} = H(\omega) = \frac{1/j\omega C}{R + j\omega L + 1/j\omega C}$$

$$= \frac{1}{j\omega + 1 - \omega^2 L}$$

Then the Fourier transform of the output signal is

$$V_0(\omega) = H(\omega)V(\omega)$$

$$= \frac{1}{(j\omega + 1 - \omega^2 L)} \frac{10}{(1 + j\omega)}$$

$$= \frac{10}{1 - \omega^2(1 + L) + j(2\omega - \omega^3 L)}$$

We will evaluate $V_0(\omega)$ at $\omega = 0$ and $\omega = 2$ and then obtain the ratio. Therefore,

$$V_0(0) = 10$$

When $\omega = 2$, we have

$$V_0(2) = \frac{10}{1 - 4(1 + L) + j(4 - 8L)}$$

and

$$|V_0(2)| = \frac{10}{[(1 - 4(1 + L))^2 + (4 - 8L)^2]^{1/2}}$$

Then the required ratio is

$$\frac{|V_0(2)|}{|V_0(0)|} = 0.01 = \frac{1}{[(1 - 4(1 + L))^2 + (4 - 8L)^2]^{1/2}}$$

which leads to the equivalent requirement that

$$[1 - 4(1 + L)]^2 + (4 - 8L)^2 = 10^{+4}$$

After completing the algebraic steps, we find it is required that $L \geq 11.43$ H. We elect to use $L = 11.5$ H. With $L = 11.5$ H, we check to find

$$\frac{|V_0(2)|}{|V_0(0)|} = \frac{1}{100.72} = 9.9 \times 10^{-3}$$

which meets the requirements.

SUMMARY

The Fourier transform provides a frequency domain description of an aperiodic time domain function. The Fourier transform can be obtained for functions with nonzero values in negative time as well as for positive time. The Fourier transform generates the steady-state response of a circuit that is excited by a sinusoidal source.

The response of a circuit can be obtained by multiplying the Fourier transform of the input source by the circuit transfer function so that

$$V_0(\omega) = H(\omega)V_s(\omega)$$

Parseval's theorem states that the energy density over a range of frequencies of a signal is the square of the magnitude of the Fourier transform of the signal. Thus, one can calculate the fraction of the total energy of a signal associated with a range of frequencies.

Furthermore, the link between the Laplace transform and the Fourier transform can be used to obtain the Fourier transform of time functions.

The magnitude spectrum for a frequency-modulated signal is shown in Figure 16-14.

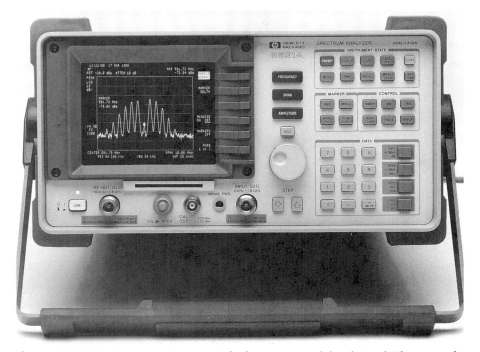

Figure 16-14 The frequency spectrum of a frequency-modulated signal. The center frequency of 501.75 MHz is the carrier signal. The HP 8591 spectrum analyzer has a frequency range from 9 kHz to 1.8 GHz. Courtesy of Hewlett-Packard Company.

TERMS AND CONCEPTS

Fourier Transform   Process of converting a time function, $f(t)$, to a frequency function, $F(\omega)$.

Inverse Fourier Transform   Process of converting a frequency function, $F(\omega)$, to a time function, $f(t)$.

Parseval's Theorem   The energy absorbed by a 1-Ω resistor is $1/2\pi$ times the integral of the energy spectrum over all frequencies from $-\infty$ to $+\infty$.

Energy Spectrum   The square of the magnitude of $V(\omega)$ written as $|V(\omega)|^2$. Also called the energy spectral density.

Spectrum   The Fourier transform, $F(\omega)$, of a signal, $f(t)$. Also called spectral density.

Unit Impulse   A pulse of infinitesimal width with the area under the pulse equal to 1.

REFERENCE   Chapter 16

Bracewell, Ronald N., "The Fourier Transform," *Scientific American,* June 1989, pp. 86–95.

PROBLEMS

Section 16-4 The Fourier Transform

P 16.4-1 Find the Fourier transform of the function

$$f(t) = -u(-t) + u(t)$$

as shown in Figure P 16.4-1. This is called the signum function.

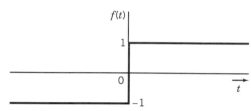

Figure P 16.4-1

P 16.4-2 Find the Fourier transform of $f(t) = Ae^{-at}u(t)$ when $a > 0$.

Answer: $F(\omega) = \dfrac{A}{a + j\omega}$

P 16.4-3 Find the Fourier transform of the waveform shown in Figure P 16.4-3.

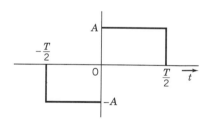

Figure P 16.4-3

P 16.4-4 Determine the Fourier transform of $f(t) = 10 \cos 50t$.

Answer: $F(\omega) = 10\pi\,\delta(\omega - 50) + 10\pi\,\delta(\omega + 50)$

P 16.4-5 Determine the Fourier transform of the pulse shown in Figure P 16.4-5.

Answer: $F(j\omega) = \dfrac{2}{\omega}(\sin \omega - \sin 2\omega) + \dfrac{j2}{\omega}(\cos \omega -$

$\cos 2\omega)$

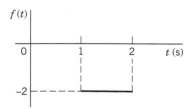

Figure P 16.4-5

P 16.4-6 Determine the Fourier transform of a signal with $f(t) = At$ between $t = 0$ and $t = B$ and $f(t) = 0$ elsewhere.

Answer: $F(j\omega) = \dfrac{A}{B}\left[\dfrac{-B}{j\omega}\, e^{-j\omega B} + \dfrac{1}{\omega^2}\, e^{-j\omega B} - \dfrac{1}{\omega^2} \right]$

P 16.4-7 Determine the Fourier transform of the waveform $f(t)$ shown in Figure P 16.4-7.

Answer: $F(j\omega) = \dfrac{2}{\omega}(\sin 2\omega - \sin \omega)$

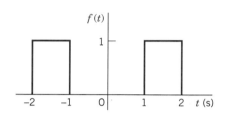

Figure P 16.4-7

Section 16-5 Fourier Transform Properties

P 16.5-1 Determine the Fourier transform of

$$f(t) = \frac{d}{dt}[e^{-at}u(t)]$$

P 16.5-2 Find the Fourier transform for the unit step function where

$$f(t) = \begin{cases} +1 & t > 0 \\ 0 & t < 0 \end{cases}$$

P 16.5-3 Use the signum function and a constant to assist in the derivation of the Fourier transform of the unit step function.

P 16.5-4 Determine the Fourier transform of the ideal delay for which the output is a replica of the input signal after a delay of T seconds.

Answer: $H(\omega) = e^{-j\omega T}$

Section 16-6 Transform Pairs

P 16.6-1 Find the Fourier transform for item 7 of Table 16-3.

P 16.6-2 Find the Fourier transform for item 6 of Table 16-3.

P 16.6-3 Find the Fourier transform for $f(t) = te^{-bt}u(t)$.

Answer: $F(j\omega) = \dfrac{1}{(j\omega + b)^2}$

Section 16-7 The Spectrum of Signals

P 16.7-1 Find the Fourier transform of $f(t) = Ae^{-at}u(t)$ when a > 0, and plot its amplitude and phase spectrum.

P 16.7-2 Find the Fourier transform of $f(t) = A \cos \omega_0 t$, and plot its amplitude and phase spectrum.

P 16.7-3 Plot the amplitude and phase spectrum of the sawtooth waveform (item 6 of Table 16-3).

P 16.7-4 Plot the amplitude and phase spectrum of $f(t)$ of P 16.6-3.

P 16.7-5 Determine the Fourier transform of

$$f(t) = \begin{cases} 1 - \dfrac{|t|}{\tau} & |t| < \tau \\ 0 & \text{otherwise} \end{cases}$$

Plot the amplitude and frequency spectrum.

Section 16-8 The Energy Spectrum

P 16.8-1 The voltage across a 50-Ω resistor is

$$v = 4te^{-t}u(t) \text{ V}$$

Find the percentage of the total energy absorbed by the resistor when the frequency range is 0 to $\sqrt{3}$ rad/s.

P 16.8-2 Find the total energy absorbed by a 1-Ω resistor with a voltage $v(t) =$ signum function (see item 10 of Table 16-3).

P 16.8-3 Find (a) the total energy absorbed by a 1-Ω resistor for a voltage $v = Ae^{-at}u(t)$ and (b) the bandwidth $\omega = BW$ required to pass 90 percent of the total energy.

P 16.8-4 (a) Find the total energy absorbed by a 1-Ω resistor for a current $i(t) = te^{-Bt}u(t)$. (b) Determine the fraction of the energy present in the band $-B$ rad/s to $+B$ rad/s.

Answer: (a) $w = 1/4B^3$
(b) 0.818

Section 16-9 Convolution and Circuit Response

P 16.9-1 Find the current $i(t)$ in the circuit of Figure P 16.9-1 when $i_s(t)$ is the signum function, so that

$$i_s(t) = \begin{cases} +40 \text{ A} & t > 0 \\ -40 \text{ A} & t < 0 \end{cases}$$

Also, sketch $i(t)$.

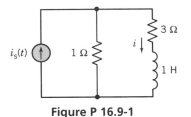

Figure P 16.9-1

P 16.9-2 Repeat Problem 16.9-1 when $i_s = 100 \cos 3t$ A.

P 16.9-3 The voltage source of Figure P 16.9-3 is $v(t) = 10 \cos 2t$ for all t. Calculate $i(t)$ using the Fourier transform.

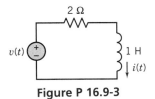

Figure P 16.9-3

P 16.9-4 Find the output voltage $v_0(t)$ using the Fourier transform for the circuit of Figure P 16.9-4 when $v(t) = e^t u(-t) + u(t)$ V.

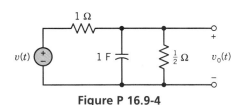

Figure P 16.9-4

P 16.9-5 The voltage source of the circuit of Figure P 16.9-5 is $v_s(t) = 15e^{-5t}$ V. Find the resistance R when it is known that the energy available in the output signal is two-thirds of the energy of the input signal.

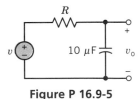

Figure P 16.9-5

P 16.9-6 The pulse signal shown in Figure P 16.9-6a is the source $v_s(t)$ for the circuit of Figure P 16.9-6b. Determine the output voltage, v_0, using the Fourier transform.

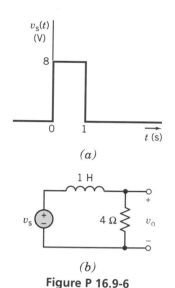

(a)

(b)

Figure P 16.9-6

Section 16-10   The Fourier Transform and the Laplace Transform

P 16.10-1   Determine the Fourier transform of $f(t)$ using its Laplace transform when

$$f(t) = Ae^{-at} \sin\beta t \, u(t)$$

Answer: $F(\omega) = \dfrac{A\beta}{\beta^2 + a^2 - \omega^2 + j2a\omega}$

P 16.10-2   Determine the Fourier transform of the ideal delay for which the output is a delayed replica of the input signal after T seconds.

Answer: $H(\omega) = e^{-j\omega T}$

P 16.10-3   A circuit is shown in Figure P 16.10-3a that acts as a low-pass filter. The input signal, v_s, is shown in Figure P 16.10-3b.

(a)   Find $V_s(\omega)$, and plot the $|V_s|^2$ as the energy spectrum.

(b)   Determine $H(j\omega) = V_0(\omega)/V_s(\omega)$, and plot $H(j\omega)$ by plotting $|H|$ and $\phi(\omega)$.

(c)   Determine $V_0(\omega)$ and $v_0(t)$.

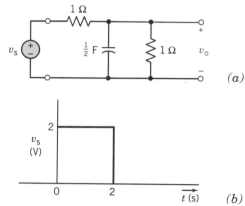

(a)

(b)

Figure P 16.10-3

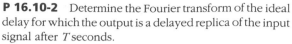

VERIFICATION PROBLEMS

VP 16-1   Verify that the magnitude of the Fourier transform of $f(t) = e^{-10t}u(t)$ is $|F(\omega)| = (100 + \omega^2)^{-1/2}$, and plot the magnitude versus ω for $0 \le \omega \le 100$.

VP 16-2   An ideal low-pass filter has an input $v(t) =$

$10e^{-20t}u(t)$ V. The filter has a cut-off frequency of 100 rad/s. Verify that the ratio of energy absorbed by a 1-Ω resistor to the input energy is 0.876.

DESIGN PROBLEM

DP 16-1   The RL circuit shown in Figure DP 16-1a is used to develop the output signal $v(t)$ shown in Figure DP 16-1b. Select a suitable current pulse source and inductance L in order to achieve the desired response.

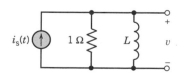

(a)

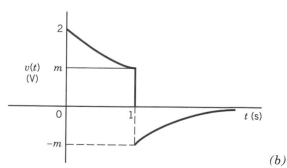

(b)

Figure DP 16-1

CHAPTER 17

TOPOLOGICAL AND SENSITIVITY ANALYSIS

Problems

PSpice Problems

Verification Problems

Design Problems

PREVIEW

Many electric circuits are complex, consisting of many nodes and sources, and it is important to choose appropriate equations in order to ease the solution for selected variables. With the growing use of computer-aided solution, it is helpful to understand the underlying methods utilized by the available computer analysis tools.

In this chapter we examine two important underlying analytical methods widely used in the analysis of complex circuits: topological and sensitivity analysis. Topological analysis is concerned with the selection of appropriate independent equations, and sensitivity analysis is concerned with the change in the circuit variables due to element variation and the impact of that change on the circuit performance.

17-1 DESIGN CHALLENGE

CURRENT SOURCE WITH A VOLTAGE CONSTRAINT

A circuit is shown in Figure 17D-1. We wish to select R_1, R_2, and R_3 so that the voltage across the current source i_s is limited to 8 V. We wish to determine the required number of node equations versus the required number of mesh equations. This will enable us to select between the node voltage method and the mesh current method. Using the appropriate method we can calculate the voltage across i_s and then select the required resistors.

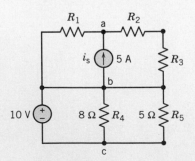

Figure 17D-1 Circuit with current source i_s.

Define the Situation, State the Assumptions, and Develop a Model

1 The current source is common to two meshes.
2 The voltage $v_{bc} = 10$ V is known.

The Goal

Determine whether to use node voltage or mesh current analysis, and then determine R_1, R_2, and R_3 so that the voltage across the current source is no more than 8 V.

In this chapter we will develop a graphical method to determine the number of equations required to obtain the voltage v_{ab}. We will return to this problem at the end of the chapter.

17-2 GRAPHS AND PARAMETER VARIATION

From the dawn of civilization, inquiring citizens have been concerned with the representation of connected lines, which we call graphs. For example, the road system of the Roman Empire could be represented by a series of cities (nodes) and a series of interconnecting lines between the cities. The drawing of the road system is a graph.

One of the early originators of graph theory was the Swiss mathematician Leonhard Euler, whose 1736 paper, "The Seven Bridges of Königsberg," was one of the first formal treatments of the subject. In Königsberg, there were two islands linked to each other and to the banks of the Pregel River by seven bridges. The problem is to start at one of the land areas and to cross all seven bridges without ever recrossing a bridge. Euler proved that there is no solution, and he established a rule that applies to any connected graph: such a traversal is possible if, and only if, at most, two points are the terminus for an odd number of lines.

During the 1930s, the engineers at the Bell Telephone Laboratories studied the poor reliability of electronic amplifiers due to variation of the circuit elements over time. It was necessary to achieve highly accurate and reliable amplification, and yet element tolerances could not readily be maintained (Bode, 1945). An amplifier's gain would vary, and the results varied from hour to hour. Thus, a formal discussion of parameter sensitivity evolved and came to fruition with the advent of computer-aided circuit analysis.

17-3 NETWORK GRAPHS

An electric circuit is determined by the elements it contains and the way in which the elements are connected. In this chapter we focus on the interconnection of the elements and not on the elements themselves. Consider the circuit shown in Figure 17-1, which consists of four nodes and five branches. Since the geometrical properties of a circuit are independent of the elements that are contained in each branch, we simply represent each branch by a line segment. Such a drawing is called a graph, and the graph of the circuit of

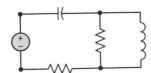

Figure 17-1 A circuit with four nodes and five branches.

Figure 17-2 The graph of the circuit of Figure 17-1.

Figure 17-1 is shown in Figure 17-2. A *graph* is a set of nodes together with a set of branches with the condition that each branch terminates at each end into a node. The topological character of the circuit is completely specified by its graph. The word *topology* refers to the science of placement of elements and is a study of the geometric configurations invariant under transformation by remapping.

Circuit topology is the study of the geometric properties of a circuit useful for discerning the underlying circuit behavior.

The circuit of Figure 17-1 is planar; that is, it can be drawn on a plane surface, with no crossovers. An example of the graph of a nonplanar circuit is shown in Figure 17-3. This circuit cannot be redrawn on a plane surface because no matter how we relocate node d by redrawing the graph, we will have one crossover. Note that one remapping of the circuit of Figure 17-3 as shown in Figure 17-4 still contains a crossover. The topological

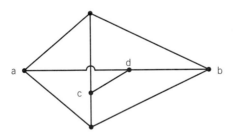

Figure 17-3 The graph of a nonplanar circuit with six nodes and nine branches.

properties of the graph are invariant under elastic deformation. The topological properties remain invariant with stretching, bending, or squeezing of the graph.

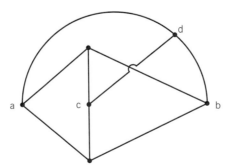

Figure 17-4 A transformation of the graph of Figure 17-3, which still results in one crossover.

A **circuit graph** displays the circuit connections of nodes and branches.

17-4 | TREES

A *tree* of a graph is any connected set of branches that connects every node to every other node directly or indirectly without forming any closed path or loop. Thus, a tree is a subgraph that contains all the nodes of a graph, but no loops. To construct a tree (1) draw all the original nodes and (2) add branches until all nodes are connected without forming any loops. Generally, a graph has many trees. For the graph of Figure 17-5a, we show the four trees in Figure 17-5b. One tree of the graph of Figure 17-2 is shown in Figure 17-6.

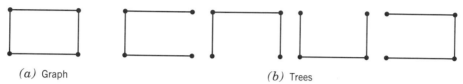

(a) Graph *(b)* Trees

Figure 17-5 *(a)* A graph and *(b)* its four trees.

The branches of the graph that are not in the chosen tree are called *links* and together with the nodes are called a *complementary tree* or *cotree*. The cotree for the chosen tree of Figure 17-6 is shown in Figure 17-7. A graph has N nodes and B branches. Therefore,

Figure 17-6 One tree of the graph of Figure 17-2.

Figure 17-7 The cotree for the tree of Figure 17-6 and the graph of Figure 17-2.

the number of nodes in any tree remains N, while the number of branches in a tree is $N - 1$. Then the number of links in the cotree is $L = B - (N - 1)$ and, therefore,

$$L = B - N + 1 \qquad (17\text{-}1)$$

For example, consider the graph of Figure 17-2, the tree of Figure 17-6, and the cotree of Figure 17-7. The graph has $N = 4$ and $B = 5$. The tree has four nodes, and it retains $N - 1 = 3$ branches. The cotree has $L = B - N + 1 = 2$ links.

Example 17-1

Let us find the graph, one tree, and its cotree for the circuit of Figure 17-8.

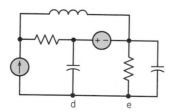

d   e

Figure 17-8 The circuit of Example 17-1.

Solution

We note that there are four nodes ($N = 4$) and seven branches ($B = 7$). We then draw the graph as shown in Figure 17-9, where node d represents both nodes d and e of Figure 17-8. We label each branch as shown for discussion purposes.

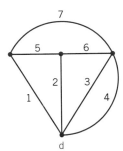

Figure 17-9 The graph of the circuit of Figure 17-8.

We draw one tree of the graph as shown in Figure 17-10a by selecting branches 2, 5, and 6. There are other trees such as branches (1, 5, and 7) and (3, 6, and 7).

The cotree for the tree of Figure 17-10a is shown in Figure 17-10b. The cotree has $L = B - N + 1 = 7 - 4 + 1 = 4$ links.

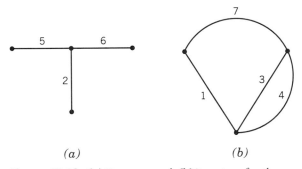

(a)   *(b)*

Figure 17-10 *(a)* One tree and *(b)* its cotree for the graph of Figure 17-9.

EXERCISE 17-1

Determine the graph, a tree, and its cotree for the circuit of Figure E 17-1. Also determine the number of links, L, of the cotree.

Answer: $L = 5$

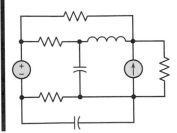

Figure E 17-1

17-5 ∥ DIRECTED BRANCHES AND CUT SETS

We have heretofore numbered the branches of a graph. If we add an assumed orientation for an element, we can represent each element with a current direction and an associated voltage as shown in Figure 17-11a. Then this element will be represented by the branch of

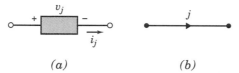

(a)   *(b)*

Figure 17-11 *(a)* The jth element with its associated voltage and current. *(b)* The jth branch of a graph representing the jth element.

a graph as shown in Figure 17-11b, where we show the assumed direction of the current by the arrow and label it with the number of the element. An example of a graph with assumed directions for each current is shown in Figure 17-12.

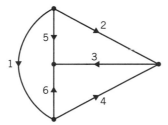

Figure 17-12 A graph with directed branches.

A *cut set* of a graph is a minimum set of elements that when cut, or removed, separates the graph into two groups of nodes. For example, two separate graphs are obtained for the graph of Figure 17-12 by selecting the cut set consisting of branches 4, 6, 5, and 2, as shown in Figure 17-13.

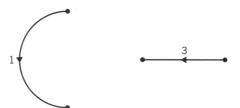

Figure 17-13 Two separate graphs created by the cut set [4, 6, 5, 2].

A *fundamental cut set* (FCS) is a cut set that cuts or contains one and only one tree branch. Thus, to find a fundamental cut set, first select a tree of the graph and then select one branch of the tree that will divide the graph into two subgraphs. The fundamental cut set for branch 5 of the selected tree is shown in Figure 17-14. Fundamental cut sets for branch 3 and branch 1 of the tree are shown in Figure 17-15.

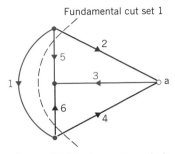

Figure 17-14 A tree (in color) and one fundamental cut set (FCS) containing branch 5 on the tree.

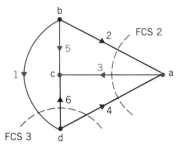

Figure 17-15 The remaining two fundamental cut sets (FCS) for branch 3 and branch 1 for the graph and tree of Figure 17-14.

The procedure for finding the fundamental cut sets is: (1) select a tree, (2) select a tree branch, and (3) divide the graph into two sets of nodes by drawing a line through the selected tree branch and appropriate cotree links while avoiding intersecting any other tree branches. Note FCS 2 yields node a and the set of nodes (b, c, d).

Example 17-2
Select a tree, and identify the fundamental cut sets of the graph of Figure 17-16.

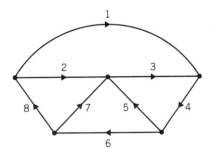

Figure 17-16 Graph for Example 17-2.

Solution
We select the tree consisting of elements 8, 7, 5, 4, as shown in color in Figure 17-17. The four fundamental cut sets are delineated in Figure 17-17. For example, fundamental cut set 3 (FCS 3) cuts branch 5 of the tree and FCS 3 contains branches [1, 3, 5, 6].

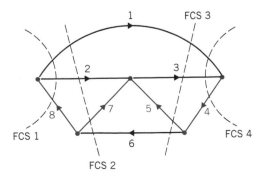

Figure 17-17 Graph, tree (in color), and four fundamental cut sets for Example 17-2.

EXERCISE 17-2

Determine the fundamental cut sets for the graph and tree (in color) shown in Figure E 17-2.

Answers: FCS 1 = [1, 2, 8]   FCS 2 = [4, 3, 2, 8]
        FCS 3 = [6, 5, 4]   FCS 4 = [8, 7, 5, 4]

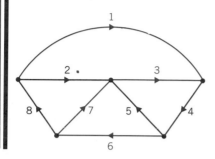

Figure E 17-2 Graph and tree (in color).

17-6 INDEPENDENT NODE VOLTAGE EQUATIONS

There are $N - 1$ independent node voltages within a circuit. We can obtain the $N - 1$ voltage equations of a circuit by using Kirchhoff's current law for either part of the circuit divided by a fundamental cut set. There are $N - 1$ fundamental cut sets, and if we obtain the current equations at each cut set we will then obtain the $N - 1$ independent equations required.

Refer to the graph and tree (in color) with the three fundamental cut sets identified as shown in Figure 17-18. Stretch the graph so that the tree (in color) lies on a horizontal line

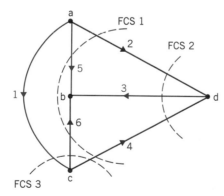

Figure 17-18 Graph, tree (in color), and three fundamental cut sets.

as shown in Figure 17-19, and assume that positive current flows to the right. Then, writing the current law for the cut sets, we have

$$\text{FCS 1:} \quad i_2 + i_5 + i_6 + i_4 = 0$$

$$\text{FCS 2:} \quad i_2 - i_3 + i_4 = 0 \tag{17-2}$$

$$\text{FCS 3:} \quad - i_1 + i_6 + i_4 = 0$$

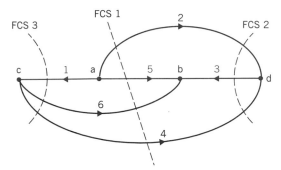

Figure 17-19 The graph of Figure 17-18 with the tree (in color) set on a horizontal line.

To find the branch voltages of the circuit, we need only find the branch voltages of the tree, and then the branch voltages of each link will be readily determined. To find the tree branch voltages we express the cut set currents as functions of these voltages.

Example 17-3
Let us consider the circuit shown in Figure 17-20 and determine the tree branch voltages, the link branch voltage, and the node voltages.

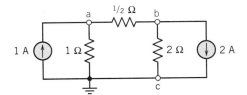

Figure 17-20 Circuit of Example 17-3.

Solution
First, we draw the graph of the circuit as shown in Figure 17-21 and select the tree (in color). We select the tree so that the branches of the tree intersect at the reference node. In

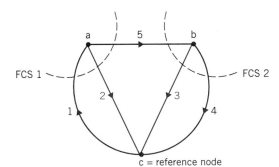

Figure 17-21 Graph, tree (in color), and fundamental cut sets of circuit of Figure 17-20.

this case, we place the current sources in the cotree. Then we identify the two fundamental cut sets as shown in Figure 17-21 and Figure 17-22. Since the node c is grounded, we set $v_c = 0$. The cut set equations are

$$\text{FCS 1:} \quad -i_1 + i_2 + i_5 = 0$$

$$\text{FCS 2:} \quad i_5 - i_3 - i_4 = 0 \tag{17-3}$$

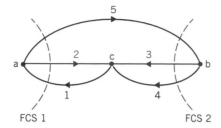

Figure 17-22 Stretched graph, tree (in color), and fundamental cut sets of circuit of Figure 17-20.

Next, we find the branch currents in terms of the node voltages as

$$i_1 = 1 \qquad i_2 = \frac{v_a}{1} \qquad i_3 = \frac{v_b}{2}$$

$$i_4 = 2 \qquad i_5 = \frac{v_a - v_b}{0.5}$$

Then, substituting into Eq. 17-3, we obtain

$$-1 + v_a + 2(v_a - v_b) = 0$$

and

$$2(v_a - v_b) - \frac{v_b}{2} - 2 = 0 \tag{17-4}$$

Rewriting Eq. 17-4, we have

$$3v_a - 2v_b = 1$$

and

$$2v_a - 2.5v_b = 2 \tag{17-5}$$

Solving Eq. 17-5, we obtain the node voltages as

$$v_a = -0.429 \text{ V} \quad \text{and} \quad v_b = -1.143 \text{ V}$$

The topological method of determining the $N - 1$ independent node voltage equations is summarized in Table 17-1.

Table 17-1
Topological Method of Obtaining $N - 1$ Independent Node Voltage Equations

Step 1	Draw a directed graph of the circuit under consideration, selecting the direction of assumed current flow to coincide for current sources.
Step 2	Select the tree of the graph of Step 1 so that current sources are in the cotree and the voltage sources are within the tree, if possible. Also, if possible, select the tree so that at least two branches of the tree are incident at the reference node.
Step 3	Draw the stretched tree on a horizontal line in color, and add the links in black.
Step 4	Draw the fundamental cut set lines on the stretched graph, and write the $N - 1$ fundamental cut set equations.
Step 5	Obtain each of the branch currents in terms of node voltages, and substitute into the fundamental cut set equations, thus obtaining the $N - 1$ independent node voltage equations.

Example 17-4

Obtain the node voltages for the circuit of Figure 17-23 with the reference node assigned as shown.

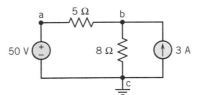

Figure 17-23 Circuit of Example 17-4.

Solution

First we draw the graph of the circuit as shown in Figure 17-24, with the appropriate reference node and noting that v_a is known to be 50 V. Then we select the tree as branches 1 and 2, as shown in Figure 17-24. Note that we include the voltage source within the tree and exclude the current source from the tree. We require only one fundamental cut set equation to determine v_b, the only unknown node voltage.

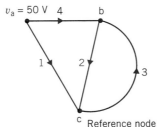

Figure 17-24 Graph and tree for Example 17-4.

We draw the stretched tree on a horizontal line, as shown in Figure 17-25, and identify the fundamental cut set at node b. Then the fundamental cut set equation is

$$i_4 - i_2 + i_3 = 0 \tag{17-6}$$

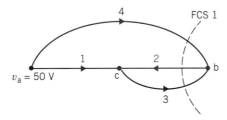

Figure 17-25 Stretched graph and fundamental cut set for Example 17-4.

The branch currents are

$$i_4 = \frac{50 - v_b}{5} \qquad i_2 = \frac{v_b}{8} \qquad i_3 = 3$$

Substituting the branch currents into Eq. 17-6, we have

$$\frac{50 - v_b}{5} - \frac{v_b}{8} + 3 = 0 \tag{17-7}$$

Solving Eq. 17-7, we find that $v_b = 40$ V.

EXERCISE 17-3

Find the graph, tree, cut sets, and node voltages for the circuit of Figure E 17-3 when $G_1 = G_3 = 1$ S and $G_2 = 0.5$ S.

Answers: $v_a = 6$ V, $v_b = 10$ V

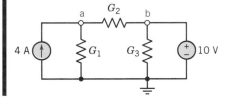

Figure E 17-3

17-7 MESH AND LOOP CURRENT EQUATIONS USING TOPOLOGICAL METHODS

We seek to determine the independent Kirchhoff voltage law equations for complex circuits. If we have selected the tree of a graph of a circuit, then we may add one link at a time, obtaining a loop equation for each link by completing the loop through the tree. Thus, there are $L = B - N + 1$ links and, therefore, L independent loop equations. If we have a planar circuit, we often may judiciously choose the tree so that each link completes a mesh and thus obtain L independent mesh equations.

For example, consider the circuit represented by the graph of Figure 17-26. We have added assumed directions for numbered branch currents, as shown for the tree and cotree

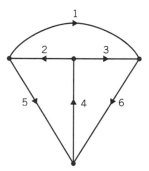

Figure 17-26 Graph of a circuit.

in Figure 17-27. This selected tree will result in three fundamental mesh KVL equations as we connect each link, in turn, to the tree. For example, we show fundamental loop 1 (FL 1)

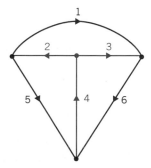

Figure 17-27 Tree (in color) and cotree for graph of Figure 17-26.

in Figure 17-28a, obtained by connecting link 1 to the tree. We sum the voltages around the loop (in clockwise direction) obtaining

$$v_1 - v_3 + v_2 = 0 \tag{17-8}$$

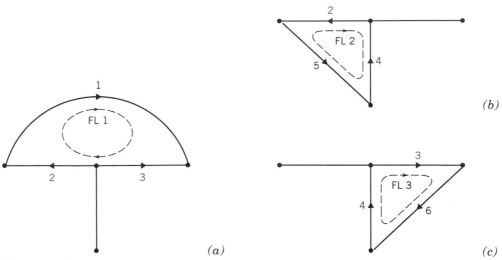

Figure 17-28 The three fundamental loops of the tree of the graph of Figure 17-27.

Similarly, we connect link 5 to obtain fundamental loop 2, as shown in Figure 17-28b, and obtain the KVL equation as

$$-v_2 - v_4 - v_5 = 0 \tag{17-9}$$

Finally, connecting link 6 to the tree, as shown in Figure 17-28c, and writing the KVL equation, we have

$$v_3 + v_6 + v_4 = 0 \tag{17-10}$$

Equations 17-8, 17-9, and 17-10 comprise the three independent mesh equations for the graph of Figure 17-26.

A summary of the topological method of determining the L independent mesh equations is provided in Table 17-2.

Table 17-2
Topological Method of Obtaining L Independent Mesh Equations

Step 1	Draw a directed graph of the circuit under consideration, selecting the direction of assumed current flow to coincide for current sources. Determine if the graph is planar. If so, proceed to Step 2; otherwise, go to Table 17-3.
Step 2	Identify the $N - 1$ meshes. Select a tree of the graph of Step 1 so that one link is removed from each mesh. Exclude all current sources from the tree.
Step 3	Add one link to the tree, creating a fundamental loop, and write a KVL equation for this fundamental loop. Repeat for each additional link until L mesh equations are obtained.

Now, let us consider the development of L independent loop equations. For example, if we choose another tree for the graph of Figure 17-26, we have three links, and therefore we will determine a set of three loop equations. For example, we will select the tree shown in Figure 17-29. We then add to the tree one link at a time to create the three fundamental loops.

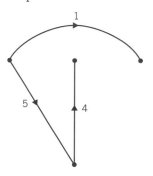

Figure 17-29 A different tree of the graph of Figure 17-26.

Adding link 2 to the tree, we have fundamental loop 1, as shown in Figure 17-30a, and the KVL equation is

$$-v_4 - v_5 - v_2 = 0 \qquad (17\text{-}11)$$

Repeating for link 3, we have fundamental loop 2, as shown in Figure 17-30b. The KVL equation is

$$v_1 - v_3 - v_4 - v_5 = 0 \qquad (17\text{-}12)$$

Finally, adding link 6 to the tree, we have fundamental loop 3 as shown in Figure 17-30c. The KVL equation is

$$v_1 + v_6 - v_5 = 0 \qquad (17\text{-}13)$$

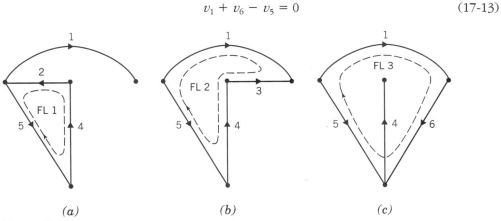

(a) (b) (c)

Figure 17-30 The three fundamental loops for the tree of Figure 17-29.

The topological approach to obtaining the L independent loop equations of a circuit is summarized in Table 17-3.

Table 17-3
Topological Method of Obtaining L Independent Loop Equations

Step 1	Draw a directed graph of the circuit under consideration, selecting the direction of assumed current flow to coincide for current sources.
Step 2	Select a tree of the graph of Step 1 while excluding current sources from the tree.
Step 3	Add one link to the tree, creating a fundamental loop, and write a KVL equation for this fundamental loop. Repeat for each additional link until L loop equations are obtained.

Example 17-5

Select a tree, and identify the four fundamental loops for the graph of the nonplanar circuit of Figure 17-4.

Solution

First, we redraw the graph and identify a tree (in color) as shown in Figure 17-31. There are four links in the cotree, and therefore, we seek four fundamental loops. If we reinsert link

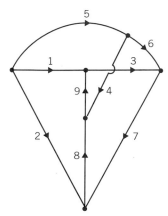

Figure 17-31 Graph for nonplanar circuit of Example 17-5.

1 into the tree, we obtain fundamental loop 1 as shown in Figure 17-32a. As we insert one link at a time, we obtain each fundamental loop as shown in Figure 17-32b, c, and d.

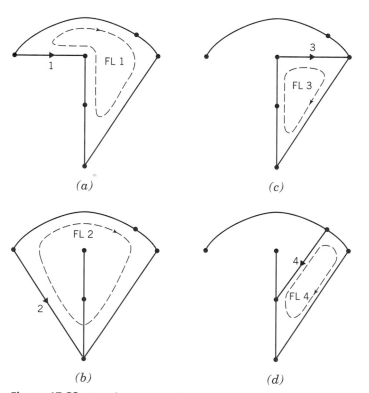

(a)

(c)

(b)

(d)

Figure 17-32 Four fundamental loops for Example 17-5.

EXERCISE 17-4

Identify a tree, and find the fundamental loops for the graph of Figure E 17-4.

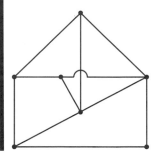

Figure E 17-4

17-8 STATE EQUATIONS USING TOPOLOGICAL METHODS

We often wish to obtain the state vector differential equation for a circuit

$$\dot{\mathbf{x}} = \mathbf{A}\mathbf{x} + \mathbf{B}\mathbf{u} \qquad (17\text{-}14)$$

where $\mathbf{x}$ is a vector consisting of the capacitor voltages and the inductor currents (see Sections 10.5 and 14.10) and $\dot{\mathbf{x}} = d\mathbf{x}/dt$. A topological method of obtaining the state variable equations using a circuit graph is summarized in Table 17-4. The goal is to obtain a set of first-order differential equations for the state variables.

Table 17-4
Topological Method of Obtaining the State Variable Equations for a Circuit

Step 1	Draw a graph of the circuit, and select the direction of assumed current flow in each element. Define the state variables as the capacitor voltages and the inductor currents. Find the number of state variables N, where N = the sum of independent capacitors and inductors.
Step 2	Select a tree so that it contains all the capacitors and none of the inductors. If possible, include the voltage sources in the tree and exclude the current sources.
Step 3	Write a fundamental cut set equation for a node of each capacitor and a fundamental loop equation for each inductor.
Step 4	Rearrange the equations of Step 3 so as to obtain N first-order differential equations.

Consider the series RLC circuit of Figure 17-33. We identify the state variables as v_c and i_L, and we seek $N = 2$ first-order differential equations. We select a tree (in color) exclud-

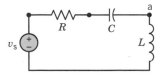

Figure 17-33 Series RLC circuit.

ing the inductor and including the capacitor and the voltage source, as shown in Figure 17-34. This enables us to obtain an equation for $C\,dv/dt$.

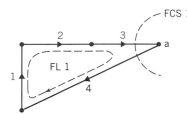

FCS 1

Figure 17-34 Graph and tree (in color) of the *RLC* circuit. FCS 1 is also shown.

The fundamental cut set KCL equation is

$$i_3 - i_4 = 0 \tag{17-15}$$

The fundamental loop KVL equation is

$$v_1 + v_2 + v_3 + v_4 = 0 \tag{17-16}$$

Since $i_3 = i_c = i_4 = i_L$, Eq. 17-15 may be written as

$$C\frac{dv_c}{dt} = i_L \tag{17-17}$$

Since $v_1 = -v_s$, $v_2 = i_2 R$, $v_3 = v_c$, and $v_4 = v_L$, we may write Eq. 17-16 as

$$-v_s + Ri_L + v_c + L\frac{di_L}{dt} = 0 \tag{17-18}$$

(We use $v_L = L\,di_L/dt$ and $i_2 = i_4 = i_L$.)

Rearranging Eqs. 17-17 and 17-18, we obtain

$$\frac{dv_c}{dt} = \frac{i_L}{C} \tag{17-19}$$

and

$$\frac{di_L}{dt} = -\frac{R}{L}i_L - \frac{1}{L}v_c + \frac{1}{L}v_s$$

Using $x_1 = v_c$, $x_2 = i_L$, and $u = v_s$, we have

$$\dot{\mathbf{x}} = \begin{bmatrix} 0 & \dfrac{1}{C} \\[2ex] -\dfrac{1}{L} & -\dfrac{R}{L} \end{bmatrix} \mathbf{x} + \begin{bmatrix} 0 \\[1ex] \dfrac{1}{L} \end{bmatrix} u \tag{17-20}$$

Example 17-6

Obtain the state variable differential equations for the circuit of Figure 17-35.

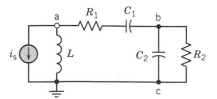

Figure 17-35 Circuit for Example 17-6.

Solution

First, we draw a graph of the circuit as shown in Figure 17-36. We choose a tree including both capacitors and excluding the inductor (elements) as shown in Figure 17-37.

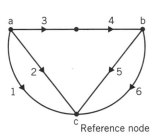

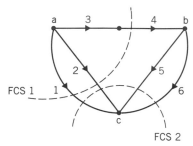

Figure 17-36 Graph of circuit for Example 17-6.

Figure 17-37 Tree (in color) and links of circuit of Example 17-6.

We choose the two fundamental cut sets in order to obtain a cut set equation for each capacitor. Then we have

$$\text{FCS 1:} \quad i_4 + i_2 + i_1 = 0 \tag{17-21}$$

$$\text{FCS 2:} \quad i_5 + i_6 + i_1 + i_2 = 0 \tag{17-22}$$

Using the fundamental loop of Figure 17-38, which incorporates the inductor, we have

$$\text{FL:} \quad v_3 + v_4 + v_5 - v_2 = 0 \tag{17-23}$$

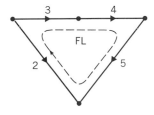

Figure 17-38 Fundamental loop of circuit of Example 17-6.

Noting that $i_4 = C_1\, dv_{c1}/dt$, $i_5 = C_2\, dv_{c2}/dt$, $v_4 = v_{c1}$, $v_5 = v_{c2}$, and $v_2 = L\, di_L/dt$, we rearrange Eqs. 17-21, 17-22, and 17-23 to obtain

$$C_1 \frac{dv_{c1}}{dt} + i_L + i_s = 0 \tag{17-24}$$

$$C_2 \frac{dv_{c2}}{dt} + \frac{v_{c2}}{R_2} + i_s + i_L = 0 \tag{17-25}$$

$$L \frac{di_L}{dt} = -R_1(i_s + i_L) + v_{c1} + v_{c2} \tag{17-26}$$

Let $v_{c1} = x_1$, $v_{c2} = x_2$, $i_L = x_3$, and $u = i_s$. Then we obtain the differential equation in matrix form as

$$\dot{\mathbf{x}} = \begin{bmatrix} 0 & 0 & \dfrac{-1}{C_1} \\[2mm] 0 & \dfrac{-1}{R_2 C_2} & \dfrac{-1}{C_2} \\[2mm] \dfrac{1}{L} & \dfrac{1}{L} & \dfrac{-R_1}{L} \end{bmatrix} \mathbf{x} + \begin{bmatrix} \dfrac{-1}{C_1} \\[2mm] \dfrac{-1}{C_2} \\[2mm] \dfrac{-R_1}{L} \end{bmatrix} u \tag{17-27}$$

EXERCISE 17-5

Obtain the state variable differential equations for the circuit of Figure E 17-5 when $x_1 = v_c$, $x_2 = i_L$, and $u = v_s$.

Partial Answer: $\dot{x}_1 = \dfrac{-1}{R_1 C} x_1 - \dfrac{1}{C} x_2 + \dfrac{1}{R_1 C} u$

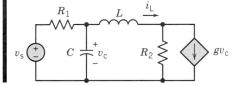

Figure E 17-5

17-9 SENSITIVITY ANALYSIS

A physical circuit is unable to perform exactly according to design in view of several practical considerations such as tolerance of circuit elements, environmental effects (temperature, humidity, radiation, interference, etc.), and aging. A good design must therefore include an analysis of the effects of element parameter variations on the circuit performance. We are usually interested in the circuit performance as represented by the output voltage v_o, the voltage gain v_o/v_{in}, the quality factor Q, or some other important performance measure. The goal is to adjust the design to minimize the effects of parameter variations.

We define the *sensitivity* of a circuit as the differential change of the gain ratio v_o/v_{in} with respect to the differential change in the parameter, p. Then, when $T = v_o/v_{in}$ we have

$$S_p^T = \frac{dT}{dp} \tag{17-28}$$

The *normalized sensitivity* is defined as

$$nS_p^T = \frac{dT}{(dp/p) \times 100} = \frac{dT \cdot p}{dp \cdot 100} \tag{17-29}$$

where $(dp/p) \times 100$ is the percentage change in the parameter.

The *relative sensitivity* is defined as

$$rS_p^T = \frac{dT/T}{dp/p} = \frac{dT}{dp} \frac{p}{T} \tag{17-30}$$

Example 17-7

The transfer function of a resistive voltage divider shown in Figure 17-39 is

$$\frac{v_o}{v_{in}} = T = \frac{R_2}{R_1 + R_2}$$

Find the required tolerance of resistor R_2 when it is required that the allowable change in the transfer function is $\Delta T/T \leq 0.0067$ (0.67%) when the nominal design value of $R_1 = 2\ \Omega$ and $R_2 = 1\ \Omega$.

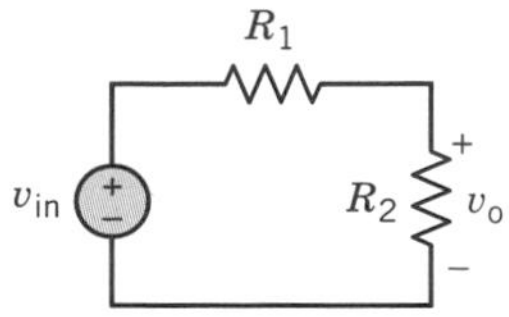

Figure 17-39 A voltage divider circuit.

Solution

The relative sensitivity is

$$rS^T_{R_2} = \frac{dT}{dR_2}\frac{R_2}{T}$$

$$= \frac{R_1}{(R_1 + R_2)^2}\frac{R_2}{T}$$

$$= \frac{R_1}{R_1 + R_2}$$

Therefore, the required tolerance is obtained from the incremental sensitivity equation as

$$\frac{\Delta T}{T} = rS^T_{R_2}\frac{\Delta R_2}{R_2}$$

and at the nominal values, we have

$$\frac{\Delta T}{T} = \frac{2}{3}\frac{\Delta R_2}{R_2}$$

Therefore, to obtain $\Delta T/T = 0.0067$, we require

$$\frac{\Delta R_2}{R_2} = \frac{3}{2}\frac{\Delta T}{T}$$

or $\Delta R_2/R_2 \le 0.01$. Thus, we require a resistor R_2 with a 1 percent tolerance.

Example 17-8

Determine the sensitivity and normalized sensitivity for R_1 for the voltage divider of Figure 17-39 when $R_1 = 3\ \Omega$ and $R_2 = 1\ \Omega$.

Solution

The sensitivity with respect to R_1 is

$$S^T_{R_1} = \frac{dT}{dR_1} = \frac{-R_2}{(R_1 + R_2)^2} = -0.0625$$

The normalized sensitivity with respect to R_1 is

$$nS^T_{R_1} = \frac{dT}{dR_1}\frac{R_1}{100} = \frac{-R_2 R_1}{(R_1 + R_2)^2 100} = \frac{-3}{16(100)} = -1.875 \times 10^{-3}$$

Finally, it may be relatively easy to generate a computer plot of the change of v_o for ± 10 percent changes in resistors. A plot of the change in v_o for several resistors of the circuit shown in Figure 17-40a is shown in Figure 17-40b. It is easy to see that the designer should focus on ensuring that R_3, R_6, R_5, and R_1 do not change while placing less emphasis on R_2 and R_4.

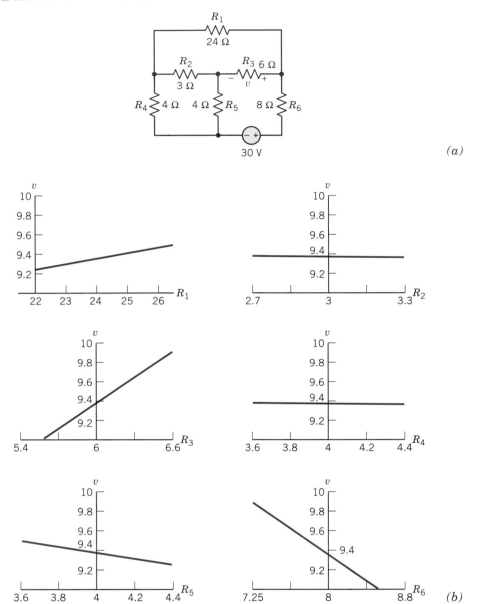

Figure 17-40 (a) A resistive circuit and (b) the sensitivity (variation) of v for each resistance.

EXERCISE 17-6

The transfer function of a low-pass filter circuit (Figure 13-6) is

$$H(s) = \frac{1}{s\tau + 1}$$

where $\tau = RC$. Find (a) the relative sensitivity rS_C^H where $H = |H(j\omega)|$ and (b) the relative sensitivity when $R = 1\ k\Omega$, $C = 1\ mF$, and $\omega = 0.5\ rad/s$.

Answers: (a) $rS_C^H = \dfrac{-(\omega\tau)^2}{1 + (\omega\tau)^2}$ (b) $rS_C^H = -0.2$

17-10 | CALCULATING SENSITIVITY USING PSpice

PSpice is able to calculate the sensitivity and normalized sensitivity for changes in circuit parameters.

A sensitivity analysis is one in which PSpice performs calculations to determine the dc sensitivity of the circuit output values to changes in element parameters. That is, it attempts to answer the question, "If I change the value of this component, what effect will it have on the output voltage or current?" In this way, sensitivity analysis can be used to "fine tune" a circuit design.

The **.SENS** statement specifies which of the outputs is to be "sensitized." PSpice will determine how a change in any individual element value will change the sensitized output. The format of the **.SENS** statement is simply

$$.SENS < output\ value > \cdots$$

The *output value* is in the same form as specified for the **.PRINT** statement. Any number of outputs may be sensitized.

As an example, consider the three-resistor circuit of Figure 17-41. By inserting the statement **.SENS** V(2,3) in the circuit description file, PSpice will calculate the change in

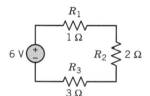

Figure 17-41 A three-resistor circuit.

the voltage between nodes 2 and 3 with respect to changes in all other circuit element parameters. Figure 17-42 contains a portion of the output file generated when this sensitivity analysis is run. Under the column labeled ELEMENT SENSITIVITY, the output shows, on a volts/unit basis (where the units are those pertinent to the element type), how

```
Sensitivity Analysis of a 3-Resistor Circuit

****        DC SENSITIVITY ANALYSIS    TEMPERATURE =   27.000 DEG C

**********************************************************************

DC SENSITIVITIES OF OUTPUT V(2,3)

            ELEMENT        ELEMENT         ELEMENT         NORMALIZED
            NAME           VALUE           SENSITIVITY     SENSITIVITY
                                           (VOLTS/UNIT)    (VOLTS/PERCENT)

            R1             1.000E+00       -3.333E-01      -3.333E-03
            R2             2.000E+00        6.667E-01       1.333E-02
            R3             3.000E+00       -3.333E-01      -1.000E-02
            Vsource        6.000E+00        3.333E-01       2.000E-02
```

Figure 17-42 Sensitivity analysis for the three-resistor circuit.

variations in each of the resistor and voltage source values will change V(2,3). Values in the column labeled NORMALIZED SENSITIVITY are calculated by multiplying the former column values by the element values and then dividing by 100 in order to obtain a percentage change.

17-11 VERIFICATION EXAMPLE

Example 17V-1

A student is studying the circuit of Figure 17V-1. His report states that the number of links is $L = 6$. Verify this result.

Figure 17V-1 Circuit under study.

Solution

A first step is to draw the graph of the circuit as shown in Figure 17V-2. This graph has four nodes, a, b, c, and d, and thus $N = 4$. The graph has nine branches ($B = 9$). Thus the

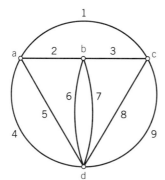

Figure 17V-2 Graph of circuit of Figure 17V-1.

number of links is $L = B - N + 1 = 9 - 4 + 1 = 6$. We can further verify this by drawing a tree as shown in Figure 17V-3 and its cotree as shown in Figure 17V-4. The number of branches of the cotree is equal to the number of links, and thus we again verify that $L = 6$.

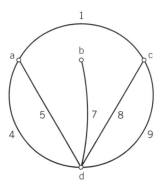

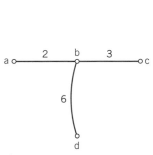

Figure 17V-3 One tree of the graph of Figure 17V-2.

Figure 17V-4 The cotree of the tree of Figure 17V-3.

17-12 DESIGN CHALLENGE SOLUTION

CURRENT SOURCE WITH A VOLTAGE CONSTRAINT

A circuit is shown in Figure 17D-1. We wish to select R_1, R_2, and R_3 so that the voltage across the current source i_s is limited to 8 V. We wish to determine the number of node equations required versus the number of mesh equations required. This will enable us to select between the node voltage method and the mesh current method. Using the appropriate method we can calculate the voltage across i_s and then select the required resistors.

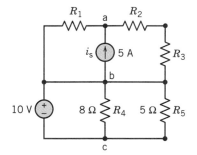

Figure 17D-1 Circuit with current source i_s.

Define the Situation, State the Assumptions, and Develop a Model

1 The current source is common to two meshes.
2 The voltage $v_{bc} = 10$ V is known.

The Goal

Determine whether to use node voltage or mesh current analysis, and then determine R_1, R_2, and R_3 so that the voltage across the current source is no more than 8 V.

Generate a Plan

1 Let $R_3 + R_2 = R_8$, where R_8 is one branch.
2 Develop a graph for the network.
3 Find a suitable tree that excludes the current source.
4 Find the number of equations required for each method.
5 Determine R_1, R_2, and R_3 so that the voltage across i_s is no more than 8 V.

Take Action Using the Plan

1 Let $R_3 + R_2 = R_8$ to obtain a branch R_8.

2 Develop a graph as shown in Figure 17D-2. This graph shows the geometry of the circuit and the central nature of the three nodes.

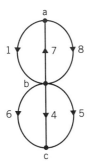

Figure 17D-2 Graph of the circuit of Figure 17D-1.

3 The graph has six branches and three nodes.

4 Draw a tree, as shown in Figure 17D-3, that excludes the current source and includes the voltage source.

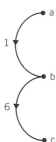

Figure 17D-3 Tree for the graph of Figure 17D-2.

5 The number of links of the graph is $L = 4$. Therefore, we need four mesh equations with four unknowns.

6 We set c as the reference node; then we have two other nodes. Therefore, we would require $N - 1 = 2$ equations. However, $v_b = v_6 = 10$ V is known, so we require only one node equation. Therefore, we select the node voltage method.

We will use the circuit diagram as shown in Figure 17D-4, where $R_{p1} = R_1 \parallel R_8$ and $R_{p2} = R_4 \parallel R_5$. At node a we have

$$\frac{v_a - 10}{R_{p1}} - 5 = 0$$

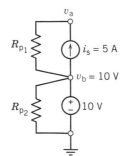

Figure 17D-4 Circuit with current source used to determine R_{p1}.

Since $\mid v_a - 10 \mid \, \leq 8$, we select $v_a = 18$ V. Then,

$$\frac{8}{R_{p1}} = 5$$

Therefore, we require $R_{p1} = 1.6\ \Omega$. One solution is to let $R_1 = 2\ \Omega$ and solve for R_8. Then, we require

$$\frac{8}{5} = \frac{2R_8}{2 + R_8}$$

and $R_8 = 8\ \Omega$.

SUMMARY

Many circuits are complex and require an orderly method for finding node or loop equations accounting for the structure of the circuit. Graphs are used to describe the geometrical properties of a circuit. Then, using a graph and a subgraph called a tree, we can develop systematic methods for determining the equations that describe a circuit.

There are $N - 1$ independent node voltage equations, where $N =$ number of nodes. We can find the $N - 1$ equations by drawing the graph of the circuit and selecting a tree where the current sources lie in the cotree and the voltage sources lie within the tree. Then we can identify and write the fundamental equations leading to the node voltage equations.

For a planar circuit, we can use mesh currents to describe the circuit behavior. We select a tree so that one link is removed from each mesh and all current sources are excluded from the tree. Then, adding one link at a time to the tree, creating a fundamental loop, we proceed to obtain L mesh equations, where L is the number of links.

For a nonplanar or complicated circuit, we can use a circuit graph and a selected tree to determine systematically the L loop equations. We add one link at a time to create a fundamental loop and thus obtain the L equations.

We can use the topological properties of a circuit to obtain the state variable differential equations. We select the capacitor voltages and inductor currents as the state variables. We select a tree incorporating the capacitors and excluding the inductors. We obtain a cut set equation for a node of each capacitor and a fundamental loop equation for each inductor.

Finally, we defined the sensitivity of a circuit as the relative change in a performance variable for a proportional change in a parameter.

TERMS AND CONCEPTS

Cotree   Links together with nodes constitute a complementary tree (cotree).

Cut Set   A minimum set of elements that, when cut, separates the graph into two parts.

Fundamental Cut Set   A cut set that contains one, and only one, tree branch.

Fundamental Loop   A loop constructed by adding a link to a path in the tree.

Graph   A set of nodes together with a set of branches. A set of interconnected branches.

Link   Branches not in a chosen tree.

Sensitivity The proportional change of a performance measure in response to a proportional change in a circuit parameter.

Topology The science of placement of elements.

Tree Any connected set of branches of a graph that connects every node to every other node directly or indirectly without forming any closed path.

REFERENCES   Chapter 17

Bode, Hendrik W., *Network Analysis and Feedback Amplifier Design,* Van Nostrand, New York, 1945.

Eslami, Mansour, "Theory of Sensitivity of Networks," *IEEE Transactions on Education,* August 1989, pp. 319–334.

PROBLEMS

Section 17-3   Network Graphs

P 17.3-1 Draw a graph of the network shown in Figure P 17.3-1.

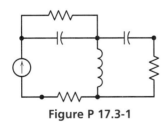

Figure P 17.3-1

P 17.3-2 Determine if the graphs of Figure P 17.3-2*a* and *b* are planar or nonplanar.

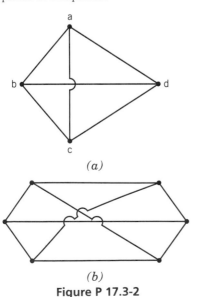

(a)

(b)

Figure P 17.3-2

Section 17-4   Trees

P 17.4-1 Consider the graph of a circuit shown in Figure P 17.4-1. Identify the trees of this graph. Determine the number of links of this graph.

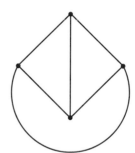

Figure P 17.4-1

P 17.4-2 Identify the 16 trees in the graph of Figure P 17.4-2.

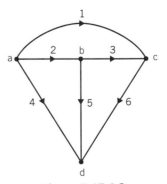

Figure P 17.4-2

Section 17-5   Directed Branches and Cut Sets

P 17.5-1   Figure P 17.5-1 shows the graph of Figure 17-12 with a selected tree (in color). Identify the fundamental cut sets of the graph.

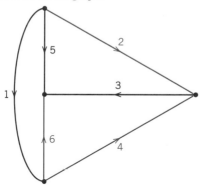

Figure P 17.5-1

P 17.5-2   Consider the bridge circuit of Figure P 17.5-2. Using node d as the reference (ground), determine the graph, select a tree, find the cut set equations, and determine v_a.

Answer: $v_a = 7.58$ V

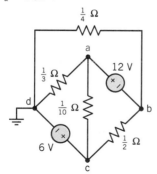

Figure P 17.5-2 A bridge circuit.

Section 17-6   Independent Node Voltage Equations

P 17.6-1   A circuit with a dependent source is shown in Figure P 17.6-1. Use node d as the reference, and use the topological method to find v_a, v_b, and v_c.

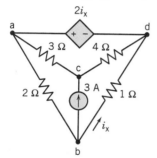

Figure P 17.6-1

P 17.6-2   The circuit of Figure P 17.6-2 has $i_s(t) = 7 \cos t$ A and $v_s(t) = 17 \cos(t + 76°)$ V. Use the topological method to find $\mathbf{I_2}$.

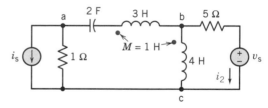

Figure P 17.6-2

P 17.6-3   Using the topological approach with node equations, find the voltage v_c and the current i_x of the circuit shown in Figure P 17.6-3.

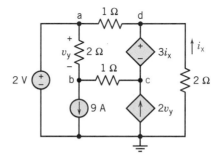

Figure P 17.6-3

Section 17-7   Mesh and Loop Current Equations Using Topological Methods

P 17.7-1   Consider the circuit of Figure P 17.5-2 and determine an appropriate graph. Identify the fundamental loops, and determine the current in the $\frac{1}{2}$-Ω resistor.

P 17.7-2   A circuit is shown in Figure P 17.7-2. (a) Determine if the circuit is planar. (b) Draw a tree, and (c) using fundamental loops, determine the current in the 2-Ω resistor. (d) Compare the number of node equations versus the number of loop equations required.

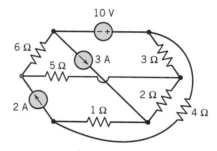

Figure P 17.7-2

P 17.7-3   For the circuit of Figure P 17.6-1, find the fundamental loop equations and determine i_x and v_c.

P 17.7-4 Repeat Problem 17.6-3 using the topological approach with loop or mesh equations.

P 17.7-5 Using topological methods for (a) mesh analysis and (b) node analysis, determine v_a in Figure P 17.7-5.

Answer: 6 V

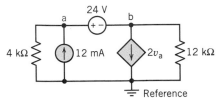

Figure P 17.7-5

P 17.7-6 A circuit is shown in Figure P 17.7-6. Choose a tree that includes the voltage source but excludes the current source, and write the loop equations. Determine the current i in the 1-Ω resistor.

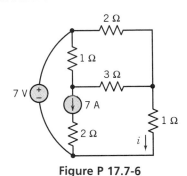

Figure P 17.7-6

Section 17-8 State Equations Using Topological Methods

P 17.8-1 Determine the state variable differential matrix equation for the circuit of Figure P 17.8-1.

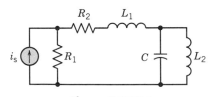

Figure P 17.8-1

P 17.8-2 Determine the state variable differential matrix equation for the circuit of Figure P 17.8-2.

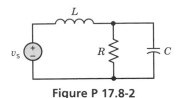

Figure P 17.8-2

P 17.8-3 Determine the state variable differential matrix equation for the circuit of Figure P 17.8-3.

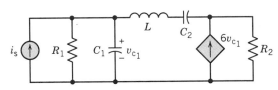

Figure P 17.8-3

Section 17-9 Sensitivity Analysis

P 17.9-1 Find the transfer function $T(j\omega)$, and evaluate the sensitivity $S_{R_1}^T$ at $R_1 = 1$ kΩ and $\omega = 0.5$ rad/s of the circuit shown in Figure P 17.9-1 when $C = 1$ mF and $R_2 = 2$ kΩ. (Note that $T = | T(j\omega) |$.)

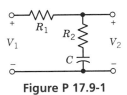

Figure P 17.9-1

P 17.9-2 The transfer function of a resistive voltage divider is

$$T = \frac{R_2}{R_1 + R_2}$$

(a) Assuming a nominal design of $R_1 = 2$ Ω and $R_2 = 1$ Ω, find $rS_{R_1}^T$ and $rS_{R_2}^T$.

(b) The maximum allowable variation in T is ± 1.5 percent. Find the required tolerances of the two resistors.

PSpice PROBLEMS

SP 17-1 Write a PSpice program and determine the sensitivities for Example 17-8.

SP 17-2 Use PSpice to determine the sensitivities of the circuit of Figure 17-39 when $R_1 = 3$ Ω and $R_2 = 1$ Ω.

VERIFICATION PROBLEMS

VP 17-1 A student notebook states that the graph of Figure VP 17-1 has eight trees. Can you verify this and name the trees?

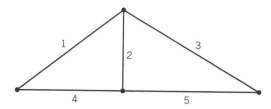

Figure VP 17-1

VP 17-2 Using the method of Section 17-6 a student analyzes the circuit of Figure VP 17-2 and obtains $v_a = 30$ V and $v_b = 27.1$ V. Verify these results.

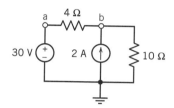

Figure VP 17-2

VP 17-3 A student report states that the circuit shown in Figure VP 17-3 has three fundamental loops. Can you verify this result and identify the fundamental loops?

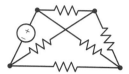

Figure VP 17-3

DESIGN PROBLEMS

DP 17-1 For the circuit in Figure DP 17-1 determine the required i_s so that the voltage across R_1 is zero. Select a tree excluding the current sources, and determine the fundamental cut sets.

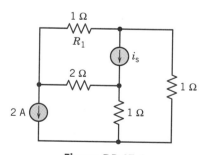

Figure DP 17-1

DP 17-2 A circuit with a nonideal op amp is shown in Figure DP 17-2. Select a tree incorporating the input voltage source and the dependent source. Determine v_o/v, and select an appropriate R_2 and R_1 so that $v_o/v = 10$ when $A = 10^4$, $R_i = 1$ MΩ, and $R_o = 10$ Ω.

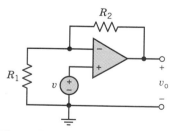

Figure DP 17-2 Op amp circuit.

CHAPTER 18

TWO-PORT AND THREE-PORT NETWORKS

Problems

PSpice Problems

Verification Problems

Design Problems

PREVIEW

Many practical circuits have two ports of access called the input and output ports. It is one purpose of this chapter to describe the parameters that can represent two-port networks. Additionally, we will study the equations that describe two special three-port passive networks called the T and Π networks. Finally, we will discuss the relationships between the six sets of parameters that describe the two-port network and obtain the equations for a network consisting of two-port networks connected in series, parallel, and cascade.

18-1 DESIGN CHALLENGE

TRANSISTOR AMPLIFIER

Figure 18D-1 shows the small signal equivalent circuit of a transistor amplifier. The data sheet for the transistor describes the transistor by specifying its h parameters to be

$$h_{ie} = 1250 \ \Omega, \quad h_{oe} = 0, \quad h_{fe} = 100, \quad \text{and} \quad h_{re} = 0$$

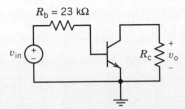

Figure 18D-1 A transistor amplifier.

The value of the resistance R_c must be between 300 Ω and 5000 Ω to ensure that the transistor will be biased correctly. The small signal gain is defined to be

$$A_v = \frac{v_o}{v_{in}}$$

The challenge is to design the amplifier so that

$$A_v = -20$$

(There is no guarantee that these specifications can be satisfied. Part of the problem is determining whether it is possible to design this amplifier so that $A_v = -20$.)

To solve this problem, we need to define h parameters that describe the transistor. We will return to this problem at the end of the chapter.

18-2 ‖ AMPLIFIERS AND FILTERS

Continued improvements in vacuum tubes and amplifier circuits made long-distance telephone lines possible, and in 1915 Alexander Graham Bell placed the first transcontinental telephone call to his famous assistant Thomas Watson. In 1921 Bell Telephone was experimenting with a 1000-mile telephone line with three voice channels, but the nonlinearity of the best tubes introduced in the numerous repeaters an intolerable amount of distortion.

Transmission of messages over long lines required the insertion of amplifiers at points along the line. These devices have two terminals for input and two terminals for output to the line. Harold S. Black, a 23-year-old engineer at Bell Laboratories, concluded that, in a rapidly growing country 4000 miles wide, a new approach would be required. First he tried to improve the amplifier tubes, but he decided that the necessary 1000-fold reduction in distortion could never be achieved that way. After hearing an inspiring talk by Charles Steinmetz, he clearly stated his problem: how to remove all the distortion from an imperfect amplifier. His first scheme was to compare the output (suitably reduced) to the input, amplify the difference (distortion) separately, and use it to cancel the distortion in the actual output. Within one day he had built a working *feed forward* amplifier.

But this 1923 invention required very precise balances and subtractions. For example, every hour on the hour somebody had to adjust the filament current to the tubes. For four years, Black struggled, and failed, to turn his idea into a practical amplifier. As he related 50 years later in the December 1977 IEEE *Spectrum:*

Then came the morning of Tuesday, August 2, 1927, when the concept of the negative feedback amplifier came to me in a flash while I was crossing the Hudson River on the Lackawanna Ferry, on my way to work. I suddenly realized that if I fed the amplifier output back to the input, in reverse phase, I would have exactly what I wanted. . . . On a page of the *New York Times,* I sketched a simple diagram . . . and the equations for amplification with feedback.

Four months later, his goal was surpassed when a 100,000-to-1 reduction of distortion was realized in a practical one-stage amplifier. Now Black's negative feedback principle is applied in practically all amplifiers.

18-3 ▌ TWO-PORT NETWORKS

Many practical circuits have just two *ports* of access, that is, two places where signals may be input or output. For example, a coaxial cable between Boston and San Francisco has two ports, one at each of those cities. The object here is to analyze such networks in terms of their terminal characteristics without particular regard to the internal composition of the network. To this end, the network will be described by relationships between the port voltages and currents.

We study two-ports and the parameters that describe them for a number of reasons. Most circuits or systems have at least two ports. We may put an input signal into one port and obtain an output signal from the other. The parameters of the two-port network completely describe its behavior in terms of the voltage and current at each port. Thus, knowing the parameters of a two-port network permits us to describe its operation when it is connected into a larger network. Two-port networks are also important in modeling electronic devices and system components. For example, in electronics, two-port networks are employed to model transistors, op amps, transformers, and transmission lines.

A two-port network is represented by the network shown in Figure 18-1. A four-terminal network is called a *two-port network* when the current entering one terminal of a pair exits the other terminal in the pair. For example, I_1 enters terminal a and exits terminal b of the input terminal pair a–b. It will be assumed in our discussion that there are no independent sources or nonzero initial conditions within the linear two-port network. Two-port networks may or may not be purely resistive and can in general be formulated in terms of the *s*-variable or the $j\omega$-variable.

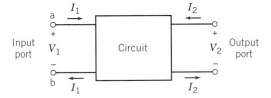

Figure 18-1 A two-port network.

A **two-port** network has two access points appearing as terminal pairs. The current entering one terminal of a pair exits the other terminal in the pair.

18-4 ▌ T-to-Π TRANSFORMATION AND TWO-PORT THREE-TERMINAL NETWORKS

Two networks that occur frequently in circuit analysis are the T and Π networks as shown in Figure 18-2. When redrawn, they can appear as the Y or delta (Δ) networks of Figure 18-3.

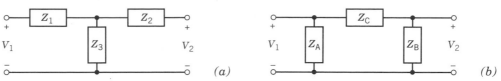

Figure 18-2 (a) T network and (b) Π network.

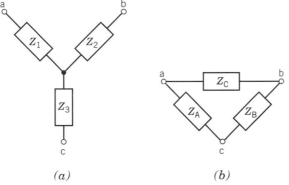

Figure 18-3 (a) Y network and (b) Δ network.

If a network has mirror-image symmetry with respect to some centerline, that is, if a line can be found to divide the network into two symmetrical halves, the network is a *symmetrical network*. The T network is symmetrical when $Z_1 = Z_2$, and the Π network is symmetrical when $Z_A = Z_B$. Furthermore, if all the impedances in either the T or Π network are equal, then the T or Π network is completely symmetrical.

Note that the networks shown in Figure 18-2 and Figure 18-3 have two access ports and three terminals. For example, one port is obtained for the terminal pair a–c and the other port is b–c.

We can obtain equations for direct transformation or conversion from a T network to a Π network, or from a Π network to a T network, by considering that for equivalence the two networks must have the same impedance when measured between the same pair of terminals. For example, at port 1 (at a–c) for the two networks of Figure 18-2, we require

$$Z_1 + Z_3 = \frac{Z_A(Z_B + Z_C)}{Z_A + Z_B + Z_C}$$

To convert a Π network to a T network, relationships for Z_1, Z_2, and Z_3 must be obtained in terms of the impedances Z_A, Z_B, and Z_C. With some algebraic effort we can show that

$$Z_1 = \frac{Z_A Z_C}{Z_A + Z_B + Z_C} \tag{18-1}$$

$$Z_2 = \frac{Z_B Z_C}{Z_A + Z_B + Z_C} \tag{18-2}$$

$$Z_3 = \frac{Z_A Z_B}{Z_A + Z_B + Z_C} \tag{18-3}$$

Similarly, we can obtain the relationships for Z_A, Z_B, and Z_C as

$$Z_A = \frac{Z_1Z_2 + Z_2Z_3 + Z_3Z_1}{Z_2} \tag{18-4}$$

$$Z_B = \frac{Z_1Z_2 + Z_2Z_3 + Z_3Z_1}{Z_1} \tag{18-5}$$

$$Z_C = \frac{Z_1Z_2 + Z_2Z_3 + Z_3Z_1}{Z_3} \tag{18-6}$$

Each T impedance equals the product of the two adjacent legs of the Π network divided by the sum of the three legs of the Π network. On the other hand, each leg of the Π network equals the sum of the possible products of the T impedances divided by the opposite T impedance.

When a T or a Π network is completely symmetrical, the conversion equations reduce to

$$Z_T = \frac{Z_\Pi}{3} \tag{18-7}$$

and

$$Z_\Pi = 3Z_T \tag{18-8}$$

where Z_T is the impedance in each leg of the T network and Z_Π is the impedance in each leg of the Π network.

Example 18-1

Find the Π form of the T circuit given in Figure 18-4a.

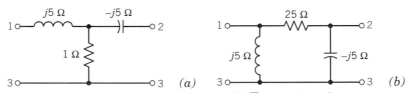

Figure 18-4 (*a*) T circuit of Example 18-1. (*b*) Π equivalent of T circuit.

Solution

The first impedance of the Π network, using Eq. 18-4, is

$$
\begin{aligned}
Z_A &= \frac{Z_1Z_2 + Z_2Z_3 + Z_3Z_1}{Z_2} \\
&= \frac{j5(-j5) + (-j5)1 + 1(j5)}{-j5} \\
&= j5\ \Omega
\end{aligned}
$$

Similarly, the second impedance using Eq. 18-5 is

$$Z_B = -j5\ \Omega$$

and the third impedance using Eq. 18-6 is

$$Z_C = 25\ \Omega$$

The Π equivalent circuit is shown in Figure 18-4b.

Example 18-2

Find the T network equivalent to the Π network shown in Figure 18-5 in the s-domain using the Laplace transform. Then, for $s = j1$, find the elements of the T network.

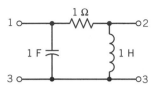

Figure 18-5 Π circuit of Example 18-2.

Solution

First, using Eq. 18-1, we have

$$Z_1 = \frac{(1)(1/s)}{s + 1 + 1/s} = \frac{1}{s^2 + s + 1}$$

Then, using Eq. 18-2, we have

$$Z_2 = \frac{1(s)}{s + 1 + 1/s} = \frac{s^2}{s^2 + s + 1}$$

Finally, the third impedance is (Eq. 18-3)

$$Z_3 = \frac{s(1/s)}{s + 1 + 1/s} = \frac{s}{s^2 + s + 1}$$

To find the elements of the T network at $s = j1$, we substitute $s = j1$ and determine each impedance. Then, we have

$$Z_1 = -j, \qquad Z_2 = j, \qquad Z_3 = 1$$

Therefore, the equivalent T network is as shown in Figure 18-6 for the value $s = j1$.

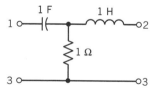

Figure 18-6 T circuit equivalent of the original Π circuit of Example 18-2 for $s = j1$.

EXERCISE 18-1

Find the T circuit equivalent to the Π circuit shown in Figure E 18-1.
Answers: $R_1 = 10\ \Omega$, $R_2 = 12.5\ \Omega$, $R_3 = 50\ \Omega$

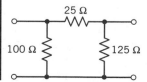

Figure E 18-1

18-5 ‖ EQUATIONS OF TWO-PORT NETWORKS

Let us consider the two-port network of Figure 18-1. By convention I_1 and I_2 are assumed to be flowing into the network as shown. The variables are V_1, V_2, I_1, and I_2. Within the two-port network two variables are independent and two are dependent and we may select a set of two independent variables from the six possible sets: (V_1, V_2), (I_1, I_2), (V_1, I_2), (I_1, V_2), (V_1, I_1), and (V_2, I_2). We will also assume linear elements.

The possibilities for independent (input) variables and the associated dependent variables are summarized in Table 18-1. The names of the associated six sets of circuit

Table 18-1
Six Circuit-Parameter Models

Independent Variables (inputs)	Dependent Variables (outputs)	Circuit Parameters
I_1, I_2	V_1, V_2	Impedance Z
V_1, V_2	I_1, I_2	Admittance Y
V_1, I_2	I_1, V_2	Hybrid g
I_1, V_2	V_1, I_2	Hybrid h
V_2, I_2	V_1, I_1	Transmission T
V_1, I_1	V_2, I_2	Inverse Transmission T'

parameters are also identified in Table 18-1. For the case of phasor transforms or Laplace transforms with the circuit of Figure 18-1, we have the familiar impedance equations where the output variables are V_1 and V_2 as follows:

$$V_1 = Z_{11}I_1 + Z_{12}I_2 \tag{18-9}$$

$$V_2 = Z_{21}I_1 + Z_{22}I_2 \tag{18-10}$$

The equations for the admittances are

$$I_1 = Y_{11}V_1 + Y_{12}V_2 \tag{18-11}$$

$$I_2 = Y_{21}V_1 + Y_{22}V_2 \tag{18-12}$$

It is appropriate, if preferred, to use lowercase letters z and y for the coefficients of Eqs. 18-9 through 18-12. The equations for the six sets of two-port parameters are summarized in Table 18-2.

For linear elements and no sources or op amps within the two-port network, we can show by the theorem of reciprocity that $Z_{12} = Z_{21}$ and $Y_{21} = Y_{12}$. One possible arrangement of a passive circuit as a T circuit is shown in Figure 18-7. Writing the two mesh equations for Figure 18-7, we can readily obtain Eqs. 18-9 and 18-10. Therefore, the circuit of Figure 18-7 can represent the impedance parameters. A possible arrangement of the admittance parameters as a Π circuit is shown in Figure 18-8.

Table 18-2
Equations for the Six Sets of Two-Port Parameters

Impedance Z	$\begin{cases} V_1 = Z_{11}I_1 + Z_{12}I_2 \\ V_2 = Z_{21}I_1 + Z_{22}I_2 \end{cases}$
Admittance Y	$\begin{cases} I_1 = Y_{11}V_1 + Y_{12}V_2 \\ I_2 = Y_{21}V_1 + Y_{22}V_2 \end{cases}$
Hybrid h	$\begin{cases} V_1 = h_{11}I_1 + h_{12}V_2 \\ I_2 = h_{21}I_1 + h_{22}V_2 \end{cases}$
Inverse hybrid g	$\begin{cases} I_1 = g_{11}V_1 + g_{12}I_2 \\ V_2 = g_{21}V_1 + g_{22}I_2 \end{cases}$
Transmission T	$\begin{cases} V_1 = AV_2 - BI_2 \\ I_1 = CV_2 - DI_2 \end{cases}$
Inverse transmission T'	$\begin{cases} V_2 = A'V_1 - B'I_1 \\ I_2 = C'V_1 - D'I_1 \end{cases}$

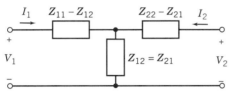

Figure 18-7 A T circuit representing the impedance parameters.

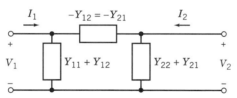

Figure 18-8 A Π circuit representing the admittance parameters.

Examining Eq. 18-9, we see that we can measure Z_{11} by obtaining

$$Z_{11} = \left. \frac{V_1}{I_1} \right|_{I_2 = 0}$$

Of course, $I_2 = 0$ implies that the output terminals are open circuited. Thus, the Z parameters are often called *open-circuit impedances*.

The Y parameters can be measured by determining

$$Y_{12} = \left. \frac{I_1}{V_2} \right|_{V_1 = 0}$$

In general, the admittance parameters are called *short-circuit admittance parameters*.

Example 18-3
Determine the admittance and the impedance parameters of the T network shown in Figure 18-9.

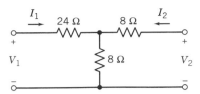

Figure **18-9** Circuit for Example 18-3.

Solution

The admittance parameters use the output terminals shorted and

$$Y_{11} = \frac{I_1}{V_1}\bigg|_{V_2=0}$$

Then, the two 8-Ω resistors are in parallel and $V_1 = 28\,I_1$. Therefore, we have

$$Y_{11} = \frac{1}{28}\ \text{S}$$

For Y_{12} we have

$$Y_{12} = \frac{I_1}{V_2}\bigg|_{V_1=0}$$

so we short-circuit the input terminals. Then we have the circuit as shown in Figure 18-10.

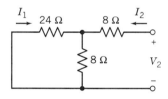

Figure **18-10** Circuit of Example 18-3 with the input terminals shorted.

Employing current division, we have

$$-I_1 = I_2\left(\frac{8}{8+24}\right)$$

and

$$I_2 = \frac{V_2}{8 + [8(24)/(8+24)]} = \frac{V_2}{14}$$

Therefore

$$Y_{12} = \frac{I_1}{V_2} = \frac{-(V_2/14)(1/4)}{V_2} = -\frac{1}{56}\ \text{S}$$

Furthermore,

$$Y_{21} = Y_{12} = -\frac{1}{56}\ \text{S}$$

Finally, Y_{22} is obtained from Figure 18-10 as

$$Y_{22} = \frac{I_2}{V_2}\bigg|_{V_1=0}$$

where

$$I_2 = \frac{V_2}{8 + [8(24)/(8 + 24)]} = \frac{V_2}{14}$$

Therefore,

$$Y_{22} = \frac{1}{14} \text{ S}$$

Thus, in matrix form we have $\mathbf{I} = \mathbf{YV}$ or

$$\begin{bmatrix} I_1 \\ I_2 \end{bmatrix} = \begin{bmatrix} \dfrac{1}{28} & -\dfrac{1}{56} \\ -\dfrac{1}{56} & \dfrac{1}{14} \end{bmatrix} \begin{bmatrix} V_1 \\ V_2 \end{bmatrix}$$

Now, let us find the impedance parameters. We have

$$Z_{11} = \frac{V_1}{I_1} \bigg|_{I_2 = 0}$$

The output terminals are open circuited, so we have the circuit of Figure 18-9. Then

$$Z_{11} = 24 + 8 = 32 \ \Omega$$

Similarly, $Z_{22} = 16 \ \Omega$ and $Z_{12} = 8 \ \Omega$. Then, in matrix form we have $\mathbf{V} = \mathbf{ZI}$ or

$$\begin{bmatrix} V_1 \\ V_2 \end{bmatrix} = \begin{bmatrix} 32 & 8 \\ 8 & 16 \end{bmatrix} \begin{bmatrix} I_1 \\ I_2 \end{bmatrix}$$

The general methods for finding the Z parameters and the Y parameters are summarized in Tables 18-3 and 18-4, respectively.

Table 18-3

Method of Obtaining the Z Parameters of a Circuit

Step IA	To determine Z_{11} and Z_{21}, connect a voltage source V_1 to the input terminals and open-circuit the output terminals.
Step IB	Find I_1 and V_2 and then $Z_{11} = V_1/I_1$ and $Z_{21} = V_2/I_1$.
Step IIA	To determine Z_{22} and Z_{12}, connect a voltage source V_2 to the output terminals and open-circuit the input terminals.
Step IIB	Find I_2 and V_1 and then $Z_{22} = V_2/I_2$ and $Z_{12} = V_1/I_2$.

Note: $Z_{12} = Z_{21}$ only when there are no dependent sources or op amps within the two-port network.

Table 18-4

Method for Obtaining the Y Parameters of a Circuit

Step IA	To determine Y_{11} and Y_{21}, connect a current source I_1 to the input terminals and short-circuit the output terminals.
Step IB	Find V_1 and I_2 and then $Y_{11} = I_1/V_1$ and $Y_{21} = I_2/V_1$.
Step IIA	To determine Y_{22} and Y_{12}, connect a current source I_2 to the output terminals and short-circuit the input terminals.
Step IIB	Find I_1 and V_2 and then $Y_{22} = I_2/V_2$ and $Y_{12} = I_1/V_2$.

Note: $Y_{12} = Y_{21}$ only when there are no dependent sources or op amps within the two-port network.

EXERCISE 18-2

Find the Z and Y parameters of the circuit of Figure E 18-2.

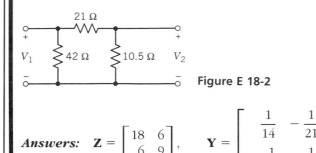

Figure E 18-2

Answers: $\mathbf{Z} = \begin{bmatrix} 18 & 6 \\ 6 & 9 \end{bmatrix}$, $\quad \mathbf{Y} = \begin{bmatrix} \dfrac{1}{14} & -\dfrac{1}{21} \\ -\dfrac{1}{21} & \dfrac{1}{7} \end{bmatrix}$

18-6   Z AND Y PARAMETERS FOR A CIRCUIT WITH DEPENDENT SOURCES

When a circuit incorporates a dependent source, it is easy to use the methods of Table 18-3 or Table 18-4 to determine the Z or Y parameters. When a dependent source is within the circuit, $Z_{21} \neq Z_{12}$ and $Y_{12} \neq Y_{21}$.

Example 18-4

Determine the Z parameters of the circuit of Figure 18-11 when $m = \frac{2}{3}$.

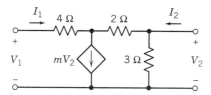

Figure 18-11 Circuit of Example 18-4.

Solution

We determine the Z parameters using the method of Table 18-3. Connect a voltage source V_1 and open-circuit the output terminals as shown in Figure 18-12a.

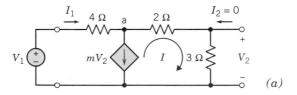

(a)

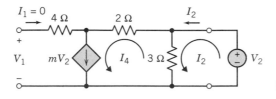

(b)

Figure 18-12 Circuit for determining *(a)* Z_{11} and Z_{21} and *(b)* Z_{22} and Z_{12}.

KCL at node a leads to

$$I_1 - mV_2 - I = 0 \qquad (18\text{-}13)$$

KVL around the outer loop is

$$V_1 = 4I_1 + 5I \qquad (18\text{-}14)$$

Furthermore, $V_2 = 3I$, so $I = V_2/3$. Substituting $I = V_2/3$ into Eq. 18-13, we have

$$I_1 = mV_2 + \frac{V_2}{3}$$

$$= (m + \tfrac{1}{3}) V_2 \qquad (18\text{-}15)$$

Therefore,

$$Z_{21} = \frac{V_2}{I_1} = 1 \ \Omega$$

Substituting $I = V_2/3$ and Eq. 18-15 into Eq. 18-14, we obtain

$$V_1 = 4I_1 + \frac{5V_2}{3}$$

$$= 4I_1 + \frac{5}{3}I_1 \qquad (18\text{-}16)$$

Therefore,

$$Z_{11} = \frac{V_1}{I_1} = \frac{17}{3} \ \Omega$$

To obtain Z_{22} and Z_{12}, we connect a voltage source V_2 to the output terminals and open-circuit the input terminals, as shown in Figure 18-12b. We can write two mesh equations for the assumed current directions shown as

$$V_1 + 5I_4 - 3I_2 = 0 \qquad (18\text{-}17)$$

and

$$V_2 + 3I_4 - 3I_2 = 0 \qquad (18\text{-}18)$$

Furthermore, $I_4 = mV_2$ so substituting into Eq. 18-18, we have

$$V_2 + 3mV_2 - 3I_2 = 0$$

or

$$V_2 = \frac{3}{3} I_2$$

Therefore,

$$Z_{22} = \frac{V_2}{I_2} = 1 \ \Omega$$

Substituting $I_4 = mV_2$ into Eq. 18-17, we have

$$V_1 + 5mV_2 = 3I_2$$

or

$$V_1 + 5mI_2 = 3I_2$$

Therefore,

$$Z_{12} = \frac{V_1}{I_2} = (3 - 5m) = -\frac{1}{3}\ \Omega$$

Then, in summary, we have

$$\mathbf{Z} = \begin{bmatrix} \dfrac{17}{3} & -\dfrac{1}{3} \\ 1 & 1 \end{bmatrix}$$

Note that $Z_{21} \neq Z_{12}$, since a dependent source is present within the circuit.

EXERCISE 18-3

Determine the Y parameters of the circuit of Figure 18-11.

Answer: $Y = \begin{bmatrix} \dfrac{1}{6} & \dfrac{1}{18} \\ -\dfrac{1}{6} & \dfrac{17}{18} \end{bmatrix}$

18-7 HYBRID AND TRANSMISSION PARAMETERS

The two-port hybrid parameter equations are based on V_1 and I_2 as the output variables, so that

$$V_1 = h_{11}I_1 + h_{12}V_2 \tag{18-19}$$

$$I_2 = h_{21}I_1 + h_{22}V_2 \tag{18-20}$$

or, in matrix form,

$$\begin{bmatrix} V_1 \\ I_2 \end{bmatrix} = \begin{bmatrix} h_{11} & h_{12} \\ h_{21} & h_{22} \end{bmatrix} \begin{bmatrix} I_1 \\ V_2 \end{bmatrix} = \mathbf{H} \begin{bmatrix} I_1 \\ V_2 \end{bmatrix} \tag{18-21}$$

These parameters are used widely in transistor circuit models. The hybrid circuit model is shown in Figure 18-13.

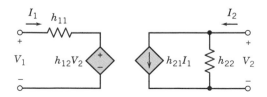

Figure 18-13 The h-parameter model of a two-port circuit.

The inverse hybrid parameter equations are

$$I_1 = g_{11}V_1 + g_{12}I_2 \tag{18-22}$$

$$V_2 = g_{21}V_1 + g_{22}I_2 \tag{18-23}$$

or, in matrix form,

$$\begin{bmatrix} I_1 \\ V_2 \end{bmatrix} = \begin{bmatrix} g_{11} & g_{12} \\ g_{21} & g_{22} \end{bmatrix} \begin{bmatrix} V_1 \\ I_2 \end{bmatrix} = \mathbf{G} \begin{bmatrix} V_1 \\ I_2 \end{bmatrix} \qquad (18\text{-}24)$$

The inverse hybrid circuit model is shown in Figure 18-14.

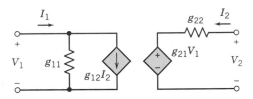

Figure 18-14 The inverse hybrid circuit (g-parameter) model.

The hybrid and inverse hybrid parameters include both impedance and admittance parameters and are thus called *hybrid*. The parameters h_{11}, h_{12}, h_{21}, and h_{22} represent the short-circuit input impedance, the open-circuit reverse voltage gain, the short-circuit forward current gain, and the open-circuit output admittance, respectively. The parameters g_{11}, g_{12}, g_{21}, and g_{22} represent the short-circuit transfer admittance, the open-circuit forward current gain, the open-circuit forward voltage gain, and the short-circuit output impedance, respectively.

The transmission parameters are written as

$$V_1 = AV_2 - BI_2 \qquad (18\text{-}25)$$

$$I_1 = CV_2 - DI_2 \qquad (18\text{-}26)$$

or, in matrix form, as

$$\begin{bmatrix} V_1 \\ I_1 \end{bmatrix} = \begin{bmatrix} A & B \\ C & D \end{bmatrix} \begin{bmatrix} V_2 \\ -I_2 \end{bmatrix} = \mathbf{T} \begin{bmatrix} V_2 \\ -I_2 \end{bmatrix} \qquad (18\text{-}27)$$

Transmission parameters are used to describe cable, fiber, and line transmission. The transmission parameters A, B, C, and D represent the open-circuit voltage ratio, the negative short-circuit transfer impedance, the open-circuit transfer admittance, and the negative short-circuit current ratio, respectively. The transmission parameters are often referred to as the *ABCD* parameters. We are primarily interested in the hybrid and transmission parameters, since they are widely used.

Example 18-5
(a) Find the h parameters for the T circuit of Figure 18-15 in terms of R_1, R_2, and R_3.
(b) Evaluate the parameters when $R_1 = 1\ \Omega$, $R_2 = 4\ \Omega$, and $R_3 = 6\ \Omega$.

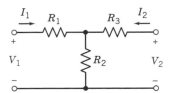

Figure 18-15 The T circuit of Example 18-5.

Solution
(a) First, we find h_{11} and h_{21} by short-circuiting the output terminals and connecting an input current source I_1, as shown in Figure 18-16a. Therefore,

$$h_{11} = \frac{V_1}{I_1}\bigg|_{V_2=0} = R_1 + \frac{R_2 R_3}{R_2 + R_3}$$

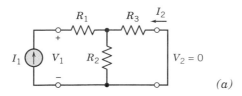

(a)

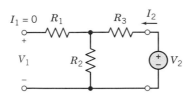

(b)

Figure 18-16 The circuits for determining (a) h_{11} and h_{21} and (b) h_{22} and h_{12}.

Then, using the current divider principle, we have

$$I_2 = \frac{-R_2}{R_2 + R_3} I_1$$

Therefore,

$$h_{21} = \left. \frac{I_2}{I_1} \right|_{V_2 = 0} = \frac{-R_2}{R_2 + R_3}$$

The next step is to redraw the circuit with $I_1 = 0$ and to connect the voltage source V_2 as shown in Figure 18-16b. Then we may determine h_{12} by using the voltage divider principle, as follows:

$$h_{12} = \left. \frac{V_1}{V_2} \right|_{I_1 = 0} = \frac{R_2}{R_2 + R_3}$$

Finally, we determine h_{22} from Figure 18-16b as

$$h_{22} = \left. \frac{I_2}{V_2} \right|_{I_1 = 0} = \frac{1}{R_2 + R_3}$$

It is a property of a passive circuit (no sources within the two-port) that $h_{12} = -h_{21}$. (b) When $R_1 = 1\ \Omega$, $R_2 = 4\ \Omega$, and $R_3 = 6\ \Omega$, we have

$$h_{11} = R_1 + \frac{R_2 R_3}{R_2 + R_3} = 3.4\ \Omega$$

$$h_{21} = \frac{-R_2}{R_2 + R_3} = -0.4$$

$$h_{12} = \frac{R_2}{R_2 + R_3} = 0.4$$

$$h_{22} = \frac{1}{R_2 + R_3} = 0.1\ S$$

EXERCISE 18-4

Find the hybrid parameter model of the circuit shown in Figure E 18-4.
Answers: $h_{11} = 0.9\ \Omega$, $h_{12} = 0.1$, $h_{21} = 4.4$, $h_{22} = 0.6$ S

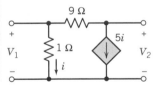

Figure E 18-4

18-8 | RELATIONSHIPS BETWEEN TWO-PORT PARAMETERS

If all the two-port parameters for a circuit exist, it is possible to relate one set of parameters to another, since the variables V_1, I_1, V_2, and I_2 are interrelated by the parameters. First, let us consider the relation between the Z parameters and the Y parameters. The matrix equation for the Z parameters is $\mathbf{V} = \mathbf{ZI}$ or

$$\begin{bmatrix} V_1 \\ V_2 \end{bmatrix} = \mathbf{Z} \begin{bmatrix} I_1 \\ I_2 \end{bmatrix} \tag{18-28}$$

Similarly, the equation for the Y parameters is $\mathbf{I} = \mathbf{YV}$ or

$$\begin{bmatrix} I_1 \\ I_2 \end{bmatrix} = \mathbf{Y} \begin{bmatrix} V_1 \\ V_2 \end{bmatrix} \tag{18-29}$$

Substituting for $\mathbf{I}$ from Eq. 18-29 into Eq. 18-28, we obtain

$$\mathbf{V} = \mathbf{ZYV}$$

or

$$\mathbf{Z} = \mathbf{Y}^{-1} \tag{18-30}$$

Thus, we can obtain the matrix $\mathbf{Z}$ by inverting the $\mathbf{Y}$ matrix. Of course, we can likewise obtain the $\mathbf{Y}$ matrix if we invert a known $\mathbf{Z}$ matrix. It is possible that a two-port network has a $\mathbf{Y}$ matrix or a $\mathbf{Z}$ matrix but not both. In other words, $\mathbf{Z}^{-1}$ or $\mathbf{Y}^{-1}$ may not exist for some networks.

If we have a known $\mathbf{Y}$ matrix, we obtain the $\mathbf{Z}$ matrix by finding the determinant of the $\mathbf{Y}$ matrix as ΔY and the adjoint of the $\mathbf{Y}$ matrix as

$$\text{adj } \mathbf{Y} = \begin{bmatrix} Y_{22} & -Y_{12} \\ -Y_{21} & Y_{11} \end{bmatrix}$$

Then

$$\mathbf{Z} = \mathbf{Y}^{-1} = \frac{\text{adj } \mathbf{Y}}{\Delta Y} \tag{18-31}$$

where $\Delta Y = Y_{11} Y_{22} - Y_{12} Y_{21}$.

The two-port parameter conversion relationships for the Z, Y, h, g, and T parameters are provided in Table 18-5.

Table 18-5
Parameter Relationships

	Z		Y		h		g		T	
Z	Z_{11}	Z_{12}	$\dfrac{Y_{22}}{\Delta Y}$	$\dfrac{-Y_{12}}{\Delta Y}$	$\dfrac{\Delta h}{h_{22}}$	$\dfrac{h_{12}}{h_{22}}$	$\dfrac{1}{g_{11}}$	$\dfrac{-g_{12}}{g_{11}}$	$\dfrac{A}{C}$	$\dfrac{\Delta T}{C}$
	Z_{21}	Z_{22}	$\dfrac{-Y_{21}}{\Delta Y}$	$\dfrac{Y_{11}}{\Delta Y}$	$\dfrac{-h_{21}}{h_{22}}$	$\dfrac{1}{h_{22}}$	$\dfrac{g_{21}}{g_{11}}$	$\dfrac{\Delta g}{g_{11}}$	$\dfrac{1}{C}$	$\dfrac{D}{C}$
Y	$\dfrac{Z_{22}}{\Delta Z}$	$\dfrac{-Z_{12}}{\Delta Z}$	Y_{11}	Y_{12}	$\dfrac{1}{h_{11}}$	$\dfrac{h_{12}}{h_{11}}$	$\dfrac{\Delta g}{g_{22}}$	$\dfrac{g_{12}}{g_{22}}$	$\dfrac{D}{B}$	$\dfrac{-\Delta T}{B}$
	$\dfrac{-Z_{21}}{\Delta Z}$	$\dfrac{Z_{11}}{\Delta Z}$	Y_{21}	Y_{22}	$\dfrac{h_{21}}{h_{11}}$	$\dfrac{\Delta h}{h_{11}}$	$\dfrac{-g_{21}}{g_{22}}$	$\dfrac{1}{g_{22}}$	$\dfrac{-1}{B}$	$\dfrac{A}{B}$
h	$\dfrac{\Delta Z}{Z_{22}}$	$\dfrac{Z_{12}}{Z_{22}}$	$\dfrac{1}{Y_{11}}$	$\dfrac{-Y_{12}}{Y_{11}}$	h_{11}	h_{12}	$\dfrac{g_{22}}{\Delta g}$	$\dfrac{g_{12}}{\Delta g}$	$\dfrac{B}{D}$	$\dfrac{\Delta T}{D}$
	$\dfrac{-Z_{21}}{Z_{22}}$	$\dfrac{1}{Z_{22}}$	$\dfrac{Y_{21}}{Y_{11}}$	$\dfrac{\Delta Y}{Y_{11}}$	h_{21}	h_{22}	$\dfrac{-g_{21}}{\Delta g}$	$\dfrac{g_{11}}{\Delta g}$	$\dfrac{-1}{D}$	$\dfrac{C}{D}$
g	$\dfrac{1}{Z_{11}}$	$\dfrac{-Z_{12}}{Z_{11}}$	$\dfrac{\Delta Y}{Y_{22}}$	$\dfrac{Y_{12}}{Y_{22}}$	$\dfrac{h_{22}}{\Delta h}$	$\dfrac{-h_{12}}{\Delta h}$	g_{11}	g_{12}	$\dfrac{C}{A}$	$\dfrac{-\Delta T}{A}$
	$\dfrac{Z_{21}}{Z_{11}}$	$\dfrac{\Delta Z}{Z_{11}}$	$\dfrac{-Y_{21}}{Y_{22}}$	$\dfrac{1}{Y_{22}}$	$\dfrac{-h_{21}}{\Delta h}$	$\dfrac{h_{11}}{\Delta h}$	g_{21}	g_{22}	$\dfrac{1}{A}$	$\dfrac{B}{A}$
T	$\dfrac{Z_{11}}{Z_{21}}$	$\dfrac{\Delta Z}{Z_{21}}$	$\dfrac{-Y_{22}}{Y_{21}}$	$\dfrac{-1}{Y_{21}}$	$\dfrac{-\Delta h}{h_{21}}$	$\dfrac{-h_{11}}{h_{21}}$	$\dfrac{1}{g_{21}}$	$\dfrac{g_{22}}{g_{21}}$	A	B
	$\dfrac{1}{Z_{21}}$	$\dfrac{Z_{22}}{Z_{21}}$	$\dfrac{-\Delta Y}{Y_{21}}$	$\dfrac{-Y_{11}}{Y_{21}}$	$\dfrac{-h_{22}}{h_{21}}$	$\dfrac{-1}{h_{21}}$	$\dfrac{g_{11}}{g_{21}}$	$\dfrac{\Delta Y}{g_{21}}$	C	D

$\Delta Z = Z_{11}Z_{22} - Z_{12}Z_{21}, \; \Delta Y = Y_{11}Y_{22} - Y_{12}Y_{21}, \; \Delta g = g_{11}g_{22} - g_{12}g_{21}, \; \Delta h = h_{11}h_{22} - h_{12}h_{21}, \; \Delta T = AD - BC.$

Example 18-6

Determine the Y and h parameters if

$$\mathbf{Z} = \begin{bmatrix} 18 & 6 \\ 6 & 9 \end{bmatrix}$$

Solution

First, we will determine the Y parameters by calculating the determinant as

$$\Delta Z = Z_{11}Z_{22} - Z_{12}Z_{21} = 18(9) - 6(6) = 126$$

Then, using Table 18-5, we obtain

$$Y_{11} = \frac{Z_{22}}{\Delta Z} = \frac{9}{126} = \frac{1}{14} \text{ S}$$

$$Y_{12} = Y_{21} = \frac{-Z_{12}}{\Delta Z} = \frac{-1}{21} \text{ S}$$

$$Y_{22} = \frac{Z_{11}}{\Delta Z} = \frac{18}{126} = \frac{1}{7} \text{ S}$$

Obtaining the h parameters by using Table 18-5, we have

$$h_{11} = \frac{\Delta Z}{Z_{22}} = \frac{126}{9} = 14 \ \Omega$$

$$h_{12} = \frac{Z_{12}}{Z_{22}} = \frac{6}{9} = \frac{2}{3}$$

$$h_{21} = \frac{-Z_{21}}{Z_{22}} = \frac{-6}{9} = \frac{-2}{3}$$

$$h_{22} = \frac{1}{Z_{22}} = \frac{1}{9} \ \text{S}$$

EXERCISE 18-5

Determine the Z parameters if the Y parameters are

$$\mathbf{Y} = \begin{bmatrix} \dfrac{2}{15} & \dfrac{-1}{5} \\[2ex] \dfrac{-1}{10} & \dfrac{2}{5} \end{bmatrix}$$

The units are siemens.
Answers: $Z_{11} = 12 \ \Omega$, $Z_{12} = 6 \ \Omega$, $Z_{21} = 3 \ \Omega$, $Z_{22} = 4 \ \Omega$

EXERCISE 18-6

Determine the T parameters from the Y parameters of Exercise 18-5.
Answers: $A = 4$, $B = 10 \ \Omega$, $C = \frac{1}{3} \ \text{S}$, $D = \frac{4}{3}$

18-9 ┃ INTERCONNECTION OF TWO-PORT NETWORKS

It is common in many circuits to have several two-port networks interconnected in parallel or in cascade. The *parallel* connection of two two-ports shown in Figure 18-17 requires that the V_1 of each two-port be equal.

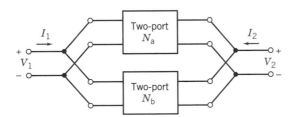

Figure 18-17 Parallel connection of two two-port networks.

Similarly, at the output port V_2 is the output voltage of both two-port networks. The defining matrix equation for network N_a is

$$\mathbf{I}_a = \mathbf{Y}_a \mathbf{V}_a \tag{18-32}$$

and for network N_b we have

$$\mathbf{I}_b = \mathbf{Y}_b \mathbf{V}_b \tag{18-33}$$

In addition, we have the total current $\mathbf{I}$ as

$$\mathbf{I} = \mathbf{I}_a + \mathbf{I}_b$$

Furthermore, since $\mathbf{V}_a = \mathbf{V}_b = \mathbf{V}$,

$$\mathbf{I} = \mathbf{Y}_a \mathbf{V} + \mathbf{Y}_b \mathbf{V} = (\mathbf{Y}_a + \mathbf{Y}_b) \, \mathbf{V} = \mathbf{YV}$$

Therefore, the Y parameters for the total network of two parallel two-ports are described by the matrix equation

$$\mathbf{Y} = \mathbf{Y}_a + \mathbf{Y}_b \tag{18-34}$$

For example,

$$Y_{11} = Y_{11a} + Y_{11b}$$

Hence, to determine the Y parameters for the total network, we add the Y parameters of each network. In general, the Y parameter matrix of the parallel connection is the sum of the Y parameter matrices of the individual two-ports.

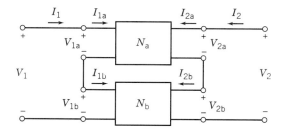

Figure 18-18 Series connection of two two-port networks.

The series interconnection of two two-port networks is shown in Figure 18-18. We will use the Z parameters to describe each two-port and the series combination. The two networks are described by the matrix equations

$$\mathbf{V}_a = \mathbf{Z}_a \mathbf{I}_a \tag{18-35}$$

and

$$\mathbf{V}_b = \mathbf{Z}_b \mathbf{I}_b \tag{18-36}$$

The terminal currents are

$$\mathbf{I} = \mathbf{I}_a = \mathbf{I}_b$$

Therefore, since $\mathbf{V} = \mathbf{V}_a + \mathbf{V}_b$, we have

$$\mathbf{V} = \mathbf{Z}_a \mathbf{I}_a + \mathbf{Z}_b \mathbf{I}_b$$
$$= (\mathbf{Z}_a + \mathbf{Z}_b)\mathbf{I} = \mathbf{ZI}$$

or

$$\mathbf{Z} = \mathbf{Z}_a + \mathbf{Z}_b \tag{18-37}$$

Therefore, the Z parameters for the total network are equal to the sum of the Z parameters for the networks.

When the output of one network is connected to the input port of the following network, as shown in Figure 18-19, the networks are said to be *cascaded*. Since the output

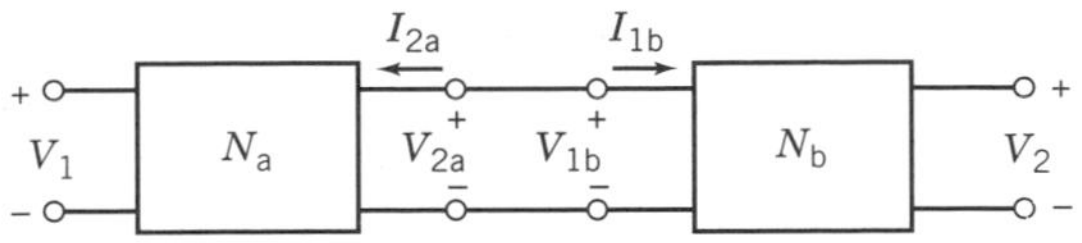

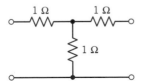

Figure 18-19 Cascade connection of two two-port networks.

variables of the first network become the input variables of the second network, the transmission parameters are utilized. The first two-port, N_a, is represented by the matrix equation

$$\begin{bmatrix} V_{1a} \\ I_{1a} \end{bmatrix} = \mathbf{T}_a \begin{bmatrix} V_{2a} \\ -I_{2a} \end{bmatrix}$$

For N_b we have

$$\begin{bmatrix} V_{1b} \\ I_{1b} \end{bmatrix} = \mathbf{T}_b \begin{bmatrix} V_{2b} \\ -I_{2b} \end{bmatrix}$$

Furthermore, we note that at the input and output, we have

$$\begin{bmatrix} V_1 \\ I_1 \end{bmatrix} = \begin{bmatrix} V_{1a} \\ I_{1a} \end{bmatrix} \quad \text{and} \quad \begin{bmatrix} V_{2b} \\ -I_{2b} \end{bmatrix} = \begin{bmatrix} V_2 \\ -I_2 \end{bmatrix}$$

At the intermediate connection, we have

$$\begin{bmatrix} V_{2a} \\ -I_{2a} \end{bmatrix} = \begin{bmatrix} V_{1b} \\ I_{1b} \end{bmatrix}$$

Therefore,

$$\begin{bmatrix} V_1 \\ I_1 \end{bmatrix} = \mathbf{T}_a \mathbf{T}_b \begin{bmatrix} V_2 \\ -I_2 \end{bmatrix}$$

and

$$\mathbf{T} = \mathbf{T}_a \mathbf{T}_b \tag{18-38}$$

Hence, the transmission parameters for the overall network are derived by matrix multiplication, observing the proper order.

All of the preceding calculations for interconnected networks assume that the interconnection does not disturb the two-port nature of the individual sub-networks.

Example 18-7

For the T network of Figure 18-20, (a) find the Z, Y, and T parameters and (b) determine the resulting parameters after connecting two two-ports in parallel and in cascade. Both two-ports are identical as in Figure 18-1.

Figure 18-20 T network of Example 18-7.

Solution

First, we find the Z parameters of the T network. Examining the network, we have

$$Z_{12} = Z_{21} = 1 \; \Omega$$
$$Z_{22} = Z_{11} = 2 \; \Omega$$

Then, using the conversion factors of Table 18-5, we find

$$\mathbf{Y} = \begin{bmatrix} \dfrac{2}{3} & \dfrac{-1}{3} \\ \dfrac{-1}{3} & \dfrac{2}{3} \end{bmatrix}$$

and

$$\mathbf{T} = \begin{bmatrix} 2 & 3 \\ 1 & 2 \end{bmatrix}$$

Two identical networks connected in parallel will have a total **Y** matrix of

$$\mathbf{Y} = \mathbf{Y}_a + \mathbf{Y}_b$$

Since $\mathbf{Y}_a = \mathbf{Y}_b$, we have

$$\mathbf{Y} = 2Y_a = \begin{bmatrix} \dfrac{4}{3} & \dfrac{-2}{3} \\ \dfrac{-2}{3} & \dfrac{4}{3} \end{bmatrix}$$

Finally, when two identical networks are connected in cascade, we have a total **T** matrix of

$$\mathbf{T} = \mathbf{T}_a \mathbf{T}_b$$

$$= \begin{bmatrix} 2 & 3 \\ 1 & 2 \end{bmatrix} \begin{bmatrix} 2 & 3 \\ 1 & 2 \end{bmatrix}$$

$$= \begin{bmatrix} 7 & 12 \\ 4 & 7 \end{bmatrix}$$

EXERCISE 18-7

Determine the total transmission parameters of the cascade connection of three two-port networks shown in Figure E 18-7.

Answers: $A = 3$, $B = 21\ \Omega$, $C = \frac{1}{6}$ S, $D = \frac{3}{2}$

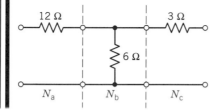

Figure E 18-7

18-10 USING PSpice TO CALCULATE TWO-PORT PARAMETERS

PSpice can be used to calculate the two-port parameters. To do this, we use a 1-A or 1-V source while using open or short circuits to impose the necessary constraints. For example, let us determine the h parameters of the circuit shown in Figure 18-21. We recall that

$$h_{11} = \frac{V_1}{I_1}\bigg|_{V_2 = 0}$$

and

$$h_{21} = \frac{I_2}{I_1}\bigg|_{V_2 = 0}$$

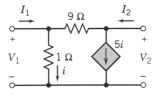

Figure 18-21 Two-port network.

Thus, we set $V_2 = 0$ by inserting a short circuit at the output and use $I_1 = 1$ A. These are the short-circuit parameters and we use the .DC PSpice statement to find V_1 and I_2. Then $h_{11} = V_1$ and $h_{21} = I_2$.

Also, recall that

$$h_{12} = \frac{V_1}{V_2}\bigg|_{I_1 = 0}$$

and

$$h_{22} = \frac{I_2}{V_2}\bigg|_{I_1 = 0}$$

We find these parameters by setting $I_1 = 0$ by open-circuiting the input terminals and inserting a source $V_2 = 1$ V at the output terminals. We then use another PSpice program to find V_1 and I_2. Then $h_{12} = V_1$ and $h_{22} = I_2$.

The circuit is redrawn to find h_{11} and h_{21} by short-circuiting the output as shown in Figure 18-22a. The dummy source VM2 is used to measure I_2, and the dummy source VM1 is used to measure i for the CCCS. The circuit to determine h_{12} and h_{22} is shown in Figure 18-22b, where $V_2 = 1$ V.

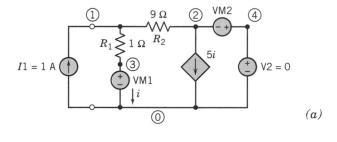

(a)

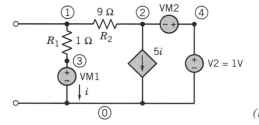

(b)

Figure 18-22 Two-port network redrawn to measure (a) h_{11} and h_{21} and (b) h_{12} and h_{22}.

Since the two circuits redrawn to find the h parameters are relatively similar, we can readily write two PSpice programs from the circuits. The program to determine h_{11} and h_{21} is shown in Figure 18-23, and the program to determine h_{12} and h_{22} is shown in Figure 18-24.

```
CALCULATE   H11   H21

I1   0   1   DC   1
R1   1   3   1
R2   1   2   9
V2   4   0   0   ;   SHORT CIRCUIT
F1   2   0   VM1   5
VM1   3   0   0   ;     MEASURE I
VM2   4   2   0   ;     MEASURE I2
.DC   I1   1   1   1
.PRINT   DC   V(1)   I(VM2)
.END
```

Figure 18-23 Program to determine h_{11} and h_{21}.

```
CALCULATE   H12   H22

R1   1   3   1
R2   1   2   9
V2   4   0   DC   1
F1   2   0   VM1   5
VM1   3   0   0
VM2   4   2   0
.DC   V2   1   1   1
.PRINT   DC   V(1)   I(VM2)
.END
```

Figure 18-24 Program to determine h_{12} and h_{22}.

After running the program of Figure 18-23, we find that $h_{11} = 0.9$ Ω and $h_{21} = 4.4$. Using the program shown in Figure 18-24, we determine $h_{12} = 0.1$ S and $h_{22} = 0.6$.

18-11 VERIFICATION EXAMPLE

Example 18V-1

The circuit shown in Figure 18V-1a was designed to have a transfer function given by

$$\frac{V_o(s)}{V_{in}(s)} = \frac{2s - 10}{s^2 + 27s + 2}$$

Does the circuit satisfy this specification?

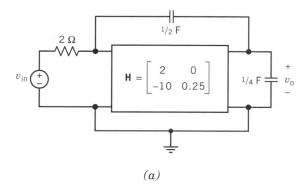

(a)

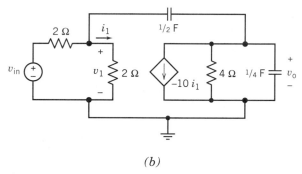

(b)

Figure 18V-1 *(a)* A circuit including a two-port network. *(b)* Using the H-parameter model to represent the two-port network.

Solution

The *h*-parameter model from Figure 18-13 can be used to redraw the circuit as shown in Figure 18V-1*b*. This circuit can be represented by node equations

$$
\begin{bmatrix} \left(1 + \dfrac{s}{2}\right) & -\dfrac{s}{2} \\ \left(-5 - \dfrac{s}{2}\right) & \left(\dfrac{3s}{4} + \dfrac{1}{4}\right) \end{bmatrix}
\begin{bmatrix} V_1(s) \\ V_o(s) \end{bmatrix}
=
\begin{bmatrix} \dfrac{V_{in}(s)}{2} \\ 0 \end{bmatrix}
$$

where $10I_1(s) = 5V_1(s)$ has been used to express the current of the dependent source in terms of the node voltages. Applying Cramer's rule gives

$$
\frac{V_o(s)}{V_{in}(s)} = \frac{\dfrac{1}{2}\left(5 + \dfrac{s}{2}\right)}{\left(1 + \dfrac{s}{2}\right)\left(\dfrac{3s}{4} + \dfrac{1}{4}\right) - \dfrac{s}{2}\left(\dfrac{s}{2} + 5\right)}
$$

$$
= \frac{2s + 20}{s^2 - 13s + 2}
$$

This is not the required transfer function, so the circuit does not satisfy the specification.

EXERCISE 18V-1

Verify that the circuit shown in Figure 18V-2 does indeed have the transfer function

$$\frac{V_o(s)}{V_{in}(s)} = \frac{2s - 10}{s^2 + 27s + 2}$$

(The circuits in Figures 18V-1a and 18V-2 differ only in the sign of h_{21} .)

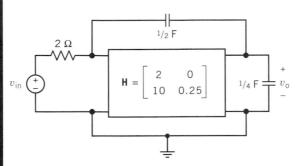

Figure 18V-2 A modified version of the circuit from Figure 18V-1a.

18-12 DESIGN CHALLENGE SOLUTION

TRANSISTOR AMPLIFIER

Figure 18D-1 shows the small signal equivalent circuit of a transistor amplifier. The data sheet for the transistor describes the transistor by specifying its h parameters to be

$$h_{ie} = 1250 \ \Omega, \quad h_{oe} = 0, \quad h_{fe} = 100, \quad \text{and} \quad h_{re} = 0$$

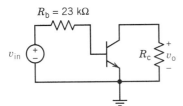

Figure 18D-1 A transistor amplifier.

The value of the resistance R_c must be between 300 Ω and 5000 Ω to ensure that the transistor will be biased correctly. The small signal gain is defined to be

$$A_v = \frac{v_o}{v_{in}}$$

The challenge is to design the amplifier so that

$$A_v = -20$$

(There is no guarantee that these specifications can be satisfied. Part of the problem is to decide whether it is possible to design this amplifier so that $A_v = -20$.)

Define the Situation

1   R_c must be between 300 Ω and 5000 Ω.
2   The transistor is represented by h parameters. Figure 18D-2a shows that the transistor can be configured to be a two-port network and represented by

h parameters. Figure 18D-2b shows an equivalent circuit for the transistor. This equivalent circuit is based on the h parameters. For this particular transistor,

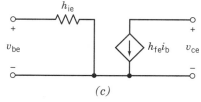

$$\begin{bmatrix} h_{ie} & h_{re} \\ h_{fe} & h_{oe} \end{bmatrix} \begin{bmatrix} i_b \\ v_{ce} \end{bmatrix} = \begin{bmatrix} v_{be} \\ i_c \end{bmatrix}$$

(a)

(b)

(c)

Figure 18D-2 (a) Using h parameters to describe a transistor. (b) An equivalent circuit. (c) A simplified equivalent circuit for $h_{re} = 0$ and $h_{oe} = 0$.

the values of the h parameters are

$$h_{ie} = 1250 \ \Omega, \quad h_{oe} = 0, \quad h_{fe} = 100, \quad \text{and} \quad h_{re} = 0$$

Since

$$\frac{1}{h_{oe}} = \infty$$

the resistor at the left side of the equivalent circuit is an open circuit. Since

$$h_{re} = 0$$

the dependent voltage source is a short circuit. Figure 18D-2c shows the equivalent circuit after these simplifications are made.

3 The voltage gain must be $A_v = -20$.

State the Goal

Select R_c so that $A_v = -20$.

Generate a Plan

Replace the transistor in Figure 18D-1 by the equivalent circuit in Figure 18D-2c. Analyze the resulting circuit to obtain a formula for the voltage gain, A_v. This formula will involve R_c. Determine the value of R_c that will make $A_v = -20$. If this value of R_c is between 300 Ω and 5000 Ω, then the amplifier design is complete. On the other hand, if this value of R_c is not between 300 Ω and 5000 Ω, then the specifications cannot be satisfied.

Take Action on the Plan

Figure 18D-3 shows the amplifier after the transistor has been replaced by the equivalent circuit. Applying Ohm's law to R_c gives

$$v_o = -R_c 100 i_b$$

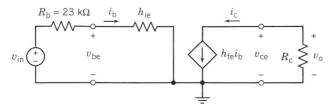

Figure 18D-3 An equivalent circuit for the transistor amplifier.

where the minus sign is due to reference directions. Next, apply KVL to the left mesh to get

$$v_{in} = 23,000 i_b + 1000 i_b$$

Then

$$A_v = \frac{v_o}{v_{in}}$$

$$= \frac{-100 R_c}{24,000}$$

Finally, set $A_v = -20$ obtaining

$$-20 = \frac{-100 R_c}{24,000}$$

Now solve for R_c to determine

$$R_c = 4800 \ \Omega$$

Since this resistance is between 300 Ω and 5000 Ω, the specifications have been satisfied.

SUMMARY

Two-port models of circuits are useful for describing the performance of a circuit in terms of its behavior at the two pairs of terminals labeled the input and output terminal pairs, respectively. The port of entry is called the input port, and the port where the signal is extracted is the output port. The two-port model is restricted to circuits where there are no independent sources and no energy storage within the two-port network.

The two-port circuit is described in terms of the input and output voltages and currents and permits the establishment of six sets of parameters called the impedance, admittance, hybrid, hybrid transmission, and inverse transmission parameters.

Because all six sets of parameters are related to the input and output voltages and currents, we may determine the relationships between the sets of parameters. Furthermore, we may use these parameter sets to describe the performance of two or more two-port networks connected in parallel, series, or cascade.

TERMS AND CONCEPTS

Completely Symmetric Network   A T or Π network where all impedances are equal.

Port   An access point consisting of a set of two terminals, where the current entering one terminal of a pair exits the other terminal in the pair.

Symmetrical Network   Two-port network that has mirror-image symmetry with respect to a centerline.

Two-Port Network   Network with two sets of access terminals called the input and output ports.

PROBLEMS

Section 18-4   T-to-Π Transformation and Two-Port Three-Terminal Networks

P 18.4-1   Determine the equivalent resistance R_{ab} of the network of Figure P 18.4-1. Use the Π-to-T transformation as one step of the reduction.
Answer: $R_{ab} = 3.2\ \Omega$

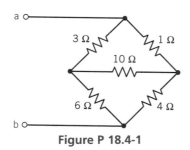

Figure P 18.4-1

P 18.4-2   Repeat Problem 18.4-1 when the 6-Ω resistance is changed to 4 Ω and the 10-Ω resistance is changed to 12 Ω.

P 18.4-3   The two-port network of Figure 18.1 has an input source V_s with a source resistance R_s connected to the input terminals so that $V_1 = V_s - I_1 R_s$ and a load resistance connected to the output terminals so that $V_2 = -I_2 R_L = I_L R_L$.   Find $R_{in} = V_1/I_1$, $A_v = V_2/V_1$, $A_i = -I_2/I_1$,   and $A_p = -V_2 I_2/V_1 I_1$ by using the Z-parameter model.

P 18.4-4   Using the Δ-to-Y transformation, determine the current I when $R_1 = 15\ \Omega$ and $R = 20\ \Omega$ for the circuit shown in Figure P 18.4-4.

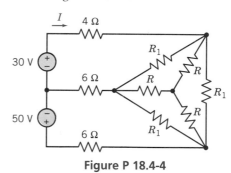

Figure P 18.4-4

P 18.4-5   Use the Y-to-Δ transformation to determine R_{in} of the circuit shown in Figure P 18.4-5.

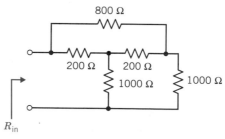

Figure P 18.4-5

Section 18-5 Equations of Two-Port Networks

P 18.5-1 Find the Y parameters and Z parameters for the two-port network of Figure P 18.5-1.

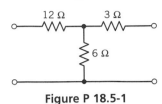

Figure P 18.5-1

P 18.5-2 Determine the Z parameters of the ac circuit shown in Figure P 18.5-2.

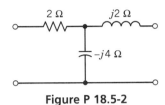

Figure P 18.5-2

P 18.5-3 Find the Y parameters of the circuit of Figure P 18.5-3 when $b = 4$, $G_1 = 2$ S, $G_2 = 1$ S, and $G_3 = 3$ S.

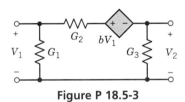

Figure P 18.5-3

P 18.5-4 Find the Y parameters for the circuit of Figure P 18.5-4.

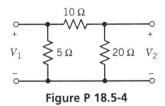

Figure P 18.5-4

P 18.5-5 Find the Y parameters of the circuit shown in Figure P 18.5-5.

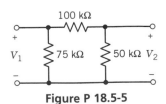

Figure P 18.5-5

P 18.5-6 Find the Z parameters for the circuit shown in Figure P 18.5-6 for sinusoidal steady-state response at $\omega = 3$ rad/s.

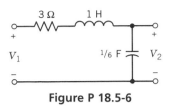

Figure P 18.5-6

P 18.5-7 Determine the impedance parameters in the s-domain (Laplace domain) for the circuit shown in Figure P 18.5-7.

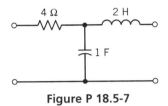

Figure P 18.5-7

P 18.5-8 Determine a two-port network that is represented by the Y parameters

$$\mathbf{Y} = \begin{bmatrix} \dfrac{s+1}{s} & -1 \\ -1 & (s+1) \end{bmatrix}$$

P 18.5-9 Find a two-port network incorporating one inductor, one capacitor, and two resistors that will give the following impedance parameters:

$$\mathbf{Z} = \frac{1}{\Delta} \begin{bmatrix} (s^2 + 2s + 2) & 1 \\ 1 & (s^2 + 1) \end{bmatrix}$$

where $\Delta = s^2 + s + 1$.

P 18.5-10 An infinite two-port network is shown in Figure P 18.5-10. When the output terminals are connected to the circuit's characteristic resistance R_o, the resistance looking down the line from each section is the same. Calculate the necessary R_o.

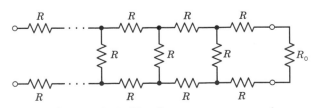

Figure P 18.5-10 Infinite two-port network.

Section 18-6 Z and Y Parameters for a Circuit with Dependent Sources

P 18.6-1 Determine the Y parameters of the circuit shown in Figure P 18.6-1.

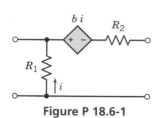

Figure P 18.6-1

P 18.6-2 An electronic amplifier has the circuit shown in Figure P 18.6-2. Determine the impedance parameters for the circuit.

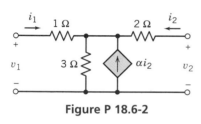

Figure P 18.6-2

P 18.6-3 (a) For the circuit shown in Figure P 18.6-3, determine the two-port Y model using impedances in the s-domain. (b) Determine the response $v_2(t)$ when a current source $i_1 = 1\,u(t)$ A is connected to the input terminals.

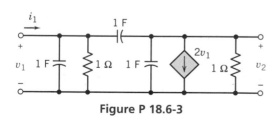

Figure P 18.6-3

P 18.6-4 One form of a heart-assist device is shown in Figure P 18.6-4a. The model of the electronic controller and pump/drive unit is shown in Figure P 18.6-4b. Determine the impedance parameters of the two-port model.

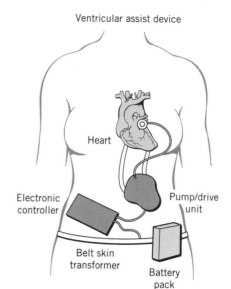

(a)

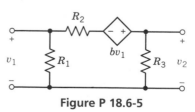

(b)

Figure P 18.6-4 (a) Heart-assist device and (b) model of controller and pump.

P 18.6-5 Determine the Y parameters for the circuit shown in Figure P 18.6-5.

Figure P 18.6-5

Partial Answer: $Y_{12} = -\dfrac{1}{R_2},\ Y_{21} = \dfrac{-(1+b)}{R_2}$

Section 18-7 Hybrid and Transmission Parameters

P 18.7-1 Find the transmission parameters of the circuit of Figure P 18.7-1.

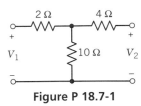

Figure P 18.7-1

P 18.7-2 An op amp circuit and its model are shown in Figure P 18.7-2. (Also see Figure 6-12.) Determine the h-parameter model of the circuit and the **H** matrix when $R_i = 100 \text{ k}\Omega$, $R_1 = R_2 = 1 \text{ M}\Omega$, $R_o = 1 \text{ k}\Omega$, and $A = 10^4$.

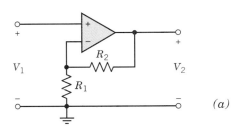

(a)

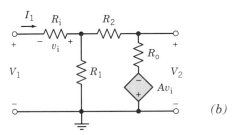

(b)

Figure P 18.7-2 (a) Op amp circuit and (b) circuit model.

P 18.7-3 Determine the h parameters for the ideal transformer of Section 12-9.

P 18.7-4 Determine the h parameters for the T circuit of Figure P 18.7-4.

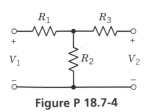

Figure P 18.7-4

P 18.7-5 A simplified model of a bipolar junction transistor is shown in Figure P 18.7-5. Determine the h parameters of this circuit.

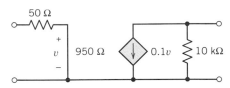

Figure P 18.7-5 Model of bipolar junction transistor.

P 18.7-6 A transformer, as shown in Figure 12-20, can be represented by the inverse hybrid circuit model shown in Figure P 18.7-6. Open-circuit measurements indicate that $V_1 = 1000$ V, $i_2 = 0$, $V_2 = 500$ V, $i_1 = 0.42$ A, and $P = 100$ W (secondary open circuit). Short-circuit measurements indicate that $i_1 = 10$ A, $V_1 = 126$ V, $P = 400$ W, and $V_2 = 0$. Determine the inverse hybrid parameters.

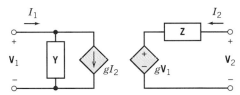

Figure P 18.7-6 Model of a transformer.

Section 18-8 Relationships Between Two-Port Parameters

P 18.8-1 Derive the relationships between the Y parameters and the h parameters by utilizing the defining equations for both parameter sets.

P 18.8-2 Determine the Y parameters if the Z parameters are (in ohms):

$$\mathbf{Z} = \begin{bmatrix} 3 & 2 \\ 2 & 6 \end{bmatrix}$$

P 18.8-3 Determine the h parameters when the Y parameters are (in siemens):

$$\mathbf{Y} = \begin{bmatrix} 0.1 & 0.1 \\ 0.4 & 0.5 \end{bmatrix}$$

P 18.8-4 A two-port has the following Y parameters: $Y_{12} = Y_{21} = -0.4$ S, $Y_{11} = 0.5$ S, and $Y_{22} = 0.6$ S. Determine the h parameters.
Answer: $h_{11} = 2 \ \Omega$, $h_{21} = -0.8$, $h_{12} = 0.8$, $h_{22} = 0.28$ S

Section 18-9   Interconnection of Two-Port Networks

P 18.9-1   Connect in parallel the two circuits shown in Figure P 18.9-1 and find the Y parameters of the parallel combination.

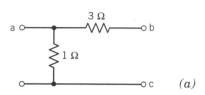

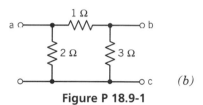

Figure P 18.9-1

P 18.9-2   For the T network of Figure P 18.9-2 find the Y and T parameters and determine the resulting parameters after the two two-ports are connected in (a) parallel and (b) cascade. Both two-ports are identical as defined in Figure 18-1.

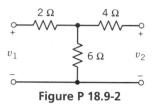

Figure P 18.9-2

P 18.9-3   Determine the Y parameters of the parallel combination of the circuits of Figure 18.9-3a and b.

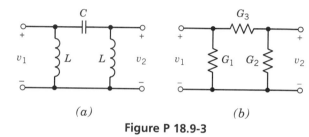

Figure P 18.9-3

VERIFICATION PROBLEMS

VP 18-1   A laboratory report concerning the circuit of Figure VP 18-1 states that $Z_{12} = 15\ \Omega$ and $Y_{11} = 24$ mS. Verify these results.

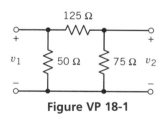

Figure VP 18-1

VP 18-2   A student report concerning the circuit of Figure VP 18-2 has determined the transmission parameters as $A = \dfrac{2(s + 10)}{s}$, $D = -A$, $C = 0.1$ S, and $B = \dfrac{-(3s^2 + 80s + 40)}{s^2}$. Verify these results when $M = 0.1$ H.

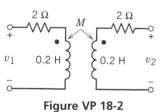

Figure VP 18-2

PSpice PROBLEMS

SP 18-1   Calculate the impedance parameters Z_{11} and Z_{21} of the circuit of Figure P 18.5-1.

SP 18-2   Calculate the admittance parameters Y_{11} and Y_{21} for the circuit of Figure SP 18-2.

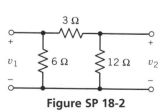

Figure SP 18-2

SP 18-3 Determine the hybrid parameters of the circuit of Figure SP 18-3.

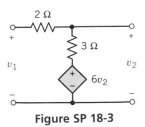

Figure SP 18-3

SP 18-4 Determine the h parameters h_{11} and h_{21} of the circuit shown in Figure SP 18-4.

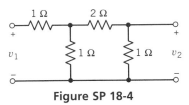

Figure SP 18-4

DESIGN PROBLEMS

DP 18-1 Select R_1 and R so that $R_{in} = 16.6\ \Omega$ for the circuit of Figure DP 18-1. A design constraint requires that both R_1 and R be less than 10 Ω.

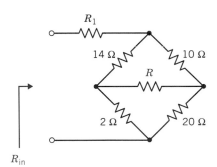

Figure DP 18-1

DP 18-2 The bridge circuit shown in Figure DP 18-2 is said to be balanced when $I = 0$. Determine the required relationship for the bridge resistances when balance is achieved.

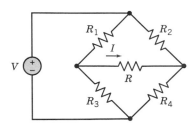

Figure DP 18-2 Bridge circuit.

DP 18-3 A hybrid model of a common-emitter transistor amplifier is shown in Figure DP 18-3. The transistor parameters are $h_{21} = 80$, $h_{11} = 45\ \Omega$, $h_{22} = 12.5\ \mu S$, and $h_{12} = 5 \times 10^{-4}$. Select R_L so that the current gain $i_2/i_1 = 79$ and the input resistance of the amplifier is less than 10 Ω.

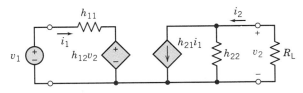

Figure DP 18-3 Model of transistor amplifier.

DP 18-4 A two-port network connected to a source v_s and a load resistance R_L is shown in Figure DP 18-4.

(a) Determine the impedance parameters of the two-port network.

(b) Select R_L so that maximum power is delivered to R_L.

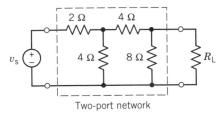

Two-port network

Figure DP 18-4

DP 18-5 (a) Determine the ABCD (transmission matrix) of the two-port networks shown in Figures DP 18-5a and DP 18-5b. (b) Using the results of (a), find the s-domain ABCD matrix of the network shown in Figure DP 18-5c. (c) Given $L_1 = (10/\pi)$ mH, $L_2 = (2.5/\pi)$ mH, $C_1 = (0.78/\pi)\,\mu F$, $C_2 = C_3 = (1/\pi)\,\mu F$, and $R_L = 100\,\Omega$, find the open-circuit voltage gain V_2/V_1 and the short-circuit current gain I_2/I_1 under sinusoidal-state conditions at the following frequencies: 2.5 kHz, 5.0 kHz, 7.5 kHz, 10 kHz, and 12.5 kHz. (Hint: Use the appropriate entries of the ABCD matrix. Also note the resonant frequencies of the circuit.)

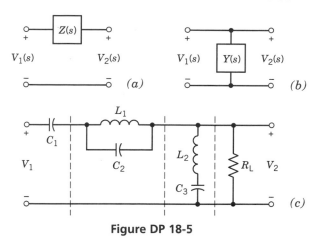

Figure DP 18-5

CHAPTER 19

BALANCED THREE-PHASE CIRCUITS

PSpice Problems

Verification Problems

Design Problems

PREVIEW

In this chapter we will consider the usefulness of a source consisting of three phase voltages of equal magnitude and frequency and where one phase voltage differs by 120° in phase from the other two. This three-phase source can be connected in two formats called the Y and Δ connections. Similarly, we will show how the load circuit can be connected in the Y or Δ format. In this chapter we consider only balanced loads.

We will then proceed to determine the current and voltage in a three-phase circuit using phasors. Finally, we will determine the power delivered to a three-phase load and demonstrate the two-wattmeter method of calculating the load power.

19-1 DESIGN CHALLENGE

POWER FACTOR CORRECTION

Figure 19D-1 shows a three-phase circuit. The capacitors are added to improve the power factor of the load. We need to determine the value of the capacitance C required to obtain a power factor of 0.9 lagging.

In this chapter we will describe three-phase circuits. In particular, we will show that balanced three-phase circuits can be analyzed by using "per-phase" equivalent circuits. This will enable us to calculate the capacitance required to correct the power factor of the load. We will return to this design problem at the end of the chapter.

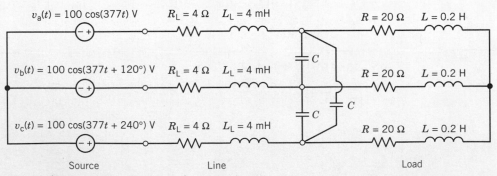

Figure 19D-1 A balanced three-phase circuit.

19-2 | TESLA AND POLYPHASE CIRCUITS

Nikola Tesla, 1856–1943, greatly influenced the adoption of ac power. He first worked for Edison, who was quoted as saying, "My personal desire would be to prohibit entirely the use of alternating currents. They are as unnecessary as they are dangerous." Tesla left Edison and within five years had obtained ten patents for ac motors employing his induction motor principle and his efficient polyphase power system. Manufacturer George Westinghouse bought these patents, and the two 3725-kW alternators that first supplied power at Niagara Falls in 1895 were based on Tesla's patents.

Tesla was awarded patent rights, after much litigation, for his invention of the polyphase-current system to run an induction motor. Polyphase (as distinguished from single-phase) alternating current comes from coils in the generator, wound to produce two or more separate circuits delivering current to the terminals in such a way that they are in sequence as the machine rotates. During the following few decades there were innumerable improvements in motors and a tremendous diversification of designs to run almost anything from a toy train to a ship.

However, because the early Westinghouse $133\frac{1}{3}$ Hz alternators, as alternating-current generators are often called, produced power at a frequency too high for efficient motor operation, 25- and 60-Hz alternators gradually became standard in the United States. At 60 Hz the human eye does not detect flicker in an incandescent lamp, while at 25 Hz a flicker is quite noticeable, especially in small lamps. So, for a time, 60 Hz was used when lighting loads predominated and 25 Hz was used when power for motors was more important. In countries other than the United States, the standard lighting frequency became 50 Hz and the power frequency became $16\frac{2}{3}$ Hz. The large Westinghouse alternators at the Columbia Exposition in Chicago in 1893 were 60-Hz machines. The voltages of alternating and direct current were also becoming standardized in the 1890s and by the end of the century, 110 V was the common American voltage for each phase of the polyphase circuits.

19-3 | THREE-PHASE VOLTAGES

The generation and transmission of electrical power are more efficient in polyphase systems employing combinations of two, three, or more sinusoidal voltages. In addition, polyphase circuits and machines possess some unique advantages. For example, the power transmitted in a three-phase circuit is constant or independent of time rather than pulsating, as it is in a single-phase circuit. Also, three-phase motors start and run much better than do single-phase motors. The most common form of polyphase system employs three balanced voltages, equal in magnitude and differing in phase by $360°/3 = 120°$.

An elementary ac generator consists of a rotating magnet and a stationary winding. The turns of the winding are spread along the periphery of the machine. The voltage generated in each turn of the winding is slightly out of phase with the voltage generated in its neighbor because it is cut by maximum magnetic flux density an instant earlier or later. The voltage produced in the first winding is $v_{aa'}$.

If the first winding were continued around the machine, the voltage generated in the last turn would be 180° out of phase with that in the first and they would cancel, producing no useful effect. For this reason, one winding is commonly spread over no more than one-third of the periphery; the other two-thirds of the periphery can hold two more windings used to generate two other similar voltages. A simplified version of three windings around the periphery of a cylindrical drum is shown in Figure 19-1a. The three

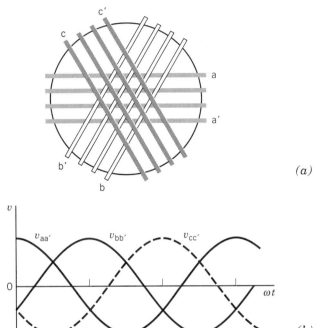

(a)

Figure 19-1 *(a)* The three windings on a cylindrical drum used to obtain three-phase voltages (end view). *(b)* Balanced three-phase voltages.

(b)

sinusoids (sinusoids are obtained with a proper winding distribution and magnet shape) generated by the three similar windings are shown in Figure 19-1b. Defining $v_{aa'}$ as the potential of terminal a with respect to terminal a′, we describe the voltages as

$$v_{aa'} = \sqrt{2}\, V \cos \omega t$$

$$v_{bb'} = \sqrt{2}\, V \cos(\omega t - 120°)$$

$$v_{cc'} = \sqrt{2}\, V \cos(\omega t - 240°) \tag{19-1}$$

where V is the effective value.

A **three-phase circuit** generates, distributes, and uses energy in the form of three voltages equal in magnitude and symmetric in phase.

The three similar portions of a three-phase system are called *phases*. Because the voltage in phase aa′ reaches its maximum first, followed by that in phase bb′ and then by that in phase cc′, we say the phase rotation is abc. This is an arbitrary convention; for any given generator the phase rotation may be reversed by reversing the direction of rotation. The six-terminal ac generator is shown in Figure 19-2.

Using phasor notation, we may write Eq. 19-1 as

$$\mathbf{V}_{aa'} = V\underline{/0°}$$

$$\mathbf{V}_{bb'} = V\underline{/-120°}$$

$$\mathbf{V}_{cc'} = V\underline{/-240°} = V\underline{/120°} \tag{19-2}$$

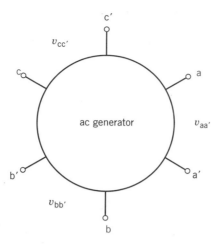

Figure 19-2 Generator with six terminals.

The three voltages are said to be *balanced* because they have identical amplitude, V, and frequency, ω, and are out of phase with each other by exactly 120°. The phasor diagram of the balanced three-phase voltages is shown in Figure 19-3. Examining Figure 19-3, we find

$$\mathbf{V}_{aa'} + \mathbf{V}_{bb'} + \mathbf{V}_{cc'} = 0 \tag{19-3}$$

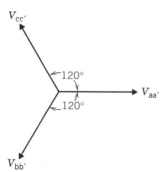

Figure 19-3 Phasor representation of the positive phase sequence of the balanced three-phase voltages.

For notational ease we henceforth use $\mathbf{V}_{aa'} = \mathbf{V}_a$, $\mathbf{V}_{bb'} = \mathbf{V}_b$, and $\mathbf{V}_{cc'} = \mathbf{V}_c$ as the three voltages.

The **positive phase** sequence is abc, as shown in Figure 19-3. The sequence acb is called the negative phase sequence, as shown in Figure 19-4.

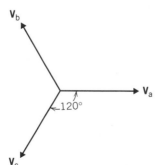

Figure 19-4 The negative phase sequence acb in the Y connection.

Often the phase voltage in the Y connection is written as

$$\mathbf{V}_a = V_p\underline{/0°}$$

where V_p is the magnitude of the phase voltage.

Referring to the generator of Figure 19-2, there are six terminals and three voltages, v_a, v_b, and v_c. We use phasor notation and assume each phase winding provides a source voltage in series with a negligible impedance. Under these assumptions, there are two ways of interconnecting the three sources, as shown in Figure 19-5. The common terminal

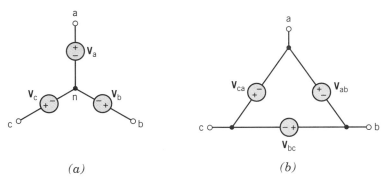

(a)   (b)

Figure 19-5 (a) Y-connected sources and (b) Δ-connected sources.

of the Y connection is called the *neutral terminal* and is labeled n. The neutral terminal may or may not be available for connection. Balanced loads result in no current in a neutral wire and thus it is often not needed.

The connection shown in Figure 19-5a is called the Y connection, and the Δ connection is shown in Figure 19-5b. The Y connection selects terminal a', b', and c' and connects them together as neutral. Then the line-to-line voltage, $\mathbf{V}_{ab}$, of the Y-connected sources is

$$\mathbf{V}_{ab} = \mathbf{V}_a - \mathbf{V}_b \tag{19-4}$$

as is evident by examining Figure 19-5a. Since $\mathbf{V}_a = V_p\underline{/0°}$ and $\mathbf{V}_b = V_p\underline{/-120°}$, we have

$$\begin{aligned}
\mathbf{V}_{ab} &= V_p - V_p(-0.5 - j0.866) \\
&= V_p(1.5 + j0.866) \\
&= \sqrt{3}\, V_p\underline{/30°}
\end{aligned} \tag{19-5}$$

This relationship is also demonstrated by the phasor diagram of Figure 19-6. Similarly,

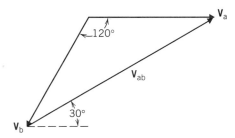

Figure 19-6 The line-to-line voltage $\mathbf{V}_{ab}$ of the Y-connected source.

$$\mathbf{V}_{bc} = \sqrt{3}\, V_p\underline{/-90°} \tag{19-6}$$

and

$$\mathbf{V}_{ca} = \sqrt{3}\, V_p\underline{/-210°} \tag{19-7}$$

Therefore, in a Y connection, the line-to-line voltage is $\sqrt{3}$ times the phase voltage and is displaced 30° in phase. The line current is equal to the phase current. We normally use V_p to remind us that it is the phase voltage.

The Y-connected three-phase voltage source has $\mathbf{V}_c = 120 \underline{/-240°}$ V. Find the line-to-line voltage $\mathbf{V}_{bc}$.
Answer: $207.8\underline{/-90°}$ V

19-4 | THE BALANCED Y-to-Y CIRCUIT

In this section, we will consider the Y-to-Y circuit connection as shown in Figure 19-7 with a neutral wire. When $\mathbf{Z}_a = \mathbf{Z}_b = \mathbf{Z}_c = \mathbf{Z}_Y$, we have

$$\mathbf{I}_A = \frac{\mathbf{V}_{an}}{\mathbf{Z}_Y}, \qquad \mathbf{I}_B = \frac{\mathbf{V}_{bn}}{\mathbf{Z}_Y} = \mathbf{I}_A\underline{/-120°}$$

and

$$\mathbf{I}_C = \frac{\mathbf{V}_{cn}}{\mathbf{Z}_Y} = \mathbf{I}_A\underline{/-240°}$$

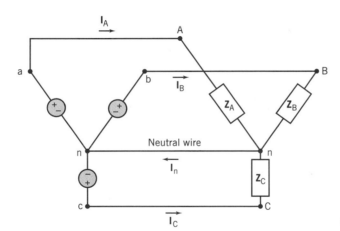

Figure 19-7 A Y-to-Y connected circuit.

Therefore, the current in the neutral wire for a balanced load is

$$\mathbf{I}_n = \mathbf{I}_A + \mathbf{I}_B + \mathbf{I}_c = 0 \qquad (19-8)$$

When each phase has the same line impedance, load impedance, and source impedance, then the circuit is balanced and $\mathbf{I}_n = 0$. Then the voltage across the neutral wire is

$$\mathbf{V}_{nn} = \mathbf{Z}_n\mathbf{I}_n = 0$$

where $\mathbf{Z}_n$ is the impedance of the neutral wire. Therefore, we can remove the neutral wire or replace it with a short circuit when calculating the line current. Then the line currents are

$$\mathbf{I}_A = \frac{\mathbf{V}_a}{\mathbf{Z}}$$

$$\mathbf{I}_B = \frac{\mathbf{V}_b}{\mathbf{Z}}$$

and

$$\mathbf{I}_C = \frac{\mathbf{V}_c}{\mathbf{Z}} \qquad (19\text{-}9)$$

where, if each phase has a source impedance $\mathbf{Z}_s$, a line impedance $\mathbf{Z}_1$, and a load impedance $\mathbf{Z}_L$, then we have

$$\mathbf{Z} = \mathbf{Z}_s + \mathbf{Z}_1 + \mathbf{Z}_L \qquad (19\text{-}10)$$

Thus, the line current in each line will be equal in amplitude and frequency and will be exactly 120° out of phase with the other two line currents.

The behavior of a Y-to-Y circuit is summarized in Table 19-1.

Example 19-1

A balanced three-phase Y system is connected to a Y load as shown in Figure 19-7 where $V_p = 114.9$ V. The source impedance is $\mathbf{Z}_s = 0.52 + j0.50$ Ω. The line impedance is $\mathbf{Z}_1 = 0.80 + j1$ Ω, and the load impedance is $\mathbf{Z}_L = 18.6 + j10$ Ω. Determine the three line currents.

Solution

The total impedance of each phase is

$$\mathbf{Z}_Y = \mathbf{Z}_s + \mathbf{Z}_1 + \mathbf{Z}_L$$
$$= 19.92 + j11.5 \ \Omega$$

Then,

$$\mathbf{I}_A = \frac{\mathbf{V}_a}{\mathbf{Z}_Y} = \frac{114.9\underline{/0°}}{23\underline{/30°}} = 5\underline{/-30°} \text{ A}$$

Because this is a balanced system, $\mathbf{I}_A$ leads $\mathbf{I}_B$ by 120° and lags $\mathbf{I}_c$ by 120°. Then

$$\mathbf{I}_B = 5\underline{/-150°} \text{ A}$$

Table 19-1
The Balanced Y-to-Y Circuit

Phase voltages	$\mathbf{V}_a = V_p\underline{/0°}$
	$\mathbf{V}_b = V_p\underline{/-120°}$
	$\mathbf{V}_c = V_p\underline{/-240°}$
Line-to-line voltages	$\mathbf{V}_{ab} = \mathbf{V}_L$
	$\quad = \sqrt{3}\, V_p\underline{/30°}$
	$\mathbf{V}_{bc} = \sqrt{3}\, V_p\underline{/-90°}$
	$\mathbf{V}_{ca} = \sqrt{3}\, V_p\underline{/-210°}$
	$V_L = \sqrt{3}\, V_p$
Currents	$\mathbf{I}_L = \mathbf{I}_p$ (line current = phase current)
	$\mathbf{I}_A = \dfrac{\mathbf{V}_a}{\mathbf{Z}_p} = \dfrac{\mathbf{V}_a}{\mathbf{Z}_Y} = I_p\underline{/-\theta} \quad$ with $\mathbf{Z}_p = Z\underline{/\theta}$
	$\mathbf{I}_B = \mathbf{I}_A\underline{/-120°}$
	$\mathbf{I}_C = \mathbf{I}_A\underline{/-240°}$

p = phase, L = line.

and

$$\mathbf{I}_C = 5\underline{/90°} \text{ A}$$

EXERCISE 19-2

The phase voltage, V_p, of a Y generator is 120 V, which is connected to a balanced Y load with the total phase impedance $\mathbf{Z}_Y = 40 + j30 \ \Omega$. Find each line current.

Answers: $I_A = 2.4\underline{/-36.9°}$, $I_B = 2.4\underline{/-156.9°}$ A, $I_C = 2.4\underline{/83.1°}$ A

19-5 ┃ THE Δ-CONNECTED SOURCE AND LOAD

The Δ-connected source is shown in Figure 19-5b. This generator connection, however, is seldom used in practice because any slight imbalance in magnitude or phase of the three-phase voltages will not result in a zero sum. The result will be a large circulating current in the generator coils that will heat the generator and depreciate the efficiency of the generator. For example, consider the condition

$$\mathbf{V}_{ab} = 120\underline{/0°}$$

$$\mathbf{V}_{bc} = 120.1\underline{/-121°}$$

$$\mathbf{V}_{ca} = 120.2\underline{/121°}$$

If the total resistance around the loop is 1 Ω, we can calculate the circulating current as

$$
\begin{aligned}
\mathbf{I} &= (\mathbf{V}_{ab} + \mathbf{V}_{bc} + \mathbf{V}_{ca})/1 \\
&= 120 + 120.1(-0.515 - j0.857) + 120.2(-0.515 + j0.857) \\
&\cong 120 - 1.03(120.15) \\
&\cong -3.75 \text{ A}
\end{aligned}
\tag{19-11}
$$

which would be unacceptable.

Therefore, we will consider only a Y-connected source as practical at the source side and consider both the Δ-connected load and the Y-connected load at the load side.

The Δ-to-Y transformation we discussed in Section 18-4 can be utilized to transform a balanced network from one form to another. Then, we have

$$\mathbf{Z}_Y = \frac{\mathbf{Z}_\Delta}{3} \tag{19-12}$$

We simply convert the Δ load to a Y load to match the Y source so that we require a calculation for a Y-to-Y circuit.

Therefore, if we have a Y-connected source and a Δ-connected load with $\mathbf{Z}_\Delta$, we convert the Δ load to a Y load with $\mathbf{Z}_Y = \mathbf{Z}_\Delta/3$. Then the line current is

$$\mathbf{I}_A = \frac{\mathbf{V}_a}{\mathbf{Z}_Y} = \frac{3\mathbf{V}_a}{\mathbf{Z}_\Delta} \tag{19-13}$$

Thus, we will consider only the Y-to-Y configuration. If the Y-to-Δ configuration is encountered, the balanced Δ load is converted to a Y-connected load equivalent, and the resulting currents and voltages are calculated.

19-6 THE Y-to-Δ CIRCUIT

Now, let us consider the Y-to-Δ circuit as shown in Figure 19-8. The goal is to calculate the line and phase currents for the load. We immediately note that the line-to-line voltages of the Y source are equal to the voltages across the Δ impedances. Thus, $\mathbf{V}_{ab}$ appears across $\mathbf{Z}_\Delta$, as shown in Figure 19-8.

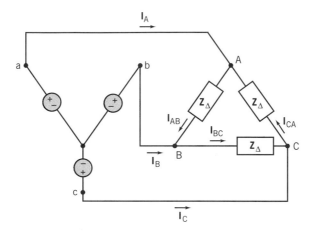

Figure 19-8 The Y-to-Δ circuit.

Examining Figure 19-8, we note that the relation between the line currents and phase currents is

$$\mathbf{I}_A = \mathbf{I}_{AB} - \mathbf{I}_{CA}$$

$$\mathbf{I}_B = \mathbf{I}_{BC} - \mathbf{I}_{AB}$$

and

$$\mathbf{I}_C = \mathbf{I}_{CA} - \mathbf{I}_{BC} \tag{19-14}$$

If $\mathbf{I}_{AB} = I\underline{/\phi^\circ}$, then $\mathbf{I}_{CA} = I\underline{/\phi + 120^\circ}$ for the ABC sequence and

$$\begin{aligned}
\mathbf{I}_A &= \mathbf{I}_{AB} - \mathbf{I}_{CA} \\
&= I\cos\phi + jI\sin\phi - I\cos(\phi + 120^\circ) - jI\sin(\phi + 120^\circ) \\
&= -2I\sin(\phi + 60^\circ)\sin(-60^\circ) + j2I\cos(\phi + 60^\circ)\sin(-60^\circ) \\
&= \sqrt{3}\,I[\sin(\phi + 60^\circ) - j\cos(\phi + 60^\circ)] \\
&= \sqrt{3}\,I[\cos(\phi - 30^\circ) + j\sin(\phi - 30^\circ)] \\
&= \sqrt{3}\,I\underline{/\phi - 30^\circ}\ \text{A}
\end{aligned} \tag{19-15}$$

Therefore,

$$|I_A| = \sqrt{3}\,|I| \tag{19-16}$$

or

$$I_L = \sqrt{3}\,I_p$$

and the line current magnitude is $\sqrt{3}$ times the phase current magnitude. This result can also be obtained from the phasor diagram shown in Figure 19-9. In a Δ connection, the line current is $\sqrt{3}$ times the phase current and is displaced -30° in phase. The line-to-line voltage is equal to the phase voltage.

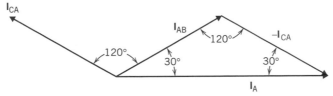

Figure 19-9 Phasor diagram for currents of a Δ load.

Example 19-2

A source provides a three-phase line-to-line voltage of 220 V and a phase rotation of abc. It is connected to a Δ-connected load $\mathbf{Z}_\Delta = 10\underline{/-50°}\ \Omega$ as shown in Figure 19-8. Determine the phase and line currents.

Solution

The phase voltages are equal to the line-to-line voltages. The phase currents are

$$\mathbf{I}_{AB} = \frac{\mathbf{V}_{AB}}{\mathbf{Z}} = \frac{220\underline{/0°}}{10\underline{/-50°}} = 22\underline{/50°}\ A$$

$$\mathbf{I}_{BC} = \frac{\mathbf{V}_{BC}}{\mathbf{Z}} = \frac{220\underline{/-120°}}{10\underline{/-50°}} = 22\underline{/-70°}\ A$$

$$\mathbf{I}_{CA} = \frac{\mathbf{V}_{CA}}{\mathbf{Z}} = \frac{220\underline{/-240°}}{10\underline{/-50°}} = 22\underline{/-190°}\ A$$

The line currents are

$$\mathbf{I}_A = \mathbf{I}_{AB} - \mathbf{I}_{CA} = 22\underline{/50°} + 22\underline{/-190°}$$
$$= 22\sqrt{3}\ \underline{/20°}\ A$$

Then

$$\mathbf{I}_B = 22\sqrt{3}\ \underline{/-100°}\ A \quad \text{and} \quad \mathbf{I}_C = 22\sqrt{3}\ \underline{/-220°}\ A$$

The current and voltage relationships for a Δ load are summarized in Table 19-2.

Table 19-2
The Current and Voltage for a Δ Load

Phase voltages	$\mathbf{V}_{AB} = \mathbf{V}_{AB}\underline{/0°} = V_p\underline{/0°}$
Line-to-line voltages	$\mathbf{V}_{AB} = \mathbf{V}_L$ (line voltage = phase voltage)
Phase currents	$\mathbf{I}_{AB} = \dfrac{\mathbf{V}_{AB}}{\mathbf{Z}_p} = \dfrac{\mathbf{V}}{\mathbf{Z}_\Delta} = I_p\underline{/-\theta}$
	with $\mathbf{Z}_p = Z\underline{/\theta}$
	$\mathbf{I}_{BC} = \mathbf{I}_{AB}\underline{/-120°}$
	$\mathbf{I}_{CA} = \mathbf{I}_{AB}\underline{/-240°}$
Line currents	$\mathbf{I}_A = \sqrt{3}\,I_p\underline{/-\theta-30°}$
	$\mathbf{I}_B = \sqrt{3}\,I_p\underline{/-\theta-150°}$
	$\mathbf{I}_C = \sqrt{3}\,I_p\underline{/-\theta+90°}$
	$I_L = \sqrt{3}\,I_p$

L = line, p = phase.

EXERCISE 19-3

A three-phase Δ-connected load, as shown in Figure 19-8, has a load of $\mathbf{Z}_\Delta = 180\underline{/-45°}\ \Omega$ in each phase. Determine the phase and line currents when the line-to-line voltage is 360 V.

Partial Answers: $\mathbf{I}_{AB} = 2\underline{/45°}$ A, $\qquad \mathbf{I}_A = 3.46\underline{/15°}$ A

19-7 THE PER-PHASE EQUIVALENT CIRCUIT

We have only two possible practical configurations, Y-to-Y and Y-to-Δ, and we can convert the latter to a Y-to-Y form. Thus, the basic balanced circuit can always be converted to the Y circuit shown in Figure 19-10. Since both the source and the load are balanced we can add a short circuit connecting the neutral of the load to the neutral of the source. The phase currents in Figure 19-10 will be balanced and the current in phase A is

$$\mathbf{I}_A = \frac{\mathbf{V}_a}{\mathbf{Z}_A} \tag{19-17}$$

where $\mathbf{Z}_A = \mathbf{Z}_Y$.

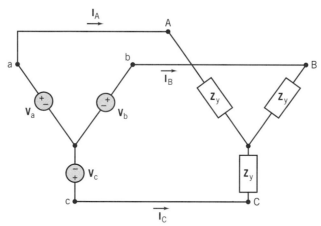

Figure 19-10 The Y-to-Y circuit obtained by converting the Y-to-Δ circuit of Figure 19-8 where $\mathbf{Z}_Y = \mathbf{Z}_\Delta/3$.

We can then construct a per-phase equivalent circuit of the balanced three-phase Y-to-Y circuit. We call the one-phase equivalent circuit a per-phase circuit because it enables us to calculate the current per phase. It follows that the current in the A-phase conductor line is simply the voltage generated in the A-phase winding of the generator divided by the total impedance in the A phase of the circuit. Thus, Eq. 19-17 describes the simple circuit in Figure 19-11, where the neutral conductor has been replaced by a perfect short circuit.

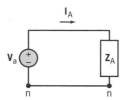

Figure 19-11 The per-phase equivalent circuit for phase a in a balanced Y-to-Y circuit.

To illustrate, let us consider the Y-to-Δ circuit of Figure 19-8. We convert the circuit of Figure 19-8 to that of Figure 19-10 and let $\mathbf{Z}_Y = \mathbf{Z}_\Delta/3$. (See Eqn. 18-7). We choose $\mathbf{V}_a = V_p\underline{/0°}$ as the reference and then

$$\mathbf{V}_b = V_p\underline{/-120°}, \qquad \mathbf{V}_c = V_p\underline{/120°}$$

where V_p is the effective value of the phase voltage. Since $\mathbf{Z}_Y = \mathbf{Z}_\Delta/3$, we know the value of $\mathbf{Z}_Y$ from the balanced load. The line current in Figure 19-11 is

$$\mathbf{I}_A = \frac{\mathbf{V}_a}{\mathbf{Z}_A} = \frac{\mathbf{V}_a}{\mathbf{Z}_Y}$$

Furthermore,

$$\mathbf{I}_B = \frac{\mathbf{V}_b}{\mathbf{Z}_Y} \quad \text{and} \quad \mathbf{I}_C = \frac{\mathbf{V}_c}{\mathbf{Z}_Y}$$

$\mathbf{I}_A$, $\mathbf{I}_B$, and $\mathbf{I}_C$ are the line currents of the Δ circuit.

Example 19-3

A three-phase Y source with a phase voltage of 120 V is connected to a Δ-connected load with $\mathbf{Z}_\Delta = 30\underline{/45°}$ Ω. Find the phase and line currents.

Solution

The equivalent Y load is

$$\mathbf{Z}_Y = \frac{\mathbf{Z}_\Delta}{3} = 10\underline{/45°} \ \Omega$$

Then the line current is (see Figure 19-11)

$$\mathbf{I}_A = \frac{120\underline{/0°}}{10\underline{/45°}} = 12\underline{/-45°} \ A$$

Therefore

$$\mathbf{I}_B = 12\underline{/-165°} \ A \quad \text{and} \quad \mathbf{I}_C = 12\underline{/-285°} \ A$$

The phase current is obtained from Figure 19-8 where

$$\mathbf{I}_{AB} = \frac{\mathbf{V}_{AB}}{\mathbf{Z}_\Delta}$$

The line-to-line voltage of the Y source becomes the phase voltage of the Δ load. Using Eq. 19-5, we have

$$\mathbf{V}_{AB} = \sqrt{3} \ V_a\underline{/30°}$$

$$= 207.8\underline{/30°} \ V$$

where V_a is the phase voltage of the Y configuration (see Figure 19-8). Then

$$\mathbf{I}_{AB} = \frac{207.8 \underline{/30°}}{30 \underline{/45°}} = 6.93 \underline{/-15°} \text{ A}$$

and, in turn,

$$\mathbf{I}_{BC} = 6.93 \underline{/-135°} \text{ A} \quad \text{and} \quad \mathbf{I}_{CA} = 6.93 \underline{/+105°} \text{ A}$$

EXERCISE 19-4

A balanced Δ load shown in Figure 19-8 with impedances $\mathbf{Z}_\Delta = 24 \underline{/-40°} \, \Omega$ is connected to a Y source with $\mathbf{V}_a = 277 \underline{/0°}$ V. Find the line current $\mathbf{I}_A$ and the phase current $\mathbf{I}_{AB}$.
Answers: $\mathbf{I}_A = 34.6 \underline{/40°}$ A, $\qquad \mathbf{I}_{AB} = 20 \underline{/70°}$ A

19-8 ‖ POWER IN A BALANCED LOAD

The total power delivered to a balanced Y-connected three-phase load is (Figure 19-10)

$$\begin{aligned} P_Y &= 3P_A \\ &= 3V_A I_A \cos \theta \end{aligned} \tag{19-18}$$

where $\cos \theta$ is the power factor, θ is the angle between the phase voltage and the phase current, and V_A and I_A are effective values.

It is easier to measure the line-to-line voltage and the line current of a circuit. Also recall that the line current equals the phase current and that the phase voltage is $V_A = V_{AB}/\sqrt{3}$ for the Y-load configuration. Therefore,

$$\begin{aligned} P &= 3 \frac{V_{AB}}{\sqrt{3}} I_A \cos \theta \\ &= \sqrt{3} \, V_{AB} I_A \cos \theta \end{aligned} \tag{19-19}$$

The power delivered to a Δ-connected load is

$$\begin{aligned} P &= 3P_{AB} \\ &= 3 V_{AB} I_{AB} \cos \theta \end{aligned} \tag{19-20}$$

Since the line-to-line voltage is equal to the phase voltage for a Δ-connected load and the phase current is related to the line current as

$$I_{AB} = I_A / \sqrt{3}$$

we have

$$\begin{aligned} P &= 3 V_A \frac{I_A}{\sqrt{3}} \cos \theta \\ &= \sqrt{3} \, V_{AB} I_A \cos \theta \end{aligned}$$

Using L as the line-to-line notation and p as the phase notation, we summarize the power delivered to a Y-connected and a Δ-connected load in Table 19-3. Thus, the power

Table 19-3

Power in a Balanced Three-Phase Circuit

	Power	
Y-Connected Load		Δ-Connected Load
$P_Y = 3V_p I_p \cos\theta$		$P = 3V_p I_p \cos\theta$
θ is the angle between phase voltage and current		
$P_Y = \sqrt{3} V_L I_L \cos\theta$		$P = \sqrt{3} V_L I_L \cos\theta$

L = line, p = phase.

delivered to the load is

$$P = \sqrt{3} V_L I_L \cos\theta \tag{19-21}$$

Remember, θ is the angle between the phase voltage and the phase current.

Example 19-4

A balanced three-phase load as shown in Figure 19-10 has $\mathbf{V}_{AB} = 440\underline{/0°}$ V and $\mathbf{I}_A = 2\underline{/-75°}$ A. Determine the power delivered to the load.

Solution

The magnitude of the line voltage is $V_L = 440$ V, and the magnitude of the line current is $I_L = 2$ A. We determine the power from

$$P = \sqrt{3} \, V_L I_L \cos\theta$$

where θ is the angle between the phase voltage and the phase current. In the Y configuration the line current equals the phase current, so we need to determine the phase voltage $\mathbf{V}_A$. Recall Eq. 19-5, which states

$$\mathbf{V}_{AB} = \sqrt{3} \, V_p\underline{/\phi + 30°}$$

when $\mathbf{V}_A = V_p\underline{/\phi}$. Therefore, we note that

$$\mathbf{V}_A = \frac{440}{\sqrt{3}} \underline{/-30°} \text{ V}$$

Since $\mathbf{I}_A = 2\underline{/-75°}$ A, we have the angle between $\mathbf{V}_A$ and $\mathbf{I}_A$ as $\theta = 45°$. Then, we obtain

$$P = \sqrt{3} \, (2)(440) \cos 45° = 1077.8 \text{ W}$$

One of the advantages of three-phase power is the smooth flow of energy to the load. Consider a balanced load with resistance R. Then the instantaneous power is

$$p(t) = \frac{v_{ab}^2}{R} + \frac{v_{bc}^2}{R} + \frac{v_{ca}^2}{R} \tag{19-22}$$

where $v_{ab} = V\cos\omega t$ and the other two phase voltages have a phase of $\pm 120°$, respectively. Furthermore,

$$\cos^2 \alpha t = (1 + \cos 2\alpha)/2$$

Therefore,

$$p(t) = \frac{V^2}{2R}[1 + \cos 2\omega t + 1 + \cos 2(\omega t - 120°) + 1 + \cos 2(\omega t - 240°)]$$

$$= \frac{3V^2}{2R} + \frac{V^2}{2R}[\cos 2\omega t + \cos(2\omega t - 240°) + \cos(2\omega t - 480°)]$$

$$= \frac{3V^2}{2R} \tag{19-23}$$

since the bracketed term is equal to zero for all time. Hence, the instantaneous power is a constant.

EXERCISE 19-5

A balanced Y-connected load has a phase voltage of 220 V and an impedance $\mathbf{Z} = 30 + j40 \ \Omega$. Determine the power absorbed by the three-phase load.
Answer: 1744 W

19-9 ‖ TWO-WATTMETER POWER MEASUREMENT

For many load configurations, for example, a three-phase motor, the phase current or voltage is unaccessible. We may wish to measure power with a wattmeter connected to each phase. However, since the phases are not available, we measure the line currents and the line-to-line voltages. A wattmeter provides a reading of $VI \cos \theta$ where V and I are the rms magnitudes and θ is the angle between $\mathbf{V}$ and $\mathbf{I}$. We choose to measure V_L and I_L, the line voltage and current respectively. We will show that two wattmeters are sufficient to read the power delivered to the three-phase load, as shown in Figure 19-12. We use cc to denote current coil and vc to denote voltage coil.

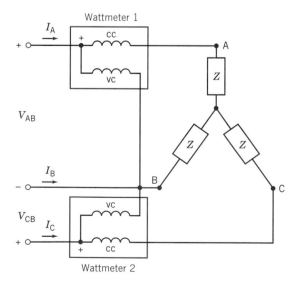

Figure 19-12 Two-wattmeter connection for a three-phase Y-connected load.

Wattmeter 1 reads

$$P_1 = V_{AB} I_A \cos \theta_1 \qquad (19\text{-}24)$$

and wattmeter 2 reads

$$P_2 = V_{CB} I_C \cos \theta_2 \qquad (19\text{-}25)$$

For the abc phase sequence for a balanced load

$$\theta_1 = \theta + 30°$$

and

$$\theta_2 = \theta - 30° \qquad (19\text{-}26)$$

where θ is the angle between the phase current and the phase voltage.

Therefore

$$P = P_1 + P_2 = 2V_L I_L \cos \theta \cos 30°$$
$$= \sqrt{3}\, V_L I_L \cos \theta \qquad (19\text{-}27)$$

which is the total average power of the three-phase circuit. The preceding derivation of Eq. 19-27 is for a balanced circuit; the result is good for any three-phase, three-wire load, even unbalanced or nonsinusoidal voltages.

Example 19-5

The two-wattmeter method is used, as shown in Figure 19-13, to measure the total power delivered to the Y-connected load when $Z = 10\underline{/45°}\ \Omega$ and the supply line-to-line voltage is 220 V. Determine the reading of each wattmeter and the total power.

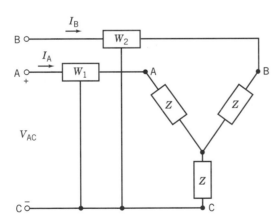

Figure 19-13 The two-wattmeter connection for Example 19-5.

Solution
The phase voltage is (see Table 19-1)

$$\mathbf{V}_A = \frac{220}{\sqrt{3}}\underline{/-30°}\ \text{V}$$

Then we obtain the line current as

$$\mathbf{I}_A = \frac{\mathbf{V}_A}{\mathbf{Z}} = \frac{220\underline{/-30°}}{10\sqrt{3}\underline{/45°}} = 12.7\underline{/-75°}\ \text{A}$$

Then the second line current is

$$\mathbf{I_B} = 12.7\underline{/-195°} \text{ A}$$

The voltage $\mathbf{V_{AB}} = 220\underline{/0°}$ V, $\mathbf{V_{CA}} = 220\underline{/+120°}$ V, and $\mathbf{V_{BC}} = 220\underline{/-120°}$ V. The first wattmeter reads

$$\begin{aligned} P_1 &= I_A V_{AC} \cos \theta_1 \\ &= 12.7(220) \cos 15° \\ &= 2698 \text{ W} \end{aligned}$$

Since $\mathbf{V_{CA}} = 220\underline{/+120°}$, then $\mathbf{V_{AC}} = 220\underline{/-60°}$. Therefore, the angle θ_1 lies between $\mathbf{V_{AC}}$ and $\mathbf{I_A}$ and is equal to $15°$. The reading of the second wattmeter is

$$\begin{aligned} P_2 &= I_B V_{BC} \cos \theta_2 \\ &= 12.7(220) \cos 75° \\ &= 723 \text{ W} \end{aligned}$$

where θ_2 is the angle between $\mathbf{I_B}$ and $\mathbf{V_{BC}}$. Therefore, the total power is

$$P = P_1 + P_2 = 3421 \text{ W}$$

We note that all of the preceding calculations assume that the wattmeter itself absorbs negligible power.

The power factor angle, θ, of a balanced three-phase system may be determined from the reading of the two wattmeters shown in Figure 19-13.

The total power is obtained from Eqs. 19-24 through 19-26 as

$$\begin{aligned} P = P_1 + P_2 &= V_L I_L [\cos(\theta + 30°) + \cos(\theta - 30°)] \\ &= V_L I_L 2 \cos \theta \cos 30° \end{aligned} \tag{19-28}$$

Similarly,

$$P_1 - P_2 = V_L I_L(-2 \sin \theta \sin 30°) \tag{19-29}$$

Dividing Eq. 19-28 by Eq. 19-29, we obtain

$$\frac{P_1 + P_2}{P_1 - P_2} = \frac{2 \cos \theta \cos 30°}{-2 \sin \theta \sin 30°} = \frac{-\sqrt{3}}{\tan \theta}$$

Therefore,

$$\tan \theta = \sqrt{3} \frac{P_2 - P_1}{P_2 + P_1} \tag{19-30}$$

where θ = power factor angle.

Example 19-6

The two wattmeters in Figure 19-12 read $P_1 = 60$ kW and $P_2 = 180$ W, respectively. Find the power factor of the circuit.

Solution

From Eq. 19-30 we have

$$\tan \theta = \sqrt{3} \frac{P_2 - P_1}{P_2 + P_1} = \sqrt{3} \frac{120}{240} = \frac{\sqrt{3}}{2} = 0.866$$

Therefore, we have $\theta = 40.9°$ and the power factor is

$$\text{pf} = \cos \theta = 0.756$$

The positive angle, θ, indicates that the power factor is lagging. If θ is negative, then the power factor is leading.

EXERCISE 19-6

The line current to a balanced three-phase load is 24 A. The line-to-line voltage is 450 V, and the power factor of the load is 0.47 lagging. If two wattmeters are connected as shown in Figure 19-12, determine the reading of each meter and the total power to the load.
Answers: $P_1 = -371$ W, $P_2 = 9162$ W, $P = 8791$ W

EXERCISE 19-7

The two wattmeters are connected as shown in Figure 19-12 with $P_1 = 60$ kW and $P_2 = 40$ kW, respectively. Determine (a) the total power and (b) the power factor.
Answers: (a) 100 kW, (b) 0.945 leading

19-10 THE BALANCED THREE-PHASE CIRCUIT AND PSpice

The analysis of three-phase balanced circuits may be facilitated by using the single-phase equivalent circuit (also called a per-phase circuit) as described in Section 19-7. Once we have arrived at a per-phase circuit, we can use PSpice to solve for the phase and line currents and voltages.

Let us consider a Y-to-Y connected system with a transmission line resistance as shown in Figure 19-14. The per-phase equivalent circuit shown in Figure 19-15 includes the neutral wire but assumes that its resistance is zero since the neutral wire is only added in the equivalent circuit. Let us consider the case where $R = 1\ \Omega$ and $\mathbf{Z} = 9 + j\omega L$ where

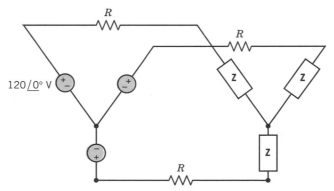

Figure 19-14 A Y-to-Y connection with a line resistance R. The circuit operates at 60 Hz.

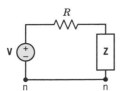

Figure 19-15 The per-phase equivalent circuit.

$f = 60$ Hz and $L = 26.5$ mH. The total impedance is $10 + j10$ and the line current is

$$\mathbf{I_A} = \frac{120\underline{/0°}}{10\sqrt{2}\underline{/45°}} = 8.5\underline{/-45°}\ \text{A}$$

To determine the line current using PSpice, we set up the PSpice program as shown in Figure 19-16. The output of the program yields the line current as IM(VA) = 8.5 and IP(VA) = 135°.

```
VA       1   0    AC     120   0
RLOSS    1   2    1
RLOAD    2   3    9
LLOAD    3   0    26.5M
.AC   LIN   1   60    60
.PRINT   AC    IM(VA)    IP(VA)
.END
```

Figure 19-16 The PSpice program for the per-phase circuit.

19-11 VERIFICATION EXAMPLES

Example 19V-1

Figure 19V-1*a* shows a balanced three-phase circuit. Computer analysis of this circuit produced the element voltages and currents tabulated in Figure 19V-1*b*. Is this computer analysis correct?

Since the three-phase circuit is balanced, it can be analyzed by using a per-phase equivalent circuit. The appropriate per-phase equivalent circuit for this example is shown in Figure 19V-2. This per-phase equivalent circuit can be analyzed by writing a single mesh equation:

$$10 = (9 + j12)\ \mathbf{I_L}(\omega)$$

or

$$\mathbf{I_L}(\omega) = 0.67e^{-j53°}\ \text{A}$$

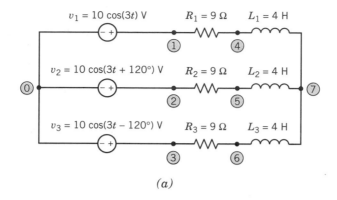

$v_1 = 10 \cos(3t)$ V $R_1 = 9\ \Omega$ $L_1 = 4$ H

$v_2 = 10 \cos(3t + 120°)$ V $R_2 = 9\ \Omega$ $L_2 = 4$ H

$v_3 = 10 \cos(3t - 120°)$ V $R_3 = 9\ \Omega$ $L_3 = 4$ H

(a)

Element	Voltage	Current
V1 1 0 10 $\underline{/0}$	10 $\underline{/0}$	0.67 $\underline{/127}$
V2 2 0 10 $\underline{/120}$	10 $\underline{/120}$	0.67 $\underline{/113}$
V3 3 0 10 $\underline{/-120}$	10 $\underline{/-120}$	0.67 $\underline{/7}$
R1 1 4 9	6 $\underline{/-53}$	0.67 $\underline{/-53}$
R2 2 5 9	6 $\underline{/67}$	0.67 $\underline{/67}$
R3 3 6 9	6 $\underline{/-173}$	0.67 $\underline{/-173}$
L1 4 7 4	8 $\underline{/37}$	0.67 $\underline{/-53}$
L2 5 7 4	8 $\underline{/157}$	0.67 $\underline{/67}$
L3 6 7 4	8 $\underline{/83}$	0.67 $\underline{/-173}$

(b)

Figure 19V-1 (a) A three-phase circuit. (b) The results of computer analysis.

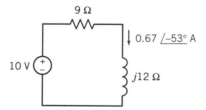

9 Ω

0.67 $\underline{/-53°}$ A

10 V

$j12$ Ω

Figure 19V-2 The per-phase equivalent circuit.

where $\mathbf{I}_L(\omega)$ is the phasor corresponding to the inductor current. The voltage across the inductor is given by

$$\mathbf{V}_L(\omega) = j12\mathbf{I}_L(\omega)$$
$$= 8e^{j37°}\ \text{V}$$

The voltage across the resistor is given by

$$\mathbf{V}_R(\omega) = 9\mathbf{I}_L(\omega)$$
$$= 6e^{-j53°}\ \text{V}$$

These currents and voltages are the same as the values given in the computer analysis for the element currents and voltages of R_1 and L_1. We conclude that the computer analysis of the three-phase circuit is correct.

Example 19V-2

A three-phase Y-connected source with a phase voltage of 120 V is connected via a line to a Y-connected load. The line impedance is $1 + 1j\ \Omega$ and the load impedance is $\mathbf{Z} = 20 + j10\ \Omega$. PSpice is used to determine the load current and voltage and the results are stated as $\mathbf{I}_A = 3\underline{/-52°}$ A and $\mathbf{V}_A = 101\underline{/0°}$ V. Verify these results.

Solution

First we draw the circuit per-phase model as shown in Figure 19V-3. We can do a simple check by using the fact that $\mathbf{V}_A = \mathbf{I}_A\mathbf{Z}$. After substituting the PSpice result for $\mathbf{I}_A$ and

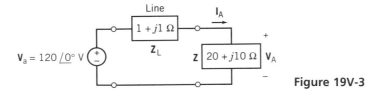

Figure 19V-3

$\mathbf{Z} = 20 + j10$, we get

$$\mathbf{V}_A = (3\underline{/-52°})(20 + j10) = 67.1\underline{/-25.4°}\ \text{V}$$

which does not equal the PSpice result $\mathbf{V}_A = 101\underline{/0°}$ V.

Let us proceed to calculate $\mathbf{I}_A$ by using $\mathbf{I}_A = \mathbf{V}_a/(\mathbf{Z} + \mathbf{Z}_L)$. Then, we have the correct load current

$$\mathbf{I}_A = \frac{120\underline{/0°}}{21 + j11} = 5.06\underline{/-27.65°}\ \text{A}$$

Thus, the correct load voltage is

$$\mathbf{V}_A = \mathbf{I}_A\mathbf{Z} = (5.06\underline{/-27.65°})(22.36\underline{/26.57°})$$
$$= 113.14\underline{/-1.1°}\ \text{V}$$

19-12 DESIGN CHALLENGE SOLUTION

POWER FACTOR CORRECTION

Figure 19D-1 shows a three-phase circuit. The capacitors are added to improve the power factor of the load. We need to determine the value of the capacitance C required to obtain a power factor of 0.9 lagging.

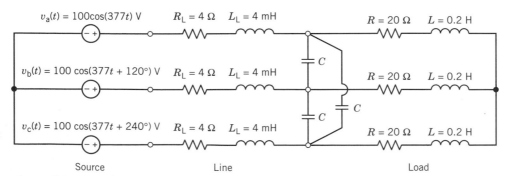

Figure 19D-1 A balanced three-phase circuit.

Define the Situation

1 The circuit is excited by sinusoidal sources all having the same frequency, 60 Hz or 377 rad/s. The circuit is at steady state. The circuit is a linear circuit. Phasors can be used to analyze this circuit.

2 The circuit is a balanced three-phase circuit. A per-phase equivalent circuit can be used to analyze this circuit.

3 The load consists of two parts. The part comprising resistors and inductors is connected as a Y. The part comprising capacitors is connected as a Δ. A Δ-to-Y transformation can be used to simplify the load.

The per-phase equivalent circuit is shown in Figure 19D-2.

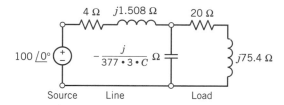

Figure 19D-2 The per-phase equivalent circuit.

State the Goal

Determine the value of C required to correct the power factor to 0.9 lagging.

Generate a Plan

Power factor correction was considered in Chapter 12. A formula was provided for calculating the reactance, X_1, needed to correct the power factor of a load

$$X_1 = \frac{R^2 + X^2}{R \tan (\cos^{-1} \text{pfc}) - X}$$

where R and X are the real and imaginary parts of the load impedance before the power factor is corrected and pfc is the corrected power factor. After this equation is used to calculate X_1, the capacitance C can be calculated from X_1. Notice that X_1 will be the reactance of the equivalent Y-connected capacitors. We will need to calculate the Y connected capacitor equivalent of the Δ connected capacitor.

Take Action on the Plan

We note that $\mathbf{Z} = R + jX = 20 + j75.4 \ \Omega$. Therefore, the reactance, X_1, needed to correct the power factor is

$$X_1 = \frac{20^2 + 75.4^2}{20 \tan(\cos^{-1} 0.9) - 75.4}$$
$$= -92.6$$

The Y connected capacitor equivalent to the Δ connected capacitor can be calculated from Eqn. 18-7, $\mathbf{Z}_Y = \mathbf{Z}_\Delta/3$. Therefore, the capacitance of the equivalent Y connected capacitor is $3C$.

Finally, since $X_1 = 1/(3C\omega)$, we have

$$C = \frac{1}{\omega \cdot 3 \cdot X_1}$$
$$= -\frac{1}{377 \cdot 3(-92.6)}$$
$$= 9.458 \ \mu\text{F}$$

SUMMARY

The generation and transmission of electrical power are more efficient in three-phase systems employing three voltages of the same magnitude and frequency and differing in phase by 120° from each other.

In this chapter we considered balanced three-phase sources in the Y connection and balanced loads in both the Y and the Δ connections. In a Y connection, the line-to-line voltage is $\sqrt{3}$ times the phase voltage and is displaced 30° in phase while the line current is equal to the phase current.

The current in the neutral wire of a Y-to-Y connection is zero and thus the wire may be removed if desired. The line current for a Y-to-Y connection is $\mathbf{V}_a/\mathbf{Z}$ for the A line, and the other two currents are displaced by ±120° from $\mathbf{I}_A$.

For a Δ load, we converted the balanced Δ load to a Y-connected load by using the relation $\mathbf{Z}_Y = \mathbf{Z}_\Delta/3$. Then we proceeded with the Y-to-Y calculation.

The line current for a Δ load is $\sqrt{3}$ times the phase current and is displaced −30° in phase. The line-to-line voltage of a Δ load is equal to the phase voltage.

The power delivered to a balanced Y-connected load is $P_Y = \sqrt{3} V_{AB} I_A \cos\theta$ where V_{AB} is the line-to-line voltage, I_A is the line current, and θ is the angle between the phase voltage and the phase current ($\mathbf{Z}_Y = Z\underline{/\theta}$).

The two-wattmeter method of measuring three-phase power delivered to a load was described. Also, we considered the usefulness of the two-wattmeter method for determining the power factor angle of a three-phase system.

TERMS AND CONCEPTS

Balanced Load   The impedance in each phase is equal.

Balanced Voltages   The phase voltages have identical amplitude and frequency and are out of phase with each other by 120°.

Delta (Δ) Connection   Three phase voltages are connected in series around a delta or loop.

Line-to-Line Voltage   Voltage between two lines in a three-phase system.

Negative Phase Sequence   Phase rotation is acb on the phasor diagram and occurs in that order with respect to time.

Per-Phase Equivalent Circuit   The equivalent circuit for a balanced three-phase system that enables the calculation of phase voltage and current.

Phase Voltage   Voltage of each phase of a three-phase system.

Phase Sequence   Sequence of phase voltages on the phasor diagram.

Positive Phase Sequence   Phase rotation is abc on the phasor diagram and occurs in that order with respect to time.

Three-Phase System   Three similar voltages of a three-phase system each displaced by 120° from its neighbor on the phasor diagram.

Two-Wattmeter Method   Method for using two wattmeters to read the total power in a three-phase load.

Y Connection   Three-phase voltages are connected to a common neutral terminal.

REFERENCES   Chapter 19

Barnes, R., and Wong, K. T., "Unbalanced and Harmonic Studies for the Channel Tunnel Railway System," *IEE Proceedings,* March 1991, pp. 41–50.

Nye, David E., *Electrifying America,* MIT Press, Cambridge, MA, 1991.

PROBLEMS

Section 19-3   Three-Phase Voltages

P 19.3-1   A balanced three-phase Y-connected load has one phase voltage

$$\mathbf{V}_c = 277\underline{/45°}$$

The phase sequence is abc. Find the line-to-line voltages $\mathbf{V}_{AB}$, $\mathbf{V}_{BC}$, and $\mathbf{V}_{CA}$. Draw a phasor diagram showing the phase and line voltages.

P 19.3-2   A three-phase system has a line-to-line voltage

$$\mathbf{V}_{BA} = 12{,}470\underline{/-35°}\ \text{V}$$

with a Y load. Find the phase voltages when the phase sequence is abc.

P 19.3-3   A three-phase system has a line-to-line voltage

$$\mathbf{V}_{ab} = 1500\underline{/30°}\ \text{V}$$

with a Y load. Determine the phase voltage.

Section 19-4   The Balanced Y-to-Y Circuit

P 19.4-1   A three-phase system has a line-to-line voltage of 208 V connected to a balanced Y-connected load $\mathbf{Z} = 12\underline{/30°}\ \Omega$.

(a) Find the phase voltages.
(b) Find the line currents and phase currents.
(c) Show the line currents and phase currents on a phasor diagram.
(d) Determine the power dissipated in the load.

P 19.4-2   A balanced three-phase Y-connected supply delivers power through a three-wire plus neutral-wire circuit in a large office building to a three-phase Y-connected load. Each transmission wire, including the neutral wire, has a 2-Ω resistance, and the balanced Y load is a

10-Ω resistance in series with 100 mH. Each supply-phase voltage is 120 V, and the circuit operates at 60 Hz. Find the line voltage and the phase current at the load.

P 19.4-3   A Y-connected source and load are shown in Figure P 19.4-3. (a) Determine the rms value of the current $i_a(t)$. (b) Determine the average power delivered to the load.

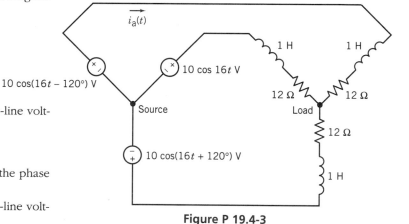

Figure P 19.4-3

Section 19-5   The Δ-Connected Source and Load

P 19.5-1   A balanced three-phase Δ-connected load has one line current

$$\mathbf{I}_B = 50\underline{/-40°}\ \text{A}$$

Find the phase currents $\mathbf{I}_{BC}$, $\mathbf{I}_{AB}$, and $\mathbf{I}_{CA}$. Draw the phasor diagram showing the line and phase currents.

P 19.5-2   A three-phase circuit has two parallel balanced Δ loads, one of 5-Ω resistors and one of 20-Ω resistors. Find the magnitude of the total line current when the line-to-line voltage is 480 V.

Section 19-6   The Y-to-Δ Circuit

P 19.6-1   The three-phase system of Problem 19.4-1 is connected to a Δ-connected load with $\mathbf{Z} = 12\underline{/30°}\ \Omega$.

Determine the line currents and calculate the power dissipated in the load.

P 19.6-2 A balanced Δ-connected load is connected by three wires, each with a 4-Ω resistance, to a Y source with a line voltage of 480 V. Find the line current I_A when $\mathbf{Z}_\Delta = 39\underline{/-40°}\ \Omega$.

P 19.6-3 The balanced circuit shown in Figure P 19.6-3 has $\mathbf{V}_{ab} = 380\underline{/30°}$ V. Determine the phase currents in the load when $\mathbf{Z} = 3 + j4\ \Omega$. Sketch a phasor diagram.

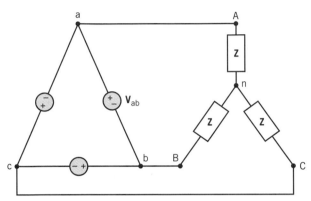

Figure P 19.6-3 A Δ-to-Y circuit.

P 19.6-4 A balanced system with a Δ-connected load has $\mathbf{Z} = 9 + j12\ \Omega$ and the line voltage is 380 V. Calculate the line currents and the phase currents. Assume $\mathbf{V}_{AB} = 380\underline{/0°}$ V.

Section 19-7   The Per-Phase Equivalent Circuit

P 19.7-1 The English Channel Tunnel rail link will be supplied at 25 kV from the United Kingdom and French grid systems. When there is a grid supply failure, each end will be capable of supplying the whole tunnel, but in a reduced operational mode.

The tunnel traction system will be a conventional catenary (overhead wire) system similar to the surface main line electric railway system of the United Kingdom and France. What makes the tunnel traction system different and unique is the high density of traction load and the end-fed supply arrangement. The tunnel traction load will be considerable. For each half tunnel the load is 120 MVA upon installation, rising to an ultimate load of 180 MVA as traffic increases (Barnes and Wong, 1991).

Assume that each line-to-line voltage is 25 kV and the three-phase system is connected to the traction motor of an electric locomotive. The motor is a Y-connected load with $\mathbf{Z} = 150\underline{/25°}\ \Omega$. Find the line currents and the power delivered to the traction motor.

P 19.7-2 A three-phase source with a line voltage of 45 kV is connected to two balanced loads. The Y-connected

load has $\mathbf{Z} = 10 + j20\ \Omega$ and the Δ load has a branch impedance of 50 Ω. The connecting lines have an impedance of 2 Ω. Determine the line current, the power delivered to the loads, and the power lost in the wires. What percentage of power is lost in the wires?

P 19.7-3 A three-phase source has a Y-connected source with $v_a = 5\cos(2t + 30°)$ connected to a three-phase Y load with $\mathbf{Z}_y = 4 + j8$. The connecting lines each have a resistance of 2 Ω. Determine the total average power delivered to the load.

Section 19-8   Power in a Balanced Load

P 19.8-1 Find the power absorbed by a balanced three-phase Y connected load when

$$\mathbf{V}_{CB} = 208\underline{/15°}\ V \quad \text{and} \quad \mathbf{I}_B = 3\underline{/110°}\ A$$

P 19.8-2 A three-phase motor delivers 20 hp operating from a 480-V line voltage. The motor operates at 85 percent efficiency and with a power factor equal to 0.8. Find the magnitude and angle of the line current for phase A.

P 19.8-3 A three-phase balanced load is fed by a line-to-line voltage of 220 V. It absorbs 1500 W at 0.8 power factor lagging. Calculate the phase impedance if it is (a) Δ connected and (b) Y connected.

P 19.8-4 A 600-V three-phase circuit has two balanced Δ loads connected to the lines. The load impedances are $40\underline{/30°}\ \Omega$ and $50\underline{/-60°}\ \Omega$, respectively. Determine the line current and the total average power.

P 19.8-5 A three-phase feeder simultaneously supplies power to two separate three-phase loads. The first total load is Δ connected and requires 39 kVA at 0.7 lagging. The second total load is Y connected and requires 15 kW at 0.21 leading. Each line has an impedance 0.038 + $j0.072\ \Omega$/phase. Calculate the line-to-line source voltage magnitude required so that the loads are supplied with 208 V line to line.

P 19.8-6 A building is supplied by a public utility at 4.16 kV. The building contains three loads connected to the three-phase lines:

(a) Δ connected, 500 kVA at 0.85 lagging.
(b) Y connected, 75 kVA at 0.0 leading.
(c) Y connected; each phase has a 150-Ω resistor parallel to a 225-Ω inductive reactance.

The utility feeder is 5 miles long with an impedance per phase of 1.69 + $j0.78\ \Omega$/mile. At what voltage must the utility supply its feeder so that the building is operating at 4.16 kV?

P 19.8-7 The diagram shown in Figure P 19.8-7 has two three-phase loads that form part of a manufacturing plant. They are connected in parallel and require 4.16 kV. Load 1 is 1.5 MVA, 0.75 lag pf, Δ connected. Load 2 is 2 MW, 0.8 lagging pf, Y connected. The feeder from the power utility's substation transformer has an impedance of $0.4 + j0.8\ \Omega$/phase. Determine

(a) The required magnitude of the line voltage at the supply.
(b) The real power drawn from the supply.
(c) The percentage of the real power drawn from the supply that is consumed by the loads.

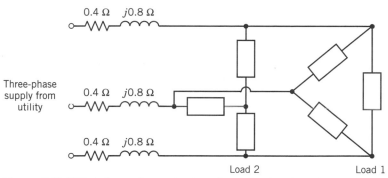

Figure P 19.8-7 A three-phase circuit with a Δ load and a Y load.

P 19.8-8 The balanced three-phase load of a large commercial building requires 480 kW at a lagging power factor of 0.8. The load is supplied by a connecting line with an impedance of $5 + j25$ mΩ for each phase. Each phase of the load has a line-to-line voltage of 600 V. Determine the line current and the line voltage at the source. Also, determine the power factor at the source. Use the line-to-neutral voltage as the reference with an angle of 0°.

Section 19-9   Two-Wattmeter Power Measurement

P 19.9-1 The two-wattmeter method is used to determine the power drawn by a three-phase 440-V motor that is a Y-connected balanced load. The motor operates at 20 hp at 74.6 % efficiency. The magnitude of the line current is 52.5 A. The wattmeters are connected in the A and C lines. Find the reading of each wattmeter. The motor has a lagging power factor.

P 19.9-2 A three-phase system has a line-to-line voltage of 4000 V and a balanced Δ-connected load with $\mathbf{Z} = 40 + j30\ \Omega$. The phase sequence is abc. Use the two wattmeters connected to lines A and C, with line B as the common for the voltage measurement. Determine the total power measurement recorded by the wattmeters.

P 19.9-3 A three-phase system with a sequence abc and a line-to-line voltage of 200 V feeds a Y-connected load with $\mathbf{Z} = 70.7\underline{/45°}\ \Omega$. Find the line currents. Find the total power by using two wattmeters connected to lines B and C.

P 19.9-4 A three-phase system with a line-to-line voltage of 208 V and phase sequence abc is connected to a Y-balanced load with impedance $10\underline{/-30°}\ \Omega$ and a balanced Δ load with impedance $15\underline{/30°}\ \Omega$. Find the line currents and the total power using two wattmeters.

P 19.9-5 The two-wattmeter method is used. The wattmeter in line A reads 920 W and the wattmeter in line C reads 460 W. Find the impedance of the balanced Δ-connected load. The circuit is a three-phase 120-V system with an ABC sequence.

P 19.9-6 Using the two-wattmeter method, determine the power reading of each wattmeter and the total power for Problem 19.6-3 when $\mathbf{Z} = 0.868 + j4.924\ \Omega$. Place the current coils in the A-to-a and C-to-c lines.

VERIFICATION PROBLEMS

VP 19-1 A Y-connected source is connected to a Y-connected load (Figure 19-7) with $\mathbf{Z} = 10 + j4\ \Omega$. The line voltage is $V_L = 416$ V. A student report states that the line current $\mathbf{I_A} = 38.63$ A and the power delivered to the load is 16.1 kW. Verify these results.

VP 19-2 A Δ load with $\mathbf{Z} = 40 + j30\ \Omega$ has a three-phase source with $V_L = 240$ V (Figure 19-8). A computer analysis program states that one phase current is $4.8\underline{/-36.9°}$ A. Verify this result.

PSpice PROBLEMS

SP 19-1 Determine the line voltage and the phase current at the load for Problem 19.4-2.

SP 19-2 Determine the line current of Problem 19.6-2 using PSpice.

SP 19-3 A circuit has a Y-to-Y connection as shown in Figure 19-14 with $R = 2\,\Omega$ and $\mathbf{Z} = 8 + j9\,\Omega$. Determine the line current and the total power delivered to the load.

DESIGN PROBLEMS

DP 19-1 A balanced three-phase Y source has a line voltage of 208 V. The total power delivered to the balanced Δ load is 1200 W with a power factor of 0.94 lagging. Determine the required load impedance for each phase of the Δ load. Calculate the resulting line current. The source is a 208-V ABC sequence.

DP 19-2 A three-phase 240-V circuit has a balanced Y load impedance $\mathbf{Z}$. Two wattmeters are connected with current coils in lines A and C. The wattmeter in line A reads 1440 W, and the wattmeter in line C reads zero. Determine the impedance required so that one wattmeter reads the total average power.

DP 19-3 A three-phase motor delivers 100 hp and operates at 80 percent efficiency with a 0.75 lagging power factor. Determine the required Δ-connected balanced set of three capacitors that will improve the power factor to 0.90 lagging. The motor operates from 480-V lines.

DP 19-4 A three-phase system has balanced conditions so that the per-phase circuit representation can be utilized as shown in Figure DP 19-4. Select the turns ratio of the step-up and step-down transformers so that the system operates with an efficiency greater than 99 percent. The load voltage is specified as 4 kV, and the load impedance is $4/3\,\Omega$.

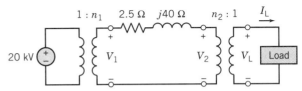

Figure DP 19-4 Per-phase diagram with two transformers.

GLOSSARY

Active Element   An element capable of supplying energy. In other words, the energy supplied by the element may be positive.

Admittance   Ratio of the phasor current $\mathbf{I}$ to the phasor voltage $\mathbf{V}$ of an element or set of elements so that $\mathbf{Y} = \mathbf{I}/\mathbf{V}$.

Apparent Power   Product of the rms voltage and the rms current. It is expressed as VI with units of voltamperes (VA).

Asymptotic Curve   A Bode plot with a straight line representing each portion of the total curve. See Bode Plot.

Automatic Control   Use of a self-directing system to control a process to achieve a desired performance.

Automation   Automatic operation or control of a process, device, or system.

Average Power Delivered by a Sinusoidal Steady-State Source $P = (V_m I_m/2) \cos \theta$, where θ is the phase angle difference between the voltage and current of the source.

Average Power of a Periodic Function   Integral of the time function over a complete period T, divided by the period:

$$P = \frac{1}{T} \int_{t_0}^{t_0 + T} p(t) \, dt$$

Balanced Three Phase Load   The impedance in each phase is equal.

Balanced Three Phase Source   The phase voltages have identical amplitude and frequency and are out of phase with each other by 120°.

Bandpass Filter   Circuit that will ideally pass unimpeded all frequencies in a selected range of frequencies and reject all frequencies outside this range.

Bandstop Filter   Circuit that will ideally reject all frequencies in a selected range of frequencies and pass unimpeded all frequencies outside this range.

Bandwidth (BW)   Range of frequencies that lie between the two frequencies where the magnitude of the gain ratio is $1/\sqrt{2}$ for $|H|/|H_0|$, where H_0 is the maximum value of the magnitude H. For a parallel or series resonant circuit $H_0 = 1$.

Bode Plot   Plot of log-gain and phase angle values on a log-frequency base for a function $\mathbf{H}(j\omega)$.

Branch   Path that connects two nodes with no intermediate nodes.

Break Frequency   Point where the asymptotic curve for the logarithmic gain exhibits a change in slope.

Buffer Amplifier   Isolating amplifier used to separate the output from the input. Usually $v_o = 1v_s$.

Capacitance   Ratio of the charge stored to the voltage difference between the two conducting plates or wires; $C = q/v$.

Capacitor   Two-terminal element whose primary purpose is to introduce capacitance into an electric circuit.

Cellular Radio   Use of transmission zones called cells served by low-power receiver-transmitters and a switching capability that enables the system to serve many radio-telephone users as they move from one cell to another.

Characteristic Equation   Equation obtained by setting $s^n = d^n/dt^n$ in the unforced differential equation; the characteristic equation is $s^2 + a_1 s + a_0 = 0$ for a second-order differential equation.

Characteristic Roots   Roots of the characteristic equation. Also called natural frequencies.

Charge   Fundamental unit of matter responsible for electric phenomena.

Circuit   Interconnection of electrical elements in a closed path.

Coefficient of Coupling   Ratio of the mutual inductance M to the square root of the product of the two inductances L_1 and L_2:

$$k = \frac{M}{\sqrt{L_1 L_2}}$$

Common Mode Rejection Ratio (CMRR)   Ratio of the gain A that multiplies v_i to the gain B that multiplies the common mode component of the input.

Compact Disk Player   Device that reads information stored on an optical disk using a laser.

Completely Symmetric Network   A T or Π network where all impedances are equal.

Complex Exponential Forcing Function $V_m e^{j\omega t}$ for a voltage source, or $I_m e^{j\omega t}$ for a current source.

Complex Fourier Coefficients   Fourier coefficients for the exponential form of the Fourier series.

Complex Frequency $s = \sigma + j\omega$

Complex Power   Sum of the average power and the reactive power, $\mathbf{S} = P + jQ$, with units of voltamperes (VA).

Computer-Aided Design   Use of computers and associated programs to design parts and processes and graphically view the resulting models.

Conductance   Real part of admittance, denoted as G. The inverse of the resistance of a resistor; $G = 1/R$ in units of siemens (S). Some engineers use units of mhos ($\mho$).

Convolution Theorem   The convolution of $h(t)$ and $f(t)$ is equal to

$$\int_0^t h(\tau) f(t - \tau)\, d\tau$$

Corner Frequency   See Break Frequency.

Cotree   Links together with nodes constitute a complementary tree. See Link.

Critically Damped   Condition that exists when the characteristic roots of a second-order circuit are equal.

Current   Time rate of change of charge, $i = dq/dt$.

Current Divider   Circuit of parallel resistors that divides the source current i_s so that

$$i_n = \frac{G_n i_s}{G_1 + G_2 + \cdots + G_n + \cdots + G_N}$$

Cut Set   A minimum set of elements that, when cut, separates the graph into two parts.

Damped Resonant Frequency $\omega_d = \sqrt{\omega_0^2 - \alpha^2}$ of the second-order circuit with the characteristic equation $s^2 + 2\alpha s + \omega_0^2 = 0$.

Decibels   Units of the logarithmic gain. See Logarithmic Gain.

Delta Connection   Three phase voltages are connected in series around a delta or loop.

Dependent Source   Source that provides a current (or a voltage) that is dependent on another variable elsewhere in the circuit.

Dielectric Constant   Property that determines the energy stored per unit volume for unit voltage difference in a capacitor.

Differentiator   Amplifier circuit that provides differentiation of the input voltage v_s.

Dot Convention   If a current goes into the dotted end of one coil, then for the other coil the reference direction of the induced voltage is positive at the dotted end.

Dual Circuits   Two circuits such that the equations describing the first circuit, with v and i interchanged and R and G interchanged, describe the second circuit.

Effective Value of a Current   The steady current that is as effective in transferring power as the average power of the varying current:

$$I_{\text{eff}} = \left(\frac{1}{T} \int_0^T i^2 \, dt \right)^{1/2}$$

We normally use $I = I_{\text{eff}} = I_{\text{rms}}$, dropping the subscript.

Efficiency of Power Transfer   Ratio of the power delivered to the load to the power supplied by the source.

Electric Field   State of a region in which charged bodies are subject to forces by virtue of their charge; the force acting on a unit positive charge.

Electric Filter   See Filter.

Electric Power System   System that generates, transmits, and distributes electric power to the end user.

Electric Signal   Voltage or current varying in time in a manner that conveys information.

Electric Vehicle   Vehicle using electric energy storage, electric controls, and electric propulsion devices.

Electricity   Physical phenomena arising from the existence and interaction of electric charge.

Electronic Computer   Device that assembles, stores, retrieves, correlates, processes, and prints information in accordance with predetermined programs.

Electronics   The engineering field and industry that uses electron devices in circuits and systems.

Energy   Capacity to perform work.

Energy Spectrum   The square of the magnitude of $V(\omega)$ written as $|V(\omega)|^2$. Also called the energy spectral density.

Energy Storage   Work performed on moving a charge resulting in energy storage in a capacitor, or work performed to establish a magnetic field resulting in energy storage in an inductor.

Engineering   The profession in which knowledge of mathematical, natural, and social sciences gained by study, experience, and practice is applied with judgment to develop ways to utilize economically the materials and forces of nature for the benefit of humankind.

Equivalent Circuit   Arrangement of circuit elements that is equivalent to another arrangement of elements. A circuit equivalent to another circuit exhibits identical characteristics (behavior) between the same terminals.

Even Function   A function with $f(t) = f(-t)$.

Exponential Solution   Assumed solution of the form $x = Ae^{st}$ where A and s are constants to be determined.

Filter   Circuit designed to provide a magnitude gain or loss over a prescribed range of frequencies.

Final Value Theorem   The value of $f(t)$ as t approaches infinity may be determined from $f(\infty) = \lim_{s \to 0} sF(s)$.

First-Order Circuit   Circuit that contains only one energy storage element, either capacitance or inductance.

First-Order Differential Equation   Differential equation where the highest order derivative term is the first derivative, i.e., dx/dt.

Forced Response   Steady-state response of a circuit to an independent source.

Fourier Series   Infinite sum of harmonically related sinusoids.

Fourier Spectrum   See Spectrum.

Fourier Transform   Process of converting a time function, $f(t)$, to a frequency function, $F(\omega)$.

Fourier Trigonometric Coefficients   Coefficients of the sinusoidal terms that in sum represent a periodic function.

Frequency   Inverse of the period T. See Period.

Frequency of a Sinusoid   Radian frequency ω in the sinusoidal waveform $x = \cos \omega t$; also ordinary frequency f, where $\omega = 2\pi f$ (f has units of hertz).

Frequency Domain   Mathematical domain where the set of possible values of a variable is expressed in terms of frequency.

Frequency Response   Frequency-dependent relation, in both gain and phase, between an input sinusoidal steady-state signal and an output sinusoidal signal.

Fundamental Cut Set   A cut set that contains one, and only one, tree branch.

Fundamental Frequency   For a periodic function of period T, the frequency $\omega_0 = 2\pi/T$.

Fundamental Loop   A loop constructed by adding a link to a path in a tree.

Gain Ratio   Ratio of a selected output signal to a designated input signal, written as $H = X_o/X_{in}$ where X may be a voltage or a current.

Gibbs Phenomenon   Regardless of the number of terms used, overshoot occurs at discontinuities when a waveform is represented by a Fourier series.

Graph   A set of nodes together with a set of branches. A set of interconnected branches.

Harmonic Frequencies   Integer multiples of the fundamental frequency.

High-Pass Filter   Filter that will ideally pass all frequencies above the cutoff frequency ω_c and reject all frequencies below the cutoff frequency.

High-Voltage Power Transmission   Use of voltages greater than 100 kV to keep low resistive losses I^2R on the transmission lines.

Homogeneity   Property that the response of an element is kv when it is subjected to an excitation ki where v is the response to an input i. The magnitude scale factor is preserved.

Ideal Operational Amplifier   The simplest model of an operational amplifier. This model is characterized by $i_1 = 0$, $i_2 = 0$ and $v_1 = v_2$.

Ideal Source   Ideal model of an actual source that assumes that the parameters of the source, such as its magnitude, are independent of other circuit variables.

Ideal Transformer   Model of a transformer with a coupling coefficient equal to unity and with the primary and secondary reactances very large compared to the impedances connected to the transformer terminals.

Impedance   Ratio of the phasor voltage V to the phasor current I for a circuit element or set of elements so that $Z = V/I$.

Impedance in the s-Domain, Z(s)   Ratio of the voltage $V(s)$ at a pair of terminals to the current $I(s)$ flowing into the positive voltage terminal, all expressed in the s-domain.

Impulse Function   A very short pulse such that its value is zero for $t \neq 0$ and the integral of the pulse is 1.

Impulse Response   Output response of a circuit when the input is a unit impulse, with no initial stored energy in the circuit.

Independent Source   Source that provides a current (or a voltage) independent of other circuit variables.

Inductance   Property of an electric device by virtue of which a time-varying current produces a voltage across the device.

Inductor   Two-terminal element consisting of a winding of N turns for introducing inductance into an electric circuit.

Information Ratio   Percentage of household expenditures for various kinds of information-related activities.

Information Society   A society in which there is an abundance in quantity and quality of information with all the necessary facilities for its distribution.

Initial Time   The time, t_0, when a new action is initiated in a circuit; usually, $t_0 = 0$.

Initial Value Theorem The initial value at time $t = 0$ may be determined from $f(0) = \lim_{s \to \infty} sF(s)$.

Instantaneous Power Product of the voltage $v(t)$ and the current $i(t)$, $p = vi$.

Integrated Circuit Combination of interconnected circuit elements inseparably on or within a continuous semiconductor.

Integrating Factor Method Method for obtaining the solution of a differential equation by multiplying the equation by an exponential factor that makes one side of the equation a perfect derivative and then integrating both sides of the equation.

Integrator Amplifier circuit that provides integration of the input voltage v_s.

Inverse Fourier Transform Process of converting a frequency function, $F(\omega)$, to a time function $f(t)$.

Inverse Laplace Transform Transform of $F(s)$ into the time domain to yield $f(t)$.

Inverting Amplifier Amplifier with a relationship of the form $v_o = -Kv_s$, where K is a positive constant, v_o is the output voltage, and v_s is the source voltage.

Kirchhoff's Current Law The algebraic sum of the currents entering a node is zero.

Kirchhoff's Voltage Law The algebraic sum of the voltages around a closed path is zero.

Laplace Transform Transform of $f(t)$ into its s-domain form $F(s)$.

Linear Element Element that satisfies the properties of superposition and homogeneity.

Line-to-Line Voltage Voltage between two lines in a three-phase system.

Link Branches not in a chosen tree.

Logarithmic Gain $20 \log_{10} H$ with units of decibels (dB).

Loop Closed path progressing from node to node and returning to the starting node; closed path around a circuit.

Low-Pass Filter Filter that will ideally pass all frequencies up to the cutoff frequency ω_c and perfectly reject all frequencies above ω_c.

Magnetic Field State produced either by current flow or by a permanent magnet that can induce voltage in a conductor when the flux linkage changes in the conductor.

Magnetic Levitation Use of magnetic field forces to raise a vehicle above the track.

Magnetic Resonance Imaging Technique in which weak radio frequency signals emitted by hydrogen nuclei in the human body are polarized by a constant magnetic field, excited by a radio frequency magnetic field, and used as a probe or input signal. (MRI)

Maximum Power Transfer for Phasor Circuits If a circuit has a Thévenin equivalent circuit with an impedance $\mathbf{Z}_T$, the maximum power is delivered to the load when the load $\mathbf{Z}_L$ is set equal to the complex conjugate of $\mathbf{Z}_T$.

Maximum Power Transfer Theorem for Resistive Circuits The maximum power delivered by a source represented by its Thévenin equivalent is attained when the load resistor R_L is equal to the Thévenin resistance R_t.

Mesh Loop that does not contain any other loops within it.

Mesh Current The current that flows around the periphery of a mesh; the current that flows through the elements constituting the mesh.

Model Representation of an element.

Mutual Inductance Property of two electric coils where a voltage in one is induced by a changing current in the other.

Natural Frequencies Roots of the characteristic equation.

Natural Response Response of an RL, RC, or RLC circuit that depends only on the nature of the circuit and not on external sources.

Negative Phase Sequence Phase rotation is acb on the phasor diagram and occurs in that order with respect to time.

Node Terminal common to two or more branches of a circuit; point where two or more elements have a common connection.

Node Voltage   Voltage from a selected node to the reference node.

Noninverting Amplifier   Amplifier with a relationship $v_o = Kv_s$, where K is a positive constant, v_o is the output voltage, and v_s is the source voltage.

Norton's Theorem   Given a linear circuit, divide it into two parts, A and B. For circuit A, determine its short-circuit current at its terminals. The equivalent circuit of A is a current source i_{sc} in parallel with a resistance R_n, where R_n is the resistance calculated with all the circuit's independent sources deactivated.

Odd Function   A function with $f(t) = -f(-t)$.

Ohm's Law   The voltage across the terminals of a resistor is related to the current into the positive terminal as $v = Ri$.

Open Circuit   Condition that exists when the current between two terminals is identically zero, irrespective of the voltage across the terminals.

Operational Amplifier   Amplifier with a high gain designed to be used with other circuit elements to perform a specified function (often called an op amp).

Operator   Differential operator s such that $s^n x = d^n x / dt^n$.

Operator Method   Method of obtaining the nth-order differential equation that uses the operator $s^n = d^n / dt^n$ and Cramer's rule to obtain the complete solution.

Optical Fibers   Tiny strands of pure glass no wider than a human hair that can carry thousands of telephone channels.

Output Offset Voltage   A small, nonzero output voltage that occurs for an operational amplifier when the voltage across the input terminals is zero.

Overdamped   Condition that exists when the roots of the characteristic equation of a second-order circuit are real and distinct.

Parallel Connection   Arrangement of resistors so that each resistor has the same voltage appearing across it.

Parallel Resonant Circuit   Circuit with a resistor, capacitor, and inductor in parallel.

Parallel *RLC* Circuit   Circuit with three parallel branches where the branches contain a resistor, a capacitor, and an inductor, respectively.

Parseval's Theorem   The energy absorbed by a 1-Ω resistor is $1/2\pi$ times the integral of the energy spectrum over all frequencies from $-\infty$ to $+\infty$.

Partial Fraction Expansion   Expansion of $F(s)$ into a series of terms.

Passive Element   Element that absorbs energy. The energy delivered to it is always nonnegative (zero or positive).

Peak Frequency, ω_p   Frequency at which a peak magnitude occurs for a pair of complex poles.

Period   Length of time, T, at which a waveform repeats itself.

Periodic Function   Function defined by the property $x(t + T) = x(t)$ so that it repeats every T seconds.

Period of Oscillation   Time for an underdamped response to proceed through one cycle of oscillation, denoted by T_d. T_d is the period of oscillation of the periodic function.

Per-Phase Equivalent Circuit   The equivalent circuit for a balanced three-phase system that enables the calculation of phase voltage and current.

Phase Sequence   Sequence of phase voltages on the phasor diagram.

Phase Shift   Phase angle ϕ associated with a variable x so that $x = X_m \sin(\omega t + \phi)$, or phase angle between a voltage and a current.

Phase Voltage   Voltage of each phase of a three-phase system.

Phasor   Complex number associated with a circuit variable—for example, the phasor voltage **V**.

Phasor Diagram   Relationship of phasors on the complex plane.

Planar Circuit   Circuit that can be drawn on a plane without branches crossing each other.

Poles   Roots of the denominator polynomial of the transfer function $H(s)$; roots of the denominator polynomial of the gain ratio $\mathbf{H}(j\omega)$.

Port   An access port consisting of a set of two terminals, where the current entering one terminal of a pair exits the other terminal in the pair.

Positive Phase Sequence   Phase rotation is abc on the phasor diagram and occurs in that order with respect to time.

Power   Time rate of change of energy, $p = dw/dt$.

Power Delivered to a Resistance $p = i^2R = v^2/R$ W.

Power Factor   Equal to $\cos \theta$, where θ is the phase angle difference between the sinusoidal steady-state voltage and current; ratio of average power to apparent power.

Primary Coil   Coil of a transformer that is normally connected to the source.

Pulse   Function of time that is zero for $t < t_0$, has magnitude M for $t_0 < t < t_1$, and is equal to zero for $t > t_1$.

Quality Factor, Q   Measure of the energy storage property in relation to the energy dissipation property of a second-order circuit. For a parallel resonant circuit $Q = R/\omega_0 L = \omega_0 RC$.

Radio Transmission   Transmission of communication messages by means of radiated electromagnetic waves other than heat or light waves.

Reactance   Imaginary part of impedance, denoted as X.

Reactive Power   Imaginary part Q of the complex power with units of voltamperes reactive (VAR).

Reference Node   Node selected as the reference for all other nodes.

Reflected Impedance   The complex impedance reflected to the primary circuit of a transformer from the secondary circuit.

Resistance   Real part of impedance, denoted as R (the units are ohms, Ω). The physical property of an element to impede current flow.

Resistivity   Ability of a material to resist the flow of current. The symbol is ρ.

Resistor   Device or element whose primary purpose is to introduce resistance R into a circuit.

Resonance   Condition in a circuit, occurring at the resonant frequency, when $\mathbf{H}(j\omega)$ becomes a real number (the circuit becomes nonreactive).

Resonant Frequency   Frequency ω_0 at which a gain ratio $\mathbf{H}(j\omega)$ is nonreactive. For a parallel resonant circuit the resonant frequency occurs when its admittance is nonreactive. $\omega_0 = 1/\sqrt{LC}$ for the parallel and the series RLC circuits.

RMS of a Periodic Waveform   Square root of the mean of the squared value. Also called the effective value. We normally use V instead of V_{rms} to denote the rms voltage, thus dropping the subscript.

Robot   Reprogrammable, multifunctional manipulator designed to move materials, parts, tools, or devices through variable programmed motions for the performance of a variety of tasks.

Secondary Coil   Coil of a transformer that is normally delivering power to the load.

Semiconductor   Electronic conductor with resistivity in the range between metals and insulators.

Sensitivity   The proportional change of a performance measure in response to a proportional change in a circuit parameter.

Sequential Switching   Action of two or more switches activated at different instants of time in a circuit.

Series Connection   Circuit of a series of elements connected so that the same current passes through each element.

Series Resonant Circuit   Circuit with a series connection of a resistor, capacitor, and inductor.

Short Circuit   Condition that exists when the voltage across two terminals is identically zero, irrespective of the current between the two terminals.

SI   Système International d'Unités; the International System of Units.

Signal   Real-valued function of time; waveform that conveys information.

Slew Rate (SR)   Maximum rate at which the output voltage of an operational amplifier can change. It is normally expressed in $V/\mu s$.

Smart House   House in which special electrical wiring and controls for appliances and safety devices are used. A single cable carries all the circuits throughout the house for easy control.

Source Transformation   Transformation of one source into another while retaining the terminal characteristics. A voltage source may be transformed to a current source and vice versa.

Spectrum   The Fourier transform, $F(\omega)$, of a signal $f(t)$. Also called spectral density.

Spectrum (Line)   A plot of the line spectra of amplitude and phase against frequency of the coefficients of the complex Fourier series.

State Variable Method   Method of identifying the state variables and obtaining a set of first-order differential equations, then proceeding to obtain the second-order differential equation in terms of one of the state variables, and then solving for the complete solution for one or more of the state variables.

State Variables   Set of variables describing the energy of the storage elements of a circuit; the capacitor voltages and the inductor currents.

Steady-State Response   Response that exists after a long time following any switch activation.

Step Response   Response of a circuit to the sudden application of a constant source when all the initial conditions of the circuit are equal to zero.

Step Voltage Source   Voltage source, v, represented by $v = Vu(t - t_0)$.

Substitution Method   Method of obtaining the second-order, or higher order, differential equation in terms of a selected variable x that substitutes equations for the desired variable in order to eliminate other variables.

Superconductivity   The phenomenon of zero-resistance electrical conduction; the resistivity of the material is equal to zero.

Supermesh   One larger "mesh" created from two meshes that have a current source in common.

Supernode   A larger "node" that includes two nodes connected by an independent or dependent voltage source.

Superposition   Property that the excitation $(i_1 + i_2)$ results in a response $(v_1 + v_2)$ where an excitation i_n results in a response v_n.

Superposition of the Average Power of Multiple Sources   The total average power delivered to a load is equal to the sum of the average power delivered by each source when no two of the sources have the same radian frequency.

Superposition Theorem   For a linear circuit containing independent sources, the voltage across (or the current through) any element may be obtained by adding algebraically all the individual voltages (or currents) caused by each independent source acting alone with all other sources set to zero.

Susceptance   Imaginary part of admittance, denoted as B.

Switched Circuit   Electric circuit with one or more switches that open or close at time t_0.

Symmetrical Network   Two-port network that has mirror-image symmetry with respect to a centerline.

System   Interconnection of electrical elements and circuits to achieve a desired objective.

Telegraph   Device for communication of messages using the activation of a switch or key to send a code.

Thévenin's Theorem   Divide a circuit into two parts, A and B, connected at a pair of terminals. Determine v_{oc} as the open-circuit voltage of A with B disconnected. Then the equivalent circuit of A is a source voltage v_{oc} in series with R_t, where R_t is the resistance seen at the terminals of circuit A when all the independent sources are deactivated.

Three-Phase Circuit   Uses energy in the form of three voltages, equal in magnitude and symmetric in phase.

Time Constant   Value in seconds, τ, in an exponential response $Ae^{-t/\tau}$. The constant τ is the time to complete 63 percent of the decay.

Topology   The science of placement of elements.

Transfer Function   Ratio of the response of a circuit to an excitation expressed as a function of complex frequency s. The initial conditions are assumed to be zero.

Transform Pair   A function in the time domain, $f(t)$, and its Laplace transform $F(s)$.

Transformation   Conversion of a set of equations from one domain to another, e.g., from the time domain to the complex frequency domain.

Transformer   Magnetic circuit with two or more multiturn coils wound on a common core.

Transient Response   Time response $x(t)$ of a circuit to a stimulus or to a response resulting from a switch activation.

Transistor   Active semiconductor device with three or more terminals.

Transponder   Active receiver-transmitter capable of receiving a signal from an originating transmitter and retransmitting the signal to one or more other receivers.

Tree   Any connected set of branches of a graph that connects every node to every other node directly or indirectly without forming any closed path.

Turns Ratio   Ratio n equal to N_2/N_1, where N_1 = number of turns in the primary coil and N_2 = turns in the secondary of an ideal transformer.

Two-Port Network   Network with two sets of access terminals called the input and output ports.

Two-Wattmeter Method   Method for using two wattmeters to read the total power in a three-phase load.

Underdamped   Condition that exists when two roots of the characteristic equation of a second-order circuit are complex conjugates.

Unit Impulse   A pulse of infinitesimal width with the area under the pulse equal to 1.

Unit Step Forcing Function   Function of time that is zero for $t < t_0$ and unity for $t > t_0$. At $t = t_0$ the magnitude changes from zero to one. The unit step is dimensionless.

Vacuum Tube   Electron tube evacuated so that its electrical characteristics are unaffected by the presence of residual gas or vapor.

Virtual Ground   Terminal in a circuit that appears to the observer to be essentially (virtually) connected to ground.

Voltage   Work or energy required to move a positive charge of 1 coulomb through an element, $v = dw/dq$.

Voltage Divider   Circuit of a series of resistors that divides the input voltage by the ratio of the resistor R_n to the total series resistance, so

$$v_n = \frac{v_s R_n}{R_1 + R_2 + \cdots + R_n + \cdots + R_N}$$

Voltage Follower   Amplifier with a voltage gain of one so that the output voltage follows the input voltage.

Y Connection   Three-phase voltages are connected to a common neutral.

Zeros   Roots of the numerator polynomial of the transfer function $H(s)$; roots of the numerator polynomial of the gain ratio $\mathbf{H}(j\omega)$.

APPENDIX B

MATRICES, DETERMINANTS, AND CRAMER'S RULE

B-1 DEFINITION OF A MATRIX

There are many situations in circuit analysis in which we have to deal with rectangular arrays of numbers. The rectangular array of numbers

$$
\mathbf{A} = \begin{bmatrix}
a_{11} & a_{12} & \cdots & a_{1n} \\
a_{21} & a_{22} & \cdots & a_{2n} \\
\cdot & \cdot & & \cdot \\
\cdot & \cdot & & \cdot \\
\cdot & \cdot & & \cdot \\
a_{m1} & a_{m2} & \cdots & a_{mn}
\end{bmatrix}
\tag{B-1}
$$

is known as a *matrix*. The numbers a_{ij} are called *elements* of the matrix, with the subscript i denoting the row and the subscript j denoting the column.

A matrix with m rows and n columns is said to be a matrix of *order* (m, n) or alternatively called an $m \times n$ (m by n) matrix. When the number of the columns equals the number of rows, $m = n$, the matrix is called a *square matrix* of order n. It is common to use boldface capital letters to denote an $m \times n$ matrix.

A matrix consisting of only one column, that is, an $m \times 1$ matrix, is known as a column matrix or, more commonly, a *column vector*. We represent a column vector with boldface lowercase letters as

$$
\mathbf{v} = \begin{bmatrix}
v_1 \\
v_2 \\
\cdot \\
\cdot \\
\cdot \\
v_m
\end{bmatrix}
\tag{B-2}
$$

When the elements of a matrix have a special relationship so that $a_{ij} = a_{ji}$, it is called a *symmetrical* matrix. Thus, for example,

$$
\mathbf{H} = \begin{bmatrix}
3 & -2 & 1 \\
-2 & 6 & 4 \\
1 & 4 & 8
\end{bmatrix}
\tag{B-3}
$$

is a symmetrical matrix of order $(3, 3)$.

944

B-2 | ADDITION AND SUBTRACTION OF MATRICES

The addition of two matrices is possible for matrices of the same order. The sum of two matrices is obtained by adding the corresponding elements. Thus, if the elements of **A** are a_{ij} and the elements of **B** are b_{ij} and if

$$\mathbf{C} = \mathbf{A} + \mathbf{B} \tag{B-4}$$

then the elements of **C** that are c_{ij} are obtained as

$$c_{ij} = a_{ij} + b_{ij} \tag{B-5}$$

For example, the matrix addition for two 3×3 matrices is as follows:

$$\mathbf{C} = \begin{bmatrix} 2 & 1 & 0 \\ 1 & -1 & 3 \\ 0 & 6 & 2 \end{bmatrix} + \begin{bmatrix} 8 & 2 & 1 \\ 1 & 3 & 0 \\ 4 & 2 & 1 \end{bmatrix} = \begin{bmatrix} 10 & 3 & 1 \\ 2 & 2 & 3 \\ 4 & 8 & 3 \end{bmatrix}$$

From the operation used for performing addition, we note that the process is commutative, that is,

$$\mathbf{A} + \mathbf{B} = \mathbf{B} + \mathbf{A} \tag{B-6}$$

Also, we note that the addition operation is associative, so that

$$(\mathbf{A} + \mathbf{B}) + \mathbf{C} = \mathbf{A} + (\mathbf{B} + \mathbf{C}) \tag{B-7}$$

To perform the operation of subtraction, we note that if a matrix **A** is multiplied by a constant α, then every element of the matrix is multiplied by this constant. Therefore, we can write

$$\alpha\mathbf{A} = \begin{bmatrix} \alpha a_{11} & \alpha a_{12} & \cdots & \alpha a_{1n} \\ \alpha a_{12} & \alpha a_{22} & \cdots & \alpha a_{2n} \\ \cdot & \cdot & & \cdot \\ \cdot & \cdot & & \cdot \\ \cdot & \cdot & & \cdot \\ \alpha a_{m1} & \alpha a_{m2} & \cdots & \alpha a_{mn} \end{bmatrix}$$

In order to carry out a subtraction operation $\mathbf{B} - \mathbf{A}$, we use $\alpha = -1$, and $-\mathbf{A}$ is obtained by multiplying each element of **A** by -1. For example,

$$\mathbf{C} = \mathbf{B} + \alpha\mathbf{A} = \mathbf{B} - \mathbf{A} = \begin{bmatrix} 2 & 1 \\ 4 & 2 \end{bmatrix} - \begin{bmatrix} 6 & 1 \\ 3 & 1 \end{bmatrix} = \begin{bmatrix} -4 & 0 \\ 1 & 1 \end{bmatrix}$$

B-3 | DETERMINANTS

The *determinant* of a matrix is a number associated with a square matrix. We define the determinant of a square matrix **A** as Δ, where

$$\Delta = \begin{vmatrix} a_{11} & a_{12} & \cdots & a_{1n} \\ a_{21} & a_{22} & \cdots & a_{2n} \\ \cdot & \cdot & & \cdot \\ \cdot & \cdot & & \cdot \\ \cdot & \cdot & & \cdot \\ a_{n1} & a_{n2} & \cdots & a_{nn} \end{vmatrix} \tag{B-8}$$

For example, consider the determinant of a 2×2 matrix

$$\Delta = \begin{vmatrix} a_{11} & a_{12} \\ a_{21} & a_{22} \end{vmatrix}$$

In this case, the determinant Δ is defined to be

$$\Delta = a_{11}a_{22} - a_{12}a_{21} \tag{B-9}$$

In the second-order, or 2×2, case of Eq. B-1, the method of obtaining Δ in Eq. B-9 uses the diagonal rule, which is

$$\Delta = \begin{vmatrix} a_{11} & a_{12} \\ a_{21} & a_{22} \end{vmatrix}$$

Thus, Δ is the difference of the product of the elements down the diagonal to the right and the product of the elements down the diagonal to the left. The determinant of a 3×3 matrix is

$$\Delta = \begin{vmatrix} a_{11} & a_{12} & a_{13} \\ a_{21} & a_{22} & a_{23} \\ a_{31} & a_{32} & a_{33} \end{vmatrix}$$
$$= (a_{11}a_{22}a_{33} + a_{12}a_{23}a_{31} + a_{13}a_{32}a_{21})$$
$$- (a_{13}a_{22}a_{31} + a_{23}a_{32}a_{11} + a_{33}a_{21}a_{12}) \tag{B-10}$$

In general, we are able to determine the determinant Δ in terms of cofactors and minors. The determinant of a submatrix of **A** obtained by deleting from **A** the ith row and the jth column is called the *minor* of the element a_{ij} and denoted as m_{ij}.

The cofactor c_{ij} is a minor with an associated sign, so that

$$c_{ij} = (-1)^{(i+j)}m_{ij} \tag{B-11}$$

As an example, consider the determinant

$$\Delta = \begin{vmatrix} 1 & -2 & 3 \\ 0 & 4 & -2 \\ 6 & -1 & -1 \end{vmatrix} \tag{B-12}$$

The minor of the element a_{11} is

$$m_{11} = \begin{vmatrix} 4 & -2 \\ -1 & -1 \end{vmatrix}$$

which is obtained by deleting the first row and the first column. The cofactor c_{11} is then

$$c_{11} = (-1)^{(1+1)}m_{11}$$
$$= m_{11} \tag{B-13}$$

The rule for evaluating the determinant Δ of a $n \times n$ matrix is

$$\Delta = \sum_{j=1}^{n} a_{ij}c_{ij} \tag{B-14}$$

for a selected value of i. Alternatively, we can obtain Δ by using the jth column, and thus

$$\Delta = \sum_{i=1}^{n} a_{ij}c_{ij} \tag{B-15}$$

for a selected value of j.

As an example, let us evaluate the determinant in Eq. B-12. It is easiest to use the summation along the first column of the matrix of Eq. B-12 since $a_{21} = 0$. Then, using Eq. B-15, we have for $j = 1$

$$\Delta = \sum_{i=1}^{3} a_{i1}c_{i1}$$
$$= a_{11}c_{11} + a_{21}c_{21} + a_{31}c_{31}$$
$$= 1(-1)^{2} \begin{vmatrix} 4 & -2 \\ -1 & -1 \end{vmatrix} + 6(-1)^{4} \begin{vmatrix} -2 & 3 \\ 4 & -2 \end{vmatrix}$$
$$= -6 + 6(-8)$$
$$= -54$$

B-4 | CRAMER'S RULE

A set of simultaneous equations

$$a_{11}x_1 + a_{12}x_2 + \cdots + a_{1n}x_n = b_1$$
$$a_{21}x_1 + a_{22}x_2 + \cdots + a_{2n}x_n = b_2$$

$$\vdots \qquad\qquad\qquad\qquad \vdots \qquad\qquad (B\text{-}16)$$

$$a_{n1}x_1 + a_{n2}x_2 + \cdots + a_{nn}x_n = b_n$$

can be written in matrix form as

$$\mathbf{Ax} = \mathbf{b} \qquad\qquad (B\text{-}17)$$

Cramer's rule states that the solution for the unknown, x_k, of the simultaneous equations of Eq. B-16 is

$$x_k = \frac{\Delta_k}{\Delta} \qquad\qquad (B\text{-}18)$$

where Δ is the determinant of $\mathbf{A}$ and Δ_k is Δ with the kth column replaced by the column vector $\mathbf{b}$.
As an example, let us consider the simultaneous equations

$$x_1 - 2x_2 + 3x_3 = 12$$
$$4x_2 - 2x_3 = -1$$
$$6x_1 - x_2 - x_3 = 0$$

In this case

$$\mathbf{A} = \begin{bmatrix} 1 & -2 & 3 \\ 0 & 4 & -2 \\ 6 & -1 & -1 \end{bmatrix}$$

and

$$\mathbf{b} = \begin{bmatrix} 12 \\ -1 \\ 0 \end{bmatrix}$$

The determinant of $\mathbf{A}$ was obtained in the preceding section as $\Delta = -54$. If we wish to obtain the unknown x_1, we have

$$x_1 = \frac{\Delta_1}{\Delta}$$

Then Δ_1 is Δ with the first column of $\mathbf{A}$ replaced by $\mathbf{b}$ so that

$$\Delta_1 = \begin{vmatrix} 12 & -2 & 3 \\ -1 & 4 & -2 \\ 0 & -1 & -1 \end{vmatrix}$$
$$= 12(-6) - 1(-1)^3(5)$$
$$= -67$$

Therefore, we have

$$x_1 = \frac{-67}{-54}$$
$$= \frac{67}{54}$$

B-5 ‖ MULTIPLICATION OF MATRICES

Matrix multiplication is defined in such a way as to assist in the solution of simultaneous linear equations. The multiplication of two matrices $\mathbf{AB}$ requires that the number of columns of $\mathbf{A}$ is equal to the number of rows of $\mathbf{B}$. Thus if $\mathbf{A}$ is of order $m \times n$ and $\mathbf{B}$ is of order $n \times q$, then the product is a matrix of order $m \times q$. The elements of a product

$$C = AB \tag{B-19}$$

are found by multiplying the ith row of $\mathbf{A}$ and the jth column of $\mathbf{B}$ and summing these products to give the element c_{ij}. That is,

$$c_{ij} = a_{i1}b_{1j} + a_{i2}b_{2j} + \cdots + a_{iq}b_{qj} = \sum_{k=1}^{q} a_{ik}b_{kj} \tag{B-20}$$

Thus we obtain c_{11}, the first element of $\mathbf{C}$, by multiplying the first row of $\mathbf{A}$ by the first column of $\mathbf{B}$ and summing the products of the elements. We should note that, in general, matrix multiplication is not commutative, that is,

$$AB \neq BA \tag{B-21}$$

Also, we will note that the multiplication of a matrix of order $m \times n$ by a column vector (order $n \times 1$) results in a column vector of order $m \times 1$.

A specific example of multiplication of a column vector by a matrix is

$$
\mathbf{x} = \mathbf{Ay} = \begin{bmatrix} a_{11} & a_{12} & a_{13} \\ a_{21} & a_{22} & a_{23} \end{bmatrix} \begin{bmatrix} y_1 \\ y_2 \\ y_3 \end{bmatrix}
$$
$$
= \begin{bmatrix} (a_{11}y_1 + a_{12}y_2 + a_{13}y_3) \\ (a_{21}y_1 + a_{22}y_2 + a_{23}y_3) \end{bmatrix} \tag{B-22}
$$

Note that $\mathbf{A}$ is of order 2×3 and $\mathbf{y}$ is of order 3×1. Therefore the resulting matrix $\mathbf{x}$ is of order 2×1, which is a column vector with two rows. There are two elements of $\mathbf{x}$, and

$$x_1 = (a_{11}y_1 + a_{12}y_2 + a_{13}y_3) \tag{B-23}$$

is the first element obtained by multiplying the first row of $\mathbf{A}$ by the first (and only) column of $\mathbf{y}$.

Another example, which the reader should verify, is

$$C = AB = \begin{bmatrix} 2 & -1 \\ -1 & 2 \end{bmatrix} \begin{bmatrix} 3 & 2 \\ -1 & -2 \end{bmatrix} = \begin{bmatrix} 7 & 6 \\ -5 & -6 \end{bmatrix} \tag{B-24}$$

For example, the element c_{22} is obtained as $c_{22} = -1(2) + 2(-2) = -6$.

COMPLEX NUMBERS

C-1 A COMPLEX NUMBER

We all are familiar with the solution of the algebraic equation

$$x^2 - 1 = 0 \tag{C-1}$$

which is $x = 1$. However, we often encounter the equation

$$x^2 + 1 = 0 \tag{C-2}$$

A number that satisfies Eq. C-2 is not a real number. We note that Eq. C-2 may be written as

$$x^2 = -1 \tag{C-3}$$

and we denote the solution of Eq. C-3 by the use of an imaginary number $j1$ so that

$$j^2 = -1 \tag{C-4}$$

and

$$j = \sqrt{-1} \tag{C-5}$$

An *imaginary number* is defined as the product of the imaginary unit j with a real number. Thus we may, for example, write an imaginary number as jb. A complex number is the sum of a real number and an imaginary number, so that

$$c = a + jb \tag{C-6}$$

where a and b are real numbers. We designate a as the real part of the complex number and b as the imaginary part and use the notation

$$\text{Re}\{c\} = a \tag{C-7}$$

and

$$\text{Im}\{c\} = b \tag{C-8}$$

949

C-2 | RECTANGULAR, EXPONENTIAL, AND POLAR FORMS

The complex number $a + jb$ may be represented on a rectangular coordinate place called a *complex plane*. The complex plane has a real axis and an imaginary axis, as shown in Figure C-1. The complex number c is the directed line identified as c with coordinates a, b. The *rectangular form* is expressed in Eq. C-6 and pictured in Figure C-1.

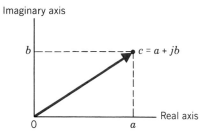

Figure C-1 Rectangular form of a complex number.

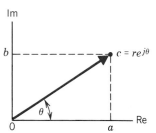

Figure C-2 Exponential form of a complex number.

An alternative way to express the complex number c is to use the distance from the origin and the angle θ, as shown in Figure C-2. The *exponential form* is written as

$$c = re^{j\theta} \tag{C-9}$$

where

$$r = (a^2 + b^2)^{1/2} \tag{C-10}$$

and

$$\theta = \tan^{-1}(b/a) \tag{C-11}$$

when $a > 0$. When $a < 0$, $\theta = 180° - \tan^{-1}(b/-a)$.

Note that $a = r\cos\theta$ and $b = r\sin\theta$. Also note that calculators give the principal value of the arc tangent and one must be sure that the final calculation is in the right quadrant.

The number r is also called the *magnitude* of c, denoted as $|c|$. The angle θ can also be denoted by the form $\underline{/\theta}$. Thus, we may represent the complex number in *polar form* as

$$c = |c| \underline{/\theta}$$
$$= r\underline{/\theta} \tag{C-12}$$

Example C-1

Express $c = 4 + j3$ in exponential and polar forms.

Solution

First, draw the complex plane diagram as shown in Figure C-3. Then find r as

$$r = (4^2 + 3^2)^{1/2} = 5$$

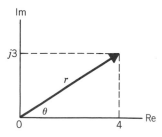

Figure C-3 Complex plane for Example C-1.

and θ as

$$\theta = \tan^{-1}(3/4) = 36.9°$$

The exponential form is then

$$c = 5e^{j.36.9°}$$

The polar form is

$$c = 5\underline{/36.9°}$$

Note that if $c = 4 - j3$, then $\theta = \tan^{-1}(-3/4) = -36.9°$. If $c = -4 + j3$, then $\theta = \tan^{-1}(-3/4)$ and a calculator will give $-36.9°$ when the correct answer is $+143.1°$. One has to check continually to ensure the angle is in the right quadrant.

C-3 | MATHEMATICAL OPERATIONS

The *conjugate* of the complex number $c = a + jb$ is denoted as c^* and is defined as

$$c^* = a - jb \tag{C-13}$$

In polar form, we have

$$c^* = r\underline{/-\theta} \tag{C-14}$$

To add or subtract two complex numbers, we add (or subtract) their real parts and their imaginary parts. Therefore, if $c = a + jb$ and $d = f + jg$, then

$$\begin{aligned} c + d &= (a + jb) + (f + jg) \\ &= (a + f) + j(b + g) \end{aligned} \tag{C-15}$$

The multiplication of two complex numbers is obtained as follows (note $j^2 = -1$):

$$\begin{aligned} cd &= (a + jb)(f + jg) \\ &= af + jag + jbf + j^2bg \\ &= (af - bg) + j(ag + bf) \end{aligned} \tag{C-16}$$

Using the exponential form, we have

$$cd = r_1e^{j\theta_1} \times r_2e^{j\theta_2} = r_1r_2e^{j(\theta_1 + \theta_2)} \tag{C-17}$$

Alternatively, we use the polar form to obtain

$$\begin{aligned} cd &= (r_1\underline{/\theta_1})(r_2\underline{/\theta_2}) \\ &= r_1r_2\underline{/\theta_1 + \theta_2} \end{aligned} \tag{C-18}$$

where

$$c = r_1\underline{/\theta_1} \quad \text{and} \quad d = r_2\underline{/\theta_2}$$

Division of one complex number by another complex number is easily obtained using the exponential form as follows:

$$\frac{c}{d} = \frac{r_1e^{j\theta_1}}{r_2e^{j\theta_2}} = (r_1/r_2)e^{j(\theta_1 - \theta_2)}$$

Alternatively, we use the polar form as

$$\frac{c}{d} = \frac{r_1 \underline{/\theta_1}}{r_2 \underline{/\theta_2}}$$

$$= \frac{r_1}{r_2} \underline{/\theta_1 - \theta_2} \tag{C-19}$$

It is easiest to add and subtract complex numbers in rectangular form and to multiply and divide them in polar form.

A few useful relations for complex numbers are summarized in Table C-1.

Table C-1
Useful Relationships for Complex Numbers

(1) $\dfrac{1}{j} = -j$

(2) $(-j)(j) = 1$

(3) $j^2 = -1$

(4) $1\underline{/\pi/2} = j$

(5) $c^k = r^k \underline{/k\theta}$

Example C-2
Find $c + d$, $c - d$, cd, and c/d when $c = 4 + j3$ and $d = 1 - j$.

Solution
First, we will express c and d in polar form as

$$c = 5\underline{/36.9°}$$

and

$$d = \sqrt{2}\underline{/-45°}$$

Then, for addition, we have

$$c + d = (4 + j3) + (1 - j)$$
$$= 5 + j2$$

For subtraction, we have

$$c - d = (4 + j3) - (1 - j)$$
$$= 3 + j4$$

For multiplication, we use the polar form to obtain

$$cd = (5\underline{/36.9°})(\sqrt{2}\underline{/-45°})$$
$$= 5\sqrt{2}\underline{/-8.1°}$$

For division, we have

$$\frac{c}{d} = \frac{5\underline{/36.9°}}{\sqrt{2}\underline{/-45°}}$$

$$= \frac{5}{\sqrt{2}}\underline{/81.9°}$$

TRIGONOMETRIC FORMULAS

$$\sin \alpha = \cos(90° - \alpha) = -\cos(90° + \alpha)$$

$$\sin \alpha = \sin(180° - \alpha)$$

$$\tan \alpha = -\tan(180° - \alpha)$$

$$\sin(\alpha \pm \beta) = \sin \alpha \cos \beta \pm \cos \alpha \sin \beta$$

$$\cos(\alpha \pm \beta) = \cos \alpha \cos \beta \mp \sin \alpha \sin \beta$$

$$\tan(\alpha \pm \beta) = (\tan \alpha \pm \tan \beta)/(1 \mp \tan \alpha \tan \beta)$$

$$\sin^2 \alpha = (1 - \cos 2\alpha)/2$$

$$\cos^2 \alpha = (1 + \cos 2\alpha)/2$$

$$\sin(-\alpha) = -\sin \alpha$$

$$\cos(-\alpha) = \cos \alpha$$

$$\sin\left(\omega t + \frac{\pi}{2}\right) = \cos \omega t$$

$$\cos\left(\omega t - \frac{\pi}{2}\right) = \sin \omega t$$

$$\sin 2\alpha = 2 \sin \alpha \cos \alpha$$

$$\cos 2\alpha = \cos^2 \alpha - \sin^2 \alpha$$

$$\sin^2 \alpha + \cos^2 \alpha = 1$$

APPENDIX E

EULER'S FORMULA

Euler's formula is

$$e^{j\theta} = \cos\theta + j\sin\theta \qquad\qquad \text{(E-1)}$$

An alternative form of Euler's formula is

$$e^{-j\theta} = \cos\theta - j\sin\theta \qquad\qquad \text{(E-2)}$$

To derive Euler's formula, let

$$f = \cos\theta + j\sin\theta$$

Differentiating, we obtain

$$\frac{df}{d\theta} = -\sin\theta + j\cos\theta$$
$$= j(\cos\theta + j\sin\theta)$$
$$= jf$$

When $f = e^{j\theta}$, we have

$$\frac{df}{d\theta} = jf$$

as required. Thus, we obtain the result, Eq. E-1.
Adding Eq. E-1 and Eq. E-2, we obtain

$$\cos\theta = \frac{1}{2}\left(e^{j\theta} + e^{-j\theta}\right) \qquad\qquad \text{(E-3)}$$

Similarly, subtracting Eq. E-2 from Eq. E-1, we obtain

$$\sin\theta = \frac{1}{2j}\left(e^{j\theta} - e^{-j\theta}\right)$$

The equivalence of the polar and rectangular forms of a complex number is a consequence of Euler's formula. To verify this, consider a complex number $\mathbf{A} = a + jb$ where $a = r\cos\theta$ and $b = r\sin\theta$. Therefore,

$$\mathbf{A} = r(\cos\theta + j\sin\theta)$$
$$= re^{j\theta}$$

by Euler's formula.

STANDARD RESISTOR COLOR CODE

Low-power resistors have a standard set of values and color-band codes as well as a tolerance. The most common types of resistors are the carbon composition and carbon film resistors.

The color code for the resistor value utilizes two digits and a multiplier digit in that order as shown in Figure F-1. A fourth band designates the tolerance. Standard values for the first two digits are listed in Table F-1.

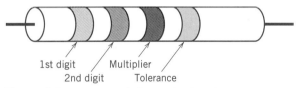

Figure F-1 Resistor with four color bands.

Table F-1
Standard Values for First Two Digits for 2% and 5% Tolerance Resistors

10	16	27	43	68
11	18	30	47	75
12	20	33	51	82
13	22	36	56	91
15	24	39	62	100

The resistance of a resistor with the four bands of color may be written as

$$R = (a \times 10 + b)m \pm \text{tolerance}$$

where a and b are the values of the first and second bands, respectively, and m is a multiplier. These resistance values are for 2% and 5% tolerance resistors as listed in Table F-1. The color code is listed in Table F-2. The multiplier and tolerance color codes are listed in Tables F-3 and F-4, respectively. Consider a resistor with the four bands yellow, violet, orange, and gold. We write the resistance as

$$R = (4 \times 10 + 7) \text{ k}\Omega \pm 5\%$$
$$= 47 \text{ k}\Omega \pm 5\%$$

Table F-2
Color Code

0	black
1	brown
2	red
3	orange
4	yellow
5	green
6	blue
7	violet
8	gray
9	white

Table F-3
Multiplier Color Code

silver	0.01
gold	0.1
black	1
brown	10
red	100
orange	1 k
yellow	10 k
green	100 k
blue	1 M
violet	10 M
gray	100 M

Table F-4
Tolerance Band Code

red	2%
gold	5%
silver	10%
none	20%

APPENDIX G

COMPUTER-AIDED ANALYSIS: PSpice

PREVIEW

This appendix is intended to serve as an introductory guide to PSpice, the circuit simulation program originally developed at the University of California at Berkeley. It is meant to provide the user with the means of learning to solve problems with this powerful computer tool. It contains sufficient detail and explanation to allow the user to advance rapidly to solving circuit analysis problems.

Many versions of PSpice are available. The software package PSpice is used throughout this chapter as a vehicle for learning how to use PSpice. The objective is to introduce the user to PSpice so that this software tool can be used to aid the circuit design and analysis cycle.

First we provide an introduction to PSpice and its capabilities and describe the types of problems that PSpice can be used to solve. We then describe how PSpice may be installed on the IBM family of personal computers. Section G-3 shows the user how to perform elementary circuit analysis. It gives the basic syntactical rules for the use of PSpice and shows how to solve simple dc circuits.

The next section deals with more sophisticated forms of dc circuit analysis including dc transfer curves and sensitivity analysis. The following sections explain how to perform time domain analysis and steady-state ac or frequency domain analysis. Finally, we show other analysis operations including temperature-dependent analysis, sensitivity analysis, analysis of circuits with mutual inductance, and Fourier analysis.

G-1 COMPUTER-AIDED ANALYSIS AND DESIGN

Engineers used mechanical calculators and slide rules for calculations in the analysis and design of circuits until the advent of digital computers in the 1950s. The world's first general-purpose electronic computer was built in 1946 by J. P. Eckert and J. W. Mauchly at the University of Pennsylvania. The nearly instantaneous response of electronic components made it possible for their ENIAC (Electronic Numerical Integrator and Computer) to multiply two 10-digit numerals in three-thousandths of a second, compared to roughly three seconds for the Mark I computer. ENIAC employed 18,000 vacuum tubes and 6000 switches to perform 5000 additions per second. It was huge, taking up the walls of a 30-foot by 50-foot room, weighing 30 tons, and requiring eighty 12-hp fans to keep the components within temperature limits.

As computers became commercially available in the 1950s and 1960s, engineers worked to reduce the size of the components and the power required to operate the computer. The price per bit of memory dropped from 0.5 cent in 1973 to 0.0005 cent in 1991 (three orders of magnitude). Over the same time period the number of gates per chip increased by three orders of magnitude and the power required to operate a computer declined by three orders of magnitude. With the reduction in size and power required, the personal computer and engineering workstations became available in the 1980s. Thus, by 1985 most engineers had sufficient computer facilities available to aid in the analysis and design of complex circuits.

In the 1970s, a mainframe computer program named SPICE was released. It was written at the University of California at Berkeley and put into the public domain.

SPICE is a general-purpose electric-circuit simulation program. The acronym SPICE stands for Simulation Program with Integrated Circuit Emphasis. The allowed components are resistors, capacitors, inductors, mutual inductances, independent dc and ac sources, dependent sources, transmission lines, diodes, and transistors.

In 1984, the first version of SPICE to run on a personal computer, named PSpice, became available. PSpice includes virtually all the features of general-purpose SPICE.

G-2 ‖ INTRODUCTION TO PSpice

Circuit analysis is an important part of the process of designing useful electric circuits. The use of computers to automate the process of circuit analysis has developed and evolved since the first digital computer became commercially available. One of the most complete circuit analysis tools is a software package known as PSpice. As the name implies, PSpice was developed to aid in the process of designing electronic circuits to be implemented as integrated circuits. This appendix will describe the use of PSpice in performing circuit analysis. Versions of PSpice are available for IBM PCs and PC compatibles as well as for Apple's line of Macintosh computers.

This appendix will attempt to describe, in general, the way in which PSpice can be used to perform circuit analysis.

The process of designing useful circuits is generally an iterative one that involves several distinct steps that may be repeated several times over. First, the circuit requirements are described in detail. This step involves deciding what the circuit must be capable of when it is finally implemented. Next, the designer will synthesize a circuit that will likely satisfy the requirements. Synthesis is usually based on experience obtained from past synthesis and analysis operations. Once the synthesis is done, an analysis of the circuit must be done to validate the fact that the circuit will meet the requirements specified in the first step. If the original specifications are met, the design process is complete. If not, modifications must be made, and the process must be started again. These modifications are generally made in the synthesis step. The designer uses the information obtained from the analysis to change the configuration or component values so that the circuit behaves closer to the desired operation. In some cases it may also be necessary to modify the original specifications if the analysis shows them to be impractical. The number of times that the design loop must be traversed depends on the ability and experience of the designer.

Circuit analysis programs like PSpice are used to perform the analysis portion of the design cycle. These programs can save much time over performing the analysis using by-hand calculations. Further, once the problem has been set up, reanalysis can be done by simply editing the PSpice circuit description file and executing it again.

Finally, because of the way in which PSpice solves a circuit analysis, there will be some circuits that PSpice will not be able to solve. That is, PSpice's solution to the problem will not always be guaranteed to converge to a unique solution. In many cases, this situation can be corrected by modifying the circuit model. Of course this must be done carefully so that the modified model still represents the circuit accurately.

The commercial version of PSpice is capable of analyzing circuits with up to approximately 130 elements and 100 nodes. The special student version of PSpice is limited to circuits consisting of approximately 10 transistor-type elements. More than 10 simple resistor, capacitor, inductor elements may be used. For most beginners' purposes, this limitation will not be detrimental.

PSpice is capable of performing three main types of analysis. It can determine the dc behavior of selected output voltages with respect to changes in input voltages. This type of analysis is usually referred to as a dc analysis. The results of a dc analysis are sometimes called the transfer characteristics of the circuit. A single-point dc analysis also determines what is called the bias-point characteristics, that is, the behavior of the circuit when only a dc voltage is applied to the circuit. In most cases, this is the starting point to introduce and apply PSpice.

A second type of analysis that is usually required in order to determine fully a circuit's behavior is called a transient analysis. Transient analyses calculate circuit voltages and currents with respect to time. This assumes that there is a time-dependent stimulus that causes an effect on the rest of the circuit. To perform a time domain circuit analysis, PSpice first calculates the bias point and then calculates the circuit response to a time-dependent change in one or more voltage or current sources.

The third main type of analysis that PSpice can perform is called an ac analysis. This analysis type is also referred to as a sinusoidal steady-state analysis. Here, voltages and currents are calculated as a function of frequency. That is, output variable changes are calculated in response to changes in the amplitude, frequency, or phase of sinusoidal input voltage or current sources. Again, the starting point for an ac analysis is the calculation of the dc bias-point.

There are several subtypes of analysis that PSpice can perform. These subtypes are generally enhanced forms of one of the three main types of analysis. First, there is a Sensitivity Analysis that can be used to determine the dc response of a circuit to changes in the values assigned to circuit elements. For example, PSpice will calculate the change in a specified output voltage with respect to changes in selected circuit elements.

Sensitivity Analysis is a type of dc analysis that shows how variations in certain component values affect the overall dc behavior of a circuit. Temperature Analysis is used to determine changes in circuit response with respect to variations in component values due to changes in temperature. PSpice's Fourier Analysis allows the user to perform a spectral analysis of a circuit, calculating the Fourier coefficients for the sinusoidal components of voltages or currents in the circuit.

G-3 || PSpice

PSpice is a commercially available version of SPICE developed by MicroSim Corporation. PSpice evolved from SPICE2, using the same input syntax and the same algorithms to perform circuit simulations. PSpice can be run on different computer hardware including a VAX, a Macintosh, an IBM PC or PC compatible, and an IBM PS/2. The remainder of this appendix will deal with running PSpice on an IBM PC or equivalent.

Installation of PSpice is quite simple and well documented on the PSpice distribution disk(s) that you receive. Exact contents of the distribution kit will depend on whether you have obtained the evaluation version of PSpice (sometimes called the student version) or whether you have purchased the full commercial version with any of several options available. Carefully read the file called README.DOC. This file contains important information on changes made to PSpice that have not yet been included in the manual. The file also contains information on the exact steps to be followed to install PSpice. These instructions are specific to installation on both floppy and hard disk drives. You will probably have to modify your CONFIG.SYS file (create one if you don't already have one) and perhaps create a new subdirectory or two if you are working with a hard disk drive. Your PSpice manual or README.DOC file will give you up-to-date instructions on how to install PSpice. Once PSpice is installed, the program is ready to run and perform circuit analysis.

When PSpice is executed on a PC, no other programs may be executed concurrently. PSpice will run on any IBM PC/XT/AT or compatible that has at least 512K of memory. It has also been run successfully on the new generation of IBM PS/2 computers. An 8087 math coprocessor is not necessary but is highly recommended. In order to run the supplemental plotting program called

PROBE, it is necessary to have a math coprocessor installed. On an IBM PC, the length of the executable file for the student version of PSpice, as stored on disk, is approximately 590K bytes. The commercial version uses approximately 470K of disk space. Because of the large size of these files, it is best to run PSpice on systems that contain a hard disk drive.

Before executing PSpice, it is necessary to create a circuit description file using a text editor. Any editor with which you are comfortable may be used as long as it produces a standard ASCII text file as its output. You may find it handy to use the filename extension .CIR for your input filenames, since PSpice will look for this extension by default.

Execution of PSpice is initiated by typing the program name, PSpice, at the DOS prompt. PSpice will ask for an input filename. You respond by typing the name of the circuit description file. If you have created this file with the extension .CIR, then there is no need to type an explicit filename extension. Next, PSpice will ask for an output filename. This will be the file where PSpice will write the results of the circuit analysis. PSpice will attach the filename extension .OUT if you do not explicitly specify one. An alternative method for initiating execution is to type the program name, PSpice, followed by the input and output filenames at the DOS prompt. The output filename may be omitted for this second initiation technique, in which case the output filename will be the same as the input filename with the .OUT extension.

Once the two filenames have been determined, PSpice will begin its analysis of the circuit description file. Messages indicating the status of the program execution like the one shown in Figure G-1 will be displayed on the screen. When PSpice is finished, it will return control to DOS as indicated by the appearance of the DOS prompt. At this point you may display the contents of the output file by printing it, using the DOS TYPE command to display it on the screen, or using a text editor to modify the file.

```
PSPICE Electrical Circuit Simulator      For the production version con-
tact:
            Demo Version
                                            MicroSim Corporation
        Copying of this program             23175 La Cadena Dr.
      is welcomed and encouraged          Laguna Hills, CA  92653
                                              (714) 770-3022

    Simulating circuit:  Edison's parallel lamp scheme
    In file  edlamp.CIR                        Writing results to file  edlamp.OUT
                          Reading and checking circuit
```

Figure G-1 Display screen showing PSpice execution status.

G-4 | CIRCUIT DESCRIPTION WITH PSpice

PSpice can be used to perform many different types of circuit analysis. In order to familiarize the user with PSpice, this section will show how a very simple circuit can be analyzed with PSpice. The rules for creating a circuit description file are given, and the results of a PSpice analysis are shown.

For a circuit analysis program to perform its function, a circuit description that can be interpreted by the program must be generated. For PSpice, the topology of a circuit is described in terms of element models and the way in which they are interconnected. This information is contained in three parameters: element names, node numbers, and element values. Syntax for providing this information is standard for all versions of PSpice.

We will start by working through the example circuit shown in Figure G-2. This is a simple circuit consisting of three resistors and a dc voltage source and is from Section 3-4. The objective is to determine the voltage across each resistor and the current flowing in the loop. The first step in creating the circuit description file is to name all of the elements. Next, each of the nodes is

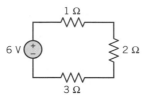

Figure G-2 Resistive circuit example with a dc source.

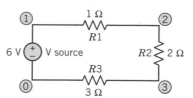

Figure G-3 Resistive circuit with PSpice element names and node numbers.

numbered. Figure G-3 is a modified version of Figure G-2 that contains node numbers and element names. Notice that the resistor names all begin with *R* and that the voltage source name begins with *V.* This is by convention and will be discussed later in this appendix. Further, nodes have been numbered starting at zero and continuing in consecutive order. Other node numbering (naming) schemes are available, and these will also be discussed later in this appendix.

A circuit description file for PSpice consists of a set of lines or statements. Each statement either describes a circuit element, is used to convey information to PSpice about the type of analysis or output that is to be generated, or is a special control statement. The first statement in each circuit description file is an example of a special control statement: a title statement. PSpice assumes that the first statement in every circuit description file is a line that gives a title to the analysis being performed. This title will be printed across the top of each output sheet when a multiple-page output is generated. Figure G-4 contains a PSpice circuit description for the example circuit of Figure G-3 with a title of "3 resistor circuit." Another important special control statement is seen at the end of the file. It is the **.END** statement that tells PSpice that it has read the entire circuit description file.

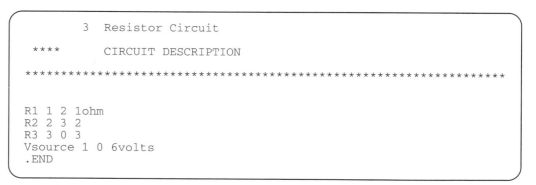

Figure G-4 Circuit description for the circuit of Figure G-3.

The general form of the statements used to describe individual elements is

$$<element\ name> <+ node> <- node> \ . \ . \ . \ <element\ value>$$

The continuation dots in this general form are used to indicate that elements with more than two terminals will have more than two nodes and may have more than one value. The value given to an element is a number that indicates the size of the device in terms of a measurement unit corresponding to that device type. For instance, resistances are always given a value in ohms whereas voltage sources are given values in volts. The second line of Figure G-4 is an element description for the 1-Ω resistor. It indicates the resistor's name as **R1**, its node connections as between node 1 and node 2, and its element value as 1 Ω. Similarly, the other two circuit resistors are described in the two subsequent statements.

The voltage source uses the statement

$$V <name> <+ node> <- node> <voltage\ value>$$

The voltage source is named **Vsource** and is connected between node 1 and node 0. Voltage source descriptions must convey polarity information to PSpice. Thus, the convention that PSpice uses in a voltage element statement is that the first node is the positive node, followed by the negative node. The value of the dc voltage source is indicated to be 6 V.

PSpice uses a passive sign convention that always assumes the positive current flows into the first indicated node, which is assumed as the positive voltage node. The statement for a current source is

$$I <name> <exit\ node> <entry\ node> <current\ value>$$

where all currents flow into the entry node.

The few lines in Figure G-4 complete the circuit description file necessary for PSpice to perform an analysis of the simple resistive circuit. If only a simple analysis is required, the next step is to submit this description to PSpice for processing.

One of the simplest analyses that PSpice is capable of performing is a single-point dc analysis. This is also referred to as an operating point analysis. An operating point analysis of the circuit in Figure G-2 may be obtained by submitting the circuit description file shown in Figure G-4 to PSpice. Since no type of circuit analysis is explicitly requested, an operating point analysis will be performed by default. An operating point analysis may be explicitly requested by including the statement **.OP** in the circuit description file.

PSpice is invoked by typing the program name, PSpice, and supplying the appropriate input and output filenames. Alternatively, one can type PS and invoke the PSpice shell, which contains a file editor and a manual. The results of this analysis example are shown in Figure G-5. The first page printed, which repeats the input circuit description and includes any syntactical error messages or warnings, is not shown in this figure. PSpice labels the result of the operating point analysis as the SMALL SIGNAL BIAS SOLUTION on the top of the results page. The voltage calculated for each node is shown in tabular format. Following the node voltage table, PSpice prints the solution for the current flowing in each voltage source in the circuit and the total dc power dissipated by the resistors. Program performance characteristics are listed after all circuit analysis solutions are printed. The performance characteristics indicate whether or not the program terminated normally and how long it took to perform the circuit analysis.

On the top of each page containing results, PSpice prints the temperature at which the analysis is performed. PSpice may perform analyses at different temperatures. This capability is primarily useful when analyzing electronic circuits. However, the element that models a resistor's behavior in PSpice can have a temperature coefficient assigned to it. That is, a resistor's resistance value can be made to vary with temperature. A discussion of temperature-dependent analysis is beyond the scope of this appendix.

Notice that the current that PSpice calculates for the voltage source is −1 amp (see Figure G-5).

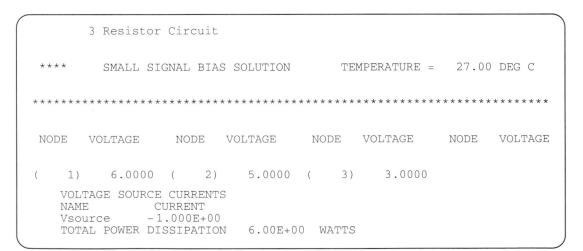

Figure G-5 Operating point analysis of the circuit of Figure G-3.

PSpice uses the convention of reporting currents as the amount of current that flows into the positive node of an element, through the element, and out the negative node. This current is usually referred to as conventional current. For elements that supply power, positive current leaves the positive node of the element. Thus, the current flowing into the positive terminal of the voltage source is minus one ampere.

If a default analysis is permitted, the only solutions are nodal voltages. Print requests for specific currents or voltages will be ignored in the .PRINT statement. If specific branch currents or potential differences across circuit elements are required, then the more detailed .DC analysis must be invoked.

Many different element models are available to represent physical devices within a PSpice analysis. Table G-1 contains a list of the basic circuit elements available in PSpice. By convention, the key letter identified in the table is used to identify a particular element to PSpice. A key letter must be the first letter of an element name. The remaining letters in an element name may be alphabetic or numeric. For PSpice, an underscore or the dollar sign may also be used within an element name. Names can be longer than 80 characters, though names longer than 20 characters are rarely used.

Table G-1
Basic Circuit Elements Available with PSpice

C	Capacitor
E	Voltage-Controlled Voltage Source
F	Current-Controlled Current Source
G	Voltage-Controlled Current Source
H	Current-Controlled Voltage Source
I	Independent Current Source
K	Inductor Coupling (Transformer)
L	Inductor
R	Resistor
T	Transmission Line
V	Independent Voltage Source
X	Subcircuit Call

Element names should be descriptive of the function of the element in the circuit if possible. Uppercase and lowercase letters may be used in names for most versions of PSpice. However, the program does not distinguish between the two types and hence, VCC is the same as Vcc, which is the same as vcc.

Nodes are the junction points where two or more circuit elements are joined. For PSpice, nodes must be numbered as positive integers. PSpice takes the node numbered 0 as the ground node. Node numbers need not be chosen in consecutive order; i.e., a continuous sequence of numbers need not be used to identify all of the nodes. PSpice allows nodes to be identified by any character string, not just a string of numbers. Throughout this appendix, numbers will be used to identify nodes in order to maintain compatibility with most versions of PSpice.

The ground node, node 0, is the node to which most other circuit voltages are referenced. For circuits in which no ground node is specified in the schematic diagram, the node with the greatest number of elements connected to it is generally specified as node 0. PSpice requires that every node in the circuit have a dc path to ground. This is necessary because of the way in which PSpice solves for the unknown voltages and currents. PSpice issues a warning message if a dc path to ground does not exist from every node. Along with this requirement, PSpice requires that each node be connected to at least two elements. This precludes having a node or wire dangling, as you might have in a lab testing environment. A very large resistor, say on the order of 100 MΩ, can be used to provide the necessary dc path to ground and circumvent the restriction on dangling nodes. For most circuits, a 100-MΩ resistor will have little or no effect on the circuit's performance.

Component values may be represented in several different forms in PSpice. Standard decimal notation may be used. For example, numbers such as 10, 10., 10.0001, − 10.0001 may be used whenever a value is needed. PSpice also accepts values in floating-point or scientific notation. In this notation, a number such as 1000 can be written as 1.0E3, which stands for 1×10^3. The letter E is used to separate the mantissa from the exponent of a base-10 floating-point number.

Because the range of values can be very large within a given PSpice circuit description, and to reduce the need for unnecessary typing and the errors that this typing might incur, the developers of PSpice included a set of abbreviations that may be used as suffixes to values. These suffixes are listed in Table G-2. For example, the value 0.004U F represents 0.000000004 F or 4E-9 F. Notice that this value could also have been written as 4N F.

Table G-2
Scale Factor Abbreviations

Letter Suffix	Multiplying Factor	Name of Suffix
T	1E12	tera
G	1E9	giga
MEG	1E6	mega
K	1E3	kilo
MIL	25.4E-6	mil
M	1E-3	milli
U	1E-6	micro
N	1E-9	nano
P	1E-12	pico
F	1E-15	femto

It should be noted that in describing component values, any letter can follow the number, but if that letter is one of the designative modifiers it will be interpreted as a modifier. In other words, you could put 14a or 14 amperes or 14A to be more descriptive of the value. Care should be used to avoid typing 10F (for 10 farads) because F is the femto multiplier.

All well-written computer software is sufficiently documented so that someone unfamiliar with the code may readily use and modify it. This should also be true for circuit models written for use with PSpice. Two methods are available for specifying comments in a PSpice circuit model. Any line beginning with an asterisk (*) is taken to be a comment line. PSpice does not interpret any of the characters in a comment line, it simply prints the characters as is in the output listing. Comments may also be included at the end of lines that contain PSpice commands or element specifications. A semicolon is used to separate a comment from a command or element specification. Of course, the comments must always follow, not precede, the command or element specification.

Another important syntactical mechanism that you will eventually need to use is the line continuation operator. As your models become more complex, you will probably generate statements that are longer than a single line. To continue a line on the next line, the succeeding line should start with a plus sign character (+). Continuation lines may be carried on repeatedly, as long as succeeding continued lines begin with the plus character.

G-5 | DC CHARACTERISTICS

Thus far, the simulations performed have calculated the behavior of a circuit when all of the circuit's sources are held at a single, individual fixed voltage or current. In many instances in the design process, it is necessary to obtain the analysis of a circuit for several different values of source voltage or current. This chapter discusses the use of a sweep analysis to determine the dc behavior of a circuit.

PSpice voltage or current sources may be swept over a range through the use of the **.DC** control statement. When the **.DC** statement is included in a simulation file, PSpice first performs the normal operating point analysis. Circuit unknowns are calculated with the single given source voltages or source currents. Next, PSpice performs the sweep analysis by stepping through a specified range of values for a particular input source. The result of using the **.DC** analysis type is the same as performing repeated **.OP** analyses where the value of a particular source is changed before each analysis. Obviously, it is much less work and much quicker to use the **.DC** statement rather than use multiple **.OP** analyses.

The syntax for the **.DC** statement is as follows:

.DC *<source name> <start value> <stop value> <increment>*

where *source name* is the name of the voltage or current source that is to be swept through a range of values. This source must be defined elsewhere in the description file by a voltage or current source statement. *Start value* and *stop value* define the range of values over which the source will be swept. The *increment* value determines the step size between succeeding analysis points. *Start value, increment,* and *stop value* are interrelated to determine the number of points for which the circuit is analyzed.

It is easily seen that the number of analysis points is (*stop value* − *start value*)/(*increment*) + 1. Depending upon the version of PSpice being used, there may be a default limit for the number of points that can be included in a **.DC** analysis. PSpice does not limit the number of iteration points that it allows. Should the **.DC** parameters be incorrectly specified and cause PSpice to initiate an overly lengthy analysis, the program can be terminated by typing a control-C (hold down Ctrl key and press the C key at the same time).

As an example of the use of the **.DC** analysis capabilities, consider the circuit shown in Figure G-6.

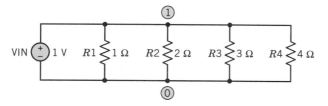

Figure G-6 Circuit diagram for Edison's parallel lamp scheme.

This circuit can be recognized as a model for Edison's parallel lamp scheme as discussed in Section 3-5. Each resistor represents an individual lamp. This scheme provides the same voltage to each lamp element, and as such, the current in individual lamps is independent of that in other lamps. The objective here is to determine the behavior of the circuit as the source voltage is increased from 0 V to 5 V. Of interest is the power dissipated in each lamp as the voltage supply increases in value. The power dissipated in a lamp can be used to determine the amount of light given off by the device. Although PSpice will not directly calculate the power supplied to an element, both the voltage across an element and the current through an element may be determined and provided as output by PSpice. The power dissipated in each element may then be calculated.

Figure G-7 contains the circuit description file for the circuit shown in Figure G-6. The **.DC** statement is used to sweep the source VIN over the range 0 to 5 V in steps of 0.2 V. Notice that the circuit description file is the same as that which would be used to calculate the operating point values, except for the addition of the **.DC** and **.PRINT** statements. The **.PRINT** statement is used to output the results of the **.DC** analysis.

PSpice presents the output of an operating point analysis automatically in a tabular form. This output displays the voltage value at each node of the circuit for the fixed driving source values. When an analysis is performed wherein the values are swept, such as the **.DC** analysis, it is convenient to provide a mechanism for choosing only the nodes of interest for output display. The **.PRINT** statement provides just such a mechanism. **.PRINT** will also be used with other analysis types that will be discussed in later sections. **.PRINT** allows you to choose which values from a

```
Edison's parallel lamp scheme


 ****      CIRCUIT DESCRIPTION


 ****************************************************************


R1 1 0 1
R2 1 0 2
R3 1 0 3
R4 1 0 4
VIN 1 0 DC 1
.DC VIN 0 5 .2
.PRINT DC I(VIN) I(R1) I(R2) I(R3) I(R4)
.PLOT DC I(VIN) I(R1) I(R2) I(R3) I(R4)
.PROBE
.END
```

Figure G-7 Circuit description for Edison's parallel lamp scheme.

particular analysis are to be displayed. By appropriately ordering the variables in the **.PRINT** statement, it is possible to order the resultant output tables for ease of value comparison.

The syntax for the **.PRINT** statement is

$$.\text{PRINT} <analysis\ type> <output\ list>$$

where the *analysis type* specifies that the variables in the succeeding list are to correspond to values calculated for a particular analysis type. To obtain output generated by a **.DC** analysis, the *analysis type* DC should be specified. This will produce a listing for each output variable at each point in the **.DC** sweep analysis. The first column of the tabular output for a **.DC** analysis will contain values of the swept source over the full range specified in the **.DC** statement. Output values can be conveniently referred back to these source values.

The *output list* is a list of the node voltages, element voltages, or element currents whose values are to be displayed. Node voltage variables for the *output list* are of the form V(n1, n2), where the voltage drop from node n1 to n2 is specified. If only a single node, n1, is included in the list, the output voltage will be referenced to ground. That is, the n2 parameter will be assumed to be a 0. Output voltages may also be specified using the form V(element). Here, the voltage across the element will be output. *Element* may be any valid element name. For example, we print the voltage across a resistor R_2 by specifying V(R2).

Element currents are specified using the form I(element) where *element* is any valid element name, including the name of a source element. The printed values of current have the direction from n1 to n2 where n1 is the first node in the list and n2 is the second. For example, current flowing in the resistor described in the statement

$$\text{R1 5 6 1Kohm}$$

will have a positive value when current flows from node 5 to node 6. That is, when the voltage at node 5 is higher than the voltage at node 6, the current flowing in R1 will be reported as positive. This convention also applies to voltage sources. As with other elements, the printed values of current will be positive for currents flowing from n1 to n2, i.e., from the positive node of the source to the negative node. This may seem to be the opposite of what is expected, but once recognized, this seeming discrepancy may be easily dealt with. For the voltage source described in the statement

$$\text{VDD 7 8 6volts}$$

the current in element VDD will be reported as positive when it flows from node 7 to node 8. Of course the voltage across the source will always have the same polarity no matter which way the current is flowing. To verify this, we refer again to the circuit in Figure G-3. Clearly, a 1-A current flows from node 0 to node 1 through the source. The input statement in Figure G-4 reads **Vsource 1 0 6volts** so that the .PRINT statement will yield the current from node 1 to 0, which is − 1 A, as shown in Figure G-5.

Several voltages and currents may be printed in a single table. The actual number will depend upon the version of PSpice being used and the default settings. Voltage and current values may be freely mixed in an output table.

The following is a set of valid .PRINT statements along with their meanings:

.PRINT V(5,2)	;print voltage between node 5 and node 2
.PRINT V(5)	;print voltage between node 5 and ground
.PRINT V(R1)	;print voltage across resistor R1
.PRINT I(R1)	;print current through resistor R1
.PRINT V(1,2) V(2,3) I(VCC)	;print voltage from node 1 to 2, the
	;voltage from node 2 to 3, and the current
	;through voltage source VCC in one table

Figure G-8 contains the output listing generated for the simulation of the circuit described in Figure G-7. This output shows the current through the source as well as through each of the resistive elements. Check to see that Kirchhoff's current law holds for each of the nodes of the circuit.

The number of digits printed in each column of the output table can be changed using the .OPTIONS statement. The syntax for this statement is

$$.OPTIONS\ [option\ name\,][option\ name = value\,]$$

where option name may be one of a list of keywords that may or may not require a *value* parameter.

```
Edison's parallel lamp scheme


****      DC TRANSFER CURVES              TEMPERATURE =   27.000 DEG C


********************************************************************
 VIN            I(VIN)        I(R1)        I(R2)        I(R3)        I(R4)

 0.000E+00     0.000E+00    0.000E+00    0.000E+00    0.000E+00    0.000E+00
 2.000E-01    -4.167E-01    2.000E-01    1.000E-01    6.667E-02    5.000E-02
 4.000E-01    -8.333E-01    4.000E-01    2.000E-01    1.333E-01    1.000E-01
 6.000E-01    -1.250E+00    6.000E-01    3.000E-01    2.000E-01    1.500E-01
 8.000E-01    -1.667E+00    8.000E-01    4.000E-01    2.667E-01    2.000E-01
 1.000E+00    -2.083E+00    1.000E+00    5.000E-01    3.333E-01    2.500E-01
 1.200E+00    -2.500E+00    1.200E+00    6.000E-01    4.000E-01    3.000E-01
 1.400E+00    -2.917E+00    1.400E+00    7.000E-01    4.667E-01    3.500E-01
 1.600E+00    -3.333E+00    1.600E+00    8.000E-01    5.333E-01    4.000E-01
 1.800E+00    -3.750E+00    1.800E+00    9.000E-01    6.000E-01    4.500E-01
 2.000E+00    -4.167E+00    2.000E+00    1.000E+00    6.667E-01    5.000E-01
 2.200E+00    -4.583E+00    2.200E+00    1.100E+00    7.333E-01    5.500E-01
 2.400E+00    -5.000E+00    2.400E+00    1.200E+00    8.000E-01    6.000E-01
 2.600E+00    -5.417E+00    2.600E+00    1.300E+00    8.667E-01    6.500E-01
 2.800E+00    -5.833E+00    2.800E+00    1.400E+00    9.333E-01    7.000E-01
 3.000E+00    -6.250E+00    3.000E+00    1.500E+00    1.000E+00    7.500E-01
 3.200E+00    -6.667E+00    3.200E+00    1.600E+00    1.067E+00    8.000E-01
 3.400E+00    -7.083E+00    3.400E+00    1.700E+00    1.133E+00    8.500E-01
 3.600E+00    -7.500E+00    3.600E+00    1.800E+00    1.200E+00    9.000E-01
 3.800E+00    -7.917E+00    3.800E+00    1.900E+00    1.267E+00    9.500E-01
 4.000E+00    -8.333E+00    4.000E+00    2.000E+00    1.333E+00    1.000E+00
 4.200E+00    -8.750E+00    4.200E+00    2.100E+00    1.400E+00    1.050E+00
 4.400E+00    -9.167E+00    4.400E+00    2.200E+00    1.467E+00    1.100E+00
 4.600E+00    -9.583E+00    4.600E+00    2.300E+00    1.533E+00    1.150E+00
 4.800E+00    -1.000E+01    4.800E+00    2.400E+00    1.600E+00    1.200E+00
 5.000E+00    -1.042E+01    5.000E+00    2.500E+00    1.667E+00    1.250E+00
```

Figure G-8 PSpice output for Edison's parallel lamp scheme using the **.PRINT** statement.

To change the number of digits printed, the keyword NUMDGTS followed by an integer value is used. The following statement changes the number of digits printed to six from the default of four:

.OPTIONS NUMDGTS = 6 WIDTH = 132

This example also demonstrates the use of the WIDTH parameter to increase the width of the printed page from the default of 80 characters per line to 132 characters per line.

In many cases, it is advantageous to get a visual perspective of the data being presented as the output of a simulation. The .PLOT statement provides a means of obtaining such a perspective by having results plotted as a part of the output generated by PSpice. Unfortunately, this means of plotting produces only what is referred to as *character-printer* quality plots. This means that each plotted data point is represented by a character, with each step on the *x*-axis being the size of a line and each step on the *y*-axis being the size of a character column. Reduced resolution is the penalty paid for the ease and speed of this plotting technique. Several other means of plotting the output of PSpice are available. One such means is a program called PROBE, available with PSpice, which will be discussed later in this section.

The syntax for using the .PLOT statement is very similar to that of the .PRINT statement. As with .PRINT, the type of analysis for which the plot is desired is specified first, followed by the variables that are to be plotted. The syntax is

.PLOT <*analysis type*> <*output list*> [<*min range*>, <*max range*>]

where as with .PRINT, *analysis type* specifies one of several different possible types of analysis that PSpice can perform and *output list* specifies which of the circuit variables are to be plotted. When more than one output variable is specified in the *output list*, each variable in the list will be plotted on the same graph. Multiple .PLOT statements will produce plots on separate axes; one axis for each .PLOT statement.

PSpice automatically calculates the range of the *y*-axis to go from the smallest to the largest value to be plotted on the *y*-axis. In order to more closely examine certain regions of an output curve, it is occasionally necessary to zoom in on a certain interval of values. This can be accomplished by using the optional *min range* and *max range* parameters. You may specify the minimum and maximum value that can be shown on the *y*-axis. The total range of the independent variable as specified on the .DC statement will always be plotted in full.

Figure G-9 shows each of the element currents calculated for Edison's parallel lamp scheme plotted on the same axis with the scale limits being automatically calculated by PSpice. The plot indicates the expected linear behavior of the resistive elements. That is, it is easy to see from the plot that the current in the resistors increases linearly as the voltage across the resistors increases. (Actually, the curves appear to have breaks in them because of the reduced precision of the printer plot.) A legend is printed at the top of the graph to indicate which symbol is related to a particular output variable. Each variable is plotted on its own scale, with the respective scale minimums and maximums printed near the legend. The symbol X is used to indicate a location where two or more plotted points intersect. Two columns of numerical data are printed alongside the plot. The first is the sequence of swept values, e.g., the dc voltage that is swept in a .DC analysis. The second column displays the numerical values of the first variable in the output list of the .PLOT statement; in this case, the value of I(VIN), the current in voltage source VIN.

A plot of the output variables for Figure G-8 can all be plotted on the same scale. This plot is obtained by using the following plot statement:

.PLOT DC I(VIN) I(R1) I(R2) I(R3) I(R4) 0 4

Limits for the *y*-axis are specified with the optional *min range* and *max range* parameters. All variables are plotted within these limits. This restriction makes for easier comparison of values. When a value is outside the specified range, a point is plotted at the edge of the plot area.

PSpice is capable of producing standard line printer plots as described above. In addition, a program called PROBE has been written by MicroSim Corporation to supplement the plotting capabilities of PSpice. PROBE uses the graphics capabilities built into a PC to produce what is called *graphics printer quality plot*. The quality of the plot will depend upon the resolution of your PC's display or printer.

```
Edison's parallel lamp scheme


****      DC TRANSFER CURVES               TEMPERATURE =   27.000 DEG C


* * * * * * * * * * * * * * * * * * * * * * * * * * * * * * * * * * * * * * * * * * * * * * * * * * * * * * * * * * * * * * * * * * * * *

LEGEND:

*:  I(VIN)
+:  I(R1)
=:  I(R2)
$:  I(R3)
0:  I(R4)

  VIN          I(VIN)
(*)----------     1.5000E+01   -1.0000E+01   -5.0000E+00    0.0000E+00    5.0000E+00
(+)----------     0.0000E+00    2.0000E+00    4.0000E+00    6.0000E+00    8.0000E+00
(=)----------     0.0000E+00    1.0000E+00    2.0000E+00    3.0000E+00    4.0000E+00
($0)---------     0.0000E+00    5.0000E-01    1.0000E+00    1.5000E+00    2.0000E+00
                - - - - - - - - - - - - - - - - - - - - - - - - - - - - - - - - - - - - - -
  0.000E+00   0.000E+00  X                .              .                    *             .
  2.000E-01  -4.167E-01  .X$              .              .                   *.             .
  4.000E-01  -8.333E-01  .   X            .              .                 *   .            .
  6.000E-01  -1.250E+00  .     X$         .              .                *    .            .
  8.000E-01  -1.667E+00  .      X $       .              .              *      .            .
  1.000E+00  -2.083E+00  .        X $     .              .             *       .            .
  1.200E+00  -2.500E+00  .         X $    .              .            *        .            .
  1.400E+00  -2.917E+00  .          X  $. .              .          *          .            .
  1.600E+00  -3.333E+00  .           X  .$.              .         *           .            .
  1.800E+00  -3.750E+00  .            X. $               .       *             .            .
  2.000E+00  -4.167E+00  .             X    $            .      *              .            .
  2.200E+00  -4.583E+00  .             .X      $         .    *                .            .
  2.400E+00  -5.000E+00  .            .   X       $      .  *                   .            .
  2.600E+00  -5.417E+00  .             .    X       $ *. .                      .            .
  2.800E+00  -5.833E+00  .             .      X       X  .                      .            .
  3.000E+00  -6.250E+00  .             .       X  *   $  .                      .            .
  3.200E+00  -6.667E+00  .             .        X*   . $ .                      .            .
  3.400E+00  -7.083E+00  .             .       *X    . $ .                      .            .
  3.600E+00  -7.500E+00  .             .      *   X   . $ .                     .            .
  3.800E+00  -7.917E+00  .             .    *     X.     $  .                   .            .
  4.000E+00  -8.333E+00  .             .   *       X       $   .               .            .
  4.200E+00  -8.750E+00  .             .  *        .X       $   .              .            .
  4.400E+00  -9.167E+00  .             . *         .   X      $. .             .            .
  4.600E+00  -9.583E+00  .             .*          .     X      .$             .            .
  4.800E+00  -1.000E+01  .            *            .       X      $             .            .
  5.000E+00  -1.042E+01  .          *.             .         X      $           .            .
                - - - - - - - - - - - - - - - - - - - - - - - - - - - - - - - - - - - - - -
```

Figure G-9 PSpice line printer plot for currents using automatic scaling for Edison's parallel lamp circuit.

To use PROBE, it is necessary to include a statement in the PSpice circuit description file so that data to be plotted by PROBE will be generated by PSpice. The data to be plotted will be put into a disk file named PROBE.DAT. When PROBE is executed, the software will check to determine that this file exists, and if not, PROBE will display an error message and terminate. The statement that must be included in the PSpice circuit description file is called the .PROBE statement. The syntax for the .PROBE statement is similar to that of the .PRINT and .PLOT statements:

.PROBE [*output list*]

The *output list* is optional and may contain a list of the node voltages or element currents that are to be put into the PROBE.DAT data file. If no parameters are contained in the *output list,* then all of the circuit voltages and currents are included in the PROBE.DAT file. This is how the **.PROBE** statement is typically used. All of the circuit variables are included in the data file, and as PROBE is run, individual variables are selected for plotting. In this way, it is unnecessary to rerun a PSpice analysis if, when a PROBE plot is done, it is suddenly recognized that an additional curve is desirable. Commands are entered via an on-screen menu, reducing the need to remember long lists of commands and their syntax. Multiple waveforms may be plotted on the same graph. Multiple graphs may be displayed on the same page. The command set provides for producing a hard copy of the graphs displayed on the screen.

Figure G-10 contains a graph generated by PROBE for the element currents in the circuit of Figure G-7. A comparison of this graph and the one shown in Figure G-9 illustrates the difference between the low-resolution line printer plot and this higher-resolution graphics printer plot.

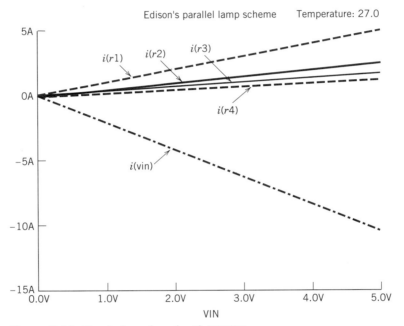

Figure G-10 Circuit data plotted with PROBE.

Subcircuits are means for reducing the amount of typing necessary to describe large circuits that contain repeated blocks. A subcircuit for PSpice is very similar to a subroutine or a macro in a programming language. A set of circuit elements may be bundled into a block structure with a defined set of nodes at the interface boundary. This subcircuit may then be "called" and connected to other circuit elements as many times as it is needed in the overall circuit. More than one subcircuit may be defined within a circuit, and each subcircuit may be called more than one time.

A subcircuit is defined using the keyword **.SUBCKT** to start a subcircuit definition and **.ENDS** to signal the end of a subcircuit definition. The syntax for subcircuit definition is

.SUBCKT <*subcircuit name*> <*interface node list*>

circuit element statements defining subcircuit

.ENDS

where the *subcircuit name* is a means by which to refer to a particular subcircuit. The *interface node list* is a list of nodes internal to the subcircuit to which external circuit elements will be attached. Node numbers contained in the subcircuit are unique to the subcircuit and may duplicate those in the external circuit. Only node 0 is an exception to this rule. Node 0 is the ground node and is common to all subcircuits and external circuits. As such, it is unnecessary to include node 0 in an interface node list.

PSpice has four basic types of dependent sources for use in circuit modeling. These sources are the *voltage-controlled voltage source* (VCVS), the *current-controlled voltage source* (CCVS), the *voltage-controlled current source* (VCCS), and the *current-controlled current source* (CCCS). Controlled sources may be linear functions of the controlling variable, or there may be a nonlinear relationship that defines how a source functions with regard to its control variable.

Linear dependent sources are syntactically specified in much the same way as independent sources. The difference is that in a dependent source, a controlling variable is specified, a multiplying or gain factor is specified, and the key letter that is used to begin the name of the element is different. The following is a list showing the syntax for each of the available dependent sources:

(VCVS)   E*name*<*element nodes*> <*control nodes*> <*gain factor*>
(CCVS)   H*name*<*element nodes*> <*control current*> <*gain factor*>
(VCCS)   G*name*<*element nodes*> <*control nodes*> <*gain factor*>
(CCCS)   F*name*<*element nodes*> <*control current*> <*gain factor*>

An example of a simple VCVS that is attached to nodes 1 and 2 and produces a voltage that is 10 times the voltage drop from node 5 to node 6 is

Esimple 1 2 5 6 10

In algebraic terms, this source would be described by the equation

$$v_1 - v_2 = 10 \times (v_5 - v_6)$$

Element nodes carry the same polarity conventions and current direction conventions as the independent sources. The first node specified is the positive node, and the positive current flows through the source from positive node to negative node. The control or sense nodes have the more positive node specified as the first node. The sense nodes have an infinite input impedance and therefore draw no current. That is, the controlled source senses the voltage across the controlling nodes but is not actually attached to those nodes. Control nodes may be the same as the element nodes.

Current-controlled sources will be a function of currents in specific branches of the circuit. The controlling branch current must be the current in an independent voltage source. This independent voltage source may have a zero or nonzero voltage value. If its value is zero, it is equivalent to a short circuit. Thus, if the controlling current is to be that which flows in a resistive branch, it is necessary to insert an independent voltage source in series with the resistor. The value of the voltage source is set to zero so that the new source does not have any effect on the branch current.

An example of a current-controlled voltage source is

Hccvs 3 4 Vdummy 5.6

where the source name is *Hccvs* and it produces a voltage drop from node 3 to node 4. The controlling current flows through the independent voltage source *Vdummy* and the gain factor is 5.6 Ω. Notice that the gain factor here has the units of resistance and is called a transresistance. This is because to produce the units of volts, the output of the source, it is necessary to multiply the sensed current in amperes by a value in ohms. The term transresistance comes from the fact that a transfer is made from one set of nodes to another.

Rules for using the VCCS and CCCS are very similar to those for the VCVS and CCVS. Outputs of the dependent current sources are specified in the same way as those of the independent current sources. Controlling voltages and currents are specified in the same way as for the VCVS and CCVS. For the CCCS, a dimensionless gain factor is provided. A transconductance is supplied for the VCCS.

Figure G-11 is the schematic for a circuit containing a CCVS. The PSpice file for describing this circuit is shown in Figure G-12. Notice that the voltage source **Vsense** is in the circuit for the sole purpose of measuring the current through the resistor R_3. A single-point solution for the circuit using the **.OP** statement is shown in Figure G-13.

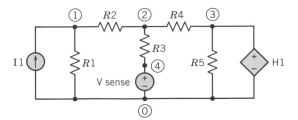

Figure G-11 Circuit with a CCVS H1.

```
Current Controlled Voltage Source Example

    ****        CIRCUIT DESCRIPTION


****************************************************************

I1 0 1 5amps
H1 3 0 Vsense 3.5
R1 1 0 10
R2 1 2 20
R3 2 4 30
R4 2 3 40
R5 3 0 50
Vsense 4 0 0volts
.OP
.END
```

Figure G-12 PSpice description of the CCVS circuit of Figure G-11.

```
Current Controlled Voltage Source Example

    ****     SMALL SIGNAL BIAS SOLUTION      TEMPERATURE =   27.000 DEG C


****************************************************************************

NODE   VOLTAGE      NODE   VOLTAGE      NODE   VOLTAGE      NODE   VOLTAGE

(    1)  -39.207  (    2)  -17.621  (    3)  -2.056  (    4)    0.0000

       VOLTAGE SOURCE CURRENTS
       NAME          CURRENT
       Vsense      -5.874E-01
       TOTAL POWER DISSIPATION   0.00E+00  WATTS
```

Figure G-13 Solution for the CCVS circuit of Figure G-11.

G-6 ‖ TIME DOMAIN ANALYSIS

Electrical circuits all have some time-dependent characteristics that must be considered during the design cycle. Even the simple flashlight driven by dc batteries has a time-dependent response. When the circuit's switch is closed, the dc source is applied to the circuit, causing a time-varying current response.

Thus far, we have discussed the use of PSpice in analyzing the performance of a circuit to which a dc source has been applied. In these analyses, it is assumed that none of the sources vary as a function of time. Further, capacitive elements were set to open circuits, and inductive elements were set to short circuits for the dc analyses. In this section, the behavior of a circuit in response to a time-varying stimulus will be discussed.

A typical circuit of interest is shown in Figure G-14. This circuit was discussed in Chapter 8. In this circuit, the switch is opened at arbitrary time $t = 0$ after being closed for a long time. If the switch has been closed for a long period of time before $t = 0$, it can be assumed that all transient effects have died out and that the circuit can be treated as a dc circuit. That is, the circuit can be analyzed with the capacitance set to an open circuit. A dc analysis provides that the voltage across the capacitor just before $t = 0$ is equal to the dc source voltage VINIT.

Once the switch is opened at $t = 0$, the circuit becomes the simple connection of a resistor in parallel with a capacitor. The initial charge stored on the capacitor will cause a current to begin flowing through the resistor at $t = 0$. Current will flow until the initial voltage is reduced to zero by discharging the capacitor through the resistor. This is termed the natural response of the circuit. It is important to be able to calculate the voltage and current in this circuit as a function of time. The transient analysis capabilities built into PSpice can perform these calculations.

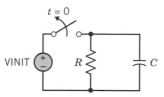

Figure G-14 *RC* circuit with dc source connected prior to $t = 0$.

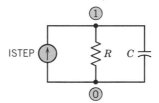

Figure G-15 Current source ISTEP supplies the initial conditions.

Unfortunately, PSpice does not have a built-in circuit element that directly models the behavior of a switch. A model for the effects of opening the switch must be built from existing element models. Figure G-15 contains the diagram of a circuit that will model this particular situation. Current source ISTEP is time-varying, changing as a step function when $t = 0$. Before $t = 0$, the current source produces sufficient current so that the appropriate initial voltage is developed across the terminals of the capacitor. At $t = 0$, the current source is set to zero, thus producing the effect of an open circuit.

In order to determine how a circuit responds as a function of time, it is necessary to have models for sources that vary as a function of time. PSpice uses a modification of the previously used dc voltage or current source element statement to produce a time-varying source description. Several different functions of time are available. The general syntax for a time-varying source is

<source name> <nodes> <time function type> <time parameters>

The *source name* and *nodes* use the same syntax as previously described for dc sources. The time function types include:

PULSE—pulse waveform that may be periodic
EXP—exponential waveform
SIN—sinusoidal waveform
PWL—piecewise linear waveform for constructing arbitrary functions of time

The PULSE input source waveform is shown in Figure G-16. The general form of the PULSE specification is

$$PULSE(<v1><v2><td><tr><tf><pw><per>)$$

where Table G-3 describes each of the parameters and their default values.

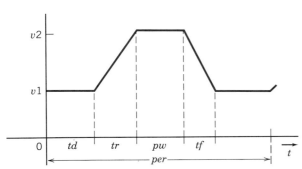

Figure G-16 Pulse waveform.

Table G-3
PULSE Waveform Specifications

Parameter	Default Value, Units
$<v1>$ initial value	none, volts or amps
$<v2>$ pulse value	none, volts or amps
$<td>$ delay time	0, seconds
$<tr>$ rise time	TSTEP, seconds
$<tf>$ fall time	TSTEP, seconds
$<pw>$ pulse width	TSTOP, seconds
$<per>$ period	TSTOP, seconds

TSTEP and TSTOP are defined as part of the .TRAN statement in the next section. Even though a user might specify rise time and fall time equal to zero in the command line, a zero value will never be implemented.

The PULSE source generates a waveform that is at $v1$ volts (or amps) at $t = 0$. After a delay time of td, the waveform makes a transition to $v2$ over a rise time interval tr. The waveform stays at $v2$ for a length of time equal to pw, then returns to $v1$, making the transition over the time interval tf. If per is specified, the waveform, starting from the beginning of tr, will repeat continuously with a period of per. The delay time is not included in the repeated waveform. Note that the value of $v1$ may be either greater or less than $v2$.

An example of a pulsed waveform current source is

$$ISPIKE\ 10\ 0\ PULSE(0\ 100\ 10N\ 1P\ 1P\ 10P\ 20N)$$

This source produces a current flowing from node 10 to node 0. The current is initially at zero amps from $t = 0$ until $t = 10$ ns (1E-9 seconds). The current then rises linearly to 100 A over the next picosecond (1E-12 seconds) and stays there for 10 ps. Next, the current falls back to zero over a 1-ps interval. The current stays at zero for the next 19.988 ns, after which the waveform repeats its transition from 0 to 100 A.

As another example of a PULSE source, consider

$$VSTEP\ 5\ 6\ PULSE(0\ 1\ 0\ 1P)$$

Only the first four PULSE parameters are specified, leaving the others to become the default values. This source produces a step waveform that changes from 0 to 1 V in 1 ps at time $t = 0$. The waveform stays at 1 V for the rest of the analysis since the pulse width parameter defaults to be equal to time TSTOP. This waveform approximates the ideal step function used in many different analysis problems.

The EXP input source is similar to the PULSE source except for the fact that it produces a waveform that varies exponentially with time. The EXP input source waveform is pictured in Figure

G-17. The general form of the EXP specification is

$$\text{EXP}(<v1><v2><td1><tc1><td2><tc2>)$$

where Table G-4 describes each of the parameters and their default values.

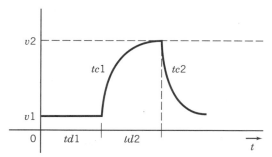

Figure G-17 Exponential waveform.

Table G-4
EXP Waveform Specifications

Parameter	Default Value, Units
$<v1>$ initial value	none, volts or amps
$<v2>$ peak value	none, volts or amps
$<td1>$ rise delay time	0, seconds
$<tc1>$ rise time constant	TSTEP, seconds
$<td2>$ fall delay time	$<td1>$ + TSTEP, seconds
$<tc2>$ fall time constant	TSTEP, seconds

The EXP source generates a waveform that is at $v1$ V (or amps) at $t = 0$. After a delay time of $td1$, the waveform makes a transition to $v2$. This transition takes the shape of an exponential waveform with a time constant of $tc1$. That is, the waveform is exponential. Fall delay $td2$ determines the length of the interval over which the rise from $v1$ to $v2$ takes place. After interval $td2$ expires, the waveform returns to $v1$ exponentially, with the constant $tc2$ specifying the exponential decay. Note that again the value of $v1$ may be either greater than or less than $v2$.

An example of an exponential waveform voltage source is

$$\text{VSAW 5 6 EXP(0 5 2M 500N 1M 1M)}$$

This source produces a voltage drop from node 5 to node 6 that is initially 0 V until $t = 2$ ms. At 2 ms, the voltage exponentially rises to 5 V at a rate defined by the time constant 500 ns. The rising exponential lasts until $t = 3$ ms, after which the voltage drops back to zero exponentially with a time constant of 1 ms.

The SIN input source waveform is pictured in Figure G-18. The general form of the SIN specification is

$$\text{SIN}(<vo><va><freq><td><df><pb>)$$

where Table G-5 describes each of the parameters and their default values.

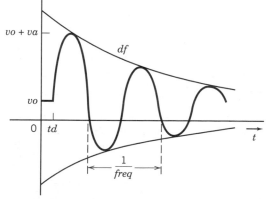

Figure G-18 Sinusoidal waveform.

Table G-5
SIN Waveform Specifications

Parameter	Default Value, Units
$<vo>$ initial offset	none, volts or amps
$<va>$ peak amplitude	none, volts or amps
$<freq>$ frequency	1/TSTOP, hertz
$<td>$ delay	0, seconds
$<df>$ damping factor	0, 1/seconds
$<pb>$ phase	0, degrees

The SIN source generates a waveform that is equal to vo V (or amps) at $t = 0$. After a delay time of td, the waveform begins to oscillate sinusoidally with an initial amplitude of va. The frequency of the sine wave is specified by *freq*. The amplitude of the waveform decays exponentially with the damping factor df specifying the decay rate. Parameter ph specifies the phase of the waveform in degrees with respect to $t = td$. An equation that describes the waveform mathematically is

$$v(t) = vo + va \times \exp(-(t - td)df) \times \sin(2\pi(freq(t - td) - ph/360))$$

A source specified with the time-varying SIN source is only for use in transient analyses. It cannot be used as a source in an ac analysis (ac analysis is explained in the next section).

An example of a sinusoidal time-varying waveform voltage source is

VRING 6 5 SIN(1 5 100K 50N 1M 90)

This source produces a voltage drop from node 6 to node 5 that is initially 1 V until $t = 50$ ns. At 50 ns, the source begins oscillating sinusoidally with a frequency of 100 KHz. The phase of the voltage is 90°, meaning that at $t = 50$ ns the voltage starts at its maximum amplitude (since sin 90° = 1) and begins declining toward zero. The magnitude of the waveform decays exponentially with an exponential damping factor of 1E-3.

The PWL (piecewise linear) input source waveform is shown in Figure G-19. The general form of the PWL specification is

PWL(<t1> <v1> <t2> <v2> . . . <tn> <vn>)

where Table G-6 describes each of the parameters and their default values.

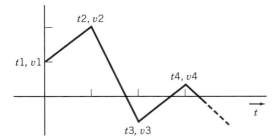

Figure G-19 Piecewise linear waveform.

Table G-6
PWL Waveform Specifications

Parameter	Default Value, Units
<tn> time at corner	none, seconds
<vn> value at corner	volts or amps

The PWL source generates a waveform that is constructed of linear segments providing either a voltage or a current. Pairs of values tn and vn specify the position of the waveform at vertices. Between vertices the waveform is specified as a waveform in the shape of a straight line connecting respective vertices. This piecewise linear waveform can be used to model arbitrary-shaped waveforms.

An example of a piecewise linear time-varying waveform voltage source is

VTRIAN 16 15 PWL(0 0 1 1 2 0 3 1 4 0)

This source produces a voltage drop from node 16 to node 15 that is initially at 0 V at $t = 0$. It rises linearly to 1 V at $t = 1$ s. Next, the voltage drops linearly to 0 at $t = 2$ s. This pattern repeats until $t = 4$ s, after which the voltage remains at 0 V. Using the PWL source, it is possible to approximate the behavior of many waveforms.

The **.TRAN** statement is used to indicate to PSpice that a transient analysis is to be performed. Further, it specifies the interval over which the analysis is to take place, the interval over which output values should be printed, and optionally a maximum value for the analysis time step. The form of the **.TRAN** statement is

.TRAN[/OP] <t step> <t stop> [<t start>][<step ceiling>]

where the optional /OP parameter, when included, causes PSpice to produce an output listing containing the operating point analysis. Although this analysis is always done before a transient analysis, it is only printed when requested. Parameter *tstep* is the size of the interval between printed output points in either .PRINT tables or .PLOTs. Parameter *tstop* specifies the time point at which the analysis (and the printing/plotting) should stop. Optional parameter *tstart* specifies the time point at which printing should start. All analysis begins at time zero, and by default, printing/plotting also begins at time zero. By specifying a value for *tstart* it is possible to avoid outputting unwanted portions of a table or plot. Optional parameter *step ceiling* specifies the maximum value that can be used as an internal time-step for the transient analysis. By default, most versions of PSpice set the maximum time step to be 1/50 of the duration of the analysis interval. In some cases, allowing the time step to become too large causes PSpice to skip over time intervals of interest, reporting little or no change in outputs over the interval. This problem can be circumvented by judicious use of the *step ceiling* parameter.

With the .TRAN statement *tstep* is the size of the interval between printed data points in either .PRINT tables or .PLOT plots and also .PROBE results. It should be noted that for any of the transient sources the value of the source at $t = 0$ is used by the PSpice program to calculate initial conditions at $t = 0$.

Output from a transient analysis can be generated in much the same way as in the .DC analysis. The .PRINT, .PLOT, and .PROBE statements are used to generate the output of selected circuit values as a function of time. Data is output at intervals as specified in the .TRAN statement. The output statements take the following forms:

.PRINT TRAN <*output values*>

and

.PLOT TRAN <*output values*>

where the *output values* specify the node voltages or element currents in the same manner as described in the previous sections. The keyword TRAN is used to specify that PSpice should output the values calculated during the transient analysis for each of the *output values* contained in the subsequent list. An example of the usage of the .PRINT and .PLOT statements is

.PRINT TRAN I(VSENS) V(3,5)

.PLOT TRAN V(2,1) V(5)

The printed and plotted output provides values as a function of time, at specific time instances. The size of the step between time instances is specified as the *tstep* parameter in the .TRAN statement.

It should be noted that it is permissible to specify more than one analysis type on both the .PRINT and .PLOT statements. For example, it is possible to request output values from both the .DC and .TRAN analyses by using a statement such as

.PLOT DC V(2,3) I(V1) TRAN V(2,3) V(4)

The .PROBE statement may also be used with PSpice to output the results of a transient analysis. As before, use of the simple .PROBE statement produces a data file containing the results of all of the circuit node voltages and element currents for use in later display operations. If a .TRAN analysis is the only analysis type specified, then only transient analysis outputs will be saved by the .PROBE statement. If any other types of analyses are included in the circuit description file, such as .DC or .AC, data generated by these analyses will also be saved in the PROBE.DAT file.

Figure G-20 contains the PSpice circuit description for the source-free *RC* circuit shown schematically in Figure G-15. This circuit is used to model the behavior of the original source-free *RC* circuit, including providing the initial condition on the capacitor *C*. The current source used to supply the initial circuit condition is a time-varying source of type PULSE. The desired initial condition is to have 5 V across C1 at $t = 0$. With the resistor R1 equal to 1 kΩ in parallel with C1, it is necessary for ISTEP to provide 5 mA of current to R1. Examination of the element statement for ISTEP shows that the source will switch at $t = 0$ from providing 5 mA to 0 mA at $t = 0^+$. The .PLOT statement causes data generated by the transient analysis to be plotted.

```
RC Circuit with an Initial Voltage on C


 ****      CIRCUIT DESCRIPTION


 ******************************************************************************


R1 1 0 1K
C1 1 0 1U
ISTEP 0 1 PULSE(5M 0 0 0 0); PROVIDES INITIAL CAPAC. VOLTAGE
.TRAN .5M 10M
.PROBE
.PRINT TRAN V(1)
.PLOT TRAN V(1)
.END
```

Figure G-20 PSpice description of the source-free circuit of Figure G-15.

Parameters for the **.**TRAN statement are chosen so that the events that take place in the circuit (i.e., the transients) can be easily viewed. If the *tstep* parameter is chosen to be too large, the transitions of interest will be seen compressed on the left side of the plot. However, if the *tstep* parameter is chosen to be too small, only a portion of the transition will be seen on the plot; most of the transition will be beyond the analysis interval. It may take one or two trial simulation runs to find the appropriate values for the **.**TRAN parameters. A reasonable rule of thumb is initially to choose the *tstep* parameter to be 1/10 of the product of the smallest R and smallest C values in the circuit. Initially choose *tstop* to be 10 times the *tstep* value. For circuits containing resistors and inductors, choose *tstep* to be 1/10 of L/R, where L and R are the smallest inductor and the largest resistor values in the circuit. After viewing the output, adjustments can be made to these parameters so that the output will provide more or less detail as required.

Details of the output generated by the **.**PLOT statement of PSpice for the *RC* circuit are shown in Figure G-21. Note that time is plotted on the vertical axis from top to bottom, while voltage amplitude is plotted horizontally increasing from left to right. As expected, the plot shows that the voltage of 5 V on the capacitor decays exponentially with a time constant equal to the product of R and C. Further details of the exponential decay can be obtained by examining the plot generated using the **.**PROBE statement. Figure G-22 contains the higher resolution plot. Although the PROBE plot provides a better picture for visualization, it does not provide the printed data that the **.**PLOT or **.**PRINT output gives. Obviously, there are circumstances in which one or the other may be more desirable.

The *RC* circuit analysis example showed how initial conditions in a circuit could be set using standard circuit element models. PSpice contains built-in facilities to perform the same operations. There are three ways in which initial conditions may be set with these built-in facilities: using the **.**IC statement, using the **.**NODESET statement, or using the optional UIC parameter in the **.**TRAN and element statements.

The **.**IC statement, which stands for initial conditions, can be used to set specific nodes to initial voltages. The syntax for **.**IC is

$$.\text{IC V}(n1) = <node\ voltage> \ . \ . \ .$$

This list of node voltages will be used in the operating point analysis that precedes the transient analysis to set particular nodes to specified voltages. At the start of the transient analysis, the initial condition generator is removed from the circuit. Note that this statement can only be used with the transient analysis type and has no effect in a **.**DC or **.**AC analysis.

A different form of this statement is found in the **.**NODESET statement. The syntax is

$$.\text{NODESET V}(n1) = <node\ voltage> \ . \ . \ .$$

```
    TIME          V(1)
(*)----------
    0.0000E+00    2.0000E+00    4.0000E+00    6.0000E+00    8.0000E+00
                  - - - - - - - - - - - - - - - - - - - - - - - - - - -
0.000E+00   5.000E+00 .                                   *          .
5.000E-04   3.935E+00 .                        *                     .
1.000E-03   2.384E+00 .                  .  *          .             .
1.500E-03   1.450E+00 .              *      .          .             .
2.000E-03   8.731E-01 .          *         .          .             .
2.500E-03   5.310E-01 .    *               .          .             .
3.000E-03   3.197E-01 .  *                 .          .             .
3.500E-03   1.944E-01 .*                   .          .             .
4.000E-03   1.171E-01 .*                   .          .             .
4.500E-03   7.120E-02 *                    .          .             .
5.000E-03   4.287E-02 *                    .          .             .
5.500E-03   2.607E-02 *                    .          .             .
6.000E-03   1.570E-02 *                    .          .             .
6.500E-03   9.548E-03 *                    .          .             .
7.000E-03   5.749E-03 *                    .          .             .
7.500E-03   3.496E-03 *                    .          .             .
8.000E-03   2.105E-03 *                    .          .             .
8.500E-03   1.280E-03 *                    .          .             .
9.000E-03   7.708E-04 *                    .          .             .
9.500E-03   4.688E-04 *                    .          .             .
1.000E-02   2.823E-04 *                    .          .             .
                  - - - - - - - - - - - - - - - - - - - - - - - - - - -
```

Figure G-21 Output of the source-free circuit of Figure G-15 using PLOT.

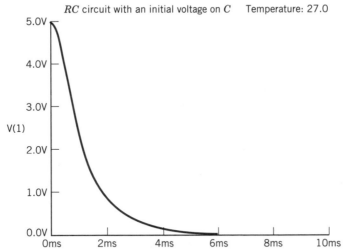

Figure G-22 Higher resolution output of V(1) for the source-free circuit of Figure G-15 using PROBE.

Statement **.NODESET** differs from **.IC** in that **.NODESET** only provides an initial node voltage guess for the simulator rather than forcing a node to a particular voltage. Effectively, **.NODESET** provides node voltages at the beginning of the operating point analysis. These values are used as the starting point in determining the actual operating point values. If no other forces in the circuit cause a change in these voltages, these become the starting point for the transient analysis. Node voltages specified in the **.IC** statement are effectively instantiated at the beginning of the operating point analysis and kept the same until the end of this analysis. Thus, conditions specified by the **.IC** statement will always be intact at the beginning of the transient analysis. This is not true when the .NODESET statement is used.

The standard form of the element statement for a capacitor or inductor is

$$<name><+ node><- node><element\ value><init.\ cond.>$$

For example, a capacitor C_1 of 1 mF can be represented as

$$C1 \quad 3 \quad 4 \quad 1M \quad IC = 1.5$$

where the initial condition is 1.5 V.

Individual elements may be set to some initial conditions using an optional parameter appended to the standard element statement. Two examples of initializing elements are given below for the C and L elements:

$$standard\ capacitor\ statement\ IC = <initial\ voltage>$$

$$standard\ inductor\ statement\ IC = <initial\ current>$$

where the units of voltage are volts and the units of current are amperes. In addition to specifying these initial conditions on the individual element statements, it is necessary to use the optional parameter UIC in the .TRAN statement. The following lines show the element statements for a capacitor and an inductor with initial conditions along with the .TRAN statement containing the necessary UIC parameter:

$$L1\ 10\ 11\ 2MH\ IC = 1MA$$

$$C1\ 12\ 13\ 1U\ IC = 2V$$

$$.TRAN[/OP]<tstep><tstop>[<tstart>][<step\ ceiling>]\ UIC$$

UIC causes PSpice to skip the operating point analysis and proceed to the transient analysis using the initial conditions specified on the element statements. We do not use the letter F following 1U in the line for C1, since F is the multiplier femto (Table G-2).

Example G-1

Determine and plot the capacitor voltage and current for the circuit shown in Figure G-23 when $R_1 = 1\ k\Omega$ and $C = 1\ \mu F$. The initial capacitor voltage is 2 V and $v_1 = 5$ V.

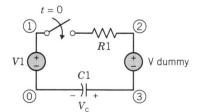

Figure G-23 Circuit diagram for a capacitor charging example.

Solution

Figure G-24 contains the circuit description file, and Figure G-25 and Figure G-26 contain PROBE plots of the voltage and current requested. Initial conditions are set using the .IC statement. The initial voltage can be seen as present at $t = 0$ with a corresponding initial current flow. After $t = 0$, the characteristic exponential increase in voltage with the corresponding current flow can be seen in the figures. We use Vdummy to obtain the capacitor current.

It is not necessary to insert the dummy source. An alternative method requests I(C1) in the .PLOT statement. Then we delete the VDUMMY line and use .PLOT TRAN V(C1) I(C1). Of course, we change the capacitor statement to C1 2 0 1U.

```
****          CIRCUIT DESCRIPTION

*************************************************************

V1 1 0 PULSE(0 5 0 1P)
R1 1 2 1K
C1 3 0 1U
VDUMMY 2 3
.IC V(3)=2
.PLOT TRAN V(3) I(VDUMMY)
.PROBE
.TRAN .2M 3M
.END
```

Figure G-24 PSpice description for Example G-1.

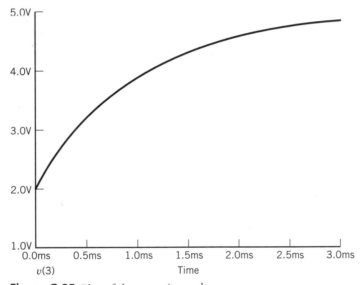

v(3)   Time

Figure G-25 Plot of the capacitor voltage.

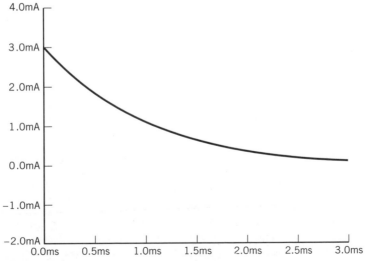

Figure G-26 Plot of the capacitor current.

G-7 | FREQUENCY DOMAIN ANALYSIS

Thus far we have covered two main types of analysis: dc analysis and time domain analysis. The third main type of analysis that is performed on a circuit is called frequency domain analysis. This type of analysis provides details on how circuits respond to stimuli of different frequencies. For PSpice, the term ac analysis is used to designate a frequency domain analysis. In an ac analysis, it is possible to observe the changes in magnitude and relative phase angle of node voltages and element currents. Unlike the transient analysis, which shows the circuit response as a function of time, the ac analysis contains no time-varying characteristics. In effect, the program assumes that each sinusoidal source has been applied for a very long time and thus calculates the steady-state response (amplitude and phase) for each frequency.

The .AC control statement specifies that the type of analysis PSpice is to perform is an ac analysis. The .AC statement further specifies the range of frequencies over which the analysis is to be performed and how the range of frequencies is to be swept. In the dc and transient analyses, the independent variables were swept linearly. That is, the increment between output points was a constant value. For the ac analysis, the independent variable frequency can be swept linearly, or it can also be swept logarithmically.

The general form of the .AC statement is

.AC *<sweep type> <n> <start freq> <end freq>*

where *n* is related to the number of points to be output. Parameter *start freq* is the frequency at which the analysis is to begin while *end freq* specifies the last frequency point to be analyzed. The *sweep type* parameter may be chosen from the following list: LIN, OCT, DEC. Choosing LIN produces a linear range of frequencies as described above. For *sweep type* LIN, parameter *n* will be the total number of points output.

Using OCT, frequencies will be swept logarithmically from *start freq* to *end freq*. That is, the frequency will be increased in octaves. Parameter *n* indicates the number of points output per octave. An octave starts at *start freq* and ends at a frequency twice the starting frequency. The next octave starts where the last octave ends, and output continues until the *end freq* is reached. Sweep type DEC works in a similar manner. DEC specifies that the frequency will be increased in decades. The *start freq* is multiplied by ten to define the first decade. Parameter *n* indicates the number of output points per decade.

The .PRINT and .PLOT statements can be used to generate tabular and graphic output. For either output statement, the organization of the data on the *x*-axis will be determined by the sweep type specified on the .AC statement. Just as with the previous analysis types, an additional parameter in the .PRINT or .PLOT statement is used to indicate what output values for the ac analysis should be displayed. The form of the output statements is

.PRINT AC *<ac output variables>*

.PLOT AC *<ac output variables>*

Various formats for the data can be provided using the ac analysis. The particular value that is to be output is specified by a suffix placed on the output variables. A list of the possible suffixes and their meanings is contained in Table G-7. An example of the use of these output suffixes is

.PLOT AC VM(1) VM(2);   plots the magnitude of the voltage at nodes 1 & 2

.PLOT AC IP(R3);   plots the phase of the current in R3

As usual, the independent variable is printed or plotted in the first column of the output. Another thing to remember is that ac, dc, and transient analyses can be mixed in the same simulation run. Outputs, of course, will be displayed in different sections of the listing.

Use of the .PROBE statement in PSpice is very much the same as for the previous two types of analysis. However, when an ac analysis is performed and the generic .PROBE (with no parameters)

Table G-7

AC Analysis Output Variable Suffixes

Suffix	Value Output
no suffix	magnitude
M	magnitude
DB	magnitude in decibels
	$(20 \times \log(\text{value}))$
P	phase
R	real part
I	imaginary part

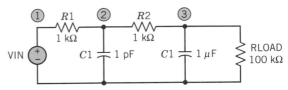

Figure G-27 Circuit diagram for a filter.

statement is used, even more data are saved in the PROBE.DAT file. This is because all of the different forms of value output are saved for possible use by the PROBE program.

Independent voltage sources for the ac analysis are specified in the same way as for the dc and transient analyses. Adding the keyword AC to an independent source statement indicates that the source is to be used in any subsequent **.AC** analysis. Two values may be specified as part of the ac source. The first value to follow the keyword AC is taken to be the magnitude of the source. If a second parameter is specified, it is taken to be the phase in degrees of the source. As discussed in the section on transient analysis sources, it is permissible to specify dc, transient, and ac values of a source all in the same statement.

An example circuit is shown in Figure G-27. This circuit acts as a low-pass filter. The PSpice description for the circuit is shown in Figure G-28. The magnitude of the source is set to 1 and the phase to zero. This is typical of single-source systems and allows the gain to be read directly from the output of the circuit. For more general circuits involving multiple sources, it may be necessary to specify different magnitudes and phases for the sources.

For the example, frequency is swept from 1 Hz to 100 kHz in octaves, with 2 output points per octave. Figure G-29 contains a **.PROBE** plot of the magnitude in decibels of the voltages at two nodes in the circuit. Examination of the figure shows that the output of the first stage (i.e., the voltage at node 2) is reduced to 50% of the input value once the frequency reaches about 1 kHz. The output of the second stage (the voltage at node 3) is reduced by 40 dB at 10 kHz.

```
RC Filter

 ****      CIRCUIT DESCRIPTION

*****************************

VIN 1 0 AC 1 0
R1 1 2 1K
C1 2 0 1P
R2 2 3 1K
C2 3 0 1U
RLOAD 3 0 100K
.PROBE
.AC OCT 12 1 100K
.PLOT AC V(2) (0,1) V(3) (0,1)
.END
```

Figure G-28 SPICE description of the filter.

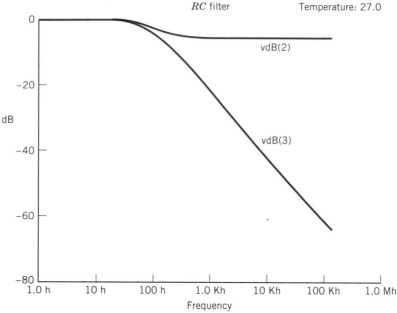

Figure G-29 Filter output plot in dB.

G-8 | MUTUAL INDUCTANCE AND OP AMPS

PSpice has the built-in capability of modeling the behavior of a mutual inductance. In its simplest form, a mutual inductance model is given as a single statement similar to a simple inductance. Syntax for the simple form model mutual inductor statement is as follows:

$$K<name> L<name> <L<name>> <coupling\ value>$$

where the inductor *names* are those of the mutual inductors (there may be more than two), the *coupling value* is a constant k where $k = M/\sqrt{L_1 L_2}$, and M is the mutual inductance in henries. The value of k will always be $0 \le k \le 1$. A coupling value of $k = 0$ indicates that there is no coupling between inductors. The "dot" convention normally follows when specifying mutual inductances is also used with PSpice. A dot is assumed at the first node listed on each of the inductor statements specified on the mutual inductance statement. Thus, the current must flow into the dot of each inductor.

As an example of the use of the mutual inductance statement, consider the problem posed in Exercise 12-11. The circuit diagram for this exercise is shown in Figure G-30. The problem is to determine the ratio of the phasor voltages V_2/V_1 when $\omega = 10$ rad/s.

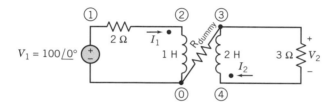

Figure G-30 Circuit diagram for the mutual inductance example. Note that the current flows into the dot of each inductor.

To solve the problem, first the value of the coupling constant is calculated to be 0.707 and the frequency of interest is 1.592 Hz. These values are used in the circuit description as shown in Figure G-31. Notice that a large resistance RDUMMY is used to tie node 3 (and consequently node 4) to the rest of the circuit. PSpice necessitates that all elements have a dc path to ground and RDUMMY

```
      Simple Circuit with Mutual Inductance


        * * * *        CIRCUIT DESCRIPTION



        * * * * * * * * * * * * * * * * * * * * * * * * * * * * * * * * * * * * * * * * * * * * * * * * * * * * *


        V1 1 0 AC 100 0
        R1 1 2 2
        R2 3 4 3
        L1 2 0 1
        L2 4 3 2
        KMUTUAL L1 L2 0.707
        * coupling value k=M/(L1*L2)^    0.5
        RDUMMY 3 0 100G
        * RDUMMY keeps nodes 3 & 4 from floating
        * no current in RDUMMY
        .AC LIN 3 1.591 1.593
        .PRINT AC VM(3,4) VP(3,4) I(RDUMMY)
        .END
```

Figure G-31 Circuit description file for the mutual inductance example.

```
          Circuit with Mutual Inductance


    * * * *        AC ANALYSIS                        TEMPERATURE =   27.000 DEG C



    * * * * * * * * * * * * * * * * * * * * * * * * * * * * * * * * * * * * * * * * * * * * * * * * * * * * * * * * * * * *


       FREQ          VM(3,4)        VP(3,4)        I(RDUMMY)


       1.591E+00     2.559E+01      1.267E+02      7.117E-15
       1.592E+00     2.558E+01      1.267E+02      2.785E-15
       1.593E+00     2.557E+01      1.266E+02      1.553E-15
```

Figure G-32 Solution for the mutual inductance example.

satisfies that need. Because of the size of the resistance, very little current flows in RDUMMY, and nodes 3 and 4 are effectively electrically isolated from the rest of the circuit. The PSpice solution for the circuit at 1.592 Hz is shown in Figure G-32.

An ideal transformer can be modeled in a PSpice program by setting L_1 and L_2 large so that ωL_1 and ωL_2 are much greater than other impedances at the frequency of interest. Also, an ideal transformer requires $k = 1$. We will use $k = 0.999999$. Another attractive approach is to use a controlled-source model of the ideal transformer as shown in Figure 12-23.

Circuits containing op amps can be readily analyzed using PSpice and the actual circuit model of the op amp, which includes the input and output resistances and the actual gain, as shown in Figure 6-16. Thus, we use a VCVS with gain A.

We can compute the input resistance, output resistance, and the gain of an op amp circuit using the transfer function statement, which is

$$.TF <output\ variable> <input\ variable>$$

For example, a circuit model of an actual (nonideal) op amp circuit is given in Figure G-33. We use the transfer function (.TF) statement as

$$.TF \quad V(3) \quad V(1)$$

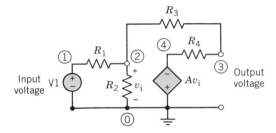

Figure G-33 Circuit model for nonideal op amp circuit. The op amp output resistance is $R_o = R_4$ and the input resistance is $R_i = R_2$.

to find the circuit gain V_3/V_1 and the circuit input resistance and output resistance. The circuit file contains the controlled source statement and calculates .TF at the .DC statement. When $R_1 = 10$ kΩ, $R_i = R_2 = 10$ MΩ, $R_3 = 50$ kΩ, $R_o = R_4 = 50$ Ω, and $A = 10^5$, the output printout is

$$V(3)/V(1) \qquad\qquad = -5.00E+00$$

$$\text{INPUT RESISTANCE AT V1} \quad = \quad 1.00E+04$$

$$\text{OUTPUT RESISTANCE AT V3} = \quad 3.03E-04$$

If we wish to model an ideal op amp, we set R_i very large, $R_o = 0$, and A very large. For the preceding example, if we desire the ratio V(3)/V(1) for an ideal op amp, we set $R_4 = R_o = 0$, $R_i = R_2 = 100$ MΩ, and $A = 10^8$.

SUMMARY

This appendix explained and illustrated the utility of PSpice as an analytical and design tool. PSpice can be used to solve many complex circuits for their dc, ac, and transient responses. This appendix focuses on PSpice from MicroSim Corporation and uses the PROBE statement to obtain higher resolution plots.

A PSpice input file or *circuit file* contains (1) title and comment statements, (2) data statements, (3) solution control statements, (4) output control statements, and (5) an end statement. PSpice is capable of performing three main types of analysis. It can determine the behavior of selected output voltages with respect to changes in input voltages. This type of analysis is usually referred to as a dc analysis. A single-point dc analysis also determines what is called the bias-point characteristics, that is, the behavior of the circuit when only a dc voltage is applied to the circuit. In most cases, this is the starting point for either of two other types of analysis.

A second type of analysis that is usually required to determine fully a circuit's behavior is called a transient analysis. Transient analyses calculate circuit voltages and currents with respect to time. This assumes that there is a time-dependent stimulus that causes an effect on the rest of the circuit. To perform a time domain circuit analysis, PSpice first calculates the bias point and then calculates the circuit response to a time-dependent change in one or more voltage or current sources.

The third main type of analysis that PSpice can perform is called an ac analysis. This analysis type is also referred to as a sinusoidal steady-state analysis. Here, voltages and currents are calculated as functions of frequency. That is, output variable changes are calculated in response to changes in the frequency or phase of sinusoidal input voltage or current sources.

A summary of important statements used for PSpice analysis is provided in Table G-8.

Table G-8
Summary of Important Statements

Function	Statement
Element description	$<element\ name><node><node><value>$
Operating point dc analysis	No statement required; occurs by default for a circuit list.
dc sweep of a voltage or current source	$.DC \left\langle \dfrac{source}{name} \right\rangle \left\langle \dfrac{start}{value} \right\rangle \left\langle \dfrac{stop}{value} \right\rangle <increment>$
Print output	$.PRINT <analysis\ type> \left[\begin{matrix} output\ list \\ of\ variables \end{matrix} \right]$
Plot output with low resolution and list numerical values	$.PLOT \left\langle \dfrac{analysis}{type} \right\rangle \left\langle \dfrac{output}{list} \right\rangle$
Higher resolution plot	$.PROBE <output\ list>$
Transient analysis for interval of time	$.TRAN <tstep><tstop><tstart>$
ac sweep of frequency; types are LIN,OCT,DEC	$.AC \left\langle \dfrac{sweep}{type} \right\rangle <n> \left\langle \dfrac{start}{freq} \right\rangle \left\langle \dfrac{end}{freq} \right\rangle$

n is related to number of points to be output

REFERENCES

Agnew, Jeremy, "Simulating Audio Transducers with SPICE," *Electronic Design,* November 7, 1991, pp. 45–47.

Monssen, Franz, *PSpice with Circuit Analysis,* Macmillan Publishing, New York, 1993.

Tuinenga, Paul W., *SPICE: A Guide to Circuit Simulation,* Prentice Hall, Englewood Cliffs, NJ, 1992.

PROBLEMS

P G-1 A model of an experimental light bulb is shown in Figure P G-1. We wish to use this model repeatedly. Develop a subcircuit PSpice description for the light when $R_1 = 0.1\ \Omega$, $L_1 = 0.1$ mH, and $C_1 = 1$ pF.

P G-2 Determine the voltage v for the circuit shown in Figure P G-2.

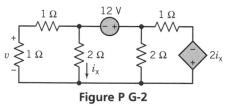

Figure P G-2

P G-3 Determine the voltage v for the circuit shown in Figure P G-3.

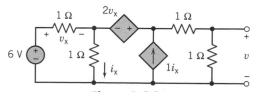

Figure P G-3

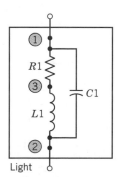

Figure P G-1 Model of an experimental light bulb.

P G-4 Determine the output voltage v when an ideal op amp is used in the circuit of Figure P G-4.

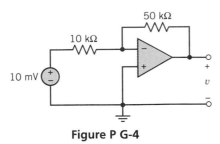

Figure P G-4

P G-5 Repeat Problem G-4 when a nonideal op amp is used. The op amp characteristics are $A = 10^5$, $R_o = 50\ \Omega$, and $R_i = 1\ M\Omega$. Compare the results with those obtained for the ideal op amp.

P G-6 For the nonideal op amp of Problem G-5, determine the input and output resistances seen by the source and a load, respectively.

P G-7 An amplifier is represented by the circuit shown in Figure P G-7. Determine v when $b = 0.1$ and $v_s = 1\ mV$.

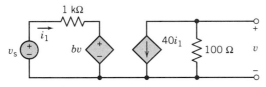

Figure P G-7 Amplifier circuit.

P G-8 An acoustic earphone can be represented by the circuit shown in Figure P G-8 (Agnew, 1991). Determine the frequency response by plotting a Bode diagram of the impedance $Z(j\omega)$.

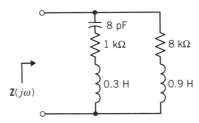

Figure P G-8 Model of acoustic earphone.

P G-9 An *RLC* circuit is shown in Figure P G-9. It is desired to select R and L so that the phase shift between input, v_1, and output, v_2, is $-90°$. Select R and L to achieve this goal and plot the Bode diagram. The input voltage operates at $\omega = 10,000$ rad/s.

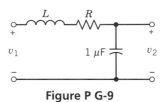

Figure P G-9

P G-10 Determine the output voltage, v, sensitivities for the resistive elements of the circuit shown in Figure P G-10. Assume an ideal op amp.

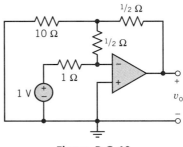

Figure P G-10

P G-11 Determine $v(t)$ for the circuit shown in Figure P G-11. Assume that the circuit is at steady state at $t = 0^-$.

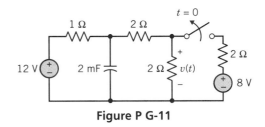

Figure P G-11

P G-12 Plot $v(t)$ for the circuit shown in Figure P G-12a. The input is the pulse shown in Figure P G-12b. Assume that the initial capacitor voltage is zero.

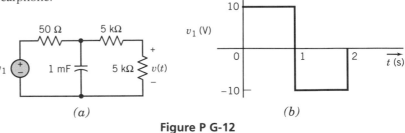

(a)   *(b)*

Figure P G-12

P G-13 Obtain a plot of the capacitor voltage, $v(t)$, for the circuit of Figure P G-13 from $t = 0$ until $t = 3$ s. Determine the voltage, v, at $t = 0.4$ s. The initial conditions are given as $i(0) = 4$ A and $v(0) = -4$ V.

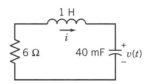

Figure P G-13

P G-14 Determine the current i for the circuit shown in Figure P G-14.

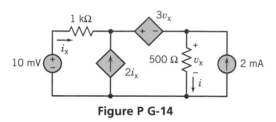

Figure P G-14

P G-15 Determine and plot the output voltage $v(t)$ of the op amp circuit when the input voltage, v_1, ranges from 1 mV to 500 mV. The parameters of the op amp shown in the circuit of Figure P G-15 are $R_i = 100$ kΩ, $R_o = 1$ kΩ, and $A = 10^5$.

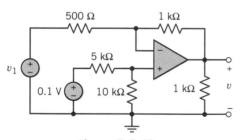

Figure P G-15

P G-16 A common-emitter bipolar junction transistor may be represented as shown in the circuit of Figure P G-16. Determine the input resistance by connecting a 1-A source to the terminals a–b.

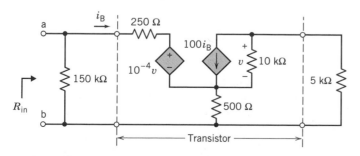

P G-17 Determine and plot the output voltage, $v(t)$, for the circuit shown in Figure P G-17 when $v_s = 11.3 \cos(\omega t + 45°)$ V and $i_s = 4 \cos \omega t$ A operating at 400 Hz.

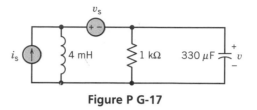

Figure P G-17

P G-18 A circuit with an ideal transformer is shown in Figure P G-18 in phasor form. Determine $\mathbf{I}_1$ when the turns ratio is $n = 10$ and $\omega = 500$ rad/s.

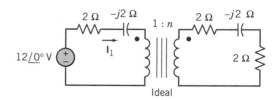

Figure P G-18

P G-19 An RLC circuit is shown in Figure P G-19. Plot the Bode diagram of $\mathbf{V}$ for the frequency range of 1000 Hz to 2000 Hz.

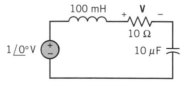

Figure P G-19

Figure P G-16 Model of a common-emitter bipolar junction transistor circuit.

P G-20 The switches of the circuit of Figure P G-20 both open at $t = 0$. Assume that the circuit is at steady state at $t = 0^-$ and determine and plot $v(t)$ for $0 < t < 25$ ms.

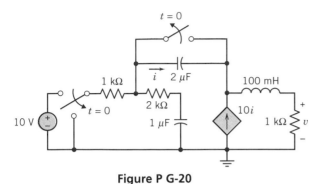

Figure P G-20

P G-21 For the circuit in Figure P G-21 determine and plot the voltage $v(t)$ for $0 < t < 2.5$ s. Find the maximum value of $|v(t)|$.

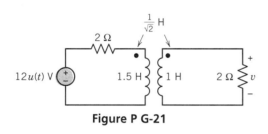

Figure P G-21

P G-22 The circuit model shown in Figure P G-22 is called the Sallen–Key configuration in honor of its inventors. The input signal is a 1-V pulse of 4 ms duration. Determine and plot the output response, $v_o(t)$.

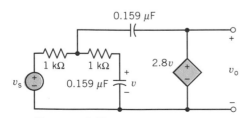

Figure P G-22 Sallen–Key circuit.

P G-23 Consider the circuit shown in Figure P G-22. Obtain a plot and printout of the magnitude and phase of $V_o(\omega)$ when the source signal is a 1-V sinusoid and the frequency is swept from 10 kHz to 500 kHz in 81 linearly spaced steps.

P G-24 Determine the steady-state ac output response v_2 of the op amp circuit shown in Figure P G-24 when $v_s = 10 \cos(1000t + 30°)$ V.

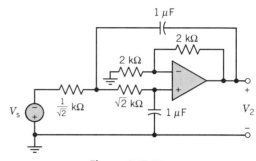

Figure P G-24

P G-25 Plot the frequency response of the circuit shown in Figure P G-25 between 10 kHz and 500 kHz when the output voltage is v. Determine the frequency when the magnitude of **V** attains a maximum.

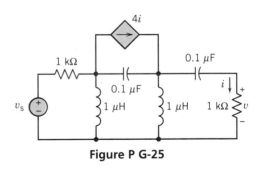

Figure P G-25

P G-26 The circuit shown in Figure P G-26 is a biquad circuit. A biquad circuit is one of the most useful circuits to the electrical engineer because it is a universal filter. It is widely available as a module from industrial sources. Assume ideal op amps.

(a) Given the element values as specified in the figure, plot the frequency response of V_2, V_3, V_4, and V_5 and comment on the type of the circuit.

(b) Can this circuit be used to eliminate the powerlike frequency (i.e., 60 Hz)? If yes, where should one take the output signal? What is the associated bandwidth?

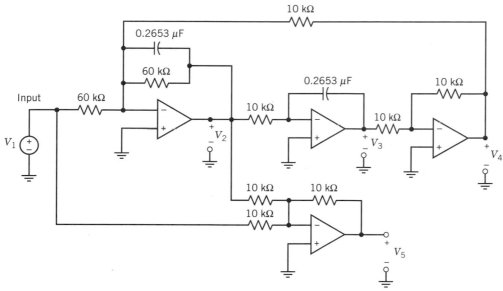

Figure P G-26 The biquad universal filter circuit.

INDEX